PREVENTION MAGAZINE'S
NUTRITION ADVISOR

By Mark Bricklin and the editors of *Prevention* Magazine

MJF BOOKS

NEW YORK

The nutritional analyses used in this book were taken from the U.S. Department of Agriculture (USDA) Nutrient Data Base for Standard Reference. Additional dietary fiber values come from the USDA's 1991 Primary Nutrient Data Set for USDA Food Consumption Surveys (PDS), Release 1: Hyattsville, Md. The "Recommended Dietary Allowances" table on page 26 is reprinted by permission of National Academy Press, Washington, D.C. From *Recommended Dietary Allowances, 10th Edition,* copyright © 1989 by the National Academy of Sciences.

Notice: this book is intended as a reference volume only, not as a medical guide or a manual for self-treatment. If you suspect that you have a medical problem, we urge you to seek competent medical help. Keep in mind that nutritional needs vary from person to person, depending on age, sex, health status, and total diet. Information here is intended to help you make informed decisions about your diet, not to substitute for any treatment that may have been prescribed by your physician.

Published by MJF Books
Fine Communications
60 West 66th Street
Two Lincoln Square
New York NY 10023

Library of Congress Catalog Card Number 94-75371
ISBN 1-56731-039-7

Printed by arrangement with Rodale Press.

This book has previously been published as *Prevention Magazine's Complete Nutrition Reference Handbook.*

Prevention is a registered trademark of Rodale Press, Inc.

MJF Books and the MJF colophon are trademarks of Fine Creative Media, Inc.

Manufactured in the United States of America

10   9   8   7   6   5   4   3   2

Prevention *Magazine's Nutrition Advisor*

Editor: Edward Claflin

Project Development and Research Director: Karen Lombardi Ingle

Author: Mark Bricklin, Editor, *Prevention* Magazine

Food Editor: Jean Rogers

Contributing Writers: Doug Dollemore, Sharon Faelten, Gale Maleskey, Ellen Michaud, Hank Nuwer, Jean Rogers, Joe Wargo

Research Chief: Ann Yermish

Nutrition Consultant: Jeffrey Blumberg, Ph.D.

Production Editor: Jane Sherman

Research/Fact-Checking Staff: Christine Dreisbach, Jewel Flegal, Jeffrey Gross, Anne Imhoff, Melissa Meyers, Paris Mihely-Muchanic, Deborah Pedron, Michele Toth

Book Designer: Denise M. Shade

Cover Designer: Acey Lee

Cover Photographer: Kurt Wilson

Illustrator: David Flaherty

Copy Editor: Sarah S. Dunn

Indexer: Ed Yeager

Office Staff: Roberta Mulliner, Karen Earl-Braymer, Julie Kehs, Mary Lou Stephen

Contents

Nutritional Reference Tables 539

Index of Foods and Dishes

Introduction

This book is the most complete and practical popular reference book on nutrition ever published.

But reference books, however complete, never tell the whole story. We need to know how to apply all that material in our daily lives and diets!

Here we do that in two ways. The reference material itself is annotated to tell you at a glance what general effect on health a food or dish is likely to have. To do that, we use a series of symbols—shown on pages 31–32. We also briefly discuss the health highlights of each food alongside its actual nutritional scorecard.

Beyond all that, though, we need a broad perspective on how all elements of nutrition *work together* to affect health for better or worse.

The first part of our book, "The ABCs of Nutrition and Health," fulfills that need. Here you will learn the basics of nutrition as interpreted by the editors of *Prevention* magazine and our many scientific advisors. The emphasis is admittedly on the practical side—for good reason. The biggest barrier to health knowledge is not lack of details, but too many, with too little guidance. Our goal in the "ABCs" is to point you in the right direction.

Immediately following the "ABCs," you'll find "A Quick and Easy Guide to This Handbook" with practical advice on how to use this book.

The reference material—supplemented if need be by advice from your dietitian or doctor—will provide all the details you'll probably need.

Read our book—*use* it—in good health!

Mark Bricklin, Editor
Prevention Magazine

The ABCs of Nutrition and Health

We suspected it all along. Somewhere along the line, something went wrong with our food, with the way we eat. From King David to Davy Crockett, our food had a certain natural vigor to it. A certain honesty that sat well with our body and its nutritional needs.

But then, in a gradual process that no one even noticed until a hundred years ago, our food began to lose that country-bred robust quality and take on the character of a fast-talking, not-to-be-trusted city slicker.

Bread was made bereft of its fibrous bran to make it soft and white. Tinned foods that had much of their nutrition boiled away became increasingly popular. Sugar became cheap and plentiful. So did milk, cream, butter, and cheese. And if you couldn't get enough variety in the cakes, pies, and desserts you made out of the white flour, cream, and sugar, you could just toss in some food additives to hit some new notes. Still not right? Spoon in some lard or pour on the oil to produce that lovely silken "mouth feel."

Beef, too, became far more plentiful than before, thanks to the western grasslands and railroads. Midwestern corn was used to produce huge tonnages of pork, ham, chops, bacon, and sausage.

More recently, our taste for gourmet food has resulted in millions of Americans eating "premium" ice creams loaded with dairy fat, croissants dripping with butter, popcorn sprayed with cheese *and* fat, and macadamia nuts—one of the fattiest nuts in the world—added to high-fat ice cream.

With all these foods to entertain the palate so universally and easily available, there came a new way of eating called "Mass Quantities." The *Prevention* Index, a national survey of health habits and attitudes, reported in the early 1990s that a record number of Americans are overweight—just about two out of three adults are above the range of weight that's considered ideal for best health. Experts on such trends say we are the fattest people on earth, perhaps the fattest nation in history.

So—*yes*. If you, too, suspected there was something wrong with our dietary habits, you were right all along. The great majority of medical experts with an interest in nutrition agree with you. They point to heart disease, high blood pressure, adult diabetes, and even some forms of cancer as diseases produced or encouraged in large part by the modern diet.

But there is more to it than that. First, let's acknowledge that food has *always* been prob-

1

lematic for people. In the days of yore, the main problem was *"Where is it?"* Scarcity, in other words. Scarcity punctuated by terrible famines. The famines in turn so lowered the strength and immunity of populations that they were often followed by epidemics that were worse than the famines.

And food wasn't necessarily wonderful in the old days. Salted fish, hardtack biscuits, cornmeal mush, and plentiful quantities of porter, ale, and gin were all popular items on the menus of history. Without refrigeration and high-speed transportation, many communities had to get through winter and early spring on moldy spuds, salt pork, molasses, and similar fare that is not quite nouvelle cuisine. During those frosty months, fresh fruits and vegetables were just about nonexistent (at least in northern settlements), so millions suffered from periodic vitamin deficiencies for months.

So let's not romanticize the diet of our forefathers into a cornucopia when, in fact, that bounty only existed at harvest time. Nor should we fail to appreciate today's engineering and business infrastructure that allows us to eat tropical fruits, fresh fish, and good vegetables (except tomatoes!) year-round.

Granted, then, that the diet of yesteryear was not perfect. But what exactly were its strengths—the qualities we should be looking for today? Put another way, what is it—*precisely*—about modern fare (combined with modern habits of daily life) that makes it so easy for us to become fat, diabetic, and prone to heart disease?

To understand why we suffer from various health problems, we have to consider factors *other* than diet that may also contribute to poor health—such as lack of exercise, environmental pollutants, and habits like alcohol consump-

tion and smoking. But various health reformers have suggested that the biggest problems are with our daily diet.

Is it lack of fiber? Even a hundred years ago, voices were raised against the white-flour syndrome. Without "roughage," as fiber used to be called, our systems would suffer, these early reformers warned.

Others said that additives such as food coloring and certain preservatives were a major hazard. Indeed, quite a few nonbeneficial and nonessential additives were eventually banned, although many remain. Are they a significant cause of chronic disease?

Some said the mere act of cooking food destroys its vitality. They advocated eating lots of raw foods and sprouts to protect health.

The processing of food—canning, especially, but also freezing and even the sterilizing of milk—was attacked by some as health robbers.

Sugar is a favorite suspect, too. It's been charged with everything from cavities and diabetes to hyperactivity and hypoglycemia.

Cholesterol, and a little later, fat, became prime suspects in due course. At first they were only associated with cholesterol-clogged arteries, but more recently they were named as major contributing factors to obesity, diabetes, and even male impotence.

Then there's salt. Although salt has been heavily used as a preservative for centuries, it was extremely scarce, except as naturally found in certain foods, for most of mankind's earthly sojourn. Today, however, we use salt as a seasoning. The question is, does it cause high blood pressure and stroke?

So, when we look at today's diet, what is it that we *lack*?

Fiber?

Vitamins?

Minerals?

Elusive sparks of life that are snuffed out by food processing?

Scientists who have been looking into these questions for the past 20 years or so have gradually painted a fairly clear picture of the strengths and weaknesses of the contemporary diet. While there are still a few faint lines and a few dark corners, the observations that follow emerge as the main features.

What's Up with Our Daily Diet?

1. People today eat too much fat. Saturated fat—from animal-source foods like meat, eggs, and cheese—is almost certainly the most harmful kind of fat.

2. People today eat too much—period. Average fat intake seems to have declined a few percentage points in recent years— from about 42 percent of total calories to just 38 percent—but our weight problem hasn't gone down at all.

3. Quite possibly, the first two points have as much negative impact on our nutritional health as everything else put together. The combination of excess fat on our plates *and* under our belts is now acknowledged as a prime contribution to clogged arteries, angina, high blood pressure, adult-onset diabetes, and most likely breast and colon cancer. Being overweight also aggravates arthritis in the hips and knees. Many of these conditions also make it more difficult to get regular exercise—a requirement every bit as important as good diet.

4. More fresh fruits and vegetables would make a dual contribution to better health. First, they would partly replace high-fat foods. Second, they are *the* richest sources of beta-carotene (the form of vitamin A found in nonanimal foods), vitamin C, and folate—all vitamins that bolster the immune system against every invader from colds to cancer.

5. To help fill the place of high-fat foods like burgers, steak, roasts, cold cuts, and cheese, turn to cereal foods and beans. While these "starchy" foods were once (and wrongly) thought to be largely devoid of nutrition—and fattening—the modern view is just the opposite. They have not only B vitamins but also important minerals like magnesium, zinc, and potassium. They have iron, too, although it's in the plant form of nonheme iron—which means it's less bioavailable than heme iron from meat. (Meat is the most efficient source.) They also have a rich amount of fiber—which helps rid the body of cholesterol and encourages healthful elimination—and they have protein, which is something essentially missing from most fruits and vegetables. Generally, these foods are far less *fattening*. They do contain calories, although those calories are less well incorporated into body fat than calories from fat.

However, when we serve cereal foods and beans with butter, syrup, bacon, pork, and similar toppings, we really add the pounds. So the modern recommendation is to regularly eat a variety of grains like whole wheat, oats, rice, and barley, along with more beans that contain protein—navy, pinto, red, black, lentils, limas, and any others that strike your fancy.

6. While it is not necessary to totally avoid canned foods or foods with added salt or lots of sugar, it's best to eat them only occasionally. Canned foods often have a great deal of added salt, while the natural potassium has been lost. That change could be harmful to people with a history of high blood pressure and perhaps to others as well. As for sugar, the danger is when it supplants real food in the diet.

Remarkably little harm has been identified as coming from small to moderate amounts of sugar in most people, though it can be a partial risk factor in hyperlipidemia (high levels of lipids in the blood) and diabetes. Additives, too, are considered relatively harmless in the big picture today: They pose less risk than food poisoning, nutritional deficiences, food contamination, and naturally occurring toxicants. In truth, there is scant evidence of harm from food additives, except to individuals who may have reactions and intolerances. Allergies to seafood, citrus, peanuts, milk, and other perfectly good foods are far more common than true allergies to food additives.

7. Eating foods that have a good supply of nutrients can boost the body's ability to defend itself against disease. For the past 40 years, physicians and researchers have been studying the way nutrients that naturally occur in foods may reduce the risk of infectious disease and cancer. Research at the U.S. Department of Agriculture's (USDA) Human Nutrition Research Center on Aging at Tufts University suggests that better immunity might slow the decline in immunity that occurs with aging.

If we stand back from the big picture to get a panoramic view, what we see is a striking confirmation by scientific inquiry of our intuitive suspicions.

If we subtract from our diet a good, thick slice of what modern technology has produced and replace it with an equally big slice of *close-to-the-earth* eating, many of our most bothersome chronic health problems begin to retreat.

The tracks of this retreat will sometimes be clear. You may begin seeing belt holes, for instance, that you haven't seen in years. You may notice different numbers on your weight scale or on the meter of a blood pressure cuff. Other changes may not have numbers attached to them, but they will be just as clear—lots more energy and stamina, for instance. Your doctor may tell you that your cholesterol is going down, that your blood sugar is no longer as high as it was a few months ago, or that your blood is a bit less "sticky."

It can be something like taking a multi-wonder-drug formula. Only it's no drug, and there's nothing at all unnatural about it. Better health is the purely *natural* product of a better diet.

In the following section, we'll tell you in greater detail just what the elements of this better diet are, and—just as important—provide a nearly foolproof way of working them into your life.

What Is "Natural Food" Anyway? Don't Ask!

Don't ask because it's the wrong question! The real question is—*what is a natural diet?*

Here's why.

Coconut cream is a perfectly natural food—simply the juice of mashed coconut. The trouble is, it happens to be loaded with saturated fat—the same fat that's high in a T-bone steak.

Sugar is natural, too. So's honey. So are butter and cheese, steak and eggs. A popcorn can call itself "all-natural" and have so much added cheese, oil, and salt that it's worse than potato chips.

On the other hand, what about nonfat frozen yogurt? Technically, it consists of natural substances, but the way milk is processed to create strawberry nonfat frozen yogurt is not the kind of thing our ancestors did in their tents. Yet it's considerably healthier than a far more natural dairy product like cream.

Actually, our body doesn't care that much about the things we do to food before we eat it. It's what it *gets* that matters. And what it doesn't get.

There is now wide agreement that the natural *overall* diet of human beings—the kind we became adapted to over the ages—is a diet that is *low* in fat, cholesterol, and salt and *high* in fiber, vitamins, and minerals (at least when food was available!).

The important thing to look after is the whole daily diet—*the total of all those pluses and minuses.*

Now, it so happens—and this is no accident—that foods very close to the way nature made them *generally* have big health bonuses.

But when you take "natural" oil, sugar, honey, coconut solids, butterfat, and chocolate and put them together in an "all-natural" candy bar or other snack item, what you get is a natural mess—a mockery of everything the word "natural" ought to imply.

Moral: Don't get hung up on the "natural food" concept. Ask yourself if what you're eating is part of a *natural diet.* If it's full of oils and sweeteners, it isn't. An occasional treat? No problem. An everyday pick-me-up? Please—try to live without it!

Dietary Fat and Your Health

Two women.

They eat exactly the same number of calories.

They exercise the same amount of time.

They have a similar basal metabolic rate—in other words, their bodies use up calories at about the same rate when they're not exercising.

But . . . the first woman habitually eats a high-fat diet. Some 43 percent of her calories come in the form of fat—just a bit higher than the national average of about 38 percent. The second woman habitually eats a low-fat diet—29 percent of her calories come from fat.

The result? Even with the same calorie intake, the same exercise habits, and the same basal metabolic rate, the first woman will weigh an average of nearly 9 pounds more than the second. Comparing men, the difference is 11 pounds.

And chances are that extra weight put on by a fatty diet is going to accumulate largely—and we mean *largely*—around the *beltline* of Mr. and Mrs. "We'll-have-the-extra-crispy-please!"

That's what new research involving hundreds of men and women showed recently in work done at Laval University in Ste. Foy, Quebec.

What happened to calorie counting? Well, it's still there—for better or worse. But there's a new way of counting calories that explains why a diet of burgers and chips is more fattening than the same number of calories from fish and veggies.

When we eat fat, only *3 percent* of its calories are consumed by metabolic processes before it's stored for eventual use as body energy fuel (assuming that day comes!). But when we eat carbohydrate calories, *23 percent* are expended before the body can turn them into flab.

That can easily work out to a gain of 10 to 15 pounds over a period of time for a person eating a high-fat diet.

The fact that there's a tendency for a disproportionate amount of that "overage" to wind up on the abdomen and trunk (as opposed to arms and legs) does not bode well for your health. People who carry their extra weight in that region have a higher chance of developing heart disease.

That research explains why some of us—who may not be eating wildly excessive calories—have become pudgy from tilting the balance toward fat. But what if we put it to a real-life test? Can we actually lose weight we've already accumulated by lowering our fat intake? Or will we simply eat more to make up for the "missing" fat? Researchers at Cornell University and the University of Götenberg, Sweden, had the same question. So they took a group of women and put them on a diet. But not the kind of diet you're used to hearing about. The women could eat as much as they wanted to. Calorie counting was out the window. Instead, these women were first given typical American fare to eat, with about 35 to 40 percent of its calories coming from fat. They munched away for 11 weeks. Then came the big switch. Now, the food they were given to eat had just 20 to 25 percent of calories from fat. Once again, they could eat all they wanted. Eleven weeks later, on this "no-diet" diet, the women had lost an average of 5.5 pounds. The weight came off because the women failed to eat enough to replace the calories that had disappeared with the fat.

The moral is as plain as the nose on your face—or perhaps some larger bodily protuberance.

The Key Is Low-Fat Cuisine

As far as diet is concerned, eating low-fat cuisine is the natural path to weight control. It is better, easier, saner, more dependable, less anguishing, and yes, simply more natural than measuring and counting every morsel of food you put into your mouth.

Incidentally, the weight loss produced by low-fat cuisine—½ pound per week—is now regarded as the ideal rate at which to peel off poundage. Research suggests that as your rate of losing exceeds that mark, the chances that you'll put it all back go up sharply. (However, for people who are obese, their doctor may okay a weight-loss goal of 1 to 2 pounds per week.)

Actually, there's another benefit of low-fat eating we didn't mention above. Besides being the most effective way to lose weight, it's also the *healthiest*.

In 1991, the research arm of *Prevention* magazine, called Medical Consensus Surveys, conducted surveys of three groups of health experts to determine what they believed were the most important steps a person could take to protect his or her health. Leading cardiologists, cancer specialists, and clinical nutritionists were all asked to rate dozens of various measures and to decide—given their respective fields of interest—*which* measures were most important.

All three groups rated "reducing fat intake" as among the very highest priorities. Why this powerful consensus? Because:

- Excess fat raises serum cholesterol.

- Excess fat increases the risk of heart attack.

- Excess fat increases the risk of adult-type diabetes.

- Excess fat appears to promote several kinds of cancer, including cancer of the breast, colon, and prostate.

- Excess fat *makes* you fat. And that, in turn, promotes high blood pressure, back pain, gout, gallstones, fatigue, and disturbed sleep. What's more, it aggravates arthritis of the weight-bearing joints and altogether makes it harder to enjoy an active life.

The toxic effect excess fat has on the human system is probably greater than the sum total harm caused by pesticides, herbicides, food additives, radon, ozone, cosmic rays, and bad vibes. Organic cheesecake, in other words, is not where it's at.

Now, perhaps you're wondering—just *how much* fat is too much? And what kind of diet must I eat to be in the low-fat range, where I can begin losing weight without trying?

Later in this chapter you'll find a precise answer to these questions, based on counting the grams of fat you consume. But you needn't get that precise, unless you're so inclined.

Instead, you can simply cut back on certain commonly eaten items that are high in fat. By "cut back," we mean eat perhaps half as much as you do now (if you're fairly average). You needn't do that all at once, but work at it gradually, over weeks, even months.

Here's what to concentrate on cutting:

- Steaks and roasts

- Hamburgers and cheeseburgers

- Hot dogs, salami, boiled ham, and cold cuts

- Butter

- Whole milk

- Eggs

- Ice cream

- Rich desserts

Those aren't the *only* foods with a hefty fat content, but they are eaten in such quantity—by many people—that they wind up delivering the biggest fat load. Nuts are high in fat; so are fried potatoes, hard cheeses, even avocados. But most people don't eat them as regularly as they do burgers, ham sandwiches, and the other foods mentioned. Of course, if *you* do, that's another story. If you consume cheese cubes as a daily appetizer, for instance, or eat pizza four times a week, you'd better cut back in the cheese department. And if breakfast is always doughnuts or croissants, you'd better change that, too.

But instead of worrying about that chocolate turtle you ate last Saturday night, you're far better off concentrating your fat-trimming energy on foods you consume as staples—like ham and eggs with buttered toast.

Besides cutting back in easy stages, you can also explore the growing world of low-fat and nonfat foods. Many are surprisingly good—particularly if you give yourself a chance to get used to the slightly different taste. And there are some fats that are good for your heart—like the omega-3 fatty acids found in many fish. (For more on omega-3's and a list of good sources, see page 582.)

Most people know that saturated fat—found for the most part in animal-source foods like lard, pork, beef, eggs, and cream—is the worst kind of fat. Both the cardiologists and nutritionists who were polled by Medical Consensus Surveys gave reducing saturated fat a slightly higher priority than reducing total fat.

The good news (and there *is* no bad news!) is that when you cut back the way we've suggested, you will *automatically* be cutting back extra hard on saturated fat. That's because the very same foods that supply most of the *total* fat in the American diet also supply the most saturated fat. So by cutting back on meat and meat products—such as cold cuts, bacon, sausage, eggs, whole milk, butter, and rich desserts—you're doing yourself a double favor. You're slashing total fat, but especially saturated fat.

When you have succeeded in cutting the fat in your diet just about in half, or close to it, you may well have halted any ongoing process of coronary vessel blockage in your body. That is the suggestion from a key piece of research by scientists at the Atherosclerosis Research Center at the University of Southern California School of Medicine.

These researchers worked with patients who'd had a bypass operation, because it's known that arterial grafts have a tendency to simply reclose with cholesterol-laden plaque after a few years. Using new inner-body imaging techniques, they examined the progress

of any new atherosclerosis and then looked at these results in relation to the diet each individual had been eating. They discovered that once total fat levels dropped below 25 percent of calories, there was no new deposition of cholesterol in 95 percent of patients. The most important measurement they found, by the way, was not just saturated fat, but *total* fat. Eating more monounsaturated fat (as from olive, peanut, or canola oil) or more polyunsaturated fat (as from corn, safflower, or sunflower oil) was no saving grace. So try cutting *all* fats and let the saturated drop way back as a kind of collateral benefit.

Rabbit Food Is Not the Answer, Doc

Some people, quite understandably, jump to the conclusion that low-fat eating means lots of bunny food—carrots, radishes, lettuce, tomatoes, cucumbers, celery, and for gourmet bunnies, arugula and radicchio. But while these foods *are* just about fat-free, hopping down the bunny road will not lead you to low-fat eating.

That's because those foods have essentially no calories and no protein, two important nutritional requirements. So, having gnawed back a big pile of the stuff, you will *still* need to eat nearly a full day's worth of food to avoid falling over from weakness. And if what you eat to fill that big void is along the lines of cheese omelets, roast beef sandwiches, and milkshakes, you haven't done a thing to reduce your fat intake.

What ultimately matters is the portion of calories you consume that come from fat. The American Heart Association recommends that you get no more than 30 percent of your calories from fat. Other health leaders, including *Prevention,* suggest even lower intakes (no more than 25 percent of calories from fat). So to do the job you're after, you must increase your intake of foods that *do* have calories (and other nutrients) but *don't* have very much fat.

Such foods include grains, grain products (like rice, corn, barley, buckwheat, oatmeal, and breakfast cereals, except those with added fat like most granolas), beans, "hearty" hard squash (like acorn), turnips, and just about all fruits. Skim and low-fat dairy products serve the same role, supplying calories (and other vital nutrients) with very little fat.

A Beginner's Guide to Cutting Fat

Now that you know the importance of eating a low-fat diet, it's time to master an easy method to tell when your fat intake is within recommended guidelines.

The standard method is to get out your calculator and figure out the percentage of calories from fat.

- First, check the label for total grams of fat per serving.

- Multiply the grams of fat by 9.

- Divide that number by the total calories per serving.

- Then multiply by 100 to come up with the *percentage* of calories from fat.

But here's another, easier way to figure

out the fat limit in your diet: Just use the "Fat Goals" table on page 10. To use the table, find your weight in the first column. The second column shows the number of calories you're probably eating every day to maintain that weight. The third column is your magic number: This is the maximum number of grams of fat you should be eating each day to ensure that you're getting no more than 25 percent of your calories from fat and to maintain your current weight. Then just count the grams of fat you eat each day and make sure the total doesn't exceed your magic number. (When you're counting up, be sure to check the serving size on the label. If you eat more than a serving, you'll obviously get more grams of fat.)

Take the example of a 140-pound woman. The table indicates that to maintain her weight, she's probably eating about 1,700 calories every day. In order to keep her fat intake below 25 percent of calories from fat, she shouldn't eat more than 47 grams of fat per day. If she's trying to lose weight, she should aim for the fat limit of her goal weight. So, if she wants to reduce her weight to 120 pounds, she should bring her fat intake to about 40 grams of fat per day.

Keep in mind that these fat limits are approximate and that the chart is for sedentary people. If you exercise vigorously, you can afford a few more grams of fat (3 grams for every extra 100 calories you burn). The chart doesn't account for age, either, and metabolism slows down with age. So it's particularly important for older people to step up their exercise and cut back on fat calories.

This approach to counting fat has a way of encouraging some very healthy attitudes toward low-fat eating. It rightly implies that it's more important to limit total fat intake than to fret about particular foods that contribute to that total. People often try to forbid high-fat foods

Fat Goals

(for limiting dietary fat to 25 percent of total calories)

Weight (lb.)	Calories (daily)	Intake Limit of Fat (g daily)	Weight (lb.)	Calories (daily)	Intake Limit of Fat (g daily)
Women			**Men**		
110	1,300	37	130	1,800	51
120	1,400	40	140	2,000	54
130	1,600	43	150	2,100	58
140	1,700	47	160	2,200	62
150	1,800	50	170	2,400	66
160	1,900	53	180	2,500	70
170	2,000	57	190	2,700	74
180	2,200	60	200	2,800	78

entirely, but it's important to look at the total picture. If you're eating small amounts—an occasional pat of butter, a few slices of lean meat—it costs just a few grams of fat. But large amounts of high-fat foods can quickly push you over your fat limit.

Also, since high-fat fare can so quickly put you over your quota, this approach encourages you to eat more of the delicious low-fat alternatives to higher-fat fare. You'll end up eating more fruits, vegetables, whole grains, and other complex carbohydrates while keeping fat intake (and weight) down—which is a good definition of a healthy diet.

Savvy Strategies to Cut Fat from Your Servings

You *can* cut fat from your diet—a lot of it. It's mostly a matter of developing shopping habits and food-preparation strategies that make for leaner meals.

Start by rethinking your plate (and we don't mean updating your china). Realize that meat, which contains a hefty chunk of fat calories, needn't be the mainstay of every meal. Consider it one of several equally important components on the dish. Cut your usual meat portion by one-third to one-half, then increase your servings of vegetables and starches accordingly. Play up your leafy green or fresh fruit garnishes, too; they're the most neglected nutritional nuggets on the plate. So eat them instead of feeding them to the garbage disposal.

At the stove, lighten up your use of butter and oil. Rely instead on no-stick spray and low-sodium broth for sautéing. To heighten flavor, team your favorite veggies with aromatics such as onions, leeks, and garlic. Experiment with different herbs and vinegars, and leave the salt-shaker in the cupboard.

The point is that you don't have to make big changes. Simple modifications can transform almost any high-fat recipe into a winning

New and Improved: Meat Loaf

When you were growing up, was meat loaf one of the standbys at your house? If so, you probably got more than enough gravy, too—gravy on the meat loaf, gravy on the mashed potatoes, gravy for the biscuits to sop up. Don't forget the butter in the mashed potatoes and the butter on the green beans almondine you might have been served. And then there was the meat loaf itself: fatty ground beef, often with ground pork and veal and a few eggs for extra measure. One plateful could easily cost you 50 to 60 grams of fat. That's more than some people should have in an entire day.

Instead:

- Use a "meat" mixture consisting of one-third cooked brown rice, one-third skinless turkey breast, and one-third extra-lean beef or sirloin. Ask your butcher to trim the meat of all visible fat before grinding it.

- For every whole egg called for in the recipe you're following, use two egg whites.

- Add flavor (without salt) and give the loaf a boost of fiber and vitamins with an assortment of aromatic vegetables. To release their flavors for better blending, sauté them in a little stock before adding to the meat mixture. Use ½ cup of finely chopped celery, carrot, onion, garlic, fresh sage, and fresh thyme per 1-pound loaf.

- Bake the meat in a covered loaf pan (use aluminum foil) to keep it from drying out. Remove foil for last 5 minutes.

- Serve boldly flavored turnip puree as a lean and surprising alternative to butter-rich mashed potatoes. Simply steam the turnips until tender, then puree with a food mill or food processor. Season with fresh chives and nutmeg. For a second puree, do the same with a combination of carrots and leeks.

- To complement the tastes and textures of the purees, serve crisp-tender steamed green beans. Sprinkle with toasted pine nuts for extra crunch.

Original Meat Loaf Dinner
(Grams of Fat per Serving)

Meat loaf with gravy: 36.7 g

Mashed potatoes with gravy: 11.3 g

Green beans almondine: 8.6 g

Total fat: 56.6 g (63% of calories)

Improved Meat Loaf Dinner
(Grams of Fat per Serving)

Meat loaf: 3.6 g

Vegetable puree: 0.6 g

Green beans with pine nuts: 2.7 g

Total fat: 6.9 g (18% of calories)

meal. To prove how easy it is, we've updated four classics to meet our guidelines. Rather than give you strict recipes, we've decided to supply you with tips so you can update *your* favorite recipes, and in the process, learn some tricks that may come in handy when you're preparing almost any meal.

The nutrient values stated for each platter reflect our specific results, modifying typical recipes found in cookbook classics such as *Joy of Cooking* and *Betty Crocker's Cookbook.* Your results may vary, of course, depending on your original recipe and the modifications you make.

New and Improved: Pork Roast

Remember the Sunday pork roast? It was tender, juicy and just dripping with fat. In fact, the fat covering the traditional shoulder roast made basting unnecessary and kept the meat moist. The main course was often accompanied by potatoes that roasted in the same pan as the pork (and oh, how they soaked up the pork fat). Maybe there was a high-calorie side dish like broccoli in cheese sauce. We won't even mention the gravy and biscuits. No wonder you felt like Porky Pig by meal's end.

Instead:

● Choose a lean cut of meat, like tenderloin, with all visible fat removed. (A 2-pound trimmed tenderloin can serve eight.) To stretch the meat, stuff it with a savory filling. Ask your butcher to make a slit in the roast to accommodate the stuffing.

● To add flavor and help tenderize the meat, rub it with a no-fat paste consisting of garlic, chilies, fresh oregano, lime juice, cinnamon, cumin, and honey. Let the meat marinate overnight in the refrigerator.

● To create a fiber-rich stuffing, combine about a cup of cooked brown rice with a cup of chopped vegetables, including corn, onions, bell peppers, and a little minced garlic. (This is enough stuffing for a 2-pound roast.) Pack the stuffing into the pocket slit in the meat and tie the roast with kitchen string.

● To keep the meat moist, roast it in an oven cooking bag. Set the roast on a rack, then put the roast and rack into the bag. Add ½ cup stock. Follow the oven bag directions.

● Bake potatoes and serve them topped with tangy nonfat yogurt and a dash of lemon-herb seasoning.

● Round out the plate with crisp-tender steamed broccoli tossed with lightly sautéed garlic slices.

● Garnish with chilled orange and apple slices. They're a refreshing contrast to the spicy pork.

Original Pork Roast Dinner
(Grams of Fat per Serving)

Roast pork shoulder: 21.3 g

Sautéed potatoes: 7.0 g

Broccoli with cheese sauce: 2.4 g

Total fat: 30.7 g (47% of calories)

Improved Pork Roast Dinner
(Grams of Fat per Serving)

Stuffed pork loin: 3.2 g

Baked potato with nonfat yogurt: 1.1 g

Sautéed broccoli: 0.2 g

Total fat: 4.5 g (9% of calories)

New and Improved: Fish Sticks

Was Friday fish day at your home? And did that mean fish sticks on the menu? Sure, they tasted great. But with nearly 14 grams of fat in just four of the skinny sticks, they gave fish a bad name. And we won't even *talk* about the gobs of tartar sauce that you smothered them with! Don't forget the lackluster (canned) stewed tomatoes and mayo-saturated cole-slaw that usually shared the plate.

Instead:

- Opt for super-lean flounder fillets. (One pound serves four.)
- To get that characteristic fish stick shape, cut each fillet in half lengthwise along the middle seam.
- Lightly dust fish with flour or cornmeal, then dip into lightly beaten egg whites. Finish off with whole wheat bread crumbs.
- Place the fish sticks on a cookie sheet coated with no-stick spray.
- Make sure the fish sticks don't touch. Bake them in a moderate oven, flipping them once, until brown on both sides, about 10 minutes.
- To make a lean tartar sauce, add some chopped sweet pickles or sweet-pickle relish, parsley, shallots, and mustard to a small container of nonfat yogurt.
- To boost the nutrition of the coleslaw, shred carrots and scallions along with the cabbage. Then dress it with your favorite vinaigrette, using just a tablespoon of oil in the recipe.
- Sprinkle fresh tomato halves with fresh herbs and a little grated Parmesan cheese. Broil for just a few minutes, until warmed through. Garnish with crunchy, nutrient-dense curly kale.

Original Fish Sticks Dinner
(Grams of Fat per Serving)

Fish sticks: 13.7 g
Tartar sauce: 8.0 g
Coleslaw: 8.7 g
Stewed tomatoes: 0.2 g
Total fat: 30.6 g (54% of calories)

Improved Fish Sticks Dinner
(Grams of Fat per Serving)

Breaded fish: 2.7 g
Lean tartar sauce: 0
Skinny coleslaw: 3.5 g
Broiled tomatoes: 0.4 g
Total fat: 6.6 g (20% of calories)

New and Improved: Steak and Fries

Times have changed since steak and fries had their heyday. Back then meat portions were large—anywhere from 12 to 18 ounces—and no one seemed to care how much fat that represented. Nor did they think twice about the heaping plates of french fries and deep-fried onion rings that kept the steak company.

Instead:

● Reduce meat servings to about 4 ounces. And choose a cut with a minimum of fat marbling, such as sirloin. Then trim all visible fat.

● To help tenderize the meat, marinate it with coarse mustard and vinegar seasoned with low-sodium soy sauce and black pepper. Let stand for about 4 hours.

● Quick-broil to keep the lean steak moist and tender.

● For extra tenderness, slice thinly across the grain.

● Add variety to your meal with snap peas steamed just long enough to make them slightly tender.

● Make low-fat oven fries by slicing baking potatoes into 6 to 8 wedges each. *Lightly* coat with olive oil. Arrange fries skin-side down on a cookie sheet. Roast them in a hot oven until tender and brown, about 30 minutes.

● Sprinkle your fries with tangy malt vinegar in place of salt.

● To make frizzled onions, heat a well-seasoned cast-iron or no-stick frying pan until hot. Add a few drops of oil, then toss in thinly sliced onion rings. Sauté on high heat, stirring, until the rings have browned nicely, about 5 minutes.

Original Steak and Fries Dinner
(Grams of Fat per Serving)

Sirloin steak: 26.1 g
French fries: 7.0 g
Fried onion rings: 18.7 g
Total fat: 51.8 g (55% of calories)

Improved Steak and Fries Dinner
(Grams of Fat per Serving)

Trimmed sirloin steak: 6.1 g
Steamed snap peas: 0.2 g
Oven fries: 1.3 g
Frizzled onions: 1.3 g
Total fat: 8.9 g (23% of calories)

Cholesterol: The Stuff of Which Heart Attacks Are Made

If you picked up a biology textbook, you might be surprised to read that cholesterol is an essential body substance. Your body uses this white, waxy, fatlike substance to manufacture hormones, digestive juices, and nerve endings, among other tasks. All animals, including humans, need cholesterol to survive.

So if cholesterol is so wonderful, why the big fuss over eating too much? Shouldn't we be taking cholesterol *supplements*? Or at least seconds on cheesecake, under doctor's orders?

Fat chance. Here's the story.

Cholesterol is so essential that your liver manufactures just enough to meet your daily needs. So even if you never ate another egg, steak, or ice cream cone for the rest of your life, you'd never run short of cholesterol. Any cholesterol you eat adds to your overall cholesterol level. That's where the trouble begins.

For more than a century, doctors have known that surplus cholesterol tends to accumulate in the coronary arteries. As cholesterol builds up, it gets tangled up with fibrous tissue and debris from the artery walls, forming plaque. If enough plaque forms, the arteries narrow to the point where blood can't reach the heart. The result? A heart attack.

The good news is that, with a few simple dietary changes, you can cut serum cholesterol and reduce your risk of heart attack by as much as 50 percent. The basic strategy: Eat more of the foods that can help lower cholesterol and less of the ones that can raise it.

Whenever you have to make a choice between saturated and unsaturated fats—go for the unsaturated. Doctors say cutting saturated fat is *more* effective in reducing blood levels of cholesterol than cutting back on dietary cholesterol. When doctors advise people to reduce saturated fat in favor of monounsaturates (like olive oil) and polyunsaturates (like corn oil), they're thinking about cholesterol as well as weight-gain considerations. Food rich in saturated fat interferes with the liver's ability to remove cholesterol from the bloodstream.

As for unsaturated fat, both monos and polys lower your total cholesterol and raise your HDL (a beneficial component of blood cholesterol). Guidelines issued by the American Heart Association (AHA) and the National Cholesterol Education Project (NCEP) recommend that we get 10 percent of our total calorie-from-fat intake from monounsaturates, 10 percent from polyunsaturates, and less than 10 percent from saturated fat.

Eat more soluble fiber. Water-soluble fiber prevalent in oat bran, oatmeal, citrus fruit, and most types of beans seems to help clean out cholesterol. Some scientists theorize that soluble fiber binds with bile, a digestive fluid that contains significant amounts of cholesterol, in a way that causes the bile to be excreted in the stool. The result: Less cholesterol circulating in the bloodstream where it can cause trouble.

Eat less cholesterol. On the average, most Americans consume about 500 milligrams of cholesterol a day, none of which they actually need. Other factors, such as smoking and elevated blood pressure, also contribute to a buildup of arterial plaque. But in general, your heart will be better off if you take steps to avoid a cholesterol pileup. In fact, among the 200 doctors polled in the Medical Consensus Survey mentioned earlier, reducing dietary cholesterol ranked just behind cutting fat in the fight against heart disease.

Specifically, the physicians advise people to reduce dietary cholesterol to 300 milligrams per day. This is the easy part. Many of the strategies discussed in this section for slashing total dietary fat and increasing your intake of fiber will automatically eliminate major sources of cholesterol, such as eggs, butter, whole milk, and fatty meats.

Keep an eye on magnesium. Magnesium is sometimes called the "forgotten" mineral because few people have severe magnesium deficiency. But many of us get *somewhat* less magnesium than we should. And that's unfortunate, because magnesium may help control cholesterol.

In studies of areas where there are high concentrations of magnesium in the water, it's been found that people have a lower buildup of plaque in their arteries. Animal studies support the high magnesium/low cholesterol link. Research suggests that there may be a lower incidence of heart disease and fewer problems with high blood pressure in people who have adequate magnesium in their diets. (Research also suggests magnesium may help a few people with arrhythmia—irregular heartbeat.)

As you'll see in the food entries, a varied diet rich in fruit, vegetables, whole grain cereals, legumes, and unprocessed flour is all you need to reach the RDA in magnesium. However, processing of foods, especially flour, can result in magnesium loss of 80 to 90 percent. (For a list of the best sources of magnesium, see page 551.)

Watch the labels. One hitch to watch out for is food labeling—specifically, products that scream "no cholesterol." Many base that claim on the fact that they contain no animal fat, the only source of cholesterol. And it's true—animals alone produce cholesterol; plants and plant oils contain none whatsoever. But *some* plant oils—like coconut oil and palm oil—are highly saturated and therefore raise cholesterol levels. So "no cholesterol" labels can be very deceiving—sort of like the fake tunnels Road Runner would paint for Wile E. Coyote. Sure, they look like they're on the up-and-up. But in reality, they can raise your cholesterol faster than you can say, "Foiled again."

"The crackers you buy may have no cholesterol, but they're made with coconut oil—all saturated fat," says William Castelli, M.D., director of the landmark Framingham Heart Study. Wily consumers can try to escape this booby trap by examining labels closely before they proceed to compare the saturated and unsaturated fat content. But many product labels don't break down the fat content. So in the course of analyzing the 1,000 foods in this book, we've designated foods in which unsaturated fats predominate, and which are also low in cholesterol or low in total fat, with a symbol indicating that they lower cholesterol. (See page 32.)

The low cholesterol symbol is a convenient shortcut to finding foods that qualify as heart-healthy.

Of course, food labels don't get heart attacks, people do. An epidemiological study (that is, a look at who gets sick and why) found that the more cholesterol people consume, the higher their risk of heart attack, regardless of their blood cholesterol levels, blood pressure, and smoking habits. Nevertheless, it's essential to get your blood levels of cholesterol checked to make sure the stuff isn't accumulating at excessive levels.

What's excessive? Here's where doctors disagree. Half the doctors polled in the survey mentioned earlier seem to think the standard set by the National Cholesterol Education Project, which recommends a total blood cholesterol level of less than 200 mg/dL (milligrams of cholesterol per deciliter of blood) is still too high. Almost 30 percent said you should shoot for under 180 mg/dL.

"About four billion people on this earth have a cholesterol level of 150, and they don't get cardiovascular disease," says Dr. Castelli. "If we pushed our cholesterol more toward this level, we could share in some of the good fortune of those four billion people."

So go ahead and take the test. (Maybe you're one of the lucky four billion!) And while you're at it, have your blood checked for HDL and LDL cholesterol and triglycerides. HDL lipoproteins act as an efficient refuse hauler, removing cholesterol from body cells and delivering it safely to the liver for disposal. In contrast, LDL lipoproteins are like HDLs' clumsy helpers. They can't grab as much cholesterol, and on the way to the liver, they tend to drop cholesterol like litterbugs.

People without heart disease should shoot for an LDL level of less than 130; folks with heart disease, less than 100. As for HDL, a level of 35 mg/dL is considered desirable.

How often should you check out your cholesterol? It depends. If your total cholesterol is under 200 (with normal HDL), and you're free of risk factors like high blood pressure and smoking, every five years is about right. But if your total falls between 200 and 240 and you have other risk factors or already show signs of heart disease, yearly checks are in order. And if your cholesterol is higher than 240, your doctor will probably want to test you more than once a year, as well as taking some action to lower your risks.

Even if your cholesterol turns out to be high, it's not too late to take corrective action. Heart-healthy food choices can keep blocked arteries from getting worse, especially if you take other positive steps such as losing excess weight and controlling your blood pressure.

The Power of Antioxidants: Vitamins That Protect Your Arteries

While the number of people dying of heart attacks has dropped almost 30 percent since doctors started monitoring cholesterol levels, one thing has remained a puzzle: Many of the people who have heart attacks *don't* have outrageously inflated cholesterol levels (over 300 mg/dL). So researchers suspect that there's more to heart disease than cholesterol alone. And this "something more" may involve vitamins C and E and beta-carotene, found in fruits, whole grains, and vegetables.

Mounting scientific evidence suggests the possibility that these vitamins act as "antioxidants." That is, they counteract what scientists call free radicals—naturally occurring substances that damage (oxidize) body tissues and blood fats.

Free radicals' effect on LDL (the harmful form of cholesterol) is similar to what happens to a steak when it sits out on the kitchen counter too long—it goes bad. That, researchers speculate, sets off a deadly process by which blood cells bloated with LDL threaten to completely block an artery.

Enter the antioxidants—vitamins and enzymes in our bodies that stem damage caused by free radicals. Scientists believe that antioxidants may be able to stop free radicals from making LDL go bad.

The protective effect of antioxidants was first observed in laboratory rabbits studied by Daniel Steinberg, M.D., professor of medicine at the University of California, San Diego. Previously, researchers had noted that heart attack victims have consistently low levels of vitamin C in their blood, as do smokers, who are at high risk for atherosclerosis. More recently, researchers found significantly lower vitamin E levels in men with angina (chest pain that signals heart disease) than in those who were symptom-free—regardless of other risk factors, such as cigarette smoking, high cholesterol, high blood pressure, and excess weight.

Scientists are now looking into whether vitamin C and other antioxidants in our food can prevent free-radical–induced heart disease in people. One researcher, for example, is exploring whether diabetic patients' blood cholesterol is, for some genetic reason, even more susceptible to attack by free radicals. That would explain why people with diabetes are at higher risk for heart disease.

Other studies are looking at people with relatively low or normal cholesterol levels who develop heart disease or have heart attacks. They may oxidize LDL differently than other people. Another provocative theory suggests that certain compounds in olive oil and canola oil that are not part of the fat itself may also work as antioxidants, independent of their recognized ability to lower cholesterol.

Antioxidants ''might give us new ways to deal with heart disease, over and above lowering cholesterol,'' says Dr. Steinberg, the brains behind the free-radical theory of heart disease.

So when it comes to heart health, vitamins C and E, beta-carotene, and certain vegetable oils may be up-and-coming superheroes in the crusade to prevent heart disease. If that turns out to be the case, increasing your intake of fruits, whole grains, and vegetables and replacing saturated fat with monounsaturates may be just as important as decreasing dietary fat and cholesterol as a way of staying heart-healthy.

Fiber and the Fabric of Health

Food fiber used to be regarded as essentially worthless. Mostly unabsorbable, even after the digestive system did its work, fiber seemed incapable of providing nutritional benefits. Until about 20 years ago, in fact, if you heard *anything* about fiber, it was usually a warning not to eat it if you had a tendency toward intestinal discomfort. Fiber, it was thought, irritated the sensitive colon. Other than that, there wasn't much to say.

Today, we have an entirely new understanding of the importance of fiber. From being a virtual nonentity, fiber has in one generation leaped to the forefront of nutritional priorities.

One way to look at fiber's newly earned role is to simply consider that when all that fat we talked about before slid into our diets, it was fiber—more than anything else—that gave way. For thousands of years, cuisine in almost every part of the world was not only low in fat but also rich in fiber. All the sustenance grains—wheat, rice, corn, millet, barley, oats, and rye—were high-fiber foods. Another staple was beans—from favas and chick-peas to lentils and kidney beans. They're even richer in fiber. So are many vegetables and fruits, seeds, and nuts.

First, white bread came into fashion, slicing out 60 percent of the fiber from whole wheat. Then onto the bread went butter, mayo, ham, roast beef, hamburger, bacon, salami, and cheese—all entirely devoid of fiber (found only in vegetable foods).

While the fat in all those foods was doing *its* work, fiber faded. And yes, it *did* have work—though we've just discovered what it does. Fiber, it turns out:

- Reduces the absorption of calories you ingest.

- Reduces your cholesterol count.

- Prevents chronic constipation and even chronic diarrhea in those so afflicted.

- Helps prevent colon cancer—the second most common killer cancer in America.

- May even help prevent a common form of kidney stones—those consisting largely of calcium.

By controlling your weight and cholesterol and protecting against colon cancer, fiber is acting directly against the influence of excess fat. Is it any wonder these three problems are so common, with so much more fat and less fiber than is natural in our diet?

The average American today is eating about 10 to 15 grams of food fiber—about half an ounce. Current recommendations are to eat just about twice that much—20 to 35 grams. It's interesting that our fat intake needs to be cut in half while our fiber intake should be doubled. Those two steps alone would go a long way to improving America's metabolism—and health.

You may be wondering just how it is that fiber has a metabolic effect if it isn't even absorbed through the intestinal wall for the most part. The answer is that a lot of action goes on in the digestive tract. Most simply put, certain forms of fiber seem to absorb or otherwise "grab hold" of cholesterol and fat and usher them out of the body before they can cause trouble. More specifically, fiber can bind up bile acids in the intestine and carry them out before they can return to the liver and be recycled into cholesterol.

Now, not all kinds of fiber do this particularly well. The best kind is fiber from beans, oats, barley, carrots, apples, prunes, and figs. Bran products made from oats, soy, and corn are also effective. These foods contain an abundance of *soluble* fiber—which sucks up the bad guys. Wheat and wheat bran are high mainly in *insoluble* fiber—great for regularity but not for putting a headlock on cholesterol.

All the experts agree that your best bet is to get your fiber from a variety of sources—all sorts of grains, vegetables, and fruits.

The idea that fiber can help protect against cancer of the colon may strike you as even more surprising than its protection against cholesterol. But the recommendation for getting 20 to 35 grams of fiber per day comes, in fact, from the National Cancer Institute. And in a recent Medical Consensus Survey by *Prevention* magazine, cancer experts said that eating more fiber was just behind eating less fat when it comes to nutritional priorities.

The current theory is that fiber either dilutes cancer-causing chemicals in the colon, ushers them out before they can have prolonged contact with the colon wall, or both. In any event, research now suggests that a diet high in fiber may prevent as many as 40 percent of all colon cancers. That translates into many thousands of lives saved each year.

But how do you get those 20 or more grams of fiber?

To be perfectly honest, it isn't easy. Many of us just don't eat enough food—because we don't do physical work—to get enough fiber without careful planning. Others find that beans and certain vegetables cause them to be gassy, so they simply avoid those foods. Others *think* they're getting fiber because they eat a salad every night featuring iceberg lettuce and a few chunks of tomato, radish, and cucumber. In truth, there's hardly any fiber at all in such a salad.

The National Research Council suggests eating six daily servings of various grain products (a piece of bread or the equivalent counts as one serving) plus five or more servings of fruits and vegetables.

To that we add these quick suggestions:

- Eat a high-fiber cereal for breakfast—one that has 5 or more grams of fiber. All-Bran has 10.

● Go for whole grain products whenever possible. Pumpernickel, by the way, is not whole grain; oatmeal is.

● Carrots are an especially low-calorie source of fiber (with a big beta-carotene bonus). Prunes and figs are higher in calories but have excellent nutritional value as well as fiber.

● To get on friendly terms with beans and other fibrous but problematic foods like cauliflower and broccoli, try using an enzyme product (like Beano) that "neutralizes" the gassiness in your stomach. We've found them quite effective.

Two cautions: Don't overdo fiber until you're used to it, and don't use fiber pills. Some of the pills begin swelling up while they're still in your throat, and a serious blockage can result.

If you want a fiber supplement to your regular foods, go for high-fiber cereal or try flaky or granulated fiber supplements like wheat bran, psyllium seed, or soy fiber. Begin with small amounts that gradually increase so your system can adjust. (As you increase fiber intake, it is important to increase your fluid intake.) You may wind up getting 10 to 15 grams of fiber from such supplements and an equal amount from your ordinary food. (For sources of fiber, see page 573.)

Potassium—A Hedge against Stroke and Circulatory Disease

Can you guess the "star player" in the following meal?

For dinner, you eat a savory Indian-style meal of wok-cooked cauliflower, potatoes, peas, okra, and chicken. This is served over a mound of steamed millet. And for dessert, you eat an orange.

Now, all those foods are high up on the *Prevention* scale of low-fat, high-fiber, vitamin-rich nutrition. But the hidden star of that culinary show is a mineral called potassium. Evidence shows that if you continue to eat that kind of meal night after night, you will have a strong shield against one of the worst enemies of health.

You don't hear much about potassium. Although it's essential to life, the lack of it isn't linked to common deficiency diseases, such as anemia is from not getting enough iron. And there is more than enough potassium in lots of foods to avoid a critical deficiency—which could lead to weakness, confusion, and even death from a wildly irregular heartbeat. If a problem does occur, it will likely be from taking certain diuretics for high blood pressure that may drain the mineral from your body or from a serious metabolic disturbance such as acute diarrhea.

The new perspective on potassium reveals another possibility. Though we may be getting enough of this mineral to easily avoid dying of potassium deficiency, many of us are not getting enough to avoid the lightning bolt of a stroke. Or high blood pressure. Or heart disease. Yes, all those common problems have been linked to a shortfall of potassium.

In 1987, an intriguing study was published in the *New England Journal of Medicine*, suggesting that if a person were to eat just a single additional serving of fresh vegetables or fruit a day, he or she might enjoy a 40 percent reduction in the risk of having a fatal stroke.

That study, which followed 859 people in

a California community for 12 years, looked at many other factors that could be involved in strokes. Smoking, for instance, just about doubles the risk. Calcium and magnesium, both of which may also help the circulatory system, were also looked at. Because potassium is found largely in foods with little fat and cholesterol, these components were also analyzed to make sure it wasn't the absence of them, and not the presence of potassium, that was responsible for different health outcomes. But when all the statistics were worked through, potassium emerged as the antistroke powerhouse, independent of all other factors.

Before this study, it had already been shown that potassium can—at least in some cases—lower blood pressure. Yet the protection afforded against strokes in this study was much higher than what would be expected from the slight fall in blood pressure produced by eating a single extra piece of fruit. There may be a protective effect from potassium that is quite independent of blood pressure, suggest the authors, Kay-Tee Khaw, M.D., and Elizabeth Barrett-Connor, M.D.

A more direct approach to this question was taken by Louis Tobian, M.D., of the University of Minnesota. Working with rats bred to develop high blood pressure, Dr. Tobian loaded some with salt (to make their problem worse) while others got the same salt treatment plus a high-potassium diet. Sure enough, the high-potassium animals had a lower mortality rate.

In another experiment, he loaded up lab rats not just with extra salt but also with a whopping dose of saturated fat and cholesterol. Again, one group was given a high-potassium diet. Result? The potassium-plus rats had 64 percent fewer cholesterol deposits on their artery walls than the animals getting only normal amounts.

Getting more potassium into your diet isn't difficult if you know where to look. Luckily, these are the same foods that are good for many other reasons. (For sources of potassium, see page 552.)

The Scoop on Salt Restraint

If you put a seat belt on when driving your car, you should put a dietary "seat belt" on your saltshaker as well. For it's likely, new research asserts, that restraining our use of salt could save even more lives than using seat restraints when driving.

In England, where the research originates, it's estimated that this simple act could save some 75,000 lives per year. In America, it could save hundreds of thousands every year. Spared, too, would be much of the enormous suffering and expense caused by stroke and heart attack.

This connection between salt, high blood pressure, and arterial disease is certainly no secret. But the new work suggests that the beneficial results of reducing our salt intake seem to be far greater than has been generally believed. The researchers, all from the Medical College of St. Bartholomew's Hospital in London, declare that "Few measures in preventive medicine are as simple and economical and yet can achieve so much."

Let's back up a few steps.

What's been known for years is that people in societies where very little salt is consumed rarely develop high blood pressure, and their pressure does not significantly rise with advancing age, as is common in our society.

Residents of Framingham, Massachusetts, for instance, were found to have an average blood pressure of 137/88, with an average salt consumption of about 9 to 10 grams per day— a typical amount for Americans.

In contrast, native communities in remote areas of South America and the Pacific Basin consume only about 1 to 3 grams of salt per day. And typical blood pressure readings there (for people in their forties, as in Framingham) are anywhere from 103/65 to 120/73.

The confusion arises when people in the same society are put on low-salt diets to see if, in fact, a significant blood pressure drop results. Often, it doesn't seem to.

But the English researchers examined some 14 of these studies and determined that many of those that failed to find the salt-to-pressure connection were incorrectly designed. Some used too few subjects; others didn't test the low-salt diet long enough (it generally takes five weeks or more to kick in).

Their own conclusion is that while some people are more sensitive to salt than others, when you look at whole populations, you find vast numbers who respond to a salt reduction.

If people in their fifties, for example, cut salt intake by about one-third (or about 3 grams), we could expect to see systolic pressure (the first number in a reading) reduced by an average of five points, with a seven-point drop in those with very high blood pressure (170 or over). Diastolic pressure (the second number) averages a decrease of half that much.

This reduction in salt intake is just about what you get from "avoiding salty foods and not adding salt in cooking or at the table," say British researchers Dr. M. R. Law and his colleagues.

You can, of course, go further than that.

You can avoid eating processed foods that have added salt. That extra step demands some label-reading savvy and probably lots of experimentation with salt substitutes like herbs and spices. But the result would be twice the reduction in salt achieved just by nailing the saltshaker to your pantry shelf.

Let's call the first strategy Plan A. If all Britons ate like that (significantly lowering their salt intake) say the authors, some 40,000 deaths a year from stroke and heart disease could be prevented. With Plan B (not salting food, avoiding salty-tasting food, and avoiding salty processed food): 75,000 lives. Since America's population is about 4.5 times that of England, we could project saving some 180,000 lives a year with Plan A and 337,500 with Plan B.

In reality, we can all work toward at least some of both plans. Use salt a lot less at home and check labels carefully to see if there's significant salt present. Look for sodium listings on those labels: If you follow plan A, you're likely to consume about 6,000 milligrams a day, and if you follow plan B, you're going to get half that much.

Since deli products don't have labels at all, it's only fair to caution you that darn near everything in that department is heavy-up in salt: cold cuts, salads, dips, and all hard cheeses except the new low-sodium variety.

Is the proffered benefit of the low-sodium lifestyle really as great as the British team suggests? Dr. Castelli tells us the issue is a hot one in medical circles. But, he concludes, "They may well be right. Even if their conclusions are somewhat exaggerated, if we *did* take salt out of our diet, we'd see a dramatic fall in blood pressure, which would lead to an even more dramatic fall in cardiovascular disease."

A Surprising Source of Sodium

If you're ready to launch an assault on sodium, an obvious place to start may be the first meal of the day, breakfast. A not-so-obvious food to watch out for may be your old breakfast standby, cereal.

"Cereal?" you ask, "But that's good for me! Grains, fiber, vitamins and minerals, low fat, low sugar—it's got it all!" And then some. But—read the label carefully, and you're liable to find your favorite boxed cereal contains more sodium per one-ounce serving than an identical amount of potato chips. We did some label reading of our own to see how some brands of cereals compare with a few typical snack foods. Here's what we discovered:

Snack Food	Sodium Content (mg)
Corn chips	164
Tortilla chips	155
Potato chips	125

Cereal	Sodium Content (mg)
Cheerios	290
Corn Flakes	290
Rice Krispies	290
Wheaties	200

While we're not suggesting that you eliminate boxed cereal from your diet (or that you eat potato chips for breakfast!), these figures do make a point: Sodium can hide where you'd least expect it—even in foods that don't taste salty. So when you're watching sodium, it's important to know what you're eating and how it fits into your total daily intake. Read those labels! That way, you'll avoid the "salt traps" that may stand between you and good health.

Calcium: The Bone Builder . . . *Plus*

"Drink plenty of milk—so you'll have strong teeth and bones."

What child hasn't heard that refrain from a parent or school nutritionist? Adults aren't always right, but in this instance, their suggestion is close to the mark. Of course, it isn't the milk itself that builds sturdier molars and tibias: It's the bone-building mineral *in* the milk—namely, calcium.

What many adults don't realize, however, is that the need for plenty of calcium in the diet doesn't end with childhood. Studies have shown that bone tissue in both children and adults is constantly "turning over." That is, minerals like calcium are continually being deposited and lost.

During childhood and adolescence, deposits exceed withdrawals as new bone is added to the skeleton. The result: heavier, denser bones. That goes on until somewhere between the ages of 25 and 35, when your bones attain what experts call "peak bone mass." You'll never have more bone than this for the rest of your life.

That being the case—why keep an eye on your calcium intake if you're an over-35 adult?

Bone Maintenance

One important reason why most doctors recommend that adults continue to get plenty of calcium is to ward off the bone-weakening effects of osteoporosis.

Osteoporosis is an exaggerated loss of bone tissue that usually begins to occur in women after menopause, when the body produces much less estrogen (a hormone that helps protect bones). This stepped-up rate of bone loss may also occur in men, but it usually occurs about ten years later than in women. Needless to say, the weaker your bones, the greater your risk of bone fractures. In fact, osteoporosis alone is responsible for 1.3 million broken bones each year—most commonly in the wrist, hip, and spinal vertebrae.

The good news is that it's never too late to take preventive steps. "Osteoporosis need not be an inevitable part of aging," according to William A. Peck, M.D., founding president of the National Osteoporosis Foundation. Many experts now believe that the best nutritional way to lower risk of osteoporosis in later life is to ensure the adequate intake of calcium when the bone's peak mass is being accumulated—that is, from childhood up to the age of 35. But there's also strong evidence that calcium helps prevent fractures. (And that it might assist in preventing bone loss after menopause as well.)

But how much is enough?

It's been estimated that one-third of middle-aged and elderly women consume less than 400 milligrams of calcium per day—far short of their needs. According to Recommended Dietary Allowance (RDA) figures, men and women of average height and weight need 1,200 milligrams per day until they're 24 years of age, and at least 800 milligrams per day if they're 25 or older. So a middle-aged man or woman who gets only 400 milligrams of calcium is taking in *half* of the RDA.

For women who are pregnant or nursing, calcium needs are significantly higher—in the range of 1,200 milligrams per day. The National Institutes of Health (NIH) Consensus Development Conference recommends 1,000 milligrams for postmenopausal women who are on estrogen, and 1,500 milligrams for those who aren't.

To help your body *use* calcium, you also need vitamin D—a nutrient added to most milk and to some cereals in the United States. Substantial amounts of vitamin D are also found in oilier fish, such as mackerel, salmon, and herring. And the body can make its own vitamin D when it's exposed to sunlight.

Since vitamin D is such an important nutrient for calcium absorption, doctors recommend that adults drink 2 cups of skim or low-fat milk every day. This is enough to provide you with sufficient vitamin D to meet the adult RDA of 200 international units (I.U.). Other alternatives are to spend some time in the sun each day to make sure you get adequate vitamin D—or to take a vitamin supplement containing 200 to 400 I.U. of vitamin D. (But don't take more than 400 I.U. per day from a supplement because high levels can be toxic.)

Finally, there's a third component that can help increase bone strength—the mineral boron. Researchers at the USDAs Human Nutrition Research Center in North Dakota found that boron may prevent you from losing too much calcium. That means it's important to have boron in your diet—which you can get from apples, pears, grapes, nuts, and leafy vegetables. But beware of boron supplements, as excessive amounts of boron might be toxic.

The Colon Cancer Link

Calcium may be of benefit in other important ways. Studies at Memorial Sloan-Kettering Cancer Center and other research institutions suggest that calcium may reduce the risk of colorectal cancer.

According to Martin Lipkin, M.D., head of the Irving Weinstein Laboratory for Gastrointestinal Cancer Prevention research at Memorial Sloan-Kettering Cancer Center, calcium may inhibit damage to the bowel lining. "Like fiber," says Dr. Lipkin, "calcium seems to bind bile acids, which prevents them from irritating the colon wall."

Although further studies are needed, it's wise to ensure that your calcium intake meets the RDA. One caution, however: If you're prone to developing kidney stones, which are mostly composed of excessive calcium salts in the urine, you may have to limit your calcium intake. So it's a good idea to check with your physician before beginning a *high-calcium* regime that exceeds the RDA intake—particularly if you have a family history of kidney stones.

Going to the Sources

But what about the three glasses of milk a day that the school nutritionist used to recommend?

Basically, it's sound advice. Milk *is* a good source of calcium—and it's also vitamin D–fortified to help your body *use* that calcium. So go ahead and drink three glasses each day. But most doctors would add a few additional suggestions:

● Drink skim milk, rather than whole, so you don't get too much fat in your diet.

● Drink those glasses of milk throughout the day. Studies show that you can absorb twice as much calcium from four 250-milligram calcium doses as from one 1,000-milligram dose.

● Look for other ways to get calcium in addition to milk. In general, many dairy products are high in calcium, but be sure to look for reduced-fat or low-fat brands. And if you're lactose intolerant, look for low-lactose products such as yogurt. Green vegetables such as broccoli, collard greens, and kale can also contribute useful amounts of calcium. For some of the other top food sources of calcium, be sure to check the table on page 549.

● Calcium supplements are sometimes recommended. If you're taking a calcium supplement, it's wise to drink lots of water—about six to eight 8-ounce glasses a day. For maximum benefit, divide the supplements throughout the day and take them with meals.

RECOMMENDED DIETARY ALLOWANCES

Vitamins

Age (years) or Condition	Weight (lb)	Height (in)	Protein (g)	Vitamin A (mcg RE)	Thiamine (mg)	Riboflavin (mg)	Niacin (mg)	Folate (mcg)
Infants								
0–0.5	13	24	13	375	0.3	0.4	5	25
0.6–1.0	20	28	14	375	0.4	0.5	6	35
Children								
1–3	29	35	16	400	0.7	0.8	9	50
4–6	44	44	24	500	0.9	1.1	12	75
7–10	62	52	28	700	1.0	1.2	13	100
Males								
11–14	99	62	45	1,000	1.3	1.5	17	150
15–18	145	69	59	1,000	1.5	1.8	20	200
19–24	160	70	58	1,000	1.5	1.7	19	200
25–50	174	70	63	1,000	1.5	1.7	19	200
51+	170	68	63	1,000	1.2	1.4	15	200
Females								
11–14	101	62	46	800	1.1	1.3	15	150
15–18	120	64	44	800	1.1	1.3	15	180
19–24	128	65	46	800	1.1	1.3	15	180
25–50	138	64	50	800	1.1	1.3	15	180
51+	143	63	50	800	1.0	1.2	13	180
Pregnant	—	—	60	800	1.5	1.6	17	400
Lactating								
1st 6 months	—	—	65	1,300	1.6	1.8	20	280
2nd 6 months	—	—	62	1,200	1.6	1.7	20	260

NOTE: The allowances, expressed as average daily intakes over time, are intended to provide for individual variations among most normal persons as they live in the United States under usual environmental stresses. Diets should be based on a variety of common foods in order to provide other nutrients for which human requirements have been less well defined. See text for detailed discussion of allowances and of nutrients tabulated.

All the Essentials of Supplements

In "The ABCs of Nutrition and Health," we spell out the fundamentals of a good diet. But some of us simply don't eat enough of the right foods to meet our requirements for key vitamins and minerals. Even for the nutrition-ally aware, factors may intervene—weight-loss plans, food restrictions or intolerances, medication interactions, pregnancy and lactation, even aging can keep us from getting the right nutrients.

The table above shows the Recommended Dietary Allowances of vitamins and minerals for everyday use. You might be able to meet these requirements with your normal daily diet.

| | | | | | **Minerals** | | | |
Vitamin B$_6$ (mg)	Vitamin B$_{12}$ (mcg)	Vitamin C (mg)	Vitamin D (mcg)	Vitamin E (mg αTE)	Calcium (mg)	Iron (mg)	Magnesium (mg)	Zinc (mg)
0.3	0.3	30	7.5	3	400	6	40	5
0.6	0.5	35	10.0	4	600	10	60	5
1.0	0.7	40	10.0	6	800	10	80	10
1.1	1.0	45	10.0	7	800	10	120	10
1.4	1.4	45	10.0	7	800	10	170	10
1.7	2.0	50	10.0	10	1,200	12	270	15
2.0	2.0	60	10.0	10	1,200	12	400	15
2.0	2.0	60	10.0	10	1,200	10	350	15
2.0	2.0	60	5.0	10	800	10	350	15
2.0	2.0	60	5.0	10	800	10	350	15
1.4	2.0	50	10.0	8	1,200	15	280	12
1.5	2.0	60	10.0	8	1,200	15	300	12
1.6	2.0	60	10.0	8	1,200	15	280	12
1.6	2.0	60	5.0	8	800	15	280	12
1.6	2.0	60	5.0	8	800	10	280	12
2.2	2.2	70	10.0	10	1,200	30	300	15
2.1	2.6	95	10.0	12	1,200	15	355	19
2.1	2.6	90	10.0	11	1,200	15	340	16

However, some experts feel that a bit of preventive insurance makes sense for the millions who may fall short in one or more areas. Others feel that even if your diet is adequate, you may be able to give yourself an edge by supplementing with those nutrients (like the antioxidants) that have been shown to reduce risks of chronic diseases like cancer, heart disease, cataracts, and osteoporosis.

Using the guidelines in the table on page 28, you can avoid the pitfall of a multiple supplement that is adequate or even high in some nutrients but pitifully low in others. Calcium, because relatively large amounts are needed and there is only so much room in a multiple nutrient tablet, is the nutrient most likely to come short. But—especially for women—calcium may be the most important supplement of all. So check your labels and see if your supplements fall within the ranges we've outlined.

Guidelines for Daily Intake of Supplements

In this table, you'll find two sets of numbers—one for the Recommended Dietary Allowance (RDA) and the other for "preventive amounts." The numbers in both columns are expressed in milligrams (mg), micrograms (mcg), retinol equivalents (RE), International Units (IU), and tocopherol equivalents (αTE). When you buy supplements, check labels: You'll find the same units of measure on the packaging, so you can compare the doses in each supplement with the guideline figures in this table.

The numbers in the RDA column simply tell you how much supplement you'd need to reach the high range of the Recommended Dietary Allowance.

In the "preventive amounts" column is a ballpark number that early research findings suggest should put you in the right range for preventive purposes. If you stay within this range, you can supplement nutrients that may be lacking in your diet without taking excessively high (and perhaps dangerous) amounts of those nutrients. For some nutrients, however, no preventive amount is shown: the dash (—) indicates that there's no known benefit in taking supplements for preventive reasons.

Nutrient	RDA*	Preventive Amount
Vitamin A[†]	1,000 RE (5,000 I.U.)	—
Beta-carotene	6 mg[‡]	15–30 mg
Thiamine (vitamin B$_1$)	1.5 mg	—
Riboflavin (vitamin B$_2$)	1.8 mg	—
Niacin	20 mg	—
Folate	200 mcg	400–800 mcg
Vitamin B$_6$	2 mg	2–10 mg
Vitamin B$_{12}$	2 mcg	2–10 mcg
Vitamin C	60 mg	100–500 mg
Vitamin D	10 mcg	10 mcg
Vitamin E	10 mg αTE (14 I.U.)	67–268 mg αTE (100–400 I.U.)
Calcium	1,200 mg	1,200–1,500 mg (women)
Iron	15 mg	—
Magnesium	400 mg	—
Zinc	15 mg	—
Selenium	70 mcg	70–100 mcg

*Represents the highest values of all ages and sex groups except pregnant and lactating women in the latest version (1989, 10th edition) of the RDA.

[†]Vitamin A is best taken in the form of beta-carotene.

[‡]6 milligrams of beta-carotene provide 100% of the RDA for vitamin A.

A Quick and Easy Guide to This Handbook

You're headed for the supermarket, ready to stock up on foods that will boost your family's health . . .

You're getting ready for a summer barbecue, determined to put together a menu that's not only sumptuous and tasty but also not overladen with fat, salt, and cholesterol . . .

Or you're rummaging in the fridge, wondering whether you have the right "fixin's" for a well-balanced meal . . .

Whatever your questions about nutrition, the following profiles of 1,000 foods should give you the answers you need—and take some of the confusion out of meal planning.

Here's what to look for in each entry:

1. Healing Strength Symbols. These symbols tell you the fortifying, blood-building, immunity-boosting, or nutritional powers of each food. The symbols are for Stronger Bones, Cancer Protection, Better Blood Pressure, Blood Building, Digestive Health, Stronger Immunity, Lower Cholesterol, and Weight Loss.

In general, where you see the healing strength symbols, you can be sure that the food has nutritional pluses that make it a standout. But there may be minuses too—so be sure to read the entire entry.

(For more information on nutritional criteria for the symbols, see the section titled "Criteria for Healing Strength Symbols" on page 31.)

2. Serving Information. Here you'll find the size of the serving, measured either by weight or by volume; whether the food is cooked or uncooked; and descriptive information about packaging, preparation, or serving. Some common measurement conversions are listed below.

3 teaspoons = 1 tablespoon
2 tablespoons = 1 fluid ounce
8 fluid ounces = 1 cup
2 cups = 1 pint
1 ounce = 28 grams

In some cases, values were not available for the cooked form, so the "Serving" is described as "raw" or "uncooked." (But please note: this does not necessarily mean that the food *should be served* raw or uncooked.)

In referring to food that has been cooked, we may use the terms "dry heat" or "moist heat." "Dry heat" includes baking, broiling, and microwaving. "Moist heat" refers to any kind of cooking that involves added water or broth—such as boiling, poaching, braising, and steaming.

3. Nutritional Profile: For each food, the amounts of total fat, saturated, monounsaturated, and polyunsaturated fats, cholesterol, sodium, protein, carbohydrate, and dietary fiber are expressed in grams (g) or milligrams (mg). The nutritional profile also tells you the number of calories in each food and the *percentage* of calories from fat. (Because there are other components of total fat besides saturated, monounsaturated, and polyunsaturated, the values given for these three types do not add up to the value for total fat.)

The nutrition data in most entries come from information provided by the U.S. Department of Agriculture (USDA). Additional information comes from food labels or from manufacturer's data. All figures have been rounded off. When there is only a small amount of one food element, it is indicated by the word "trace." The notation N/A indicates that information about food composition was not available at the time this book was published.

4. Chief Nutrients: Here are listed nutrients that meet or exceed a minimum standard. The "minimum" is the equivalent of 5 percent of the Recommended Dietary Allowance (RDA) for a 25- to 50-year-old male. (For a complete table that specifies RDAs for all sexes and ages, see page 26.)

Since 1974, the USDA has been using retinol equivalents (RE) instead of interna-

tional units (I.U.) as the unit of measurement for vitamin A requirements and as the measure of vitamin A found in food. So, in all entries you'll find vitamin A expressed in (micrograms of) RE. The RDA, however, is still given as a percentage. (Since the daily RE of vitamin A for males aged 25 to 50 is 1,000, 250 RE of vitamin A equals 25 percent of the RDA.) Vitamin E is expressed in mg αTE, or milligrams of alpha-tocopherol, the most easily absorbed form of vitamin E. For all other vitamins and minerals, the amounts are expressed in milligrams (mg) or micrograms (mcg—one-millionth of a gram).

5. Quick-Reference Headings: Is this a food with super-high nutritional power? Or does it have drawbacks that outweigh the benefits?

To find out, look at the heading of **Caution** that appears in the entry. If it comes first, this food has some definite minuses: It could raise your blood pressure, increase levels of sodium or cholesterol in the blood, or lead to weight gain. And if you're currently taking medication or you know you react adversely to certain allergens, be sure to take note of *Drug interaction* and *Allergy alert* listed under **Caution**.

Under the **Strengths** heading, you'll find more information about the preventive and healing benefits of the components found in each food. In addition, other nutrients and beneficial substances may be described under the **Strengths** heading. In the "Fish and Seafood" entries, for instance, you can find out whether the fish contains fair, good, or excellent amounts of heart-healthy omega-3's.

Further information about each food is included under other headings. You'll find specific dietary recommendations under **Eat with**.

For some background about where the food comes from or how it was introduced into our diet, check **Curiosity, Origin,** and **History.** And if you're unfamiliar with the food, pay particular attention to the information under **Description** and **Selection.**

Some tips to help the cook are included in **Storage, Serve,** and **Preparation.**

If you decide this food is not for you, be sure to look for other suggestions under **A Better Idea.** Here you'll also find recommendations for food substitutes that will help you maintain a healthy diet.

Criteria for Healing Strength Symbols

To determine whether a food should have a healing strength symbol, our researchers weighed the pluses against the minuses in each nutritional profile.

Here are the reasons why some foods have the healing strength symbols, while others don't.

 Stronger Bones. Calcium and vitamin D are the most important bone-builders, but high sodium in the diet can detract from their benefits. A food that provides at least 20 percent of the RDA of calcium or vitamin D gets the stronger bones symbol—but only if it's *not* high in sodium. (But keep in mind, for overall health you'll want to avoid food sources that are too high in fat like whole milk and hard cheeses.)

 Cancer Protection. Research has shown that there is lower risk of some types of cancer among those who maintain low-fat, high-fiber diets. Also, cancer protection may be enhanced by eating low-fat foods that are high in beta-carotene or vitamin C.

In general, any food that's low in fat and has more than 2 grams of fiber per serving qualifies for the cancer protection symbol.

The symbol also appears when a food that is low in fat has at least 20 percent of the RDA of vitamin A or C.

All cruciferous (cabbage-type) and allium (onion-type) vegetables get the cancer protection symbol because research has shown that they help prevent certain kinds of cancer.

 Better Blood Pressure. Any food that's not high in sodium can help lower blood pressure if it has beneficial amounts of potassium, magnesium, or calcium. If the sodium amounts to less than 200 milligrams per serving, and the food has either potassium, magnesium, or calcium (at least 20 percent of the RDA), the entry rates a better blood pressure symbol.

 Blood Building. Iron and vitamin C are the ideal duo for healthy blood. If a food is generally low to moderate in fat and has 20 percent or more of the RDA of iron, vitamin C, folate, or vitamin B_{12}, you'll find a blood building symbol on the entry. (Vitamin C helps convert nonheme iron—which comes from fruit and vegetables—into an absorbable form of iron that builds red blood cells.)

 Digestive Health. The digestive health symbol appears if one serving provides at least 2 grams of fiber. A minimum of 20 grams per day is recommended by the National Cancer Institute.

 Stronger Immunity. Vitamins A, E, B$_6$, C, and folate—and the minerals iron and zinc—help to increase the body's immunity to certain kinds of disease. Any food that contains more than 20 percent of the RDA of any one of these nutrients is eligible for the stronger immunity symbol.

 Lower Cholesterol. As we've already noted, artery-clogging cholesterol can become a problem if you eat foods that are predominant in saturated fat rather than monounsaturated or polyunsaturated fat. For this reason, none of the foods that are predominantly saturated earn a lower cholesterol symbol, even though they might have little or no cholesterol.

So, to qualify for lower cholesterol, a food that is predominantly monounsaturated or polyunsaturated must also have at least one of these cholesterol-lowering features:

- 2 grams of fiber

- Less than 20 milligrams of cholesterol per 3½-ounce serving

- Less than 30 percent of its calories from fat (no more than 25 percent is ideal), and no more than 150 mg of cholesterol per 3½-ounce serving

Weight Loss. Caution, dieters: When you see the weight loss symbol on an entry, remind yourself that it only applies to a specific serving size. Needless to say, if you go overboard on serving size you're probably headed for weight *gain* instead of weight *loss*.

As you might expect, any food that is too high in fat (more than 50 percent of its calories from fat) or calories (more than 250 calories in 3½ ounces) is automatically disqualified for a weight loss symbol. So you'll find the weight loss symbol on foods that get less than 30 percent of their calories from fat or have less than 50 calories per 3½ ounces.

Research has shown that plenty of fiber in the diet can contribute to weight loss if intake of fat and calories is reasonable. So you'll find the weight loss symbol on low-fat or low-calorie foods that have more than 2 grams of fiber per serving. But note: Even a low-fat, high-fiber food (like dried apricots) is disqualified if the total *calories* are too high.

A Few More Tips about Using This Book

First of all, be sure to check the chart on page 26 to determine the Recommended Dietary Allowances for someone of your age and sex.

Also, in the back of the book, beginning on page 539, you'll find tables that include extensive nutritional information about many kinds of foods. If you can't find the food you're looking for among the 1,000 entries, be sure to check this last section too.

And keep in mind, some foods have a variety of names. Chick-peas, for example, are also known as garbanzo beans; club soda is also called seltzer water. Be sure to look in the index if you can't find an entry by the name you know.

1,000 Foods Analyzed and Rated

Quick, think of a meal!

Let's suppose your taste buds are tending toward flavorful home-cooked chicken, broccoli, sweet potato, and (well, why not?) some cranberry sauce on the side. Perhaps a fresh garden salad with French dressing? And then you find yourself contemplating the prospect of a Boston cream pie for dessert (should you or shouldn't you?), and a cup of coffee afterward.

A satisfying prospect—but what's the nutritional makeup of that tempting meal? And what's the potential cost in terms of fat, calories, and cholesterol?

Well, if you want to know, you've come to the right place.

On the following pages, you'll find all the information you need to make choices about every one of the foods in that dreamed-up meal—and practically any other meal you can imagine. Food by food, you'll find complete nutritional tables with easy-to-read healing strength symbols, plus descriptions of a thousand foods and beverages. In a glance, you can even figure out whether you want to pour French dressing on your salad—and if so, how much. Have no more doubt about the total fat cost of that slice of Boston cream pie: Just turn to "Desserts" and look it up.

One piece of advice before you begin, however: If you haven't done so already, be sure to read the section titled "A Quick and Easy Guide to This Handbook" on page 29, where you'll find important tips about reading these entries.

Beans, Peas, and Soy Products
BEANS AND PEAS

ADZUKI BEANS

Serving: ½ *cup, boiled*

Calories	147
Fat	0.1 g
Saturated	trace
Monounsaturated	N/A
Polyunsaturated	N/A
Calories from fat	1%
Cholesterol	0
Sodium	9 mg
Protein	8.7 g
Carbohydrate	28.5 g
Dietary fiber	N/A

CHIEF NUTRIENTS

NUTRIENT	AMOUNT	% RDA
Folate	139.3 mcg	70
Iron	2.3 mg	23
Magnesium	59.8 mg	17
Potassium	611.8 mg	16
Zinc	2.0 mg	14

Strengths: Adzuki beans offer healthy doses of anemia-fighting iron, nerve-soothing B vitamins, and immunity-boosting zinc—nutrients that dieters may need on their menu. Because the beans are digested slowly, they can suppress your appetite for hours. And adzukis are an excellent source of complex carbohydrates for diabetics.

Eat with: Vitamin C–rich foods (to boost iron absorption) and rice, grains, or pasta (to form complete protein).

Caution: Beans are moderately high in purines and should be eaten sparingly by anyone with gout. Unfortunately, these beans do give some people gas. *Allergy alert:* Legumes (including beans) sometimes trigger migraine headaches in sensitive individuals.

Origin: China and Japan.

Description: Small, red, sweet-flavored beans.

Selection: Found in health-food stores and oriental grocery shops.

BLACK BEANS

Serving: ½ *cup, boiled*

Calories	114
Fat	0.5 g
Saturated	trace
Monounsaturated	trace
Polyunsaturated	trace
Calories from fat	4%
Cholesterol	0
Sodium	1 mg
Protein	7.6 g
Carbohydrate	20.4 g
Dietary fiber	6.1 g

CHIEF NUTRIENTS

NUTRIENT	AMOUNT	% RDA
Folate	128.0 mcg	64
Iron	1.8 mg	18
Magnesium	60.2 mg	17
Thiamine	0.2 mg	14
Potassium	305.3 mg	8

Strengths: Ideal for dieters, black beans are digested slowly and can suppress your appetite for hours. They're packed with fiber, hefty amounts of anemia-fighting iron, and nerve-soothing B vitamins. (These nutrients are essential if you're on a weight-loss diet.) Plus, beans are exceptionally high in folate. (People who take antacids regularly may run short of this B vitamin.)

Eat with: Vitamin C–rich foods to boost iron absorption. Rice or other grains (as in black beans and rice) to form complete protein. For added color, texture, *and* nutrition, add black beans to any three-bean salad.

Caution: Beans are moderately high in purines and should be eaten sparingly by anyone with gout. And beans give some people gas. *Allergy alert:* Legumes sometimes trigger migraine headaches in sensitive individuals.

Description: Sweet-flavored, with black skin and cream-colored flesh. Also called turtle beans.

BROADBEANS

Serving: ½ cup, boiled

Calories	94
Fat	0.3 g
Saturated	trace
Monounsaturated	trace
Polyunsaturated	trace
Calories from fat	3%
Cholesterol	0
Sodium	4 mg
Protein	6.5 g
Carbohydrate	16.7 g
Dietary fiber	4.6 g

CHIEF NUTRIENTS

NUTRIENT	AMOUNT	% RDA
Folate	88.5 mcg	44
Iron	1.3 mg	13
Magnesium	36.6 mg	10
Potassium	227.8 mg	6
Zinc	0.9 mg	6
Thiamine	0.1 mg	5

Strengths: Like most other beans, broadbeans are high in fiber and folate, with respectable amounts of iron and other minerals. And they're low in fat.

Eat with: Peppers or other vitamin C–rich foods: They help your body utilize the nonheme iron in broadbeans.

Caution: Due to an inherited metabolic disorder, some people of African, Asian, or Mediterranean ancestry who eat broadbeans (often called fava beans) develop dizziness, nausea, vomiting, high fever, and other symptoms of hemolytic anemia (favism). Most people can eat fava beans safely, though. *Drug interaction:* Broadbeans contain an enzyme that is converted to dopamine, which is similar to tyramine and must be avoided by anyone taking MAO-inhibitor drugs. *Allergy alert:* May trigger migraine headaches in sensitive people.

Description: A tan, flat bean resembling a very large lima bean. The pod is inedible.

Preparation: Broadbeans should always be cooked, with the skin removed before serving.

CHICK-PEAS, Canned

Serving: ½ cup

Calories	143
Fat	1.4 g
Saturated	0.1 g
Monounsaturated	0.3 g
Polyunsaturated	0.6 g
Calories from fat	9%
Cholesterol	0
Sodium	359 mg
Protein	5.9 g
Carbohydrate	27.1 g
Dietary fiber	7.0 g

CHIEF NUTRIENTS

NUTRIENT	AMOUNT	% RDA
Folate	80.2 mcg	40
Vitamin B_6	0.6 mg	29
Iron	1.6 mg	16
Magnesium	34.8 mg	10
Zinc	1.3 mg	9
Vitamin C	4.6 mg	8
Potassium	206.4 mg	6

Strengths: Chick-peas get high marks for immunity-boosting folate and for iron. And they're a higher-than-average source of fiber and vitamin B_6. (Women who are pregnant or taking oral contraceptives need to pay special attention to B_6 in their diet.) They supply complex carbohydrates, which makes them preferred fare for diabetics.

Eat with: Rice or other grains (like couscous) to form complete protein. If you accompany chick-peas with vitamin C–rich foods like green peppers, your body makes better use of their nonheme iron.

Caution: Like other legumes, chick-peas are moderately high in purines and should be eaten sparingly by anyone with gout. *Drug interaction:* People taking levodopa are often warned to avoid excessive intake of beans and other foods that are high in vitamin B_6.

Description: Round, irregularly shaped, buff-colored legumes, chick-peas are also called garbanzo beans or ceci.

Preparation: Cooking chick-peas thoroughly can reduce their tendency to produce gas.

COWPEAS

Serving: 1/2 cup, boiled

Calories	100
Fat	0.5 g
Saturated	trace
Monounsaturated	trace
Polyunsaturated	trace
Calories from fat	4%
Cholesterol	0
Sodium	3 mg
Protein	6.7 g
Carbohydrate	17.9 g
Dietary fiber	8.3 g

CHIEF NUTRIENTS

NUTRIENT	AMOUNT	% RDA
Folate	178.8 mcg	89
Iron	2.2 mg	22
Magnesium	45.6 mg	13
Thiamine	0.2 mg	11
Zinc	1.1 mg	7
Potassium	239.1 mg	6

Strengths: Packed with fiber, cowpeas are digested slowly and can suppress your appetite for hours. These legumes pack hefty amounts of anemia-fighting iron and nerve-soothing B vitamins. They're extraordinarily rich in folate. (People who take antacids regularly may run short of this B vitamin.) With an abundance of good-quality vegetable protein, cowpeas are especially beneficial to vegetarians.

Eat with: Rice (as in hopping John, a traditional Southern dish) to form complete protein. If you also eat vitamin C–rich foods like sweet potatoes or collard greens, you'll boost iron absorption.

Caution: On a salt- or fat-restricted diet? If so, omit salt pork or bacon from recipes using cowpeas. Cowpeas, like other legumes, are moderately high in purines; anyone with gout should eat sparingly. *Allergy alert:* Legumes (including cowpeas) sometimes trigger migraine headaches in sensitive individuals.

Description: Beige, with a black "eye." Also called black-eyed peas.

EDIBLE-PODDED PEAS

Serving: 1/2 cup, boiled

Calories	34
Fat	0.2 g
Saturated	trace
Monounsaturated	trace
Polyunsaturated	trace
Calories from fat	5%
Cholesterol	0
Sodium	3 mg
Protein	2.6 g
Carbohydrate	5.6 g
Dietary fiber	2.2 g

CHIEF NUTRIENTS

NUTRIENT	AMOUNT	% RDA
Vitamin C	38.3 mg	64
Iron	1.6 mg	16
Folate	23.3 mcg	12
Thiamine	0.1 mg	7
Magnesium	20.8 mg	6
Vitamin B$_6$	0.1 mg	6
Potassium	192.0 mg	5

Strengths: A respectable source of fiber, edible-podded peas offer good digestive health, colon cancer protection, and relief of constipation. Like other legumes, pea pods have folate (needed by pregnant women for healthy babies). The super vitamin C content helps boost absorption of nonheme iron in the edible pods, and that's added insurance against anemia. Gram for gram, nutrient values for raw pea pods are slightly higher than for cooked.

Caution: Peas contain substances that bind to iron, zinc, and other minerals, so you may not get their full benefit. Since peas are moderately high in purines, they should be eaten sparingly by anyone with gout. *Allergy alert:* This vegetable may trigger hives or other allergic reactions in sensitive individuals.

Selection: Also called snow peas or sugar snap peas.

Preparation: Superfast cooking—especially microwave cooking— helps save nutrients.

FRENCH BEANS

Serving: ½ cup, boiled

Calories	111
Fat	0.7 g
Saturated	trace
Monounsaturated	trace
Polyunsaturated	trace
Calories from fat	5%
Cholesterol	0
Sodium	5 mg
Protein	6.1 g
Carbohydrate	20.7 g
Dietary fiber	N/A

CHIEF NUTRIENTS

NUTRIENT	AMOUNT	% RDA
Folate	64.2 mcg	32
Magnesium	48.2 mg	14
Iron	0.9 mg	9
Potassium	318.2 mg	8
Calcium	52.4 mg	7
Thiamine	0.1 mg	7

Strengths: Just ½ cup of French beans supplies nearly one-third of a day's folate requirements—a good choice for pregnant women, who need adequate amounts of this B vitamin. French beans also offer some potassium and a respectable amount of magnesium, mighty minerals that help keep your blood flowing freely and easily. These beans have some anemia-fighting iron, too.

Eat with: Lean meat or vitamin C–rich foods (like tomatoes or peppers) to boost absorption of nonheme iron.

Curiosity: French beans were introduced to England by French Huguenots.

Description: A French bean is any young green string bean with an edible pod. (When a green bean is cut lengthwise into thin strips, it's "french cut.")

Selection: The young pods are most tender and may be eaten whole. Longer, stringier pods need to be cut and destringed.

Preparation: Steam, do not boil, to save nutrients. A few drops of lemon juice bring out the flavor.

GREAT NORTHERN BEANS

Serving: ½ cup, boiled

Calories	104
Fat	0.4 g
Saturated	trace
Monounsaturated	trace
Polyunsaturated	trace
Calories from fat	3%
Cholesterol	0
Sodium	2 mg
Protein	7.3 g
Carbohydrate	18.6 g
Dietary fiber	4.8 g

CHIEF NUTRIENTS

NUTRIENT	AMOUNT	% RDA
Folate	89.9 mcg	45
Iron	1.9 mg	19
Magnesium	44.0 mg	13
Potassium	344.1 mg	9
Thiamine	0.1 mg	9
Calcium	59.8 mg	7
Zinc	0.8 mg	5

Strengths: Packed with fiber, beans are digested slowly and can suppress your appetite for hours. If you eat ½ cup, you'll satisfy well over one-third of the day's folate requirements—a good choice for pregnant women who need adequate amounts of this B vitamin. The beans have a decent amount of anemia-fighting iron and some immunity-boosting zinc, as well as some potassium and magnesium. They're also a better-than-average source of vegetable protein.

Eat with: Green peppers, tomatoes, or other vitamin C–rich foods to boost absorption of nonheme iron.

Caution: Beans are moderately high in purines and should be eaten sparingly by anyone with gout. And they do give some people gas.

Curiosity: By the year 2000, scientists expect to develop beans that don't cause flatulence.

Preparation: Soak and cook beans thoroughly to reduce their tendency to create intestinal gas.

Serve: In recipes calling for white beans (such as many baked bean recipes).

GREEN PEAS

Serving: ¹/₂ cup, boiled

Calories	67
Fat	0.2 g
Saturated	trace
Monounsaturated	trace
Polyunsaturated	trace
Calories from fat	2%
Cholesterol	0
Sodium	2 mg
Protein	4.3 g
Carbohydrate	12.5 g
Dietary fiber	2.4 g

CHIEF NUTRIENTS

NUTRIENT	AMOUNT	% RDA
Folate	50.6 mcg	25
Vitamin C	11.4 mg	19
Thiamine	0.2 mg	14
Iron	1.2 mg	12
Magnesium	31.2 mg	9
Niacin	1.6 mg	9
Vitamin B₆	0.2 mg	9
Riboflavin	0.1 mg	7
Potassium	216.8 mg	6
Zinc	1.0 mg	6

Strengths: Green peas are a valuable source of B vitamins, especially folate and thiamine. (B-vitamin deficits may trigger skin and nerve problems.) They have barely any fat, with useful amounts of protein, fiber, and complex carbohydrates.

Eat with: Lean meat, a rich source of heme iron that helps boost absorption of nonheme iron in peas and other legumes.

Caution: Peas contain substances that bind to iron, zinc, and other minerals, interfering to some degree with their use. They're moderately high in purines and should be eaten sparingly by anyone with gout. *Drug interaction:* People taking levodopa are often warned to avoid excessive intake of peas and other foods high in vitamin B₆. *Allergy alert:* May trigger hives or other allergic reactions in sensitive individuals.

Preparation: Cooking peas in the microwave saves nutrients.

KIDNEY BEANS

Serving: ¹/₂ cup, boiled

Calories	112
Fat	0.4 g
Saturated	trace
Monounsaturated	trace
Polyunsaturated	trace
Calories from fat	4%
Cholesterol	0
Sodium	2 mg
Protein	7.6 g
Carbohydrate	20.1 g
Dietary fiber	6.9 g

CHIEF NUTRIENTS

NUTRIENT	AMOUNT	% RDA
Folate	114.1 mcg	57
Iron	2.6 mg	26
Magnesium	39.6 mg	11
Potassium	354.6 mg	10
Thiamine	0.1 mg	9
Vitamin B₆	0.1 mg	6
Zinc	0.9 mg	6

Strengths: Packed with fiber, legumes like kidney beans are digested slowly and can suppress your appetite for hours. Supplying over one-half the day's folate requirements in a ¹/₂-cup serving, they're a good choice for pregnant women (who need adequate amounts of folate for healthy babies). Like other beans, the kidney variety is a good source of iron. They also offer respectable amounts of potassium and magnesium.

Eat with: Tomatoes and other vitamin C–rich foods to boost iron absorption. In combination with rice or other grains, kidney beans form complete protein.

Caution: Beans are moderately high in purines and should be eaten sparingly by anyone with gout. And go slowly if you're prone to flatulence. *Allergy alert:* Legumes (including beans) sometimes trigger migraine headaches in sensitive individuals.

Preparation: To reduce gas-inducing properties, soak for at least 3 hours, then discard the soaking water and cook thoroughly in fresh water.

LENTILS

Serving: ½ cup, boiled

Calories	115
Fat	0.4 g
Saturated	trace
Monounsaturated	trace
Polyunsaturated	trace
Calories from fat	3%
Cholesterol	0
Sodium	2 mg
Protein	8.9 g
Carbohydrate	19.9 g
Dietary fiber	5.2 g

CHIEF NUTRIENTS

NUTRIENT	AMOUNT	% RDA
Folate	179.0 mcg	89
Iron	3.3 mg	33
Thiamine	0.2 mg	11
Magnesium	35.6 mg	10
Potassium	365.3 mg	10
Vitamin B$_6$	0.2 mg	9
Zinc	1.3 mg	8
Niacin	1.1 mg	6

Strengths: A super-extraordinary source of folate, lentils are an excellent choice for women who are pregnant or taking birth control pills. Lentils also offer decent amounts of other nerve-soothing vitamins, with respectable amounts of potassium and magnesium. Like other legumes, lentils are rich in iron and protein. Packed with fiber, lentils can help relieve constipation and reduce risk of hemorrhoids. (Although lentils are rich in nonheme iron, the presence of phytate hinders absorption.)

Eat with: Grains (like corn, rice, or wheat) to form complete protein. The addition of lean meat and vitamin C–rich fruits and vegetables helps boost absorption of nonheme iron.

Caution: Moderately high in purines—should be eaten sparingly by anyone with gout. Lentils may give some people gas. *Allergy alert:* This legume sometimes triggers migraine headaches.

History: Lentils have long been used as a protein source in India and the Middle East.

LIMA BEANS

Serving: ½ cup, boiled

Calories	108
Fat	0.4 g
Saturated	trace
Monounsaturated	trace
Polyunsaturated	trace
Calories from fat	3%
Cholesterol	0
Sodium	2 mg
Protein	7.3 g
Carbohydrate	19.6 g
Dietary fiber	6.8 g

CHIEF NUTRIENTS

NUTRIENT	AMOUNT	% RDA
Folate	78.1 mcg	39
Iron	2.3 mg	23
Potassium	477.5 mg	13
Magnesium	40.4 mg	12
Thiamine	0.2 mg	10
Vitamin B$_6$	0.2 mg	8
Zinc	0.9 mg	6

Strengths: Although somewhat lower in folate than baby limas, the larger beans supply over one-third of a day's requirements. (Either kind is a good choice for pregnant women, who need adequate amounts of folate for healthy babies.) Lima beans also offer decent amounts of other nerve-soothing B vitamins. Packed with fiber, they're digested slowly and can suppress your appetite for hours. They're also a good source of protein, magnesium, and potassium.

Eat with: Lean meat or vitamin C–rich foods (to boost iron absorption). With grains like corn or rice, they form complete protein.

Caution: Moderately high in purines, beans should be eaten sparingly by anyone with gout. Limas, especially, are notorious for their gas-producing potential: Eat in moderation if you're troubled by flatulence. *Allergy alert:* Beans sometimes trigger migraine headaches in sensitive individuals.

Curiosity: Scientists hope to develop non-gas-forming beans in the near future.

Selection: Lima beans are also called Fordhook beans (and sometimes butter beans in the South).

LIMA BEANS, *Baby*

Serving: $^{1}/_{2}$ *cup, boiled*

Calories	115
Fat	0.4 g
Saturated	trace
Monounsaturated	trace
Polyunsaturated	trace
Calories from fat	3%
Cholesterol	0
Sodium	3 mg
Protein	7.3 g
Carbohydrate	21.2 g
Dietary fiber	6.6 g

CHIEF NUTRIENTS

NUTRIENT	AMOUNT	% RDA
Folate	136.4 mcg	68
Iron	2.2 mg	22
Magnesium	48.2 mg	14
Potassium	364.9 mg	10
Thiamine	0.2 mg	10
Zinc	0.9 mg	6

Strengths: If you eat a hearty serving of baby lima beans, you'll get nearly two-thirds of your folate requirements for the day. So they're a good choice for pregnant women, who need adequate amounts of this B vitamin for healthy babies. Lima beans also offer decent amounts of thiamine, a nerve-soothing B vitamin, and anemia-fighting iron. And they're a good source of protein, magnesium, and potassium. A respectable amount of fiber helps relieve constipation and prevent hemorrhoids.

Eat with: Lean meat or vitamin C–rich foods (to boost iron absorption). Also with grains (like corn and rice) to form complete protein.

Caution: Beans are moderately high in purines and should be eaten sparingly by anyone with gout. And if you're troubled by flatulence, proceed with caution. *Allergy alert:* Legumes (including beans) sometimes trigger migraine headaches in sensitive individuals.

MUNG BEANS

Serving: $^{1}/_{2}$ *cup, boiled*

Calories	106
Fat	0.4 g
Saturated	trace
Monounsaturated	trace
Polyunsaturated	trace
Calories from fat	3%
Cholesterol	0
Sodium	2 mg
Protein	7.1 g
Carbohydrate	19.3 g
Dietary fiber	0.8 g

CHIEF NUTRIENTS

NUTRIENT	AMOUNT	% RDA
Folate	160.4 mcg	80
Iron	1.4 mg	14
Magnesium	48.5 mg	14
Thiamine	0.2 mg	11
Potassium	268.7 mg	7
Zinc	0.9 mg	6

Strengths: Mung beans supply nearly an entire day's folate requirement—which means they're a top choice for pregnant women, who need adequate amounts of this B vitamin for healthy babies. They're a decent source of thiamine (another nerve-soothing B vitamin) and anemia-fighting iron. Protein, magnesium, and potassium are also present.

Eat with: Lean meat and vitamin C–rich foods like green peppers or broccoli in a stir-fry to boost iron absorption. With grains (like rice), they form complete protein.

Curiosity: Cellophane noodles are made from mung bean flour.

Description: Mung beans are small with yellow flesh and green, yellow, or black skin. They're tender and slightly sweet when cooked.

NAVY BEANS

Serving: ½ cup, boiled

Calories	129
Fat	0.5 g
Saturated	trace
Monounsaturated	trace
Polyunsaturated	trace
Calories from fat	4%
Cholesterol	0
Sodium	0.9 mg
Protein	7.9 g
Carbohydrate	23.9 g
Dietary fiber	4.9 g

CHIEF NUTRIENTS

NUTRIENT	AMOUNT	% RDA
Folate	127.3 mcg	64
Iron	2.3 mg	23
Magnesium	53.7 mg	15
Thiamine	0.2 mg	12
Potassium	334.9 mg	9
Calcium	63.7 mg	8
Vitamin B$_6$	0.2 mg	8
Zinc	1.0 mg	6

Strengths: Some studies have shown that men who eat navy beans and other high-fiber foods can lower their cholesterol as much as 60 points. An excellent source of complex carbohydrates, navy beans are good for diabetics. They supply nearly two-thirds of a day's folate requirements as well as healthy doses of thiamine and anemia-fighting iron. A useful source of vegetable protein, too.

Eat with: Lean meat or vitamin C–rich foods (to boost iron absorption). Or with grains (like rice) to form complete protein.

Caution: Beans are moderately high in purines and should be eaten sparingly by anyone with gout. And navy beans do give some people gas. *Allergy alert:* Legumes (including beans) sometimes trigger migraine headaches in sensitive individuals.

Curiosity: The name stuck because the U.S. Navy has used this bean as a staple since the mid-1800s.

Preparation: To reduce gas-inducing properties, soak for at least 3 hours, then discard the soaking water and cook thoroughly in fresh water.

PINK BEANS

Serving: ½ cup, boiled

Calories	125
Fat	0.4 g
Saturated	trace
Monounsaturated	trace
Polyunsaturated	trace
Calories from fat	3%
Cholesterol	0
Sodium	2 mg
Protein	7.6 g
Carbohydrate	23.4 g
Dietary fiber	3.7 g

CHIEF NUTRIENTS

NUTRIENT	AMOUNT	% RDA
Folate	141.4 mcg	71
Iron	1.9 mg	19
Magnesium	54.6 mg	16
Thiamine	0.2 mg	15
Potassium	426.7 mg	11
Vitamin B$_6$	0.2 mg	8
Calcium	43.7 mg	5
Zinc	0.8 mg	5

Strengths: Supplying more than two-thirds of a day's folate requirements, pink beans (like other bean varieties) are a good choice for pregnant women: This B vitamin helps nurture healthy babies. Beans are a good low-fat source of protein and complex carbohydrates (good for diabetics) and they have anemia-fighting iron.

Eat with: Lean meat or vitamin C–rich foods (to boost iron absorption). And with grains (like rice) to form complete protein.

Caution: Beans are moderately high in purines and should be eaten sparingly by anyone with gout. And they give some people gas. *Drug interaction:* People taking levodopa are often warned to avoid excessive intake of beans and other foods high in vitamin B$_6$. *Allergy alert:* Beans may trigger migraine headaches in sensitive individuals.

Serve: In chili con carne or other Mexican dishes (instead of pinto beans).

PINTO BEANS

Serving: ½ cup, boiled

Calories	117
Fat	0.4 g
Saturated	trace
Monounsaturated	trace
Polyunsaturated	trace
Calories from fat	3%
Cholesterol	0
Sodium	2 mg
Protein	7.0 g
Carbohydrate	21.8 g
Dietary fiber	3.4 g

CHIEF NUTRIENTS

NUTRIENT	AMOUNT	% RDA
Folate	146.2 mcg	73
Iron	2.2 mg	22
Magnesium	46.8 mg	13
Potassium	397.8 mg	11
Thiamine	0.2 mg	11
Vitamin B_6	0.1 mg	7
Zinc	0.9 mg	6
Calcium	40.8 mg	5

Strengths: Pinto beans, in a ½-cup serving, supply more than two-thirds of a day's folate requirements, along with other nerve-soothing B vitamins in smaller amounts. They have good amounts of iron and contain other valuable minerals. Some studies have shown that men who eat pinto beans and other high-fiber foods can lower their cholesterol as much as 60 points. These beans are also a good low-fat source of protein and complex carbohydrates.

Eat with: Lean meat or vitamin C–rich foods (to boost iron absorption). And with grains (like rice) to form complete protein.

Caution: Beans are moderately high in purines and should be eaten sparingly by anyone with gout. And they give some people gas. *Drug interaction:* If you're taking levodopa, you'll want to avoid foods like beans that are high in vitamin B_6.

Description: Pale pink beans with reddish-brown streaks. (*Pinto* is Spanish for "painted.")

Serve: In chili con carne or other Mexican dishes.

SNAP BEANS, *Green*

Serving: ½ cup, boiled

Calories	22
Fat	0.2 g
Saturated	trace
Monounsaturated	trace
Polyunsaturated	trace
Calories from fat	7%
Cholesterol	0
Sodium	2 mg
Protein	1.2 g
Carbohydrate	4.9 g
Dietary fiber	1.8 g

CHIEF NUTRIENTS

NUTRIENT	AMOUNT	% RDA
Folate	20.7 mcg	10
Vitamin C	6.0 mg	10
Iron	0.8 mg	8
Potassium	185.4 mg	5

Strengths: A nice low-calorie source of fiber, green snap beans are practically fat-free, with a little immunity-boosting folate, iron, and vitamin C. And they're not as likely to cause gas as navy or kidney beans (which are mature beans).

Caution: Green beans contain modest amounts of oxalates, which should be restricted by those with calcium oxalate kidney stones.

Description: A long, slender bean pod. Also called string beans.

Selection: Should be firm, crisp, and tender. Choose the ones that snap readily when broken.

Preparation: Steam, boil, or stir-fry.

Serve: As a side dish or vegetable. These beans go well in soups, stews, or casseroles.

SNAP BEANS, *Yellow*

Serving: ½ cup, boiled

Calories	22
Fat	0.2 g
Saturated	trace
Monounsaturated	trace
Polyunsaturated	trace
Calories from fat	7%
Cholesterol	0
Sodium	2 mg
Protein	1.2 g
Carbohydrate	4.9 g
Dietary fiber	1.8 g

CHIEF NUTRIENTS

NUTRIENTS	AMOUNT	% RDA
Folate	20.7 mcg	10
Vitamin C	6.0 mg	10
Iron	0.8 mg	8
Potassium	185.4 mg	5

Strengths: A nice low-calorie side dish, yellow snap beans are practically fat-free, with a bit of immunity-boosting folate, iron, and vitamin C. They're not as likely to cause gas as mature beans (such as navy or kidney beans).

Caution: Contain modest amounts of oxalates, which should be restricted by those with calcium oxalate kidney stones.

Description: A pale yellow variety of green bean; also called wax beans.

Selection: Should be firm, crisp, and tender, and should snap readily when broken.

Preparation: Steam, boil, or stir-fry quickly to conserve nutrients.

Serve: As a vegetable side dish, or in three-bean salad.

SOYBEANS

Serving: ½ cup, boiled

Calories	149
Fat	7.7 g
Saturated	1.1 g
Monounsaturated	1.7 g
Polyunsaturated	4.4 g
Calories from fat	47%
Cholesterol	0
Sodium	0.9 mg
Protein	14.3 g
Carbohydrate	8.5 g
Dietary fiber	N/A

CHIEF NUTRIENTS

NUTRIENT	AMOUNT	% RDA
Iron	4.4 mg	44
Folate	46.3 mcg	23
Magnesium	74.0 mg	21
Riboflavin	0.3 mg	15
Potassium	442.9 mg	12
Calcium	87.7 mg	11
Vitamin B$_6$	0.2 mg	10
Thiamine	0.1 mg	9
Zinc	1.0 mg	7

Strengths: Soybeans supply twice as much protein as most other legumes. Their higher-than-average amount of magnesium may help to stave off migraine headaches in some people. Some research has indicated that a soybean-rich diet may help prevent or slow the spread of breast cancer. Soybeans have many pluses: They provide useful amounts of nerve-soothing B vitamins (especially folate), significant calcium, and a little zinc.

Eat with: Grains (like corn and rice) to form complete protein.

Caution: Soybeans contain substances that bind some minerals, limiting the body's use of them. Soybeans are moderately high in purines and should be eaten sparingly by anyone with gout. *Drug interaction:* People taking levodopa are often warned to avoid excessive intake of vitamin B$_6$–rich legumes like soybeans. *Allergy alert:* Soybeans are a common food allergen, yet they're found in many processed foods!

Serve: In chili, beans and rice, hummus, or vegetable pâté.

SPLIT PEAS

Serving: ½ cup, boiled

Calories	116
Fat	0.4 g
Saturated	trace
Monounsaturated	trace
Polyunsaturated	trace
Calories from fat	3%
Cholesterol	0
Sodium	2 mg
Protein	8.2 g
Carbohydrate	20.7 g
Dietary fiber	3.1 g

CHIEF NUTRIENTS

NUTRIENT	AMOUNT	% RDA
Folate	63.6 mcg	32
Iron	1.3 mg	13
Thiamine	0.2 mg	13
Magnesium	35.3 mg	10
Potassium	354.8 mg	9
Zinc	1.0 mg	7

Strengths: A good source of protein, fiber, and complex carbohydrates, with barely a trace of fat, split peas are rich in folate. (People who take antacids on a regular basis may have increased needs for folate. So do women who are pregnant or taking birth control pills.) They also supply some magnesium and potassium.

Eat with: Lean meat, a rich source of heme iron that helps boost absorption of nonheme iron in legumes. Rice and other grains to form complete protein.

Caution: Peas contain substances that bind to iron, zinc, and other minerals, interfering to some degree with their use. Moderately high in purines, peas should be eaten sparingly by anyone with gout. *Drug interaction:* People taking levodopa are often warned to avoid overeating peas and other foods high in vitamin B_6. *Allergy alert:* Hives or other allergic reactions may occur in sensitive individuals.

Preparation: Split peas must always be cooked thoroughly.

WINGED BEANS

Serving: ½ cup, boiled

Calories	126
Fat	5.0 g
Saturated	0.7 g
Monounsaturated	1.9 g
Polyunsaturated	1.3 g
Calories from fat	36%
Cholesterol	0
Sodium	11 mg
Protein	9.1 g
Carbohydrate	12.9 g
Dietary fiber	N/A

CHIEF NUTRIENTS

NUTRIENT	AMOUNT	% RDA
Iron	3.7 mg	37
Thiamine	0.3 mg	17
Calcium	122.1 mg	15
Magnesium	46.4 mg	13
Zinc	1.2 mg	8
Riboflavin	0.1 mg	7
Potassium	240.8 mg	6

Strengths: Higher in protein than other legumes, winged beans supply a little bit of zinc, some magnesium and potassium, and a considerable amount of iron. Like soybeans, winged beans supply calcium.

Eat with: Lean meat or vitamin C–rich foods (to boost iron absorption).

Origin: Southeast Asia.

Description: Thin, green, rectangular pods, about 2 inches long, which flare into ruffled ridges, or "wings," like maple seeds. Winged beans have a texture something like a starchy green bean. You'll be most likely to find these in specialty produce markets.

Preparation: Steam; or simmer for 10 minutes in very little water.

Serve: As a vegetable dish, like green beans. You can cut them into thin slices to go in stir-fries. Or cook, chill, and marinate in dressing.

YARD-LONG BEANS

Serving: ½ cup, boiled

Calories	102
Fat	0.4 g
Saturated	trace
Monounsaturated	trace
Polyunsaturated	trace
Calories from fat	4%
Cholesterol	0
Sodium	4 mg
Protein	7.1 g
Carbohydrate	18.1 g
Dietary fiber	1.6 g

CHIEF NUTRIENTS

NUTRIENT	AMOUNT	% RDA
Folate	125.3 mcg	63
Magnesium	84.3 mg	24
Iron	2.3 mg	23
Thiamine	0.2 mg	12

Strengths: An interesting change of pace for vegetable lovers, yard-long beans are virtually fat-free, with plenty of texture. They contain about the same amount of calories, protein, and nutrients as snap beans. Their immunity-building folate and iron are accompanied by a little magnesium. (Some thiazide diuretics can increase requirements for magnesium.)

Curiosity: One yard-long bean equals 5 or 6 string beans.

Description: Closely related to cowpeas (black-eyed peas). The grayish-green, pencil-thin pods may reach 3 feet in length, although they're usually picked at 18 inches or less. Not as stringy as green beans.

Selection: Available in oriental grocery shops and markets. Look for small (that is, younger) flexible pods, with immature seeds.

Preparation: Cut into 2-inch lengths and steam briefly, like green beans. Don't overcook—they'll get mushy.

Serve: As a side dish, or stir-fried with pork, ginger, and chili peppers.

SOY PRODUCTS

MISO

Serving: ½ cup

Calories	284
Fat	8.4 g
Saturated	1.2 g
Monounsaturated	1.9 g
Polyunsaturated	4.7 g
Calories from fat	27%
Cholesterol	0
Sodium	5,033 mg
Protein	16.3 g
Carbohydrate	38.6 g
Dietary fiber	7.5 g

CHIEF NUTRIENTS

NUTRIENT	AMOUNT	% RDA
Iron	3.8 mg	38
Zinc	4.6 mg	31
Folate	45.5 mcg	23
Riboflavin	0.4 mg	21
Magnesium	58.0 mg	17
Vitamin B$_6$	0.3 mg	15

Caution: Extraordinarily high in sodium. *Drug interaction:* Contains tyramine, an amino acid derivative that can raise blood pressure to dangerously high levels in anyone taking MAO inhibitors. *Allergy alert:* Contains soybeans, a common food allergen. Also, miso may trigger headaches in people sensitive to tyramines. (But small amounts, eaten occasionally, may not cause a problem; and serving size is usually smaller than ½ cup.)

Strengths: A superb source of protein and fiber, with significant amounts of B vitamins and minerals. Research suggests that frequent consumption of miso may be one reason why there's a lower incidence of breast cancer among Japanese women.

Description: Miso is a bean paste made from soybeans and rice (or barley) along with salt and water. This paste is inoculated with *Aspergillus oryzae,* a mold that triggers fermentation. A mainstay in Japanese cuisine, used to flavor soups, sauces, marinades, dips, and entrées.

A Better Idea: Natto, if you're on a low-sodium diet.

NATTO

Serving: ½ cup

Calories	187
Fat	9.7 g
Saturated	1.4 g
Monounsaturated	2.1 g
Polyunsaturated	5.5 g
Calories from fat	47%
Cholesterol	0
Sodium	6 mg
Protein	15.6 g
Carbohydrate	12.6 g
Dietary fiber	4.8 g

CHIEF NUTRIENTS

NUTRIENT	AMOUNT	% RDA
Iron	7.6 mg	76
Magnesium	101.2 mg	29
Calcium	191.0 mg	24
Vitamin C	11.4 mg	19
Zinc	2.7 mg	18
Potassium	641.5 mg	17
Riboflavin	0.2 mg	10
Thiamine	0.1 mg	9
Vitamin B$_6$	0.1 mg	6

Strengths: High in protein, rich in fiber, with very little sodium, natto contains fat that is mostly unsaturated. Like miso, natto contains generous amounts of iron and other minerals, as well as nice amounts of B vitamins, which are hard to get on a vegetarian diet.

Caution: *Drug interaction:* Fermented soy products like natto contain tyramine, an amino acid derivative that can raise blood pressure to dangerously high levels in anyone taking MAO inhibitors. *Allergy alert:* Contains soybeans, a common food allergen. May trigger headaches in people sensitive to tyramines. (But small amounts, eaten occasionally, may not cause a problem.)

Description: A fermented Japanese condiment with a strong, cheeselike flavor. Made from whole cooked soybeans that have been inoculated with *Bacillus subtilis,* a helpful bacterium.

Serve: Over rice or noodles, or in soups and salads. (Note: Serving size is usually smaller than ½ cup.)

SOYBEAN SPROUTS

Serving: ½ cup, raw

Calories	45
Fat	2.4 g
Saturated	0.3 g
Monounsaturated	0.3 g
Polyunsaturated	1.3 g
Calories from fat	47%
Cholesterol	0
Sodium	5 mg
Protein	4.6 g
Carbohydrate	3.9 g
Dietary fiber	N/A

CHIEF NUTRIENTS

NUTRIENT	AMOUNT	% RDA
Folate	60.1 mcg	30
Vitamin C	5.4 mg	9
Thiamine	0.1 mg	8
Iron	0.7 mg	7
Magnesium	25.2 mg	7

Strengths: The most nutritious of all sprouts, soybean sprouts are cholesterol-free and very low in sodium. Unlike other beans, they contain complete animal-quality protein. Soybean sprouts also are an ample source of immunity-boosting folate, and they also have some vitamin C, thiamine, iron, and magnesium. The vitamin C enhances the absorption of nonheme iron.

Caution: Beans contain purines and should be eaten sparingly by anyone with gout. *Allergy alert:* Some people may develop hives as an allergic reaction to beans.

Description: As a bean sprouts, so does its vitamin content, particularly vitamin C. Sprouts usually are eaten soon after they begin to grow but can be consumed even after the first leaves form.

Preparation: Steam or simmer sprouts for a few minutes. Uncooked bean sprouts and seeds contain a substance that inhibits one of the enzymes the body uses to digest protein.

TEMPEH

Serving: ½ cup

Calories	165
Fat	6.4 g
Saturated	0.9 g
Monounsaturated	1.4 g
Polyunsaturated	3.6 g
Calories from fat	35%
Cholesterol	0
Sodium	5 mg
Protein	15.7 g
Carbohydrate	14.1 g
Dietary fiber	N/A

CHIEF NUTRIENTS

NUTRIENT	AMOUNT	% RDA
Vitamin B$_{12}$	0.8 mcg	42
Folate	43.2 mcg	22
Niacin	3.8 mg	20
Iron	1.9 mg	19
Magnesium	58.1 mg	17
Vitamin B$_6$	0.3 mg	13
Calcium	77.2 mg	10
Zinc	1.5 mg	10
Potassium	304.6 mg	8
Thiamine	0.1 mg	7
Vitamin A	57.3 RE	6

Strengths: High in protein—and one of the few vegetables containing substantial amounts of vitamin B$_{12}$. Tempeh also supplies significant amounts of other B vitamins, plus vitamin A and a number of valuable minerals.

Caution: *Drug interaction:* May contain tyramine, an amino acid derivative that can raise blood pressure to dangerously high levels in anyone taking MAO inhibitors. *Allergy alert:* Contains soybeans, a common food allergen. Also may contain tyramine, which can trigger headaches or light-headedness in people sensitive to other fermented foods, like aged cheese.

Origin: Indonesia.

Description: Compact tempeh cakes or patties are made from cooked fermented soybeans laced with a non-harmful mold that gives them their flavor. Available fresh, canned, frozen, or dehydrated.

Preparation: Tempeh is usually fried or roasted, but it may also be steamed or cooked in soup. Sometimes eaten as a vegetarian "burger."

TOFU, *Firm*

Serving: ¼ block (about 3 oz.)

Calories	118
Fat	7.1 g
Saturated	1.0 g
Monounsaturated	1.6 g
Polyunsaturated	4.0 g
Calories from fat	54%
Cholesterol	0
Sodium	11 mg
Protein	12.8 g
Carbohydrate	3.5 g
Dietary fiber	N/A

CHIEF NUTRIENTS

NUTRIENT	AMOUNT	% RDA
Iron	8.5 mg	85
Magnesium	76.1 mg	22
Calcium	166.1 mg	21
Folate	23.7 mcg	12
Thiamine	0.1 mg	9
Zinc	1.3 mg	9

Strengths: An excellent source of vegetable protein, firm tofu is slightly higher in vitamins and most minerals than regular tofu. It also has 34% more protein, and it's cholesterol-free. Unlike tempeh, tofu is not fermented, so it's acceptable for those who are sensitive to fermented foods.

Eat with: Vitamin C–rich foods, to enhance absorption of iron. Animal foods, rich in heme iron, will also improve absorption of nonheme iron.

Caution: *Allergy alert:* Contains soybeans, a common food allergen.

Description: Comes in creamy-white, soft, cheeselike cakes of curdled soy milk made from ground cooked soybeans.

Storage: Keeps 1 week in a container of fresh water in the refrigerator. Change water daily. May be frozen up to 3 months, but freezing changes the texture, making it slightly chewier.

Serve: May be sliced, diced, or mashed and used in soups, stir-fries, casseroles, and sandwiches.

TOFU, *Regular*

Serving: ¼ block (about 4 oz.)

Calories	88
Fat	5.5 g
Saturated	0.8 g
Monounsaturated	1.2 g
Polyunsaturated	3.1 g
Calories from fat	57%
Cholesterol	0
Sodium	8 mg
Protein	9.4 g
Carbohydrate	2.2 g
Dietary fiber	1.4 g

CHIEF NUTRIENTS

NUTRIENT	AMOUNT	% RDA
Iron	6.2 mg	62
Magnesium	120.0 mg	34
Calcium	122.0 mg	15
Folate	17.4 mcg	9
Thiamine	0.1 mg	6
Zinc	0.1 mg	6

Strengths: An excellent source of vegetable protein, tofu is cholesterol-free. High in iron, rich in magnesium, it has a respectable amount of calcium and some zinc. Supplies some fiber, a little folate, and most of its fat is unsaturated. Unlike tempeh, tofu is not fermented, so it's acceptable to those who are sensitive to fermented foods.

Eat with: Vitamin C–rich foods to enhance absorption of iron. Animal foods, rich in heme iron, will also improve absorption of nonheme iron.

Caution: *Allergy alert:* Derived from soybeans, a common food allergen.

Description: Regular tofu comes in soft, creamy-white, custardlike cakes of curdled soy milk made from ground cooked soybeans; also called soybean curd or bean curd.

Selection: Some tofu is calcium-fortified.

Storage: Tofu is highly perishable. Store in fresh water in the refrigerator for up to 1 week, changing water daily.

Serve: May be sliced, diced, or mashed and used in soups, stir-fries, casseroles, and sandwiches.

Beverages
CARBONATED BEVERAGES

CLUB SODA

Serving: 1 can (12 fl. oz.)

Calories	0
Fat	0
Saturated	0
Monounsaturated	0
Polyunsaturated	0
Calories from fat	0
Cholesterol	0
Sodium	75 mg
Protein	0
Carbohydrate	0
Dietary fiber	0

CHIEF NUTRIENTS

(None of the nutrients meet or exceed 5% of the RDA.)

Strengths: This delightful, zingy beverage is completely devoid of fat, cholesterol, sugar, or alcohol. Also, club soda contains small amounts of sodium bicarbonate which, because it's alkaline, can help neutralize an acid stomach and relieve indigestion.

Eat with: Fruit juice, to concoct your own bubbly, custom-made, low-calorie, hip-whittling soda. Also tastes great with a twist of lemon or lime.

Caution: If you've been advised to abridge your sodium intake, be aware that club soda metes out a moderate amount of sodium.

Curiosity: Sparkling water was the basis for the first soda pop. In the 1840s, flavors were added to seltzer water, a naturally effervescent water that takes its name from the town of Nieder Selters in the Weisbaden region of Germany.

Description: Club soda is water that has been highly charged with carbon dioxide, a gas that renders this beverage wonderfully effervescent. It's also called soda water or seltzer water.

COLA

Serving: 1 can (12 fl. oz.)

Calories	152
Fat	trace
Saturated	trace
Monounsaturated	0
Polyunsaturated	0
Calories from fat	0
Cholesterol	0
Sodium	15 mg
Protein	0
Carbohydrate	38.5 g
Dietary fiber	0

CHIEF NUTRIENTS

(None of the nutrients meet or exceed 5% of the RDA.)

Caution: Contains calories, 9 tsp. of sugar, and little else. Excessive consumption may contribute to excessive weight gain and tooth decay. Cola contains 37 mg of caffeine, which makes some people jumpy when consumed in large quantities. Cola beverages contain oxalates, which should be avoided by those who tend to develop oxalate-containing kidney stones. *Allergy alert:* Some ingredients used in cola may trigger hives, a rash around the mouth, swollen lips, or other reactions in sensitive people. Some allergists advise people who are allergic to aspirin to avoid soft drinks, including colas.

History: Created by a druggist in Atlanta in May 1886, cola was originally sold as a medicinal cure for headaches and hangovers.

Description: Cola formulas are heavily guarded industrial secrets, but most are based on kola nut extract.

COLA, Low-Cal, with Aspartame

Serving: 1 can (12 fl. oz.)

Calories	4
Fat	0
Saturated	0
Monounsaturated	0
Polyunsaturated	0
Calories from fat	0
Cholesterol	0
Sodium	21 mg
Protein	0.4 g
Carbohydrate	0.4 g
Dietary fiber	0

CHIEF NUTRIENTS

(None of the nutrients meet or exceed 5% of the RDA.)

Caution: The caffeine in cola makes some people jumpy when it's consumed in large quantities. Cola beverages contain oxalates, which should be avoided by those who tend to develop oxalate-containing kidney stones. People with phenylketonuria (PKU) tend to accumulate phenylalanine and must avoid the aspartame in low-calorie cola. *Allergy alert:* Some ingredients may trigger hives, a rash around the mouth, swollen lips, or other reactions in sensitive people. People allergic to aspirin are sometimes advised to avoid soft drinks, including colas. Aspartame may trigger or worsen migraine headaches.

Strength: Practically calorie-free; can help heavy soda drinkers lose weight.

Curiosity: The aspartame in diet colas is an artificial sweetener derived from phenylalanine, an amino acid.

COLA, Low-Cal, with Saccharin

Serving: 1 can (12 fl. oz.)

Calories	0
Fat	0
Saturated	0
Monounsaturated	0
Polyunsaturated	0
Calories from fat	0
Cholesterol	0
Sodium	57 mg
Protein	0
Carbohydrate	0.4 g
Dietary fiber	0

CHIEF NUTRIENTS

(None of the nutrients meet or exceed 5% of the RDA.)

Strength: The calorie, fat, and sugar content of this beverage is zero—so it's the choice of many dieters.

Caution: Diet sodas like this contain some sodium. Take that into account if you're trying to limit your daily intake. And like regular cola drinks, low-cal cola- and pepper-type drinks contain some caffeine, which might cause jumpiness or nervousness in some people. *Allergy alert*: Some people get hives and other allergic reactions from sodas.

Curiosity: This diet drink contains saccharin, an artificial sweetener once suspected of causing bladder cancer. However, the National Cancer Institute has concluded that consuming saccharin is safe, even for diabetics, who tend to rely more heavily on artificially sweetened beverages.

Description: A bubbly, carbonated, cola-flavored drink containing sodium, saccharin, and other ingredients.

A Better Idea: For a low-cal beverage free of artificial colors or other additives, splash some fruit juice or a twist of lemon into a tall glass of cool club soda.

CREAM SODA

Serving: 1 can (12 fl. oz.)

Calories	189
Fat	0
Saturated	0
Monounsaturated	0
Polyunsaturated	0
Calories from fat	0
Cholesterol	0
Sodium	45 mg
Protein	0
Carbohydrate	49.3 g
Dietary fiber	0

CHIEF NUTRIENTS

(None of the nutrients meet or exceed 5% of the RDA.)

Caution: Considerably higher in calories than other soft drinks, cream soda is not for weight watchers. And it's fairly high in sugar, which may contribute to tooth decay. *Allergy alert:* Many allergists recommend that people who are allergic to aspirin avoid artificially flavored soft drinks, which may contain salicylate, an active ingredient in aspirin.

Curiosity: Cream soda's characteristic flavor comes from vanilla, derived from the pods of climbing orchids native to tropical American forests. Artificial vanilla is made from wood pulp by-products that have been treated with chemicals.

GINGER ALE

Serving: 1 can (12 fl. oz.)

Calories	124
Fat	0
Saturated	0
Monounsaturated	0
Polyunsaturated	0
Calories from fat	0
Cholesterol	0
Sodium	26 mg
Protein	0
Carbohydrate	31.8 g
Dietary fiber	0

CHIEF NUTRIENTS

NUTRIENT	AMOUNT	% RDA
Iron	0.7 mg	7

Caution: Contains 7 tsp. of sugar per serving. Excessive consumption may contribute to weight gain and tooth decay. *Allergy alert:* May contain citric acid, a potential problem for people allergic to citrus. Many allergists recommend that people who are allergic to aspirin avoid artificially flavored soft drinks in general, as they may contain salicylate, an active ingredient in aspirin.

Strengths: Ginger ale has a few calories less than cola, lemon-lime soda, or other soft drinks. A glass or two may help to relieve an upset stomach. As a bonus, you get a little iron.

Description: A carbonated, sweetened soft drink flavored with ginger root oil or extract (or both) and lemon or lime oil (or both), plus caramel coloring and citric acid.

GRAPE SODA

Serving: 1 can (12 fl. oz.)

Calories	160
Fat	0
Saturated	0
Monounsaturated	0
Polyunsaturated	0
Calories from fat	0
Cholesterol	0
Sodium	56 mg
Protein	0
Carbohydrate	41.7 g
Dietary fiber	0

CHIEF NUTRIENTS

(None of the nutrients meet or exceed 5% of the RDA.)

Caution: Contains moderate amounts of sodium. Grape soda is fairly high in sugar, so excessive consumption may contribute to weight gain and tooth decay. *Allergy alert:* Many allergists recommend that people sensitive to aspirin avoid grape juice, which contains naturally occurring salicylate, an active ingredient in aspirin.

Description: A sweetened, carbonated soft drink flavored with oil of cognac, methyl anthranilate (artificial grape flavoring), and sometimes even some real grape juice.

A Better Idea: For more nutritious, low-calorie, additive-free grape soda, combine carbonated water and undiluted frozen grape juice in proportions of about 3 to 1.

LEMON-LIME SODA

Serving: 1 can (12 fl. oz.)

Calories	148
Fat	0
Saturated	0
Monounsaturated	0
Polyunsaturated	0
Calories from fat	0
Cholesterol	0
Sodium	40 mg
Protein	0
Carbohydrate	38.3 g
Dietary fiber	0

CHIEF NUTRIENTS

(None of the nutrients meet or exceed 5% of the RDA.)

Caution: Contains approximately 9 tsp. of sugar, so excessive consumption may contribute to weight gain and tooth decay. *Allergy alert:* May contain citric acid, a potential problem for people allergic to citrus. Many allergists recommend that people who are allergic to aspirin avoid artificially flavored soft drinks (including citrus-flavored drinks), which may contain salicylate (also found in aspirin).

Description: A sweetened, carbonated soft drink flavored with lemon and lime oil and citric acid.

Curiosity: 7-Up, a popular version of this soda, was named after a card game, seven-up—after unsuccessful attempts to market it under six other names. Its first name was Bib-label Lithiated Lemon-Lime soda.

A Better Idea: For more nutritious, low-calorie, additive-free lemon-lime soda, combine carbonated water with undiluted frozen lemon and lime juices in proportions of about 3 to 1.

NONCOLA, *Low-Cal, with Saccharin*

Serving: 1 can (12 fl. oz.)

Calories	0
Fat	0
Saturated	0
Monounsaturated	0
Polyunsaturated	0
Calories from fat	0
Cholesterol	0
Sodium	57 mg
Protein	0
Carbohydrate	0.4 g
Dietary fiber	0

CHIEF NUTRIENTS

(None of the nutrients meet or exceed 5% of the RDA.)

Strengths: Low-cal soda's big draw is that it contains no fat, sugar, or calories, making it a popular recreational beverage among dieters.

Caution: Diet sodas do contain some sodium, so if you're trying to limit your daily sodium intake, you probably shouldn't drink more than a couple of diet sodas a day. A yellow light to the caffeine-sensitive: Low-cal soft drinks may contain some caffeine. The amount varies from product to product, but it's something to keep in mind if caffeine makes you jumpy.

Description: This carbonated beverage contains saccharin as a sweetener. Though once suspected of causing bladder cancer, saccharin has been found safe by the National Cancer Institute.

A Better Idea: You can make a low-cal beverage free of additives: Just add some fruit juice to club soda.

ORANGE SODA

Serving: 1 can (12 fl. oz.)

Calories	179
Fat	0
Saturated	0
Monounsaturated	0
Polyunsaturated	0
Calories from fat	0
Cholesterol	0
Sodium	45 mg
Protein	0
Carbohydrate	46.0 g
Dietary fiber	0

CHIEF NUTRIENTS

(None of the nutrients meet or exceed 5% of the RDA.)

Caution: One of the higher-calorie soft drinks, orange soda has enough sugar to contribute to weight gain and tooth decay. *Allergy alert:* May contain citric acid, a potential problem for people allergic to citrus. Many allergists recommend that people who are allergic to aspirin avoid soft drinks that may contain salicylate, an active ingredient in aspirin.

Description: A carbonated, sweetened soft drink flavored with oil of orange, citric acid, and sometimes real orange juice.

A Better Idea: For a more nutritious, low-calorie orange soda, combine carbonated water with undiluted frozen orange juice in proportions of about 3 to 1.

PEPPER-TYPE SODA

Serving: 1 can (12 fl. oz.)

Calories	151
Fat	0.4 g
Saturated	0.3 g
Monounsaturated	N/A
Polyunsaturated	N/A
Calories from fat	2%
Cholesterol	0
Sodium	37 mg
Protein	0
Carbohydrate	38.2 g
Dietary fiber	0

CHIEF NUTRIENTS

(None of the nutrients meet or exceed 5% of the RDA.)

Caution: This soda is fairly high in sugar, so avoid it if you want to keep your weight down and prevent tooth decay. Contains some caffeine, which makes some people jumpy. *Allergy alert:* If you are allergic to aspirin, you may be allergic to artificially flavored soft drinks: They contain salicylate, an active ingredient in aspirin.

Description: A carbonated, sweetened soft drink flavored with cherry bark extract, oil of bitter almond, and lemon and lime oils.

ROOT BEER

Serving: 1 can (12 fl. oz.)

Calories	152
Fat	0
Saturated	0
Monounsaturated	0
Polyunsaturated	0
Calories from fat	0
Cholesterol	0
Sodium	48 mg
Protein	0
Carbohydrate	39.2 g
Dietary fiber	0

CHIEF NUTRIENTS

(None of the nutrients meet or exceed 5% of the RDA.)

Caution: This traditional hot-summer's-day soda is fairly high in sugar, which may contribute to weight gain and tooth decay. *Allergy alert:* The citric acid in this beverage could be a problem for people allergic to citrus. Many allergists recommend that people who are allergic to aspirin avoid artificially flavored soft drinks that contain oil of wintergreen (methyl salicylate) or other sources of salicylate (an active ingredient in aspirin). Also, people who are allergic to eggs should be aware that eggs are frequently used to make root beer foamy.

Origin: Created in the mid-1800s by Philadelphia pharmacist Charles Hires, who fermented a blend of sugar and yeast with various roots, herbs, and barks.

Description: A carbonated, sweetened soft drink flavored with oil of wintergreen, vanillin, nutmeg, cloves, anise oil, and eucalyptus oil. It also contains citric acid and caramel coloring.

TONIC WATER

Serving: 1 can (12 fl. oz.)

Calories	124
Fat	0
Saturated	0
Monounsaturated	0
Polyunsaturated	0
Calories from fat	0
Cholesterol	0
Sodium	15 mg
Protein	0
Carbohydrate	32.2 g
Dietary fiber	0

CHIEF NUTRIENTS

(None of the nutrients meet or exceed 5% of the RDA.)

Caution: *Allergy alert:* Many allergists recommend that people who are allergic to aspirin avoid artificially flavored soft drinks, which may contain salicylate (an active ingredient in aspirin).
Strengths: Tonic contains a few calories less than cola, lemon-lime soda, or other soft drinks, and it's low in sodium.
Description: A carbonated, lightly sweetened beverage, tonic is flavored with a small amount of quinine (a bitter alkaloid), lemon, and lime. It's also called quinine water. In parts of New England, however, any kind of carbonated soft drink may be referred to as tonic water.
Serve: Customarily mixed with gin to make a gin and tonic cocktail.

FRUIT AND VEGETABLE JUICES/DRINKS

APPLE JUICE, *Unsweetened, Canned/Bottled*

Serving: 8 fl. oz.

Calories	117
Fat	0.3 g
Saturated	trace
Monounsaturated	trace
Polyunsaturated	trace
Calories from fat	2%
Cholesterol	0
Sodium	7 mg
Protein	0.2 g
Carbohydrate	29.0 g
Dietary fiber	0.3 g

CHIEF NUTRIENTS

NUTRIENTS	AMOUNT	% RDA
Iron	0.9 mg	9
Potassium	295.1 mg	8

Strengths: Very low in fat and sodium; no cholesterol. A nice source of iron for healthy blood, and potassium for the normal functioning of nerves and muscles.
Eat with: Whole grain breads or cereals to add fiber that the juice lacks.
Caution: Most brands of apple juice contain very little fiber. For better cholesterol control, healthy digestion, and weight loss, choose whole apples instead. *Allergy alert:* Those allergic to aspirin may react to the natural salicylate in apples.
History: Basic preparation methods haven't changed since colonial times: Apples are ground, pressed to extract the juice, then strained to remove large particles. If the resultant cider is further clarified, juice results.
Storage: Pasteurizing the juice prevents it from fermenting into hard cider; vacuum-packed juice stays fresh for a year without refrigeration.
Serve: Frozen! Apple juice makes naturally sweet frozen pops.

APRICOT NECTAR, Canned

Serving: 8 fl. oz.

Calories	141
Fat	0.2 g
Saturated	trace
Monounsaturated	trace
Polyunsaturated	trace
Calories from fat	1%
Cholesterol	0
Sodium	8 mg
Protein	0.9 g
Carbohydrate	36.1 g
Dietary fiber	1.5 g

CHIEF NUTRIENTS

NUTRIENT	AMOUNT	% RDA
Vitamin A	331.3 RE	33
Iron	1.0 mg	10
Potassium	286.1 mg	8

Strengths: Apricot nectar has a negligible amount of fat and no cholesterol—all the better to help keep serum cholesterol levels down. A very good source of vitamin A and especially beta-carotene, which help prevent certain types of cancer. Vitamin A and iron also help boost immunity to various diseases. There's very little sodium and a nice amount of potassium in apricot nectar, and that's a boon if you have high blood pressure linked to salt sensitivity. More fiber than many other juices because it contains quite a bit of pulp.

Eat with: Other fruits rich in vitamin C (which helps the iron be absorbed more readily). Makes a nice thick syrup for whole grain waffles and pancakes; heat briefly and add some berries for extra vitamin C.

Caution: Be aware that nectar drinks often contain added corn syrup or other sweeteners; read labels. *Allergy alert:* Those allergic to aspirin may react to the natural salicylate in apricots.

CARROT JUICE, Canned

Serving: 6 fl. oz.

Calories	74
Fat	0.3 g
Saturated	trace
Monounsaturated	trace
Polyunsaturated	trace
Calories from fat	3%
Cholesterol	0
Sodium	53 mg
Protein	1.8 g
Carbohydrate	17.1 g
Dietary fiber	1.5 g

CHIEF NUTRIENTS

NUTRIENT	AMOUNT	% RDA
Vitamin A	4,738.0 RE	474
Vitamin C	15.6 mg	26
Vitamin B$_6$	0.4 mg	20
Potassium	537.3 mg	14
Thiamine	0.2 mg	11
Iron	0.9 mg	9
Magnesium	25.8 mg	7
Calcium	44.2 mg	6
Riboflavin	0.1 mg	6

Strengths: Carrot juice is a nutritional powerhouse, with over 400% of the RDA for vitamin A. As a source of vitamin C, a glass of this hearty juice also helps fight infection, heal wounds, and boost calcium and iron absorption. Carrot juice is also a good source of vitamin B$_6$, which reduces excretion of oxalic acid and helps prevent kidney stones from forming.

Eat with: Foods, such as spinach, that are high in oxalates.

Selection: Vegetable juice cocktail often has carrot juice as one ingredient. Either vegetable juice or pure carrot juice makes a good chaser to a cocktail, since they replenish some of the B$_6$ destroyed by alcohol.

CITRUS FRUIT JUICE DRINK *from Concentrate*

Serving: 8 fl. oz.

Calories	114
Fat	0
Saturated	0
Monounsaturated	0
Polyunsaturated	trace
Calories from fat	0
Cholesterol	0
Sodium	7 mg
Protein	0.7 g
Carbohydrate	28.5 g
Dietary fiber	0.2 g

CHIEF NUTRIENTS

NUTRIENT	AMOUNT	% RDA
Vitamin C	67.2 mg	112
Iron	2.8 mg	28
Potassium	277.8 mg	7

Strengths: A terrific source of vitamin C, this drink supplies 112% of the RDA. In addition, the drink contains some potassium, which may help protect against high blood pressure. A significant amount of iron helps the body fight off infections.
Caution: Some fruits contain oxalates, which should be restricted by people with calcium-oxalate stones. *Allergy alert:* Citrus may cause allergic reactions such as hives and headaches in some people.
Description: The drink may be made from a mixture of several types of citrus juice, including orange, lime, lemon, and occasionally grapefruit.
Selection: Choose products that are labeled ''100% fruit juice'' to ensure that you're getting good nutritional value.
A Better Idea: For fresher juice and even higher nutrition, make your own citrus drinks.

CRANBERRY JUICE COCKTAIL, *Bottled*

Serving: 8 fl. oz.

Calories	144
Fat	0.3 g
Saturated	trace
Monounsaturated	N/A
Polyunsaturated	N/A
Calories from fat	2%
Cholesterol	0
Sodium	5 mg
Protein	0
Carbohydrate	36.5 g
Dietary fiber	0.3 g

CHIEF NUTRIENTS

NUTRIENT	AMOUNT	% RDA
Vitamin C	89.7 mg	149

Strengths: Cranberry juice cocktail is so high in vitamin C because it is supplemented with ascorbic acid. It's ideal for women who take birth control pills because they tend to need extra C. The juice is believed by many to help alleviate urinary tract infections, possibly by preventing harmful bacteria from anchoring on bladder walls. Further research is needed, but in the meantime, many doctors say this remedy can't hurt.
Eat with: Any iron-rich food to take advantage of vitamin C's ability to enhance iron absorption.
Caution: Most brands contain added sweetener. Cranberries contain some oxalic acid, which should be restricted by those with calcium-oxalate stones.
Selection: Products labeled cranberry juice cocktail must contain at least 25% cranberry juice. It is possible to buy unsweetened and artificially sweetened juice, both of which are considerably lower in calories than sugared blends.

FRUIT PUNCH JUICE DRINK *from Concentrate*

Serving: 8 fl. oz.

Calories	124
Fat	0.5 g
Saturated	trace
Monounsaturated	trace
Polyunsaturated	trace
Calories from fat	4%
Cholesterol	0
Sodium	12 mg
Protein	0.3 g
Carbohydrate	30.3 g
Dietary fiber	0.2 g

CHIEF NUTRIENTS

NUTRIENT	AMOUNT	% RDA
Vitamin C	13.9 mg	23
Riboflavin	0.2 mg	9
Iron	0.6 mg	6
Potassium	191.0 mg	5

Caution: Although called fruit punch, this drink is mostly composed of water and various forms of sugars, including fructose, corn syrup, and sucrose.

Strengths: Fruit punch contains vitamin C, which helps in the absorption of iron. It has some riboflavin, potassium, and also some iron, which is essential for healthy red blood cells. The fruit used to make this drink also may contain oxalates which should be restricted by people with calcium oxalate stones. *Allergy alert:* Fresh fruit can cause allergic reactions, such as hives, in some people.

A Better Idea: Instead of fruit punch, why not drink pure fruit juices that have more vitamin C and don't contain sugar?

GRAPE JUICE, *Unsweetened, Canned/Bottled*

Serving: 8 fl. oz.

Calories	154
Fat	0.2 g
Saturated	trace
Monounsaturated	trace
Polyunsaturated	trace
Calories from fat	1%
Cholesterol	0
Sodium	8 mg
Protein	1.4 g
Carbohydrate	37.9 g
Dietary fiber	trace

CHIEF NUTRIENTS

NUTRIENT	AMOUNT	% RDA
Potassium	334.0 mg	9
Vitamin B$_6$	0.2 mg	8
Magnesium	25.3 mg	7
Iron	0.6 mg	6
Riboflavin	0.1 mg	5

Strengths: No fat or sodium to speak of. Has some potassium for possible protection against stroke and control of high blood pressure. Vitamin C is added to some brands, which helps the iron in the juice be better absorbed.

Caution: *Allergy alert:* Those allergic to aspirin may react to the natural salicylate in grapes.

History: The production of grape juice in the U.S. began in 1869 when some churches needed a substitute for wine to serve in communion services.

Description: Most of the grape juice produced in this country comes from Concord and other dark-skinned grapes. To make sure you're getting 100% juice, read the label carefully.

Storage: To safeguard vitamin C, store in a tightly closed jar with little air space at the top. Transfer to progressively smaller containers as you use the juice.

GRAPE JUICE DRINK, *Canned*   ⓒ ☀ ✵Ⅱ✵ ⬇ 🔟

Serving: 8 fl. oz.

Calories	125
Fat	0
Saturated	0
Monounsaturated	0
Polyunsaturated	0
Calories from fat	0
Cholesterol	0
Sodium	3 mg
Protein	0.3 g
Carbohydrate	32.3 g
Dietary fiber	0

CHIEF NUTRIENTS

NUTRIENT	AMOUNT	% RDA
Vitamin C	40.1 mg	67

Caution: This product is a watered-down version of pure juice and contains added sugar in some form. Since this drink has a lower concentration of extract, it contains fewer nutrients than pure juice. Might have enough oxalic acid to cause problems for those with calcium-oxalate stones. *Allergy alert:* Depending on the amount of juice in the drink, it may trigger allergic reactions in those sensitive to the natural salicylate in grapes.

Strengths: Has been enriched with vitamin C, which contributes some measure of protection against certain types of cancer and helps build immunity.

A Better Idea: To avoid the added sugar that comes in prepackaged grape drink, dilute pure grape juice yourself. (Try adding sparkling water for a refreshing drink). That way, you'll avoid the added sugar that comes in prepackaged grape drink, and you'll get more nutrients.

GRAPEFRUIT JUICE, *Unsweetened, Canned*   ⓒ ☀ ✵Ⅱ✵ ⬇ 🔟

Serving: 8 fl. oz.

Calories	94
Fat	0.3 g
Saturated	trace
Monounsaturated	trace
Polyunsaturated	trace
Calories from fat	2%
Cholesterol	0
Sodium	2 mg
Protein	1.3 g
Carbohydrate	22.1 g
Dietary fiber	N/A

CHIEF NUTRIENTS

NUTRIENT	AMOUNT	% RDA
Vitamin C	72.1 mg	120
Folate	25.7 mcg	13
Potassium	377.9 mg	10
Magnesium	24.7 mg	7
Thiamine	0.1 mg	7

Strengths: A wonderful source of vitamin C for healthy skin, gums, and blood vessels, plus immunity to certain diseases. With a good amount of potassium (and little sodium), grapefruit juice may help control high blood pressure. Its magnesium, working along with the potassium, acts as a muscle relaxant. Be aware that red and pink grapefruit juices have considerably more beta-carotene than white, so they're more valuable for possible protection against certain cancers.

Caution: The citrus acid in grapefruit juice may erode tooth enamel; sip the juice through a straw. Also, citrus may cause intestinal gas. *Allergy alert:* Allergy to citrus fruits may trigger bed-wetting in some children. May also cause allergic headaches. Those allergic to citrus peel may react to the juice (which may contain some peel).

Storage: To safeguard vitamin C, keep the juice cold and in a tightly closed glass jar. If the juice is stored this way and used within 2 weeks, little vitamin C will be lost.

LEMONADE, *White or Pink, from Concentrate* ⬇ ▥

Serving: 8 fl. oz.

Calories	99
Fat	0.1 g
Saturated	trace
Monounsaturated	N/A
Polyunsaturated	trace
Calories from fat	1%
Cholesterol	0
Sodium	7 mg
Protein	0.3 g
Carbohydrate	26.0 g
Dietary fiber	1.0 g

CHIEF NUTRIENTS

NUTRIENT	AMOUNT	% RDA
Vitamin C	9.7 mg	16

Caution: This drink is one way to get vitamin C, but there's a high-calorie cost: Lemonade contains lots of sweetener to counteract the acidity of lemon juice. Although the lemon juice is highly diluted, large quantities could cause problems for those with calcium-oxalate stones. Too much could also sensitize teeth and lead to enamel erosion; use a straw when drinking lemonade. *Allergy alert:* If you make fresh lemonade, squeezing lemons could provoke contact dermatitis if you're sensitive to the lemon oil in the peel; wear plastic gloves.

Strengths: A decent amount of vitamin C. No fat or cholesterol; very little sodium. Pink lemonade gets its color from grape juice, which adds just a touch of vitamin A to the product.

A Better Idea: To make your own low-sugar, low-calorie lemonade, mix lemon juice with sparkling water. To get the most juice from a lemon, make sure it's at room temperature before squeezing (or microwave it on high for about 30 seconds). It also helps to roll the lemon around on the counter under the palm of your hand before cutting it to break open juice sacs.

LEMONADE *from Powder* ⬇ ▥

Serving: 8 fl. oz.

Calories	97
Fat	0
Saturated	0
Monounsaturated	0
Polyunsaturated	N/A
Calories from fat	0%
Cholesterol	0
Sodium	12 mg
Protein	0
Carbohydrate	25.3 g
Dietary fiber	0

CHIEF NUTRIENTS

NUTRIENT	AMOUNT	% RDA
Vitamin C	7.9 mg	13
Calcium	67.0 mg	8

Caution: The main ingredient in powdered lemonade is sugar; the store-bought mix is less than 5% fruit juice. So if you crave a healthy fruit drink, you probably should quench your thirst with something else.

Strengths: Powdered lemonade has just about 100 calories per serving and does have a good amount of vitamin C. Some calcium helps improve bones and teeth. *Allergy alert:* Although this product is made with little fruit juice, some people who are allergic to citrus may have reactions such as hives and headaches.

Curiosity: Prospectors who could afford them ate lemons to combat scurvy during the California gold rush in 1849. The California lemon industry began soon after.

A Better Idea: Make your own lemonade, or try citrus drinks that contain less sugar, and more fruit juice and vitamin C.

LIMEADE *from Concentrate*

Serving: 8 fl. oz.

Calories	101
Fat	0.1 g
Saturated	N/A
Monounsaturated	N/A
Polyunsaturated	trace
Calories from fat	1%
Cholesterol	0
Sodium	5 mg
Protein	0.1 g
Carbohydrate	27.2 g
Dietary fiber	N/A

CHIEF NUTRIENTS

NUTRIENT	AMOUNT	% RDA
Vitamin C	6.7 mg	11

Caution: Limeade is a high-calorie way to get vitamin C, because it contains lots of sweetener to counteract the acidity of lime juice. Although the juice is highly diluted, too much could sensitize teeth and lead to enamel erosion; use a straw when drinking. *Allergy alert:* Squeezing limes for homemade limeade could provoke contact dermatitis if you're sensitive to oil in the peel; wear plastic gloves.

Strengths: A decent amount of vitamin C. No fat or cholesterol; very little sodium.

A Better Idea: To make your own low-sugar, low-calorie limeade, mix lime juice with sparkling water. To get the most juice from a lime, make sure it's at room temperature before squeezing (or microwave it on high for about 30 seconds). It also helps to roll the fruit around on the counter under the palm of your hand before cutting to break open juice sacs.

ORANGE DRINK, *Canned*

Serving: 8 fl. oz.

Calories	126
Fat	0
Saturated	0
Monounsaturated	0
Polyunsaturated	0
Calories from fat	0
Cholesterol	0
Sodium	40 mg
Protein	0
Carbohydrate	32.0 g
Dietary fiber	0.2 g

CHIEF NUTRIENTS

NUTRIENT	AMOUNT	% RDA
Vitamin C	84.5 mg	141
Iron	0.7 mg	7

Caution: Not the best way to get your vitamin C: Orange drink is a diluted and sweetened form of orange juice, and it's much higher in sodium than pure juice. If the juice content is high, the cautions that apply to pure juice are relevant to the drink as well. Orange juice contains some oxalic acid, which should be restricted by those with calcium-oxalate stones; it may cause intestinal gas, and it's hard on tooth enamel. To help protect your teeth, use a straw when sipping orange drink. *Allergy alert:* Some people have allergic reactions to orange peel, which may be present in orange drink. Those allergic to grass pollen or aspirin may react to oranges in any form.

Strengths: Lots of vitamin C (due to extra supplementation). Orange drink is a surprising source of iron. It has no fat or cholesterol, and its caloric content is comparable to that of pure juice.

A Better Idea: If you want diluted orange juice, you're better off mixing real juice with water or seltzer rather than buying prepackaged juice. You'll get more for your money, and you won't be consuming extra sweeteners.

ORANGE JUICE

Serving: 8 fl. oz.

Calories	112
Fat	0.5 g
Saturated	trace
Monounsaturated	trace
Polyunsaturated	trace
Calories from fat	4%
Cholesterol	0
Sodium	2 mg
Protein	1.7 g
Carbohydrate	25.8 g
Dietary fiber	0.2 g

CHIEF NUTRIENTS

NUTRIENT	AMOUNT	% RDA
Vitamin C	124.0 mg	207
Folate	75.1 mcg	38
Thiamine	0.2 mg	15
Potassium	496.0 mg	13
Magnesium	27.3 mg	8
Niacin	1.0 mg	5

Strengths: Fresh-squeezed orange juice provides a whopping dose of vitamin C and plenty of folate (both of which benefit women taking birth control pills). Nice amounts of other B vitamins plus magnesium promote healthy skin, nerves, and muscles. There's some fiber for cholesterol control, although the exact amount varies according to how much pulp is in the juice.

Eat with: Anything—and the vitamin C in orange juice boosts absorption of iron from plant foods.

Caution: Since oranges contain some oxalic acid, which should be restricted by those with calcium-oxalate stones, orange juice may be problematic. It may also cause intestinal gas. It's better to drink orange juice with a straw, because citrus juice is hard on tooth enamel. *Allergy alert:* Some people have allergic reactions to orange peel, which may be present in the juice. Those allergic to grass pollen or aspirin may react to oranges in any form.

Curiosity: As orange juice loses vitamin C, it also loses flavor, so trust your taste buds for judging the juice's nutritional value.

PAPAYA NECTAR, *Canned*

Serving: 8 fl. oz.

Calories	143
Fat	0.4 g
Saturated	trace
Monounsaturated	trace
Polyunsaturated	trace
Calories from fat	2%
Cholesterol	0
Sodium	13 mg
Protein	0.4 g
Carbohydrate	36.3 g
Dietary fiber	1.5 g

CHIEF NUTRIENTS

NUTRIENT	AMOUNT	% RDA
Vitamin C	7.5 mg	13
Iron	0.9 mg	9

Strengths: Barely any fat, not much sodium, no cholesterol. Papaya nectar is sure to contain some beta-carotene (the deep color is a tipoff) for an extra measure of protection from some types of cancer. This is a surprising source of iron, with enough vitamin C to help enhance absorption.

Selection: The official FDA standard of identity for papaya nectar demands that the juice contain at least 33% (by weight) papaya. But the remainder may be comprised of water and corn syrup or other sweeteners. So read labels to be aware of what you're buying.

Storage: When using canned papaya nectar, be sure to transfer leftover juice to a glass or other nonmetal container for storage. The juice could react with any lead in the can.

Serve: With whole grain waffles and pancakes: It makes a nice, thick low-fat topping. Heat briefly and add some blueberries, raspberries, or chopped oranges for extra vitamin C and a few nuts or seeds for additional fiber.

PASSION FRUIT JUICE

Serving: 2 fl. oz.

Calories	31
Fat	trace
Saturated	N/A
Monounsaturated	N/A
Polyunsaturated	N/A
Calories from fat	1%
Cholesterol	0
Sodium	4 mg
Protein	0.2 g
Carbohydrate	8.4 g
Dietary fiber	N/A

CHIEF NUTRIENTS

NUTRIENT	AMOUNT	% RDA
Vitamin C	18.4 mg	31
Niacin	0.9 mg	5
Potassium	171.7 mg	5
Riboflavin	0.1 mg	5

Strengths: Low in calories, fat, and sodium; no cholesterol. A very good source of vitamin C for stronger immunity and possible protection against certain types of cancer. Passion fruit juice has some B vitamins for healthy skin and nerves. A portion of potassium (combined with a minimum of sodium) helps protect against high blood pressure and stroke.

Selection: If you're buying passion fruit juice, check labels for water, sweeteners, and other additives.

Serve: As a nice marinade for poultry, seafood, and meat. (Try seasoning the juice with a little soy sauce, rice vinegar, ginger, and garlic.) To make an intriguingly different dressing for green or fruit salads, mix with a little oil, lime juice, orange juice, and honey or hot pepper sauce.

A Better Idea: You can make your own juice. Here's how: Cut the fruits in half crosswise and spoon the flesh and seeds into a nonaluminum sieve; press with the back of a spoon to extract all the juice. If the resulting juice is too tart, dilute with a little water and a touch of sweetener.

PEACH NECTAR, *Canned*

Serving: 8 fl. oz.

Calories	135
Fat	trace
Saturated	trace
Monounsaturated	trace
Polyunsaturated	trace
Calories from fat	0.3%
Cholesterol	0
Sodium	17 mg
Protein	0.7 g
Carbohydrate	34.7 g
Dietary fiber	1.5 g

CHIEF NUTRIENTS

NUTRIENT	AMOUNT	% RDA
Vitamin C	13.2 mg	22
Vitamin A	64.7 RE	7

Strengths: Peach nectar has only trace amounts of fat, no cholesterol, and it's quite low in sodium. And peach nectar has fiber because it contains a good bit of pulp. One cup has almost a quarter of the RDA for vitamin C. Some vitamin A enhances immunity and possibly prevents certain cancers, especially of the breast, lungs, and colon. Some brands are enriched with extra vitamin C.

Caution: Many nectars are made from fruit concentrate that is then diluted with water and sweetened with sugar or corn syrup. Read labels if you'd rather avoid such sweeteners. *Allergy alert:* Those allergic to aspirin may react to the natural salicylate in peaches.

Storage: When using canned peach nectar, be sure to transfer leftover juice to a glass or other nonmetal container for storage. Otherwise, there could be a chemical reaction between the juice and the can.

PINEAPPLE JUICE, *Unsweetened, Canned*

Serving: 8 fl. oz.

Calories	140
Fat	0.2 g
Saturated	trace
Monounsaturated	trace
Polyunsaturated	trace
Calories from fat	1%
Cholesterol	0
Sodium	3 mg
Protein	0.8 g
Carbohydrate	34.5 g
Dietary fiber	N/A

CHIEF NUTRIENTS

NUTRIENT	AMOUNT	% RDA
Vitamin C	26.8 mg	45
Folate	57.8 mcg	29
Vitamin B_6	0.2 mg	12
Magnesium	32.5 mg	9
Potassium	335.0 mg	9
Thiamine	0.1 mg	9
Iron	0.7 mg	7
Calcium	42.5 mg	5

Strengths: Pineapple juice has lots of vitamin C to enhance immunity and provide some protection against certain types of cancer. (Some brands are also supplemented with extra C.) The juice has a nice complement of vitamins and minerals, including various B vitamins for energy. It's very low in fat and sodium, with no cholesterol and only a moderate amount of calories. It even has some iron and calcium for strong blood and bones.

Eat with: Any iron-containing food to take advantage of vitamin C's ability to help enhance absorption of the mineral. Add to baked goods instead of sugar, as a natural sweetener.

Caution: *Allergy alert:* Contains bromelin, an enzyme that can cause dermatitis in some people.

Curiosity: The bromelin in pineapple juice breaks down protein, so marinating meat in the juice can help tenderize it.

History: Columbus found pineapple growing on the island of Guadaloupe in 1493. The original Indian name meant "exquisite fruit."

PRUNE JUICE, *Canned*

Serving: 8 fl. oz.

Calories	182
Fat	0.1 g
Saturated	trace
Monounsaturated	trace
Polyunsaturated	trace
Calories from fat	0.4%
Cholesterol	0
Sodium	10 mg
Protein	1.6 g
Carbohydrate	44.7 g
Dietary fiber	2.6 g

CHIEF NUTRIENTS

NUTRIENT	AMOUNT	% RDA
Iron	3.0 mg	30
Vitamin B_6	0.6 mg	28
Potassium	706.6 mg	19
Vitamin C	10.5 mg	18
Niacin	2.0 mg	11
Riboflavin	0.2 mg	11
Magnesium	35.8 mg	10

Strengths: Made from the water extract of dried prunes, the juice is high in fiber, which helps improve digestion and control cholesterol. Prune juice is a rich source of iron, with plentiful vitamin C built in to enhance its absorption. Good amounts of potassium and magnesium (and not very much sodium) may help control high blood pressure in some individuals. And there's quite a bit of vitamin B_6 to help fight infection.

Eat with: This concentrated source of nutrients complements low-fat cheeses or meats, which provide protein the juice lacks.

Caution: A bit high in calories. (And some brands contain added sweetener; check labels.) Excess consumption may cause intestinal gas. *Drug interaction:* People taking levodopa should not drink excessive amounts of prune juice because the B_6 may interfere with the drug. *Allergy alert:* Those allergic to aspirin may react to the natural salicylate in prunes.

TOMATO JUICE, *Canned*
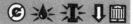

Serving: 6 fl. oz.

Calories	31
Fat	0.1 g
Saturated	trace
Monounsaturated	trace
Polyunsaturated	trace
Calories from fat	3%
Cholesterol	0
Sodium	657 mg
Protein	1.4 g
Carbohydrate	7.7 g
Dietary fiber	1.5 g

CHIEF NUTRIENTS

NUTRIENT	AMOUNT	% RDA
Vitamin C	33.3 mg	56
Folate	36.2 mcg	18
Iron	1.1 mg	11
Potassium	400.4 mg	11
Vitamin A	101.9 RE	10
Vitamin B$_6$	0.2 mg	10
Magnesium	20.0 mg	6
Niacin	1.2 mg	6
Thiamine	0.1 mg	6

Strengths: Tomato juice is a whopping good supplier of vitamin C, with more than half your daily quota in a single 6-oz. glass. It's also a decent source of vitamin A, iron, folate, and vitamin B$_6$. And it has a fair amount of other B vitamins that help prevent anemia and skin problems and boost immunity. All in all, tomato juice is a filling, low-calorie meal-starter for people trying to control their weight and appetite.

Eat with: Beans, greens, and other vegetable sources of iron to take advantage of the high level of iron-enhancing vitamin C.

Caution: Look for low-sodium varieties if you're on a sodium-restricted diet. *Allergy alert:* Tomatoes are a common food allergen and may trigger hives, headaches, mouth itching, or other reactions in sensitive people. Also, if you are allergic to salicylate in aspirin, you might want to avoid tomato juice, since tomatoes also contain salicylate.

Description: Contains the juice and pureed pulp of tomatoes.

Selection: If you hear a "plop" when you start pouring, you can be sure the juice is thick and rich.

VEGETABLE JUICE COCKTAIL, *Canned*

Serving: 6 fl. oz.

Calories	35
Fat	0.2 g
Saturated fat	trace
Monounsaturated fat	trace
Polyunsaturated fat	trace
Calories from fat	4%
Cholesterol	0
Sodium	664 mg
Protein	1.2 g
Carbohydrate	8.3 g
Dietary fiber	1.5 g

CHIEF NUTRIENTS

NUTRIENT	AMOUNT	% RDA
Vitamin C	50.4 mg	84
Vitamin A	212.9 RE	21
Folate	38.4 mcg	19
Vitamin B$_6$	0.3 mg	13
Potassium	351.3 mg	9
Iron	0.8 mg	8
Niacin	1.3 mg	7
Magnesium	20.0 mg	6

Strengths: For an easy way to add an extra serving of vegetables to your diet, just quaff some vegetable juice cocktail. It's filling yet low in calories, with a bit of fiber to help aid digestion. A terrific source of vitamin C, this drink also provides a nice assortment of B vitamins. It has three times as much immunity-boosting vitamin A as pink grapefruit juice, along with some blood pressure–lowering potassium.

Eat with: Beans and iron-rich vegetables: The high vitamin C content aids iron absorption.

Caution: High in sodium; look for low-sodium brands.

Curiosity: Invented in 1933 by W. G. Peacock, Sr., of Evanston, Illinois.

Description: Usually contains tomato juice, along with carrot juice, celery juice, and other seasonings.

Storage: To protect the vitamin C content, store in a tightly closed glass container and refrigerate.

HOT BEVERAGES

COCOA, *Homemade*

Serving: 8 fl. oz.

Calories	218
Fat	9.1 g
Saturated	5.6 g
Monounsaturated	2.7 g
Polyunsaturated	0.3 g
Calories from fat	37%
Cholesterol	33 mg
Sodium	123 mg
Protein	9.1 g
Carbohydrate	25.8 g
Dietary fiber	3.8 g

CHIEF NUTRIENTS

NUTRIENT	AMOUNT	% RDA
Vitamin B_{12}	0.9 mcg	44
Calcium	298.3 mg	37
Riboflavin	0.4 mg	26
Magnesium	55.5 mg	16
Potassium	479.8 mg	13
Vitamin A	85.0 RE	9
Iron	0.8 mg	8
Zinc	1.2 mg	8
Thiamine	0.1 mg	7
Folate	12.3 mcg	6
Vitamin B_6	0.1 mg	6

Caution: Moderate sodium content. *Allergy alert:* Some allergists suspect that cocoa consumption is somehow linked to multiple sclerosis, although further study is necessary.

Strengths: A very good source of calcium for stronger bones and vitamin B_{12} to promote the normal growth of nerve fibers. In addition, fiber-rich cocoa contains a good amount of riboflavin to help build energy, as well as magnesium and potassium, which may help lower blood pressure.

Description: Cocoa comes from seeds of the cacao tree, which grows in South America's Amazon/Orinoco river basin.

History: Cacao, which means "bitter water" in Aztec, was corrupted to "cocoa" during the 1700s. Cocoa beans were used as money by Spanish explorers during the slave-trading era. Not until the 20th century did a mix of cocoa powder, sugar, and water—hot cocoa—become a favorite beverage.

COFFEE, *Brewed*

Serving: 6 fl. oz.

Calories	4
Fat	0
Saturated	0
Monounsaturated	0
Polyunsaturated	0
Calories from fat	0
Cholesterol	0
Sodium	4 mg
Protein	0.2 g
Carbohydrate	0.7 g
Dietary fiber	0

CHIEF NUTRIENTS

(None of the nutrients meet or exceed 5% of the RDA.)

Caution: Contains about 102 mg of caffeine per serving. Too much caffeine—300 mg or more—overstimulates the system, causing increased heart rate, insomnia, nervousness, headache, irritability, diarrhea, and frequent urination. Tolerance varies from person to person. Drinking coffee within 1 hour after a meal can cut iron absorption by 40% due to the presence of bitter, iron-binding substances called polyphenols. Coffee also contains substances that break down thiamine, rendering the B vitamin inactive.

Curiosity: Three strong cups of coffee can be used as an emergency asthma medicine, since one of caffeine's effects is to relax the bronchial passages.

Origin: Native to Ethiopia, ground beans mixed with fat were first used by African tribal warriors as a food to heighten aggressiveness before going to battle. A local monk brewed coffee after he noticed that goats that ate coffee beans were particularly sprightly.

COFFEE, *Instant*

Serving: 6 fl. oz.

Calories	4
Fat	0
Saturated	0
Monounsaturated	0
Polyunsaturated	0
Calories from fat	0
Cholesterol	0
Sodium	5 mg
Protein	0.2 g
Carbohydrate	0.7 g
Dietary fiber	0

CHIEF NUTRIENTS

(None of the nutrients meet or exceed 5% of the RDA.)

Caution: Contains a little over half as much caffeine as brewed coffee. Tolerance varies from person to person, but caffeinated coffee may cause increased heart rate, insomnia, nervousness, and headache in some people. Drinking coffee within 1 hour after a meal can cut iron absorption by 40%. Bitter, iron-binding polyphenols are the substances that reduce absorption. Other substances in coffee break down thiamine, rendering the B vitamin inactive. May aggravate ulcers.

Strengths: Practically calorie-free, with no fat.

Description: A dehydrated coffee extract.

COFFEE, *Decaffeinated, Instant*

Serving: 6 fl. oz.

Calories	4
Fat	0
Saturated	0
Monounsaturated	0
Polyunsaturated	0
Calories from fat	0
Cholesterol	0
Sodium	5 mg
Protein	0.2 g
Carbohydrate	0.7 g
Dietary fiber	0

CHIEF NUTRIENTS

(None of the nutrients meet or exceed 5% of the RDA.)

Caution: May aggravate ulcers. Drinking coffee within 1 hour after a meal can cut iron absorption by 40% due to the presence of bitter, iron-binding substances called polyphenols. Coffee also contains substances that break down thiamine, rendering the B vitamin inactive.

Strengths: With practically no calories, instant decaffeinated has less than 2 mg of caffeine per serving, compared to 102 mg for brewed and 57 mg for instant coffee. A welcome alternative for those who feel caffeine makes them jittery or keeps them awake at night.

Description: In the decaffeination process, caffeine is extracted from unroasted green coffee beans. This process removes some of coffee's oils and waxes; it also affects the flavor.

TEA, Brewed

Serving: 6 fl. oz.

Calories	2
Fat	0
Saturated	0
Monounsaturated	0
Polyunsaturated	0
Calories from fat	0
Cholesterol	0
Sodium	5 mg
Protein	0
Carbohydrate	0.5 g
Dietary fiber	0

CHIEF NUTRIENTS

(None of the nutrients meet or exceed 5% of the RDA.)

Strengths: Brewed tea contains about half as much caffeine as instant coffee, with no fat or cholesterol and too few calories to mention.

Caution: Tea is a diuretic that increases water loss through the kidneys. So don't count on drinking tea to up your daily intake of water. Also, tea contains tannins and other iron-binding substances, which tend to interfere with absorption of iron. It also contains 35 mg or so of caffeine, which is something to keep in mind if you're caffeine sensitive. *Allergy alert:* Some allergists advise those who are sensitive to aspirin to avoid tea.

History: Tea is native to China, where the Chinese relied on tea leaves to flavor water that tasted flat after being boiled for purification.

Preparation: Pour boiling water over tea leaves and let steep for 3 to 5 minutes.

TEA, Herbal

Serving: 6 fl. oz.

Calories	2
Fat	0
Saturated	0
Monounsaturated	0
Polyunsaturated	0
Calories from fat	0
Cholesterol	0
Sodium	2 mg
Protein	0
Carbohydrate	0.4 g
Dietary fiber	0

CHIEF NUTRIENTS

(None of the nutrients meet or exceed 5% of the RDA.)

Strengths: Except for maté, herbal tea contains no caffeine, so it's a welcome alternative beverage for those who want to avoid caffeine.

Caution: *Allergy alert:* Individuals who are allergic to flowering weeds or herbs or certain fruits may not tolerate certain herbal teas.

Curiosity: Reputedly, mint tea is a digestive aid. Rosemary tea is supposedly a balm for headaches. And tansy tea is allegedly helpful for alleviating gout.

OTHER BEVERAGES

CAROB BEVERAGE, *Powdered*

Serving: 3 tsp., made w/1 cup milk

Calories	195
Fat	8.2 g
Saturated	5.1 g
Monounsaturated	2.4 g
Polyunsaturated	0.3 g
Calories from fat	38%
Cholesterol	33 mg
Sodium	133 mg
Protein	8.2 g
Carbohydrate	22.5 g
Dietary fiber	N/A

CHIEF NUTRIENTS

NUTRIENT	AMOUNT	% RDA
Vitamin B_{12}	0.9 mcg	44
Calcium	291.8 mg	36
Riboflavin	0.4 mg	23
Magnesium	33.3 mg	10
Potassium	368.6 mg	10
Vitamin A	76.8 RE	8
Iron	0.7 mg	7
Folate	12.3 mcg	6
Thiamine	0.1 mg	6

Caution: This drink is moderate in calories and fat. Those on salt-free diets will probably want to avoid carob beverage because of its sodium content.

Strengths: Rich in vitamin B_{12} and calcium, this beverage is good for healthy blood and helps build strong bones. It has a respectable supply of riboflavin to give you an energy boost. A good amount of potassium may help the body get rid of some of the sodium, and, along with magnesium, may help promote better blood pressure levels.

Origin: Carob may have originated in Syria. However, it has long thrived on the rocky, otherwise barren hillsides surrounding the Mediterranean.

History: Some people believe that during John the Baptist's fast in the desert, the ''locusts'' he ate were actually pods from carob trees.

Description: Carob pods contain a sweet pulp that is 50% sugar. Those who can become accustomed to its taste will enjoy a more nutritious alternative to chocolate.

A Better Idea: Mix carob powder with skim milk if you're calorie conscious.

CHOCOLATE MILK

Serving: 8 fl. oz.

Calories	208
Fat	8.5 g
Saturated	5.3 g
Monounsaturated	2.5 g
Polyunsaturated	0.3 g
Calories from fat	37%
Cholesterol	31 mg
Sodium	149 mg
Protein	7.9 g
Carbohydrate	25.9 g
Dietary fiber	3.8 g

CHIEF NUTRIENTS

NUTRIENT	AMOUNT	% RDA
Vitamin B_{12}	0.8 mcg	42
Calcium	280.3 mg	35
Riboflavin	0.4 mg	24
Potassium	417.3 mg	11
Magnesium	32.6 mg	9
Vitamin A	72.5 RE	7
Zinc	1.0 mg	7
Folate	11.8 mcg	6

Caution: Anyone being treated for kidney stones should avoid chocolate, since it contains oxalate. *Drug interaction:* Chocolate contains tyramine, which may cause dangerous elevations in blood pressure in people taking MAO inhibitors. *Allergy alert*: Canker sores, headaches, tension-fatigue syndrome, and hives have been linked to consumption of chocolate.

Strengths: Preliminary research has found that people who are lactose intolerant may be able to tolerate chocolate milk without suffering ill effects. The drink is rich in fiber, vitamin B_{12}, and calcium, and it has a good dose of potassium to help nerve function. It also has riboflavin for energy, vitamin A for healthy skin, and nice supplies of folate, magnesium, and zinc.

History: Instant-style chocolate milk—powdered, sweetened chocolate added to plain milk—was introduced to the market after World War II.

A Better Idea: Mix cocoa with nonfat milk—or buy ready-made low-fat chocolate milk in your supermarket—and give your arteries a kinder, gentler beverage.

EGGNOG

Serving: 4 fl. oz.

Calories	171
Fat	9.5 g
Saturated	5.7 g
Monounsaturated	2.8 g
Polyunsaturated	0.4 g
Calories from fat	50%
Cholesterol	75 mg
Sodium	69 mg
Protein	4.8 g
Carbohydrate	17.2 g
Dietary fiber	0

CHIEF NUTRIENTS

NUTRIENT	AMOUNT	% RDA
Vitamin B_{12}	0.6 mcg	29
Calcium	165.1 mg	21
Riboflavin	0.2 mg	14
Vitamin A	101.6 RE	10
Magnesium	23.5 mg	7
Potassium	209.8 mg	6

Caution: High in calories, fat, and cholesterol, eggnog may also contain alcohol, so check the label. Home recipes often call for uncooked eggs, which may harbor salmonella bacteria and could cause food poisoning. For safety, use egg substitute in place of fresh eggs. Or prepare a cooked custard from eggs, sugar, and milk, then proceed with the recipe. (Note that commercial eggnog is pasteurized, which kills the bacteria.) *Allergy alert:* Likely to trigger reactions in those allergic to milk or eggs.

Strengths: Good to very good amounts of many of the nutrients prominent in milk, including calcium for strong bones, vitamin A for good vision and healthy skin, and B vitamins for nerves. But all things considered, plain milk is a healthier choice, and eggnog should be reserved for special occasions.

Curiosity: The word *nog* is an Old English term for ale, indicating that alcohol has long been an integral part of the beverage. In early times, red wine was often the spirit of choice in eggnog.

MALTED MILK BEVERAGE, *Powdered*

Serving: 1 Tbsp., made w/1 cup milk

Calories	236
Fat	9.8 g
Saturated	6.0 g
Monounsaturated	2.8 g
Polyunsaturated	0.6 g
Calories from fat	37%
Cholesterol	37 mg
Sodium	223 mg
Protein	10.3 g
Carbohydrate	27.3 g
Dietary fiber	N/A

CHIEF NUTRIENTS

NUTRIENT	AMOUNT	% RDA
Vitamin B_{12}	1.0 mcg	52
Calcium	355.1 mg	44
Riboflavin	0.6 mg	35
Magnesium	53.0 mg	15
Potassium	530.0 mg	14
Folate	21.7 mcg	11
Thiamine	0.2 mg	10
Vitamin A	95.4 RE	10
Vitamin B_6	0.2 mg	10
Zinc	1.1 mg	8

Caution: If high sodium is a concern that has led to "salt talks" with your doctor, it's unlikely that malted milk should be your choice. When combined with whole milk, this high-sodium powder becomes a beverage that's moderately high in fat. *Allergy alert:* People allergic to wheat, barley, or eggs should avoid malt beverages that may contain all these allergens.

Strengths: Low in cholesterol, malted milk beverage gives you a stupendous supply of vitamin B_{12} to perk up your blood. It's very rich in riboflavin to convert food into energy, and it supplies calcium for stronger bones. Also a good source of magnesium and potassium, as well as a respectable source of thiamine and vitamin B_6 for energy. Some vitamin A contributes to better eyesight.

History: A porridge-and-malt mixture used to be the prescription of choice for the young or sick. It was believed that the natural malt helped build stronger bodies and combat disease.

Description: Malt is a grain (usually barley) that's rich in starch. The commercial form is a sweet-tasting powder that dissolves in milk.

SPORTS DRINK, Bottled

Serving: 8 fl. oz.

Calories	60
Fat	0
Saturated	0
Monounsaturated	0
Polyunsaturated	0
Calories from fat	0
Cholesterol	N/A
Sodium	96 mg
Protein	0
Carbohydrate	15.2 g
Dietary fiber	N/A

CHIEF NUTRIENTS

(None of the nutrients meet or exceed 5% of the RDA.)

Strengths: One cup of a sports drink (or ''thirst-quencher'') delivers a powerful boost to your body. The beauty of these low-calorie drinks is that most of them contain no fat, yet provide carbohydrates for energy. While helping to replace electrolytes, a thirst-quencher lives up to its name, subduing your thirst during and after exercise. The key elements are glucose polymers, derivatives of complex carbohydrates. Unlike simple carbohydrates, such as sugar, that release energy in surges, glucose polymers do the same job at more moderate rate. In essence, the drink can mimic simple sugars and help deliver energy quickly, but it also has some of the long-lasting effects of complex carbohydrates.

History: Sports drinks were first used in the 1960s by the University of Florida football team.

Curiosity: A loss of just 2 to 4% of your body weight through sweating can cause a 5 to 10% drop in athletic performance and impair concentration and judgment.

TEA, Iced, Unsweetened, Instant

Serving: 8 fl. oz.

Calories	2
Fat	0
Saturated	0
Monounsaturated	0
Polyunsaturated	0
Calories from fat	0
Cholesterol	0
Sodium	7 mg
Protein	0
Carbohydrate	0.5 g
Dietary fiber	0

CHIEF NUTRIENTS

(None of the nutrients meet or exceed 5% of the RDA.)

Strengths: Tea contains so few calories, they're hardly worth mentioning, and its sodium content is almost nil.

Caution: Instant tea contains about 30 mg of caffeine—about half as much as a cup of instant coffee. So if caffeine affects your nerves, don't drink more than a couple of glasses of instant iced tea at a sitting. Also, tea tends to run high in substances called tannins, which interfere with absorption of iron—a potential problem if you drink lots of tea and consume very little iron-rich food. *Allergy alert:* Those who are sensitive to aspirin may react to the salicylate found in tea.

History: Iced tea was first consumed in 1904, at the St. Louis World's Fair.

Description: Instant tea consists of dehydrated brewed tea that dissolves quickly in cold water. It often contains sugar or sugar substitutes and other flavorings such as cinnamon or lemon.

Preparation: To prevent tea from becoming cloudy, prepare with distilled water.

WATER, *Bottled*

Serving: 8 fl. oz.

Calories	0
Fat	0
Saturated	0
Monounsaturated	0
Polyunsaturated	0
Calories from fat	0
Cholesterol	0
Sodium	2 mg
Protein	0
Carbohydrate	0
Dietary fiber	0

CHIEF NUTRIENTS

(None of the nutrients meet or exceed 5% of the RDA.)

Strengths: Bottled water is a convenient alternative to tap water if your home water supply tastes "off" or contains unacceptable amounts of undesirable contaminants, such as bacteria or nitrate. Some bottled water manufacturers add calcium and magnesium to improve the taste; both are healthful minerals.

Caution: Some bottled waters may be higher in sodium than the example given here. Check labels if you need to monitor your sodium intake for medical reasons.

History: A bottle of mineral water has long been a normal part of the meal for many people in Italy, Spain, Germany, and other parts of Europe. Originally, it was bottled for health reasons. But the custom has also caught on in the U.S., where municipal water is safe to drink.

Description: Bottled water is a generic term for spring water, sparkling water, mineral water, distilled water, or plain tap water that comes in a bottle.

WATER, *Municipal*

Serving: 8 fl. oz.

Calories	0
Fat	0
Saturated	0
Monounsaturated	0
Polyunsaturated	0
Calories from fat	0
Cholesterol	0
Sodium	7 mg
Protein	0
Carbohydrate	0
Dietary fiber	0

CHIEF NUTRIENTS

(None of the nutrients meet or exceed 5% of the RDA.)

Strengths: Virtually every cell in the body needs water to survive. It transports nutrients to all parts of the body via the blood, and serves as the medium for thousands of life-supporting chemical reactions that are constantly taking place inside our bodies. Even tissues that are not thought of as "watery"—like the brain and muscles—consist of about three-quarters water. And bone is more than 20% water. The average person needs to consume 2½ to 3 quarts of water per day whether he or she is thirsty or not. But people who live in hot climates or whose jobs or whose hobbies involve strenuous exertion need to consume more than that. So drink up!

Caution: Tap water may pick up undesirable metals, such as lead, from some pipes used in homes built or remodeled before 1985. Hot water picks up lead more easily, so you might want to consider using only cold water for drinking and cooking. For the purest water possible from the tap, run the faucet first before filling your glass.

Curiosity: It's almost impossible to drink too much water, since the body is very efficient at getting rid of what it doesn't need.

Breads, Rolls, and Muffins
BREADS AND ROLLS

BAGEL, *Egg* ⇩

Serving: 1 (about 2 oz.)

Calories	163
Fat	1.4 g
Saturated	0.5 g
Monounsaturated	N/A
Polyunsaturated	N/A
Calories from fat	8%
Cholesterol	8 mg
Sodium	198 mg
Protein	6.0 g
Carbohydrate	31.0 g
Dietary fiber	1.2 g

CHIEF NUTRIENTS

NUTRIENT	AMOUNT	% RDA
Iron	1.5 mg	15
Thiamine	0.2 mg	14
Niacin	1.9 mg	10
Riboflavin	0.2 mg	9

Strengths: Egg bagels are low in fat, and even though they contain whole eggs, they're low in cholesterol. Made with enriched flour, they offer a good amount of blood-building iron, along with a healthy dose of niacin and thiamine and some riboflavin, all B-complex vitamins essential for energy.

Caution: With almost 200 mg of sodium per serving, egg bagels border on high levels of salt.

Origin: A traditional Jewish food; authentic versions are most often found in delicatessens. Bagels are traditionally eaten with lox (smoked salmon), cream cheese, and sliced white onion.

History: The first printed mention of the word bagel is found in the Community Regulations of Kracow, Poland, for the year 1610. The regulations stated that the food was to be given as a gift to women in childbirth.

A Better Idea: To improve on the lox, bagel, and cream cheese combination, use low-fat cream cheese or cottage cheese.

BAGEL, *Water* ⇩

Serving: 1 (about 2 oz.)

Calories	163
Fat	1.4 g
Saturated	0.2 g
Monounsaturated	N/A
Polyunsaturated	N/A
Calories from fat	8%
Cholesterol	0
Sodium	198 mg
Protein	6.0 g
Carbohydrate	31.0 g
Dietary fiber	1.2 g

CHIEF NUTRIENTS

NUTRIENT	AMOUNT	% RDA
Iron	1.5 mg	15
Thiamine	0.2 mg	14
Niacin	1.9 mg	10
Riboflavin	0.2 mg	9

Strengths: Water bagels are low in fat and cholesterol. Made with enriched flour, they offer a good amount of blood-building iron, along with niacin, riboflavin, and thiamine, B-complex vitamins that aid energy production.

Caution: With almost 200 mg of sodium per serving, water bagels contain moderately high amounts of salt. Eating them with lox (smoked salmon) and cream cheese adds more fat and sodium.

Origin: A Jewish food. The word bagel is derived from the Yiddish word *beygel* (from the German *Beugel*), a round loaf of bread.

Description: All bagels are boiled in water, then baked, but water bagels are chewier than egg bagels.

A Better Idea: If you're fond of cream cheese on your bagels, try replacing it with low-fat cream cheese, cottage cheese, or part-skim ricotta.

BISCUIT, *Baking Powder, Homemade*

Serving: 1 (1 oz.)

Calories	103
Fat	4.8 g
Saturated	1.2 g
Monounsaturated	N/A
Polyunsaturated	N/A
Calories from fat	41%
Cholesterol	0.3 mg
Sodium	175 mg
Protein	2.1 g
Carbohydrate	12.8 g
Dietary fiber	0.5 g

CHIEF NUTRIENTS

NUTRIENT	AMOUNT	% RDA
Iron	0.7 mg	7
Thiamine	0.1 mg	5

Caution: A biscuit is high in fat compared to most bread, and its sodium is moderately high. A single homemade biscuit gets over 40% of its calories from fat and has 175 mg of sodium. Prepared mixes have about 15% less fat but nearly 100 mg more sodium per biscuit. *Allergy alert:* Some people are allergic to wheat flour.

Strengths: The enriched white flour used in most biscuits lends a bit of blood-building iron, along with some thiamine for metabolism.

History: The first biscuits resembled hockey pucks—they were bone-dry, tooth-breaking hardtack, made for long sea voyages

Description: A good biscuit is crispy-brown on the outside, flaky-tender inside—a testament to its fat content. A concoction of flour, fat, baking powder, salt, and water, it is either rolled out and cut into rounds or dropped from a spoon onto a baking sheet.

A Better Idea: For more fiber, substitute whole wheat flour, rye flour, or some cornmeal for part of the white flour. Use margarine or oil instead of butter.

BOSTON BROWN BREAD

Serving: 1 piece (about 1½ oz.)

Calories	95
Fat	0.6 g
Saturated	trace
Monounsaturated	trace
Polyunsaturated	trace
Calories from fat	6%
Cholesterol	0.5 mg
Sodium	113 mg
Protein	2.5 g
Carbohydrate	20.5 g
Dietary fiber	2.1 g

CHIEF NUTRIENTS

NUTRIENT	AMOUNT	% RDA
Iron	1.0 mg	10
Calcium	40.5 mg	5

Strengths: Some versions of Boston brown bread contain more whole wheat and rye flour than others—but most offer a good share of fiber. The blood-boosting iron and calcium in this bread are essential for strong bones.

Caution: *Allergy alert:* May contain eggs, which cause an allergic reaction in some people. Wheat flour is also a common allergen.

History: Boston brown bread was well known among the Puritans, who traditionally served it on Sundays along with Boston baked beans.

Description: Similar to a steamed pudding, Boston brown bread is a heavy, moist, dark bread made with rye flour, coarsely milled whole wheat (graham) flour, white cornmeal, and molasses, along with buttermilk or sour milk. It may also contain raisins and eggs. Commercial brands may be made of white flour with a little bran and coloring added.

Serve: With Boston baked beans.

BREAD STICK, Vienna-Type

Serving: 1 (1¼ oz.)

Calories	106
Fat	1.1 g
Saturated	0.2 g
Monounsaturated	N/A
Polyunsaturated	N/A
Calories from fat	9%
Cholesterol	1 mg
Sodium	548 mg
Protein	3.3 g
Carbohydrate	20.3 g
Dietary fiber	1.1 g

CHIEF NUTRIENTS

(None of the nutrients meet or exceed 5% of the RDA.)

Caution: A bread stick is pretty much a nutritional nothing; about the only thing it offers is sodium, a whopping 547 mg per stick. Definitely to be avoided if you're watching salt levels in your diet. It's also high in calories.

Strengths: Low in fat and cholesterol, a bread stick has a decent amount of fiber.

Origin: Bread sticks are said to have come from a town called Turin in the mountainous region of northern Italy. Napoleon, who was very fond of them, called them *les petits batons de Turin* ("those little Turinese sticks").

Description: A long, thin, hard-baked wheat flour stick often sprinkled with coarse salt or sesame or poppy seeds. Commonly served in Italian restaurants.

A Better Idea: If you're trying to reduce your sodium intake, a slice of plain bread is a better choice.

CORNBREAD, Homemade

Serving: 1 slice (about 1½ oz.)

Calories	93
Fat	3.2 g
Saturated	0.8 g
Monounsaturated	N/A
Polyunsaturated	N/A
Calories from fat	31%
Cholesterol	35 mg
Sodium	283 mg
Protein	3.3 g
Carbohydrate	13.1 g
Dietary fiber	1.2 g

CHIEF NUTRIENTS

NUTRIENT	AMOUNT	% RDA
Calcium	54.0 mg	7
Riboflavin	0.1 mg	5

Strengths: Cornbread offers a bit of bone-strengthening calcium, along with some riboflavin that contributes to healthy blood and good energy levels. Cornbread is a decent source of fiber.

Caution: Because cornbread is usually made with oil, butter, or meat drippings (and often milk and eggs besides), it gets a large percentage of calories from fat. Even when it's made with vegetable oil, nearly one-third of its calories come from fat. *Allergy alert:* Eggs and milk are both common allergens.

Curiosity: Johnnycake, thought to be the precursor of the pancake, was basically a flat cornbread.

Description: A quick bread made with whole-ground yellow cornmeal, wheat flour, vegetable shortening, baking powder, eggs, milk, and sometimes sugar.

Preparation: For an authentic taste and texture, use stone-ground cornmeal; bake it in a heavy, preheated, well-greased pan to give it a rich brown crust. Dress up cornbread by adding chopped green chili peppers or shredded cheddar cheese and green onions to the batter.

Serve: With traditional southern fare—yams, cooked turnip greens, or country ham.

CRACKED WHEAT BREAD ⇩

Serving: 1 slice (about 1 oz.)

Calories	66
Fat	0.6 g
Saturated	trace
Monounsaturated	trace
Polyunsaturated	trace
Calories from fat	8%
Cholesterol	0.5 mg
Sodium	132 mg
Protein	2.2 g
Carbohydrate	13.0 g
Dietary fiber	1.3 g

CHIEF NUTRIENTS

NUTRIENT	AMOUNT	% RDA
Iron	0.7 mg	7
Thiamine	0.1 mg	6

Strengths: Contains some thiamine, which helps convert carbohydrates to energy; and also some iron. Low in fat, with almost no cholesterol, cracked wheat bread offers a nice amount of fiber.

Caution: Watch out for the sodium if you're on a low-salt diet. *Allergy alert*: Some people are allergic to wheat.

Description: This coarse-texture, chewy bread is made from wheat kernels that have been broken into fragments.

Selection: Many commercial brands of cracked wheat bread are mostly white flour, so check labels.

Storage: Whole grain breads keep best in the refrigerator.

Preparation: For homemade bread, soak the cracked wheat for about ½ hour or so to soften it before adding the ingredients. Add ¼ cup of wheat bran for extra fiber.

Serve: Good for sandwiches, and cracked wheat bread makes great toast!

CROISSANT

Serving: 1 (about 2 oz.)

Calories	235
Fat	12.0 g
Saturated	3.5 g
Monounsaturated	6.7 g
Polyunsaturated	1.4 g
Calories from fat	46%
Cholesterol	13 mg
Sodium	452 mg
Protein	5.0 g
Carbohydrate	27.0 g
Dietary fiber	N/A

CHIEF NUTRIENTS

NUTRIENT	AMOUNT	% RDA
Iron	2.1 mg	21
Thiamine	0.2 mg	11
Riboflavin	0.1 mg	8
Niacin	1.3 mg	7

Caution: A croissant is crammed with calories and saturated fat, and its sodium is also very high. *Allergy alert:* Many of the ingredients in this buttery bread—including milk, eggs, wheat, and yeast—are common allergens.

Strengths: A croissant does contain a good amount of iron, which helps combat infection and assists the body in converting beta-carotene to usable vitamin A—and also boosts the body's supply of B vitamins for better energy.

Description: The butter-rich croissant is made with layers of puff pastry or yeast dough, shaped into a crescent, and sometimes filled with cheese or chocolate.

Origin: In the 1680s, Austrian bakers alerted their government to an impending attack by Turkish troops. In honor of this event they created a pastry in the shape of the crescent on the Turkish flag and gave it the name *croissant*—crescent in French.

ENGLISH MUFFIN ⬇

Serving: 1 (about 2 oz.), plain, toasted

Calories	154
Fat	1.3 g
Saturated	N/A
Monounsaturated	N/A
Polyunsaturated	N/A
Calories from fat	7%
Cholesterol	0
Sodium	414 mg
Protein	5.1 g
Carbohydrate	30.0 g
Dietary fiber	1.5 g

CHIEF NUTRIENTS

NUTRIENT	AMOUNT	% RDA
Iron	1.8 mg	18
Thiamine	0.2 mg	16
Calcium	105.0 mg	13
Niacin	2.4 mg	13
Riboflavin	0.2 mg	12
Potassium	364.0 mg	10
Folate	20.7 mcg	5

Strengths: Low in fat and cholesterol-free, a plain English muffin offers a good amount of blood-building iron, along with niacin, thiamine, folate, and riboflavin, B-complex vitamins that enhance energy production. It also has a good amount of bone-strengthening calcium and some potassium.

Description: A round, rather flat, yeast-raised roll, sprinkled with cornmeal and baked on a greased griddle. It may also contain malted barley, vinegar, and farina.

Preparation: Purists insist that English muffins be split with a fork and gently pulled in half, never cut with a knife. In fact, fork-splitting produces a surface of peaks and craters that toasts up crisp and provides lots of pockets for butter and jelly.

Serve: With tea.

A Better Idea: There's no doubt that a "dry" English muffin is pitiful to eat. If you're trying to go easy on the fat, spread low-sugar preserves or jam on your muffin.

FRENCH OR VIENNA BREAD ⬇

Serving: 1 slice (1¼ oz.)

Calories	102
Fat	1.0 g
Saturated	0.2 g
Monounsaturated	N/A
Polyunsaturated	N/A
Calories from fat	9%
Cholesterol	1 mg
Sodium	203 mg
Protein	3.2 g
Carbohydrate	19.4 g
Dietary fiber	0.8 g

CHIEF NUTRIENTS

NUTRIENT	AMOUNT	% RDA
Iron	1.0 mg	10
Thiamine	0.1 mg	9
Niacin	1.2 mg	6
Riboflavin	0.1 mg	5

Strengths: Because it's usually made with enriched white flour, French bread is a good source of blood-building iron. It also offers fair amounts of niacin, riboflavin, and thiamine—B-complex vitamins that perform many functions in the body, including the production of energy. Contains little fat and almost no cholesterol.

Eat with: Fruit or grain to add more fiber.

Caution: Each slice is moderately high in sodium, so be wary if you're on a low-salt diet. *Allergy alert:* Some people are allergic to wheat.

Description: A light, crusty, yeast-raised bread made with white flour, water, salt, and usually a bit of sugar and shortening. To make a crisp crust, the loaf is brushed or sprayed with water during baking.

Serve: Before meals with cheese; thin diagonal slices make a great foundation for sandwiches.

ITALIAN BREAD ⇩

Serving: 1 slice (about 1 oz.)

Calories	83
Fat	0.2 g
Saturated	trace
Monounsaturated	trace
Polyunsaturated	0
Calories from fat	3%
Cholesterol	0.3 mg
Sodium	176 mg
Protein	2.7 g
Carbohydrate	16.9 g
Dietary fiber	0.8 g

CHIEF NUTRIENTS

NUTRIENT	AMOUNT	% RDA
Iron	0.8 mg	8
Thiamine	0.1 mg	8
Niacin	1.0 mg	5

Strengths: Italian bread is usually made with enriched wheat flour, so it offers some iron, thiamine, and niacin, all important for good energy levels. It contains almost no fat or cholesterol.
Caution: Has a moderate amount of sodium, so watch how much you eat if you're on a low-salt diet. *Allergy alert:* Some people are allergic to wheat flour.
Description: Italian bread is similar to French bread, but the loaves are often shorter and thicker. It's a light, crusty, yeast-raised bread made with white flour, water, salt, and usually a bit of sugar and shortening. The crisp crust is sometimes sprinkled with sesame seeds.
Selection: A good fresh loaf should be crisp but squeezable and should smell irresistible.
Preparation: Wrap the bread in foil and rewarm it in the oven before serving.
Serve: With cold meats, salads, and vegetables or with coffee and preserves.

PITA BREAD ⇩

Serving: 1 pocket (about 1¼ oz.)

Calories	105
Fat	0.6 g
Saturated	trace
Monounsaturated	trace
Polyunsaturated	trace
Calories from fat	5%
Cholesterol	0
Sodium	215 mg
Protein	4.0 g
Carbohydrate	20.6 g
Dietary fiber	0.4 g

CHIEF NUTRIENTS

NUTRIENT	AMOUNT	% RDA
Thiamine	0.2 mg	11
Iron	0.9 mg	9
Niacin	1.4 mg	7

Strengths: Made with enriched flour, pita bread contains some blood-building iron. It also has niacin and thiamine, which are important for many body functions, including energy production.
Caution: A moderate amount of sodium; not recommended for people on low-salt diets. *Allergy alert:* Some people are allergic to wheat flour.
Description: Circles of twice-risen dough are baked for a short time in a hot oven. The circles separate, creating a hollow center.
Selection: Delicious whole wheat pita bread, which contains more fiber, is available at some stores.
Preparation: Wrap in foil and warm in the oven before serving.
Serve: Pita bread is usually sliced lengthwise, and the resulting pocket is filled with broiled lamb kabobs, chopped fresh vegetables, cooked chick-peas, or shredded cheese. Or it can be cut into wedges and dipped into baba ghanoush or hummus.

POPOVER, *Homemade*

Serving: 1 (about 1½ oz.)

Calories	90
Fat	3.7 g
Saturated	1.3 g
Monounsaturated	N/A
Polyunsaturated	N/A
Calories from fat	37%
Cholesterol	59 mg
Sodium	88 mg
Protein	3.5 g
Carbohydrate	10.3 g
Dietary fiber	N/A

CHIEF NUTRIENTS

NUTRIENT	AMOUNT	% RDA
Iron	0.8 mg	8
Riboflavin	0.1 mg	7
Thiamine	0.1 mg	5

Strengths: Because a popover is made with enriched wheat flour and whole eggs, it offers a bit of riboflavin, thiamine, and blood-building iron.

Caution: Almost 40% of a popover's calories are from fat. *Allergy alert:* Popovers contain eggs and milk, both common allergens.

Curiosity: Portland popover pudding consists of popovers flavored with meat drippings, garlic, and frequently herbs.

Description: A popover is a crisp, crusty, muffin-shaped, tasty puff of flour with a hollow center. When it bakes, it pops over the sides of the cup-shaped indentation of the muffin tin.

Preparation: A popover is often filled with pieces of cooked egg, fish, meat, or (for dessert) jelly.

A Better Idea: For less fat and more fiber, use a rice, millet, or bulgur pilaf as the base for your meal.

PUMPERNICKEL BREAD

Serving: 1 slice (about 1 oz.)

Calories	79
Fat	0.4 g
Saturated	trace
Monounsaturated	trace
Polyunsaturated	trace
Calories from fat	4%
Cholesterol	0.3 mg
Sodium	182 mg
Protein	2.9 g
Carbohydrate	17.0 g
Dietary fiber	1.9 g

CHIEF NUTRIENTS

NUTRIENT	AMOUNT	% RDA
Iron	0.9 mg	9
Thiamine	0.1 mg	6

Strengths: Pumpernickel bread has a fair amount of blood-building iron, along with some thiamine, a B-complex vitamin essential for plenty of energy. A good source of fiber, it is very low in fat and cholesterol.

Caution: Has a moderate amount of sodium. *Allergy alert:* Some people are allergic to wheat flour, an ingredient in pumpernickel bread.

Curiosity: In German, a "pumpernickel" is a dolt or fool.

Origin: One food writer has speculated that the word is a corruption of the phrase *pain pour Nicole* ("bread for Nicole"); Napoleon's horse, Nicole, was said to have been fond of black bread.

Description: A dark bread made from rye and wheat flour, along with added molasses for color and flavor. Sometimes seasoned.

Selection: Real pumpernickel bread is fairly heavy, chewy, and fragrant, with a thick crust. Poor imitations are light, soft, and fairly odor-free.

RAISIN BREAD ⇩

Serving: 1 slice (about 1 oz.)

Calories	66
Fat	0.7 g
Saturated	trace
Monounsaturated	trace
Polyunsaturated	trace
Calories from fat	10%
Cholesterol	0.8 mg
Sodium	91 mg
Protein	1.7 g
Carbohydrate	13.4 g
Dietary fiber	1.0 g

CHIEF NUTRIENTS

NUTRIENT	AMOUNT	% RDA
Iron	0.7 mg	7
Thiamine	0.1 mg	7

Strengths: Raisin bread contains some blood-building iron, along with a bit of thiamine—a B-complex vitamin important for energy production and other functions.

Caution: Some bakeries may soak their raisins in liquor before adding them to the dough. Check labels or ask about preparation if you want to avoid alcohol. *Allergy alert:* Some raisin breads are made with eggs, a common allergen. And wheat may cause allergies in some people.

Description: Raisin bread is usually a white-flour or egg-dough bread dotted with raisins and flavored with cinnamon. Other ingredients may include walnuts, butter, brown sugar, or honey.

A Better Idea: For more fiber and nutrition, make your own raisin bread, substituting oat bran or whole wheat flour for part of the white flour.

ROLL OR BUN, *Homemade*

Serving: 1 (about 1¼ oz.)

Calories	119
Fat	3.1 g
Saturated	0.9 g
Monounsaturated	N/A
Polyunsaturated	N/A
Calories from fat	23%
Cholesterol	12 mg
Sodium	98 mg
Protein	2.9 g
Carbohydrate	19.6 g
Dietary fiber	0.7 g

CHIEF NUTRIENTS

NUTRIENT	AMOUNT	% RDA
Iron	1.1 mg	11
Thiamine	0.1 mg	8
Riboflavin	0.1 mg	7
Niacin	1.2 mg	6

Strengths: Because it is made with enriched wheat flour, a homemade roll or bun offers a good amount of blood-building iron. It also has a bit of niacin, riboflavin, and thiamine—B vitamins essential for plenty of energy. A plain roll or bun is low in fat and moderate in cholesterol.

Caution: Fancied-up versions could contain hidden butter. *Allergy alert:* Some rolls and buns contain milk or eggs, which are common allergens.

Curiosity: The Parker House roll—a plain, puffy yeast roll—was created about 1855 by a chef at the Parker House hotel in Boston. Because of its purselike appearance, it was also called a pocketbook roll.

Description: Rolls and buns come in all shapes and sizes: cloverleaf, crescent, twist, fan-tan, and plain round. They're made with a yeast-raised dough consisting of flour and milk, along with a bit of salt, sugar, shortening, and sometimes eggs.

A Better Idea: Get additional fiber and nutrients by substituting ½ cup of wheat bran for flour, or using part–whole wheat flour.

RUSKS

Serving: 5 pieces (about 1½ oz.)

Calories	182
Fat	3.8 g
Saturated	1.1 g
Monounsaturated	N/A
Polyunsaturated	N/A
Calories from fat	19%
Cholesterol	4 mg
Sodium	107 mg
Protein	6.0 g
Carbohydrate	30.9 g
Dietary fiber	N/A

CHIEF NUTRIENTS

NUTRIENT	AMOUNT	% RDA
Iron	0.6 mg	6
Riboflavin	0.1 mg	6

Strengths: Rusks have a bit of iron and riboflavin, both essential for high energy. They're quite low in fat and cholesterol and moderate in sodium. Dieters enjoy them because of their satisfying crunchiness.

Caution: *Allergy alert:* Rusks contain eggs and milk, common allergens.

Description: Also called zwieback or twice-baked bread, a rusk is a slice of yeast bread that has been baked until it is dry, crisp, and golden brown. The bread may be slightly sweet, seasoned with mace, cinnamon, and nutmeg, and may be topped with a lemon glaze.

Selection: Rusks are found in most grocery stores, often in the baby food section.

Storage: If stored in an airtight container, rusks will keep indefinitely.

Preparation: Rusks can be recrisped in a warm oven.

Serve: To teething babies, who love to chew on rusks.

RYE BREAD, *American*

Serving: 1 slice (about 1 oz.),

Calories	61
Fat	0.3 g
Saturated	trace
Monounsaturated	trace
Polyunsaturated	trace
Calories from fat	4%
Cholesterol	0.3 mg
Sodium	139 mg
Protein	2.3 g
Carbohydrate	13.0 g
Dietary fiber	1.6 g

CHIEF NUTRIENTS

NUTRIENT	AMOUNT	% RDA
Iron	0.7 mg	7
Thiamine	0.1 mg	5

Strengths: Most commercial brands of rye bread are made with enriched white flour and provide some iron and a little thiamine, which is essential for high energy levels.

Caution: Each slice is moderate in sodium, so proceed with caution if you're watching your salt intake. *Allergy alert:* Contains wheat flour, which is a common allergen.

Origin: Rye bread comes from Germany, Russia, and Scandinavia, where it is considered a cherished staple.

History: Available before World War II, the traditional bread was often seasoned with caraway seeds.

Description: Traditional rye bread, also called black bread, was dense, heavy, and sour-tasting. Today, many commercial rye breads contain ⅓ rye flour and ⅔ white flour and are flavored with acid and colored with caramel.

Selection: Delicatessens and small bakeries are most likely to offer authentic rye bread.

Serve: With lean roast beef and mustard, sliced turkey, or whitefish.

SOURDOUGH BREAD

Serving: 1 slice (1 oz.)

Calories	70
Fat	1.0 g
Saturated	N/A
Monounsaturated	N/A
Polyunsaturated	N/A
Calories from fat	13%
Cholesterol	0
Sodium	140 mg
Protein	3.0 g
Carbohydrate	12.0 g
Dietary fiber	0.8 g

CHIEF NUTRIENTS

(Other nutrient data unavailable.)

Strengths: Sourdough bread is low in fat and contains no cholesterol. If it's made with enriched wheat flour, it offers a bit of blood-building iron, along with B-complex vitamins thiamine, riboflavin, and niacin, essential for energy production.
Caution: Commercial sourdough bread may be high in sodium. *Allergy alert:* Some people are allergic to wheat.
Description: A chewy, sour-tasting bread leavened with a starter made by mixing together flour, water (or potato water), and sugar and letting it ferment in a warm place for 2 or 3 days. When it begins to smell sour, it's ready for bread-making.
Serve: On a patio with a view of the Golden Gate Bridge (San Francisco is known for its superior sourdough bread).

SPOONBREAD

Serving: 1 cup

Calories	468
Fat	27.4 g
Saturated	8.7 g
Monounsaturated	N/A
Polyunsaturated	N/A
Calories from fat	53%
Cholesterol	293 mg
Sodium	1,157 mg
Protein	16.1 g
Carbohydrate	40.6 g
Dietary fiber	N/A

CHIEF NUTRIENTS

NUTRIENT	AMOUNT	% RDA
Calcium	230.4 mg	29
Riboflavin	0.4 mg	25
Iron	2.4 mg	24
Thiamine	0.2 mg	15
Potassium	316.8 mg	8
Niacin	1.0 mg	5

Caution: More than half the calories are from fat, and the sodium level is sky-high. If you're watching your weight or your salt intake, steer clear of spoonbread. *Allergy alert:* Contains eggs and milk, both common allergens.
Strengths: Contains a healthy portion of calcium. This soufflé-soft cornbread also offers good shares of iron and riboflavin, both important for healthy blood. And it has a good amount of thiamine and some niacin and potassium.
Description: Spoonbread is a puddinglike bread so soft it must be eaten with a spoon or fork. Made with ground cornmeal (either yellow or white) and vegetable shortening, and baked as a casserole. It can also be made with rice or millet, and may include buttermilk, chopped green chilis, shredded cheddar cheese, and bits of ham or bacon. It may be served with a tomato or mushroom sauce.
Preparation: Most recipes produce a crusty spoonbread with a soft center. To keep the crust soft, baste with a few tablespoons of milk from time to time while the bread is baking.
Serve: In moderation, as a side dish.

WHITE BREAD, *Soft Crumb* ⇩

Serving: 1 slice (about 1 oz.)

Calories	68
Fat	0.8 g
Saturated	trace
Monounsaturated	trace
Polyunsaturated	trace
Calories from fat	11%
Cholesterol	0.8 mg
Sodium	127 mg
Protein	2.2 g
Carbohydrate	12.6 g
Dietary fiber	0.5 g

CHIEF NUTRIENTS

NUTRIENT	AMOUNT	% RDA
Iron	0.7 mg	7
Thiamine	0.1 mg	7

Strengths: Made with vitamin- and mineral-enriched white flour, white bread offers some blood-building iron and some thiamine, a B-complex vitamin essential for energy production. It's low in fat and has almost no cholesterol.

Caution: As in other bread, the sodium level is moderate, so be aware of how much you're eating, especially if you're on a low-salt diet. *Allergy alert:* Contains wheat flour and may be made with eggs; both are common allergens.

Curiosity: During the Middle Ages, the more privileged ate their meals on square slabs of thick tough bread, called trenchers, that served as plates.

History: In ancient times, the making of white flour was a tedious process; white bread was a gourmet food only the nobility could afford.

Description: An extremely light bread with a soft brown crust and a smooth, soft center. Commercial white bread is made with white flour, yeast, milk, a bit of sugar, shortening, salt, and sometimes eggs.

A Better Idea: Whole grain breads provide more fiber.

WHOLE WHEAT BREAD, *Soft Crumb* ♺ 🕐 ⇩ 🗑

Serving: 1 slice (about 1 oz.)

Calories	67
Fat	0.7 g
Saturated	trace
Monounsaturated	trace
Polyunsaturated	trace
Calories from fat	10%
Cholesterol	0.8 mg
Sodium	148 mg
Protein	2.6 g
Carbohydrate	13.8 g
Dietary fiber	2.1 g

CHIEF NUTRIENTS

NUTRIENT	AMOUNT	% RDA
Iron	0.8 mg	8
Thiamine	0.1 mg	5

Strengths: Commercial whole wheat bread often contains more white flour than whole wheat. It offers some iron and thiamine, both essential for healthy blood and high energy levels. It's low in fat and cholesterol.

Caution: Has sodium—so be cautious if you're on a low-salt diet. *Allergy alert:* Contains wheat, a common allergen.

Curiosity: Bread making is often the subject of reverence and awe. In Turkey, children are taught to kiss bread that has fallen to the floor.

Description: Commercial whole wheat bread is made with whole wheat and white flours, yeast, sweetener, shortening, and salt.

Selection: Read the label on the loaf. If whole wheat is the main ingredient, it should be listed first on the label.

Storage: Like other whole grain products, whole wheat bread can get moldy. If you can't use it within a day or two, keep it in a plastic bag in the refrigerator.

MUFFINS

BLUEBERRY, *Homemade*

Serving: 1 (about 1½ oz.)

Calories	112
Fat	3.7 g
Saturated	1.1 g
Monounsaturated	N/A
Polyunsaturated	N/A
Calories from fat	30%
Cholesterol	33 mg
Sodium	253 mg
Protein	2.9 g
Carbohydrate	16.8 g
Dietary fiber	1.4 g

CHIEF NUTRIENTS

NUTRIENT	AMOUNT	% RDA
Iron	0.8 mg	8
Riboflavin	0.1 mg	6
Thiamine	0.1 mg	6

Strengths: One homemade blueberry muffin made with enriched white flour and vegetable shortening provides fiber, plus some iron, riboflavin, and thiamine. Although 30% of its calories are from fat, a blueberry muffin contains about one-third less fat and half the calories of a Danish pastry or a doughnut.
Caution: Don't load up your muffin with butter or margarine if you intend to keep it low-fat. *Allergy alert:* Muffins usually contain milk and eggs, so beware if you're allergic to either.
Curiosity: The word muffin may come from Low German *Muffe,* meaning cake.
Description: A homemade blueberry muffin is a fragrant, tender, slightly sweet, cakelike bread leavened with baking powder or baking soda.
Serve: With fruit butters or flavored yogurt, warm from the oven. Or wrap in foil and reheat for about 5 minutes in a preheated 450°F oven.
A Better Idea: For more fiber, try using half cornmeal and half whole wheat flour. Or add wheat bran to your recipe.

BRAN, *Homemade*

Serving: 1 (about 1½ oz.)

Calories	104
Fat	3.9 g
Saturated	1.2 g
Monounsaturated	N/A
Polyunsaturated	N/A
Calories from fat	34%
Cholesterol	41 mg
Sodium	179 mg
Protein	3.1 g
Carbohydrate	17.2 g
Dietary fiber	3.0 g

CHIEF NUTRIENTS

NUTRIENT	AMOUNT	% RDA
Iron	1.6 mg	16
Niacin	1.7 mg	9
Calcium	56.8 mg	7
Riboflavin	0.1 mg	6

Strengths: A homemade bran muffin made with enriched white flour and vegetable shortening provides a good helping of fiber and blood-boosting iron, along with some calcium, important for strong bones. Its riboflavin and niacin are two B-complex vitamins that are essential for many body functions, including energy production and the formation of red blood cells.
Caution: Although one bran muffin contains less fat than a doughnut or a pastry, the percentage of calories from fat (34%) is higher than desirable. *Allergy alert:* Don't eat if you have allergies to milk or eggs: Most muffins are made with both.
Description: A bran muffin is light to dark brown, with a crusty top and a soft, cakelike interior. It's made with wheat flour, wheat bran, baking soda or powder, sugar, oil, and usually milk and eggs.
Serve: Hot from the oven, with fruit butters, preserves, or flavored yogurts. Or rewarm before serving.
A Better Idea: Some bran muffin recipes tell you exactly how much fiber and fat are in each muffin: Look for at least 2 g of fiber per muffin, and no more than 30% of calories from fat.

CORN, *Homemade*

Serving: 1 (about 1½ oz.)

Calories	115
Fat	4.1 g
Saturated	1.2 g
Monounsaturated	N/A
Polyunsaturated	N/A
Calories from fat	32%
Cholesterol	22 mg
Sodium	198 mg
Protein	2.9 g
Carbohydrate	17.0 g
Dietary fiber	1.0 g

CHIEF NUTRIENTS

NUTRIENT	AMOUNT	% RDA
Calcium	44.8 mg	6
Iron	0.6 mg	6
Thiamine	0.1 mg	5

Strengths: A homemade corn muffin made with enriched whole grain cornmeal and vegetable shortening provides some fiber and a bit of iron and thiamine, both essential for production of energy. Plus, it offers a little calcium.

Caution: Although it contains less fat and a bit more fiber than a doughnut or a pastry, over 30% of its calories come from fat. *Allergy alert:* Muffins usually contain milk and eggs, both common allergens.

Preparation: Use coarsely ground whole cornmeal for an authentic, slightly gritty texture. The corn muffin is particularly adaptable to seasoning; add a little chili powder, oregano, or marjoram for a savory alternative.

Serve: With strawberry jam or apricot preserves.

A Better Idea: Substitute ½ cup of corn bran for cornmeal to boost fiber.

PLAIN, *Homemade*

Serving: 1 (about 1½ oz.)

Calories	118
Fat	4.0 g
Saturated	1.0 g
Monounsaturated	N/A
Polyunsaturated	N/A
Calories from fat	31%
Cholesterol	21 mg
Sodium	176 mg
Protein	3.1 g
Carbohydrate	16.9 g
Dietary fiber	N/A

CHIEF NUTRIENTS

NUTRIENT	AMOUNT	% RDA
Iron	0.9 mg	9
Riboflavin	0.1 mg	7
Calcium	41.6 mg	5

Strengths: A plain muffin is usually made with enriched white flour and vegetable shortening, which offers some iron and riboflavin. Made from a typical recipe, a plain muffin gets about 31% of its calories from fat, which is less than a pastry or a doughnut.

Caution: *Allergy alert:* Like cake, muffins usually contain milk and eggs, two common allergens.

Description: Muffins are usually made from white flour, baking powder or baking soda, sugar, milk, eggs, oil, and salt.

Preparation: Batter that's beaten too vigorously results in muffins that look like they've been tunneled by termites. When dough is mixed to the right consistency, it will pour from the spoon in ribbons—but it should still break in coarse globs.

Serve: Spread with apple butter or yogurt, accompanied by a glass of skim milk.

A Better Idea: There are many ways to add fiber and nutrients to plain muffins—include mashed bananas and walnuts, cranberries and grated orange rind, or wheat or oat bran.

Candies and Sweets
CANDIES

ALMONDS, *Chocolate-Coated*

Serving: 7 (about 1 oz.)

Calories	159
Fat	12.2 g
Saturated	2.1 g
Monounsaturated	N/A
Polyunsaturated	N/A
Calories from fat	69%
Cholesterol	0.3 mg
Sodium	17 mg
Protein	3.4 g
Carbohydrate	11.1 g
Dietary fiber	2.4 g

CHIEF NUTRIENTS

NUTRIENT	AMOUNT	% RDA
Riboflavin	0.2 mg	9
Iron	0.8 mg	8
Calcium	56.8 mg	7

Caution: Almonds are high in fat and calories. Although they contain a bit of iron, one study has demonstrated that almonds can markedly inhibit iron absorption. *Drug interaction:* Chocolate contains a substance that may cause dangerous elevations in blood pressure in people taking MAO inhibitors. *Allergy alert:* Chocolate may cause allergic headaches in some people. People allergic to salicylate in aspirin should also avoid foods that contain this substance, including almonds.

Eat with: Foods high in vitamin C to boost the absorption of iron.

Strengths: Chocolate-coated almonds contain some riboflavin for increased energy. They also have a bit of calcium.

Curiosity: The almond tree is related to the rose.

History: California produces nearly 50% of the world's almonds; they were introduced to the state in the mid-19th century.

Preparation: If you decide to make your own chocolate-covered almonds, toast them first to add flavor and crunch before dipping in chocolate.

ALMONDS, *Sugar-Coated*

Serving: 8 (1 oz.)

Calories	128
Fat	5.2 g
Saturated	0.4 g
Monounsaturated	N/A
Polyunsaturated	N/A
Calories from fat	37%
Cholesterol	0
Sodium	6 mg
Protein	2.2 g
Carbohydrate	19.7 g
Dietary fiber	1.3 g

CHIEF NUTRIENTS

NUTRIENT	AMOUNT	% RDA
Iron	0.5 mg	5

Caution: High in calories with moderate fat, sugar-coated almonds are not recommended for anyone on a weight-loss diet. *Allergy alert:* People allergic to the salicylate in aspirin should also avoid almonds, which contain the same allergen.

Eat with: Foods high in vitamin C to improve iron absorption.

Strengths: This confection has some fiber and a bit of iron for better blood. However, according to a recent study, almonds can inhibit absorption of iron by the body.

Curiosity: Sweet almonds are the only variety legally sold in the U.S. Bitter almonds, sold overseas, contain lethal prussic acid, which is eliminated when the nuts are toasted.

Description: Almonds are the seed of the almond tree (*Prunus amygdalus*). If not harvested, the mature fruit of the almond becomes as green and leathery as a turtle shell—and about as tasty.

BUTTERSCOTCH

Serving: 4 pieces (1 oz.)

Calories	111
Fat	1.0 g
Saturated	0.5 g
Monounsaturated	N/A
Polyunsaturated	N/A
Calories from fat	8%
Cholesterol	3 mg
Sodium	18 mg
Protein	0
Carbohydrate	26.5 g
Dietary fiber	0

CHIEF NUTRIENTS

(None of the nutrients meet or exceed 5% of the RDA.)

Caution: High in calories, butterscotch candy ought to be avoided by those watching their weight. *Allergy alert:* Read labels if you're sensitive to corn, since most commercial brands are made with corn syrup.

Origin: As the name implies, butterscotch candy probably originated in Scotland.

Description: The old-fashion butterscotch hard candy was made from butter, brown sugar, and lemon juice. Most commercial candy produced today contains sugar, dark corn syrup, butter, and salt.

Preparation: Butterscotch candy is made by boiling a sugar syrup to a point where the mixture contains very little moisture. Butter is added to give the candy a smoother texture.

CANDY CORN

Serving: 20 pieces (about 1 oz.)

Calories	102
Fat	0.6 g
Saturated	trace
Monounsaturated	trace
Polyunsaturated	trace
Calories from fat	5%
Cholesterol	0
Sodium	59 mg
Protein	trace
Carbohydrate	25.1 g
Dietary fiber	N/A

CHIEF NUTRIENTS

(None of the nutrients meet or exceed 5% of the RDA.)

Caution: Because candy corn is high in calories and sugar, this treat is not recommended for weight watchers or diabetics. Those suspicious of artificial colorings won't like what they read on a candy corn label; the product contains Red No. 40, Blue No. 1, and Yellows No. 5 and 6. *Allergy alert:* Corn syrup is used in cooking—so avoid this candy if you're sensitive to corn. Also, if you're allergic to eggs, read labels before eating candy corn: Many brands are made with egg whites.

Description: The sweet paste that makes up candy corn consists of sugar, corn syrup, water, and flavoring (often there's also cream of tartar and egg whites); the mixture is very similar to cake icing.

CARAMEL, *Plain or Chocolate*

Serving: 1 oz.

Calories	112
Fat	2.9 g
Saturated	1.6 g
Monounsaturated	N/A
Polyunsaturated	N/A
Calories from fat	23%
Cholesterol	0.6 mg
Sodium	63 mg
Protein	1.1 g
Carbohydrate	21.5 g
Dietary fiber	0.3 g

CHIEF NUTRIENTS

NUTRIENT	AMOUNT	% RDA
Calcium	41.4 mg	5

Caution: High in calories and saturated fat, caramel should be avoided if you're trying to lose weight. *Drug interaction:* Those taking MAO inhibitors should avoid chocolate caramel. *Allergy alert:* Corn syrup is also used, so this candy is off-limits to anyone who is allergic to corn. Chocolate caramel may cause allergic headaches in some people. Since caramel is made with butter and milk, avoid it if you're lactose intolerant.

History: The candy was first mentioned by the *Philadelphia Times* in 1884.

Description: A chewy confection formed when sugar is boiled to a temperature of about 245°F. Chocolate caramel is made by adding unsweetened chocolate during the boiling process.

Preparation: Caramel gets its characteristic taste and coloring during cooking. If you're making a batch, be sure to watch it closely: Caramel turns black and bitter if it's overcooked.

CHEWING GUM

Serving: 1 stick

Calories	5
Fat	0
Saturated	0
Monounsaturated	0
Polyunsaturated	0
Calories from fat	0
Cholesterol	0
Sodium	0
Protein	0
Carbohydrate	1.6 g
Dietary fiber	0

CHIEF NUTRIENTS

(None of the nutrients meet or exceed 5% of the RDA.)

Caution: Chewing gum made with sugar has been blamed for many cases of tooth decay. *Allergy alert:* Many brands of gum are made with corn syrup, which means they're off-limits to people who are sensitive to corn. Some people are allergic to the artificial colors or flavors in gum.

History: Arab sugar traders first tried mixing certain kinds of acacia gum with sugar, producing what came to be known as gum arabic. Early doctors mixed some medicines with gum so that the medication would dissolve slowly in the patient's mouth. Gum as we know it was the creation of New York inventor Thomas Adams in the 1870s. After experimenting with a sassafras-flavored gum, he introduced licorice to produce a chewing gum called Black Jack.

A Better Idea: Save your teeth by chewing sugarless gum.

CHOCOLATE, Milk

Serving: 1 oz.

Calories	146
Fat	9.0 g
Saturated	5.1 g
Monounsaturated	N/A
Polyunsaturated	N/A
Calories from fat	56%
Cholesterol	6 mg
Sodium	26 mg
Protein	2.2 g
Carbohydrate	15.9 g
Dietary fiber	0.8 g

CHIEF NUTRIENTS

NUTRIENT	AMOUNT	% RDA
Calcium	63.8 mg	8
Riboflavin	0.1 mg	6

Caution: High in calories and fat, chocolate shouldn't even be sniffed by those on a diet. Over half the calories are from fat, much of which is saturated. Those who easily develop canker sores or kidney stones should also resist milk chocolate. *Drug interaction:* Dangerous elevations in blood pressure may result if you eat chocolate while taking MAO inhibitors. *Allergy alert:* Chocolate is a common allergen.

Curiosity: Believing that chocolate was an aphrodisiac, Montezuma—king of the Aztecs—supposedly drank 50 unsweetened cups of cocoa daily. During the 17th century, the Catholic Church rallied against chocolate, believing it to be the drink of sorcerers.

CHOCOLATE, Semisweet

Serving: 1 oz.

Calories	142
Fat	10.0 g
Saturated	5.6 g
Monounsaturated	N/A
Polyunsaturated	N/A
Calories from fat	63%
Cholesterol	0
Sodium	0.6 mg
Protein	1.2 g
Carbohydrate	16.0 g
Dietary fiber	1.8 g

CHIEF NUTRIENTS

NUTRIENT	AMOUNT	% RDA
Iron	0.7 mg	7

Caution: High in calories, semisweet chocolate is off-limits for those watching their weight. (Sweet and bittersweet chocolate are also high in calories.) Also, people who are susceptible to kidney stones should avoid chocolate. *Drug interaction:* Chocolate contains a substance that may cause dangerous elevations in blood pressure in those taking MAO inhibitors. *Allergy alert:* Chocolate is a common allergen. People susceptible to canker sores or migraine headaches should avoid it.

Selection: Available in 8-oz. squares and other forms, semisweet chocolate is often melted to use for dipping candy. True to its name, semisweet has less sugar than sweet chocolate. Milk chocolate, of course, contains milk.

Storage: The best temperature for keeping chocolate is 78°F.

Preparation: By varying the amounts of sugar and other ingredients, bakers create sweet, bittersweet, and semisweet chocolate.

COCONUT, Chocolate-Coated

Serving: 1 oz.

Calories	123
Fat	4.9 g
Saturated	2.9 g
Monounsaturated	N/A
Polyunsaturated	N/A
Calories from fat	36%
Cholesterol	0.3 mg
Sodium	55 mg
Protein	0.8 g
Carbohydrate	20.2 g
Dietary fiber	N/A

CHIEF NUTRIENTS

(None of the nutrients meet or exceed 5% of the RDA.)

Caution: High in calories and saturated fat, chocolate-covered coconut should be considered off-limits for weight watchers. *Drug interaction:* For people taking MAO inhibitors, it's advisable to avoid chocolate: It contains a substance that may cause blood pressure to jump to a dangerous level. *Allergy alert:* Chocolate may cause allergic headaches in some people. And anyone allergic to corn should read labels; some brands of chocolate-coated coconut are made with corn oil.

History: The popular commercial chocolate bar with a coconut center—Mounds—was produced in 1922.

FUDGE, Chocolate

Serving: 1 oz.

Calories	112
Fat	3.4 g
Saturated	1.2 g
Monounsaturated	N/A
Polyunsaturated	N/A
Calories from fat	27%
Cholesterol	0.3 mg
Sodium	53 mg
Protein	0.8 g
Carbohydrate	21.0 g
Dietary fiber	0.4 g

CHIEF NUTRIENTS

(None of the nutrients meet or exceed 5% of the RDA.)

Caution: High in calories, chocolate fudge has a moderate amount of saturated fat. *Allergy alert:* Fudge is often made from corn syrup, which can cause problems for those allergic to corn. People with allergies to milk also need to be wary.

Origin: Tradition says fudge was invented by students at an unnamed prestigious women's school during the late 19th century. Whatever its true origins, it became a popular treat at Smith, Vassar, and other Seven Sister schools.

Curiosity: The term "fudge" originally meant playing a hoax. According to rumor, coeds used to trick school officials into letting them stay up late to make candy; that's probably how the candy got it name. Later, "oh, fudge!" became an expletive used by those too refined to cuss.

Description: A tasty confection often made with sugar, butter or cream, and corn syrup; a hybrid of creams and caramels.

Preparation: Sugar, corn syrup, and milk are heated to 238°F, then butter is added and the mixture is allowed to cool.

FUDGE, *Chocolate, Chocolate-Coated*

Serving: 1 oz.

Calories	120
Fat	4.5 g
Saturated	1.5 g
Monounsaturated	N/A
Polyunsaturated	N/A
Calories from fat	33%
Cholesterol	0.6 mg
Sodium	64 mg
Protein	1.1 g
Carbohydrate	20.5 g
Dietary fiber	0.5 g

CHIEF NUTRIENTS

(None of the nutrients meet or exceed 5% of the RDA.)

Caution: Moderately high in saturated fat and high in calories—weight watchers beware. *Allergy alert:* People with an allergy to corn should avoid fudge, which is often made from corn syrup. People allergic to milk also need to avoid this confection.

Curiosity: The word "fudge" may derive from the Old English word *fadge,* meaning "to fit pieces together."

Description: A tasty confection made with sugar, butter or cream, and corn syrup; a hybrid of creams and caramels. Because the concoction is about 10% water, it has a creamy consistency.

FUDGE, *Vanilla*

Serving: 1 oz.

Calories	111
Fat	3.1 g
Saturated	0.8 g
Monounsaturated	N/A
Polyunsaturated	N/A
Calories from fat	25%
Cholesterol	0.6 mg
Sodium	58 mg
Protein	0.8 g
Carbohydrate	20.9 g
Dietary fiber	0

CHIEF NUTRIENTS

(None of the nutrients meet or exceed 5% of the RDA.)

Caution: High in calories. Vanilla can cause swollen lips in some people. *Allergy alert:* People allergic to corn should be advised that fudge is often made from corn syrup.

Strengths: Of all the fudges, vanilla is the lowest in calories. It's also low in fat and cholesterol.

Curiosity: Large doses of vanilla are poisonous.

Origin: Vanilla comes from the Spanish *vainilla,* "little sheath." The ancient Aztecs first derived vanilla from the pod of an orchid known to scientists today as *Vanilla planifolia.*

GINGER ROOT, *Crystallized, Candied*

Serving: 1 oz.

Calories	95
Fat	0.1 g
Saturated	0
Monounsaturated	N/A
Polyunsaturated	N/A
Calories from fat	1%
Cholesterol	0
Sodium	17 mg
Protein	0.1 g
Carbohydrate	24.4 g
Dietary fiber	N/A

CHIEF NUTRIENTS

NUTRIENT	AMOUNT	% RDA
Iron	5.9 mg	59
Potassium	739.2 mg	20
Vitamin C	11.2 mg	19
Niacin	2.0 mg	10
Calcium	64.4 mg	8
Riboflavin	0.1 mg	6

Strengths: Ginger root is a remarkable storehouse of iron, which may be particularly needed if someone suffers chronic fatigue, headaches, lack of energy, or irritability—all common signs of anemia. Vitamin C helps the body absorb iron, and potassium helps flush sodium from the system. Ginger also has a little calcium to help fight osteoporosis.

Caution: Ginger root is high in calories and sodium, and it has a little fat.

Curiosity: Although called a root, ginger actually is an underground stem that can sprout new roots and shoots.

Description: Ginger root is cooked in a sugar syrup, then coated with sugar.

Serve: Ginger in this form usually is used as a confection or added to desserts. Candied ginger has been served as an after-meal digestive aid.

GUMDROPS

Serving: 1 oz.

Calories	100
Fat	trace
Saturated	trace
Monounsaturated	trace
Polyunsaturated	trace
Calories from fat	N/A
Cholesterol	0
Sodium	10 mg
Protein	0
Carbohydrate	25.0 g
Dietary fiber	1.7 g

CHIEF NUTRIENTS

(None of the nutrients meet or exceed 5% of the RDA.)

Caution: Gumdrops are high in calories and therefore taboo if you're dieting. *Allergy alert:* People who have allergies to corn should avoid gumdrops; they contain corn syrup and modified corn starch. Artificial flavoring, preservatives, and coloring are used in preparing most commercial gumdrops.

HARD CANDY

Serving: 1 oz.

Calories	108
Fat	0.3 g
Saturated	0
Monounsaturated	N/A
Polyunsaturated	N/A
Calories from fat	3%
Cholesterol	0
Sodium	9 mg
Protein	0
Carbohydrate	27.2 g
Dietary fiber	0

CHIEF NUTRIENTS

NUTRIENT	AMOUNT	% RDA
Iron	0.5 mg	5

Caution: A single ounce of hard candy is high in calories. The corn syrup and sugar often used to sweeten candy are bad for teeth. *Allergy alert:* People allergic to corn may have an adverse reaction to the corn syrup in hard candy.

Strengths: Hard candy is low in saturated fat, cholesterol, and sodium.

History: During the 19th century, hard candies such as peppermint sticks and horehound candy were favorites with people who had a sweet tooth.

Description: Hard candy is a boiled mixture of syrup and sugar—with or without flavoring, fillings, and coloring. Some common forms include mints, lollipops, and butterscotch candies.

Preparation: To improve flavor, candymakers add citric acid to the basic mixture.

JELLY BEANS

Serving: 10 (1 oz.)

Calories	103
Fat	0.1 g
Saturated	0
Monounsaturated	trace
Polyunsaturated	trace
Calories from fat	1%
Cholesterol	0
Sodium	3 mg
Protein	0
Carbohydrate	26.1 g
Dietary fiber	N/A

CHIEF NUTRIENTS

(None of the nutrients meet or exceed 5% of the RDA.)

Caution: With more than 10 calories per jelly bean, this little candy is high in calories, so it's not for people watching their weight—unless they can practice restraint. *Allergy alert:* Since corn syrup is a primary ingredient in jelly beans, people allergic to corn may suffer a reaction.

Curiosity: During his presidency, Ronald Reagan made headlines after confessing a weakness for jelly beans. Unlike old-style jelly beans, which came in mundane flavors like licorice and orange, exotic new candies are available in flavors such as chocolate fudge-mint, piña colada, and pink lemonade.

History: Soon after the invention of jelly beans, the first ad for them ran in a Chicago paper on July 5, 1905. They were 9¢ per pound.

Description: Jelly beans are sugary, chewy, egg-shaped candies. They're made of boiled sugar and artificial flavors set with either gelatin or pectin.

MARSHMALLOW

Serving: 1 large

Calories	23
Fat	0
Saturated	0
Monounsaturated	0
Polyunsaturated	0
Calories from fat	0
Cholesterol	0.1 mg
Sodium	3 mg
Protein	0.1 g
Carbohydrate	5.8 g
Dietary fiber	N/A

CHIEF NUTRIENTS

(None of the nutrients meet or exceed 5% of the RDA.)

Caution: The calorie count for marshmallows can add up quickly if you eat too many. *Allergy alert:* Those who are allergic to corn should avoid marshmallows, which are made with corn syrup. Since some marshmallows are made with egg whites, read labels if you have an allergy to eggs; find marshmallows that are made with gelatin instead.

Strengths: Better than most candies, marshmallows have no fat, and they're low in cholesterol and sodium.

Curiosity: Before corn syrup was used to make marshmallows, the candy's flavoring was derived from a sweetened extract that came from the roots of the marshmallow plant (*Althaea officinalis*). Until recently, the root itself was boiled and eaten in some parts of Britain, and the syrup was used as a cough remedy.

History: Although their exact origin is unknown, marshmallows were eaten during the 1880s.

Description: Ingredients include sugar, corn syrup, gelatin, and sometimes egg whites. Vanilla is often added as a flavoring.

MINTS, *Chocolate-Coated*

Serving: 12 miniature (about 1 oz.)

Calories	115
Fat	2.9 g
Saturated	0.9 g
Monounsaturated	N/A
Polyunsaturated	N/A
Calories from fat	23%
Cholesterol	0.3 mg
Sodium	52 mg
Protein	0.5 g
Carbohydrate	22.6 g
Dietary fiber	N/A

CHIEF NUTRIENTS

(None of the nutrients meet or exceed 5% of the RDA.)

Caution: High in calories, with a moderate amount of sodium in each ounce, mints are inadvisable for those who are weight conscious or on a salt-restricted diet. *Drug interaction*: Chocolate contains a substance that may cause dangerous elevations in blood pressure in people taking MAO inhibitors. *Allergy alert:* Chocolate may cause allergic headaches in some people. Occasionally, people get hives after eating mints. This candy may contain corn syrup, which should be avoided by those sensitive to corn products.

Curiosity: One commercial mint candy—Junior Mints—was named after a Broadway play, *Junior Miss*.

Storage: Immediately after mints are made, they are stored for a while in an airtight container. This process gives them a more desirable softness and smoothness.

Preparation: Like many other candies, the sugar in chocolate-coated mints is cooked at a temperature of 236°F.

MINTS, *Plain*

Serving: 1 oz.

Calories	102
Fat	0.6 g
Saturated	trace
Monounsaturated	trace
Polyunsaturated	trace
Calories from fat	5%
Cholesterol	0
Sodium	59 mg
Protein	trace
Carbohydrate	25.1 g
Dietary fiber	N/A

CHIEF NUTRIENTS

(None of the nutrients meet or exceed 5% of the RDA.)

Caution: High in calories, with 59 mg of salt in a single ounce, uncoated mints should be consumed sparingly by the weight conscious and those on a salt-restricted diet. *Allergy alert:* Some people have been known to get hives after eating mints. The corn syrup used to make mints should be avoided by those sensitive to corn.

Curiosity: The zingy taste in some mints comes from peppermint, which surprisingly is not a species of mint—it's a hybrid of water mint and spearmint.

Description: A mint is a type of creamy candy classified as a fondant. It's made from a cooked syrup of water and sugar, with flavoring added.

Preparation: Fondants are made from sugar that has been heated to a temperature of 236°F.

NOUGAT AND CARAMEL CANDY, *Chocolate-Coated*

Serving: 1 oz.

Calories	116
Fat	3.9 g
Saturated	1.2 g
Monounsaturated	N/A
Polyunsaturated	N/A
Calories from fat	30%
Cholesterol	1 mg
Sodium	48 mg
Protein	1.1 g
Carbohydrate	20.4 g
Dietary fiber	0.4 g

CHIEF NUTRIENTS

(None of the nutrients meet or exceed 5% of the RDA.)

Caution: Weight watchers beware: This candy is high in calories. *Drug interaction:* Chocolate contains a substance that may cause dangerous elevations in blood pressure in people taking MAO inhibitors. *Allergy alert:* The nuts in nougat have been known to cause hives. People who are sensitive to corn or eggs should avoid this popular candy because both corn syrup and eggs are ingredients. And the chocolate coating may cause allergic headaches in some people.

History: Nougat probably was introduced to Europe by the Arabs during the Middle Ages. The word "nougat" comes from the Latin *nux* (nut). One of the most famous chocolate-coated candy bars made of nougat and caramel, Milky Way, debuted in 1923.

Description: Nougat is a confection made with sugar or honey and roasted nuts; it ranges in texture from chewy to hard. Caramel is a thick brown or golden confection.

PEANUT BARS

Serving: 1 oz.

Calories	144
Fat	9.0 g
Saturated	2.0 g
Monounsaturated	N/A
Polyunsaturated	N/A
Calories from fat	56%
Cholesterol	0
Sodium	3 mg
Protein	4.9 g
Carbohydrate	13.2 g
Dietary fiber	1.7 g

CHIEF NUTRIENTS

NUTRIENT	AMOUNT	% RDA
Niacin	2.6 mg	14
Thiamine	0.1 mg	8
Iron	0.5 mg	5

Caution: Peanut bars are very unkind to your arteries because they're high in fat and calories. And peanuts tend to inhibit absorption of iron. Since the peanuts in peanut bars may get rancid, the fresher the candy, the better. *Allergy alert:* Made with corn syrup, so individuals sensitive to corn should pass.

Eat with: Foods high in vitamin C to help improve absorption of iron.

Strengths: Peanut bars have a good amount of niacin and a bit of thiamine for energy. A bit of iron helps to create strong blood.

PEANUT BRITTLE

Serving: 1 oz.

Calories	118
Fat	2.9 g
Saturated	0.6 g
Monounsaturated	N/A
Polyunsaturated	N/A
Calories from fat	22%
Cholesterol	0
Sodium	9 mg
Protein	1.6 g
Carbohydrate	22.7 g
Dietary fiber	0.5 g

CHIEF NUTRIENTS

NUTRIENT	AMOUNT	% RDA
Iron	0.6 mg	6

Caution: High in calories, peanut brittle is undesirable for weight watchers. Some researchers have found that nuts can inhibit the body's absorption of iron. Peanuts also can give you gas. Make sure the peanut brittle you eat is fresh—the peanuts tend to get rancid much more quickly than they do in other types of candies. *Allergy alert:* Since peanut brittle is made with corn syrup, people who are corn sensitive should avoid it.

Eat with: Foods high in vitamin C to help iron absorption.

Strength: This confection has a bit of iron.

History: Peanut brittle became a sticky hit late in the 1800s.

Preparation: Peanut brittle is perhaps the simplest candy to make. It *is* brittle because it has a very low moisture content—after the ingredients are cooked, moisture is about 2%. The sugar caramelizes in cooking, which gives the candy its dark-to medium-brown color.

PEANUTS, *Chocolate-Coated*

Serving: 12 (about 1 oz.)

Calories	157
Fat	11.6 g
Saturated	3.0 g
Monounsaturated	N/A
Polyunsaturated	N/A
Calories from fat	66%
Cholesterol	0.3 mg
Sodium	17 mg
Protein	4.6 g
Carbohydrate	11.0 g
Dietary fiber	1.6 g

CHIEF NUTRIENTS

NUTRIENT	AMOUNT	% RDA
Niacin	2.1 mg	11
Thiamine	0.1 mg	7

Caution: Chocolate-coated peanuts are high in fat and calories, making them an undesirable snack for those watching their waistline. Researchers say that eating peanuts can deter the absorption of iron by the body. *Drug interaction:* Chocolate and peanuts contain a substance that may cause a dangerous elevation in blood pressure in people who take MAO inhibitors. *Allergy alert:* Both chocolate and peanuts may cause headaches in some sensitive people.

Strengths: They have fiber, plus a good amount of protein to build a better body. Also a good supply of niacin and a bit of thiamine for more energy.

Curiosity: The classic Southern label for peanuts—goobers—is derived from African culture. *Nguba* was the African word for certain edible roots and legumes. Brought to America by slaves, the word became goober.

A Better Idea: Nearly all the nutrition comes from the peanuts in this confection. If you hunger for the taste of peanuts, have some plain, rather than coated with chocolate.

RAISINS, *Chocolate-Coated*

Serving: 30 (about 1 oz.)

Calories	119
Fat	4.8 g
Saturated	2.7 g
Monounsaturated	N/A
Polyunsaturated	N/A
Calories from fat	36%
Cholesterol	3 mg
Sodium	18 mg
Protein	1.5 g
Carbohydrate	19.7 g
Dietary fiber	1.3 g

CHIEF NUTRIENTS

NUTRIENT	AMOUNT	% RDA
Iron	0.7 mg	7
Calcium	42.6 mg	5

Caution: High in saturated fat, chocolate-covered raisins should be avoided by those on a weight-loss regimen. To prevent choking, chop raisins before feeding them to children 4 and under. It's advisable to brush your teeth after eating raisins, since they enhance decay. Raisins can also cause flatulence in some people. *Drug interaction:* A substance in chocolate may cause dangerous jumps in blood pressure in people taking MAO inhibitors. *Allergy alert:* Those who are allergic to aspirin should avoid raisins: They contain salicylate, the allergen found in aspirin.

Strengths: Chocolate-coated raisins have some fiber, plus a bit of calcium for strong teeth and a fair amount of iron.

VANILLA CREAMS, *Chocolate-Coated*

Serving: 1 oz.

Calories	122
Fat	4.8 g
Saturated	1.4 g
Monounsaturated	N/A
Polyunsaturated	N/A
Calories from fat	35%
Cholesterol	0.6 mg
Sodium	51 mg
Protein	1.1 g
Carbohydrate	19.7 g
Dietary fiber	N/A

CHIEF NUTRIENTS

(None of the nutrients meet or exceed 5% of the RDA.)

Caution: Vanilla creams are a bit high in saturated fat, and each ounce packs 122 calories. Plus, there's hidden salt in these treats. *Allergy alert*: People allergic to corn should beware: Vanilla creams are often made with corn syrup.

History: The precursor of this confection was a so-called French-style candy with tantalizing cream on the inside. It was introduced in England during the Great Exhibition of 1851. Cream fillings became so popular when they were introduced in the U.S. that some 380 factories were built to turn out "penny candies."

Description: The "cream" in vanilla creams is created by first cooking a combination of sugar, water, and cream of tartar—a paste known as fondant. The scrumptious paste is dipped in chocolate after the mixture cools.

SWEETENERS AND TOPPINGS

ALMOND PASTE

Serving: 1 oz.

Calories	127
Fat	7.7 g
Saturated	0.7 g
Monounsaturated	5.0 g
Polyunsaturated	1.6 g
Calories from fat	55%
Cholesterol	0
Sodium	3 mg
Protein	3.4 g
Carbohydrate	12.4 g
Dietary fiber	N/A

CHIEF NUTRIENTS

NUTRIENT	AMOUNT	% RDA
Magnesium	73.6 mg	21
Riboflavin	0.2 mg	12
Iron	0.9 mg	9
Calcium	65.3 mg	8
Folate	15.8 mcg	8
Potassium	183.8 mg	5

Strengths: Low in sodium, with no cholesterol, almond paste is high in monounsaturated fat, which can help control serum cholesterol. It has nice amounts of folate and iron, and fair to good amounts of blood pressure regulators (magnesium, calcium, potassium). It's also a source of iron. If unblanched nuts are used in the paste, the fiber content is higher than in paste made from blanched almonds.

Caution: Too high in fat and calories for frequent consumption, almond paste contains sugar and it's high in oxalic acid, which should be restricted by those with calcium-oxalate stones. *Allergy alert:* Nuts are highly allergenic, often causing hives, headaches, and other reactions. Those allergic to aspirin may react to the natural salicylate in almonds.

Description: Ground almonds and sugar blended into a pliable mixture; it's less creamy than almond butter but coarser than marzipan. Almond extract is often added for an extra boost of flavor.

HONEY

Serving: 1 Tbsp.

Calories	64
Fat	0
Saturated	0
Monounsaturated	0
Polyunsaturated	0
Calories from fat	0
Cholesterol	0
Sodium	1 mg
Protein	0.1 g
Carbohydrate	17.3 g
Dietary fiber	trace

CHIEF NUTRIENTS

(None of the nutrients meet or exceed 5% of the RDA.)

Caution: Do not feed honey to infants under 1 year old—bacteria, which can grow in unsterile products, can cause infant botulism. The same bacteria, however, will usually not affect older children and adults. Honey derived from mountain laurel or rhododendron can contain toxins that will knock you for a loop; symptoms resembling drunkenness or heart attack can last up to 24 hours. Honey from oleander trees can be poisonous.

Curiosity: Medical researchers in Africa have shown that honey can help clean burns and wounds, reducing the possibility of infection. Indians in New England referred to honey bees as ''English flies'' or ''white man's flies.''

Description: The flavor of honey may vary, depending on the type of nectar collected by bees. Hundreds of varieties of honey are consumed in the U.S. Colors of honey range from white to black, purple, or green. Clover honey is generally light in color, while buckwheat honey is dark.

MOLASSES, *Dark*

Serving: 2 Tbsp.

Calories	95
Fat	0
Saturated	0
Monounsaturated	0
Polyunsaturated	0
Calories from fat	0
Cholesterol	0
Sodium	15 mg
Protein	0
Carbohydrate	24.6 g
Dietary fiber	0

CHIEF NUTRIENTS

NUTRIENT	AMOUNT	% RDA
Iron	2.5 mg	25
Calcium	118.9 mg	15
Potassium	435.9 mg	12

Caution: *Allergy alert:* Occasionally, molasses triggers a reaction in susceptible people.

Strengths: Dark molasses has a high-intensity supply of iron, with more than one-quarter of the RDA in a single 1-oz. portion. The dark molasses is more potent because it contains less sugar and is more concentrated.

Curiosity: Distilled molasses is fermented to make rum.

History: Molasses was the predominant sweetener in the American colonies during the 1700s. Many colonists refused to pay the British tax on this luxury item—among the first acts of rebellion leading to the Revolutionary War.

Description: Molasses is the residue left during the refining of sugar cane or sugar beets. Dark or blackstrap molasses is what's left after the third (last) extraction of sugar cane. Other grades of molasses that contain more sugar (such as ''medium''), are also available.

Serve: Molasses is an ingredient in Boston baked beans and shoo-fly pie.

NONDAIRY WHIPPED TOPPING

Serving: 2 Tbsp.

Calories	30
Fat	2.4 g
Saturated	2.0 g
Monounsaturated	0.2 g
Polyunsaturated	0.1 g
Calories from fat	72%
Cholesterol	0
Sodium	2 mg
Protein	0.1 g
Carbohydrate	2.2 g
Dietary fiber	0

CHIEF NUTRIENTS

(None of the nutrients meet or exceed 5% of the RDA.)

Caution: Many people who are trying to cut their fat intake or avoid milk use nondairy topping, but it doesn't have the desired effect. Even if you limit a portion to 2 Tbsp., the topping is still high in calories. Fully 72% of those calories come from fat—and the fat is often highly saturated because it's made from coconut and other hydrogenated oils. Worst of all, nondairy whipped topping contains various additives that you might wish to avoid. Seemingly identical products may contain some cream—or even egg yolks. *Allergy alert:* Some brands have caseinate, a milk protein that people who are allergic to milk or are lactose intolerant may want to avoid.

Eat with: Restraint.

A Better Idea: You can whip skim milk or evaporated skim milk to make an airy low-fat topping. Here's how: Chill some milk in the freezer until ice crystals just begin to form (about 20 minutes). Then whip using chilled beaters and serve immediately.

SUGAR, *Brown*

Serving: 1 tsp.

Calories	11
Fat	0
Saturated	0
Monounsaturated	0
Polyunsaturated	0
Calories from fat	0
Cholesterol	0
Sodium	0.9 mg
Protein	0
Carbohydrate	2.9 g
Dietary fiber	N/A

CHIEF NUTRIENTS

(None of the nutrients meet or exceed 5% of the RDA.)

Caution: Bacteria flourish in the mouth when you load up on any sweets—resulting in tooth decay. *Allergy alert:* Sugar has been suspected of causing allergic tension-fatigue syndrome.

History: The Persians grew sugar 400 years before Christ's birth.

Description: Brown sugar is a combination of white sugar and molasses. The sugar may be derived either from sugar beets or sugar cane; nutrient values are the same in either case. The darker the sugar, the more molasses it contains.

Storage: Hardened brown sugar can be resoftened by placing it with an apple wedge in a plastic bag and sealing tightly for 1 to 2 days.

SUGAR, *Granulated*

Serving: 1 tsp.

Calories	15
Fat	0
Saturated	0
Monounsaturated	0
Polyunsaturated	0
Calories from fat	0
Cholesterol	0
Sodium	trace
Protein	0
Carbohydrate	4.0 g
Dietary fiber	0

CHIEF NUTRIENTS

(None of the nutrients meet or exceed 5% of the RDA.)

Caution: Sweeteners invite bacteria to thrive in the mouth, causing tooth decay. If that isn't enough to steer you away, consider this—it's very high in calories. *Allergy alert:* Sugar is often a suspect when allergic tension-fatigue syndrome occurs.
History: Although Arabic countries have grown sugar since 400 B.C., it was not eaten in Europe until the 9th century. Early sugar came in huge chunks and had to be chipped away before it could be ground with a mortar and pestle. When sugar was rare, it was an expensive luxury—much as saffron is today.
Description: Also known as white sugar, granulated sugar is highly refined beet or cane sugar. It's the most commonly used form for table use and cooking.

SUGAR, *Powdered*

Serving: 1 tsp.

Calories	10
Fat	0
Saturated	0
Monounsaturated	0
Polyunsaturated	0
Calories from fat	0
Cholesterol	0
Sodium	trace
Protein	0
Carbohydrate	2.5 g
Dietary fiber	0

CHIEF NUTRIENTS

(None of the nutrients meet or exceed 5% of the RDA.)

Caution: Be careful if you eat many sugary snacks during a day. The sugar that's a residue from sweet snacks—especially from sticky candy—can cause cavities. Also high in calories. *Allergy alert:* Sugar is often a suspect when allergic tension-fatigue syndrome occurs.
History: In the 9th century the Moors conquered the Iberian Peninsula; they brought sugar to the Western world.
Description: Powdered sugar (also known as confectioners' sugar) is derived from sugar beets or sugar cane. It is simply granulated sugar that has been crushed until it's fine and powdery. Cornstarch is added.
Serve: Use powdered sugar in icings, candies, and fine dustings for desserts.

SYRUP, *Maple*

Serving: 1 Tbsp.

Calories	50
Fat	0
Saturated	0
Monounsaturated	0
Polyunsaturated	0
Calories from fat	0
Cholesterol	0
Sodium	2 mg
Protein	0
Carbohydrate	12.8 g
Dietary fiber	0

CHIEF NUTRIENTS

(None of the nutrients meet or exceed 5% of the RDA.)

Caution: Use sparingly; syrup's high calorie count could be troublesome for people with weight problems, especially if it's poured liberally over French toast or pancakes.

Strengths: A delicious, delicate natural sweetener that can enhance the flavor of a wide variety of foods.

Curiosity: It takes 20 to 80 gallons of sap to obtain a single gallon of maple syrup. A grove of sugar, black, or red maple trees is called a ''sugarbush'' by tappers. The trees need to grow 35 to 60 years before they reach tapping size. In Vermont, sap boilers sometimes pour the boiled syrup onto clean snow, making a tasty taffy. The maple syrup industry exists only in North America.

Origin: North American Indians were the first to boil the sap into sweet, tasty maple syrup.

Serve: A light touch of real maple syrup can add flavor to healthy foods like baked acorn squash.

SYRUP, *Pancake*

Serving: 1 Tbsp.

Calories	50
Fat	0
Saturated	0
Monounsaturated	0
Polyunsaturated	0
Calories from fat	0
Cholesterol	0
Sodium	0.4 mg
Protein	0
Carbohydrate	12.8 g
Dietary fiber	0

CHIEF NUTRIENTS

(None of the nutrients meet or exceed 5% of the RDA.)

Caution: High in calories, so should be used sparingly by people watching their weight. *Allergy alert:* People allergic to corn should be aware that some pancake syrups contain corn syrup.

Strength: A good sweetener that can provide more flavor if used judiciously.

History: Today, blends of maple syrup and less expensive syrups that are flavored with maple extracts have largely replaced maple syrup.

A Better Idea: If you must use syrup, real maple syrup gives more flavor.

WHIPPED CREAM TOPPING

Serving: 2 Tbsp.

Calories	19
Fat	1.7 g
Saturated	1.0 g
Monounsaturated	0.5 g
Polyunsaturated	0.1 g
Calories from fat	78%
Cholesterol	6 mg
Sodium	10 mg
Protein	0.2 g
Carbohydrate	0.9 g
Dietary fiber	0

CHIEF NUTRIENTS

(None of the nutrients meet or exceed 5% of the RDA.)

Caution: This product generally includes real cream plus sugar, stabilizers, emulsifiers, and gas propellant—all of which you may wish to avoid. Even if you limit your serving to 2 Tbsp., whipped cream topping has a high proportion of calories and fat. Fully 78% of the calories come from fat—and most of it is saturated. And since it's tempting to consume large amounts of fluffy topping, it's easy to overindulge. *Allergy alert:* Milk can cause a wide range of health problems, including migraines, hives, and other allergic reactions. And it's often off-limits for those with lactose intolerance.

Selection: In addition to whipped cream topping, there are also canned aerosol products, called ''dessert toppings.'' These toppings contain no cream and are based instead on hydrogenated vegetable oils—hardly an improvement.

A Better Idea: Whip skim milk with instant nonfat dry milk for an airy low-fat topping.

Condiments, Preserves, and Spreads
CONDIMENTS

CATSUP ⬇ 🏛

Serving: 1 Tbsp.

Calories	16
Fat	0.1 g
Saturated	trace
Monounsaturated	trace
Polyunsaturated	trace
Calories from fat	3%
Cholesterol	0
Sodium	178 mg
Protein	0.2 g
Carbohydrate	4.1 g
Dietary fiber	0.2 g

CHIEF NUTRIENTS

(None of the nutrients meet or exceed 5% of the RDA.)

Strengths: Barely a smidgen of fat, with a negligible calorie count. As condiments go, catsup is lower in fat than mayonnaise, and somewhat lower in sodium than mustard.

Caution: Catsup is moderately high in sodium, so look for low-sodium varieties if you're on a sodium-restricted diet. *Allergy alert:* Because tomatoes are a common food allergen, they may cause mouth itching or other reactions in sensitive people. People who are allergic to aspirin may want to avoid tomatoes, which contain salicylate, the allergen found in aspirin.

Curiosity: By law, catsup must pour no faster than 14 cm— that's about 5½"—in 30 seconds.

Description: A concentrated tomato product. May also contain vinegar, sugar, corn syrup, or other sweeteners, as well as onions, garlic, salt, spices, or other flavorings.

CHOWCHOW, *Sour* 🔆 🔀 ⬇ 🏛

Serving: ½ cup

Calories	35
Fat	1.6 g
Saturated	0
Monounsaturated	N/A
Polyunsaturated	N/A
Calories from fat	40%
Cholesterol	0
Sodium	1,606 mg
Protein	1.7 g
Carbohydrate	4.9 g
Dietary fiber	N/A

CHIEF NUTRIENTS

NUTRIENT	AMOUNT	% RDA
Iron	3.1 mg	31
Vitamin C	8.4 mg	14
Potassium	240.0 mg	6

Caution: This combination of pickled vegetables assaults you with sodium; in the pickling process, the vegetables are cooked in boiling salted water. *Allergy alert:* People allergic to aspirin may have trouble with chowchow, which contains the same allergen, salicylate.

Strengths: This mixture of chopped veggies and a sweet 'n' hot blend of spices and vinegar has a couple things going for it besides good taste. First, there's no cholesterol. Also, sour chowchow is rich in iron, has a good amount of vitamin C, and contains a dab of potassium.

Selection: Some chowchow may include cucumbers, green tomatoes, peppers, beans, and celery in addition to the standard cauliflower and onions.

Preparation: Chowchow can be made at home—it's especially good using fresh-grown cucumbers and other fresh garden vegetables.

Serve: As an appetizer or side dish.

CHOWCHOW, *Sweet*

Serving: ½ cup

Calories	142
Fat	1.1 g
Saturated	0
Monounsaturated	N/A
Polyunsaturated	N/A
Calories from fat	7%
Cholesterol	0
Sodium	646 mg
Protein	1.8 g
Carbohydrate	33.1 g
Dietary fiber	1.8 g

CHIEF NUTRIENTS

NUTRIENT	AMOUNT	% RDA
Iron	1.8 mg	18
Vitamin C	7.4 mg	12
Potassium	245.0 mg	7

Caution: Less salty than sour chowchow, the sweet variety still packs quite a bit of sodium because it's cooked in boiling salted water. Sweet chowchow has 4 times the calories of sour—142 in ½ cup. *Allergy alert:* Anyone sensitive to aspirin should avoid chowchow, which also contains the allergen salicylate.
Strengths: Cholesterol-free, chowchow has good amounts of vitamin C and iron, and a fair amount of potassium.
History: Both sour and sweet chowchow may have been introduced to the U.S. by Chinese railroad laborers during the 19th century.
Description: Contains cauliflower and onions.
Preparation: Cook it to taste with your favorite veggies. Look in the "Relish" section of cookbooks.
Serve: As an appetizer, a side dish, or a topping for hamburgers.

HORSERADISH, *Prepared*

Serving: 1 tsp.

Calories	2
Fat	trace
Saturated	0
Monounsaturated	N/A
Polyunsaturated	N/A
Calories from fat	5%
Cholesterol	0
Sodium	5 mg
Protein	0.1 g
Carbohydrate	0.5 g
Dietary fiber	0.1 g

CHIEF NUTRIENTS

(None of the nutrients meet or exceed 5% of the RDA.)

Caution: *Allergy alert:* Can cause skin eruptions in people who are allergic to allyl isothiocyanate, an oil present in horseradish.
Strengths: Horseradish adds extra zing to any dish, especially if you use an extra-hot variety. Depending on its strength, it can also be an effective way to clear sinus passages.
History: The perennial horseradish plant hails from eastern Europe and western Asia. The Roman scholar Pliny thought it cured asthma. Horseradish is a bitter herb that is traditionally served at the Jewish Passover meal.
Selection: White horseradish is preserved with vinegar; the red kind contains beet juice.
Serve: With meats, egg dishes, and chicken, tuna, or salmon salads. Mixed with nonfat yogurt, horseradish creates a tangy sauce for seafood or cold meats.

HORSERADISH, *Raw*

Serving: 1 tsp.

Calories	4
Fat	trace
Saturated	0
Monounsaturated	N/A
Polyunsaturated	N/A
Calories from fat	3%
Cholesterol	0
Sodium	0.4 mg
Protein	0.2 g
Carbohydrate	1.0 g
Dietary fiber	0.1 g

CHIEF NUTRIENTS

NUTRIENT	AMOUNT	% RDA
Vitamin C	4.1 mg	7

Caution: *Allergy alert:* Individuals allergic to the allyl isothiocyanate oil that is present in horseradish could develop a skin rash.

Strengths: Valued for stimulating the appetite. Depending on how strong it is, horseradish may also clear the sinuses.

History: This root vegetable has been cultivated for two millennia and was used for medicinal purposes for hundreds of years. Cooks began using it in prepared foods during the 17th century.

Description: A pungent, edible root, horseradish is a sharp-tasting member of the mustard family. It's harvested in early spring, late fall, or early winter—just after the ground thaws or just before it freezes. The edible portion of the root is grated.

Storage: Fresh horseradish, available in many supermarkets, should be stored in a plastic bag in the refrigerator.

Serve: Sprinkled on meats and salads; for a really unusual treat, add a little to applesauce. For best flavor, grate horseradish just before serving.

LEMON JUICE

Serving: 1 Tbsp.

Calories	4
Fat	0
Saturated	0
Monounsaturated	0
Polyunsaturated	0
Calories from fat	0
Cholesterol	0
Sodium	trace
Protein	trace
Carbohydrate	1.3 g
Dietary fiber	trace

CHIEF NUTRIENTS

NUTRIENT	AMOUNT	% RDA
Vitamin C	7.0 mg	12

Strengths: For just 4 skinny calories, you can get 12% of your daily requirement of vitamin C in a single tablespoon. Although it has a trace of sodium, lemon juice imparts a robust flavor—so adding the juice may satisfy a craving for sodium.

Caution: Although a few tablespoons of lemon juice contain just a small amount of oxalic acid, larger quantities could cause problems for those with calcium-oxalate stones. Excessive consumption of pure lemon juice may sensitize teeth and lead to enamel erosion. *Allergy alert:* Squeezing lemons could provoke contact dermatitis; if you're sensitive to oil in the peel, wear plastic gloves.

Preparation: To get the most juice from a lemon, make sure it's at room temperature before squeezing (or microwave it with the oven set on high for about 30 seconds).

Serve: Mixed with sparkling water to get a low-cal lemonade. And with all types of salads, fish, poultry, and vegetables. To make a fine no-fat, low-cal salad dressing, simply squeeze lemon juice over greens.

OLIVES, Black ⇩

Serving: 5 large

Calories	25
Fat	2.4 g
Saturated	0.3 g
Monounsaturated	1.8 g
Polyunsaturated	0.2 g
Calories from fat	84%
Cholesterol	0
Sodium	192 mg
Protein	0.2 g
Carbohydrate	1.4 g
Dietary fiber	0.7 g

CHIEF NUTRIENTS

NUTRIENT	AMOUNT	% RDA
Iron	0.8 mg	8

Strengths: Moderate in calories. Although a high percentage of those calories come from fat, a serving of black olives still has less than 3 g of fat. Because most of it is the monounsaturated form, olive oil is beneficial for controlling high blood cholesterol. Despite some popular misconceptions, olives contain no cholesterol. Black olives, while not exactly low in sodium, have considerably less than green ones, as long as they aren't cured in salt. They contain a surprising amount of iron, which can help prevent anemia.

Eat with: Foods rich in vitamin C to help the iron be absorbed better. (Try adding olives to a salad containing green peppers, broccoli, or asparagus.)

Caution: Sodium and fat levels can add up if you eat large quantities of olives, and olives do contribute some calories.

History: The olive tree has been considered sacred since at least the 17th century B.C.

Description: Black olives are mature green olives that get their color either naturally from being ripened on the tree, or chemically from curing.

OLIVES, Green, Unstuffed

Serving: 5 large

Calories	27
Fat	2.9 g
Saturated	0.3 g
Monounsaturated	N/A
Polyunsaturated	N/A
Calories from fat	99%
Cholesterol	0
Sodium	552 mg
Protein	0.3 g
Carbohydrate	0.3 g
Dietary fiber	0.6 g

CHIEF NUTRIENTS

(None of the nutrients meet or exceed 5% of the RDA.)

Strengths: Green olives are the unripened version of black olives and have the same general strengths—low in calories and high in monounsaturated fat, which is good for controlling high blood cholesterol. Some protein and dietary fiber are present.

Caution: Green olives have more than twice as much sodium as the black kind.

History: Ancient people considered the olive tree sacred, and olives were cultivated by Syrians and Palestinians as early as 4,000 years before Christ's birth. Today, most olives come from Italy and Spain, although California also raises a respectable crop.

Selection: Olives are always green before they ripen, whereas ripe olives may be either green or black. Bottled olives usually have been processed in some way to alleviate the bitter flavor—and they're packed in oil or brine. Spanish olives may be pitted and stuffed with pimientos, almonds, onions, or jalapeños.

PICKLES, *Dill*   ⬇ 🗑

Serving: 1 (about 2 oz.)

Calories	12
Fat	0.1 g
Saturated	trace
Monounsaturated	N/A
Polyunsaturated	trace
Calories from fat	9%
Cholesterol	0
Sodium	833 mg
Protein	0.4 g
Carbohydrate	2.7 g
Dietary fiber	0.8 g

CHIEF NUTRIENTS

(None of the nutrients meet or exceed 5% of the RDA.)

Strengths: Aside from the fact that dill pickles have a little bit of fiber, they're nutritional zeros. But they are virtually free of fat and calories. If you're trying to lose weight, you might choose to eat a pickle before a meal or snack. The jolt to your taste buds may blunt your appetite.

Caution: Dill pickles are extraordinarily high in sodium. So if you're trying to restrict your sodium intake, look for low-sodium varieties that may contain as little as 12 mg of sodium. *Allergy alert:* Pickles may trigger hives or other allergic reactions in people who are allergic to molds. Also, many allergists recommend that those who are allergic to aspirin avoid pickles; they contain salicylate, a basic component of aspirin.

PICKLES, *Sour*   ⬇ 🗑

Serving: 1 (about 1 oz.)

Calories	4
Fat	0.1 g
Saturated	trace
Monounsaturated	N/A
Polyunsaturated	trace
Calories from fat	16%
Cholesterol	0
Sodium	423 mg
Protein	0.1 g
Carbohydrate	0.8 g
Dietary fiber	0.5 g

CHIEF NUTRIENTS

(None of the nutrients meet or exceed 5% of the RDA.)

Strengths: Sour pickles don't have much going for them nutritionally, but they may help you lose weight. If you eat a pickle when you get a food craving, you just might take the edge off your appetite at the cost of just a few calories. Besides, pickles have virtually no fat.

Caution: Like dill pickles, sour pickles are extremely high in sodium due to the brine in which they are preserved. Look for low-sodium varieties that may contain as little as 6 mg per pickle. *Allergy alert:* Pickles may trigger hives or other allergic reactions in people who are allergic to molds. If you are allergic to the salicylate in aspirin, you'll probably want to avoid pickles, since they have the same ingredient.

PICKLES, *Sweet*   ⬇ 🏛

Serving: 1 (about 1 oz.)

Calories	41
Fat	0.1 g
Saturated	trace
Monounsaturated	N/A
Polyunsaturated	trace
Calories from fat	2%
Cholesterol	0
Sodium	329 mg
Protein	0.1 g
Carbohydrate	11.1 g
Dietary fiber	0.4 g

CHIEF NUTRIENTS

(None of the nutrients meet or exceed 5% of the RDA.)

Caution: Sweet pickles are higher in calories than dill or sour pickles. They're also high in sodium—though they contain less than dill or sour. However, if you want to restrict your sodium intake, you can find low-sodium varieties that may contain as little as 6 mg of sodium. *Allergy alert:* Pickles may trigger hives or other allergic reactions in people who are allergic to molds. Many allergists recommend that those who are allergic to aspirin avoid pickles, which contain salicylate, a basic component of aspirin.

RELISH, *Sour Pickle*   ⬇ 🏛

Serving: 1 Tbsp.

Calories	3
Fat	0.1 g
Saturated	trace
Monounsaturated	N/A
Polyunsaturated	N/A
Calories from fat	43%
Cholesterol	0
Sodium	203 mg
Protein	0.1 g
Carbohydrate	0.4 g
Dietary fiber	0.3 g

CHIEF NUTRIENTS

(None of the nutrients meet or exceed 5% of the RDA.)

Caution: Eating sour relish is like licking a salt block—it's packed with sodium. If you make your own pickle relish, it's essential to use proper canning procedures to kill the micro-organisms that could cause spoilage. Note: Canker sores may be aggravated by condiments like relish. *Allergy alert*: Like pickles and sauerkraut, relish contains vinegar—so those with mold allergies should avoid it.

History: The word *relish* was first used in the late 18th century; it comes from a Middle English term for "a taste." In early America, relishes were made with a large variety of garden vegetables. Cranberries, peaches, and pears were also preserved as relishes.

Selection: Many kinds of relish are available. Or you can make your own varieties of relish with vegetables or fruit. May be canned immediately or refrigerated for up to 2 weeks.

RELISH, *Sweet Pickle*   ⬇ ▥

Serving: 1 Tbsp.

Calories	21
Fat	0.1 g
Saturated	0
Monounsaturated	N/A
Polyunsaturated	N/A
Calories from fat	4%
Cholesterol	0
Sodium	107 mg
Protein	0.1 g
Carbohydrate	5.1 g
Dietary fiber	0.3 g

CHIEF NUTRIENTS

(None of the nutrients meet or exceed 5% of the RDA.)

Caution: The sweet variety of relish has about half the salt of the sour kind. Still, there's enough sodium in a single tablespoon to make it taboo for those on salt-restricted diets. If you're canning your own sweet relish, use proper procedures to avoid food spoilage. *Allergy alert:* Relish contains vinegar, which is likely to aggravate symptoms in those allergic to mold.

History: In the late 1800s, Henry Heinz offered the American people a sweet pepper relish in bottled form. It became a popular general-store item, the all-American way to enhance the flavors of meats, fish, and poultry.

Selection: Besides the standard sweet pickle relish, there are many other kinds.

Preparation: By pickling and canning your own fresh vegetables in summer, you can have homemade relishes throughout the winter.

Serve: A tomato-apple relish is particularly good as a side dish with a main meal—or add it to salad fixings out of season when home-grown veggies aren't available.

VINEGAR, *Cider*   ⬇ ▥

Serving: 1 Tbsp.

Calories	2
Fat	0
Saturated	0
Monounsaturated	0
Polyunsaturated	0
Calories from fat	0
Cholesterol	0
Sodium	0.2 mg
Protein	0
Carbohydrate	0.9 g
Dietary fiber	0

CHIEF NUTRIENTS

(None of the nutrients meet or exceed 5% of the RDA.)

Strengths: Cider vinegar adds zip to food, with practically no sodium, fat, cholesterol, or calories.

Caution: *Allergy alert:* Individuals who are allergic to molds, yeasts, and other fermented foods may develop headaches or have other reactions to vinegar.

Description: Cider vinegar is a pungent, sour, acidic solution made from the fermented juice of crushed pressed apples.

Storage: Store vinegar in a cool, dark place.

Serve: Sprinkle a few drops of plain cider vinegar on lightly steamed asparagus, broccoli, cauliflower, cabbage, brussels sprouts, green beans, or other foods high in vitamins, minerals, fiber, or other healthful components.

VINEGAR, *Distilled* ⬇ 🏛

Serving: 1 Tbsp.

Calories	2
Fat	0
Saturated	0
Monounsaturated	0
Polyunsaturated	0
Calories from fat	0
Cholesterol	0
Sodium	0.2 mg
Protein	0
Carbohydrate	0.8 g
Dietary fiber	0

CHIEF NUTRIENTS

(None of the nutrients meet or exceed 5% of the RDA.)

Strengths: Virtually free of fat, calories, sodium, and cholesterol.
Caution: *Allergy alert:* Vinegar may trigger hives or other reactions in those who are sensitive to molds or yeasts.
Curiosity: Vinegar comes from the Latin *vinum,* meaning wine, plus *acer,* meaning sharp.
History: Until about the 17th century, vinegar was a by-product of the wine and beer industry. Then vinegar-making grew into a separate industry.
Description: Distilled vinegar is made from grain alcohol diluted with water to a strength of about 5% acidity.
Storage: Store in an airtight container in a cool, dark place. Keeps well for up to 6 months after opening.
Serve: Distilled white vinegar is used as a flavoring, especially in salad dressings, and as a pickling solution. Whisk vinegar together with oil and herbs to make a zingy, reduced-fat dressing for steamed fish or vegetables.

PRESERVES

JAMS ⬇

Serving: 1 Tbsp.

Calories	54
Fat	trace
Saturated	0
Monounsaturated	N/A
Polyunsaturated	N/A
Calories from fat	0.3%
Cholesterol	0
Sodium	2 mg
Protein	0.1 g
Carbohydrate	14.0 g
Dietary fiber	0.4 g

CHIEF NUTRIENTS

(None of the nutrients meet or exceed 5% of the RDA)

Caution: A single tablespoon of jam is high in calories, and it has only a trace of a few nutrients. *Allergy alert:* People allergic to aspirin should avoid jam because it contains salicylate. If you are allergic to glutens, you may have an adverse reaction to some of the fruit used in jam or preserves.
Curiosity: In some Mediterranean countries, it's considered hospitable to serve guests a glass of water with a spoonful of very sweet jam on the side.
History: Jams and preserves have long been a sweet treat in Mediterranean countries. By the 16th century, recipes for quince and orange marmalades were common in England.
Description: Both jams and preserves are made by boiling fruit and sweetener (pectin or sugar) to a thick consistency before canning; however, only preserves contain fruit chunks. Store-bought jams and preserves usually are made with commercial pectin and contain a high proportion of sugar.
A Better Idea: Look for recipes for less-sweet preserves that use the natural pectin in fruit rather than commercial pectins.

JELLIES   ⇩

Serving: 1 Tbsp.

Calories	49
Fat	trace
Saturated	0
Monounsaturated	N/A
Polyunsaturated	N/A
Calories from fat	0.4%
Cholesterol	0
Sodium	3 mg
Protein	trace
Carbohydrate	12.7 g
Dietary fiber	0.5 g

CHIEF NUTRIENTS

(None of the nutrients meet or exceed 5% of the RDA.)

Caution: With nearly 50 calories in each tablespoon of jelly, it's high in calories. *Allergy alert:* People allergic to aspirin should stay away from jelly because it contains the same allergen, salicylate.

History: Jelly comes from the Latin *gelare,* which means "to freeze."

Description: This sweet product is made by boiling a mixture of fruit juice, sugar, and sometimes pectin. In addition to the many kinds of fruit jellies, there are also herb jellies—made with extracts of marjoram, cloves, and other aromatics.

A Better Idea: Commercial jellies, by law, must have 45% fruit—and the rest is usually sugar. You can reduce sweeteners dramatically by canning your own low-sugar jelly. Look for canning recipes that combine fruits naturally high in pectin, such as apples, with other fruits that are low in pectin.

MARMALADE   ⇩

Serving: 1 Tbsp.

Calories	51
Fat	trace
Saturated	0
Monounsaturated	trace
Polyunsaturated	trace
Calories from fat	0.4%
Cholesterol	0
Sodium	3 mg
Protein	0.1 g
Carbohydrate	14.0 g
Dietary fiber	1.0 g

CHIEF NUTRIENTS

(None of the nutrients meet or exceed 5% of the RDA.)

Caution: Contains a high 51 calories per tablespoon. *Allergy alert:* People allergic to aspirin should avoid marmalade because it contains salicylate. Those allergic to glutens may have trouble with the fruit in marmalade.

History: The word marmalade first came from the Greek word *melimēlon* meaning "honey apple," and then from *marmelada,* Portuguese for "quince jam." Recipes for quince, guava, and other fruit marmalades appeared in 16th- and 17th-century cookbooks.

Description: Marmalade is a type of preserve that generally contains portions of citrus fruit rind and pulp. Marmalade is often made with Seville oranges.

Preparation: You can make your own marmalade with a variety of fruits—oranges, lemons, limes, and grapefruit.

SPREADS

ALMOND BUTTER, *Unsalted* ⬇

Serving: 1 Tbsp.

Calories	101
Fat	9.5 g
Saturated	0.9 g
Monounsaturated	6.1 g
Polyunsaturated	2.0 g
Calories from fat	84%
Cholesterol	0
Sodium	2 mg
Protein	2.4 g
Carbohydrate	3.4 g
Dietary fiber	N/A

CHIEF NUTRIENTS

NUTRIENT	AMOUNT	% RDA
Magnesium	48.0 mg	14
Iron	0.6 mg	6
Riboflavin	0.1 mg	6
Calcium	43.0 mg	5
Folate	10.4 mcg	5

Strengths: Almond butter is low in sodium (if the product is unsalted—but you need to check the label to find out). Although high in fat, most of it is the monounsaturated kind, which can help control serum cholesterol. Almond butter has no cholesterol. Its immunity nutrients are iron and folate; blood pressure regulators include magnesium, calcium, and a bit of potassium. The amount of fiber in the butter depends on whether it was made from blanched or unblanched nuts.

Eat with: Bananas, apples, and other foods rich in vitamin C to enhance absorption of the iron.

Caution: Too high in fat and calories for heavy consumption, almond butter contains oxalic acid, which should be restricted by those with calcium-oxalate stones. *Allergy alert:* Nuts are highly allergenic, often causing hives, headaches, and other reactions. Those allergic to aspirin may react to the natural salicylate in almonds.

Description: A creamy paste similar in texture and appearance to peanut butter. May be made from raw or roasted nuts; both salted and unsalted brands are available.

APPLE BUTTER ⬇ 🗎

Serving: 1 Tbsp.

Calories	33
Fat	0.1 g
Saturated	0
Monounsaturated	N/A
Polyunsaturated	N/A
Calories from fat	4%
Cholesterol	0
Sodium	0.4 mg
Protein	0.1 g
Carbohydrate	8.2 g
Dietary fiber	0.2 g

CHIEF NUTRIENTS

(None of the nutrients meet or exceed 5% of the RDA.)

Caution: Moderately high in calories.

Strengths: Low in fat. Adds some potassium to your diet to give a little boost to nerve and muscle functions.

Origin: This Pennsylvania Dutch treat goes back to Revolutionary War days.

History: Colonists occasionally added apple cider, orange peel, and quince to give zing to their apple butter.

Description: Tasty brown preserves made of cooked apples, sugar, spices (cloves, ginger, cinnamon), and apple cider.

Preparation: 3 pounds of fresh apples make about 2 cups of apple butter.

A Better Idea: To control the amount of sugar in your apple butter, make your own. Use the Jonathan variety of apples for delicious results. If you need to add sweetener, use a sugar substitute instead of sugar.

CASHEW BUTTER, Unsalted ⬇

Serving: 1 Tbsp.

Calories	94
Fat	7.9 g
Saturated	1.6 g
Monounsaturated	4.7 g
Polyunsaturated	1.3 g
Calories from fat	76%
Cholesterol	0
Sodium	2 mg
Protein	2.8 g
Carbohydrate	4.4 g
Dietary fiber	N/A

CHIEF NUTRIENTS

NUTRIENT	AMOUNT	% RDA
Magnesium	41.3 mg	12
Iron	0.8 mg	8
Zinc	0.8 mg	6
Folate	10.9 mcg	5

Strengths: Unsalted cashew butter is low in sodium, so check the label to make sure the product is salt-free. It has no cholesterol, and the fat is mostly monounsaturated, which can help control serum cholesterol. Its immunity nutrients include iron, zinc, and folate. Plentiful magnesium enhances the normal functioning of nerves and muscles.

Eat with: Foods rich in vitamin C to improve absorption of the nonheme iron.

Caution: Too high in fat and calories for heavy consumption, cashews also contain oxalic acid, which should be restricted by those with calcium-oxalate stones. Certain components of nuts may inhibit iron absorption in those on no-meat, low-C diets. *Allergy alert:* Nuts are highly allergenic, often causing hives, headaches, and other reactions.

Description: A creamy paste similar in texture and appearance to peanut butter. It may be made from raw or roasted nuts and is available salted and unsalted.

Serve: On slices of apples, bananas, peppers, or carrots.

DEVILED HAM, Canned

Serving: 2 oz.

Calories	199
Fat	18.3 g
Saturated	6.6 g
Monounsaturated	N/A
Polyunsaturated	N/A
Calories from fat	83%
Cholesterol	37 mg
Sodium	670 mg
Protein	7.9 g
Carbohydrate	0
Dietary fiber	0

CHIEF NUTRIENTS

NUTRIENT	AMOUNT	% RDA
Thiamine	0.1 mg	5

Caution: When you have the urge to chew the fat, you probably couldn't find a better place to start than with canned deviled ham. More than 80% of its calories are from fat, and its cholesterol count is somewhat less than ideal, too. The sodium in one serving is high—not recommended for those on a low-salt diet. Gout sufferers will want to shun deviled ham because it contains purines, insoluble uric acids. *Drug interaction:* People taking MAO-inhibitor antidepressants should avoid tyramine, which is found in meats that have been aged, dried, or pickled. The interaction could raise blood pressure considerably. *Allergy alert:* Pork products may trigger allergic reactions such as hives and headaches in some people. Those allergic to molds should stay away from deviled ham.

Strengths: Canned deviled ham is a source of protein, and it contains some thiamine for more energy.

Description: Processed from cured ham and ham fat seasoned with hot spices, this deviled-meat product is often used as a sandwich spread or as a dip.

HUMMUS

Serving: ⅓ cup

Calories	140
Fat	6.9 g
Saturated	1.0 g
Monounsaturated	2.9 g
Polyunsaturated	2.6 g
Calories from fat	44%
Cholesterol	0
Sodium	200 mg
Protein	4.0 g
Carbohydrate	16.5 g
Dietary fiber	N/A

CHIEF NUTRIENTS

NUTRIENT	AMOUNT	% RDA
Folate	48.7 mcg	24
Vitamin B$_6$	0.3 mg	17
Iron	1.3 mg	13
Vitamin C	6.5 mg	11
Magnesium	23.8 mg	7
Zinc	0.9 mg	6
Calcium	41.0 mg	5
Thiamine	0.1 mg	5

Strengths: Although nearly half the calories in hummus come from fat, very little is saturated—so it's a fairly decent alternative to cheese dips and spreads. Hummus supplies significant amounts of folate and vitamin B$_6$. (Women who are taking oral contraceptives may need more of both these B vitamins.) In addition, it offers immunity-boosting iron and vitamin C and contains some fiber (exact quantities not available).

Caution: Sodium content may be a bit high for those on sodium-restricted diets. Hummus contains chick-peas, which, like other beans, are moderately high in purines and should be eaten sparingly by anyone with gout.

Description: A thick Middle Eastern spread made from mashed chickpeas and tahini (sesame seed paste), hummus is seasoned with lemon juice, olive oil, and garlic. It's sometimes called hummus bi tahini.

Serve: As an appetizer, with whole wheat pita bread cut into triangles and toasted.

MARGARINE, *Soft*

Serving: 2 tsp.

Calories	67
Fat	7.6 g
Saturated	1.3 g
Monounsaturated	2.7 g
Polyunsaturated	3.3 g
Calories from fat	100%
Cholesterol	0
Sodium	101 mg
Protein	0.1 g
Carbohydrate	trace
Dietary fiber	0

CHIEF NUTRIENTS

NUTRIENT	AMOUNT	% RDA
Vitamin A	93.9 RE	9
Vitamin E	0.7 mg αTE	7

Caution: High in fat, a small dab of soft margarine has more calories and sodium than is advisable. Trans fatty acids in margarine may raise cholesterol as much as butter does. *Allergy alert:* Margarine often contains corn, soy, or milk products as well as salicylate, all of which may cause allergic reactions.

Strengths: No cholesterol. Depending on the brand, soft margarine may have less saturated fat than regular margarine. So check the labels. It's soft when cold—easy to spread in a thin layer—so you'll be able to eat less in each serving. Contains vitamins A and E, both of which enhance immunity—but fruits, vegetables, and cereals are healthier sources.

History: Considered a threat by the dairy industry, margarine was heavily taxed until 1950. By law, it also had to be lard white (except for a few states where it was allowed to be dyed pink).

A Better Idea: Light or diet margarines—these products are often lower in fat and calories than regular margarine. Better yet, try jelly on bread and muffins. It's fat-free and cholesterol-free!

MARGARINE, Stick

Serving: 2 tsp.

Calories	68
Fat	7.6 g
Saturated	1.5 g
Monounsaturated	3.4 g
Polyunsaturated	2.4 g
Calories from fat	100%
Cholesterol	0
Sodium	89 mg
Protein	0.1 g
Carbohydrate	0.1 g
Dietary fiber	0

CHIEF NUTRIENTS

NUTRIENT	AMOUNT	% RDA
Vitamin A	93.8 RE	9
Vitamin E	0.9 mg αTE	9

Caution: Margarine is high in fat and has an overabundance of calories and sodium. Contrary to popular opinion, margarine is not lower in calories and fat than butter. Trans fatty acids in margarine may raise body cholesterol as much as butter does. *Allergy alert:* Margarine often contains corn, soy, or milk products as well as salicylate, all of which may cause allergic reactions.

Strengths: No cholesterol. Contains vitamins A and E, both of which enhance immunity; other foods like fruits, vegetables, and cereals are healthier sources of these vitamins.

Origin: Invented in France in 1869 at the behest of Napoleon III. Early types were made from animal fat colored with natural dyes.

A Better Idea: Diet or reduced-fat spreads, especially the soft or liquid types. These products often contain unhydrogenated vegetable oil, making them lower in saturated fat than regular margarine.

MAYONNAISE

Serving: 1 Tbsp.

Calories	99
Fat	11.0 g
Saturated	1.6 g
Monounsaturated	3.1 g
Polyunsaturated	5.7 g
Calories from fat	98%
Cholesterol	8 mg
Sodium	78 mg
Protein	0.2 g
Carbohydrate	0.4 g
Dietary fiber	0

CHIEF NUTRIENTS

NUTRIENT	AMOUNT	% RDA
Vitamin E	2.9 mg αTE	29

Caution: Nearly all the calories in mayonnaise come from fat—and in any food, nearly all fat calories are converted to body fat. *Allergy alert:* This popular sandwich spread and salad binder contains egg, a common food allergen. Some mayo also contains small amounts of monosodium glutamate (MSG), a common headache trigger.

Strengths: Mayonnaise supplies a generous helping of vitamin E, a nutrient that protects polyunsaturated fatty acids and other nutrients from destructive oxidative reactions. As an antioxidant, vitamin E may also help protect against some of the effects of lead and other environmental pollutants.

Description: A creamy condiment made from any one of various vegetable oils, egg yolks, vinegar, or other acidifying ingredients, plus seasonings.

A Better Idea: You can prepare "slim mayo" with less fat and sodium and fewer calories using soft tofu, white vinegar, lemon, and seasonings. "Lite" mayo is also available in most supermarkets.

MUSTARD, *Prepared, Brown*

Serving: 1 Tbsp.

Calories	14
Fat	1.0 g
Saturated	0.2 g
Monounsaturated	N/A
Polyunsaturated	N/A
Calories from fat	62%
Cholesterol	0
Sodium	196 mg
Protein	0.9 g
Carbohydrate	0.8 g
Dietary fiber	N/A

CHIEF NUTRIENTS

(None of the nutrients meet or exceed 5% of the RDA.)

Caution: Mustard is high in sodium, so any frank served with mustard combines to make an overload—especially for salt-sensitive people who are watching their blood pressure. *Allergy alert:* May cause canker sores or hives in some people.

History: Mustard seeds have been around since 3000 B.C., and the Chinese ancients believed they were an aphrodisiac. The French claim that mustard's place in culinary history began when the Duke of Burgundy's chef came up with a spicy yellow sauce to kill the taste of rancid meat.

Description: Brown mustard seeds are the main ingredient in Chinese and European mustard. English mustard is a blend of white and brown mustard seeds.

Selection: A wide range of tasty mustards is available, imported from all parts of the globe. Varieties include Dijon brands from France, mild Dusseldorf, sweet Bavarian and Swedish, and extremely hot Chinese mustards.

Storage: Mustard tends to lose its sharpness if it's stored too long.

Serve: Put a light coating of hot mustard on poached or broiled fish. (Note: Serving size is usually less than 1 Tbsp.)

MUSTARD, *Prepared, Yellow*

Serving: 1 Tbsp.

Calories	11
Fat	0.7 g
Saturated	trace
Monounsaturated	N/A
Polyunsaturated	N/A
Calories from fat	53%
Cholesterol	0
Sodium	188 mg
Protein	0.7 g
Carbohydrate	1.0 g
Dietary fiber	0.4 g

CHIEF NUTRIENTS

(None of the nutrients meet or exceed 5% of the RDA.)

Caution: High in sodium. *Allergy alert*: Mustard has been identified as one of the foods that may cause hives in some people.

Curiosity: More than 700 million pounds is consumed worldwide each year—making it the most-favored seasoning in the world.

History: One theory of how mustard got its name is that it's from the Latin *must*—the term for the juice of crushed, unfermented grapes; during the Middle Ages, wine was mixed with mustard paste for flavor.

Description: Commercial yellow mustard is a condiment made of sugar, vinegar, turmeric, and white wine—and of course ground mustard seeds.

Preparation: You can make your own mustard using dry mustard powder, vinegar, and spices to taste. If you're on a low-sodium diet, omit salt.

PÂTÉ, *Chicken Liver, Canned*

Serving: 1 oz.

Calories	57
Fat	3.7 g
Saturated	1.1 g
Monounsaturated	1.5 g
Polyunsaturated	0.7 g
Calories from fat	59%
Cholesterol	111 mg
Sodium	109 mg
Protein	3.8 g
Carbohydrate	1.9 g
Dietary fiber	0

CHIEF NUTRIENTS

NUTRIENT	AMOUNT	% RDA
Vitamin B$_{12}$	2.3 mcg	115
Folate	91.0 mcg	46
Iron	2.6 mg	26
Riboflavin	0.4 mg	24
Niacin	2.1 mg	11
Vitamin A	61.5 RE	6

Strengths: A fabulous source of vitamin B$_{12}$, chicken liver pâté gets high marks for a generous amount of folate. It's a significant source of heme iron, the form best absorbed by the body, which helps insure against iron-deficiency anemia.

Eat with: Whole grain crackers, raw vegetables like carrots, and other high-fiber foods to offset the high percentage of calories from fat.

Caution: Extremely high in cholesterol; keep portions to 1 oz. or less. *Drug interaction:* Chicken livers contain tyramine, which may trigger dangerously high increases in blood pressure in people taking MAO inhibitors.

PÂTÉ, *Goose Liver, Canned, Smoked*

Serving: 1 oz.

Calories	131
Fat	12.4 g
Saturated	4.1 g
Monounsaturated	7.3 g
Polyunsaturated	0.2 g
Calories from fat	85%
Cholesterol	43 mg
Sodium	198 mg
Protein	3.2 g
Carbohydrate	1.3 g
Dietary fiber	0

CHIEF NUTRIENTS

NUTRIENT	AMOUNT	% RDA
Vitamin B$_{12}$	2.7 mcg	133
Vitamin A	283.5 RE	28
Iron	1.6 mg	16
Folate	17.0 mcg	9

Strengths: A whopping source of vitamin B$_{12}$, goose liver pâté has a decent amount of resistance-building vitamin A. The amount of absorbable iron is a redeeming factor if you're trying to insure against iron-deficiency anemia. The pâté also contributes some folate for added protection against anemia.

Eat with: High-fiber crackers and sliced raw vegetables instead of the traditional bits of buttered toast.

Caution: *Very* fatty, so keep portions to a modest size of 1 oz. or less. *Drug interaction:* Pâté contains tyramine and may cause dangerous increases in blood pressure in those taking MAO inhibitors. *Allergy alert:* Pâté may also contain eggs, which are an allergen that can affect some people.

Description: A spread made from the enlarged liver of a goose that has been force-fed and fattened, it's also called pâté de foie gras (*foie gras* is French for "fat liver").

PEANUT BUTTER

Serving: 2 Tbsp.

Calories	188
Fat	16.0 g
Saturated	3.1 g
Monounsaturated	7.6 g
Polyunsaturated	4.6 g
Calories from fat	76%
Cholesterol	0
Sodium	153 mg
Protein	7.9 g
Carbohydrate	6.6 g
Dietary fiber	1.9 g

CHIEF NUTRIENTS

NUTRIENT	AMOUNT	% RDA
Niacin	4.2 mg	22
Magnesium	50.2 mg	14
Folate	25.0 mcg	13
Potassium	230.7 mg	6
Vitamin B$_6$	0.1 mg	6
Iron	0.5 mg	5
Zinc	0.8 mg	5

Strengths: The nutritional profile of peanut butter is roughly the same as for peanuts. It provides a good amount of protein and it's fair to high in B vitamins, with some minerals. Fiber per serving is more if you choose chunk style, less if you choose smooth—but either way you're getting a good amount of fiber.

Caution: Peanuts contain phytates and other substances that inhibit absorption of iron from plant foods unless vitamin C is consumed with the meal. If you're trying to avoid excess sodium, look for unsalted brands. And don't allow children under age 4 to eat spoonfuls of peanut butter—they could choke. *Allergy alert:* Peanuts are a potent and common food allergen.

Curiosity: Developed in 1890 and promoted as a health food at the 1904 St. Louis World's Fair.

Description: A blend of ground shelled peanuts, vegetable oil (often hydrogenated), and salt. May also contain sugar and additives to smooth texture and prevent separation.

Serve: In a sandwich or on crackers—with jelly, of course! Also good with bananas.

SANDWICH SPREAD, *Chicken or Turkey Salad*

Serving: 2 oz.

Calories	113
Fat	7.6 g
Saturated	2.0 g
Monounsaturated	1.8 g
Polyunsaturated	3.4 g
Calories from fat	61%
Cholesterol	17 mg
Sodium	214 mg
Protein	6.6 g
Carbohydrate	4.2 g
Dietary fiber	0

CHIEF NUTRIENTS

NUTRIENT	AMOUNT	% RDA
Vitamin B$_{12}$	0.2 mcg	11

Caution: The percentage of calories from fat in turkey or chicken spread is double the maximum considered acceptable by the American Heart Association. The sodium level is on the high side, too.

Eat with: Foods containing potassium, like potatoes or beets, to help counteract the potential effects of sodium on blood pressure.

Strength: The vitamin B$_{12}$ in these sandwich spreads is good for healthy nerves.

Description: Contains at least 50% chicken or turkey, plus mayonnaise and seasonings.

A Better Idea: Sliced chicken or turkey supplies twice as much protein and less than half as much fat as chicken or turkey sandwich spread. To improve vastly on the prepackaged spread, make your own with lean ground poultry, fat-free mayonnaise, finely chopped celery, and a little grated onion.

SANDWICH SPREAD, *Ham Salad*

Serving: 2 oz.

Calories	123
Fat	8.8 g
Saturated	2.9 g
Monounsaturated	4.1 g
Polyunsaturated	1.5 g
Calories from fat	65%
Cholesterol	21 mg
Sodium	517 mg
Protein	4.9 g
Carbohydrate	6.0 g
Dietary fiber	0

CHIEF NUTRIENTS

NUTRIENT	AMOUNT	% RDA
Vitamin B$_{12}$	0.2 mcg	22
Thiamine	0.2 mg	16
Niacin	1.2 mg	6
Vitamin C	3.4 mg	6

Caution: As with other prepared sandwich spreads, ham salad is high in sodium. The percentage of calories from fat is a warning signal to dieters—it's far above the level recommended by the American Heart Association. Those who have a problem with eating ham—because of allergies or drug interactions—are likely to have the same problems with ham spread.

Eat with: Foods containing potassium, like oranges or sliced tomatoes, to help counteract the effects of sodium on blood pressure.

Strengths: This sandwich spread is a minor source of vitamin C, derived primarily from sodium ascorbate, an additive used in some cured meats. It also supplies decent amounts of thiamine and vitamin B$_{12}$ and a little niacin.

Description: Contains at least 50% ham, plus mayonnaise and seasonings.

A Better Idea: Sliced ham supplies twice as much protein and about half as much fat as ham spread, making it a smarter choice nutritionally. To improve on store-bought ham spread, make your own ham salad with ground lean ham, fat-free mayonnaise, finely chopped celery, and a little grated onion.

SANDWICH SPREAD, *Pork or Beef Salad*

Serving: 2 oz.

Calories	133
Fat	9.8 g
Saturated	3.4 g
Monounsaturated	4.3 g
Polyunsaturated	1.5 g
Calories from fat	67%
Cholesterol	22 mg
Sodium	574 mg
Protein	4.3 g
Carbohydrate	6.8 g
Dietary fiber	0

CHIEF NUTRIENTS

NUTRIENT	AMOUNT	% RDA
Vitamin B$_{12}$	0.6 mcg	32
Thiamine	0.1 mg	7
Niacin	1.0 mg	5

Caution: High in sodium, pork and beef spreads yield about 67% of calories from fat. That means they're even fattier than chicken and turkey spreads—with more than twice the amount of calories from fat recommended by the American Heart Association.

Eat with: Foods containing potassium, like honeydew melon or romaine lettuce and sliced tomatoes, to help counteract the potential effects of sodium on blood pressure.

Strengths: To their credit, both of these salad spreads are high in vitamin B$_{12}$ for healthy nerves. They also have small amounts of other B vitamins.

Description: They contain at least 50% pork or beef (or both), plus mayonnaise and seasoning.

A Better Idea: Choose sliced beef or ham, which supplies twice as much protein and half as much fat as sandwich spread.

TAHINI

Serving: 1 Tbsp.

Calories	85
Fat	7.9 g
Saturated	1.1 g
Monounsaturated	3.0 g
Polyunsaturated	3.5 g
Calories from fat	84%
Cholesterol	0
Sodium	0.1 mg
Protein	2.5 g
Carbohydrate	2.5 g
Dietary fiber	1.3 g

CHIEF NUTRIENTS

NUTRIENT	AMOUNT	% RDA
Thiamine	0.2 mg	15
Magnesium	49.4 mg	14
Zinc	1.5 mg	10
Iron	0.9 mg	9
Folate	13.7 mcg	7

Strengths: Tahini has no cholesterol, and if the product is unsalted, no sodium either. The fat is mostly unsaturated. There's some iron, zinc, and folate for enhanced immunity to infections. And the thiamine and magnesium can help keep nerves nice and steady.

Caution: Too high in fat for everyday use. Tahini that is made from unhulled seeds contains some calcium oxalate, which may be of concern to those prone to calcium-oxalate stones.

Curiosity: The magic words "open sesame" from the tale of Ali Baba are thought to be a reference to the characteristic bursting of the sesame pods when the seeds ripen.

Description: Tahini resembles peanut butter, although it's generally creamier and thinner in consistency. It may be made from hulled or unhulled seeds and comes in raw, toasted, salted, and unsalted varieties.

Serve: With eggplant (as in baba ghanoush), chick-peas (as in hummus), and other foods predominant in Middle Eastern, Greek, and oriental cuisines.

Dairy Products and Eggs
BUTTER

SALTED

Serving: 2 tsp.

Calories	68
Fat	7.7 g
Saturated	4.8 g
Monounsaturated	2.2 g
Polyunsaturated	0.3 g
Calories from fat	100%
Cholesterol	21 mg
Sodium	78 mg
Protein	0.1 g
Carbohydrate	trace
Dietary fiber	0

CHIEF NUTRIENTS

NUTRIENT	AMOUNT	% RDA
Vitamin A	71.3 RE	7

Caution: Very high in fat, much of it saturated. Eating a lot of butter has been associated with higher blood pressure, serum cholesterol, and glucose levels—not to mention weight. A small serving is full of calories, sodium, and cholesterol, and there's no fiber to help counterbalance the fat. *Allergy alert:* Some people who are allergic to milk may have problems with butter, but in general small amounts are well-tolerated.

Eat with: Restraint.

Strength: Contains a little vitamin A, which is vital for good vision, healthy skin, and strong immunity. But fruits and vegetables are far more desirable sources of the vitamin.

A Better Idea: Unsalted butter contains only 1 mg of sodium per 2 tsp. For cooking, use nonstick spray. For table use, try butter-flavor granules.

WHIPPED SALTED

Serving: 2 tsp.

Calories	45
Fat	5.1 g
Saturated	3.2 g
Monounsaturated	1.5 g
Polyunsaturated	0.2 g
Calories from fat	100%
Cholesterol	14 mg
Sodium	52 mg
Protein	0.1 g
Carbohydrate	0
Dietary fiber	0

CHIEF NUTRIENTS

(None of the nutrients meet or exceed 5% of the RDA.)

Caution: Very high in fat, much of it saturated. Consumption of whipped butter has been associated with higher blood pressure, serum cholesterol, and glucose levels. A small dab is high in calories, sodium, and cholesterol, with no fiber to help counterbalance the fat. *Allergy alert:* Some people who are allergic to milk may have problems with whipped butter, but in general small amounts are well-tolerated.

Strength: Whipped butter has had air beaten into it, which increases its volume. The net effect is that, teaspoon for teaspoon, whipped butter contains one-third less fat, calories, cholesterol, and sodium than regular butter. That should not, however, be taken as a license to eat more of the whipped variety.

A Better Idea: Unsalted whipped butter contains less than 1 mg of sodium per 2 tsp. For cooking, use nonstick spray. For table use, try butter-flavor granules.

CHEESES

AMERICAN

Serving: 1 oz.

Calories	105
Fat	8.8 g
Saturated	5.5 g
Monounsaturated	2.5 g
Polyunsaturated	0.3 g
Calories from fat	75%
Cholesterol	26 mg
Sodium	401 mg
Protein	6.2 g
Carbohydrate	0.5 g
Dietary fiber	0

CHIEF NUTRIENTS

NUTRIENT	AMOUNT	% RDA
Calcium	172.3 mg	22
Vitamin B$_{12}$	0.2 mcg	10
Vitamin A	81.2 RE	8
Riboflavin	0.1 mg	6
Zinc	0.8 mg	6

Caution: American cheese is high in fat, much of it saturated, and it has lots of sodium, which can contribute to high blood pressure in salt-sensitive people. *Drug interaction:* May cause problems for those sensitive to tyramine or those taking MAO-inhibitor antidepressants. *Allergy alert:* May affect those sensitive to casein.

Strengths: Cheese is a good source of calcium for strong teeth and bones, and it contains high-quality protein that's easily digested. Certain cheeses, such as American, contain small amounts of fatty acids that may prevent some types of cancer, and they have some B vitamins for healthy skin and nerves. It's somewhat high in cholesterol, so avoid eating a lot. Because most cured cheeses are low in lactose, they are suitable for people with lactose intolerance.

Description: A processed product made of both fresh and aged cheeses.

A Better Idea: Look for reduced-fat versions.

BLUE

Serving: 1 oz.

Calories	99
Fat	8.1 g
Saturated	5.2 g
Monounsaturated	2.2 g
Polyunsaturated	0.2 g
Calories from fat	73%
Cholesterol	21 mg
Sodium	391 mg
Protein	6.0 g
Carbohydrate	0.7 g
Dietary fiber	0

CHIEF NUTRIENTS

NUTRIENT	AMOUNT	% RDA
Calcium	147.7 mg	18
Vitamin B$_{12}$	0.3 mcg	17
Riboflavin	0.1 mg	6
Vitamin A	63.8 RE	6
Folate	10.2 mcg	5

Caution: Much of the fat in blue cheese is saturated, and the cheese is also high in sodium, which can contribute to high blood pressure in salt-sensitive people. *Drug interaction:* May cause problems for those sensitive to tyramine or taking MAO inhibitors. *Allergy alert:* Blue cheese has a bluish mold, *Penicillium roquefortii,* which could cause problems for those allergic to it. (However, this mold doesn't usually affect those who are allergic to the derivative antibiotic, penicillin.)

Strengths: Like all cheeses, the blue variety is a good source of calcium for strong teeth and bones. It contains high-quality protein that's easily digested, and certain components of the cheese may help protect against cavities as well. In addition, its fatty acids may help protect against certain types of cancer. However, other lower-fat dairy foods are better sources. Low in lactose, it's suitable for people with lactose intolerance.

Description: Blue cheese is a term for Roquefort-type cheeses that are *not* produced in Roquefort, France. Called Stilton in England and Gorgonzola in Italy.

BRICK

Serving: 1 oz.

Calories	104
Fat	8.3 g
Saturated	5.3 g
Monounsaturated	2.4 g
Polyunsaturated	0.2 g
Calories from fat	72%
Cholesterol	26 mg
Sodium	157 mg
Protein	6.5 g
Carbohydrate	0.8 g
Dietary fiber	0

CHIEF NUTRIENTS

NUTRIENT	AMOUNT	% RDA
Calcium	188.6 mg	24
Vitamin B_{12}	0.4 mcg	18
Vitamin A	84.6 RE	8
Riboflavin	0.1 mg	6

Caution: All brick cheese is high in fat, and much of it is saturated. Although the sodium amount is moderate, in salt-sensitive people it may be enough to contribute to high blood pressure. *Drug interaction:* May cause problems for those sensitive to tyramine or those taking MAO-inhibitor antidepressants.

Strengths: This is a good source of calcium for strong teeth and bones, and its high-quality protein is easily digested. Most cured cheeses are low in lactose, making them suitable for people with lactose intolerance. Although there's a good amount of vitamin B_{12}, you'd be better off getting it from lower-fat dairy foods. The vitamin A enhances immunity, although the amount is small.

Curiosity: The name of this native Wisconsin cheese is said to have come from the fact that bricks were once used to weigh down the curds and press out the whey.

Description: A semisoft cheese with a mild flavor. Be aware that it can take on Limburger qualities if it ages too long.

BRIE

Serving: 1 oz.

Calories	93
Fat	7.8 g
Saturated	4.9 g
Monounsaturated	2.2 g
Polyunsaturated	0.2 g
Calories from fat	75%
Cholesterol	28 mg
Sodium	176 mg
Protein	5.8 g
Carbohydrate	0.1 g
Dietary fiber	0

CHIEF NUTRIENTS

NUTRIENT	AMOUNT	% RDA
Vitamin B_{12}	0.5 mcg	23
Folate	18.2 mcg	9
Riboflavin	0.2 mg	9
Calcium	51.5 mg	6
Vitamin A	51.0 RE	5

Caution: Brie is high in saturated fat, and it has a moderate amount of sodium, which can contribute to high blood pressure in salt-sensitive people. A good amount of vitamin B_{12} contributes to healthy nerves and blood, but you'd be better off getting it from lower-fat dairy foods. *Drug interaction:* May cause problems for those sensitive to tyramine or those taking MAO-inhibitor antidepressants.

Strengths: Although Brie has some calcium for strong teeth and bones, it's not as high in this mineral as many other cheeses. On the plus side, its high-quality protein is easily digested, and certain components of the cheese may help protect against cavities. Low in lactose, Brie is often suitable for people who have lactose intolerance.

Description: Similar to Camembert, Brie is characterized by an edible white rind and a creamy, soft interior that oozes when the cheese is at the peak of ripeness. French Brie is considered the world's best.

CAMEMBERT

Serving: 1 oz.

Calories	84
Fat	6.8 g
Saturated	4.3 g
Monounsaturated	2.0 g
Polyunsaturated	1.0 g
Calories from fat	73%
Cholesterol	20 mg
Sodium	236 mg
Protein	5.5 g
Carbohydrate	0.1 g
Dietary fiber	0

CHIEF NUTRIENTS

NUTRIENT	AMOUNT	% RDA
Vitamin B$_{12}$	0.4 mcg	18
Calcium	108.5 mg	14
Folate	17.4 mcg	9
Riboflavin	0.1 mg	8

Caution: High in fat and sodium, Camembert should be avoided if you're on a low-fat diet or guarding against high blood pressure. *Drug interaction:* May cause problems for those sensitive to tyramine or those taking MAO-inhibitor drugs.

Strengths: A good source of calcium, cheese contains high-quality protein that's easily digested. Because it's similar to Brie, Camembert may be equally effective in helping to fight cavities. Most cured cheeses are low in lactose, which makes them suitable for people with lactose intolerance. A good amount of vitamin B$_{12}$ contributes to healthy nerves and blood, but you'd be better off getting it from lower-fat dairy foods.

History: Legend has it that Camembert was named by Napoleon after the Norman village where he first tasted it.

Description: Like Brie, this cheese has a white rind and smooth, creamy interior. It's best when it feels plump and soft to the touch. A wheel of Camembert is about 4½ inches in diameter.

CARAWAY

Serving: 1 oz.

Calories	105
Fat	8.2 g
Saturated	5.2 g
Monounsaturated	2.3 g
Polyunsaturated	0.2 g
Calories from fat	70%
Cholesterol	26 mg
Sodium	193 mg
Protein	7.1 g
Carbohydrate	0.9 g
Dietary fiber	0

CHIEF NUTRIENTS

NUTRIENT	AMOUNT	% RDA
Calcium	188.5 mg	24
Riboflavin	0.1 mg	8
Vitamin A	80.9 RE	8
Zinc	0.8 mg	5

Caution: Like all cheeses, this one is high in saturated fat, and it contains enough sodium to contribute to high blood pressure in salt-sensitive people.

Eat with: Restraint.

Strengths: A good source of calcium for strong teeth and bones, caraway cheese contains high-quality protein that's easily digested. (Coincidentally, the caraway seeds present in this cheese are a longtime remedy for upset stomach.) Most cured cheeses are low in lactose, which makes them suitable for people with lactose intolerance. Caraway cheese has some riboflavin, which aids in blood formation, plus a little zinc for increased immunity and faster wound healing.

Description: An aged cheese, the caraway variety is considered ''spiced''—in the same category as those cheeses containing cumin, pepper, cloves, sage, or other aromatics. The spices are added during the cheese-making process so they're evenly distributed throughout the final product.

CHEDDAR

Serving: 1 oz.

Calories	113
Fat	9.3 g
Saturated	5.9 g
Monounsaturated	2.6 g
Polyunsaturated	1.3 g
Calories from fat	74%
Cholesterol	29 mg
Sodium	174 mg
Protein	7.0 g
Carbohydrate	0.4 g
Dietary fiber	0

CHIEF NUTRIENTS

NUTRIENT	AMOUNT	% RDA
Calcium	202.0 mg	25
Vitamin B$_{12}$	0.2 mcg	12
Riboflavin	0.1 mg	6
Zinc	0.9 mg	6

Caution: Cheddar cheese is high in fat, and in salt-sensitive people—about one-third of the population—its moderate amount of sodium can contribute to high blood pressure in salt-sensitive people. *Drug interaction:* May cause problems for those sensitive to tyramine or those taking MAO-inhibitor antidepressants. *Allergy alert:* May affect those sensitive to casein.

Eat with: Restraint.

Strengths: A good source of calcium for strong teeth and bones, Cheddar contains high-quality protein that's easily digested. Certain components of the cheese may help prevent cavities; others may protect against some types of cancer. Low in lactose, it's therefore suitable for people with lactose intolerance.

Origin: Cheddar, England, probably in the 16th century.

Description: Firm cheese ranging from mild to sharp in flavor and from white to orange in color. The deep color comes from a natural dye called annatto.

A Better Idea: Look for reduced-fat versions.

CHESHIRE

Serving: 1 oz.

Calories	108
Fat	8.6 g
Saturated	5.5 g
Monounsaturated	2.4 g
Polyunsaturated	0.2 g
Calories from fat	71%
Cholesterol	29 mg
Sodium	196 mg
Protein	6.5 g
Carbohydrate	1.3 g
Dietary fiber	0

CHIEF NUTRIENTS

NUTRIENT	AMOUNT	% RDA
Calcium	180.0 mg	23
Vitamin B$_{12}$	0.2 mcg	12
Vitamin A	68.6 RE	7
Zinc	0.8 mg	5

Caution: Cheshire is high in fat, primarily the saturated kind, and has a moderate amount of sodium. *Drug interaction:* May cause problems for those sensitive to tyramine or those taking MAO-inhibitor antidepressants.

Strengths: A good source of calcium for strong teeth and bones, Cheshire cheese contains high-quality protein that's easily digested. Its low lactose content makes it suitable for people who have lactose intolerance, and its B vitamins, although in small amounts, aid in healthy blood formation and have some benefits for nerves.

Origin: First made in Cheshire County, England, around the time of Queen Elizabeth I.

Description: A firm, somewhat crumbly cheese resembling Cheddar, Cheshire comes mainly in white and red varieties. (The white is actually pale yellow; the red is apricot colored.) A third type, blue Cheshire, resembles Stilton and might be inadvisable for anyone who reacts badly to Roquefort or blue cheese.

Serve: In Welsh rarebit. Cheshire is often the cheese of choice for this dish.

COLBY

Serving: 1 oz.

Calories	110
Fat	9.0 g
Saturated	5.7 g
Monounsaturated	2.6 g
Polyunsaturated	0.3 g
Calories from fat	73%
Cholesterol	27 mg
Sodium	169 mg
Protein	6.7 g
Carbohydrate	0.7 g
Dietary fiber	0

CHIEF NUTRIENTS

NUTRIENT	AMOUNT	% RDA
Calcium	191.7 mg	24
Vitamin B$_{12}$	0.2 mcg	12
Vitamin A	77.0 RE	8
Riboflavin	0.1 mg	6
Zinc	0.9 mg	6

Caution: High in saturated fat, Colby is a moderate source of sodium, which can contribute to high blood pressure in some people who are salt sensitive. *Drug interaction:* May cause problems for those sensitive to tyramine or those taking MAO-inhibitor antidepressants.

Strengths: A good source of calcium for strong teeth and bones, cheese contains high-quality protein that's easily digested. Because Colby is a close relative of Cheddar, it's possible that certain of Cheddar's strengths apply; these include cavity-fighting properties and some cancer protection. Low in lactose, so it's suitable for people with lactose intolerance, this cheese has a good amount of vitamin B$_{12}$ for healthy blood formation, and it can also help nerves. However, other sources of the vitamin are better bets because they're lower in fat.

Description: A mild, American-made type of Cheddar that's softer and moister than aged Cheddars (and therefore doesn't keep as long).

A Better Idea: Look for reduced-fat versions.

COTTAGE CHEESE, *Creamed*

Serving: 1/2 cup

Calories	117
Fat	5.1 g
Saturated	3.2 g
Monounsaturated	1.5 g
Polyunsaturated	0.2 g
Calories from fat	39%
Cholesterol	17 mg
Sodium	457 mg
Protein	14.1 g
Carbohydrate	3.0 g
Dietary fiber	0

CHIEF NUTRIENTS

NUTRIENT	AMOUNT	% RDA
Vitamin B$_{12}$	0.7 mcg	35
Riboflavin	0.2 mg	11
Calcium	67.8 mg	8
Folate	13.8 mcg	7
Vitamin A	54.2 RE	5

Caution: Higher in fat than other types of cottage cheese, the creamed variety is still reasonable, considering the serving size. But it's quite high in sodium. *Allergy alert:* May affect those sensitive to whey.

Strengths: Fresh cheeses, being low in tyramine, should be safe for those who are sensitive to this amine compound. It's a benefit to have some calcium present, although cottage cheeses have less than aged cheeses. Creamed cottage cheese has a very good amount of vitamin B$_{12}$, which contributes to healthy nerves and blood, but lower-fat dairy foods might be better sources. Its protein can provide a supplement for those cutting back on meat.

Description: Generally mild, this type of cottage cheese comes in small, medium, and large curds. Creamed cottage cheese has had from 4 to 8% cream added.

A Better Idea: Lower-fat versions are preferable. If creamed is all you have, try placing it in a strainer and rinsing with cold water to remove some sodium and fat.

COTTAGE CHEESE, *Dry Curd*

Serving: ¹/₂ cup

Calories	96
Fat	0.5 g
Saturated	trace
Monounsaturated	trace
Polyunsaturated	trace
Calories from fat	4%
Cholesterol	8 mg
Sodium	14 mg
Protein	19.5 g
Carbohydrate	2.1 g
Dietary fiber	0

CHIEF NUTRIENTS

NUTRIENT	AMOUNT	% RDA
Vitamin B_{12}	0.9 mcg	47
Riboflavin	0.2 mg	9
Folate	16.7 mcg	8

Strengths: Super-low in fat, dry curd cottage cheese is also low in calories and cholesterol. Fresh cheeses such as this should be safe for those sensitive to tyramine. It's rich in vitamin B_{12}, which contributes to healthy nerves and blood.

Caution: Not a good source of calcium, so if you're seeking this mineral, look elsewhere. *Allergy alert:* May affect those sensitive to whey.

Storage: Quite perishable once the container is opened. To keep air exposure of the curds to a minimum, store the container upside down.

Serve: With any meal. Works well in casseroles such as macaroni and cheese as a total or partial replacement for higher-fat cheeses. For a spread to go on bagels or muffins, season with herbs, spices or fruit puree.

Preparation: If you want a smooth substitute for sour cream, puree the cottage cheese, adding a little milk to facilitate blending, and squeeze in some fresh lemon juice for tang.

COTTAGE CHEESE, *Low-Fat (1%)*

Serving: ¹/₂ cup

Calories	82
Fat	1.2 g
Saturated	0.7 g
Monounsaturated	0.3 g
Polyunsaturated	trace
Calories from fat	13%
Cholesterol	5 mg
Sodium	459 mg
Protein	14.0 g
Carbohydrate	3.1 g
Dietary fiber	0

CHIEF NUTRIENTS

NUTRIENT	AMOUNT	% RDA
Vitamin B_{12}	0.7 mcg	36
Riboflavin	0.2 mg	11
Calcium	68.8 mg	9
Folate	14.0 mcg	7

Strengths: Not only is 1% cottage cheese low in fat, it also has a reasonable amount of calories and low cholesterol. And because it's a fresh cheese, it's safe for those sensitive to tyramine. A very good amount of vitamin B_{12} contributes to healthy nerves and blood. It contains a good amount of calcium as well, but not as much as many other dairy foods.

Caution: *Allergy alert:* May affect those sensitive to whey.

Storage: Cottage cheese tends to be quite perishable once the container has been opened. To keep air exposure to a minimum, try storing the container upside down (make sure the lid's on tight!)

Serve: With any meal. Makes a nice dieter's lunch with fresh fruit and low-fat crackers.

Preparation: Puree, as a smooth substitute for sour cream. Add a little milk to facilitate blending and a bit of lemon juice for tang.

COTTAGE CHEESE, *Low-Fat (2%)*

Serving: ½ cup

Calories	101
Fat	2.2 g
Saturated	1.4 g
Monounsaturated	0.6 g
Polyunsaturated	0.1 g
Calories from fat	19%
Cholesterol	9 mg
Sodium	459 mg
Protein	15.5 g
Carbohydrate	4.1 g
Dietary fiber	0

CHIEF NUTRIENTS

NUTRIENT	AMOUNT	% RDA
Vitamin B_{12}	0.8 mcg	40
Riboflavin	0.2 mg	12
Calcium	77.4 mg	10
Folate	14.8 mcg	7

Strengths: Even though 2% is slightly higher in fat than 1% cottage cheese, it's still quite low in fat, calories, and cholesterol. A fresh cheese such as this should be safe for those sensitive to tyramine. It is rich in vitamin B_{12}, which contributes to healthy nerves and blood, and it has a good amount of calcium for stronger bones and teeth—though the high concentration of sodium is a drawback.

Caution: *Allergy alert:* May affect those sensitive to whey.

Storage: Cottage cheese tends to be quite perishable once the container is opened. To prevent the curds from being exposed to air, try storing the container upside down with the lid on *tight!*

Serve: With any meal. Makes a nice dieter's lunch with fresh berries or melon and rice cakes. And pureed cottage cheese is a smooth substitute for sour cream. It's excellent on baked potatoes or Mexican foods.

CREAM CHEESE

Serving: 1 oz.

Calories	98
Fat	9.8 g
Saturated	6.2 g
Monounsaturated	2.8 g
Polyunsaturated	0.4 g
Calories from fat	90%
Cholesterol	31 mg
Sodium	83 mg
Protein	2.1 g
Carbohydrate	0.7 g
Dietary fiber	0

CHIEF NUTRIENTS

NUTRIENT	AMOUNT	% RDA
Vitamin A	122.4 RE	12
Vitamin B_{12}	0.1 mcg	6

Caution: Cream cheese is high in fat, and much of it is saturated. It also has some cholesterol—so excess consumption of cream cheese just isn't a good idea. *Allergy alert:* May affect those sensitive to whey.

Strengths: Fresh cheeses such as this should be safe for those sensitive to tyramine.

Description: Legally, pure cream cheese must contain at least 33% butterfat (by weight).

A Better Idea: Look for reduced-fat versions; some contain as little as 2 g of fat per oz. Be aware that Neufchâtel cheese, sometimes sold as light cream cheese, isn't much lower in fat than the original. Other alternatives: nonfat yogurt cheese or pureed low-fat cottage cheese. A creamier substitute (with fewer calories, less cholesterol and half the fat of cream cheese) is 1 cup dry curd cottage cheese pureed with ¼ cup margarine.

EDAM

Serving: 1 oz.

Calories	100
Fat	7.8 g
Saturated	4.9 g
Monounsaturated	2.3 g
Polyunsaturated	0.2 g
Calories from fat	70%
Cholesterol	25 mg
Sodium	270 mg
Protein	7.0 g
Carbohydrate	0.4 g
Dietary fiber	0

CHIEF NUTRIENTS

NUTRIENT	AMOUNT	% RDA
Calcium	204.7 mg	26
Vitamin B$_{12}$	0.4 mcg	22
Vitamin A	70.8 RE	7
Zinc	1.1 mg	7
Riboflavin	0.1 mg	6

Caution: High in saturated and monounsaturated fats, Edam is also high in sodium. *Drug interaction:* May cause problems for those sensitive to tyramine or taking MAO inhibitors. *Allergy alert:* May affect those sensitive to casein.

Strengths: A good source of calcium for strong teeth and bones, Edam cheese also contains high-quality protein that's easily digested. Certain components of the cheese may help protect against cavities. With low lactose, it's suitable for people with lactose intolerance. A good amount of vitamin B$_{12}$ helps produce healthy nerves and blood, but you'd be better off getting the vitamin from lower-fat foods.

Description: A mellow, savory cheese with a pale yellow interior and red or yellow paraffin coating. Although Edam is made from part-skim milk, the finished cheese is no lower in fat than other cheeses.

FETA

Serving: 1 oz.

Calories	74
Fat	6.0 g
Saturated	4.2 g
Monounsaturated	1.3 g
Polyunsaturated	0.2 g
Calories from fat	73%
Cholesterol	25 mg
Sodium	313 mg
Protein	4.0 g
Carbohydrate	1.2 g
Dietary fiber	0

CHIEF NUTRIENTS

NUTRIENT	AMOUNT	% RDA
Vitamin B$_{12}$	0.5 mcg	24
Calcium	137.9 mg	17
Riboflavin	0.2 mg	14
Vitamin B$_6$	0.1 mg	6
Zinc	0.8 mg	5

Caution: Feta cheese is very high in sodium because it's cured and stored in a salty whey brine—so avoid it if you have salt-sensitive high blood pressure. And it's also high in fat (much of it saturated). *Drug interaction:* May cause problems for those sensitive to tyramine or taking MAO inhibitors. *Allergy alert:* May affect those sensitive to whey.

Strengths: A good source of calcium for strong teeth and bones, though the high concentration of sodium makes this a less-than-ideal source. Feta cheese contains high-quality protein that's easily digested. Low in lactose, it's suitable for the lactose intolerant. A good amount of vitamin B$_{12}$ plus some other B vitamins boosts healthy nerves and blood, but you'd be better off getting the vitamins from lower-fat foods.

Description: A classic Greek cheese that's white and crumbly with a tangy, salty flavor. Texture can range from soft to semi-dry. Feta is traditionally made from ewe's milk, but goat's milk and cow's milk are also used.

A Better Idea: Soak in cold water to remove some salt.

FONTINA

Serving: 1 oz.

Calories	109
Fat	8.7 g
Saturated	5.4 g
Monounsaturated	2.4 g
Polyunsaturated	0.5 g
Calories from fat	72%
Cholesterol	32 mg
Sodium	224 mg
Protein	7.2 g
Carbohydrate	0.4 g
Dietary fiber	0

CHIEF NUTRIENTS

NUTRIENT	AMOUNT	% RDA
Vitamin B_{12}	0.5 mcg	24
Calcium	154.0 mg	19
Vitamin A	81.2 RE	8
Zinc	1.0 mg	7

Caution: Fontina is high in fat, and its large dose of sodium can contribute to high blood pressure in salt-sensitive people. *Drug interaction:* May cause problems for those sensitive to tyramine or those taking MAO-inhibitor antidepressants.

Eat with: Restraint.

Strengths: A good source of calcium for strong teeth and bones, though the high concentration of sodium is a drawback. Fontina contains high-quality protein that's easily digested. Most cured cheeses are low in lactose, so they're suitable for people with lactose intolerance. A good amount of vitamin B_{12} in fontina contributes to healthy nerves and blood, but you'd be better off getting the vitamin from lower-fat foods.

Origin: This cheese comes from Valle d'Aosta, Italy.

Description: A cow's-milk cheese with a golden brown rind and a pale yellow interior dotted with tiny holes, fontina has a mild, nutty flavor. It melts easily and smoothly, making it a favorite for fondue.

FROMAGE BLANC

Serving: 1 oz.

Calories	19
Fat	0.5 g
Saturated fat	trace
Monounsaturated fat	trace
Polyunsaturated fat	trace
Calories from fat	24%
Cholesterol	1 mg
Sodium	20 mg
Protein	1.4 g
Carbohydrate	2.1 g
Dietary fiber	0

CHIEF NUTRIENTS

(Other nutrient data unavailable.)

Strength: Fromage blanc made with skim milk is a low-fat alternative to cream cheese or sour cream.

Caution: Like other dairy products, fromage blanc may cause digestive problems in those with lactose intolerance.

Description: Pronounced *froh-MAHZH BLAHNGK*, this is an extremely soft, fresh, creamy cheese, with the consistency of sour cream.

Selection: Fromage blanc is available in some supermarkets, but be sure to look for the kind made with skim milk. (Whole-milk varieties are much higher in fat and calories.) You can also make your own using a special starter culture available from some mail-order companies.

Serve: With strawberries, blueberries, apricots, nectarines, or other fruit. Fromage blanc is also an excellent filling for cheese blintzes.

GJETOST

Serving: 1 oz.

Calories	130
Fat	8.3 g
Saturated	5.4 g
Monounsaturated	2.2 g
Polyunsaturated	0.3 g
Calories from fat	57%
Cholesterol	26 mg
Sodium	168 mg
Protein	2.7 g
Carbohydrate	11.9 g
Dietary fiber	0

CHIEF NUTRIENTS

NUTRIENT	AMOUNT	% RDA
Vitamin B_{12}	0.7 mcg	34
Riboflavin	0.4 mg	23
Calcium	112.0 mg	14
Potassium	394.5 mg	11
Vitamin A	76.7 RE	8
Magnesium	19.6 mg	6
Thiamine	0.1 mg	6

Caution: Much of the fat in gjetost cheese is saturated, and it also carries a moderate amount of sodium, which can contribute to high blood pressure in some people. May be problematic for those with lactose intolerance. *Drug interaction:* May cause problems for those sensitive to tyramine or those taking MAO inhibitors. *Allergy alert:* May affect those sensitive to whey.

Strengths: A good source of calcium for strong teeth and bones, gjetost has more magnesium and potassium than usually found in cheese. It contains high-quality protein that's easily digested. Gjetost cheese is a very rich source of vitamin B_{12}, and it has good amounts of riboflavin. However, you'd be better off getting these vitamins from lower-fat foods.

Origin: A Norwegian cheese originally made from goat's milk (as indicated by the ''Gje-'').

Description: Pronounced *YEHT-ohst.* Commonly made from a combination of goat's- and cow's-milk whey that's slowly cooked until the milk sugars caramelize. This cheese is faintly sweet with a dark color, and its texture ranges from fudgelike to the consistency of stiff peanut butter.

GOUDA

Serving: 1 oz.

Calories	100
Fat	7.7 g
Saturated	4.9 g
Monounsaturated	2.2 g
Polyunsaturated	0.2 g
Calories from fat	69%
Cholesterol	32 mg
Sodium	229 mg
Protein	7.0 g
Carbohydrate	0.6 g
Dietary fiber	0

CHIEF NUTRIENTS

NUTRIENT	AMOUNT	% RDA
Calcium	195.9 mg	24
Vitamin B_{12}	0.4 mcg	22
Zinc	1.1 mg	7
Riboflavin	0.1 mg	5

Caution: High in saturated fat, Gouda also has lots of sodium, which can contribute to high blood pressure in salt-sensitive people. *Drug interaction:* May cause problems for those sensitive to tyramine or those taking MAO-inhibitor anti-depressants.

Strengths: A good source of calcium for strong teeth and bones. Certain components of the cheese may help prevent cavities; others may give a degree of cancer protection. The high-quality protein is easily digested, and a good amount of vitamin B_{12} promotes healthy nerves and blood. As with most cured cheeses that are low in lactose, Gouda is suitable for people with lactose intolerance.

Description: A semisoft Dutch cheese similar to Edam but often a little creamier and higher in fat. Gouda can be made from whole or part-skim milk and aged from a few weeks to over a year.

A Better Idea: For sources of B_{12}, try lower-fat dairy foods.

GRUYÈRE

Serving: 1 oz.

Calories	116
Fat	9.1 g
Saturated	5.3 g
Monounsaturated	2.8 g
Polyunsaturated	0.5 g
Calories from fat	71%
Cholesterol	31 mg
Sodium	94 mg
Protein	8.4 g
Carbohydrate	0.1 g
Dietary fiber	0

CHIEF NUTRIENTS

NUTRIENT	AMOUNT	% RDA
Calcium	283.1 mg	35
Vitamin B$_{12}$	0.5 mcg	23
Vitamin A	84.3 RE	8
Zinc	1.1 mg	7

Caution: Gruyère is high in saturated fat. *Drug interaction:* May cause problems for those sensitive to tyramine or those taking MAO-inhibitor drugs. *Allergy alert:* May affect those sensitive to casein.

Strengths: A very good source of calcium for strong teeth and bones, cheese also has certain components that may give a degree of cancer protection. Its high-quality protein is easily digested, and a good amount of vitamin B$_{12}$ contributes to healthy nerves. Gruyère is lower in sodium than many other cheeses. As with other cured cheeses, it's also low in lactose, which makes it suitable for people with lactose intolerance.

Description: A Swiss cheese named for the valley where it originated, Gruyère has a rich, nutty flavor. Its firm, pale yellow interior has medium-size holes. True Gruyère is not the same as the processed cheese sold in little foil-wrapped triangles.

LIMBURGER

Serving: 1 oz.

Calories	92
Fat	7.6 g
Saturated	4.7 g
Monounsaturated	2.4 g
Polyunsaturated	0.1 g
Calories from fat	75%
Cholesterol	25 mg
Sodium	224 mg
Protein	5.6 g
Carbohydrate	0.1 g
Dietary fiber	0

CHIEF NUTRIENTS

NUTRIENT	AMOUNT	% RDA
Calcium	139.1 mg	17
Vitamin B$_{12}$	0.3 mcg	15
Vitamin A	88.5 RE	9
Folate	16.1 mcg	8
Riboflavin	0.1 mg	8

Caution: High in saturated fat, Limburger has lots of sodium, which can contribute to high blood pressure in salt-sensitive people. *Drug interaction:* May cause problems for those sensitive to tyramine or those taking MAO-inhibitor antidepressants.

Strengths: A good source of calcium for strong teeth and bones, though the amount of sodium counts against it. Cheese contains high-quality protein that's easily digested. Several B vitamins contribute to healthy nerves and blood, but you'd be better off getting them from lower-fat foods. A cured cheese low in lactose, Limburger is suitable for people with lactose intolerance.

Description: A very strong-smelling cheese that's definitely an acquired taste. Ranges in color from yellow to reddish-brown, with a yellow, pasty interior. Generally imported from Germany, it continues to ripen in transit, accentuating its flavor and odor.

MONTEREY JACK

Serving: 1 oz.

Calories	105
Fat	8.5 g
Saturated	5.3 g
Monounsaturated	2.5 g
Polyunsaturated	0.3 g
Calories from fat	73%
Cholesterol	25 mg
Sodium	150 mg
Protein	6.9 g
Carbohydrate	0.2 g
Dietary fiber	0

CHIEF NUTRIENTS

NUTRIENT	AMOUNT	% RDA
Calcium	209.0 mg	26
Vitamin B$_{12}$	0.2 mcg	12
Vitamin A	70.8 RE	7
Riboflavin	0.1 mg	6
Zinc	0.8 mg	6

Caution: High in saturated fat, Monterey Jack has a moderate amount of sodium, which can contribute to high blood pressure in salt-sensitive people. *Drug interaction:* May cause problems for those sensitive to tyramine or those taking MAO-inhibitor antidepressants.

Strengths: A good source of calcium for strong teeth and bones, cheese also has certain components that may help protect against cavities. The high-quality protein is easily digested, and some B vitamins help healthy skin and nerves. Monterey Jack is low in lactose, so it's suitable for people with lactose intolerance.

Origin: Monterey, California; hence the name.

Description: Can be made from whole, part-skim, or skim milk, so fat content can vary. The cheese made from skim or part-skim milk is hard. Soft Monterey Jack is made from whole milk. The unaged form is reminiscent of Muenster, while the aged form resembles Cheddar.

A Better Idea: Look for reduced-fat versions of this cheese.

MOZZARELLA, *Part-Skim*

Serving: 1 oz.

Calories	71
Fat	4.5 g
Saturated	2.8 g
Monounsaturated	1.3 g
Polyunsaturated	0.1 g
Calories from fat	56%
Cholesterol	16 mg
Sodium	130 mg
Protein	6.8 g
Carbohydrate	0.8 g
Dietary fiber	0

CHIEF NUTRIENTS

NUTRIENT	AMOUNT	% RDA
Calcium	180.8 mg	23
Vitamin B$_{12}$	0.2 mcg	12
Zinc	0.8 mg	5

Strengths: Lower in fat than many other cheeses, part-skim mozzarella is not too high in cholesterol as long as you don't eat a lot. A good source of calcium for strong teeth and bones, cheese in general contains high-quality protein that's easily digested and some B vitamins for healthy blood formation and nerves. Certain components of cheese may help protect against cavities.

Caution: Some mozzarellas, especially those touted as cholesterol-free, may contain more fat than the regular cheese. They have a moderate amount of sodium—somewhat more than regular mozzarella. *Allergy alert:* May affect those sensitive to whey. Imitation mozzarellas may be problematic if you're sensitive to casein.

Curiosity: String cheese is a form of mozzarella; it is prepared at a higher temperature than mozzarella so it is usually lower in fat.

Serve: With summer's best tomatoes and fresh basil.

MOZZARELLA, *Whole-Milk*

Serving: 1 oz.

Calories	79
Fat	6.1 g
Saturated	3.7 g
Monounsaturated	1.8 g
Polyunsaturated	0.2 g
Calories from fat	69%
Cholesterol	22 mg
Sodium	104 mg
Protein	5.4 g
Carbohydrate	0.6 g
Dietary fiber	0

CHIEF NUTRIENTS

NUTRIENT	AMOUNT	% RDA
Calcium	144.8 mg	18
Vitamin B$_{12}$	0.2 mcg	9
Vitamin A	67.5 RE	7

Caution: Higher in fat than part-skim mozzarella. *Allergy alert:* May affect those sensitive to whey.

Strengths: Certain components of the cheese may help protect against cavities. Even when made from whole milk, mozzarella is not too high in cholesterol, and it's a source of calcium for strong teeth and bones. The high-quality protein is easily digested, while some vitamin B$_{12}$ contributes to healthy nerves and blood formation.

Curiosity: String cheese is a form of mozzarella that is dried and pulled into strings. It generally has a lower fat content than regular mozzarella.

Origin: Southern Italy, where it is made from water-buffalo milk.

Selection: The very fresh varieties found in many Italian markets and cheese shops may be more suitable than longer-cured mozzarella for those sensitive to tyramine or those taking MAO inhibitors. Most of the mozzarellas sold in the U.S. are very fresh.

A Better Idea: Look for part-skim and other reduced-fat versions of this cheese.

MUENSTER

Serving: 1 oz.

Calories	103
Fat	8.4 g
Saturated	5.4 g
Monounsaturated	2.4 g
Polyunsaturated	0.2 g
Calories from fat	73%
Cholesterol	27 mg
Sodium	176 mg
Protein	6.6 g
Carbohydrate	0.3 g
Dietary fiber	0

CHIEF NUTRIENTS

NUTRIENT	AMOUNT	% RDA
Calcium	200.8 mg	25
Vitamin B$_{12}$	0.4 mcg	21
Vitamin A	88.5 RE	9
Riboflavin	0.1 mg	5
Zinc	0.8 mg	5

Caution: High in fat—much of it saturated—Muenster has a moderate amount of sodium, which can contribute to high blood pressure in salt-sensitive people. *Drug interaction:* May cause problems for those sensitive to tyramine or those taking MAO-inhibitor antidepressants.

Strengths: Along with calcium for strong teeth and bones, cheese has certain components that may help protect against cavities. The high-quality protein is easily digested, and a good amount of vitamin B$_{12}$ contributes to healthy nerves. Most cured cheeses are low in lactose, making them suitable for people with lactose intolerance.

Description: The best Muensters come from Europe and have red or orange rinds and a smooth, yellow interior with small holes. Flavor can range from mild to mellow. American versions are often quite bland.

NEUFCHÂTEL

Serving: 1 oz.

Calories	73
Fat	6.6 g
Saturated	4.1 g
Monounsaturated	1.9 g
Polyunsaturated	0.2 g
Calories from fat	81%
Cholesterol	21 mg
Sodium	112 mg
Protein	2.8 g
Carbohydrate	0.8 g
Dietary fiber	0

CHIEF NUTRIENTS

NUTRIENT	AMOUNT	% RDA
Vitamin A	73.9 RE	7

Caution: High in saturated fat, Neufchâtel has a moderate amount of sodium, which can contribute to high blood pressure in salt-sensitive people. *Allergy alert:* May affect those sensitive to whey.

Strengths: Neufchâtel, closely related to cream cheese, may share certain properties that help prevent cancer. Fresh cheeses such as this should be safe for those sensitive to tyramine. A bit of vitamin A enhances immunity.

Description: The Neufchâtel sold in this country is nearly identical to cream cheese but slightly lower in fat, calories, and cholesterol. The French version, from the town of Neufchâtel, is a soft, mild, white, unripened cheese that turns more pungent with aging.

A Better Idea: Nonfat yogurt cheese or diet cream cheese: These dairy products may contain as little as 2 g of fat per oz.

PARMESAN, *Grated*

Serving: 1 Tbsp.

Calories	23
Fat	1.5 g
Saturated	1.0 g
Monounsaturated	0.4 g
Polyunsaturated	trace
Calories from fat	59%
Cholesterol	4 mg
Sodium	93 mg
Protein	2.1 g
Carbohydrate	0.2 g
Dietary fiber	0

CHIEF NUTRIENTS

NUTRIENT	AMOUNT	% RDA
Calcium	68.8 mg	9

Strengths: Certain fatty acids in Parmesan may protect against some types of cancer. A nice source of calcium for strong teeth and bones, this cheese is very flavorful, so a little goes a long way. One Tbsp. has only a moderate amount of fat, is low in calories and cholesterol, and contains high-quality protein that's easily digested. Low in lactose, Parmesan is suitable for people with lactose intolerance.

Caution: Considering the serving size, Parmesan has quite a bit of sodium, which can contribute to high blood pressure in salt-sensitive people. *Drug interaction:* May cause problems for those sensitive to tyramine or those taking MAO-inhibitor antidepressants. *Allergy alert:* May affect those sensitive to casein.

Serve: With salad, soup, pasta, or any casserole that needs a little flavor lift.

PORT DU SALUT

Serving: 1 oz.

Calories	98
Fat	7.9 g
Saturated	4.7 g
Monounsaturated	2.6 g
Polyunsaturated	0.2 g
Calories from fat	72%
Cholesterol	34 mg
Sodium	150 mg
Protein	6.7 g
Carbohydrate	0.2 g
Dietary fiber	0

CHIEF NUTRIENTS

NUTRIENT	AMOUNT	% RDA
Calcium	181.9 mg	23
Vitamin B$_{12}$	0.4 mcg	21
Vitamin A	104.2 RE	10

Caution: Port du Salut is high in saturated fat. Although the sodium amount is moderate, in salt-sensitive people it may be enough to contribute to high blood pressure. *Drug interaction:* May cause problems for those sensitive to tyramine or those taking MAO-inhibitor antidepressants.

Strengths: A good source of calcium for strong teeth and bones, cheese has high-quality protein that's easily digested. In addition, certain components of cheese may help protect against cavities. A good amount of vitamin B$_{12}$ for healthy nerves and blood. Most cured cheeses are low in lactose, making them suitable for people with lactose intolerance.

Origin: First made at the Trappist Monastery of Port du Salut in France.

Description: A semisoft cheese with a mild flavor, yellow interior, satiny texture, and orange-colored rind.

PROVOLONE

Serving size: 1 oz.

Calories	98
Fat	7.5 g
Saturated	4.8 g
Monounsaturated	2.1 g
Polyunsaturated	0.2 g
Calories from fat	68%
Cholesterol	19 mg
Sodium	245 mg
Protein	7.2 g
Carbohydrate	0.6 g
Dietary fiber	0

CHIEF NUTRIENTS

NUTRIENT	AMOUNT	% RDA
Calcium	211.7 mg	26
Vitamin B$_{12}$	0.4 mcg	21
Vitamin A	73.9 RE	7
Zinc	0.9 mg	6
Riboflavin	0.1 mg	5

Caution: High in fat, provolone has lots of sodium, which can contribute to high blood pressure in salt-sensitive people. *Drug interaction:* May cause problems for those sensitive to tyramine or those taking MAO-inhibitor antidepressants. *Allergy alert:* May affect those sensitive to whey.

Strengths: A good source of calcium for strong teeth and bones, cheese contains high-quality protein that's easily digested. A good amount of vitamin B$_{12}$ contributes to healthy nerves and blood, but you'd be better off getting the vitamin from lower-fat foods. Low in lactose, this cheese is suitable for people with lactose intolerance, and it's not too high in cholesterol as long as you don't eat a lot.

Description: A stretched-curd cheese similar in texture to mozzarella, with a mild, smoky flavor. If aged longer than 2 to 3 months, color and flavor intensify.

A Better Idea: Look for reduced-fat versions.

RICOTTA, *Part-Skim*

Serving: ¹/₂ cup

Calories	171
Fat	9.8 g
Saturated	6.1 g
Monounsaturated	2.9 g
Polyunsaturated	0.3 g
Calories from fat	52%
Cholesterol	38 mg
Sodium	155 mg
Protein	14.1 g
Carbohydrate	6.4 g
Dietary fiber	0

CHIEF NUTRIENTS

NUTRIENT	AMOUNT	% RDA
Calcium	337.3 mg	42
Vitamin B$_{12}$	0.4 mcg	18
Riboflavin	0.2 mg	14
Vitamin A	140.1 RE	14
Zinc	1.7 mg	11
Folate	16.2 mcg	8
Iron	0.6 mg	6
Magnesium	18.3 mg	5

Strengths: A very good source of calcium for strong teeth and bones, this fresh cheese should be safe for those sensitive to tyramine. Contains high-quality protein that's easily digested. Various B vitamins improve healthy skin and nerves, and it's not too high in sodium, considering the serving size. A surprising source of iron.

Eat with: Foods that could use a calcium boost, such as pasta, breads, and other grain products.

Caution: Although made partly from skim milk, this ricotta is not really low in fat. *Allergy alert:* May affect those sensitive to whey.

Curiosity: Technically, ricotta is a cheese by-product: It's produced from whey left over from making mozzarella and provolone.

Description: Similar to cottage cheese, ricotta is higher in fat and tends to have a smoother texture.

A Better Idea: Substitute low-fat or dry curd cottage cheese, which can be pureed or pressed through a sieve if a finer texture is desired.

RICOTTA, *Whole-Milk*

Serving: ¹/₂ cup

Calories	216
Fat	16.1 g
Saturated	10.3 g
Monounsaturated	4.5 g
Polyunsaturated	0.5 g
Calories from fat	67%
Cholesterol	63 mg
Sodium	104 mg
Protein	14.0 g
Carbohydrate	3.8 g
Dietary fiber	0

CHIEF NUTRIENTS

NUTRIENT	AMOUNT	% RDA
Calcium	256.7 mg	32
Vitamin B$_{12}$	0.4 mcg	21
Vitamin A	166.2 RE	17
Riboflavin	0.2 mg	14
Zinc	1.4 mg	10
Folate	15.1 mcg	8

Caution: Quite high in fat and cholesterol, considering the serving size. *Allergy alert:* May affect those sensitive to whey.

Strengths: A very good source of calcium for strong teeth and bones, but not as high as part-skim ricotta. A fresh cheese, ricotta should be safe for those sensitive to tyramine. Its high-quality protein is easily digested, and various B vitamins benefit healthy skin and nerves. A good amount of vitamin A and some folate and zinc help improve immunity.

A Better Idea: Part-skim ricotta is preferable to the kind made from whole milk. Even better are extra-light versions. If you can't find any that are significantly lower in fat, opt for low-fat or dry curd cottage cheese. When baked into lasagna or other casseroles, they're perfectly fine substitutes.

ROMANO, *Grated*

Serving: 1 Tbsp.

Calories	19
Fat	1.4 g
Saturated	0.9 g
Monounsaturated	0.4 g
Polyunsaturated	trace
Calories from fat	63%
Cholesterol	5 mg
Sodium	60 mg
Protein	1.6 g
Carbohydrate	0.2 g
Dietary fiber	0

CHIEF NUTRIENTS

NUTRIENT	AMOUNT	% RDA
Calcium	53.2 mg	7

Caution: Romano is high in saturated fat and has a moderate amount of sodium. This hard cheese is usually grated and sprinkled over food like Parmesan, so you'll probably be using only a small amount for flavoring. *Drug interaction:* May cause problems for those sensitive to tyramine or those taking MAO inhibitors. *Allergy alert:* May affect those sensitive to casein.

Strengths: Cheese may help prevent cavities, and some components may offer a degree of cancer protection. Romano provides some calcium for strong teeth and bones. Low in lactose, it's suitable for the lactose intolerant.

ROQUEFORT

Serving: 1 oz.

Calories	103
Fat	8.6 g
Saturated	5.4 g
Monounsaturated	2.4 g
Polyunsaturated	0.4 g
Calories from fat	75%
Cholesterol	25 mg
Sodium	507 mg
Protein	6.0 g
Carbohydrate	0.6 g
Dietary fiber	0

CHIEF NUTRIENTS

NUTRIENT	AMOUNT	% RDA
Calcium	185.3 mg	23
Riboflavin	0.2 mg	9
Vitamin B$_{12}$	0.2 mcg	9
Vitamin A	83.7 RE	8
Folate	13.7 mcg	7

Caution: Roquefort is high in fat and very high in sodium, which can contribute to high blood pressure. *Drug interaction:* May cause problems for those sensitive to tyramine or those taking MAO-inhibitor antidepressants. *Allergy alert:* Roquefort contains a bluish-green mold, *Penicillium roquefortii,* which may cause problems for those allergic to it. However, those allergic to the antibiotic penicillin usually are not affected by the *Penicillium* in the cheese.

Strengths: Cheese may help prevent cavities, and certain components may give a degree of protection against some types of cancer. Along with calcium for strong teeth and bones, Roquefort contains high-quality protein that's easily digested. Some B vitamins improve nerves and blood and help to boost energy. Most cured cheeses are low in lactose, so they're suitable for people with lactose intolerance.

SWISS

Serving: 1 oz.

Calories	105
Fat	7.7 g
Saturated	5.0 g
Monounsaturated	2.0 g
Polyunsaturated	0.3 g
Calories from fat	66%
Cholesterol	26 mg
Sodium	73 mg
Protein	8.0 g
Carbohydrate	1.0 g
Dietary fiber	0

CHIEF NUTRIENTS

NUTRIENT	AMOUNT	% RDA
Calcium	269.1 mg	34
Vitamin B_{12}	0.5 mcg	24
Vitamin A	70.8 RE	7
Zinc	1.1 mg	7
Riboflavin	0.1 mg	6

Caution: Swiss cheese sometimes provokes histamine poisoning, a mild form of "food intoxication" that produces symptoms such as a bright red rash, flushing, sweating, nausea, and hives. Symptoms usually subside in a few hours without treatment; antihistamine therapy provides prompt relief. *Drug interaction:* May cause problems for those sensitive to tyramine or those taking MAO-inhibitor drugs. *Allergy alert:* May affect those sensitive to casein.

Strengths: Certain components of cheese may help protect against cavities. Swiss contains less sodium than most other cheeses, and it's a very good source of calcium for strong teeth and bones. Its high-quality protein is easily digested. A good amount of vitamin B_{12} contributes to healthy nerves and blood, but you'd be better off getting it from lower-fat dairy foods. Most cured cheeses are low in lactose, which makes them suitable for people with lactose intolerance.

A Better Idea: Look for reduced-fat Swiss.

TILSIT

Serving: 1 oz.

Calories	95
Fat	7.3 g
Saturated	4.7 g
Monounsaturated	2.0 g
Polyunsaturated	0.2 g
Calories from fat	69%
Cholesterol	29 mg
Sodium	211 mg
Protein	6.8 g
Carbohydrate	0.5 g
Dietary fiber	0

CHIEF NUTRIENTS

NUTRIENT	AMOUNT	% RDA
Vitamin B_{12}	0.6 mcg	30
Calcium	196.0 mg	25
Vitamin A	81.5 RE	8
Zinc	1.0 mg	7
Riboflavin	0.1 mg	6

Caution: High in saturated fat, Tilsit also has lots of sodium, which can contribute to high blood pressure in salt-sensitive people. *Drug interaction:* May cause problems for those sensitive to tyramine or those taking MAO-inhibitor antidepressants.

Strengths: A good source of calcium for strong teeth and bones. The high-quality protein is easily digested, and a very good amount of vitamin B_{12} contributes to healthy nerves and blood. As with most cured cheeses, it's low in lactose, making it suitable for people with lactose intolerance.

Origin: Tilsit, East Prussia. Dutch immigrants accidentally created the cheese while trying to make Gouda.

Description: A medium-firm cheese with a pale yellow interior, irregular eyes or cracks, and a dark rind. The mild flavor intensifies with age.

YOGURT CHEESE, *Low-Fat*   ⬇ 🗑

Serving: 1 oz.

Calories	30
Fat	0.6 g
Saturated	trace
Monounsaturated	trace
Polyunsaturated	trace
Calories from fat	18%
Cholesterol	0
Sodium	1 mg
Protein	2.2 g
Carbohydrate	2.5 g
Dietary fiber	0

CHIEF NUTRIENTS

(Other nutrient data unavailable.)

Strengths: Yogurt cheese contains just a fraction of the calories of cream cheese, and practically no fat. Since it contains the digestive enzymes found in yogurt, this cheese allows for easy digestion even if you have trouble with the lactose in milk and other dairy products. (In fact, it has been shown that eating some form of yogurt every day can help reduce the symptoms of lactose intolerance.)

Description: Lusciously smooth and creamy.

Selection: Whether you buy yogurt cheese or make your own, look for brands that contain live cultures.

Preparation: You can make yogurt cheese at home. Line a large strainer with cheesecloth and spoon in 4 cups of plain nonfat yogurt made without stabilizers or emulsifiers. Refrigerate and let drain overnight. You'll get 1½ to 2 cups of yogurt cheese.

Serve: As a dip for crisp raw vegetables or a topping for baked potatoes. Yogurt cheese also makes a delicious spread for bread, crackers, or bagels.

CREAM AND NONDAIRY CREAMER

HALF-AND-HALF

Serving: 2 Tbsp.

Calories	39
Fat	3.5 g
Saturated	2.2 g
Monounsaturated	1.0 g
Polyunsaturated	0.1 g
Calories from fat	79%
Cholesterol	11 mg
Sodium	12 mg
Protein	0.9 g
Carbohydrate	1.3 g
Dietary fiber	0

CHIEF NUTRIENTS

NUTRIENT	AMOUNT	% RDA
Vitamin B_{12}	0.1 mcg	5

Caution: Seventy-nine percent of the calories in half-and-half come from fat, much of it saturated; that's just too high for frequent consumption. This dairy product, which is half milk and half cream, also has some cholesterol and sodium, which can add up quickly. Some half-and-half contains emulsifiers, stabilizers, or other ingredients; check labels if you want to avoid them. *Allergy alert:* Milk can cause a wide range of health problems, including migraines, hives, and other allergic reactions. And some people have lactose intolerance, an inability to digest cow's milk.

Eat with: Restraint.

Strengths: Contains a small amount of vitamin B_{12}, which contributes to healthy nerves and blood, although you'd be better off getting the vitamin from low-fat dairy foods.

HEAVY CREAM

Serving: 2 Tbsp.

Calories	103
Fat	11.0 g
Saturated	6.9 g
Monounsaturated	3.2 g
Polyunsaturated	0.4 g
Calories from fat	97%
Cholesterol	41 mg
Sodium	11 mg
Protein	0.6 g
Carbohydrate	0.8 g
Dietary fiber	0

CHIEF NUTRIENTS

NUTRIENT	AMOUNT	% RDA
Vitamin A	125.3 RE	13

Caution: Cream is too high in fat for frequent consumption. The sodium and cholesterol in heavy cream add up quickly if it's consumed in large quantities. Like other dairy products, heavy cream contributes no fiber to the diet. May contain emulsifiers, stabilizers, or other ingredients; check labels if you want to avoid them. *Allergy alert:* Milk can cause a wide range of health problems, including migraines, hives, and other allergic reactions. And of course cream is ill-advised for those with lactose intolerance—an inability to digest cow's milk. If you have high cholesterol or weight problems, try to cut cream out of your diet entirely.

Strength: This dairy product does contain a good amount of vitamin A, which contributes to healthy vision and immunity and may prevent certain types of cancer. Still, you're better off getting A from lower-fat sources.

LIGHT CREAM

Serving: 2 Tbsp.

Calories	59
Fat	5.8 g
Saturated	3.6 g
Monounsaturated	1.7 g
Polyunsaturated	0.2 g
Calories from fat	89%
Cholesterol	20 mg
Sodium	12 mg
Protein	0.8 g
Carbohydrate	1.1 g
Dietary fiber	0

CHIEF NUTRIENTS

NUTRIENT	AMOUNT	% RDA
Vitamin A	54.6 RE	5

Caution: Even light cream is too high in fat. Although it has half the cholesterol of heavy cream, there's just as much sodium. Like other dairy products, it contributes no fiber to the diet. *Allergy alert:* Milk can cause a wide range of health problems, including migraines, hives, and other allergic reactions. And cream is probably off-limits if you have lactose intolerance, an inability to digest cow's milk.

Curiosity: The word *cream* comes from the Greek *chriein*, meaning ''to anoint,'' and is a reference to the product's large amount of oil (fat).

Description: Based on its saturated fat content, light cream falls between half-and-half and heavy cream. It's sometimes called coffee cream or table cream.

NONDAIRY CREAMER, *Powdered*

Serving: 1 tsp.

Calories	11
Fat	0.7 g
Saturated	trace
Monounsaturated	trace
Polyunsaturated	trace
Calories from fat	58%
Cholesterol	0
Sodium	4 mg
Protein	0.1 g
Carbohydrate	1.1 g
Dietary fiber	0

CHIEF NUTRIENTS

(None of the nutrients meet or exceed 5% of the RDA.)

Caution: Used by many people who are trying to cut their fat intake or avoid milk, powdered nondairy creamer falls short on both counts. For the serving size, it's high in calories—higher than half-and-half and light cream. And 58% of those calories come from fat, which is often highly saturated, because it's derived from coconut and other hydrogenated oils. Furthermore, nondairy products contain various additives that you might wish to avoid, so check labels. *Allergy alert:* Some brands have caseinate, a milk protein that you may want to avoid if you are allergic to milk or are lactose intolerant.

A Better Idea: Use regular milk—even whole milk—to gain nutrients the artificial products lack. If lactose intolerance is a problem, opt for lactose-reduced milk or milk with added lactase.

SOUR CREAM

Serving: 2 Tbsp.

Calories	62
Fat	6.0 g
Saturated	3.8 g
Monounsaturated	1.7 g
Polyunsaturated	0.2 g
Calories from fat	88%
Cholesterol	13 mg
Sodium	15 mg
Protein	0.9 g
Carbohydrate	1.2 g
Dietary fiber	0

CHIEF NUTRIENTS

NUTRIENT	AMOUNT	% RDA
Vitamin A	56.1 RE	6

Caution: With 88% of calories coming from fat (much of it saturated), sour cream is not for frequent consumption. Its cholesterol and sodium can add up if you overindulge. Contributes no fiber to the diet. *Drug interaction:* May cause problems for those taking MAO-inhibitor antidepressants. *Allergy alert:* Milk can cause a wide range of health problems, including migraines, hives, and other allergic reactions. However, lactose content is reduced by the fermentation process, so sour cream may be eaten by some people who have a reaction to high-lactose foods.

Strength: Contains some vitamin A, but not enough to justify frequent use.

A Better Idea: Look for reduced-fat sour cream. Even better, substitute plain yogurt for part or all of the sour cream called for in recipes.

EGGS AND EGG SUBSTITUTES

EGG SUBSTITUTE, *Frozen*

Serving: ¼ cup

Calories	25
Fat	0
Saturated	0
Monounsaturated	0
Polyunsaturated	0
Calories from fat	0
Cholesterol	0
Sodium	80 mg
Protein	5.0 g
Carbohydrate	1.0 g
Dietary fiber	0

CHIEF NUTRIENTS

NUTRIENT	AMOUNT	% RDA
Vitamin B_{12}	0.9 mcg	45
Folate	32.0 mcg	16
Iron	1.1 mg	11
Riboflavin	0.2 mg	10
Vitamin A	91.0 RE	9
Thiamine	0.1 mg	6

Strengths: Lower in calories than regular eggs, the substitute has no fat or cholesterol. Its vitamins include lots of B_{12} for healthy blood and nerves. Among a broad spectrum of immunity-enhancing nutrients are folate, iron, and vitamin A. The product is pasteurized, so salmonella is not a concern if you have a recipe calling for uncooked eggs.

Caution: *Allergy alert:* Egg substitutes contain mostly egg whites, which could trigger reactions in anyone allergic to them. Corn, soy, or milk products may also be present.

Selection: Nutrient profile will vary according to brand, so check labels.

Serve: With all sorts of fruits, vegetables, grains, and meats. Appropriate in omelets, casseroles, desserts, baked goods, and other dishes that call for beaten whole eggs.

EGG SUBSTITUTE, *Liquid*

Serving: ¼ cup

Calories	60
Fat	2.0 g
Saturated	<1 g
Monounsaturated	N/A
Polyunsaturated	1.0 g
Calories from fat	30%
Cholesterol	0
Sodium	90 mg
Protein	6.0 g
Carbohydrate	3.0 g
Dietary fiber	0

CHIEF NUTRIENTS

NUTRIENT	AMOUNT	% RDA
Vitamin B_{12}	0.5 mcg	24
Folate	32.0 mcg	16
Riboflavin	0.3 mg	15
Vitamin A	121.0 RE	12
Iron	1.1 mg	11
Calcium	40.0 mg	5

Strengths: Lower in calories and fat than regular eggs, the liquid substitute has no cholesterol. Vitamins include a good amount of B_{12} for healthy blood and nerves. In its broad spectrum of immunity-enhancing nutrients are folate, iron, and vitamin A. The product is pasteurized, so salmonella is not a concern, even when making foods that customarily require uncooked eggs.

Caution: *Allergy alert:* Egg substitute is made mostly from egg whites, which could trigger reactions in anyone allergic to them. Corn or milk products may also be present.

Selection: Nutrients vary according to brand, so check labels. In general, egg substitutes are cholesterol-free and lower in fat than eggs.

Serve: Appropriate for French toast, frittatas, casseroles, crêpes, meat loaves, and other dishes that call for beaten whole eggs. Eat with all sorts of high-fiber fruits, vegetables, and grains.

144

EGG WHITE

⬇ 🏛

Serving: 1 large, uncooked

Calories	17
Fat	0
Saturated	0
Monounsaturated	0
Polyunsaturated	0
Calories from fat	0
Cholesterol	0
Sodium	55 mg
Protein	3.5 g
Carbohydrate	0.3 g
Dietary fiber	0

CHIEF NUTRIENTS

NUTRIENT	AMOUNT	% RDA
Riboflavin	0.2 mg	9

Strengths: Very low in calories, with no fat and no cholesterol, egg white has high-quality protein. The white is much less likely than the yolk to harbor dangerous salmonella. Even so, food scientists discourage consumption of raw whites—as in certain types of meringues and salad dressings.

Eat with: Fiber-rich foods.

Caution: *Allergy alert:* May provoke allergic reactions.

Preparation: When separating whites from yolks, use an egg separator device rather than passing the yolk from shell half to shell half. (Bacteria on shells can contaminate whites or yolks.)

Serve: You can make zero-cholesterol omelets and frittatas using egg whites and a variety of vegetables, such as peppers, potatoes, asparagus, carrots, or broccoli. In many recipes—including casseroles and baked goods—you can successfully substitute 2 egg whites for each whole egg, thereby eliminating a good bit of cholesterol.

EGG YOLK

Serving: 1 large, uncooked

Calories	59
Fat	5.1 g
Saturated	1.6 g
Monounsaturated	2.0 g
Polyunsaturated	0.7 g
Calories from fat	78%
Cholesterol	213 mg
Sodium	7 mg
Protein	2.8 g
Carbohydrate	0.3 g
Dietary fiber	0

CHIEF NUTRIENTS

NUTRIENT	AMOUNT	% RDA
Vitamin B$_{12}$	0.5 mcg	26
Folate	24.2 mcg	12
Vitamin A	96.9 RE	10
Iron	0.6 mg	6
Riboflavin	0.1 mg	6

Caution: Quite high in cholesterol, egg yolk contains more than two-thirds the recommended daily limit of 300 mg. Salmonella may be present in fresh yolks, rendering them unsuitable for recipes in which the eggs are not thoroughly cooked, such as homemade mayonnaise and eggnog. (The bacteria are killed at 160°F.) *Allergy alert:* Eggs are a common allergen, often triggering hives, headaches, and other reactions.

Eat with: Restraint.

Strengths: Egg yolks have good amounts of B vitamins, along with vitamin A, for enhanced immunity and healthy skin, nerves, and vision. They also have iron and riboflavin.

Curiosity: Many tempera paints, which artists have used for centuries, have an egg yolk base.

A Better Idea: In many baked goods, casseroles, and quiches, replace yolks with an equal number of whites or an equivalent amount of egg substitute.

FRIED EGG

Serving size: 1 large, fried in butter

Calories	92
Fat	6.9 g
Saturated	1.9 g
Monounsaturated	2.8 g
Polyunsaturated	1.3 g
Calories from fat	68%
Cholesterol	211 mg
Sodium	162 mg
Protein	6.2 g
Carbohydrate	0.6 g
Dietary fiber	0

CHIEF NUTRIENTS

NUTRIENT	AMOUNT	% RDA
Vitamin B$_{12}$	0.4 mcg	21
Riboflavin	0.2 mg	14
Vitamin A	114.1 RE	11
Folate	17.5 mcg	9
Iron	0.7 mg	7

Caution: All eggs are high in cholesterol, with more than two-thirds the recommended daily limit of 300 mg. Salmonella may be present in fresh eggs, so it's important to cook them properly. For over-easy fried eggs, that means at least 3 minutes on one side and 2 on the other. For sunny-side-up eggs, 4 minutes in a covered frying pan should render them safe. *Allergy alert:* Eggs are a common allergen, often triggering hives, headaches, and other reactions.

Strengths: Eggs have high-quality protein, along with good amounts of A and B vitamins. So they enhance immunity and healthy skin, nerves, and vision. Some iron helps build strong blood.

HARD-BOILED EGG

Serving: 1 large

Calories	78
Fat	5.3 g
Saturated	1.6 g
Monounsaturated	2.0 g
Polyunsaturated	0.7 g
Calories from fat	62%
Cholesterol	212 mg
Sodium	62 mg
Protein	6.3 g
Carbohydrate	0.6 g
Dietary fiber	0

CHIEF NUTRIENTS

NUTRIENT	AMOUNT	% RDA
Vitamin B$_{12}$	0.6 mcg	28
Riboflavin	0.3 mg	15
Folate	22.0 mcg	11
Vitamin A	84.0 RE	8
Iron	0.6 mg	6

Caution: Quite high in cholesterol, a hard-boiled egg has more than two-thirds the recommended daily limit of 300 mg. Calories and fat are the same as in uncooked eggs. To kill salmonella, it's imperative to cook eggs in the shell until the yolk is firm—a minimum of 7 minutes in boiling water. *Allergy alert:* Eggs are a common allergen, often triggering hives, headaches, and other reactions.

Strengths: A hard-boiled egg has high-quality protein, good amounts of several B vitamins, and a fair amount of vitamin A for enhanced immunity and healthy skin, nerves, and vision. Some iron helps build strong blood.

Curiosity: The dark ring that sometimes appears around the yolk is the result of iron and sulfur compounds that form when an egg is cooked too long and at too high a heat. It's unsightly but harmless.

Preparation: Very fresh eggs are difficult to peel after boiling. Cool the cooked eggs thoroughly in cold water. Then tap the shell gently and peel the eggs under cold running water.

POACHED EGG

Serving: 1 large

Calories	75
Fat	5.0 g
Saturated	1.5 g
Monounsaturated	1.9 g
Polyunsaturated	0.7 g
Calories from fat	60%
Cholesterol	212 mg
Sodium	140 mg
Protein	6.2 g
Carbohydrate	0.6 g
Dietary fiber	0

CHIEF NUTRIENTS

NUTRIENT	AMOUNT	% RDA
Vitamin B$_{12}$	0.4 mcg	20
Riboflavin	0.2 mg	13
Vitamin A	95.0 RE	10
Folate	17.5 mcg	9
Iron	0.7 mg	7

Caution: Quite high in cholesterol, a poached egg has more than two-thirds the recommended daily limit of 300 mg. Cook properly to kill salmonella—at least 5 minutes in boiling water. *Allergy alert:* Eggs are a common allergen, often triggering hives, headaches, and other reactions.

Eat with: Restraint.

Strengths: Eggs have high-quality protein. Good amounts of A and B vitamins enhance immunity and healthy skin, nerves, and vision. Also contain folate and iron.

Preparation: For best results, use very fresh eggs, which hold their shape well. Or use special egg-poaching devices.

Serve: For occasional consumption, pair the eggs with whole grain toast and fiber-rich fruits and vegetables. Try lightly sautéed potatoes and a fruit cup containing melons and berries.

RAW EGG

Serving: 1

Calories	75
Fat	5.0 g
Saturated	1.6 g
Monounsaturated	1.9 g
Polyunsaturated	0.7 g
Calories from fat	61%
Cholesterol	213 mg
Sodium	63 mg
Protein	6.3 g
Carbohydrate	0.6 g
Dietary fiber	0

CHIEF NUTRIENTS

NUTRIENT	AMOUNT	% RDA
Vitamin B$_{12}$	0.5 mcg	25
Riboflavin	0.3 mg	15
Folate	23.5 mcg	12
Vitamin A	95.5 RE	10
Iron	0.7 mg	7

Caution: Quite high in cholesterol, one egg has more than two-thirds the recommended daily limit of 300 mg. Salmonella may be present in fresh eggs, rendering them unsuitable for recipes in which the eggs are not thoroughly cooked, such as home-made mayonnaise and eggnog. (The bacteria are killed at 160°F.) *Allergy alert:* Eggs are a common allergen, often triggering hives, headaches, and other reactions.

Strengths: Low in calories, eggs contain high-quality protein. Several B vitamins, along with vitamin A, enhance immunity and improve the health of skin, nerves, and vision. Some iron helps build strong blood.

Curiosity: The shell color is determined by the breed of chicken. Brown eggs and white eggs have the same taste and nutrient content.

Storage: More quality is lost in a day at room temperature than in a week in the refrigerator. Refrigerate in cartons so the eggs don't absorb odors from other foods.

A Better Idea: Egg substitute is cholesterol-free and pasteurized, so salmonella is not a danger. And the substitute is generally lower in fat and calories than whole eggs.

SCRAMBLED EGG

Serving: 1

Calories	100
Fat	7.3 g
Saturated	2.2 g
Monounsaturated	2.9 g
Polyunsaturated	1.3 g
Calories from fat	66%
Cholesterol	211 mg
Sodium	168 mg
Protein	6.7 g
Carbohydrate	1.3 g
Dietary fiber	0

CHIEF NUTRIENTS

NUTRIENT	AMOUNT	% RDA
Vitamin B$_{12}$	0.5 mcg	23
Riboflavin	0.3 mg	15
Vitamin A	117.0 RE	12
Folate	18.0 mcg	9
Iron	0.7 mg	7
Calcium	42.6 mg	5

Caution: The high amount of cholesterol is more than two-thirds the recommended daily limit of 300 mg. Salmonella may be present in fresh eggs, so cook properly. For scrambled eggs, that means at least 1 minute in an electric frying pan set at 250°F. *Allergy alert:* Eggs are a common allergen, often triggering hives, headaches, and other reactions.

Strengths: Along with high-quality protein, eggs have several A and B vitamins for healthy skin, nerves, and vision. Some iron helps build strong blood. In scrambled eggs you get a calcium boost from the butter and the 1⅓ Tbsp. of milk used to prepare the eggs.

A Better Idea: To slash cholesterol, use only 1 yolk with 2, 3, or 4 whites. Better yet, opt for egg substitute, which is ideal for scrambling and contributes no cholesterol to the meal. Cook the eggs in a nonstick pan to avoid adding extra fat.

MILK

BUTTERMILK

Serving: 8 fl. oz.

Calories	99
Fat	2.2 g
Saturated	1.3 g
Monounsaturated	0.6 g
Polyunsaturated	0.1 g
Calories from fat	20%
Cholesterol	9 mg
Sodium	257 mg
Protein	8.1 g
Carbohydrate	11.7 g
Dietary fiber	0

CHIEF NUTRIENTS

NUTRIENT	AMOUNT	% RDA
Calcium	285.2 mg	36
Vitamin B$_{12}$	0.5 mcg	27
Riboflavin	0.4 mg	22
Potassium	370.7 mg	10
Magnesium	26.8 mg	8
Zinc	1.0 mg	7
Folate	12.3 mcg	6
Thiamine	0.1 mg	5

Strengths: Low in fat and calories, buttermilk is almost on a par with 1% low-fat milk. It's rich in bone-building calcium and has lots of B vitamins for healthy skin, blood, and nerves. Plus, it's low in cholesterol.

Caution: Some brands are considerably higher in fat than others. If no nutrient data is given, check the label for cream, whole milk, or butter. Buttermilk is always higher in sodium than regular milk. *Drug interaction:* May cause problems for those taking MAO-inhibitor drugs. *Allergy alert:* Milk can cause a wide range of allergic reactions and is ill-advised for those with lactose intolerance; however, the fermentation process does reduce lactose content.

Curiosity: Named buttermilk because it was the low-fat fluid left after butter was churned.

Description: A refreshing, tangy beverage, good by itself or with savory foods.

Serve: With meats, sandwiches or casseroles. An excellent substitute for sour cream on baked potatoes, with far fewer calories from fat (20%). Also good in baked goods, mashed potatoes, and noodle dishes.

LOW-FAT MILK *(1%)*

Serving: 8 fl. oz.

Calories	102
Fat	2.6 g
Saturated	1.6 g
Monounsaturated	0.8 g
Polyunsaturated	0.1 g
Calories from fat	23%
Cholesterol	10 mg
Sodium	123 mg
Protein	8.0 g
Carbohydrate	11.7 g
Dietary fiber	0

CHIEF NUTRIENTS

NUTRIENT	AMOUNT	% RDA
Vitamin B_{12}	0.9 mcg	45
Calcium	300.1 mg	38
Riboflavin	0.4 mg	24
Vitamin A	144.0 RE	14
Magnesium	33.7 mg	10
Potassium	380.9 mg	10
Thiamine	0.1 mg	7
Folate	12.4 mcg	6
Zinc	1.0 mg	6

Strengths: Nicely low in fat, a cupful of 1% milk has a reasonable amount of calories and cholesterol. And it has a very good amount of vitamin B_{12}, contributing to healthy nerves and blood, along with beneficial amounts of calcium for strong bones. Although not extra-low in sodium, low-fat milk contains plenty of calcium and good amounts of magnesium and potassium to help ensure healthy blood pressure. There's a good amount of vitamin A and some zinc for stronger immunity, and nice amounts of some B vitamins. Generally, vitamin D is added to low-fat milk: It enhances calcium absorption and is crucial in the formation of bones and teeth. By consistently drinking low-fat instead of regular milk, there's a good chance you may lower cholesterol levels and prevent certain types of cancer.

Caution: *Allergy alert:* Milk can cause a wide range of health problems, including migraines, hives, and other allergic reactions. It's often off-limits for those with lactose intolerance. (Choosing lactose-reduced milk or milk with added lactase may overcome the problem.)

LOW-FAT MILK *(2%)*

Serving: 8 fl. oz.

Calories	121
Fat	4.7 g
Saturated	2.9 g
Monounsaturated	1.4 g
Polyunsaturated	0.2 g
Calories from fat	35%
Cholesterol	18 mg
Sodium	122 mg
Protein	8.1 g
Carbohydrate	11.7 g
Dietary fiber	0

CHIEF NUTRIENTS

NUTRIENT	AMOUNT	% RDA
Vitamin B_{12}	0.9 mcg	45
Calcium	296.7 mg	37
Riboflavin	0.4 mg	24
Vitamin A	139.1 RE	14
Magnesium	33.4 mg	10
Potassium	376.7 mg	10
Thiamine	0.1 mg	7
Folate	12.4 mcg	6
Zinc	1.0 mg	6

Strengths: Moderate in fat, 2% milk is low in calories and cholesterol for its serving size. A very good amount of vitamin B_{12} contributes to healthy nerves and blood, and there's lots of calcium for strong bones. Although it's not especially low in sodium, low-fat milk contains a generous amount of calcium, as well as respectable amounts of magnesium and potassium, to help ensure healthy blood pressure. Another benefit is a good amount of vitamin A and some zinc for stronger immunity. Nice amounts of some B vitamins, and a good amount of riboflavin, are usually enhanced with added vitamin D, which helps calcium absorption and is crucial in the formation of bones and teeth. If you regularly drink low-fat instead of regular milk, you may lower cholesterol levels as well as help prevent certain types of cancer.

Caution: *Allergy alert:* Milk can cause a wide range of health problems, including migraines, hives, and other allergic reactions. And it's often off-limits for those with lactose intolerance. (Choosing lactose-reduced milk or milk with added lactase may overcome the problem.)

MOTHER'S MILK

Serving: 8 fl. oz.

Calories	171
Fat	10.8 g
Saturated	4.9 g
Monounsaturated	4.1 g
Polyunsaturated	1.1 g
Calories from fat	57%
Cholesterol	34 mg
Sodium	42 mg
Protein	2.5 g
Carbohydrate	17.0 g
Dietary fiber	0

CHIEF NUTRIENTS

NUTRIENT	AMOUNT	% RDA
Vitamin C	12.3 mg	21
Vitamin A	157.4 RE	16
Calcium	79.2 mg	10
Folate	12.8 mcg	6
Vitamin B$_{12}$	0.1 mcg	6
Riboflavin	0.1 mg	5

Strengths: For infants, mother's milk is generally considered superior to cow's milk. Many doctors advocate breast-feeding as a way to prevent development of food allergies in early infancy. (However, the evidence for this is contradictory.) Very easily digested, human milk promotes the growth of harmless organisms in the infant's digestive tract that inhibit the growth of harmful microbes. It also provides immunity from certain infectious diseases because of these beneficial organisms and because the infant gets antibodies from the mother. Contains good amounts of vitamins A and C and calcium.

Caution: Breast milk is not the perfect food for extended use, as it is deficient in iron, copper, and vitamin D. After 4 months of breast-feeding, supplementation may be needed. *Allergy alert:* Nursing mothers who are highly allergic may need to monitor their diet carefully to avoid allergenic foods. Some research suggests that some babies are sensitized to such foods when there are allergenic proteins in the mother's milk.

SKIM MILK

Serving: 8 fl. oz.

Calories	86
Fat	0.4 g
Saturated	0.3 g
Monounsaturated	0.1 g
Polyunsaturated	trace
Calories from fat	5%
Cholesterol	4 mg
Sodium	126 mg
Protein	8.4 g
Carbohydrate	11.9 g
Dietary fiber	0

CHIEF NUTRIENTS

NUTRIENT	AMOUNT	% RDA
Vitamin B$_{12}$	0.9 mcg	47
Calcium	302.3 mg	38
Riboflavin	0.3 mg	20
Vitamin A	149.4 RE	15
Potassium	405.7 mg	11
Magnesium	27.8 mg	8
Zinc	1.0 mg	7
Folate	12.7 mcg	6
Thiamine	0.1 mg	6
Vitamin B$_6$	0.1 mg	5

Strengths: Very low in fat and cholesterol, skim milk is also low in calories for the serving size. Its vitamin B$_{12}$ contributes to healthy nerves and blood, while lots of calcium helps strengthen bones. Magnesium and potassium help protect against high blood pressure. A good amount of vitamin A, along with some zinc, builds stronger immunity. Yielding nice amounts of some B vitamins, skim milk generally has added vitamin D, which enhances calcium absorption and is crucial in the formation of bones and teeth. Frequent consumption of skim milk may help lower cholesterol levels and blood pressure as well as prevent certain types of cancer.

Caution: *Allergy alert:* Milk can cause a wide range of health problems, including migraines, hives, and other allergic reactions. And it's often off-limits for those with lactose intolerance. (Choosing lactose-reduced milk or milk with added lactase may overcome the problem.)

WHOLE MILK

Serving: 8 fl. oz.

Calories	157
Fat	8.9 g
Saturated	5.6 g
Monounsaturated	2.6 g
Polyunsaturated	0.3 g
Calories from fat	51%
Cholesterol	35 mg
Sodium	119 mg
Protein	8.0 g
Carbohydrate	11.4 g
Dietary fiber	0

CHIEF NUTRIENTS

NUTRIENT	AMOUNT	% RDA
Vitamin B$_{12}$	0.9 mcg	44
Calcium	290.4 mg	36
Riboflavin	0.4 mg	23
Potassium	368.4 mg	10
Magnesium	32.7 mg	9
Vitamin A	83.0 RE	8
Zinc	0.9 mg	6

Caution: Since whole milk is much higher in fat than other milks, the low-fat kind is always a better choice. *Allergy alert:* Milk can cause a wide range of health problems, including migraines, hives, and other allergic reactions. And it's usually undesirable for those with lactose intolerance.

Strengths: Calories and cholesterol are reasonable, if you drink only about 1 cup. Whole milk has a very good amount of vitamin B$_{12}$, which contributes to healthy nerves. The calcium is good for strong bones, while its magnesium and potassium help ensure healthy blood pressure. A fair amount of vitamin A and some zinc lend stronger immunity. Milk has good amounts of some B vitamins, and it generally has added vitamin D, which enhances calcium absorption and is crucial in the formation of bones and teeth.

YOGURT

LOW-FAT YOGURT

Serving: 1 cup

Calories	144
Fat	3.5 g
Saturated	2.3 g
Monounsaturated	1.0 g
Polyunsaturated	0.1 g
Calories from fat	22%
Cholesterol	14 mg
Sodium	159 mg
Protein	11.9 g
Carbohydrate	16.0 g
Dietary fiber	0

CHIEF NUTRIENTS

NUTRIENT	AMOUNT	% RDA
Vitamin B$_{12}$	1.3 mcg	64
Calcium	414.5 mg	52
Riboflavin	0.5 mg	29
Potassium	530.7 mg	14
Folate	25.4 mcg	13
Zinc	2.0 mg	13
Magnesium	39.6 mg	11
Thiamine	0.1 mg	7

Strengths: The fat content of low-fat yogurt is about on a par with that of low-fat (1%) milk. And sometimes those with lactose intolerance can eat yogurt even though they can't drink milk. Yogurt brands with active cultures may help reduce the number of recurrent yeast infections. Yogurt is a super source of calcium for strong bones and teeth; the calcium, along with beneficial amounts of magnesium and potassium, helps ensure healthy blood pressure. Excellent amounts of vitamin B$_{12}$ contribute to healthy nerves and blood.

Caution: *Allergy alert:* Milk can cause a wide range of health problems, including migraines, hives and other allergic reactions.

History: It's believed the ancient Greeks and Romans may have eaten yogurt.

Serve: Flavor with herbs for a sour cream substitute; add fruit for a tangy dessert.

LOW-FAT YOGURT, *Fruit-Flavored*

Serving: 1 cup

Calories	225
Fat	2.6 g
Saturated	1.7 g
Monounsaturated	0.7 g
Polyunsaturated	0.1 g
Calories from fat	10%
Cholesterol	10 mg
Sodium	121 mg
Protein	9.0 g
Carbohydrate	42.3 g
Dietary fiber	N/A

CHIEF NUTRIENTS

NUTRIENT	AMOUNT	% RDA
Vitamin B_{12}	1.0 mcg	49
Calcium	313.9 mg	39
Riboflavin	0.4 mg	22
Potassium	402.0 mg	11
Folate	19.3 mcg	10
Zinc	1.5 mg	10
Magnesium	30.1 mg	9
Thiamine	0.1 mg	5

Strengths: Low-fat fruit-flavored yogurt tends to be acceptable for those with lactose intolerance. Brands with active cultures may help reduce the number of recurrent yeast infections. The high calcium is necessary for strong bones and teeth, and it is also associated with healthier blood pressure. Very good amounts of vitamin B_{12} contribute to healthy nerves and blood. There are also commendable amounts of other B vitamins. Although not extra-low in sodium, all yogurts contain magnesium and potassium, which (along with calcium) help ensure healthy blood pressure.

Caution: Fruit-flavored yogurt may be much higher in calories than plain yogurt if sugar or other high-calorie sweeteners are added. However, you can buy many brands of ''light'' fruit yogurt that are sweetened with aspartame. *Allergy alert:* Milk can cause a wide range of allergy problems.

Serve: With pancakes, waffles, angel food cake. To cut calories per serving, dilute with some plain nonfat yogurt and add chopped fresh fruit.

NONFAT YOGURT

Serving: 1 cup

Calories	127
Fat	0.4 g
Saturated	trace
Monounsaturated	trace
Polyunsaturated	trace
Calories from fat	3%
Cholesterol	4 mg
Sodium	174 mg
Protein	13.0 g
Carbohydrate	17.4 g
Dietary fiber	0

CHIEF NUTRIENTS

NUTRIENT	AMOUNT	% RDA
Vitamin B_{12}	1.4 mcg	70
Calcium	452.0 mg	56
Riboflavin	0.5 mg	31
Potassium	578.6 mg	15
Zinc	2.2 mg	15
Folate	27.7 mcg	14
Magnesium	43.4 mg	12
Thiamine	0.1 mg	7
Vitamin B_6	0.1 mg	6

Strengths: Lower in fat than an equal amount of skim milk, nonfat yogurt is acceptable to some people who generally have a lactose intolerance. Brands with active cultures may help reduce the number of recurrent yeast infections. It's a super source of calcium for strong bones and teeth. An excellent amount of vitamin B_{12} contributes to healthy nerves and blood; there are commendable amounts of other B vitamins as well. Yogurt does have sodium, but the fabulous amount of calcium, plus respectable amounts of magnesium and potassium, helps ensure healthy blood pressure.

Caution: *Allergy alert:* Milk can cause a wide range of health problems, including migraines, hives, and other allergic reactions.

Serve: Yogurt is an excellent sour cream substitute. If you drain the excess whey through a cheesecloth, you can make a tangy low-fat spread for bagels, muffins, or quick breads.

WHOLE-MILK YOGURT

Serving: 1 cup

Calories	139
Fat	7.4 g
Saturated	4.8 g
Monounsaturated	2.0 g
Polyunsaturated	0.2 g
Calories from fat	48%
Cholesterol	29 mg
Sodium	105 mg
Protein	7.9 g
Carbohydrate	10.6 g
Dietary fiber	0

CHIEF NUTRIENTS

NUTRIENT	AMOUNT	% RDA
Vitamin B_{12}	0.8 mcg	42
Calcium	274.0 mg	34
Riboflavin	0.3 mg	19
Potassium	350.9 mg	9
Zinc	1.3 mg	9
Folate	16.8 mcg	8
Magnesium	26.3 mg	8
Vitamin A	68.1 RE	7

Strengths: Yogurt, more often than milk, tends to be acceptable for those with lactose intolerance. Brands with active cultures may help reduce the number of recurrent yeast infections. Whole-milk yogurt has plenty of calcium for strong bones and teeth, in combination with some magnesium and potassium to help ensure healthy blood pressure. Very good amounts of vitamin B_{12} contribute to healthy nerves.

Caution: Whole-milk yogurt contains quite a bit more fat than low-fat yogurt, while its vitamin and mineral values are consistently lower. *Allergy alert:* Milk can cause a wide range of health problems, including migraines, hives, and other allergic reactions.

Serve: Flavor with herbs for a sour cream substitute; add fruit for a tangy dessert.

A Better Idea: Whenever possible, diet watchers are better off choosing lower-fat varieties.

Desserts
CAKES

ANGEL FOOD *from Mix* ⬇

Serving: 1 slice (about 2 oz.)

Calories	137
Fat	0.1 g
Saturated	0
Monounsaturated	N/A
Polyunsaturated	N/A
Calories from fat	1%
Cholesterol	0
Sodium	77 mg
Protein	3.0 g
Carbohydrate	31.5 g
Dietary fiber	0.4 g

CHIEF NUTRIENTS

NUTRIENT	AMOUNT	% RDA
Calcium	50.4 mg	6

Strengths: Angel food cake is extremely low in fat, with no cholesterol.

Origin: Some say angel food cake was first made by thrifty Pennsylvania Dutch bakers who developed the recipe to use up leftover egg whites.

Description: A very light, spongelike cake made with lots of whipped egg whites, sugar, and flour, with small amounts of salt, flavoring, and cream of tartar (to keep the egg whites fluffed). It's usually baked in a tube pan.

Selection: Angel food cake mixes come in many flavors, including classic lemon, strawberry, and chocolate. Not to be confused with sponge cake, which has much more fat.

Storage: Wrap in an airtight plastic bag to seal in moisture—angel food cake can be stored this way for several days.

Serve: With sliced fresh or frozen strawberries or peaches, along with a scoop of nonfat frozen yogurt.

BOSTON CREAM

Serving: 1 slice (about 2½ oz.)

Calories	208
Fat	6.5 g
Saturated	2.0 g
Monounsaturated	N/A
Polyunsaturated	N/A
Calories from fat	28%
Cholesterol	59 mg
Sodium	128 mg
Protein	3.5 g
Carbohydrate	34.4 g
Dietary fiber	N/A

CHIEF NUTRIENTS

NUTRIENT	AMOUNT	% RDA
Iron	0.9 mg	9
Calcium	46.2 mg	6
Riboflavin	0.1 mg	6
Thiamine	0.1 mg	6

Strengths: Lower in fat than most other pies, Boston cream is moderate in sodium and cholesterol. It offers a bit of blood-pumping iron and bone-bolstering calcium, along with riboflavin and thiamine, B vitamins that aid energy production.

History: Several New England restaurants would like to claim Boston cream pie as their own creation—but no one knows for sure.

Description: Traditionally called a pie, this is really a sponge cake or yellow layer cake made with enriched flour and vegetable shortening; it has a thick custard filling. It's topped with a chocolate glaze.

Preparation: In her 1930 *Boston Cooking School Cook Book*, Fannie Farmer says to bake this cake in a heavy frying pan.

A Better Idea: Use a "light" yellow cake mix and a reduced-calorie pudding mix. Add sliced bananas to the filling for extra nutrition and flavor.

CHEESECAKE

Serving: 1 slice (about 3 oz.)

Calories	257
Fat	16.3 g
Saturated	N/A
Monounsaturated	N/A
Polyunsaturated	N/A
Calories from fat	57%
Cholesterol	N/A
Sodium	189 mg
Protein	4.6 g
Carbohydrate	24.3 g
Dietary fiber	1.8 g

CHIEF NUTRIENTS

NUTRIENT	AMOUNT	% RDA
Vitamin B$_{12}$	0.4 mcg	21
Riboflavin	0.1 mg	7
Vitamin C	4.3 mg	7
Calcium	48.0 mg	6

Caution: In commercial cheesecake, almost 60% of the calories come from fat, making it a real waist-expander.

Strengths: This cake is rich in vitamin B$_{12}$, a nutrient important for proper function of the nervous system and red blood cell formation. *Allergy alert:* The eggs are likely to be a problem for anyone who's allergic to them.

History: Various forms of cheescake have been popular for centuries; the first mention of such a dessert was in 1440.

Description: Jewish, or New York, cheesecake is made with a smooth cream cheese. Italian cheesecake is made with a ricotta cheese filling and may contain toasted pine nuts, almonds, and citron. Both types may have a pastry or graham cracker crust.

Storage: Cheesecake should be refrigerated.

A Better Idea: You can make a fairly tasty, lower-fat cheesecake using part-skim ricotta cheese, yogurt, or even silky-textured tofu.

COFFEE CAKE *from Mix*

Serving: 1 slice (about 2½ oz.)

Calories	232
Fat	6.9 g
Saturated	2.0 g
Monounsaturated	N/A
Polyunsaturated	N/A
Calories from fat	27%
Cholesterol	45 mg
Sodium	310 mg
Protein	4.5 g
Carbohydrate	37.7 g
Dietary fiber	1.8 g

CHIEF NUTRIENTS

NUTRIENT	AMOUNT	% RDA
Iron	1.2 mg	12
Riboflavin	0.2 mg	9
Thiamine	0.1 mg	9
Niacin	1.3 mg	7
Calcium	43.9 mg	5

Caution: With 310 mg of sodium per serving, coffee cake is high in salt. It's also high in calories—so be cautious if you're trying to lose weight.

Strengths: A good source of fiber, depending on the recipe. Because coffee cake is made with enriched flour, a serving offers a good share of blood-building iron, along with some niacin, riboflavin, and thiamine, B-complex vitamins that help convert food into energy. Made with milk, coffee cake has a tad of bone-strengthening calcium. It does qualify as a low-fat food, with fewer than 30% of calories from fat.

Description: A rich, sweet, cakelike bread that comes in an infinite number of forms. Made with milk, eggs, enriched flour, and yeast or baking powder, it may contain fruit, nuts, or a cream cheese, poppyseed, or jam filling. It may be glazed or topped with crumbs.

Serve: Slightly warm. You may want to wrap the cake in foil and reheat in the oven before serving.

A Better Idea: Look for new lower-fat coffee cake mixes.

DEVIL'S FOOD *from Mix, with Chocolate Icing*

Serving: 1 slice (about 2½ oz.)

Calories	234
Fat	8.5 g
Saturated	3.3 g
Monounsaturated	N/A
Polyunsaturated	N/A
Calories from fat	33%
Cholesterol	33 mg
Sodium	181 mg
Protein	3.0 g
Carbohydrate	40.2 g
Dietary fiber	1.5 g

CHIEF NUTRIENTS

NUTRIENT	AMOUNT	% RDA
Iron	1.4 mg	14
Riboflavin	0.1 mg	6
Calcium	40.7 mg	5

Caution: Devil's food is very high in calories, and over a third of them come from fat.

Strengths: Enriched flour adds a good amount of blood-building iron and some riboflavin and calcium to devil's food cake. Plus, it offers some fiber.

Curiosity: This cake may have been dubbed "devil's food" because of its rich, delicious taste. To a moralist, such pleasure *must* be the work of the devil.

Origin: The first devil's food cake recipe appeared in 1905.

Description: A moist, reddish-brown chocolate cake made with eggs, water, and enriched flour, devil's food is the exact opposite of light, white angel food cake.

A Better Idea: Even devil's food cake mix now comes in a reduced-fat version. Top it with lower-fat chocolate icing or a scoop of light whipped cream and you've got a delicious dessert.

FRUITCAKE, *Dark*

Serving: 1 slice (about 1½ oz.)

Calories	163
Fat	6.6 g
Saturated	1.4 g
Monounsaturated	N/A
Polyunsaturated	N/A
Calories from fat	36%
Cholesterol	19 mg
Sodium	68 mg
Protein	2.1 g
Carbohydrate	25.7 g
Dietary fiber	1.6 g

CHIEF NUTRIENTS

NUTRIENT	AMOUNT	% RDA
Iron	1.2 mg	12
Potassium	213.3 mg	6

Caution: Fruitcake is high in calories, and more than a third of those calories come from fat.

Strengths: Enriched flour, dried fruit, and molasses give dark fruitcake a good amount of blood-pumping iron, along with some potassium, important for healthy blood pressure.

Curiosity: "Nutty as a fruitcake" refers to an eccentric person.

Description: A yellow butter cake with fruit, made with vegetable shortening and enriched flour. Dark fruitcakes generally include ingredients such as molasses or brown sugar and a dark liquor such as bourbon. Prunes, dates, walnuts, and raisins may be added.

Storage: Keep in the refrigerator or freezer.

A Better Idea: If you don't like fruitcake, perhaps it's because you've eaten only the sticky, sweet, citron-laden, store-bought versions. Try making your own. You can use dried apricots, dates, raisins, and even pumpkin and sunflower seeds.

FRUITCAKE, *Light*

Serving: 1 slice (about 1½ oz.)

Calories	167
Fat	7.1 g
Saturated	1.6 g
Monounsaturated	N/A
Polyunsaturated	N/A
Calories from fat	38%
Cholesterol	3 mg
Sodium	83 mg
Protein	2.6 g
Carbohydrate	24.7 g
Dietary fiber	1.6 g

CHIEF NUTRIENTS

NUTRIENT	AMOUNT	% RDA
Iron	0.8 mg	8

Caution: Fruitcake is high in calories.

Strength: Provides a bit of iron, essential for healthy blood.

Curiosity: Some fruitcakes have been known to keep for 25 years or more. They are usually soaked with liquor, wrapped in layers of liquor-soaked cheesecloth, then packed in powdered sugar in an airtight tin.

Description: A pound cake with fruit, made with vegetable shortening and enriched flour. Light fruitcake is often sweetened with corn syrup or white sugar. It may be studded with almonds, dried apricots, and golden raisins.

A Better Idea: You can make your own healthy-tasting light fruitcake from whole wheat flour, apricots, dates, raisins, and pumpkin or sunflower seeds. Avoid using dried candied fruits or glacé fruits, which are oversweetened.

GINGERBREAD *from Mix*

Serving: 1 slice (2¼ oz.)

Calories	174
Fat	4.3 g
Saturated	1.1 g
Monounsaturated	N/A
Polyunsaturated	N/A
Calories from fat	22%
Cholesterol	1 mg
Sodium	192 mg
Protein	2.0 g
Carbohydrate	32.2 g
Dietary fiber	N/A

CHIEF NUTRIENTS

NUTRIENT	AMOUNT	% RDA
Iron	1.2 mg	12
Calcium	56.7 mg	7
Riboflavin	0.1 mg	7
Thiamine	0.1 mg	6

Strengths: Gingerbread is low in fat and cholesterol. Because it is made with enriched flour, it offers a good amount of blood-building iron, along with some riboflavin and thiamine, B-complex vitamins essential for energy, and some calcium for stronger bones.

Curiosity: Parkin, a gingerbread cake made with oatmeal and English molasses, is served on Guy Fawkes Day in England. It's a full-bodied, dark, moist cake with a wonderful texture.

History: Gingerbread dates back to the Middle Ages, when fair ladies presented the honey-spice cake as a favor to dashing knights before they entered tournaments. In those days, gingerbread was intricately shaped and decorated, sometimes with gold leaf.

Description: A cake made with powdered ginger, molasses, wheat flour, eggs, and oil or butter. It sometimes contains cinnamon, brown sugar, or honey.

Serve: Warm, with a light lemon sauce.

A Better Idea: Make your own gingerbread using whole wheat flour. Or add some wheat germ to a white-flour recipe for added nutritional value.

PINEAPPLE UPSIDE-DOWN, *Homemade*

Serving: 1 slice (2½ oz.)

Calories	221
Fat	8.5 g
Saturated	N/A
Monounsaturated	N/A
Polyunsaturated	N/A
Calories from fat	35%
Cholesterol	N/A
Sodium	167 mg
Protein	2.4 g
Carbohydrate	34.9 g
Dietary fiber	N/A

CHIEF NUTRIENTS

NUTRIENT	AMOUNT	% RDA
Iron	1.1 mg	11
Thiamine	0.1 mg	7
Calcium	50.4 mg	6
Vitamin C	3.5 mg	6
Vitamin A	54.4 RE	5

Caution: Pineapple upside-down cake has a moderate amount of fat and sodium and is high in calories.

Strengths: Contains a good amount of blood-building iron and some thiamine, plus a tad of calcium, which is good for bones. The pineapple contributes a bit of vitamins C and A, which both help protect against certain types of cancer.

Description: A yellow cake with a glazed fruit topping.

Selection: You can choose among tasty variations of pineapple upside-down cake, including ginger-pear, apple-cranberry, apricot-prune, and peach-pecan.

Storage: Keep refrigerated.

Serve: Warm—plain or with a dab of low-fat whipped topping.

A Better Idea: For less fat and fewer calories, stick with angel food cake with sliced fruit.

POUND

Serving: 1 slice (about 1 oz.)

Calories	119
Fat	5.4 g
Saturated	1.4 g
Monounsaturated	N/A
Polyunsaturated	N/A
Calories from fat	41%
Cholesterol	44 mg
Sodium	52 mg
Protein	1.9 g
Carbohydrate	15.9 g
Dietary fiber	N/A

CHIEF NUTRIENTS

NUTRIENT	AMOUNT	% RDA
Iron	0.6 mg	6

Caution: A cake that's fairly high in fat and very high in calories. About 40% of calories are from fat. And "old-fashioned" pound cake has significantly more fat than regular pound cake. Both of these butter-and-egg–packed cakes also contain moderate amounts of cholesterol.

Strength: Because it's made with enriched flour, pound cake offers a bit of blood-building iron.

Curiosity: Its name derives from the traditional weight of the ingredients—1 lb. of flour, 1 lb. of butter, 1 lb. of sugar, and 1 lb. of eggs—although these measurements are generally not followed in most modern recipes.

Description: Made with vegetable shortening and enriched flour.

Storage: This cake can be tightly wrapped and refrigerated for several weeks, or it can be frozen.

A Better Idea: Look for lower-fat versions of pound cake, now available in most supermarkets. Or stick with lower-fat cakes such as angel food.

SPONGE

Serving: 1 slice (about 2¼ oz.)

Calories	196
Fat	3.8 g
Saturated	1.2 g
Monounsaturated	N/A
Polyunsaturated	N/A
Calories from fat	17%
Cholesterol	162 mg
Sodium	110 mg
Protein	5.0 g
Carbohydrate	35.7 g
Dietary fiber	N/A

CHIEF NUTRIENTS

NUTRIENT	AMOUNT	% RDA
Iron	1.3 mg	13
Vitamin A	90.0 RE	9
Riboflavin	0.1 mg	8
Thiamine	0.1 mg	6

Caution: Sponge cake is high in cholesterol because it's jam-packed with whole eggs.

Strengths: Low in total fat. The eggs and enriched flour in this cake add a good amount of blood-building iron, along with some riboflavin and thiamine, B-complex vitamins that aid energy production. It also contains a decent amount of vitamin A for healthy skin and eyesight.

History: In the 16th century, Catherine de Medici brought along a sponge cake recipe as part of her dowry when she married Henry II and moved to France. French chefs quickly adapted the recipe and made it a classic of their own.

Description: A light, airy cake that gets its lift from eggs beaten to a foam—hence its generic name, "foam cake." Sponge cakes include genoises, jelly rolls, chiffon cakes, and tortes. The cakes are usually moistened with flavored syrups and filled with mousse or Bavarian cream.

A Better Idea: Angel food cake, which does not contain egg yolks. It also has less fat and no cholesterol.

WHITE *from Mix, with Chocolate Icing*

Serving: 1 slice (about 2½ oz.)

Calories	249
Fat	7.6 g
Saturated	2.8 g
Monounsaturated	N/A
Polyunsaturated	N/A
Calories from fat	27%
Cholesterol	1 mg
Sodium	161 mg
Protein	2.8 g
Carbohydrate	44.6 g
Dietary fiber	N/A

CHIEF NUTRIENTS

NUTRIENT	AMOUNT	% RDA
Calcium	70.3 mg	9
Iron	0.9 mg	9
Riboflavin	0.1 mg	7
Thiamine	0.1 mg	6

Caution: Very high in calories, and a bit high in saturated fat. Almost one-third of the calories come from fat.

Strengths: Enriched flour adds some blood-building iron and a bit of riboflavin and thiamine, both important for good energy levels, along with a tad of bone-strengthening calcium.

Description: A white, light, fine-textured layer cake made with flour, baking powder, salt, butter, sugar, egg whites, milk, and vanilla extract. This basic cake graciously accommodates a variety of fillings and flavorings, including dates and nuts.

Selection: Packaged cake mix may be as much as two years old if the store has a slow turnover. Check dates on the packaging and choose the freshest mix.

Storage: Keep in a cool place.

A Better Idea: Use low-fat cake mix and icing, now available in supermarkets.

YELLOW *from Mix, with Chocolate Icing*

Serving: 1 slice (about 2½ oz.)

Calories	233
Fat	7.8 g
Saturated	2.8 g
Monounsaturated	N/A
Polyunsaturated	N/A
Calories from fat	30%
Cholesterol	33 mg
Sodium	157 mg
Protein	2.8 g
Carbohydrate	39.7 g
Dietary fiber	0.6 g

CHIEF NUTRIENTS

NUTRIENT	AMOUNT	% RDA
Iron	1.0 mg	10
Calcium	62.8 mg	8
Riboflavin	0.1 mg	6
Thiamine	0.1 mg	5

Caution: High in calories, and saturated fat is a bit high. 30% of the calories are from fat.

Strengths: Offers a good amount of blood-building iron, along with riboflavin and thiamine, B vitamins essential for energy production. A bit of calcium helps keep bones strong.

Curiosity: The largest cake ever created was a 90,000-lb. monster made from Duncan Hines yellow cake mix by an Austin, Texas chef. Topped with 30,000 lbs. of vanilla icing, it took 24 hours to assemble and 32 hours to bake.

Description: A light, fine-textured yellow layer cake made with enriched flour, baking powder, salt, butter, sugar, eggs, milk, and vanilla extract. Often used with part chocolate batter to make marble cake. May have a lemon or raspberry jam filling.

A Better Idea: Look for low-fat cake mix and icing, now available in supermarkets.

COOKIES

BROWNIE *with Nuts, Homemade*

Serving: 1 (about ¾ oz.)

Calories	93
Fat	6.0 g
Saturated	2.0 g
Monounsaturated	N/A
Polyunsaturated	N/A
Calories from fat	58%
Cholesterol	22 mg
Sodium	74 mg
Protein	1.3 g
Carbohydrate	10.1 g
Dietary fiber	0.5 g

CHIEF NUTRIENTS

(None of the nutrients meet or exceed 5% of the RDA.)

Caution: Very high in saturated fat and calories. *Allergy alert:* Recipes call for use of wheat, chocolate, and eggs, which are common allergens.

Strength: Has a small amount of fiber.

Curiosity: The brownie was first mentioned in print in the 1897 edition of the Sears, Roebuck and Co. catalog.

Description: A rich, chewy, cakelike cookie, usually flavored with chocolate.

Preparation: The size of the baking pan can affect the texture of brownies. In general, a pan that's too big will produce dry, brittle brownies; a pan too small will yield cakelike brownies that are not chewy.

A Better Idea: If you bake your own brownies, you can substitute carob for chocolate. Look for recipes that include high-nutrition ingredients, such as whole wheat flour, dates, carrots, and apples.

CHOCOLATE CHIP

Serving: 5 (about 2 oz.)

Calories	247
Fat	11.0 g
Saturated	3.4 g
Monounsaturated	N/A
Polyunsaturated	N/A
Calories from fat	40%
Cholesterol	20 mg
Sodium	211 mg
Protein	2.8 g
Carbohydrate	36.6 g
Dietary fiber	1.4 g

CHIEF NUTRIENTS

NUTRIENT	AMOUNT	% RDA
Iron	1.5 mg	15
Thiamine	0.1 mg	6
Riboflavin	0.1 mg	5

Caution: This, the most popular type of cookie in America, is high in calories and loaded with saturated fat. A serving of 5 commercial chocolate chip cookies amounts to more than 200 calories. About 40 to 50% of those calories are derived from fat—and much of that fat is the saturated kind that may increase blood cholesterol levels. *Allergy alert:* Chocolate can cause allergic reactions in some people and is known to contribute to heartburn. The chocolate chips in these cookies also contain a substance called tyramine that may cause headaches in sensitive individuals. They're often made with milk, eggs, and flour, substances known to cause allergic reactions.

Strengths: Chocolate chip cookies are a good source of iron, a mineral that boosts your body's immune system. They also contain fair amounts of riboflavin and thiamine, two B vitamins that help convert carbohydrates and other foods into energy.

A Better Idea: If you're making chocolate chip cookies, substitute margarine or vegetable shortening for butter, and egg whites or egg substitute for whole eggs.

COCONUT BARS

Serving: 5 (about 1½ oz.)

Calories	222
Fat	11.0 g
Saturated	4.2 g
Monounsaturated	N/A
Polyunsaturated	N/A
Calories from fat	45%
Cholesterol	49 mg
Sodium	67 mg
Protein	2.8 g
Carbohydrate	28.8 g
Dietary fiber	1.8 g

CHIEF NUTRIENTS

NUTRIENT	AMOUNT	% RDA
Iron	1.4 mg	14
Thiamine	0.1 mg	6

Caution: Commercial varieties are sometimes made with partially hydrogenated vegetable oil or coconut oil. Both kinds of oil are very high in saturated fat and may increase blood cholesterol levels. As with all cookies, read the label carefully. Many manufacturers scatter several types of sugar throughout their list of ingredients. These sugars, when added together, often constitute the main ingredient—so coconut bars are very high in calories and are deficient in nutrients. *Allergy alert:* Cookies are commonly made with milk and flour, substances that may cause allergic reactions in some people.

Strengths: Coconut bars are a good source of iron, a nutrient that contributes to a strong immune system and the synthesis of collagen. They also contain a moderate amount of carbohydrates and a bit of thiamine, a B vitamin that helps convert carbohydrates into energy.

A Better Idea: For a more nutritious dessert, substitute fresh fruit for coconut bars.

FIG BARS

Serving: 4 (about 2 oz.)

Calories	200
Fat	3.1 g
Saturated	0.9 g
Monounsaturated	N/A
Polyunsaturated	N/A
Calories from fat	14%
Cholesterol	22 mg
Sodium	141 mg
Protein	2.2 g
Carbohydrate	42.2 g
Dietary fiber	2.6 g

CHIEF NUTRIENTS

NUTRIENT	AMOUNT	% RDA
Iron	1.2 mg	12
Thiamine	0.1 mg	7
Riboflavin	0.1 mg	6
Calcium	43.7 mg	5

Caution: Although a better choice than many cookies, fig bars are high in calories—200 in a 2-oz. serving. They also have moderate amounts of sodium and cholesterol. *Allergy alert:* Cookies are commonly made with eggs, milk, and flour, substances that may cause allergic reactions in some people.

Strengths: Fig bars get only about 14% of their calories from fat and contain significantly less saturated fat than many other cookies. Like most other cookies, fig bars have a good amount of iron and fair amounts of thiamine and riboflavin. They're also high in fiber.

Curiosity: Figs were an ancient symbol of peace and prosperity.

History: Fig Newtons—first made in 1892—were named after Newton, Massachusetts.

A Better Idea: In the summer, try eating fresh figs. They are available from June to October. And dried figs are a better alternative to cookies the rest of the year.

GINGERSNAPS

Serving: 5 (about 1¼ oz.)

Calories	147
Fat	3.1 g
Saturated	0.8 g
Monounsaturated	N/A
Polyunsaturated	N/A
Calories from fat	19%
Cholesterol	14 mg
Sodium	200 mg
Protein	1.9 g
Carbohydrate	27.9 g
Dietary fiber	0.3 g

CHIEF NUTRIENTS

NUTRIENT	AMOUNT	% RDA
Iron	1.1 mg	11
Thiamine	0.1 mg	7

Caution: Gingersnaps are high in sodium, which makes them off-limits for those on salt-restricted diets. *Allergy alert:* Gingersnaps are commonly made with eggs and flour, substances that may cause allergic reactions in some people.

Strengths: Lower in calories and saturated fat than many other cookies, gingersnaps boast a good amount of iron and some thiamine. In addition, they have high amounts of carbohydrates.

History: Ginger was used to make gingerbread in the Middle Ages. The term *gingersnap,* first printed in 1805, probably comes from the Middle Dutch word, *snappen,* meaning to seize quickly.

Description: Small, crisp cookies made from ginger and molasses, often cut into human shapes and sold as gingerbread people.

A Better Idea: Overall, gingersnaps aren't a bad choice, but you'll probably want another type of cookie, such as ladyfingers, if you're on a salt-restricted diet.

LADYFINGERS

Serving: 4 (about 1½ oz.)

Calories	158
Fat	3.4 g
Saturated	1.1 g
Monounsaturated	N/A
Polyunsaturated	N/A
Calories from fat	19%
Cholesterol	157 mg
Sodium	31 mg
Protein	3.4 g
Carbohydrate	28.4 g
Dietary fiber	0.3 g

CHIEF NUTRIENTS

NUTRIENT	AMOUNT	% RDA
Iron	1.3 mg	13
Riboflavin	0.2 mg	9
Thiamine	0.1 mg	7
Vitamin A	57.2 RE	6

Strengths: Ladyfingers are a low-fat, low-sodium alternative to many cookies on the dessert tray. They contain good amounts of iron, which is a component of hemoglobin, the oxygen-carrying protein in red blood cells. Fair amounts of riboflavin and thiamine are also present, and these two types of B vitamins help convert carbohydrates and other foods into energy.

Caution: *Allergy alert:* Ladyfingers are commonly made with eggs and flour, substances that may cause allergic reactions in some people.

Curiosity: Ladyfingers were first referred to by John Keats in his 1820 poem "The Cap and Bells" in which an emperor requests "some lady's-fingers nice in Candy wine."

Description: A light sponge cake biscuit resembling a large finger.

Serve: With fresh fruit and low-fat yogurt for a tasty blend of flavors.

MACAROONS

Serving: 2 (about 1½ oz.)

Calories	181
Fat	8.8 g
Saturated	6.1 g
Monounsaturated	N/A
Polyunsaturated	N/A
Calories from fat	44%
Cholesterol	41 mg
Sodium	13 mg
Protein	2.0 g
Carbohydrate	25.1 g
Dietary fiber	0.8 g

CHIEF NUTRIENTS

(None of the nutrients meet or exceed 5% of the RDA.)

Caution: Macaroons are very high in saturated fat, which may increase blood cholesterol levels. Also, they're not the best choice if you're watching your waistline; two macaroons contain more than 180 calories. *Allergy alert:* Macaroons traditionally are made of almond paste or ground almonds; this nut can cause allergic reactions such as swelling of the lips. Like other cookies, they are also likely to be made with eggs and sometimes milk. These substances may cause allergic reactions in some people.

Strengths: The cookies contain moderate amounts of carbohydrates, and they're low in sodium—so you might choose this dessert if you're on a low-salt diet.

A Better Idea: If baking your own, try substituting high-fiber bran flakes for almonds or coconut.

MARSHMALLOW *with Coconut*

Serving: 4 (about 2½ oz.)

Calories	294
Fat	9.5 g
Saturated	5.8 g
Monounsaturated	N/A
Polyunsaturated	N/A
Calories from fat	29%
Cholesterol	55 mg
Sodium	150 mg
Protein	2.9 g
Carbohydrate	52.1 g
Dietary fiber	0.1 g

CHIEF NUTRIENTS

NUTRIENT	AMOUNT	% RDA
Iron	1.2 mg	12
Thiamine	0.1 mg	6
Riboflavin	0.1 mg	5

Caution: Marshmallow cookies are high in calories and saturated fat, which tends to increase blood cholesterol levels. This tempting dessert also has a moderate amount of sodium. *Allergy alert:* Marshmallow cookies often contain corn syrup and eggs, and the chocolate-covered kind have, of course, been dipped in chocolate. Those substances may cause allergic reactions in some people.

Strengths: Marshmallow cookies have a good amount of iron, which contributes to healthy blood and gives the immune system a boost. They also contain some riboflavin and thiamine, two B vitamins that help convert carbohydrates and other foods into energy.

Selection: Cookie for cookie, coconut-covered marshmallow cookies are higher in calories, saturated fat, cholesterol, and sodium than the chocolate-covered variety. The percentage of calories from fat is about the same in both.

A Better Idea: Forget the fancy cookie notion and get back to basics. Start a fire, get a stick, and roast marshmallows over the open flame. Plain marshmallows have only 19 calories apiece and no fat.

MOLASSES

Serving: 2 (about 2 oz.)

Calories	274
Fat	6.9 g
Saturated	1.7 g
Monounsaturated	N/A
Polyunsaturated	N/A
Calories from fat	23%
Cholesterol	25 mg
Sodium	251 mg
Protein	4.2 g
Carbohydrate	49.4 g
Dietary fiber	0.8 g

CHIEF NUTRIENTS

NUTRIENT	AMOUNT	% RDA
Iron	2.3 mg	23
Thiamine	0.2 mg	11
Niacin	1.6 mg	9
Riboflavin	0.1 mg	8

Caution: Molasses is a concentrated sweetener, so you're getting a big dollop of calories with every bite of molasses cookie. *Allergy alert:* Molasses can occasionally cause allergic reactions. These cookies may contain eggs and flour, substances that can cause allergies in some people.

Strengths: Molasses cookies have a good amount of iron, which is needed for good blood circulation. They also have favorable amounts of thiamine, niacin, and riboflavin.

Curiosity: Molasses played a role in the slave trade. The sweetener was shipped from the West Indies to New England; there it was made into rum.

History: Molasses was first mentioned in a 1582 Portuguese manuscript about the East Indies. The word is derived from the Portuguese word *melaco,* which means honey.

A Better Idea: Fig bars and ladyfingers have fewer calories.

OATMEAL *with Raisins*

Serving: 4 (about 2 oz.)

Calories	235
Fat	8.0 g
Saturated	2.1 g
Monounsaturated	N/A
Polyunsaturated	N/A
Calories from fat	31%
Cholesterol	20 mg
Sodium	84 mg
Protein	3.2 g
Carbohydrate	38.2 g
Dietary fiber	1.2 g

CHIEF NUTRIENTS

NUTRIENT	AMOUNT	% RDA
Iron	1.7 mg	17
Thiamine	0.2 mg	10
Riboflavin	0.1 mg	6
Niacin	1.0 mg	5
Potassium	192.4 mg	5

Caution: If you're munching on these tempting confections, you're getting about 30% of calories from fat. *Allergy alert:* Eggs and flour are common ingredients in these cookies, and may cause allergic reactions in some people.

Strengths: Raisins boost the iron content. These cookies also provide a good dose of thiamine and some ribovflavin, two B vitamins that help convert carbohydrates and other foods into energy. In addition, they have a bit of potassium, a mineral that helps flush sodium from the body to lower blood pressure.

A Better Idea: If baking your own, try using oat bran; this is the high-fiber outer casing of oat grain that has been shown to help fight high cholesterol.

PEANUT BUTTER, *Homemade*

Serving: 4 (about 1¾ oz.)

Calories	245
Fat	14.0 g
Saturated	4.0 g
Monounsaturated	5.8 g
Polyunsaturated	2.8 g
Calories from fat	51%
Cholesterol	22 mg
Sodium	142 mg
Protein	4.0 g
Carbohydrate	28.0 g
Dietary fiber	0.9 g

CHIEF NUTRIENTS

NUTRIENT	AMOUNT	% RDA
Iron	1.1 mg	11
Niacin	1.9 mg	10

Caution: High in fat and very high in calories, these treats also have moderate amounts of sodium and cholesterol. *Allergy alert:* Peanuts, eggs, and flour are substances that may cause allergic reactions such as headaches and fatigue in some people.

Strengths: Homemade peanut butter cookies have a good level of iron, a nutrient essential for healthy red blood cells. They also have a significant amount of niacin, a B vitamin that helps body cells maintain their structure. Moderate amounts of protein and carbohydrates are also present.

Curiosity: In the 1890s, George Washington Carver suggested to farmers in the South that they plant peanuts to replace cotton that was being devastated by the boll weevil.

History: Peanuts are native to South America; the Incas may have been the first to cultivate them.

PEANUT BUTTER SANDWICH

Serving: 4 (about 1¾ oz.)

Calories	232
Fat	9.4 g
Saturated	2.0 g
Monounsaturated	N/A
Polyunsaturated	N/A
Calories from fat	36%
Cholesterol	19 mg
Sodium	85 mg
Protein	4.9 g
Carbohydrate	32.8 g
Dietary fiber	N/A

CHIEF NUTRIENTS

NUTRIENT	AMOUNT	% RDA
Iron	1.4 mg	14
Niacin	2.3 mg	12
Thiamine	0.2 mg	11
Riboflavin	0.2 mg	7

Caution: Moderate in fat, these cookies are very high in calories. Four peanut butter sandwich cookies have about 232 calories, and 36% of those calories come from fat. There's also a moderate amount of sodium. *Allergy alert:* Some people suffer from skin rashes and itching if they eat or smell peanut butter. In addition, these cookies are commonly made from milk and flour, other substances that may cause allergic reactions in some people.

Strengths: Peanut butter sandwich cookies are a moderately good source of protein, iron, thiamine, and niacin. They also have some riboflavin, a B vitamin that helps convert food into energy. They are more nutritious and have less fat than vanilla- or chocolate-filled sandwich cookies.

Curiosity: Prior to the arrival of European explorers, there were no peanuts in North America.

A Better Idea: For dessert, have a single cookie with fruit; that way, you'll get the peanut-buttery sweetness along with vitamin C and fiber.

SHORTBREAD

Serving: 5 (about 1¼ oz.)

Calories	187
Fat	8.7 g
Saturated	2.2 g
Monounsaturated	N/A
Polyunsaturated	N/A
Calories from fat	42%
Cholesterol	15 mg
Sodium	23 mg
Protein	2.7 g
Carbohydrate	24.4 g
Dietary fiber	N/A

CHIEF NUTRIENTS

NUTRIENT	AMOUNT	% RDA
Iron	1.1 mg	11
Thiamine	0.2 mg	11
Niacin	1.2 mg	6
Riboflavin	0.1 mg	6

Caution: Shortbread cookies are loaded with butter and are a bit high in saturated fat (the type most likely to increase blood cholesterol). *Allergy alert:* Shortbread cookies are commonly made with wheat flour, a substance that may cause allergic reactions in some people.

Strengths: Shortbread cookies are a good source of thiamine and are equally strong in iron, a mineral that improves the body's ability to fight off infections. Along with some riboflavin, they also have some niacin, a nutrient that helps hold cells together. Five shortbread cookies have about 186 calories, but they're fine if you can be satisfied with one or two occasionally.

History: The original cookies were patterned after a Scottish Yule cake, the bannock, which was usually made of barley meal and oatmeal. They were traditionally eaten at Christmas and Hogmanay, the Scottish New Year's Eve. Today, shortbread is a year-round favorite.

A Better Idea: Gingersnaps and ladyfingers have less fat.

SUGAR, *Homemade*

Serving: 5 (about 1½ oz.)

Calories	170
Fat	6.1 g
Saturated	3.2 g
Monounsaturated	N/A
Polyunsaturated	N/A
Calories from fat	32%
Cholesterol	30 mg
Sodium	188 mg
Protein	2.4 g
Carbohydrate	26.8 g
Dietary fiber	N/A

CHIEF NUTRIENTS

NUTRIENT	AMOUNT	% RDA
Iron	0.8 mg	8
Thiamine	0.1 mg	6
Riboflavin	0.1 mg	5

Caution: Where there's sugar, there are empty calories. Where there's butter, there's ugly fat. Unfortunately, these cookies have both sugar and butter, with a significant amount of sodium tossed in. Nearly a third of the 170 calories in five cookies come from fat; since half of that fat is saturated, it is likely to increase blood cholesterol. *Allergy alert:* These cookies are commonly made with eggs and flour, ingredients that can cause allergic reactions in some people.

Strengths: Sugar cookies have some iron, riboflavin, and thiamine. However, other cookies are richer in these nutrients.

A Better Idea: If you do bake these cookies, use margarine instead of butter to cut down on saturated fat.

SUGAR WAFERS

Serving: 5 (about 1¾ oz.)

Calories	230
Fat	9.2 g
Saturated	2.3 g
Monounsaturated	N/A
Polyunsaturated	N/A
Calories from fat	36%
Cholesterol	19 mg
Sodium	90 mg
Protein	2.3 g
Carbohydrate	34.9 g
Dietary fiber	0.2 g

CHIEF NUTRIENTS

NUTRIENT	AMOUNT	% RDA
Iron	1.0 mg	10
Thiamine	0.1 mg	7
Niacin	1.0 mg	5
Riboflavin	0.1 mg	5

Caution: Sometimes made with partially hydrogenated soybean oil, a type of saturated fat that may increase blood cholesterol levels, sugar wafers are also high in calories. Overall, this cookie gets about 36% of its calories from fat. *Allergy alert:* These cookies may be made with milk, eggs, and flour, substances that cause allergic reactions in some people.

Strengths: They don't contain spinach, but sugar wafers do have a good amount of iron, a catalyst that helps the body process vitamin A. They also have fair amounts of riboflavin and thiamine, two B vitamins that help convert carbohydrates and other foods into energy.

Description: Two thin, wafflelike crackers surrounding a creamy filling.

A Better Idea: A 2-oz. serving of fig bars has fewer calories and less fat.

VANILLA OR CHOCOLATE SANDWICH

Serving: 4 ovals (about 2 oz.)

Calories	297
Fat	13.5 g
Saturated	3.7 g
Monounsaturated	N/A
Polyunsaturated	N/A
Calories from fat	41%
Cholesterol	23 mg
Sodium	290 mg
Protein	2.9 g
Carbohydrate	41.6 g
Dietary fiber (vanilla)	0.9 g
Dietary fiber (chocolate)	1.7 g

CHIEF NUTRIENTS

NUTRIENT	AMOUNT	% RDA
Iron	1.2 mg	12
Thiamine	0.1 mg	9
Niacin	1.3 mg	7
Riboflavin	0.1 mg	6

Caution: More than 40% of the classic sandwich cookie's calories are fat. Oval types, being larger, also have more calories (297 per 4-cookie serving) than round ones (198 per 4-cookie serving). Both vanilla and chocolate sandwich cookies contain a moderate amount of cholesterol. *Allergy alert:* These cookies may be made with milk and wheat flour, substances that cause allergic reactions in some people.

Strengths: A good amount of iron is the sandwich cookie's best selling point. This creamy treat also has some thiamine, riboflavin, and niacin (a nutrient essential for the formation of red blood cells). Tastes great, but you might ask yourself if the fat is worth it.

Serve: Great for dipping in coffee, tea, or milk.

A Better Idea: If you really crave a sandwich cookie, try one with a more nutritious filling, such as peanut butter.

VANILLA WAFERS

Serving: 10 (about 1½ oz.)

Calories	185
Fat	6.4 g
Saturated	1.6 g
Monounsaturated	N/A
Polyunsaturated	N/A
Calories from fat	31%
Cholesterol	16 mg
Sodium	101 mg
Protein	2.2 g
Carbohydrate	29.8 g
Dietary fiber	0.7 g

CHIEF NUTRIENTS

NUTRIENT	AMOUNT	% RDA
Iron	1.0 mg	9
Thiamine	0.1 mg	7
Riboflavin	0.1 mg	5

Caution: Vanilla wafers have a moderate amount of sodium. About 32% of this cookie's calories come from fat. *Allergy alert:* Some wafers are made with milk, eggs, and wheat flour, substances that may cause allergic reactions.

Strengths: Though they contain few nutrients, these cookies do have a fair amount of iron and some thiamine and riboflavin, a B vitamin that contributes to the formation of red blood cells in bone marrow.

History: Aztecs used the vanilla bean, and Spanish explorers took it back to Europe from Mexico. Vanilla was once considered an aphrodisiac and was so scarce that it was only available to royalty.

A Better Idea: Eat an apple.

DOUGHNUTS

CAKE-TYPE

Serving: 1 (2 oz.)

Calories	227
Fat	10.8 g
Saturated	2.7 g
Monounsaturated	N/A
Polyunsaturated	N/A
Calories from fat	43%
Cholesterol	35 mg
Sodium	291 mg
Protein	2.7 g
Carbohydrate	29.8 g
Dietary fiber	0.8 g

CHIEF NUTRIENTS

NUTRIENT	AMOUNT	% RDA
Iron	1.2 mg	12
Thiamine	0.1 mg	8
Riboflavin	0.1 mg	7
Niacin	1.0 mg	5

Caution: Doughnuts are usually deep-fried, not baked, and they get almost half their calories from fat. They are also high in sugar and calories. Sodium is steep, so avoid them if you're on a low-salt diet. *Allergy alert:* Wheat flour causes an allergic reaction in some people.

Strengths: Because they are made with enriched flour, doughnuts provide a good amount of iron and some thiamine, riboflavin, and niacin, all essential for healthy blood and good energy levels. Doughnuts also have some fiber.

Curiosity: Legend has it that a Rockport, Maine, fisherman made the first doughnuts with holes: He poked out the centers so he could slip the doughnuts over the spokes of his ship's wheel. Now and then, he would grab a doughnut for a snack.

Description: Soft, fine-textured doughnuts raised with baking powder, not yeast. They are often flavored.

A Better Idea: Hearty whole grain muffins with less fat and more fiber can satisfy a sweet tooth without the high-fat factor.

YEAST OR RAISED

Serving: 1 (1½ oz.)

Calories	174
Fat	11.2 g
Saturated	2.8 g
Monounsaturated	N/A
Polyunsaturated	N/A
Calories from fat	58%
Cholesterol	11 mg
Sodium	98 mg
Protein	2.7 g
Carbohydrate	15.8 g
Dietary fiber	0.9 g

CHIEF NUTRIENTS

NUTRIENT	AMOUNT	% RDA
Iron	0.9 mg	9
Thiamine	0.1 mg	7
Riboflavin	0.1 mg	5

Caution: Because they are deep-fried, yeast doughnuts get almost 60% of their calories from fat. *Allergy alert:* Some people are allergic to the wheat in flour.

Strengths: The enriched wheat flour in doughnuts provides some blood-building iron, along with riboflavin and thiamine—B-complex vitamins important for energy production.

Curiosity: A "sinker" is one term for a heavy, heavy doughnut.

Origin: German. The Pennsylvania Dutch traditionally served doughnuts called "fastnachts" as the last indulgence before Lent. On Fastnacht Day, the last person to arrive at the breakfast table was labeled a lazy fastnacht.

Description: Raised doughnuts are made with yeast and allowed to rise at least once before being fried. They come in rings, squares, and twists. Sometimes they are filled with jelly or cream and glazed or dusted with granulated sugar.

A Better Idea: Whole grain muffins make great breakfast and snack fare with less fat and sugar.

169

FROZEN DESSERTS

FROZEN YOGURT, *Nonfat, All Flavors* ⬇ 🏛

Serving: ½ cup

Calories	100
Fat	0
Saturated	0
Monounsaturated	0
Polyunsaturated	0
Calories from fat	0
Cholesterol	0
Sodium	60 mg
Protein	4.0 g
Carbohydrate	21.0 g
Dietary fiber	N/A

CHIEF NUTRIENTS

NUTRIENT	AMOUNT	% RDA
Calcium	150.0 mg	19
Riboflavin	0.2 mg	10

Strengths: Frozen nonfat yogurt contains no fat or cholesterol and a moderate amount of calories. (Nonfat frozen yogurt has significantly fewer calories than regular frozen yogurt.) The nonfat variety is a good source of bone-bolstering calcium and a fair source of energy-enhancing riboflavin.
Eat with: Fresh sliced fruit, which may add only 10 calories.
Caution: A syrup topping can add 75 calories or more.
Selection: Compare calories and fat as listed on the labels of store-bought brands. Even nonfat yogurts have a wide range of calories per ½-cup serving. In general, frozen yogurts usually contain significantly lower levels of beneficial live culture cells than regular yogurt.

FROZEN YOGURT, *Regular, Chocolate* 🏛

Serving: ½ cup

Calories	140
Fat	4.0 g
Saturated	N/A
Monounsaturated	N/A
Polyunsaturated	N/A
Calories from fat	26%
Cholesterol	N/A
Sodium	65 mg
Protein	3.0 g
Carbohydrate	24.0 g
Dietary fiber	N/A

CHIEF NUTRIENTS

NUTRIENT	AMOUNT	% RDA
Calcium	100.0 mg	13
Vitamin B_{12}	0.2 mcg	12
Riboflavin	0.1 mg	8

Strengths: Frozen yogurt is low in fat and moderate in calories. It's a good source of bone-strengthening calcium and vitamin B_{12}, important for healthy nerves and blood. It's also a fair source of riboflavin, a B-complex vitamin that aids energy production. A serving contains less than half the calories of premium ice cream.
Caution: Beware of syrup toppings; they can add lots of calories.
Description: Tastes remarkably like ice milk; doesn't have the tangy taste of regular yogurt.
Selection: Compare calories and fat by reading labels; frozen yogurt ranges from 95 to 160 calories per ½-cup serving, depending on the brand. Rich-sounding flavors like cheesecake and Dutch chocolate tend to be a bit higher in fat and calories than vanilla or fruit flavors.
Serve: With sliced fresh fruit.

FRUIT ICE

Serving: 1 cup

Calories	247
Fat	0
Saturated	0
Monounsaturated	N/A
Polyunsaturated	N/A
Calories from fat	0
Cholesterol	0
Sodium	0
Protein	0.8 g
Carbohydrate	62.9 g
Dietary fiber	0

CHIEF NUTRIENTS

(None of the nutrients meet or exceed 5% of the RDA.)

Strengths: Fruit ice contains few nutrients but can be a real thirst quencher on a hot, humid day. It has no fat, no sodium, and no cholesterol, and it has a trace of protein.
Caution: A 1-cup serving does have a noticeable 247 calories.
Allergy alert: The fruit and berry flavorings may cause allergic reactions in some people.
Description: Ices, also known as Italian ices, are made with water, sugar, and fruit flavorings such as mint, watermelon, raspberry, strawberry, and orange.

FRUIT JUICE BAR

Serving: 1 (about 1³/₄ oz.)

Calories	42
Fat	0
Saturated	0
Monounsaturated	0
Polyunsaturated	0
Calories from fat	0
Cholesterol	0
Sodium	4 mg
Protein	0.6 g
Carbohydrate	10.1 g
Dietary fiber	N/A

CHIEF NUTRIENTS

NUTRIENT	AMOUNT	% RDA
Vitamin C	5.0 mg	8

Strengths: Fruit juice bars make a decent alternative to chocolate bars—they have no fat or cholesterol, and they're relatively low in calories. Plus you get a little vitamin C to help boost immunity.
Caution: *Allergy alert:* If you have asthma, beware: Sulfites have been known to cause severe allergic reactions in sensitive individuals. Also, if you're allergic to berries, oranges, or other fruit, you'll probably be allergic to bars made from the concentrated fruit juice.
Description: Fruit juice bars are made from juice concentrates, locust bean, and guar gums (used as thickeners), with added sulfites (as preservatives). You can choose from a variety of flavors, including banana, cherry, grape, raspberry, orange, pineapple, and piña colada.

ICE CREAM, *French Vanilla, Soft-Serve*

Serving: ½ cup

Calories	188
Fat	11.3 g
Saturated	6.8 g
Monounsaturated	3.3 g
Polyunsaturated	0.5 g
Calories from fat	54%
Cholesterol	77 mg
Sodium	77 mg
Protein	3.5 g
Carbohydrate	19.1 g
Dietary fiber	0

CHIEF NUTRIENTS

NUTRIENT	AMOUNT	% RDA
Vitamin B_{12}	0.5 mcg	25
Calcium	117.9 mg	15
Riboflavin	0.2 mg	13
Vitamin A	99.5 RE	10
Zinc	1.0 mg	7

Caution: Contains a lot of fat, most of it saturated, and quite a few calories. French vanilla is generally higher in cholesterol than regular ice cream due to the presence of egg yolks in the mixture. *Allergy alert:* May be problematic for those allergic to milk or eggs. And it may contain salicylate from artificial coloring or flavoring: Should be avoided by those with aspirin allergy.

Eat with: Restraint.

Strengths: Has appreciable vitamin B_{12} for healthy nerves and blood, calcium for strong bones, and vitamin A for good vision and enhanced immunity.

History: Thomas Jefferson is credited with bringing French-style ice cream to America. He even had an ice cream machine at Monticello.

A Better Idea: Look for ice milk, especially the new fat-free varieties. Or serve low-fat frozen yogurt or fruit ice instead. Top with lots of fresh fruit to add fiber: You'll eat less and still feel full.

ICE CREAM, *Vanilla (10% Fat)*

Serving: ½ cup

Calories	135
Fat	7.2 g
Saturated	4.5 g
Monounsaturated	2.1 g
Polyunsaturated	0.3 g
Calories from fat	48%
Cholesterol	30 mg
Sodium	58 mg
Protein	2.4 g
Carbohydrate	15.9 g
Dietary fiber	0

CHIEF NUTRIENTS

NUTRIENT	AMOUNT	% RDA
Vitamin B_{12}	0.3 mcg	16
Calcium	87.9 mg	11
Riboflavin	0.2 mg	10
Vitamin A	66.5 RE	7

Caution: Contains quite a bit of fat—most of it the unhealthy saturated variety. Vanilla ice cream is a little heavy on calories, even in a small serving. *Allergy alert:* May be a problem for those allergic to milk. May contain salicylate from artificial coloring or flavoring, which should be avoided by those sensitive to aspirin.

Strengths: Has good amounts of vitamin B_{12} and riboflavin for healthy nerves and blood, as well as calcium for strong bones. Ice cream even has some vitamin A, which helps improve vision.

Curiosity: George Washington was so enamored of ice cream that he spent at least $200 on the confection during the summer of 1790.

A Better Idea: Try the new fat-free varieties of ice milk, or low-fat frozen yogurt. Top with a big helping of berries, crushed pineapple, or other fresh fruit to add fiber and vitamins.

ICE CREAM, *Vanilla (16% Fat)*

Serving: ½ cup

Calories	175
Fat	11.9 g
Saturated	7.4 g
Monounsaturated	3.4 g
Polyunsaturated	0.4 g
Calories from fat	61%
Cholesterol	44 mg
Sodium	54 mg
Protein	2.1 g
Carbohydrate	16.0 g
Dietary fiber	0

CHIEF NUTRIENTS

NUTRIENT	AMOUNT	% RDA
Vitamin B_{12}	0.3 mcg	14
Vitamin A	109.5 RE	11
Calcium	75.6 mg	9
Riboflavin	0.1 mg	8

Caution: This is a "rich" ice cream, containing quite a bit of fat—most in the unhealthy saturated form that can raise cholesterol. It's also high in calories. Ice cream labeled "premium" may contain even *more* fat and calories! *Allergy alert:* May cause problems for those allergic to milk. Some brands may contain salicylate from artificial coloring and flavorings, which should be avoided by those sensitive to aspirin.

Strengths: Has plenty of vitamin B_{12} for healthy nerves and blood, calcium for strong bones, and vitamin A for enhanced immunity.

Storage: High-fat varieties of ice cream can easily absorb other food odors and flavors. After you open a carton, wrap it well in plastic or foil before returning it to the freezer.

A Better Idea: Select low-fat or nonfat ice milk for dessert. You can serve these treats piled high with your favorite fresh fruit to add fiber.

ICE MILK, *Vanilla*

Serving: ½ cup

Calories	92
Fat	2.8 g
Saturated	1.8 g
Monounsaturated	0.8 g
Polyunsaturated	0.1 g
Calories from fat	28%
Cholesterol	9 mg
Sodium	52 mg
Protein	2.6 g
Carbohydrate	14.5 g
Dietary fiber	0

CHIEF NUTRIENTS

NUTRIENT	AMOUNT	% RDA
Vitamin B_{12}	0.4 mcg	22
Calcium	88.0 mg	11
Riboflavin	0.2 mg	10

Strengths: Ice milk is usually much lower in fat and cholesterol than ice cream. Good amounts of vitamin B_{12} and riboflavin help improve healthy skin and nerves; its calcium helps build strong bones.

Eat with: Lots of fresh fruit to add fiber and other nutrients—especially vitamins A and C. Bananas, berries, and cherries are good high-C choices. Apricots, peaches, and mangoes add beta-carotene for a measure of cancer protection.

Caution: The calories and sodium in ice milk can add up if you really indulge. *Allergy alert:* May cause problems for those allergic to milk. Some brands might contain salicylate from artificial coloring or flavorings—a concern for those sensitive to aspirin.

Description: According to FDA regulations, ice milk contains 2 to 7% milkfat, compared with a minimum of 10% for ice cream. Lower-fat and nonfat versions cannot be labeled "ice milk."

Selection: Look for fat-free varieties.

ICE MILK, *Vanilla, Soft-Serve*

Serving: ½ cup

Calories	112
Fat	2.3 g
Saturated	1.4 g
Monounsaturated	0.7 g
Polyunsaturated	0.1 g
Calories from fat	19%
Cholesterol	7 mg
Sodium	81 mg
Protein	4.0 g
Carbohydrate	19.2 g
Dietary fiber	0

CHIEF NUTRIENTS

NUTRIENT	AMOUNT	% RDA
Vitamin B_{12}	0.7 mcg	34
Calcium	137.2 mg	17
Riboflavin	0.3 mg	16
Potassium	206.2 mg	6

Strengths: Soft-serve ice milk is higher in calories and sodium than regular ice milk, but somewhat lower in saturated fat and cholesterol. It has good amounts of vitamin B_{12} and riboflavin for healthy skin and nerves, plus calcium for strong bones. In fact, it is slightly higher in these nutrients than hardened ice milk. But nutrient composition may vary according to brand.

Eat with: Angel food cake, which is fat-free. Top with plenty of raspberries or blackberries for a hefty dose of fiber. A sprinkling of toasted wheat germ adds other nutrients, including other B vitamins, along with magnesium, iron, and zinc.

Caution: Calories and sodium can add up if you really indulge. *Allergy alert:* May be problematic for those allergic to milk. And it may contain salicylate from artificial coloring and flavoring—a concern for those sensitive to aspirin.

Selection: Look for fat-free varieties or those with the lowest amounts of fat, cholesterol, and calories.

SHERBET, *Orange*

Serving: ½ cup

Calories	135
Fat	1.9 g
Saturated	1.2 g
Monounsaturated	0.6 g
Polyunsaturated	0.1 g
Calories from fat	13%
Cholesterol	7 mg
Sodium	44 mg
Protein	1.1 g
Carbohydrate	29.4 g
Dietary fiber	N/A

CHIEF NUTRIENTS

NUTRIENT	AMOUNT	% RDA
Calcium	51.7 mg	6

Strengths: Typically lower in fat and cholesterol than ice cream, sherbet is as high in calories. Has a little calcium, thanks to the milk content. May contain some fiber, depending on whether actual fruit is used rather than fruit juice only. Sherbet is fairly low in sodium.

Eat with: Plenty of fresh fruit to add fiber and to counterbalance the sherbet's sweetness.

Caution: *Allergy alert:* Although sherbet contains only a small amount of milk, it may be enough to trigger an allergic reaction. May contain egg whites, which could affect those with egg allergy.

Origin: Can be traced back to a popular Middle Eastern drink called *charbet*, which was made from sweetened fruit juice and water.

Serve: With cottage cheese, sliced fruit, and date-nut bread for a light lunch. Or eat a tiny portion as a refreshing, low-cal palate cleanser between the courses of a large meal.

TOFUTTI, *All Flavors*   ⬇

Serving: ½ cup

Calories	217
Fat	12.0 g
Saturated	2.0 g
Monounsaturated	3.0 g
Polyunsaturated	7.0 g
Calories from fat	53%
Cholesterol	0
Sodium	105 mg
Protein	2.6 g
Carbohydrate	22.4 g
Dietary fiber	N/A

CHIEF NUTRIENTS

(None of the nutrients meet or exceed 5% of the RDA.)

Strengths: Tofutti contains zero cholesterol, compared with 44 g for regular ice cream. A nondairy product, Tofutti is also lactose-free, so it's easier for people with lactose intolerance to digest. It's great for kids with milk allergies who feel left out if they can't eat ice cream like their friends do.

Caution: Regular Tofutti contains more fat and calories than premium ice cream (although most of the fat is unsaturated).

Description: A frozen dessert containing corn oil, soybean oil, various gums, sweeteners, flavorings, and sometimes tofu. It's available in 7 flavors.

A Better Idea: Lite Lite Tofutti has only 90 calories and is virtually fat-free. It comes in 5 different flavors and tastes a lot like frozen yogurt.

PIES

APPLE

Serving: 1 slice (about 4⅛ oz.)

Calories	302
Fat	13.1 g
Saturated	3.4 g
Monounsaturated	N/A
Polyunsaturated	N/A
Calories from fat	39%
Cholesterol	0
Sodium	355 mg
Protein	2.6 g
Carbohydrate	45.0 g
Dietary fiber	1.9 g

CHIEF NUTRIENTS

NUTRIENT	AMOUNT	% RDA
Iron	1.2 mg	12
Thiamine	0.1 mg	9
Niacin	1.2 mg	6
Riboflavin	0.1 mg	5

Caution: Pie crust is moderately high in sodium, and it also contains a high amount of fat, even when made with vegetable shortening. Apple pie has about 6% more fat and 70 more calories per serving than a slice of devil's food cake with chocolate icing.

Strengths: This pie offers a decent amount of blood-building iron, along with some niacin, riboflavin, and thiamine, B-complex vitamins that aid energy production. It also offers some fiber.

Curiosity: Researchers have discovered that apple-spice scent, like relaxation exercises, can produce a calming of brainwave patterns.

Preparation: Use apples that are firm and tart, such as greenings, Jonathans, Cortlands, or Granny Smiths. Make your crust with enriched flour and vegetable shortening.

Serve: Traditionally served warm, sometimes accompanied by cheddar cheese, heavy cream, or ice cream—additions that increase fat and calories.

A Better Idea: Boost fiber by using unpeeled apples and oat bran crumb topping. Serve with a glass of skim milk, not ice cream or cheese.

BANANA CREAM, *Homemade*

Serving: 1 slice (about 4½ oz.)

Calories	285
Fat	12.0 g
Saturated	3.8 g
Monounsaturated	N/A
Polyunsaturated	N/A
Calories from fat	38%
Cholesterol	40 mg
Sodium	252 mg
Protein	6.0 g
Carbohydrate	40.0 g
Dietary fiber	N/A

CHIEF NUTRIENTS

NUTRIENT	AMOUNT	% RDA
Riboflavin	0.2 mg	13
Calcium	86.0 mg	11
Iron	1.0 mg	10
Thiamine	0.1 mg	7
Vitamin A	66.0 RE	7
Niacin	1.0 mg	5

Caution: Banana cream pie is high in sodium, largely because of salt in the crust. It's also moderate in calories—and about 38% of those come from fat. (However, it has fewer calories than apple pie.) *Allergy alert:* Contains eggs and milk and sometimes cornstarch or wheat flour—all common allergens.

Strengths: Banana cream pie offers appreciable amounts of blood-building iron and bone-strengthening calcium. There's some vitamin A for better vision and healthier skin. Its niacin, riboflavin, and thiamine aid energy production. The thiamine and niacin also help maintain healthy nerves.

Description: A vanilla cream pie containing sliced bananas. The cream is made with sugar, milk or heavy cream, eggs, flour or cornstarch, and butter. It may be topped with meringue or whipped cream.

Storage: Keep this pie refrigerated.

A Better Idea: Create a leaner version of banana cream pie using a low-calorie vanilla pudding mix and a low-fat whipped topping.

BLACKBERRY

Serving: 1 slice (about 4⅛ oz.)

Calories	287
Fat	13.0 g
Saturated	3.2 g
Monounsaturated	N/A
Polyunsaturated	N/A
Calories from fat	41%
Cholesterol	0
Sodium	316 mg
Protein	3.1 g
Carbohydrate	40.6 g
Dietary fiber	N/A

CHIEF NUTRIENTS

NUTRIENT	AMOUNT	% RDA
Iron	1.4 mg	14
Thiamine	0.1 mg	9
Vitamin C	4.7 mg	8
Niacin	1.3 mg	7
Riboflavin	0.1 mg	6

Caution: Blackberry pie is moderate in calories and fat; it's high in sodium, with 316 mg per slice.

Strengths: Offers a good share of blood-boosting iron, along with some niacin, riboflavin, and thiamine, B-complex vitamins that aid energy production. Thiamine and niacin also help maintain healthy nerves. Blackberry pie also has a bit of vitamin C, which may boost immune function and protect against cancer. Plus, blackberries are a decent source of fiber.

Curiosity: The nursery rhyme "Sing a Song of Sixpence" with its "four and twenty blackbirds baked in a pie" refers to medieval chefs' attempts to outdo each other with outrageous pies. Not only did they tether live birds in prebaked pie shells, they sometimes used rabbits, frogs, and turtles.

Serve: Warm, fresh from the oven.

BLUEBERRY

Serving: 1 slice (about 4¹/₈ oz.)

Calories	286
Fat	12.7 g
Saturated	3.2 g
Monounsaturated	N/A
Polyunsaturated	N/A
Calories from fat	40%
Cholesterol	0
Sodium	474 mg
Protein	2.8 g
Carbohydrate	41.2 g
Dietary fiber	N/A

CHIEF NUTRIENTS

NUTRIENT	AMOUNT	% RDA
Iron	1.5 mg	15
Thiamine	0.1 mg	9
Niacin	1.3 mg	7
Riboflavin	0.1 mg	6
Vitamin C	3.5 mg	6

Caution: Moderate in fat and calories, and high in sodium. When salt is included in both the filling and the crust, each slice of blueberry pie has a whopping 474 mg of sodium. So it's not for anyone on a low-salt diet.

Strengths: Blueberry pie has a decent amount of blood-building iron, along with some niacin, riboflavin, and thiamine, B-complex vitamins essential for producing energy. Since raw blueberries have a good amount of fiber, there may be some in blueberry pie, though figures are not available. It also has a little vitamin C, which may offer cancer protection.

Description: The usual ingredients are blueberries, sugar, lemon juice, and cinnamon baked in a crust of enriched flour, water, and vegetable shortening, often topped with either a crust or crumbs.

Selection: If you are choosing blueberries for a homemade pie, choose ones that are firm, uniform in size, and indigo blue with a silvery tinge.

Serve: As part of a picnic lunch, in a cool shady grove. Warning: You may have company, since bears also love blueberries.

CHERRY

Serving: 1 slice (about 4¹/₈ oz.)

Calories	308
Fat	13.3 g
Saturated	3.5 g
Monounsaturated	N/A
Polyunsaturated	N/A
Calories from fat	39%
Cholesterol	0
Sodium	359 mg
Protein	3.1 g
Carbohydrate	45.3 g
Dietary fiber	0.9 g

CHIEF NUTRIENTS

NUTRIENT	AMOUNT	% RDA
Iron	1.2 mg	12
Vitamin A	103.8 RE	10
Thiamine	0.1 mg	9
Niacin	1.2 mg	6
Riboflavin	0.1 mg	6

Caution: Moderate in fat, and high in sodium and calories.

Strengths: Cherry pie offers a respectable amount of blood-building iron, and a good amount of vitamin A to help maintain healthy blood and nerves. It also has nice amounts of thiamine, niacin, and riboflavin—B vitamins that help maintain healthy nerves and blood.

Description: Usually made with sour cherries and sugar. A squirt of almond extract or kirsch, a cherry-flavored liqueur, is added to enhance the flavor, and tapioca is sometimes used to thicken the cherry juice. The piecrust is made with enriched flour and vegetable shortening.

Selection: If making your own pie, choose brightly colored, shiny, plump cherries—or use pitted canned sour cherries.

A Better Idea: Most fruit pie recipes call for the fruit filling to be dotted with butter; for less fat and fewer calories, skip that step. For less sodium, omit salt from the crust. Use only a light crumb topping, not a second crust. And taste the fruit filling as you add the sugar, using only as much sweetener as you need.

CHOCOLATE CREAM, *Homemade*

Serving: 1 slice (about 3½ oz.)

Calories	264
Fat	15.1 g
Saturated	N/A
Monounsaturated	N/A
Polyunsaturated	N/A
Calories from fat	51%
Cholesterol	N/A
Sodium	273 mg
Protein	4.6 g
Carbohydrate	29.5 g
Dietary fiber	0.2 g

CHIEF NUTRIENTS

NUTRIENT	AMOUNT	% RDA
Vitamin B$_{12}$	0.4 mg	18
Calcium	84.0 mg	11
Iron	1.1 mg	11
Riboflavin	0.2 mg	10
Magnesium	25.0 mg	7
Thiamine	0.1 mg	7
Vitamin A	52.8 RE	5

Caution: Chocolate cream pie has a higher percentage of calories from fat than many other pies. *Allergy alert:* Not for those who are sensitive to food allergens; this pie is a veritable minefield of ingredients that cause allergic reactions, including chocolate, milk, eggs, wheat flour, and alcohol-based flavorings.

Strengths: Offers satisfactory amounts of blood-building iron as well as riboflavin. It has a bit of thiamine, a B-complex vitamin that aids in the production of energy. In addition, a good amount of vitamin B$_{12}$ contributes to proper nerve function, and a tad of magnesium helps maintain proper blood pressure. There's some vitamin A for healthy skin.

Description: A very rich and creamy pie often made with milk, egg yolks, butter, sugar, and chocolate.

A Better Idea: If you must have this pie, try making it using a low-calorie chocolate pudding mix and a low-fat whipped topping.

CUSTARD

Serving: 1 slice (about 4 oz.)

Calories	249
Fat	12.7 g
Saturated	4.3 g
Monounsaturated	N/A
Polyunsaturated	N/A
Calories from fat	46%
Cholesterol	120 mg
Sodium	327 mg
Protein	7.0 g
Carbohydrate	26.7 g
Dietary fiber	N/A

CHIEF NUTRIENTS

NUTRIENT	AMOUNT	% RDA
Calcium	109.4 mg	14
Riboflavin	0.2 mg	14
Iron	1.1 mg	11
Thiamine	0.1 mg	7
Vitamin A	52.4 RE	5

Caution: Custard pie weighs in heavily, with 46% of its 249 calories coming from fat. It is moderate in both calories and cholesterol.

Strengths: Offers a decent amount of blood-boosting iron. It also has a good supply of riboflavin to aid in energy production, and some thiamine, a B-complex vitamin essential for healthy nerves. The milk or cream in custard pie provides a good dose of calcium for stronger bones and enough protein to help build body tissue. And it has some vitamin A for healthier skin.

Description: Made of eggs or egg yolks, milk or cream, sugar, salt, and vanilla or rum flavoring—and the piecrust is made with enriched flour and vegetable shortening. Bananas, coconut, pineapple, or other fruit—or chocolate—may be added. Coconut custard pie has about 2% more fat than regular custard, and about 1½ g more saturated fat per serving.

Storage: Custard pie must be refrigerated; unfortunately, refrigeration tends to make the custard watery.

A Better Idea: Try making your own lower-fat filling with skim milk and egg substitute.

LEMON CHIFFON

Serving: 1 slice (about 3 oz.)

Calories	254
Fat	10.2 g
Saturated	2.8 g
Monounsaturated	N/A
Polyunsaturated	N/A
Calories from fat	36%
Cholesterol	137 mg
Sodium	211 mg
Protein	5.7 g
Carbohydrate	35.5 g
Dietary fiber	N/A

CHIEF NUTRIENTS

NUTRIENT	AMOUNT	% RDA
Iron	1.2 mg	12
Riboflavin	0.1 mg	5
Thiamine	0.1 mg	5

Caution: Contains a moderate amount of fat and is high in cholesterol, calories, and sodium.

Strengths: Lemon chiffon pie offers a good amount of blood-building iron, along with some riboflavin and thiamine, B-complex vitamins that enhance energy production.

Curiosity: The word "chiffon," from the French meaning, also refers to pieces of sheer, delicate ribbon or fabric.

Description: An airy, fluffy, sweet pie made with stiffly beaten egg whites, sugar, egg yolks, lemon juice, grated lemon peel, and sometimes gelatin. The crust is made with enriched flour and vegetable shortening. The pie filling is cooked in a double boiler, then chilled. Lime, raspberry, strawberry, and chocolate are chiffon variations.

A Better Idea: It's possible to make a leaner version of this pie using drained low-fat yogurt and less sugar in the filling. For the leanest version, omit the crust.

LEMON MERINGUE

Serving: 1 slice (about 3¾ oz.)

Calories	268
Fat	10.7 g
Saturated	3.2 g
Monounsaturated	N/A
Polyunsaturated	N/A
Calories from fat	36%
Cholesterol	98 mg
Sodium	296 mg
Protein	3.9 g
Carbohydrate	39.6 g
Dietary fiber	1.3 g

CHIEF NUTRIENTS

NUTRIENT	AMOUNT	% RDA
Iron	1.1 mg	11
Riboflavin	0.1 mg	6
Vitamin C	3.2 mg	5

Caution: Lemon meringue pie contains moderate amounts of fat and cholesterol and is high in calories and sodium.

Strengths: Offers a good amount of iron and a bit of riboflavin, essential for healthy blood and good energy production. The lemon juice adds some vitamin C, which aids in cancer protection.

Description: A rich, sweet-tart, deeply colored custard pie made with lemon juice, cornstarch, sugar, egg yolks, and butter. The piecrust is made with enriched flour and vegetable shortening. The meringue topping, which should emerge from the oven lightly browned, contains whipped egg whites and sugar.

A Better Idea: Satisfy your "lemon sweet tooth" with plain lemon custard sprinkled with a crunchy cereal such as Grape-Nuts.

MINCE

Serving: 1 slice (about 4⅛ oz.)

Calories	320
Fat	13.6 g
Saturated	3.6 g
Monounsaturated	N/A
Polyunsaturated	N/A
Calories from fat	38%
Cholesterol	1 mg
Sodium	529 mg
Protein	3.0 g
Carbohydrate	48.6 g
Dietary fiber	N/A

CHIEF NUTRIENTS

NUTRIENT	AMOUNT	% RDA
Iron	2.0 mg	20
Thiamine	0.1 mg	8
Niacin	1.2 mg	6
Potassium	210.0 mg	6
Riboflavin	0.1 mg	6

Caution: Very high in sodium, with over 500 mg per serving. In one slice of mince pie you'll get about 3½ g of saturated fat from the crust and from the suet in the filling.

Strengths: Mince pie offers a good share of blood-building iron. It also offers some niacin, riboflavin, and thiamine, B-complex vitamins needed to produce energy. The dried fruit adds a bit of potassium, which helps maintain healthy nerves and muscles.

Curiosity: In colonial America, these pies were made in the fall and sometimes frozen through the winter.

Description: A traditional dessert for Christmas holidays, mince pie has a filling made with raisins, citron, apple, sugar, spices, and suet. Sometimes finely chopped beef is added, hence the name "mincemeat." The mixture is simmered, then baked in a pie shell made from enriched flour and vegetable shortening. It may also contain sherry, brandy, rum, lemon juice, or chopped almonds.

A Better Idea: For less saturated fat, make mince pie without suet.

PEACH

Serving: 1 slice (about 4⅛ oz.)

Calories	301
Fat	12.6 g
Saturated	3.1 g
Monounsaturated	N/A
Polyunsaturated	N/A
Calories from fat	38%
Cholesterol	0
Sodium	316 mg
Protein	3.0 g
Carbohydrate	45.1 g
Dietary fiber	N/A

CHIEF NUTRIENTS

NUTRIENT	AMOUNT	% RDA
Vitamin A	172.3 RE	17
Iron	1.4 mg	14
Niacin	1.8 mg	9
Thiamine	0.1 mg	9
Riboflavin	0.1 mg	7
Vitamin C	3.5 mg	6

Caution: Peach pie is high in sodium. It's also high in calories, with a moderate amount of fat.

Strengths: There's a good amount of vitamin A, which plays a vital role in vision and is essential for healthy skin. There's also a good amount of blood-building iron. Some niacin, riboflavin, and thiamine also provide a boost: These B-complex vitamins aid in the production of energy. Peaches offer some vitamin C and a bit of fiber.

Description: The filling includes sliced peaches, sugar, cinnamon, lemon juice, and margarine or butter. The piecrust is made with enriched flour and vegetable shortening.

Preparation: Use freestone peaches, which are easy to prepare. Before peeling, blanch the peaches by dropping them into boiling water for about 2 minutes; remove to a bowl of cold water, then drain when cool. The skins should slip off.

A Better Idea: For less fat and sodium and fewer calories, have a bowlful of freshly sliced peaches with a very light sprinkling of confectioner's sugar.

PECAN

Serving: 1 slice (about 3¾ oz.)

Calories	431
Fat	23.6 g
Saturated	3.3 g
Monounsaturated	N/A
Polyunsaturated	N/A
Calories from fat	49%
Cholesterol	65 mg
Sodium	228 mg
Protein	5.3 g
Carbohydrate	52.8 g
Dietary fiber	3.6 g

CHIEF NUTRIENTS

NUTRIENT	AMOUNT	% RDA
Iron	3.4 mg	34
Thiamine	0.2 mg	15
Riboflavin	0.1 mg	7
Calcium	48.4 mg	6

Caution: With a waistline-stretching 431 calories per serving, pecan pie is a weight watcher's nemesis. Nearly half its calories are from fat. Pecan pie is also high in sodium.

Strengths: A very good source of blood-building iron, with 34% of the RDA in each slice. Pecan pie is a good source of thiamine, and it provides some riboflavin; these B-complex vitamins aid in energy production. Plus, this pie offers a tad of bone-strengthening calcium.

Description: Pecan pie is a typical Southern dessert, rich in eggs, butter, sugar, and pecans. The crust is made from enriched flour and vegetable shortening.

PUMPKIN

Serving: 1 slice (about 4 oz.)

Calories	241
Fat	12.8 g
Saturated	4.5 g
Monounsaturated	N/A
Polyunsaturated	N/A
Calories from fat	48%
Cholesterol	70 mg
Sodium	244 mg
Protein	4.6 g
Carbohydrate	27.9 g
Dietary fiber	3.1 g

CHIEF NUTRIENTS

NUTRIENT	AMOUNT	% RDA
Vitamin A	563.2 RE	56
Iron	1.0 mg	10
Riboflavin	0.2 mg	9
Calcium	58.1 mg	7
Thiamine	0.1 mg	7

Caution: Nearly half the 241 calories are from fat, and pumpkin pie has moderate cholesterol. It's high in sodium.

Strengths: A super-high source of vitamin A from the pumpkin filling, pumpkin pie helps skin, nerves, and blood—and the A may also help reduce the risk of certain types of cancer. Pumpkin pie offers a good amount of blood-building iron. It also has riboflavin and thiamine, B-complex vitamins that aid energy production, and a little bone-boosting calcium.

Curiosity: Pumpkin pie was much beloved by the nation's early settlers, who adopted the ungainly pumpkin gourd from Indians.

Description: The fragrant, orange-brown custard pie is made with eggs, evaporated milk, pureed pumpkin, sugar, and spices. The crust is made with enriched flour and vegetable shortening.

A Better Idea: Make a leaner version of pumpkin pie using evaporated skim milk instead of whole milk, and egg whites or egg substitute.

RAISIN

Serving: 1 slice (about 4⅛ oz.)

Calories	319
Fat	12.6 g
Saturated	3.1 g
Monounsaturated	N/A
Polyunsaturated	N/A
Calories from fat	36%
Cholesterol	0
Sodium	336 mg
Protein	3.1 g
Carbohydrate	50.7 g
Dietary fiber	N/A

CHIEF NUTRIENTS

NUTRIENT	AMOUNT	% RDA
Iron	1.9 mg	19
Thiamine	0.1 mg	9
Niacin	1.3 mg	7
Potassium	226.6 mg	6
Riboflavin	0.1 mg	6

Caution: Raisin pie is high in calories and sodium and has a moderate amount of fat. *Allergy alert:* Some recipes for raisin pie include eggs, cornstarch, sour cream, or rum flavoring—all common allergens.

Strengths: The raisins and enriched wheat flour in this pie provide a moderate amount of blood-building iron; the raisins also offer some potassium and fiber. Both raisins and flour contribute niacin, riboflavin, and thiamine, B-complex vitamins that enhance energy production.

Description: Ingredients include raisins, sugar, butter, flour, lemon rind, and lemon juice; these may be combined with eggs, egg yolks, or cornstarch. The piecrust is made with enriched flour and vegetable shortening. The pie can be topped with crumbs, a lattice crust, or with meringue. Variations include raisin-apricot, raisin-cranberry, and raisin-rum sour cream.

A Better Idea: Reduce calories, fat, and sodium by skipping the bottom crust. Add some apples and treat this dessert more like a crisp.

RHUBARB

Serving: 1 slice (about 4⅛ oz.)

Calories	299
Fat	12.6 g
Saturated	3.1 g
Monounsaturated	N/A
Polyunsaturated	N/A
Calories from fat	38%
Cholesterol	0
Sodium	319 mg
Protein	3.0 g
Carbohydrate	45.1 g
Dietary fiber	N/A

CHIEF NUTRIENTS

NUTRIENT	AMOUNT	% RDA
Iron	1.7 mg	17
Calcium	75.5 mg	9
Thiamine	0.1 mg	9
Riboflavin	0.1 mg	8
Niacin	1.3 mg	7
Vitamin C	3.5 mg	6
Potassium	187.6 mg	5

Caution: Rhubarb pie is high in calories and sodium, and contains a moderate amount of fat.

Strengths: It offers a good share of blood-boosting iron, and the rhubarb contributes some vitamin C. Niacin, thiamine and riboflavin, all B-complex vitamins, aid energy production.

Curiosity: In some regions, rhubarb is known as *pieplant,* an indication of its popularity as a pie filling.

History: Rhubarb plants may have been brought over to America by 18th-century Russian fur traders.

Description: A pie made of cubed rhubarb stalks combined with enough sugar to counteract the sourness. The piecrust is made with enriched flour and vegetable shortening. Variations include strawberry-rhubarb and rhubarb custard.

Selection: Select crisp stalks with fresh-looking, blemish-free leaves. Do not use the leaves; they contain oxalic acid and are poisonous.

STRAWBERRY

Serving: 1 slice (about 3¹⁄₄ oz.)

Calories	184
Fat	7.4 g
Saturated	1.8 g
Monounsaturated	N/A
Polyunsaturated	N/A
Calories from fat	36%
Cholesterol	0
Sodium	180 mg
Protein	1.8 g
Carbohydrate	28.7 g
Dietary fiber	N/A

CHIEF NUTRIENTS

NUTRIENT	AMOUNT	% RDA
Vitamin C	23.3 mg	39
Iron	1.1 mg	11

Caution: Strawberry pie contains a moderate amount of fat. There's some sodium from the salt that's added to the crust. This pie is also moderately high in calories.

Strengths: One cup of strawberries provides 141% of the RDA of vitamin C—so it's not surprising that strawberry pie is rich in this vitamin.

Origin: The strawberry is related to the rose and has grown wild for centuries in America and Europe.

Description: The filling of this fragrant, deep-red pie is composed of strawberries, sugar, flour, lemon juice, and a bit of cinnamon. The crust is made with enriched flour and vegetable shortening. It's sometimes topped with crumbs instead of crust. If the fruit is very juicy, tapioca is added as a thickener. Variations include strawberries with blueberries, rhubarb, or bananas.

A Better Idea: For a lot less fat, serve fresh berries over angel food cake.

SWEET POTATO

Serving: 1 slice (about 4 oz.)

Calories	243
Fat	12.9 g
Saturated	4.6 g
Monounsaturated	N/A
Polyunsaturated	N/A
Calories from fat	48%
Cholesterol	62 mg
Sodium	249 mg
Protein	5.1 g
Carbohydrate	27.0 g
Dietary fiber	N/A

CHIEF NUTRIENTS

NUTRIENT	AMOUNT	% RDA
Vitamin A	547.2 RE	55
Iron	1.1 mg	11
Riboflavin	0.2 mg	11
Calcium	78.7 mg	10
Vitamin C	4.6 mg	8
Thiamine	0.1 mg	7
Potassium	185.8 mg	5

Caution: This gussied-up 'tater is high in sodium, and moderate in fat and calories.

Strengths: Because sweet potatoes are a fabulous source of vitamin A, a slice of sweet potato pie offers some of that vitamin's protection against cancer—and it helps improve skin, nerves, and red blood cells. It is also a good source of blood-building iron; riboflavin and thiamine aid energy production, and there's enough vitamin C to contribute to cancer protection.

Curiosity: Sweet potatoes are a member of the morning-glory family of plants.

Description: A moist orange custard pie made with milk, eggs, sweet potatoes, sugar, and spices. The crust is made with enriched flour and vegetable shortening.

A Better Idea: It's possible to make a tasty, leaner version of this Southern classic. Use evaporated skim milk instead of whole milk; replace whole eggs with egg whites or egg substitute. If you're watching sodium and fat in your diet, leave the crust on your plate.

PUDDINGS

BREAD with Raisins

Serving: ½ cup

Calories	248
Fat	8.1 g
Saturated	3.8 g
Monounsaturated	N/A
Polyunsaturated	N/A
Calories from fat	29%
Cholesterol	90 mg
Sodium	266 mg
Protein	7.4 g
Carbohydrate	37.6 g
Dietary fiber	N/A

CHIEF NUTRIENTS

NUTRIENT	AMOUNT	% RDA
Calcium	144.4 mg	18
Iron	1.5 mg	15
Riboflavin	0.3 mg	15
Potassium	284.9 mg	8
Thiamine	0.1 mg	6

Caution: If you're on a heart-healthy diet, beware: Bread pudding contains more fat, calories, and cholesterol than chocolate pudding. And nearly half the fat is saturated. *Allergy alert:* Bread pudding contains wheat, milk, and eggs, all common food allergens.

Strengths: This dessert yields appreciable amounts of bone-strengthening calcium, thanks to the milk, along with some riboflavin and anemia-fighting iron.

History: Frugal cooks concocted bread pudding as a way to use up stale bread.

Description: Made of cubed bread, milk, eggs, sugar, vanilla, spices, and raisins.

A Better Idea: If you prepare bread pudding with skim milk and an egg substitute, you can cut the fat and calorie content considerably.

CHOCOLATE, Homemade

Serving: ½ cup

Calories	192
Fat	6.1 g
Saturated	3.4 g
Monounsaturated	N/A
Polyunsaturated	N/A
Calories from fat	29%
Cholesterol	14 mg
Sodium	73 mg
Protein	4.0 g
Carbohydrate	33.4 g
Dietary fiber	N/A

CHIEF NUTRIENTS

NUTRIENT	AMOUNT	% RDA
Calcium	124.8 mg	16
Riboflavin	0.2 mg	11
Iron	0.7 mg	7
Potassium	222.3 mg	6

Caution: If you're trying to keep your cuisine lean, bear in mind that ½ cup of chocolate pudding has about as much fat as ice cream. *Allergy alert:* This dessert contains milk, corn, and chocolate, all known allergens.

Strengths: Customarily made with milk, chocolate pudding offers a good amount of bone-building calcium, with smaller amounts of riboflavin, iron, and potassium. Dessert lovers, take note: With just 29% of calories from fat, chocolate pudding is far less sinful than chocolate mousse, which derives 72% of its calories from fat.

Description: A popular comfort food for children and grown-ups alike, chocolate pudding is a milk-based dessert that needs to cool and thicken before it's ready for serving. So be patient.

A Better Idea: To reduce the fat and calories, prepare homemade pudding with skim milk instead of whole milk.

CORN

Serving: 1/2 cup

Calories	136
Fat	6.7 g
Saturated	3.2 g
Monounsaturated	2.2 g
Polyunsaturated	0.9 g
Calories from fat	44%
Cholesterol	125 mg
Sodium	69 mg
Protein	5.5 g
Carbohydrate	16.0 g
Dietary fiber	N/A

CHIEF NUTRIENTS

NUTRIENT	AMOUNT	% RDA
Thiamine	0.5 mg	34
Folate	31.6 mcg	16
Riboflavin	0.2 mg	9
Vitamin B6	0.2 mg	8
Iron	0.7 mg	7
Niacin	1.2 mg	7
Calcium	50.0 mg	6
Vitamin B12	0.1 mcg	6
Vitamin C	3.5 mg	6
Magnesium	18.8 mg	5
Potassium	201.3 mg	5

Caution: Although corn pudding has fewer calories than bread pudding or rice pudding, the percentage of calories from fat is still higher than the 30% considered acceptable by the American Heart Association. So if you're on a heart-healthy diet, keep portions modest. *Allergy alert:* Corn, milk, and eggs are common allergens.

Strengths: Thanks to corn, the thiamine content of this dessert really stands out. That's a benefit if you customarily take antacids, since they can impair thiamine absorption. Among the array of other useful vitamins and minerals in corn pudding is folate, a B vitamin that can help boost immunity.

Description: This pudding is usually made with yellow corn, whole milk, eggs, sugar, butter, salt, and pepper.

A Better Idea: You can reap the nutritional benefits of corn pudding with far less fat and cholesterol if you use skim milk, egg substitute, and diet margarine instead of whole milk, eggs, and butter.

CUSTARD, *Baked*

Serving: 1/2 cup

Calories	152
Fat	7.3 g
Saturated	3.4 g
Monounsaturated	N/A
Polyunsaturated	N/A
Calories from fat	43%
Cholesterol	139 mg
Sodium	105 mg
Protein	7.2 g
Carbohydrate	14.7 g
Dietary fiber	0

CHIEF NUTRIENTS

NUTRIENT	AMOUNT	% RDA
Calcium	148.4 mg	19
Riboflavin	0.6 mg	15
Iron	0.5 mg	5
Potassium	193.5 mg	5

Caution: Ending your dinner with baked custard adds moderate amounts of fat, sodium, cholesterol, and calories to the meal. So keep portions small—1/2 cup or so—if you're on a special diet. *Allergy alert:* Custard contains milk and eggs, both common food allergens.

Eat with: Fresh raspberries or blueberries for added fiber.

Strengths: Made with milk, this sweet, puddinglike dessert contains a respectable amount of bone-building calcium. You also get some riboflavin, a B vitamin that aids in the formation of red blood cells, plus a small amount of anemia-fighting iron.

Storage: Always store custard in a covered container in the refrigerator. Leftover custard is a fertile medium for bacterial growth, and when it spoils, it doesn't always look or smell bad.

RICE *with Raisins*

Serving: ½ cup

Calories	193
Fat	4.1 g
Saturated	2.2 g
Monounsaturated	N/A
Polyunsaturated	N/A
Calories from fat	19%
Cholesterol	15 mg
Sodium	94 mg
Protein	4.8 g
Carbohydrate	35.4 g
Dietary fiber	N/A

CHIEF NUTRIENTS

NUTRIENT	AMOUNT	% RDA
Calcium	129.9 mg	16
Riboflavin	0.2 mg	11
Potassium	234.5 mg	6
Iron	0.5 mg	5

Strengths: Topping off a meal with rice pudding contributes some useful calcium, a mineral valued for its role in building strong bones and keeping blood pressure levels under control. And while rice pudding has a fair amount of calories, the amounts of sodium and fat are quite reasonable. In fact, the percentage of calories from fat is well below the benchmark of 30% considered acceptable by the American Heart Association. Also, rice and rice-based desserts are excellent alternatives for those who are allergic to wheat or gluten.

Caution: *Allergy alert:* Rice pudding contains milk and eggs, both common food allergens.

Description: This is a hearty baked dessert made from milk, eggs, sugar, rice, and raisins, often seasoned with vanilla and cinnamon or nutmeg.

A Better Idea: Use brown rice—it has more fiber than white rice.

TAPIOCA

Serving: ½ cup

Calories	111
Fat	4.2 g
Saturated	2.0 g
Monounsaturated	N/A
Polyunsaturated	N/A
Calories from fat	34%
Cholesterol	80 mg
Sodium	129 mg
Protein	4.1 g
Carbohydrate	14.1 g
Dietary fiber	N/A

CHIEF NUTRIENTS

NUTRIENT	AMOUNT	% RDA
Calcium	86.6 mg	11
Riboflavin	0.2 mg	9

Strengths: Tapioca is a starch extracted from the root of the cassava plant. For those allergic to grains, this may be a welcome alternative to desserts thickened with wheat or corn. And if you're a dessert-lover on a diet, take note: Tapioca pudding has a fairly reasonable amount of fat—only a bit more than rice pudding, with even fewer calories.

Caution: Tapioca pudding contains a moderate amount of cholesterol and a noticeable amount of sodium, factors to keep in mind if you're watching your cholesterol levels or monitoring your blood pressure. *Allergy alert:* Tapioca pudding is customarily made with milk and eggs, two common food allergens.

VANILLA, *Homemade*

Serving: ½ cup

Calories	142
Fat	5.0 g
Saturated	2.7 g
Monounsaturated	N/A
Polyunsaturated	N/A
Calories from fat	32%
Cholesterol	18 mg
Sodium	83 mg
Protein	4.5 g
Carbohydrate	20.2 g
Dietary fiber	N/A

CHIEF NUTRIENTS

NUTRIENT	AMOUNT	% RDA
Calcium	149.2 mg	19
Riboflavin	0.2 mg	12

Strengths: This is one of the more reasonable desserts you can select as far as fat, calories, sodium, and cholesterol are concerned. Vanilla pudding also offers a fairly useful amount of calcium, a mineral that helps build stronger bones and also helps keep your blood pressure from soaring.

Caution: *Allergy alert:* Contains corn, milk, and eggs, ingredients that can trigger allergies in sensitive individuals.

Description: This plain-and-simple treat is made from sugar, cornstarch, milk, egg yolks, and margarine or butter, flavored with vanilla.

Preparation: To keep a "skin" from forming on the top of the pudding, cover the warm surface with waxed paper and remove it after the pudding has cooled.

A Better Idea: You can reduce the fat and calories if you prepare vanilla pudding with skim milk and an egg substitute instead of the whole milk and egg yolks called for in most recipes.

OTHER DESSERTS

APPLE BROWN BETTY

Serving: ½ cup

Calories	163
Fat	3.8 g
Saturated	1.6 g
Monounsaturated	N/A
Polyunsaturated	N/A
Calories from fat	21%
Cholesterol	9 mg
Sodium	165 mg
Protein	1.7 g
Carbohydrate	32.1 g
Dietary fiber	N/A

CHIEF NUTRIENTS

NUTRIENT	AMOUNT	% RDA
Iron	0.8 mg	8
Thiamine	0.1 mg	5

Strengths: In the world of sweet, high-calorie desserts, an apple brown betty is a relatively benign, almost healthy treat (although a bit high in saturated fat). It is low in cholesterol, low in fat, and reasonable in the percentage of calories from fat. Its iron aids in the delivery of oxygen to cell tissues, while thiamine gives a small boost to the body's ability to convert food into energy.

Caution: Has a moderate amount of sodium.

History: Betties, which date back to colonial times, are baked puddings made from alternating layers of fruit and buttered bread crumbs.

Description: Apple brown betty includes bread crumbs, melted butter, and sliced apples sprinkled with cinnamon, ground cloves, nutmeg, and lemon juice.

A Better Idea: Use whole wheat bread crumbs and light margarine. To keep your apple brown betty low in fat and calories, avoid serving with whipped cream.

CHARLOTTE RUSSE

Serving: 4 oz.

Calories	326
Fat	16.6 g
Saturated	8.3 g
Monounsaturated	N/A
Polyunsaturated	N/A
Calories from fat	46%
Cholesterol	225 mg
Sodium	49 mg
Protein	6.7 g
Carbohydrate	38.2 g
Dietary fiber	N/A

CHIEF NUTRIENTS

NUTRIENT	AMOUNT	% RDA
Vitamin A	168.7 RE	17
Iron	1.1 mg	11
Riboflavin	0.2 mg	10
Thiamine	0.1 mg	9
Calcium	52.4 mg	7

Caution: The cholesterol in one high-calorie, high–saturated fat serving of charlotte russe comes close to meeting the maximum daily amount recommended by the American Heart Association. *Allergy alert:* Dairy products used in the filling are common allergens.

Strengths: This dessert has a good amount of vitamin A, which is essential for proper vision and healthy skin. It also has iron, which is important in building stronger blood and helping the body to fight infection. Good amounts of riboflavin and some thiamine can help boost energy. Charlotte russe is low in sodium and has some calcium for strong bones.

Origin: A charlotte russe is said to have been first prepared for the Russian czar Alexander.

Description: A charlotte is any molded dessert filled with layers of gelatin-based custard or whipped cream. A baking mold is lined with ladyfingers, then filled with a Bavarian cream mixture.

CREAM PUFF *with Custard Filling*

Serving: 1 (about 4½ oz.)

Calories	303
Fat	18.1 g
Saturated	5.6 g
Monounsaturated	N/A
Polyunsaturated	N/A
Calories from fat	54%
Cholesterol	187 mg
Sodium	108 mg
Protein	8.5 g
Carbohydrate	26.7 g
Dietary fiber	N/A

CHIEF NUTRIENTS

NUTRIENT	AMOUNT	% RDA
Calcium	105.3 mg	13
Riboflavin	0.2 mg	13
Iron	0.9 mg	9
Vitamin A	91.0 RE	9

Caution: The cream puff is a nutritional lightweight that plays the heavy when it comes to fat. One custard cream puff also has more than half the recommended daily maximum of cholesterol. *Allergy alert:* Cream puff fillings often contain eggs and other dairy products, which are common allergens.

Strengths: The dairy-based filling in a cream puff has a good amount of calcium, which contributes to strong bones and teeth and can help ward off osteoporosis. It also has a good amount of riboflavin which helps convert food into energy.

Description: Cream puffs are small, hollowed-out pastry balls filled with custard or whipped cream. The specially prepared pastry, called choux, is made by mixing flour with boiling water and butter, then beating in eggs. The eggs make the pastry puff into irregular domes during baking.

ÉCLAIR

Serving: 1 (about 3½ oz.)

Calories	239
Fat	13.6 g
Saturated	4.4 g
Monounsaturated	N/A
Polyunsaturated	N/A
Calories from fat	51%
Cholesterol	136 mg
Sodium	82 mg
Protein	6.2 g
Carbohydrate	23.2 g
Dietary fiber	N/A

CHIEF NUTRIENTS

NUTRIENT	AMOUNT	% RDA
Calcium	80.0 mg	10
Riboflavin	0.2 mg	9
Iron	0.7 mg	7
Vitamin A	68.0 RE	7

Caution: These sweet pastries are crammed with fat. *Allergy alert:* Ingredients in the cream filling often include eggs or other dairy products, which are common allergens.

Strengths: The custard filling supplies a good amount of calcium for sturdy bones. Éclairs also have riboflavin, one of the B vitamins contributing to formation of red blood cells, and iron, a mineral that helps with production of antibodies. Its vitamin A is a precursor to beta-carotene, which may help protect against some cancers.

Description: Elongated cream puffs with icing on top. Like cream puffs, éclairs are made of choux pastry—a very sticky dough that's made with flour, boiling water, butter, and eggs. When baked, the dough puffs up; then the chef splits the pastry along its side and adds a custard filling and chocolate frosting.

Storage: Must be kept refrigerated. Egg fillings can spoil easily, leading to growth of harmful bacteria.

GELATIN DESSERT *from Powder*

Serving: ½ cup, made w/water

Calories	72
Fat	0
Saturated	0
Monounsaturated	0
Polyunsaturated	0
Calories from fat	0
Cholesterol	0
Sodium	55 mg
Protein	2.0 g
Carbohydrate	17.0 g
Dietary fiber	0

CHIEF NUTRIENTS

(None of the nutrients meet or exceed 5% of the RDA.)

Caution: Moderate in calories and sodium. *Allergy alert:* Some forms of gelatin contain salicylate, a substance that is present in aspirin and causes allergic reactions in some people.

Curiosity: Gelatin desserts such as Jell-O have been nicknamed ''nervous pudding'' or ''shimmy'' because they quiver when shaken.

History: The patent for a gelatin dessert was issued in 1845. At the turn of the century the product caught the nation's fancy when Orator Woodward packaged this dessert as a product he labeled Jell-O.

Description: Gelatin by itself is a thickening agent. The protein in gelatin comes primarily from scraps of white connective tissue such as animal skins and ossein (the protein matrix of bone).

Selection: Most packaged gelatin is artificially flavored. Sugar-free varieties are available.

A Better Idea: You can make your own flavored gelatin from fruit juice. Bring juice to a boil, add unflavored gelatin, then chill.

PRUNE WHIP

Serving: ¹/₂ cup, cold

Calories	101
Fat	0.1 g
Saturated	0
Monounsaturated	trace
Polyunsaturated	trace
Calories from fat	2%
Cholesterol	0
Sodium	107 mg
Protein	2.9 g
Carbohydrate	24.0 g
Dietary fiber	N/A

CHIEF NUTRIENTS

NUTRIENT	AMOUNT	% RDA
Iron	0.9 mg	8
Potassium	188.5 mg	5
Riboflavin	0.1 mg	5

Strengths: As desserts go, prune whip is among the lowest in fat and calories, making it a dieter's delight. It has no cholesterol either. As a bonus, you get some iron and a small amount of potassium.

Eat with: Prune whip would be quite appropriate served after a rich meal to help limit total fat and calories.

Caution: *Allergy alert:* Prune whip contains eggs, a common food allergen.

Description: This airy dessert is made from cooked prunes, egg whites, and sugar, often seasoned with lemon juice or rind.

Preparation: Use prunes that are slightly soft and flexible.

TOASTER PASTRY

Serving: 1 (about 1³/₄ oz.)

Calories	195
Fat	5.7 g
Saturated	N/A
Monounsaturated	N/A
Polyunsaturated	N/A
Calories from fat	26%
Cholesterol	N/A
Sodium	229 mg
Protein	1.9 g
Carbohydrate	35.2 g
Dietary fiber	0.1 g

CHIEF NUTRIENTS

NUTRIENT	AMOUNT	% RDA
Folate	40.0 mcg	20
Iron	2.0 mg	20
Calcium	96.0 mg	12
Niacin	2.1 mg	11
Thiamine	0.2 mg	11
Riboflavin	0.2 mg	10
Vitamin B₆	0.2 mg	10

Caution: High in sodium, toaster pastries should be avoided by people on low-salt diets. And the calories are also excessive, so you'll want to avoid these treats if you're trying to lose weight. *Allergy alert:* If you're allergic to glutens, be warned that many toaster pastries are made with wheat. This ingredient gives some people migraines and others an allergic reaction known as tension-fatigue syndrome. Those sensitive to soybeans should also avoid toaster pastries because they may contain soybean oil.

Strengths: These pastries have plenty of folate to build immunity and a rich amount of iron to give you healthier blood. They have lots of thiamine, riboflavin, and calcium, and a decent amount of vitamin B₆ to help fight infection.

A Better Idea: Plan breakfasts with high-fiber cereals and fresh fruit to start the day right.

Fast Food
BREAKFAST FOODS

BISCUIT with Egg and Bacon

Serving: 1 (about 5¼ oz.)

Calories	458
Fat	31.1 g
Saturated	9.9 g
Monounsaturated	13.3 g
Polyunsaturated	5.7 g
Calories from fat	61%
Cholesterol	353 mg
Sodium	999 mg
Protein	17.0 g
Carbohydrate	28.6 g
Dietary fiber	N/A

CHIEF NUTRIENTS

NUTRIENT	AMOUNT	% RDA
Vitamin B_{12}	1.0 mcg	52
Iron	3.7 mg	37
Calcium	189.0 mg	24
Folate	30.0 mcg	15
Riboflavin	0.2 mg	14
Niacin	2.4 mg	13

Caution: A bacon and egg biscuit is excessively high in cholesterol, surpassing the recommended daily limit of 300 mg. It also is very high in sodium and calories. More than half of the total calories come from fat; protein and carbohydrate counts are moderate. *Allergy alert:* Eggs and pork are common allergens.

Strengths: A fast food biscuit with egg and bacon contains more than half the recommended amount of vitamin B_{12}. The iron in this breakfast is needed for stronger blood, and it helps prevent anemia. A quarter of the daily requirement of calcium helps strengthen bones.

A Better Idea: More healthy alternatives are available, supplying equivalent nutrients without high cholesterol, sodium, and calories. A glass of skim milk is a low-fat source of calcium and vitamin B_{12}. For a leaner diet, switch to an English muffin. And skip the bacon if you're on a low-salt plan or want to reduce the sodium in your diet.

BISCUIT with Egg and Ham

Serving: 1 (about 7 oz.)

Calories	442
Fat	27.0 g
Saturated	8.4 g
Monounsaturated	11.3 g
Polyunsaturated	5.2 g
Calories from fat	55%
Cholesterol	300 mg
Sodium	1,382 mg
Protein	20.4 g
Carbohydrate	30.3 g
Dietary fiber	N/A

CHIEF NUTRIENTS

NUTRIENT	AMOUNT	% RDA
Vitamin B_{12}	1.2 mcg	60
Iron	4.6 mg	46
Thiamine	0.7 mg	45
Riboflavin	0.6 mg	35
Calcium	220.8 mg	28
Vitamin A	240.0 RE	24

Caution: A little more than half the calories in a biscuit with egg and ham come from fat. The cholesterol count in this loaded-up breakfast matches the American Heart Association's recommended daily maximum of 300 mg, and sodium is sky-high. Even though this breakfast combo is a source of protein and carbohydrates, consider the caloric cost. *Allergy alert:* Eggs and pork are common allergens.

Strengths: A good amount of vitamin B_{12} in the ham and egg biscuit help the body to form red blood cells, and thiamine boosts energy. Vitamin A is a key to good eyesight and promotes night vision.

A Better Idea: Always try to replace a biscuit with an English muffin, which is lower in fat. If you wish for a better option, use leaner meat or omit the ham altogether to help reduce sodium. Better yet, try a bowl of a nutrient-enriched, high-fiber cereal with some skim milk and a banana.

BISCUIT *with Egg and Sausage*

Serving: 1 (about 6½ oz.)

Calories	581
Fat	38.7 g
Saturated	15.0 g
Monounsaturated	16.4 g
Polyunsaturated	4.5 g
Calories from fat	60%
Cholesterol	302 mg
Sodium	1,141 mg
Protein	19.2 g
Carbohydrate	41.2 g
Dietary fiber	N/A

CHIEF NUTRIENTS

NUTRIENT	AMOUNT	% RDA
Vitamin B$_{12}$	1.4 mcg	69
Iron	4.0 mg	40
Thiamine	0.5 mg	33
Riboflavin	0.5 mg	26
Folate	39.6 mcg	20
Calcium	154.8 mg	19
Niacin	3.6 mg	19
Vitamin A	163.8 RE	16

Caution: In any breakfast combination with an egg and a cured pork product, calories, cholesterol, and sodium are all too high. A biscuit with egg and sausage takes the prize: Of all biscuit combinations, it is the highest in calories. And a full 60% of those calories come from fat. *Allergy alert:* Eggs and pork are common allergens.

Strengths: Good amounts of thiamine and riboflavin in the egg and sausage biscuit help the body convert food into energy. Has vitamin B$_{12}$ for healthy blood and nerves.

Description: The nutritional value of sausage depends on how it is made. The casing could be stuffed with anything from animal by-products and fats to pork, veal, or chicken. Ingredients will vary depending on brand. All usually have a high salt content.

A Better Idea: Substitute an English muffin for the biscuit, and skip the sausage. Better yet, try an omelet with leeks or broccoli; both vegetables protect against cancer.

BISCUIT *with Egg, Cheese, and Bacon*

Serving: 1 (about 5 oz.)

Calories	477
Fat	31.4 g
Saturated	11.4 g
Monounsaturated	14.2 g
Polyunsaturated	3.5 g
Calories from fat	59%
Cholesterol	261 mg
Sodium	1,260 mg
Protein	16.3 g
Carbohydrate	33.4 g
Dietary fiber	N/A

CHIEF NUTRIENTS

NUTRIENT	AMOUNT	% RDA
Vitamin B$_{12}$	1.1 mcg	53
Iron	2.6 mg	26
Riboflavin	0.4 mg	25
Calcium	164.2 mg	21
Thiamine	0.3 mg	20
Folate	37.4 mcg	19
Vitamin A	165.6 RE	17
Niacin	2.3 mg	12

Caution: Cheese raises the amount of fat in the biscuit combination: In this tasty but risky fast food breakfast handful, more than 50% of the calories come from fat. The cholesterol in one serving nearly fulfills the American Heart Association's recommended allotment for the entire day, and sodium is very high. *Allergy alert:* Eggs, dairy products, and pork are common allergens.

Strengths: Significant amounts of riboflavin and thiamine help the body create energy from food; vitamin B$_{12}$ helps maintain healthy nerves and blood. Calcium builds bones, and iron aids in carrying oxygen to cells.

Description: Depending on where you buy this breakfast, the cheese plopped onto it may be an imitation food product—or it might contain some natural dairy cheese.

A Better Idea: Skip the bacon, nix the cheese, order an English muffin instead of the biscuit, and get the eggs à la carte. Even pancakes with a little butter on them are lower in cholesterol and fat than eggs and bacon.

CROISSANT *with Egg and Cheese*

Serving: 1 (about 4½ oz.)

Calories	368
Fat	24.7 g
Saturated	14.1 g
Monounsaturated	7.5 g
Polyunsaturated	1.4 g
Calories from fat	60%
Cholesterol	216 mg
Sodium	551 mg
Protein	12.8 g
Carbohydrate	24.3 g
Dietary fiber	N/A

CHIEF NUTRIENTS

NUTRIENT	AMOUNT	% RDA
Vitamin B$_{12}$	0.8 mcg	39
Calcium	243.8 mg	30
Vitamin A	255.3 RE	26
Iron	2.2 mg	22
Riboflavin	0.4 mg	22
Folate	36.8 mcg	18
Thiamine	0.2 mg	13
Zinc	1.8 mg	12

Caution: Alarms should go off if you're on the watch for calories, sodium, and cholesterol. More than half the calories in an egg and cheese croissant come from fat. Although this food has moderate protein and carbohydrates, cholesterol and sodium levels are exceptionally high. *Allergy alert:* Eggs and dairy products are common allergens.

Strengths: Good amounts of riboflavin, vitamin B$_{12}$, and iron. Plus, the vitamin A in this breakfast can aid in healing wounds and combating infection.

History: The French word *croissant* means "crescent"; the pastry was named by Austrian bakers to celebrate the 1686 victory over Turkey, whose flag bore a crescent.

Description: A croissant is basically buttered layers of yeast dough.

A Better Idea: Croissants are higher in fat than any other breakfast bread, with twice the fat of biscuits and 6 times what you would get from English muffins. For a leaner way to start the day, move to muffins, or order dry whole wheat toast.

CROISSANT *with Egg, Cheese, and Bacon*

Serving: 1 (about 4½ oz.)

Calories	413
Fat	28.4 g
Saturated	15.4 g
Monounsaturated	9.2 g
Polyunsaturated	1.8 g
Calories from fat	62%
Cholesterol	215 mg
Sodium	889 mg
Protein	16.2 g
Carbohydrate	23.7 g
Dietary fiber	N/A

CHIEF NUTRIENTS

NUTRIENT	AMOUNT	% RDA
Vitamin B$_{12}$	0.9 mcg	43
Thiamine	0.4 mg	23
Iron	2.2 mg	22
Riboflavin	0.3 mg	20
Calcium	150.9 mg	19
Folate	34.8 mcg	17
Zinc	1.9 mg	13
Niacin	2.2 mg	12
Vitamin A	120.0 RE	12

Caution: Add bacon, and you add fat, calories, and sodium. You also reduce the amount of vitamin A and calcium found in the plain cheese and egg croissant. This croissant-wrapped breakfast is higher in saturated fat than the equivalent made with a biscuit. In fact, well over half the calories come from fat. The cholesterol count is also high according to American Heart Association standards. *Allergy alert:* Eggs, pork, and milk products are common allergens.

Strength: Although a poor means of getting protein, bacon slightly boosts the amount of this muscle-building nutrient.

A Better Idea: At least bounce the bacon, which adds little nutrition. Instead, munch on an English muffin or toast and have a cup of nonfat yogurt on the side.

CROISSANT with Egg, Cheese, and Ham

Serving: 1 (about 5½ oz.)

Calories	474
Fat	33.6 g
Saturated	17.5 g
Monounsaturated	11.4 g
Polyunsaturated	2.4 g
Calories from fat	64%
Cholesterol	213 mg
Sodium	1,081 mg
Protein	18.9 g
Carbohydrate	24.2 g
Dietary fiber	N/A

CHIEF NUTRIENTS

NUTRIENT	AMOUNT	% RDA
Vitamin B$_{12}$	1.0 mcg	50
Thiamine	0.5 mg	35
Iron	2.1 mg	21
Vitamin C	11.4 mg	19
Calcium	144.4 mg	18
Folate	36.5 mcg	18
Riboflavin	0.3 mg	18
Niacin	3.2 mg	17

Caution: Add ham to any egg and cheese combination and you'll get more fat and much more sodium. The percentage of calories from fat is high, and sodium levels are in the danger zone for anyone on a low-salt diet. The cholesterol count is borderline high according to standards recommended by the American Heart Association. *Allergy alert:* Eggs, pork, and dairy products are common allergens.

Strengths: The protein in ham is a little better in quality than what's found in bacon. And the vitamin B$_6$ is needed by the body to break down amino acids into energy.

A Better Idea: Trade the croissant for an English muffin. If you must have a pork product for breakfast, try to find a lean slice of ham at the breakfast bar—preferably with reduced sodium. Choose a bowl of cereal or oatmeal and garnish it with a favorite fruit to add some fiber to your meal.

CROISSANT with Egg, Cheese, and Sausage

Serving: 1 (about 5¾ oz.)

Calories	523
Fat	38.2 g
Saturated	18.2 g
Monounsaturated	14.3 g
Polyunsaturated	3.0 g
Calories from fat	66%
Cholesterol	216 mg
Sodium	1,115 mg
Protein	20.3 g
Carbohydrate	24.7 g
Dietary fiber	N/A

CHIEF NUTRIENTS

NUTRIENT	AMOUNT	% RDA
Thiamine	1.0 mg	66
Vitamin B$_{12}$	0.9 mcg	45
Iron	3.0 mg	30
Niacin	4.0 mg	21
Folate	38.4 mcg	19
Riboflavin	0.3 mg	19
Calcium	144.0 mg	18
Zinc	2.1 mg	14

Caution: Having a fast food breakfast that includes sausage instead of ham increases calories, cholesterol, and sodium. The increase in nutrients is negligible. All things considered, the combination of croissant, egg, cheese, and sausage is a quadruple overload—the most unhealthy of the croissant combinations. The percentage of calories from fat is a few points shy of 70% and the levels of sodium and cholesterol should set off alarm bells if you're trying to eat healthy. *Allergy alert:* Eggs, pork, and dairy products are common allergens.

Strengths: Provides more than half the daily requirement of thiamine, which helps convert carbohydrates into energy. Also has generous amounts of vitamin B$_{12}$ and iron for healthier blood.

Description: Sausage is the big unknown in this breakfast. Fat and animal by-products often are used in addition to meat.

A Better Idea: Forget sausage exists. Order some pancakes with low-fat cottage cheese on the side.

DANISH PASTRY, *Cheese*

Serving: 1 (about 3¼ oz.)

Calories	353
Fat	24.6 g
Saturated	5.1 g
Monounsaturated	15.6 g
Polyunsaturated	2.4 g
Calories from fat	63%
Cholesterol	20 mg
Sodium	319 mg
Protein	5.8 g
Carbohydrate	28.7 g
Dietary fiber	1.3 g

CHIEF NUTRIENTS

NUTRIENT	AMOUNT	% RDA
Iron	1.9 mg	19
Thiamine	0.3 mg	17
Niacin	2.6 mg	13
Riboflavin	0.2 mg	12
Vitamin B₁₂	0.2 mcg	12
Calcium	70.1 mg	9
Folate	14.6 mcg	7

Caution: Of all the Danish variations, the one with cheese contains the most calories *and* the highest percentage of calories from fat. It is high in sodium, and diabetics should beware: Even plain Danish has sugar in the dough. *Allergy alert:* Eggs and dairy products are common allergens.

Strengths: The cholesterol level is considered moderate, and the cheese adds a little protein and calcium. This version of the Danish is a little lower in sodium than the fruit or cinnamon type.

Description: A cheese Danish is made with a butter-rich dough. Although other kinds of Danish have different fillings, the caloric contents are all in the same ballpark.

A Better Idea: Two plain English muffins and ½ cup of low-fat cottage cheese would be a more filling, more nutritious alternative with about the same number of calories. Or try a couple of waffles without butter and just a trace of syrup.

DANISH PASTRY, *Cinnamon*

Serving: 1 (about 3 oz.)

Calories	349
Fat	16.7 g
Saturated	3.5 g
Monounsaturated	10.6 g
Polyunsaturated	1.7 g
Calories from fat	43%
Cholesterol	27 mg
Sodium	326 mg
Protein	4.8 g
Carbohydrate	46.9 g
Dietary fiber	1.3 g

CHIEF NUTRIENTS

NUTRIENT	AMOUNT	% RDA
Iron	1.8 mg	18
Thiamine	0.3 mg	17
Niacin	2.2 mg	12
Riboflavin	0.2 mg	11
Vitamin B₁₂	0.2 mcg	11
Folate	14.1 mcg	7

Caution: Over 40% of a cinnamon Danish's calories are from fat. There is a moderate amount of cholesterol, and the sodium content is considered high. A spice Danish is not a significant source of vitamins, although it contains economical amounts of iron and B vitamins. Butter comprises about 13% of the ingredients. Sugar makes up another 5% in the dough alone, so diabetics should beware. *Allergy alert:* Eggs and dairy products are common allergens.

Strength: Borderline high in carbohydrates, which is just fine if you need the energy to play several sets of tennis and burn off the Danish's calories.

Description: The butter-rich Danish dough is sprinkled heavily with cinnamon, then rolled and folded several times.

A Better Idea: For something more nutritious and lower in fat, try an English muffin either plain or with just a glazing of jelly or applesauce. A bowl of oatmeal or shredded wheat has a lot fewer calories and would be an all-around better way to start the day.

DANISH PASTRY, *Fruit*

Serving: 1 (about 3¼ oz.)

Calories	335
Fat	15.9 g
Saturated	3.3 g
Monounsaturated	10.1 g
Polyunsaturated	1.6 g
Calories from fat	43%
Cholesterol	19 mg
Sodium	333 mg
Protein	4.8 g
Carbohydrate	45.1 g
Dietary fiber	1.9 g

CHIEF NUTRIENTS

NUTRIENT	AMOUNT	% RDA
Thiamine	0.3 mg	19
Iron	1.4 mg	14
Riboflavin	0.2 mg	12
Vitamin B$_{12}$	0.2 mcg	12
Niacin	1.8 mg	9
Folate	15.0 mcg	8

Caution: A fruit Danish is high in sodium and borderline high in fat, and its cholesterol is not exactly ideal. Some 43% of calories come from fat. Butter is almost 13% of the recipe, and at least 5% is sugar. *Allergy alert:* Eggs and dairy products are common allergens.

Strengths: A fruit Danish is more modest in cholesterol and marginally lower in calories than a cheese or cinnamon Danish. The thiamine in a Danish will help the body generate energy from food.

A Better Idea: Although the fruit variety is the best choice among the Danishes, a better breakfast selection would be anything that offered more protein, more carbohydrates, and less fat—a couple of pancakes with a glass of skim milk, or oat cereal with banana slices and skim milk.

ENGLISH MUFFIN with *Egg, Cheese, and Canadian Bacon*

Serving: 1 (about 5¼ oz.)

Calories	383
Fat	19.8 g
Saturated	9.1 g
Monounsaturated	6.8 g
Polyunsaturated	2.1 g
Calories from fat	46%
Cholesterol	234 mg
Sodium	784 mg
Protein	19.8 g
Carbohydrate	31.5 g
Dietary fiber	1.6 g

CHIEF NUTRIENTS

NUTRIENT	AMOUNT	% RDA
Vitamin B$_{12}$	0.8 mcg	40
Iron	3.3 mg	33
Thiamine	0.5 mg	32
Riboflavin	0.5 mg	31
Calcium	207.3 mg	26
Folate	43.8 mcg	22
Niacin	3.9 mg	21
Vitamin A	157.7 RE	16
Zinc	1.8 mg	12
Magnesium	33.6 mg	10

Strengths: Canadian bacon, which comes from the eye of the loin on the middle of a pig's back, is much leaner than traditional bacon and more closely related to ham. English muffins are six times lower in fat than croissants and half as fatty as biscuits, so any comparable English muffin combo is lower in fat and calories than any other fast food biscuit or croissant combination. Good amounts of some B vitamins help the body convert food to fuel.

Caution: An ''undesirable'' for cholesterol and calories, an English muffin crammed with egg, cheese, and Canadian bacon definitely is high in sodium. Somewhat less than half of its calories are from fat. *Allergy alert:* Eggs, pork, and dairy products are common allergens.

Curiosity: The British have no idea why an English muffin is called English. But it does have a counterpart in Great Britain—the once popular tea muffin. It arrived in the U.S. via Samuel Bath Thomas, an Englishman who brought over his mother's recipe in 1880.

Description: The muffin itself is made from a soft yeast dough that can include flour, malted barley vinegar, and farina.

ENGLISH MUFFIN with Egg, Cheese, and Sausage

Serving: 1 (about 6 oz.)

Calories	487
Fat	30.9 g
Saturated	12.4 g
Monounsaturated	12.8 g
Polyunsaturated	3.3 g
Calories from fat	57%
Cholesterol	274 mg
Sodium	1,135 mg
Protein	21.7 g
Carbohydrate	31.0 g
Dietary fiber	1.6 g

CHIEF NUTRIENTS

NUTRIENT	AMOUNT	% RDA
Vitamin B_{12}	1.4 mcg	69
Thiamine	0.8 mg	56
Iron	3.5 mg	35
Riboflavin	0.5 mg	29
Folate	54.5 mcg	27
Calcium	196.4 mg	25
Niacin	4.5 mg	23
Vitamin A	171.6 RE	17
Zinc	2.4 mg	16

Caution: When you order sausage instead of Canadian bacon, you're immediately adding 11 g of fat and 100 calories to your morning muffin combination. The percentage of calories from fat climbs to almost 60%, and the cholesterol content also rises. Sodium is very high, so avoid this food if you're on a low-salt diet. *Allergy alert:* Eggs, pork, and dairy products are common allergens.

Strengths: With sausage instead of Canadian bacon, the amounts of thiamine and vitamin B_{12} increase, boosting the body's ability to process carbohydrates and helping to maintain the nervous system. Folate, niacin, riboflavin, and calcium come in high amounts, and there is a very good amount of iron.

A Better Idea: Get a plain English muffin and eat it along with some low-fat yogurt, cottage cheese, or an à la carte omelet. A little jelly on the muffin would be preferable to a pat of butter or margarine.

FRENCH TOAST with Butter

Serving: 2 slices (about 5 oz.)

Calories	356
Fat	18.8 g
Saturated	7.8 g
Monounsaturated	7.1 g
Polyunsaturated	2.5 g
Calories from fat	47%
Cholesterol	116 mg
Sodium	513 mg
Protein	10.3 g
Carbohydrate	36.1 g
Dietary fiber	4.2 g

CHIEF NUTRIENTS

NUTRIENT	AMOUNT	% RDA
Thiamine	0.6 mg	39
Riboflavin	0.5 mg	29
Niacin	3.9 mg	21
Iron	1.9 mg	19
Vitamin B_{12}	0.4 mcg	18
Folate	29.7 mcg	15
Vitamin A	145.8 RE	15

Caution: Almost half the calories in French toast come from fat. High in sodium, it contains a moderate level of cholesterol. But a few excess pats of butter will send the cholesterol level over the top. *Allergy alert:* Eggs and dairy products are common allergens.

Strengths: French toast is a source of carbohydrates for body fuel. Some iron and several B vitamins, especially thiamine and riboflavin, help the body more efficiently convert those carbohydrates into energy.

Curiosity: Unlike Canadian bacon and English muffin, French toast did originate in the country for which it is named.

Origin: In France, French toast is called *pain perdu*—"lost bread"—because it is a way of using stale bread.

A Better Idea: Make your own. Beat together an egg or egg substitute and a little skim milk. Dip bread into that mixture and brown both sides on an ungreased nonstick griddle. Use apple butter or pour on a little apple juice instead of syrup.

FRENCH TOAST STICKS

Serving: 5 (about 5 oz.)

Calories	478
Fat	29.1 g
Saturated	4.7 g
Monounsaturated	12.7 g
Polyunsaturated	9.9 g
Calories from fat	55%
Cholesterol	75 mg
Sodium	499 mg
Protein	8.3 g
Carbohydrate	49.1 g
Dietary fiber	2.3 g

CHIEF NUTRIENTS

NUTRIENT	AMOUNT	% RDA
Folate	134.0 mcg	67
Iron	3.0 mg	30
Niacin	3.0 mg	16
Riboflavin	0.3 mg	15
Thiamine	0.2 mg	15
Vitamin B₆	0.3 mg	13
Calcium	77.6 mg	10

Caution: More than half the calories in French toast sticks come from fat. They're high in sodium, with a moderate level of cholesterol.

Strengths: Supplies one-third the RDA for iron and two-thirds the RDA for folate—a B vitamin necessary for healthy blood and a strong immune system. (Folate is especially important to pregnant women and growing children.) French toast sticks are lower in cholesterol than regular French toast that is served with butter on top.

A Better Idea: As a rule, less-processed foods are better—and that's certainly true in this case. Regular French toast is lower in calories and fat than the stick version.

PANCAKES *with Butter and Syrup*

Serving: 3 (about 8¼ oz.)

Calories	520
Fat	14.0 g
Saturated	5.9 g
Monounsaturated	5.3 g
Polyunsaturated fat	2.0 g
Calories from fat	24%
Cholesterol	58 mg
Sodium	1,104 mg
Protein	8.3 g
Carbohydrate	90.9 g
Dietary fiber	N/A

CHIEF NUTRIENTS

NUTRIENT	AMOUNT	% RDA
Riboflavin	0.6 mg	33
Iron	2.6 mg	26
Thiamine	0.4 mg	26
Niacin	3.4 mg	18
Folate	34.8 mcg	17
Calcium	127.6 mg	16
Magnesium	48.7 mg	14
Vitamin E	1.4 mg αTE	14
Vitamin B₁₂	0.2 mcg	12

Strengths: Pancakes are composed almost entirely of carbohydrates. For the farmer in the field or the runner on a track, they load up the body for a strenuous workout. They're a good source of iron, and they also have some B vitamins to help convert food into energy.

Caution: Certainly not the breakfast of choice for weight watchers. Pancakes with butter and syrup are borderline high in calories and the sodium level is very high. *Allergy alert:* Eggs and dairy products are common allergens.

Curiosity: Every time they land on the griddle, pancakes seem to get a different name—Indian cakes, buckwheat cakes, hoe cakes, slapjacks, flapjacks, flatcars, sweatpads, griddle cakes, flannel cakes, and—particularly in New England and Rhode Island—johnnycakes.

History: American Indians made an early version of the pancake from cornmeal. Dutch settlers made their cakes from buckwheat flour.

A Better Idea: When eating out, cut calories by eliminating the butter and going light with the syrup. Better yet, make pancakes at home with more nutritious whole wheat flour.

CHICKEN, PIZZA, AND SALADS

FRIED CHICKEN *Drumsticks or Thighs*

Serving: 2 pieces (about 5¼ oz.)

Calories	431
Fat	26.7 g
Saturated	7.1 g
Monounsaturated	10.9 g
Polyunsaturated	6.3 g
Calories from fat	56%
Cholesterol	166 mg
Sodium	755 mg
Protein	30.1 g
Carbohydrate	15.7 g
Dietary fiber	N/A

CHIEF NUTRIENTS

NUTRIENT	AMOUNT	% RDA
Vitamin B$_{12}$	0.8 mcg	42
Niacin	7.2 mg	38
Riboflavin	0.4 mg	25
Zinc	3.2 mg	22
Vitamin B$_6$	0.3 mg	17
Iron	1.6 mg	16
Potassium	445.5 mg	12
Magnesium	37.0 mg	11

Caution: Call this all-American favorite *chicken à la fat:* Once poultry is breaded and deep-fried, it's brimming with fat and cholesterol. As a take-out food, it's not a bit better than burgers. In fact, fried chicken may be worse than a burger since the batter soaks up large amounts of fat and it's highly salted. Two fried drumsticks contain about as much fat as a double cheeseburger.

Eat with: Fruit or a salad to add vitamin C and some fiber.

Strengths: Fried chicken does have high-quality protein, and it also supplies handsome amounts of hard-to-get vitamins, along with a bonus of vitamin B$_{12}$. There's a commendable amount of anemia-fighting iron and good doses of magnesium and potassium. Dark-meat chicken is a good source of zinc, a mineral that boosts immunity and helps burns and scrapes heal fast.

A Better Idea: Peel off the skin and get rid of extra fat, calories, and sodium.

FRIED CHICKEN *Wings or Breasts*

Serving: 2 pieces (about 5¾ oz.)

Calories	494
Fat	29.5 g
Saturated	7.8 g
Monounsaturated	12.2 g
Polyunsaturated	6.8 g
Calories from fat	54%
Cholesterol	148 mg
Sodium	975 mg
Protein	35.7 g
Carbohydrate	19.6 g
Dietary fiber	N/A

CHIEF NUTRIENTS

NUTRIENT	AMOUNT	% RDA
Niacin	12.0 mg	63
Vitamin B$_{12}$	0.7 mcg	34
Vitamin B$_6$	0.6 mg	29
Riboflavin	0.3 mg	17
Iron	1.5 mg	15
Potassium	565.6 mg	15
Magnesium	37.5 mg	11
Thiamine	0.2 mg	10
Zinc	1.6 mg	10

Caution: Chicken may be naturally lower in fat than hamburger, but when batter-fried, it loses its nutritional edge. Breaded fried chicken is high in fat and calories, an important consideration for those with diabetes, heart disease, or a weight problem. And it contains a whopping amount of sodium, a real problem for anyone who is trying to stay on a salt-restricted diet. *Allergy alert:* Chicken breading usually contains wheat, a common food allergen.

Eat with: Fruit, or a green salad with low-fat, reduced-sodium dressing.

Strengths: Fried chicken is a great source of B vitamins (and has an excellent helping of niacin), essential for healthy skin and nerves. It also provides a useful amount of iron for rich red blood and tip-top immunity, along with worthwhile portions of zinc, potassium, and magnesium.

A Better Idea: For a chicken treat that's far lower in fat, bake at home and remove the skin.

FRIED CHICKEN PIECES

Serving: 6 pieces (about 3½ oz.)

Calories	290
Fat	17.7 g
Saturated	5.6 g
Monounsaturated	8.7 g
Polyunsaturated	2.3 g
Calories from fat	55%
Cholesterol	61 mg
Sodium	543 mg
Protein	16.9 g
Carbohydrate	15.5 g
Dietary fiber	N/A

CHIEF NUTRIENTS

NUTRIENT	AMOUNT	% RDA
Niacin	6.9 mg	36
Vitamin B$_6$	0.3 mg	16
Vitamin B$_{12}$	0.3 mcg	16
Iron	1.3 mg	13
Riboflavin	0.1 mg	8
Potassium	250.9 mg	7
Zinc	1.1 mg	7
Folate	11.2 mcg	6

Caution: Don't be fooled into thinking chicken is automatically leaner than a hamburger. The total fat in 6 breaded and fried pieces is 17 g, while a regular hamburger contains about 12 g. Breaded chicken contains more sodium than regular fried chicken, and ''crispy'' chicken pieces soak up even more oil than those with regular coating. *Allergy alert:* When fried chicken is breaded, it usually contains wheat and corn, two common food allergens.

Eat with: Nonfried, nutritious fare, like a salad, low-fat milk, and fruit.

Strengths: Fried chicken pieces are rich in niacin and reasonably high in other B vitamins. You also get a good amount of iron and small amounts of other minerals.

Description: Small chunks of fried chicken meat, usually made of combined, processed pieces of boneless chicken.

A Better Idea: Eat chicken pieces as an occasional treat, if at all. To save calories, dip in cocktail sauce instead of honey, hot mustard, barbecue, or sweet-and-sour sauce.

GARDEN SALAD

Serving: 1½ cups, w/out dressing

Calories	33
Fat	0.1 g
Saturated	trace
Monounsaturated	trace
Polyunsaturated	trace
Calories from fat	4%
Cholesterol	0
Sodium	54 mg
Protein	2.6 g
Carbohydrate	6.7 g
Dietary fiber	N/A

CHIEF NUTRIENTS

NUTRIENT	AMOUNT	% RDA
Vitamin C	48.0 mg	80
Folate	76.6 mcg	38
Vitamin A	236.0 RE	24
Iron	1.3 mg	13
Potassium	356.0 mg	9
Vitamin B$_6$	0.2 mg	9
Magnesium	22.8 mg	7
Niacin	1.1 mg	6

Strengths: A tossed garden salad is a excellent source of vitamin C, a nutrient that's absent in most other fast food items. Salad also supplies an appreciable amount of vitamin A for healthy skin and better vision. Along with folate, vegetable salad has smaller amounts of other vitamins and minerals that are in short supply on fast food menus. As a bonus, you get a little fiber. A vegetable salad with low-fat dressing is a great alternative to french fries—higher in nutrients, with less fat and fewer calories.

Description: Values given here are for a salad made with lettuce, cabbage, cucumber, green pepper, tomatoes, radishes, and carrots.

GARDEN SALAD *with Cheese and Egg*

Serving: 1½ cups, w/out dressing

Calories	102
Fat	5.8 g
Saturated	3.0 g
Monounsaturated	1.8 g
Polyunsaturated	0.5 g
Calories from fat	51%
Cholesterol	98 mg
Sodium	119 mg
Protein	8.8 g
Carbohydrate	4.8 g
Dietary fiber	N/A

CHIEF NUTRIENTS

NUTRIENT	AMOUNT	% RDA
Folate	84.6 mcg	42
Vitamin C	9.8 mg	16
Vitamin B$_{12}$	0.3 mcg	15
Calcium	99.8 mg	12
Vitamin A	115.0 RE	12
Potassium	371.1 mg	10
Riboflavin	0.2 mg	10
Iron	0.7 mg	7

Strengths: When you order this salad, you get respectable amounts of vitamins A and C and other nutrients that are in short supply on fast food menus. Sodium, cholesterol, and calories are not unreasonably high when compared to other fast food side dishes. If you decide to add a dressing, be sure it's a low-fat variety to keep fat and calorie count down to reasonable levels.

Caution: Contains a small amount of fat—about equal to ½ Tbsp. of butter.

Description: Values shown here are based on a salad made of lettuce, tomato, egg, cheese, celery, cucumber, and radishes.

A Better Idea: A tossed salad with chicken has half as much fat as salad with cheese and egg and twice as much protein.

GARDEN SALAD *with Chicken*

Serving: 1½ cups, w/out dressing

Calories	105
Fat	2.2 g
Saturated	0.6 g
Monounsaturated	0.7 g
Polyunsaturated	0.6 g
Calories from fat	19%
Cholesterol	72 mg
Sodium	209 mg
Protein	17.4 g
Carbohydrate	3.7 g
Dietary fiber	N/A

CHIEF NUTRIENTS

NUTRIENT	AMOUNT	% RDA
Folate	67.6 mcg	34
Niacin	5.9 mg	31
Vitamin C	17.4 mg	29
Vitamin B$_6$	0.4 mg	22
Potassium	446.9 mg	12
Iron	1.1 mg	11
Vitamin A	95.9 RE	10
Vitamin B$_{12}$	0.2 mcg	10

Strengths: Low in fat and calories, high in protein, a vegetable salad with chicken (and without dressing) may represent the best nutritional bargain of all fast food fare. Chicken is an ideal source of low-fat, high-quality protein. Very good amounts of folate and niacin, plus plentiful vitamin B$_6$, help the body process food into energy. Potassium keeps nerve and muscle reactions sharp, while significant amounts of vitamins C and A work to protect the immune system and possibly reduce the chance of certain cancers.

Caution: The sodium content is high, which can raise blood pressure levels in salt-sensitive individuals.

A Better Idea: When offered the salty, fatty croutons and high-calorie dressings, it's best to pass. A couple of tablespoons of salad dressing, for example, can add several hundred calories, quickly ruining one of fast food's best bargains.

GARDEN SALAD with Turkey, Ham, and Cheese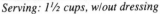

Serving: 1½ cups, w/out dressing

Calories	267
Fat	16.1 g
Saturated	8.2 g
Monounsaturated	5.2 g
Polyunsaturated	1.4 g
Calories from fat	54%
Cholesterol	140 mg
Sodium	743 mg
Protein	26.0 g
Carbohydrate	4.7 g
Dietary fiber	N/A

CHIEF NUTRIENTS

NUTRIENT	AMOUNT	% RDA
Folate	101.6 mcg	51
Vitamin B_{12}	0.9 mcg	43
Niacin	6.0 mg	31
Calcium	234.7 mg	29
Vitamin C	16.3 mg	27
Thiamine	0.4 mg	26
Riboflavin	0.4 mg	23
Vitamin B_6	0.4 mg	21
Zinc	3.1 mg	21
Iron	2.0 mg	20
Magnesium	48.9 mg	14

Caution: This chef's-type salad isn't quite as damaging to your waistline and blood pressure as pasta and seafood salad. Nevertheless, you should be aware that it does contain a considerable amount of fat, and half of that fat is saturated. The sodium count is significant, and both the meats—turkey and ham—are high in purines, so eat sparingly if you have gout. *Allergy alert:* Contains egg, a common food allergen. Tyramine in ham and aged cheeses may raise blood pressure in some sensitive individuals.

Strengths: This filling entrée offers an awesome amount of folate, which is a major boost to red blood cells. It's strong in B vitamins, especially the B_{12} that assists the nervous system. One helping of this salad is high in calcium and other important minerals and vitamins. If you use a dressing, be sure it's low-calorie to keep total fat intake of the meal reasonable.

Description: The salad consists of lettuce, tomato, turkey, ham, cheese, egg, celery, cucumber, radishes, and carrots.

PIZZA, *Cheese*

Serving: 1 slice (about 2¼ oz.)

Calories	140
Fat	3.2 g
Saturated	1.5 g
Monounsaturated	1.0 g
Polyunsaturated	0.5 g
Calories from fat	21%
Cholesterol	9 mg
Sodium	336 mg
Protein	7.7 g
Carbohydrate	20.5 g
Dietary fiber	N/A

CHIEF NUTRIENTS

NUTRIENT	AMOUNT	% RDA
Folate	58.6 mcg	29
Vitamin B_{12}	0.3 mcg	17
Calcium	116.6 mg	15
Niacin	2.5 mg	13
Thiamine	0.2 mg	12

Strengths: High in carbohydrates and protein for energy and muscle, low in fat and cholesterol. When prepared with the proper ingredients, pizza just may be, despite the apparent contradiction in terms, a healthy fast food. The crust contains B vitamins (in particular, folate) to bolster the immune system. Cheese offers calcium to encourage sturdy bones.

Caution: High in sodium; has a moderate number of calories. *Allergy alert:* Cheese and tomatoes are common allergens.

Curiosity: Though now an American diet staple, pizza originally could be found only in the Naples region of Italy and was considered a peasant food.

History: American soldiers returning home from battle in World War II popularized pizza as a meal.

A Better Idea: Fast food pizza seems to be as healthy as its homemade and frozen counterparts. Always look for pizza topped with mozzarella made partly from skim milk, and avoid high-fat toppings like pepperoni.

PIZZA, *Cheese, Meat, and Vegetable*   ⬇ 🗑

Serving: 1 slice (about 2¾ oz.)

Calories	184
Fat	5.4 g
Saturated	1.5 g
Monounsaturated	2.5 g
Polyunsaturated	0.9 g
Calories from fat	26%
Cholesterol	21 mg
Sodium	382 mg
Protein	13.0 g
Carbohydrate	21.3 g
Dietary fiber	N/A

CHIEF NUTRIENTS

NUTRIENT	AMOUNT	% RDA
Vitamin B$_{12}$	0.4 mcg	18
Iron	1.5 mg	15
Thiamine	0.2 mg	14
Calcium	101.1 mg	13
Folate	26.9 mcg	13
Niacin	0.2 mg	10
Riboflavin	0.2 mg	10
Magnesium	18.2 mg	5
Potassium	178.5 mg	5

Strengths: High in carbohydrates, moderate in protein, and low in fat. Moderate amounts of calcium, iron, and several B vitamins. The addition of meat and vegetables to cheese pizza provides a small amount of magnesium, which keeps calcium in tooth enamel and thus prevents tooth decay. Some extra potassium helps to remove sodium from the body and can help lower blood pressure.

Caution: High in sodium. Higher in percentage of calories from fat than a plain cheese pizza. *Allergy alert:* Cheese and tomatoes are common allergens.

A Better Idea: Extra toppings, carefully chosen, can make a pizza even more pleasing—to the palate and the waistline. Stay away from high-fat meat toppings to reduce unnecessary calories and sodium. Top the pie with fresh mushrooms, green peppers, and onions.

PIZZA, *Pepperoni*   ☀ ⬌ ⬇

Serving: 1 slice (about 2½ oz.)

Calories	181
Fat	7.0 g
Saturated	2.2 g
Monounsaturated	3.1 g
Polyunsaturated	1.2 g
Calories from fat	35%
Cholesterol	14 mg
Sodium	267 mg
Protein	10.1 g
Carbohydrate	19.9 g
Dietary fiber	N/A

CHIEF NUTRIENTS

NUTRIENT	AMOUNT	% RDA
Folate	52.5 mcg	26
Niacin	3.1 mg	16
Riboflavin	0.2 mg	14
Iron	0.9 mg	9
Thiamine	0.1 mg	9
Vitamin B$_{12}$	0.2 mcg	9
Calcium	64.6 mg	8

Strengths: High in protein and moderate in carbohydrates and fat (if you can eat just one slice). The traditional pizza with pepperoni fulfills a little more than a quarter of the body's daily need for folate, which is important for tissue growth and metabolizing amino acids. It also adds good amounts of niacin and riboflavin for healthy skin, as well as other B vitamins, calcium, and iron.

Caution: High in sodium and calories. Pepperoni pizza derives one-third of its calories from fat—more than a plain cheese pizza. *Allergy alert:* Cheese and tomatoes are common allergens.

A Better Idea: Pepperoni is a half-dollar–sized disk of fat and sodium; 80% of its calories are from fat, and it can boast of virtually no vitamins or minerals. Order the pizza plain or with fresh mushrooms. Or at least use a napkin or paper towel to sop up those little puddles of fat and grease that form in pepperoni when it's cooked.

SALAD, *Pasta and Seafood*

Serving: 1½ cups

Calories	379
Fat	20.9 g
Saturated	2.6 g
Monounsaturated	4.8 g
Polyunsaturated	9.1 g
Calories from fat	49%
Cholesterol	50 mg
Sodium	1,572 mg
Protein	16.4 g
Carbohydrate	32.0 g
Dietary fiber	N/A

CHIEF NUTRIENTS

NUTRIENT	AMOUNT	% RDA
Vitamin B$_{12}$	1.7 mcg	86
Vitamin A	638.0 RE	64
Vitamin C	38.4 mg	64
Folate	100.1 mcg	50
Iron	3.2 mg	32
Niacin	3.5 mg	19
Thiamine	0.3 mg	19
Vitamin B$_6$	0.3 mg	17
Potassium	600.5 mg	16
Magnesium	50.0 mg	14
Riboflavin	0.2 mg	12
Zinc	1.7 mg	11

Caution: Watch out—this pasta and seafood salad offers more fat than 3 doughnuts and more calories than a cheeseburger with mayo. Also, it's so high in sodium that you end up eating the equivalent of ⅔ tsp. of salt—a major dietary faux pas if you're watching your blood pressure.

Strengths: You get a fair amount of high-quality protein, along with excellent amounts of vitamins A, B$_{12}$, C, and folate; ample iron is accompanied by a smattering of other vitamins and minerals. Cholesterol is moderate.

Description: A typical seafood and pasta salad contains lettuce, macaroni, salad dressing, pollock, sweet peppers, carrots, celery, crab, turbot, olives, and onions.

A Better Idea: You might want to substitute salad with chicken; though it doesn't have as many nutrients, it contains only a fraction of the fat, calories, and sodium.

DESSERTS AND SHAKES

FRUIT PIE, *Fried*

Serving: 1 (about 3 oz.)

Calories	266
Fat	14.4 g
Saturated	6.5 g
Monounsaturated	5.8 g
Polyunsaturated	1.2 g
Calories from fat	49%
Cholesterol	13 mg
Sodium	325 mg
Protein	2.4 g
Carbohydrate	33.1 g
Dietary fiber	0

CHIEF NUTRIENTS

NUTRIENT	AMOUNT	% RDA
Iron	0.9 mg	9
Thiamine	0.1 mg	7

Caution: This fruit-flavored pastry is high in calories and moderately high in fat (almost half of it is saturated). It also contains a fair amount of sodium, with hardly any nutrients to speak of. *Allergy alert:* Contains wheat, corn, and other potential allergens.

Description: May contain a trace of apple, cherry, or lemon, plus typical pie ingredients—sugar, corn syrup, cornstarch, margarine, citric acid, flour, shortening, oil, and various stabilizers and flavorings.

A Better Idea: If your heart is set on having dessert, why not opt for the calories in a good home-style fruit pie? Better still, stop at a roadside produce stand and buy some apples, cherries, or other "fast fruit" to go. You'll get some heart-healthy soluble fiber with hardly any fat or sodium and relatively few calories.

SHAKE, *Chocolate*

Serving: 10 fl. oz.

Calories	359
Fat	10.5 g
Saturated	6.6 g
Monounsaturated	3.0 g
Polyunsaturated	0.4 g
Calories from fat	26%
Cholesterol	37 mg
Sodium	275 mg
Protein	9.6 g
Carbohydrate	58.0 g
Dietary fiber	0

CHIEF NUTRIENTS

NUTRIENT	AMOUNT	% RDA
Vitamin B$_{12}$	1.0 mcg	48
Riboflavin	0.7 mg	41
Calcium	319.8 mg	40
Potassium	566.0 mg	15
Magnesium	48.1 mg	14
Thiamine	0.2 mg	11
Iron	0.9 mg	9
Zinc	1.2 mg	8

Caution: A typical shake contains about as many calories as a 12-oz. can of regular soda, with about as much fat as a tablespoon of butter. A chocolate shake is also high in sodium and moderately high in cholesterol. Though some fast food chains now offer low-fat shakes made with skim milk, the calorie count is still high. *Allergy alert:* May contain milk, a common food allergen.

Strengths: A chocolate shake is rich in calcium and B vitamins. It also offers valuable amounts of potassium and magnesium. But given the steep calorie count, you'd be much better off just ordering a glass of low-fat milk.

Curiosity: When milkshakes first appeared in 1889, they're said to have contained whiskey. But by the 1900s, the whiskey was being omitted from most shakes.

Description: A chocolate shake is usually made of milk, ice cream, and flavored syrup.

SHAKE, *Strawberry*

Serving: 10 fl. oz.

Calories	320
Fat	7.9 g
Saturated	4.9 g
Monounsaturated	N/A
Polyunsaturated	N/A
Calories from fat	22%
Cholesterol	31 mg
Sodium	235 mg
Protein	9.6 g
Carbohydrate	53.5 g
Dietary fiber	0

CHIEF NUTRIENTS

NUTRIENT	AMOUNT	% RDA
Vitamin B$_{12}$	0.9 mcg	44
Calcium	319.8 mg	40
Riboflavin	0.6 mg	32
Potassium	515.1 mg	14
Magnesium	36.8 mg	11
Thiamine	0.1 mg	9
Vitamin A	82.1 RE	8
Zinc	1.0 mg	7

Caution: While a strawberry shake has a tad less fat than the chocolate or vanilla variety, it's still a dieter's nightmare, with 320 calories and quite a bit of sugar. Some fast food menus offer shakes made from skim milk, but the calorie count is still high. *Allergy alert:* May contain milk, a common food allergen.

Strengths: A strawberry shake is rich in bone-building calcium, riboflavin, and vitamin B$_{12}$ for healthy skin and nerves.

Curiosity: According to the *Guinness Book of World Records,* the largest strawberry milkshake ever made consisted of 117½ gallons of vanilla ice cream, 60 gallons of milk, and 17 gallons of strawberry flavoring. (No calorie count available.)

A Better Idea: A glass of low-fat milk would give you the same amount of bone-building calcium as a regular-size shake, with one-third the fat and calories.

SHAKE, *Vanilla*

Serving: 10 fl. oz.

Calories	314
Fat	8.5 g
Saturated	5.3 g
Monounsaturated	2.4 g
Polyunsaturated	0.3 g
Calories from fat	24%
Cholesterol	31 mg
Sodium	232 mg
Protein	10.0 g
Carbohydrate	50.7 g
Dietary fiber	0

CHIEF NUTRIENTS

NUTRIENT	AMOUNT	% RDA
Vitamin B_{12}	1.0 mcg	51
Calcium	345.3 mg	43
Riboflavin	0.5 mg	31
Potassium	492.4 mg	13
Magnesium	34.0 mg	10
Thiamine	0.1 mg	9
Vitamin A	90.6 RE	9
Vitamin B_6	0.2 mg	8

Caution: Most of the fat in a shake is saturated—and therefore unkind to your arteries. Shakes are high in calories, too—so the more you drink, the more you risk adding inches to your hips and waistline. Since shakes are also high in sodium, you might say a vanilla shake is a milk-based meanie. *Allergy alert:* If you're allergic to milk, shakes could be a problem.

Strengths: A vanilla shake is rich in bone-building calcium, vitamin B_{12}, and riboflavin, with smaller amounts of other valuable nutrients.

Curiosity: In some parts of the country, a shake is called a frappé (French for "chill," which refers to a frozen mixture of fruit and ice).

Description: A vanilla shake usually consists of milk, ice cream, and flavoring blended to a thick, creamy consistency.

A Better Idea: Order a glass of low-fat milk instead. If you must have a shake, make one with skim milk and ice milk.

SUNDAE, *Caramel*

Serving: 1 (about 5½ oz.)

Calories	304
Fat	9.3 g
Saturated	4.5 g
Monounsaturated	3.0 g
Polyunsaturated	1.0 g
Calories from fat	27%
Cholesterol	25 mg
Sodium	195 mg
Protein	7.3 g
Carbohydrate	49.3 g
Dietary fiber	0

CHIEF NUTRIENTS

NUTRIENT	AMOUNT	% RDA
Vitamin B_{12}	0.6 mcg	30
Calcium	189.1 mg	24
Riboflavin	0.3 mg	17
Vitamin E	0.9 mg αTE	9
Magnesium	27.9 mg	8
Potassium	317.8 mg	8
Folate	12.4 mcg	6
Zinc	0.8 mg	5

Caution: Surprisingly, a caramel sundae has more fat and calories than the notoriously sinful hot fudge sundae. It's also a bit high in sodium, which is a drawback if you're on a sodium-restricted diet. Like milk, ice cream contains lactose, a type of carbohydrate that's poorly digested by many adults. *Allergy alert:* Ice cream sundaes could be a problem for those allergic to milk.

Eat with: Restraint. Ice cream cannot be considered a mainstay of a healthful diet.

Strengths: A caramel sundae is rich in bone-building calcium and vitamin B_{12} (for healthy skin and nerves). You also get smaller amounts of zinc and folate, nutrients that may enhance the immune response to certain diseases. (However, the high calorie count and lack of fiber make a sundae a less-than-ideal source of vitamins and minerals.)

A Better Idea: If it's available, order a container of low-fat yogurt instead of a sundae. You'll get much more bone-building calcium, with fewer calories and less fat.

SUNDAE, *Hot Fudge*

Serving: 1 (about 5½ oz.)

Calories	284
Fat	8.6 g
Saturated	5.0 g
Monounsaturated	2.3 g
Polyunsaturated	0.8 g
Calories from fat	27%
Cholesterol	21 mg
Sodium	182 mg
Protein	5.6 g
Carbohydrate	47.7 g
Dietary fiber	0

CHIEF NUTRIENTS

NUTRIENT	AMOUNT	% RDA
Vitamin B$_{12}$	0.7 mcg	33
Calcium	207.0 mg	26
Riboflavin	0.3 mg	18
Potassium	395.0 mg	11
Magnesium	33.2 mg	9
Vitamin B$_6$	0.1 mg	7
Iron	0.6 mg	6
Zinc	1.0 mg	6

Caution: If you're curbing your intake of fat, sodium, and calories, save this dessert for special occasions only. Double caution: The worst problem with a hot fudge sundae is that many people can't be contented with just an occasional one. And a hot fudge *habit* can be extraordinarily hard to break. Like other milk products, an ice cream sundae contains lactose, a carbohydrate that many adults digest poorly.

Strengths: This sweet indulgence offers significant amounts of bone-building calcium and B vitamins that keep skin and nerves healthy. You also get a little iron and zinc, nutrients that help ward off infections.

A Better Idea: Order a low-calorie frozen yogurt and sprinkle with fresh berries instead of hot fudge sauce, nuts, and whipped cream.

SUNDAE, *Strawberry*

Serving: 1 (about 5½ oz.)

Calories	268
Fat	7.9 g
Saturated	3.8 g
Monounsaturated	2.7 g
Polyunsaturated	1.0 g
Calories from fat	26%
Cholesterol	21 mg
Sodium	92 mg
Protein	6.3 g
Carbohydrate	44.7 g
Dietary fiber	0

CHIEF NUTRIENTS

NUTRIENT	AMOUNT	% RDA
Vitamin B$_{12}$	0.6 mcg	32
Calcium	160.7 mg	20
Riboflavin	0.3 mg	16
Folate	18.4 mcg	9
Vitamin E	0.8 mg αTE	8
Magnesium	24.5 mg	7
Potassium	270.8 mg	7
Vitamin A	58.1 RE	6

Caution: A strawberry sundae isn't quite as damaging to your waistline as a hot fudge sundae—but it's close. Since ice cream contains lactose, avoid it if you are lactose-intolerant. *Allergy alert:* Milk is a common food allergen.

Strengths: There's a generous serving of bone-building calcium in the ice cream, as well as small amounts of the resistance-boosting nutrients vitamins A and E and folate.

History: In the late 1800s, when the ''sundae'' was first created, church leaders were opposed to people drinking carbonated beverages on Sunday. Ice cream was allowed, however; thus ice cream with toppings got the name sundae.

A Better Idea: To avoid adverse effects on your heart and arteries, choose low-fat frozen yogurt or a small dish of sherbet instead. Beneficial types of bacteria in yogurt may also help digest the lactose in that product.

MEXICAN FOODS

BURRITO, Bean

Serving: 2 (about 7¾ oz.)

Calories	447
Fat	13.5 g
Saturated	6.9 g
Monounsaturated	4.7 g
Polyunsaturated	1.2 g
Calories from fat	27%
Cholesterol	4 mg
Sodium	985 mg
Protein	14.1 g
Carbohydrate	71.4 g
Dietary fiber	N/A

CHIEF NUTRIENTS

NUTRIENT	AMOUNT	% RDA
Folate	117.2 mcg	59
Vitamin B_{12}	1.1 mcg	55
Iron	4.5 mg	45
Thiamine	0.6 mg	42
Riboflavin	0.6 mg	36
Magnesium	86.8 mg	25
Niacin	4.1 mg	21
Potassium	653.2 mg	17

Strengths: Bean burritos are high in complex carbohydrates and very low in cholesterol. They contain impressive amounts of iron and several B vitamins to convert food into energy and maintain healthy red blood cells. One serving supplies one-quarter of the RDA of magnesium. Although the USDA does not give a measurement for fiber, most beans do contain fiber.
Caution: Fast food bean burritos are high in sodium, and more than one-quarter of their calories come from fat.
Curiosity: The name translates to ''little donkey.''
Description: The filling consists of mashed beans, often with ground beef, chicken, vegetables, or cheese, rolled inside a crepelike soft tortilla made of either corn or wheat flour.
A Better Idea: Make burritos at home. Refried beans, even from a can, are extremely low in fat and high in some B vitamins. (But they can contain lard, so check the label.)

BURRITO, Bean, Cheese, and Beef

Serving: 2 (about 7¼ oz.)

Calories	331
Fat	13.3 g
Saturated	7.2 g
Monounsaturated	4.5 g
Polyunsaturated	1.0 g
Calories from fat	36%
Cholesterol	125 mg
Sodium	990 mg
Protein	14.6 g
Carbohydrate	39.7 g
Dietary fiber	N/A

CHIEF NUTRIENTS

NUTRIENT	AMOUNT	% RDA
Vitamin B_{12}	1.1 mcg	55
Riboflavin	0.7 mg	42
Iron	3.7 mg	37
Folate	61.0 mcg	31
Niacin	3.9 mg	20
Thiamine	0.3 mg	20
Calcium	131.0 mg	16
Zinc	2.4 mg	16

Strengths: This variation on the burrito is high in complex carbohydrates, with moderate percentages of calories from protein and fat. It's strong in B vitamins, which help convert food into energy and assist in maintaining red blood cells. It meets more than one-third of the daily requirement of iron, which facilitates the transport of oxygen to the cells and helps the body fight infection.
Caution: The cholesterol count is moderate, and the level of sodium is high. *Allergy alert:* Wheat and milk products are common allergens.
A Better Idea: Have a plain bean burrito with some fresh salad on the side. As soon as you add cheese and beef to a burrito order, you're asking for about 30 times more cholesterol.

BURRITO, *Beef*

Serving: 2 (about 8 oz.)

Calories	524
Fat	20.8 g
Saturated	10.5 g
Monounsaturated	7.4 g
Polyunsaturated	0.9 g
Calories from fat	36%
Cholesterol	64 mg
Sodium	1,492 mg
Protein	26.6 g
Carbohydrate	58.5 g
Dietary fiber	N/A

CHIEF NUTRIENTS

NUTRIENT	AMOUNT	% RDA
Vitamin B_{12}	2.0 mcg	98
Iron	6.1 mg	61
Riboflavin	0.9 mg	54
Niacin	6.5 mg	34
Zinc	4.7 mg	32
Magnesium	81.4 mg	23
Folate	39.6 mcg	20
Potassium	739.2 mg	20

Caution: Extremely high in sodium, a beef burrito gets more than 30% of its calories from fat.

Strengths: The level of cholesterol is rated moderate by the American Heart Association, and a beef burrito is a good source of protein. It offers virtually an entire day's supply of vitamin B_{12}, which is needed for a smoothly functioning nervous system. Copious iron helps the body fight infection, and zinc contributes to brain function—particularly behavior and learning performance.

A Better Idea: Because it is much lower in cholesterol and fat, a burrito filled with beans would be a healthier alternative. For a better source of vitamin B_{12} and zinc, eat a handful of peanuts.

CHIMICHANGA, *Beef*

Serving: 1 (about 6¼ oz.)

Calories	425
Fat	19.7 g
Saturated	8.5 g
Monounsaturated	8.1 g
Polyunsaturated	1.1 g
Calories from fat	42%
Cholesterol	9 mg
Sodium	910 mg
Protein	19.6 g
Carbohydrate	42.8 g
Dietary fiber	N/A

CHIEF NUTRIENTS

NUTRIENT	AMOUNT	% RDA
Vitamin B_{12}	1.5 mcg	76
Iron	4.5 mg	45
Riboflavin	0.6 mg	38
Thiamine	0.5 mg	33
Zinc	5.0 mg	33
Niacin	5.8 mg	30
Magnesium	62.6 mg	18
Folate	31.3 mcg	16

Caution: A high percentage of the calories in a beef chimichanga come from saturated fat, so it's not recommended for anyone prone to high blood pressure. This food is also very high in sodium.

Strengths: Moderate in carbohydrates and protein, a beef chimichanga also is very low in cholesterol. It contains three-quarters of the recommended daily amount of vitamin B_{12} to protect the nervous system, and supplies almost half of the body's iron needs.

Curiosity: The name is a Mexican version of thingamajig or whatchamacallit.

Description: Chimichangas are burritos that have been fried or deep-fried.

A Better Idea: Deep-six the deep-fried food and choose oven-cooked burritos when you can. A plain bean burrito has little more than half the calories.

CHIMICHANGA, *Beef and Cheese*

Serving: 1 (about 6½ oz.)

Calories	443
Fat	23.4 g
Saturated	11.2 g
Monounsaturated	9.4 g
Polyunsaturated	0.7 g
Calories from fat	48%
Cholesterol	51 mg
Sodium	957 mg
Protein	20.1 g
Carbohydrate	39.3 g
Dietary fiber	N/A

CHIEF NUTRIENTS

NUTRIENT	AMOUNT	% RDA
Vitamin B$_{12}$	1.3 mcg	65
Riboflavin	0.9 mg	51
Iron	3.8 mg	38
Calcium	237.9 mg	30
Niacin	4.7 mg	25
Thiamine	0.4 mg	25
Zinc	3.4 mg	22
Magnesium	60.4 mg	17

Caution: Almost half of the calories in a beef and cheese chimichanga come from fat. It's very high in sodium, and the cholesterol count is considered moderate. *Allergy alert:* Wheat and dairy products such as cheese are common allergens.

Strengths: Because it contains meat, a beef chimichanga offers good amounts of several B vitamins to ensure proper blood formation and assist in the conversion of food to energy. Beef is also rich in heme iron—a form of iron that's more easily absorbed than the kind found in plants. Cheese adds calcium to strengthen bones.

A Better Idea: Steer away from this deep-fried version of a burrito. Not only is it higher in calories than either a beef or bean burrito, but more of those calories come from fat. Choose a plain bean burrito for nearly half the calories and fat.

ENCHILADA, *Cheese*

Serving: 1 (about 6 oz.)

Calories	319
Fat	18.8 g
Saturated	10.6 g
Monounsaturated	6.3 g
Polyunsaturated	0.8 g
Calories from fat	53%
Cholesterol	44 mg
Sodium	784 mg
Protein	9.6 g
Carbohydrate	28.5 g
Dietary fiber	N/A

CHIEF NUTRIENTS

NUTRIENT	AMOUNT	% RDA
Calcium	324.4 mg	41
Vitamin B$_{12}$	0.8 mcg	38
Riboflavin	0.4 mg	25
Vitamin B$_6$	0.4 mg	20
Vitamin A	185.8 RE	19
Folate	34.2 mcg	17
Zinc	2.5 mg	17
Magnesium	50.5 mg	14

Caution: Fat, most of it saturated, provides more than half of a cheese enchilada's calories. This high-sodium dish also contains a moderate amount of cholesterol. *Allergy alert:* Cheese, wheat, and corn are common allergens.

Strengths: An enchilada with cheese satisfies more than one-third of the daily requirement of vitamin B$_{12}$, which helps protect the nervous system. This food also contains almost half of the recommended daily amount of calcium, which is required to absorb B$_{12}$.

Description: An enchilada (the name means "filled with chili") is a rolled corn or wheat tortilla stuffed with any of various mixtures. It's usually baked, then topped with sour cream or a tomato sauce and shredded cheese.

A Better Idea: Choose corn flour tortillas if you can, since wheat flour tortillas are usually made with a small amount of lard. Or try a chicken- or seafood-filled version, which will provide lower-fat protein. Skip sour cream.

ENCHILADA, *Cheese and Beef*

Serving: 1 (about 6¾ oz.)

Calories	323
Fat	17.6 g
Saturated	9.1 g
Monounsaturated	6.2 g
Polyunsaturated	1.4 g
Calories from fat	49%
Cholesterol	40 mg
Sodium	1,319 mg
Protein	11.9 g
Carbohydrate	30.5 g
Dietary fiber	N/A

CHIEF NUTRIENTS

NUTRIENT	AMOUNT	% RDA
Folate	192.0 mcg	96
Vitamin B$_{12}$	1.0 mcg	51
Iron	3.1 mg	31
Calcium	228.5 mg	29
Magnesium	82.6 mg	24
Riboflavin	0.4 mg	24
Zinc	2.7 mg	18
Potassium	574.1 mg	15

Caution: Nearly half the calories come from fat—mainly the saturated kind, which is known to raise serum cholesterol levels. Also, the sodium content is extremely high.

Strengths: One serving provides almost an entire day's requirement of folate, a B vitamin essential for regeneration of red blood cells and a strong immune system. An excellent amount of vitamin B$_{12}$ will maintain nerves and help to release folate that the body stores. And the magnesium may help keep blood pressure levels healthy.

Description: Every enchilada is a rolled tortilla with some kind of stuffing and topping. Beef and cheese enchiladas are usually baked. Toppings might include sour cream or a tomato sauce and shredded cheese.

A Better Idea: Replace the beef with chicken or fish for lower-fat protein.

FRIJOLES, *Cheese*

Serving: 1 cup

Calories	225
Fat	7.8 g
Saturated	4.1 g
Monounsaturated	2.6 g
Polyunsaturated	0.7 g
Calories from fat	31%
Cholesterol	37 mg
Sodium	882 mg
Protein	11.4 g
Carbohydrate	28.7 g
Dietary fiber	N/A

CHIEF NUTRIENTS

NUTRIENT	AMOUNT	% RDA
Folate	111.9 mcg	56
Vitamin B$_{12}$	0.7 mcg	34
Calcium	188.7 mg	24
Magnesium	85.2 mg	24
Iron	2.2 mg	22
Riboflavin	0.3 mg	19
Potassium	604.5 mg	16
Zinc	1.7 mg	12

Strengths: Frijoles with cheese are a good source of complex carbohydrates and protein. This Mexican food is moderate in fat, although some of it may come from lard. In an 8-oz. serving, you'll get more than half the RDA of folate, which the body needs to boost its immune system and to break down the amino acids in protein.

Caution: High in sodium. Beans don't have cholesterol, but the cheese melted inside this Mexican food can be moderate to high in it. The net result is a fast food that ranks high in the cholesterol category.

Description: *Frijole* is the Mexican word for "bean." The dish consists of pinto beans that have been mashed, then fried, usually in lard.

A Better Idea: To cut calories as much as possible, don't add any sour cream or guacamole. If you're buying canned or frozen frijoles, look for ones made without lard.

NACHOS, *Cheese*

Serving: 6–8 (about 4 oz.)

Calories	346
Fat	19.0 g
Saturated	7.8 g
Monounsaturated	8.0 g
Polyunsaturated	2.2 g
Calories from fat	49%
Cholesterol	18 mg
Sodium	816 mg
Protein	9.1 g
Carbohydrate	36.3 g
Dietary fiber	N/A

CHIEF NUTRIENTS

NUTRIENT	AMOUNT	% RDA
Vitamin B_{12}	0.8 mcg	41
Calcium	272.3 mg	34
Riboflavin	0.4 mg	22
Magnesium	55.4 mg	16
Iron	1.3 mg	13
Thiamine	0.2 mg	13
Zinc	1.8 mg	12
Vitamin B_6	0.2 mg	10

Caution: A handful of nachos dipped in cheese is very high in sodium. Half the calories come from fat. A serving of nachos with cheese has about the same fat and calorie content as two fudge brownies—and more than a slice of banana cream pie. *Allergy alert:* Dairy products and corn are common allergens. **Strengths:** Low in cholesterol. Nachos with cheese provide more than a third of the RDA of calcium—aiding the body in absorbing vitamin B_{12}, which is necessary for healthy blood and nerves. **Description:** Nachos are fried tortilla chips with cheese. **A Better Idea:** Nix the nachos. If you're hungry for a Mexican appetizer, order a plain tortilla and spoon on a little salsa. Or get a salad and use the salsa as a dressing

NACHOS, *Supreme*

Serving: 6–8 (about 9 oz.)

Calories	569
Fat	30.7 g
Saturated	12.5 g
Monounsaturated	11.0 g
Polyunsaturated	5.7 g
Calories from fat	49%
Cholesterol	20 mg
Sodium	1,800 mg
Protein	19.8 g
Carbohydrate	55.8 g
Dietary fiber	N/A

CHIEF NUTRIENTS

NUTRIENT	AMOUNT	% RDA
Vitamin B_{12}	1.0 mcg	51
Calcium	385.1 mg	48
Vitamin A	469.2 RE	47
Riboflavin	0.7 mg	41
Iron	2.8 mg	28
Magnesium	96.9 mg	28
Zinc	3.7 mg	24
Vitamin B_6	0.4 mg	21

Caution: This appetizer has some very unappealing qualities. The amount of sodium is sky-high; it's like swallowing almost a teaspoon of salt. About half the calories come from fat. A hot fudge sundae has less fat and fewer calories. **Strengths:** Nutrient-rich. An excellent source of vitamin B_{12} for healthy nerves. One serving provides about half the RDA of calcium and vitamin A for bone growth and strength and good vision. This dish also is a rich source of riboflavin, which helps in the production of red blood cells in bone marrow. **Description:** Nachos topped with cheese, beans, ground beef, and green pepper. **A Better Idea:** Bag the nachos and order a low-fat, nonfried tortilla to eat along with a bowl of frijoles.

TACO

Serving: 1 small (about 2½ oz.)

Calories	369
Fat	20.6 g
Saturated	11.4 g
Monounsaturated	6.6 g
Polyunsaturated	1.0 g
Calories from fat	50%
Cholesterol	56 mg
Sodium	802 mg
Protein	20.7 g
Carbohydrate	26.7 g
Dietary fiber	N/A

CHIEF NUTRIENTS

NUTRIENT	AMOUNT	% RDA
Vitamin B$_{12}$	1.0 mcg	52
Calcium	220.6 mg	28
Riboflavin	0.4 mg	26
Zinc	3.9 mg	26
Iron	2.4 mg	24
Magnesium	70.1 mg	20
Niacin	3.2 mg	17
Vitamin A	147.1 RE	15

Caution: Half the calories come from fat, most of which is the saturated kind known to raise cholesterol levels. *Allergy alert:* Cheese, wheat, corn, beans, and tomatoes are common allergens.

Strengths: Several B vitamins in a taco help the body produce red blood cells and assist in the process of changing food into energy. Zinc aids in metabolizing proteins and amino acids, while calcium strengthens the bones.

Curiosity: The word *taco* translates literally into "wad" or "plug," but in the broader definition it means a light snack.

Description: A crisp-fried U-shaped tortilla filled with ground meat or beans and topped with lettuce, tomato, cheese, and onion, sometimes with a Mexican sauce.

A Better Idea: Order a taco filled with beans instead of meat to boost your meal's fiber content. Ask about a soft taco shell that is not fried. Steer clear of fat-filled sour cream or guacamole toppings.

TACO SALAD

Serving: 1½ cups

Calories	279
Fat	14.8 g
Saturated	6.8 g
Monounsaturated	5.2 g
Polyunsaturated	1.8 g
Calories from fat	48%
Cholesterol	44 mg
Sodium	762 mg
Protein	13.2 g
Carbohydrate	23.6 g
Dietary fiber	N/A

CHIEF NUTRIENTS

NUTRIENT	AMOUNT	% RDA
Vitamin B$_{12}$	0.6 mcg	32
Calcium	192.1 mg	24
Iron	2.3 mg	23
Riboflavin	0.4 mg	21
Folate	39.6 mcg	20
Zinc	2.7 mg	18
Magnesium	51.5 mg	15
Niacin	2.5 mg	13

Caution: Even though lettuce may make up 40% of the ingredients, almost half a taco salad's calories come from fat. It's high in sodium and has a moderate amount of cholesterol. *Alergy alert:* Cheese and other dairy products are common allergens.

Strengths: A taco salad is rich in vitamin B$_{12}$ and provides moderate amounts of other B vitamins to aid the body in deriving energy from food. It also contains about a quarter of the daily requirement of iron, which helps the body fight infection.

Description: Generally, the salad has the same ingredients as the taco, though it's delivered on a flat or bowl-shaped shell.

A Better Idea: Lower the fat content by cutting back on cheese and reducing the dollops of high-calorie guacamole, chili sauce, or dressing. To eliminate more fat and add fiber, order beans instead of ground beef as the main filling.

TOSTADA, *Bean and Cheese*

Serving: 1 (about 5 oz.)

Calories	223
Fat	9.9 g
Saturated	5.4 g
Monounsaturated	3.1 g
Polyunsaturated	0.8 g
Calories from fat	40%
Cholesterol	30 mg
Sodium	543 mg
Protein	9.6 g
Carbohydrate	26.5 g
Dietary fiber	N/A

CHIEF NUTRIENTS

NUTRIENT	AMOUNT	% RDA
Folate	74.9 mcg	37
Vitamin B_{12}	0.7 mcg	35
Calcium	210.2 mg	26
Iron	1.9 mg	19
Riboflavin	0.3 mg	19
Magnesium	59.0 mg	17
Zinc	1.9 mg	13
Potassium	403.2 mg	11

Caution: A full 40% of the calories in a bean and cheese tostada come from fat. Also, this dish is not for those on a low-salt diet. *Allergy alert:* Cheese, wheat, corn, beans, and tomatoes are common allergens.

Strengths: This combination is lower in fat, sodium, and cholesterol than any beef and cheese tostada. Beans offer a good source of complex carbohydrates and fiber, which can relieve constipation and may protect against colon cancer. A bean and cheese tostada supplies over one-third the RDA of vitamin B_{12}, which is needed to assist the growth and maintenance of the nervous system. Calcium helps the body absorb vitamin B_{12}. The cholesterol count is reasonable enough—at the low end of the moderate range.

Description: A crisp-fried tortilla topped with a meat or bean filling, tomatoes, lettuce, cheese, and a sauce.

TOSTADA, *Bean, Beef, and Cheese*

Serving: 1 (about 8 oz.)

Calories	333
Fat	16.9 g
Saturated	11.5 g
Monounsaturated	3.5 g
Polyunsaturated	0.6 g
Calories from fat	46%
Cholesterol	74 mg
Sodium	871 mg
Protein	16.1 g
Carbohydrate	29.7 g
Dietary fiber	N/A

CHIEF NUTRIENTS

NUTRIENT	AMOUNT	% RDA
Vitamin B_{12}	1.1 mcg	57
Folate	96.8 mcg	48
Riboflavin	0.5 mg	29
Iron	2.5 mg	25
Calcium	189.0 mg	24
Zinc	3.2 mg	21
Magnesium	67.5 mg	19
Vitamin A	173.3 RE	17

Caution: Very high in sodium. Almost half the calories in this tostada come from fat. The amount of saturated fat is twice what you would get from a bean and cheese tostada, and the cholesterol count is moderate. *Allergy alert:* Cheese, wheat, corn, beans, and tomatoes are common allergens.

Strengths: The dish supplies about half the folate and vitamin B_{12} needed daily: Both nutrients assist proper formation of blood in the body. A good amount of calcium contributes to strong bones; a good amount of iron helps carry oxygen to cell tissues. Vitamin A assists in the healing of wounds and helps reinforce the immune system.

A Better Idea: Beef is only a minor ingredient in the recipe for a bean, beef, and cheese tostada, yet its sodium and fat are high. For better all-around blood and body health, opt for the bean tostada.

TOSTADA, *Beef and Cheese*

Serving: 1 (about 5¾ oz.)

Calories	315
Fat	16.4 g
Saturated	10.4 g
Monounsaturated	3.3 g
Polyunsaturated	1.0 g
Calories from fat	47%
Cholesterol	41 mg
Sodium	897 mg
Protein	19.0 g
Carbohydrate	22.8 g
Dietary fiber	N/A

CHIEF NUTRIENTS

NUTRIENT	AMOUNT	% RDA
Vitamin B$_{12}$	1.2 mcg	59
Riboflavin	0.6 mg	32
Iron	2.9 mg	29
Calcium	216.8 mg	27
Zinc	3.7 mg	25
Magnesium	63.6 mg	18
Potassium	572.1 mg	15

Caution: The sodium level is very high, and the cholesterol count is moderate. Almost half the calories in a beef and cheese tostada come from fat. *Allergy alert:* Cheese, wheat, corn, beans, and tomatoes are common food allergens.

Strengths: This version of the tostada is packed with vitamin B$_{12}$ and is high in riboflavin, nutrients that assist the body's transformation of food into energy. The presence of magnesium and potassium may help lower blood pressure.

A Better Idea: While meat offers some important vitamins and minerals, it also multiplies the calories from fat. Weight watchers especially will want to order the plain bean version of this tasty Mexican food.

SANDWICHES

CHEESEBURGER, *Plain*

Serving: 1 (3½ oz.)

Calories	319
Fat	15.2 g
Saturated	6.5 g
Monounsaturated	5.8 g
Polyunsaturated	N/A
Calories from fat	43%
Cholesterol	50 mg
Sodium	500 mg
Protein	14.8 g
Carbohydrate	31.8 g
Dietary fiber	N/A

CHIEF NUTRIENTS

NUTRIENT	AMOUNT	% RDA
Vitamin B$_{12}$	0.1 mcg	49
Thiamine	0.4 mg	27
Iron	2.4 mg	24
Riboflavin	0.4 mg	24
Niacin	3.7 mg	19

Caution: The American Heart Association recommends that only 30% of calories come from fat, but a fast food cheeseburger is up in the 40% range. Not recommended for anyone on a heart-healthy diet. And double cheeseburgers or cheeseburgers with bacon are fattier yet.

Eat with: A green salad with fresh vegetables to contribute nutrition and fiber to the meal.

Strengths: A cheeseburger offers a decent amount of high-quality protein. It's a good source of anemia-fighting iron and various B vitamins and minerals. But just eat *one*. If you have a double-decker appetite, pile on the lettuce, onions, and tomato.

A Better Idea: Patronize fast food outlets that offer lean hamburgers. Or order broiled chicken breast and a salad with low-fat dressing or another reduced-fat entrée.

CHEESEBURGER *with Condiments*

Serving: 1 (5½ oz.)

Calories	359
Fat	19.8 g
Saturated	9.2 g
Monounsaturated	7.2 g
Polyunsaturated	1.5 g
Calories from fat	50%
Cholesterol	52 mg
Sodium	976 mg
Protein	17.8 g
Carbohydrate	28.1 g
Dietary fiber	N/A

CHIEF NUTRIENTS

NUTRIENT	AMOUNT	% RDA
Vitamin B$_{12}$	1.2 mcg	62
Niacin	6.4 mg	34
Iron	2.7 mg	27
Calcium	181.7 mg	23
Thiamine	0.3 mg	21
Zinc	2.6 mg	17
Riboflavin	0.2 mg	14
Magnesium	26.1 mg	7
Potassium	229.5 mg	6

Caution: If you're watching your blood pressure, take heed: Condiments add quite a bit of sodium to a cheeseburger. Note, too, that ordering a cheeseburger with lots of regular mayo increases the fat count considerably.

Strengths: Decked out with fixings, a cheeseburger gives you a little more resistance-boosting iron, zinc, and protein than you get in a plain cheeseburger. You also get a fair amount of bone-building calcium and a little magnesium and potassium.

Description: Values shown here are based on a cheeseburger garnished with catsup, mustard, mayonnaise-style dressing, pickles, onions, lettuce, and tomatoes.

A Better Idea: If you ask the server to hold the mayo, you save fat and calories.

CHEESE STEAK

Serving: 1 (9¾ oz.)

Calories	519
Fat	18.6 g
Saturated	8.1 g
Monounsaturated	3.6 g
Polyunsaturated	3.0 g
Calories from fat	32%
Cholesterol	87 mg
Sodium	1,303 mg
Protein	35.3 g
Carbohydrate	48.8 g
Dietary fiber	2.2 g

CHIEF NUTRIENTS

NUTRIENT	AMOUNT	% RDA
Vitamin B$_{12}$	3.3 mcg	167
Iron	4.8 mg	48
Niacin	6.3 mg	33
Zinc	5.0 mg	33
Riboflavin	0.5 mg	31
Thiamine	0.5 mg	30
Calcium	204.0 mg	26

Caution: Extremely high in sodium, a cheese steak supplies more than a whole day's salt quota if you're on a low-sodium diet. A cheese steak contains a moderate amount of calories and fat, but if it's cooked on a greasy grill and smothered in grilled onions, it can contain lots more. "Philly" cheese steaks are about twice as large as the normal version.

Strengths: A cheese steak offers plenty of nutrition, but at a calorie cost. It contains 167% of the RDA of vitamin B$_{12}$, a nutrient essential for a smooth-functioning nervous system and for healthy blood. It's a rich source of blood-boosting iron and supplies good amounts of B-complex vitamins that help produce energy. It's also a good source of bone-bolstering calcium and a very good source of zinc, a mineral needed for strong immunity and wound healing.

Origin: Supposedly created at Pat's restaurant in Philadelphia in 1930.

Description: Thinly sliced steak, lightly grilled, with American cheese. Served in an Italian roll.

CHICKEN FILLET, *Plain*

Serving: 1 (6½ oz.)

Calories	515
Fat	29.5 g
Saturated	8.5 g
Monounsaturated	10.4 g
Polyunsaturated	8.4 g
Calories from fat	51%
Cholesterol	60 mg
Sodium	957 mg
Protein	24.1 g
Carbohydrate	38.7 g
Dietary fiber	N/A

CHIEF NUTRIENTS

NUTRIENT	AMOUNT	% RDA
Iron	4.7 mg	47
Niacin	6.8 mg	36
Thiamine	0.3 mg	22
Vitamin B$_{12}$	0.4 mcg	19
Folate	29.1 mcg	15
Vitamin C	8.9 mg	15
Riboflavin	0.2 mg	14
Zinc	1.9 mg	12

Caution: With lots of sodium and more than half of the calories derived from fat, chicken fillet sandwiches are in the same high salt and fat ballpark as deluxe hamburgers. Some chains use chicken that has been breaded and deep-fried, a cooking method that escalates the threat to your waistline and heart health.

Strengths: This sandwich is rich in easy-to-absorb heme iron that boosts resistance and helps fight anemia. You also get plenty of niacin and a good amount of thiamine, along with respectable amounts of other valuable B vitamins. Chicken is an excellent source of high-quality protein.

Serve: With a plain baked potato or a salad with low-fat topping.

A Better Idea: Skip the mayo. Also scout around for lean versions of this entrée that are offered by some chains—your arteries and waistline will thank you.

CORNDOG

Serving: 1 (6¼ oz.)

Calories	460
Fat	18.9 g
Saturated	5.2 g
Monounsaturated	9.1 g
Polyunsaturated	3.5 g
Calories from fat	37%
Cholesterol	79 mg
Sodium	973 mg
Protein	16.8 g
Carbohydrate	55.8 g
Dietary fiber	N/A

CHIEF NUTRIENTS

NUTRIENT	AMOUNT	% RDA
Iron	6.2 mg	62
Riboflavin	0.7 mg	41
Folate	59.5 mcg	30
Niacin	4.2 mg	22
Vitamin B$_{12}$	0.4 mcg	22
Thiamine	0.3 mg	19
Calcium	101.5 mg	13

Caution: Like many other vendor foods, corndogs are packed with hefty amounts of salt and calories, with more fat than you probably need—so you don't want to eat them often. *Allergy alert:* May trigger a reaction in people allergic to wheat, corn, milk, or ingredients in cured meats. Since a corndog is mostly hot dog, all the cautions for frankfurters apply to corndogs as well.

Strengths: An excellent source of anemia-fighting iron. One corndog provides significant amounts of protein for lean muscle mass and a good supply of B vitamins for healthy nerves and skin, along with calcium for stronger bones.

Origin: A popular vendor item, corndogs were created for the Texas State Fair by Neil Fletcher in 1942.

Description: A frankfurter coated with a heavy batter of cornmeal, flour, milk, egg, sugar, and baking powder, then deep-fried in shortening. Often served on a stick rather than in a roll. (Turkey franks are sometimes used instead of beef franks.)

EGG AND CHEESE

Serving: 1 (about 5¼ oz.)

Calories	340
Fat	19.4 g
Saturated	6.6 g
Monounsaturated	8.3 g
Polyunsaturated	2.6 g
Calories from fat	51%
Cholesterol	291 mg
Sodium	804 mg
Protein	15.6 g
Carbohydrate	25.9 g
Dietary fiber	N/A

CHIEF NUTRIENTS

NUTRIENT	AMOUNT	% RDA
Vitamin B$_{12}$	1.1 mcg	57
Riboflavin	0.6 mg	34
Iron	3.0 mg	30
Calcium	224.8 mg	28
Folate	36.5 mcg	18
Vitamin A	181.0 RE	18
Thiamine	0.3 mg	17
Niacin	2.1 mg	11
Zinc	1.7 mg	11

Caution: An egg and cheese sandwich is high in cholesterol and sodium—not a prudent choice if you're pursuing a heart-healthy diet. This sandwich also gives you nearly 7 g of saturated fat—about as much as you'd get from a cheeseburger or an order of onion rings.

Eat with: Low-fat, low-sodium, high-fiber side dishes like salads—or have some fruit juice with your meal.

Strengths: An egg and cheese sandwich is extremely rich in vitamin B$_{12}$, and has respectable amounts of riboflavin, folate, and other valuable B vitamins. You also get commendable amounts of vitamin A, iron, and zinc—nutrients that team up to defend against disease. There's a good supply of calcium to help build stronger bones and teeth.

A Better Idea: Skip the sandwich altogether, and order a baked potato with low-fat dressing.

FISH *with Tartar Sauce*

Serving: 1 (about 5½ oz.)

Calories	431
Fat	22.8 g
Saturated	5.2 g
Monounsaturated	7.7 g
Polyunsaturated	8.3 g
Calories from fat	48%
Cholesterol	55 mg
Sodium	615 mg
Protein	16.9 g
Carbohydrate	41.0 g
Dietary fiber	N/A

CHIEF NUTRIENTS

NUTRIENT	AMOUNT	% RDA
Vitamin B$_{12}$	1.1 mcg	54
Iron	2.6 mg	26
Folate	44.2 mcg	22
Thiamine	0.3 mg	22
Niacin	3.4 mg	18
Riboflavin	0.2 mg	13
Calcium	83.7 mg	10
Magnesium	33.2 mg	9

Caution: Fish starts out very lean and healthful, but after it has been dipped in batter, breaded, and deep-fried, seafood turns into a fatty, high-calorie entrée. Deep-frying destroys its heart-healthy omega-3 fatty acids, and any slathering of tartar sauce contributes heavily to the fat and calorie load. Note the sodium content in case you're trying to limit your salt intake.

Eat with: Coleslaw for fiber and nonfat milk for calcium.

Strengths: Fish gives you a substantial amount of vitamin B$_{12}$, which you might need in quantity if you shy away from meat, chicken, and dairy products. It's also a good source of folate and heme iron, which fight anemia. An array of other B vitamins help to maintain healthy skin and nerves.

Description: Fast food fish sandwiches are usually made with cod or other white North Atlantic fish.

A Better Idea: If possible, choose grilled or broiled fish, prepared without oil or butter and served with lemon juice instead of mayo or tartar sauce.

HAM AND CHEESE

Serving: 1 (about 5¼ oz.)

Calories	352
Fat	15.5 g
Saturated	6.4 g
Monounsaturated	6.7 g
Polyunsaturated	1.4 g
Calories from fat	40%
Cholesterol	58 mg
Sodium	771 mg
Protein	20.7 g
Carbohydrate	33.3 g
Dietary fiber	N/A

CHIEF NUTRIENTS

NUTRIENT	AMOUNT	% RDA
Folate	71.5 mcg	36
Iron	3.2 mg	32
Riboflavin	0.5 mg	28
Vitamin B$_{12}$	0.5 mcg	27
Thiamine	0.3 mg	21
Calcium	129.9 mg	16
Niacin	2.7 mg	14
Vitamin B$_6$	0.2 mg	10

Caution: With nearly 40% of its calories derived from fat, a ham and cheese sandwich is a slightly better choice than a breaded fish or chicken sandwich. Still, it gets more calories from fat than what the American Heart Association would like to see. In fact, a ham and cheese sandwich has as many grams of fat as a slice of chocolate cream pie. The sodium content may put this off-limits to those on salt-restricted diets. *Allergy alert:* Ham and cheese are both common allergens.

Eat with: A low-sodium, high-fiber side dish such as a crisp garden salad.

Strengths: If your diet is low in anemia-fighting folate and iron—or if you're trying to enhance your intake of B vitamins—it could make sense to have a ham and cheese sandwich for lunch. You also get some useful bone-building calcium.

HAM, EGG, AND CHEESE

Serving: 1 (about 5 oz.)

Calories	347
Fat	16.3 g
Saturated	7.4 g
Monounsaturated	5.7 g
Polyunsaturated	1.7 g
Calories from fat	42%
Cholesterol	246 mg
Sodium	1,005 mg
Protein	19.3 g
Carbohydrate	31.0 g
Dietary fiber	N/A

CHIEF NUTRIENTS

NUTRIENT	AMOUNT	% RDA
Vitamin B$_{12}$	1.2 mcg	62
Riboflavin	0.6 mg	33
Iron	3.1 mg	31
Thiamine	0.4 mg	29
Calcium	211.6 mg	26
Niacin	4.2 mg	22
Folate	42.9 mcg	21
Vitamin A	148.7 RE	15
Zinc	2.0 mg	13

Caution: This combo is much higher in cholesterol and sodium than a ham and cheese sandwich. It's also higher in fat—much of it saturated—creating a barnyard-size array of threats to your blood pressure and arteries. Since saturated fat raises blood cholesterol, this is a sandwich you'll probably want to pass up. *Allergy alert:* Ham, egg, and cheese are common allergens.

Strengths: While it may pose a triple threat to your heart and blood vessels, at least a ham, egg, and cheese sandwich has some things to offer your blood—an abundance of heme iron and plentiful folate, along with B$_{12}$ and other B vitamins. You also get a decent amount of immunity-boosting zinc. (Still, the high fat content makes it a shabby vehicle for these nutrients.)

A Better Idea: Even a burger is preferable. And some raw vegetables and leafy greens from the salad bar would be best of all.

HAMBURGER, *McLean Deluxe*

Serving: 1 (about 7 oz.)

Calories	320
Fat	10.0 g
Saturated	4.0 g
Monounsaturated	5.0 g
Polyunsaturated	1.0 g
Calories from fat	28%
Cholesterol	60 mg
Sodium	670 mg
Protein	22.0 g
Carbohydrate	35.0 g
Dietary fiber	N/A

CHIEF NUTRIENTS

NUTRIENT	AMOUNT	% RDA
Niacin	7.0 mg	37
Thiamine	0.4 mg	25
Iron	3.6 mg	24
Riboflavin	0.3 mg	20
Calcium	150.0 mg	19
Vitamin A	100.0 RE	10

Strengths: The McLean Deluxe, introduced by McDonald's in 1991, finally presents an alternative to McD's high-fat, high-calorie burgers. The McLean is relatively low in calories and fat, thanks to the addition of a seaweed extract called carrageenan to the meat. Plenty of niacin and a good amount of thiamine contribute to the body's formation of red blood cells and the conversion of food to energy. A healthy dose of iron boosts the body's infection-fighting ability.

Caution: New and improved, yes. Low in sodium, no. The McLean Deluxe still has more sodium than you can shake a saltshaker at. Also, the beef is high in purines such as uric acid, which can irritate gout sufferers. *Allergy alert:* Beef can be a food allergen.

History: Carrageenan, the key to the low-fat burger, long has been used as a thickening agent in processed foods.

Description: The McLean Deluxe comes on McDonald's famous sesame seed bun with all the traditional fixin's—lettuce, tomato, mustard, ketchup, pickles, and onions.

HAMBURGER, *Plain*

Serving: 1 (about 3¼ oz.)

Calories	275
Fat	11.8 g
Saturated	4.1 g
Monounsaturated	5.5 g
Polyunsaturated	0.9 g
Calories from fat	39%
Cholesterol	35 mg
Sodium	387 mg
Protein	12.3 g
Carbohydrate	30.5 g
Dietary fiber	N/A

CHIEF NUTRIENTS

NUTRIENT	AMOUNT	% RDA
Vitamin B$_{12}$	0.9 mcg	45
Iron	2.4 mg	24
Thiamine	0.3 mg	22
Niacin	3.7 mg	20
Riboflavin	0.3 mg	16
Folate	25.2 mcg	13
Zinc	2.0 mg	13

Strengths: With under 12 g of fat—most of it unsaturated—a plain broiled burger may be the best choice among all the hot sandwiches on a typical fast food menu. It gives you handsome amounts of anemia-fighting iron, along with lots of vitamin B$_{12}$, which is not available from plant foods. Burgers provide a good amount of zinc, too, for faster wound healing.

Eat with: A glass of low-fat milk or juice and a salad. These accompaniments have more vitamins and minerals than the french fries and soda that people usually order with burgers.

Caution: Double hamburgers with sauce can have 3 times as much fat as a plain hamburger.

A Better Idea: Take advantage of newly introduced extra-lean burgers.

HAMBURGER with Condiments

Serving: 1 (about 4 oz.)

Calories	279
Fat	13.5 g
Saturated	4.1 g
Monounsaturated	5.3 g
Polyunsaturated	N/A
Calories from fat	43%
Cholesterol	26 mg
Sodium	504 mg
Protein	12.9 g
Carbohydrate	27.3 g
Dietary fiber	N/A

CHIEF NUTRIENTS

NUTRIENT	AMOUNT	% RDA
Vitamin B_{12}	0.9 mcg	44
Iron	2.6 mg	26
Niacin	3.7 mg	19
Thiamine	0.2 mg	15
Zinc	2.1 mg	14
Riboflavin	0.2 mg	12
Folate	18.7 mcg	9
Calcium	62.7 mg	8
Magnesium	22.0 mg	6

Caution: If you garnish your burger with standard toppings, like catsup, mustard, mayo, pickles, onions, lettuce, and tomato, the sodium and fat contents edge upward. And if you add fries and a shake to the meal, you're really putting some stress on your waistline, arteries, and blood pressure.

Eat with: A nice salad with low-fat dressing and a glass of juice to round out the meal.

Strengths: Except for the sodium and fat content, the nutritional profile for a hamburger with condiments is roughly the same as that of a plain burger. A remarkable amount of vitamin B_{12} helps to promote sound nerves and healthier blood, and a good quantity of iron helps your body fight infection. Plus, there's niacin to aid in energy production and some calcium to contribute to stronger bones and teeth.

A Better Idea: If you order a burger, pile on the vegetable toppings, but skip the mayo and high-sodium garnishes.

HOT DOG

Serving: 1 (3½ oz.)

Calories	242
Fat	14.5 g
Saturated	5.1 g
Monounsaturated	6.9 g
Polyunsaturated	1.7 g
Calories from fat	54%
Cholesterol	44 mg
Sodium	670 mg
Protein	10.4 g
Carbohydrate	18.0 g
Dietary fiber	N/A

CHIEF NUTRIENTS

NUTRIENT	AMOUNT	% RDA
Vitamin B_{12}	0.5 mcg	26
Iron	2.3 mg	23
Niacin	3.7 mg	19
Riboflavin	0.3 mg	16
Thiamine	0.2 mg	16
Folate	29.4 mcg	15
Zinc	2.0 mg	13

Caution: A plain, unadorned hot dog is high in sodium. If you then add catsup, mustard, pickle relish, or sauerkraut, the salt level skyrockets. A hot dog also contains a moderate amount of cholesterol and a lot of fat—more than you'd get from a pat of butter. *Allergy alert:* Hot dogs may contain curing agents and nonmeat fillers such as nonfat dry milk, cereal, or soy protein—all potential allergy triggers.

Eat with: Self-control.

Strengths: Hot dogs lend valuable iron and vitamin B_{12}, with lesser amounts of zinc and other B vitamins.

ROAST BEEF

Serving: 1 (about 5 oz.)

Calories	346
Fat	13.8 g
Saturated	3.6 g
Monounsaturated	6.8 g
Polyunsaturated	1.7 g
Calories from fat	36%
Cholesterol	51 mg
Sodium	792 mg
Protein	21.5 g
Carbohydrate	33.4 g
Dietary fiber	N/A

CHIEF NUTRIENTS

NUTRIENT	AMOUNT	% RDA
Vitamin B$_{12}$	1.2 mcg	61
Iron	4.2 mg	42
Niacin	5.9 mg	31
Thiamine	0.4 mg	25
Zinc	3.4 mg	23
Folate	40.3 mcg	20
Riboflavin	0.3 mg	18

Strengths: Roast beef sandwiches tend to be relatively low in fat compared to hamburgers, fried chicken, fish sandwiches, and other popular fast food menu items. In fact, in some chain restaurants a roast beef sandwich with no sauce may be the leanest item on the menu. Nutrientwise, this sandwich is rich in iron for rich red blood. And it boasts handsome amounts of immunity-boosting zinc and an array of valuable B vitamins.

Caution: This sandwich is fairly high in sodium, so be careful to select low-sodium foods for the rest of the meal—and the rest of the day.

Serve: With sliced tomatoes, onions, and lettuce.

A Better Idea: Choose the regular- or junior-size roast beef sandwich over the deluxe sandwich and you'll save considerable calories. To spare yourself added fat, skip the mayonnaise or mayo-horseradish sauce.

STEAK

Serving: 1 (about 7¼ oz.)

Calories	459
Fat	14.1 g
Saturated	3.8 g
Monounsaturated	5.4 g
Polyunsaturated	3.4 g
Calories from fat	28%
Cholesterol	73 mg
Sodium	798 mg
Protein	30.3 g
Carbohydrate	52.0 g
Dietary fiber	N/A

CHIEF NUTRIENTS

NUTRIENT	AMOUNT	% RDA
Vitamin B$_{12}$	1.6 mcg	79
Iron	5.2 mg	52
Folate	89.8 mcg	45
Niacin	7.3 mg	38
Zinc	4.5 mg	30
Thiamine	0.4 mg	27
Riboflavin	0.4 mg	22
Vitamin B$_6$	0.4 mg	19
Magnesium	49.0 mg	14

Strengths: As nutritional profiles go, this sandwich isn't half bad: You get a ton of anemia-fighting iron and lots of folate, awesome amounts of vitamin B$_{12}$, and fine amounts of niacin, thiamine, and other nutrients needed for healthy skin and nerves. Steak sandwiches are rich in immunity-boosting zinc, too—all for under 30% of calories from fat.

Eat with: Coleslaw, to make up for missing fiber, vitamins, and minerals. (If you have a hearty appetite, don't ruin a good thing by ordering fries!)

Caution: With nearly 500 calories, steak sandwiches aren't exactly diet food. Split one with a friend, and you'll reduce the fat, sodium, cholesterol, and calories that you consume.

Description: Values given here are for ground beef with tomato, lettuce, mayonnaise, and salt on a roll.

A Better Idea: Ask the server to hold the salt and mayo, and you've got yourself a halfway decent fast food alternative to burgers and fries.

SUBMARINE with Cold Cuts

Serving: 1 (about 8 oz.)

Calories	456
Fat	18.6 g
Saturated	6.8 g
Monounsaturated	8.2 g
Polyunsaturated	2.3 g
Calories from fat	37%
Cholesterol	36 mg
Sodium	1,651 mg
Protein	21.8 g
Carbohydrate	51.1 g
Dietary fiber	N/A

CHIEF NUTRIENTS

NUTRIENT	AMOUNT	% RDA
Thiamine	1.0 mg	67
Vitamin B$_{12}$	1.1 mcg	55
Riboflavin	0.8 mg	47
Niacin	5.5 mg	29
Folate	54.7 mcg	27
Iron	2.5 mg	25
Vitamin C	12.3 mg	21
Zinc	2.6 mg	17
Vitamin A	79.8 RE	8

Caution: Subs made with Italian cold cuts contain a considerable amount of fat and a veritable blizzard of sodium. *Allergy alert:* Contains cheese, salami, and ham—a potential problem for those with allergies to cured meats or dairy products.

Strengths: This hefty bundle of lunch meat and bread is conspicuously high in B vitamins. It yields a good amount of zinc, some vitamin A, and is high in vitamin C, all resistance-building nutrients.

Curiosity: Submarine sandwiches are also called grinders, heros, hoagies, poor boys, torpedos, and Italian sandwiches, depending on which part of the country you live in.

A Better Idea: You can shave off a fair amount of fat and sodium if you order a roast beef sub with no mayo or salt instead of cold cuts. If you have a weakness for Italian hoagies, split one, give away half, and round out the meal with fruit or nonfat yogurt.

SUBMARINE with Roast Beef

Serving: 1 (about 7¾ oz.)

Calories	410
Fat	13.0 g
Saturated	7.1 g
Monounsaturated	1.8 g
Polyunsaturated	2.6 g
Calories from fat	28%
Cholesterol	73 mg
Sodium	845 mg
Protein	29.0 g
Carbohydrate	44.0 g
Dietary fiber	N/A

CHIEF NUTRIENTS

NUTRIENT	AMOUNT	% RDA
Vitamin B$_{12}$	1.8 mcg	91
Niacin	6.0 mg	31
Zinc	4.4 mg	29
Iron	2.8 mg	28
Thiamine	0.4 mg	27
Riboflavin	0.4 mg	24
Folate	45.4 mcg	23

Strengths: Loaded with more protein, iron, zinc, vitamin B$_{12}$, and niacin than a sub with cold cuts, the roast beef cousin has one-third less fat and about half as much sodium. You also get an appreciable amount of magnesium.

Eat with: Coleslaw or a salad with low-fat dressing to provide more fiber.

Caution: Although a roast beef sub is lower in sodium than subs made with cold cuts, it's still fairly high—a potential problem if you're on a low-salt diet to prevent high blood pressure. This sandwich has a fair amount of cholesterol and saturated fat, which are on the "Unwanted" list for people on heart-healthy diets. So don't make a steady diet of roast beef subs.

A Better Idea: Order half a sub, with no added oil or salt, and use horseradish or mustard instead of mayo. You'll still get a decent amount of protein, with less fat, cholesterol, and calories.

SUBMARINE *with Tuna Salad*

Serving: 1 (about 9 oz.)

Calories	584
Fat	28.0 g
Saturated	5.3 g
Monounsaturated	13.4 g
Polyunsaturated	7.3 g
Calories from fat	43%
Cholesterol	49 mg
Sodium	1,293 mg
Protein	29.7 g
Carbohydrate	55.4 g
Dietary fiber	N/A

CHIEF NUTRIENTS

NUTRIENT	AMOUNT	% RDA
Vitamin B_{12}	1.6 mcg	81
Niacin	11.3 mg	60
Thiamine	0.5 mg	31
Folate	56.3 mcg	28
Iron	2.6 mg	26
Magnesium	79.4 mg	23

Caution: Dieters, take heed: Don't assume that opting for fish lets you off the hook caloriewise. With 28 g of fat, this tuna sandwich contains nearly 600 calories—more than the calories in a double cheeseburger. And it has a higher-than-average amount of sodium.

Strengths: Luckily, most of the fat in tuna salad is monounsaturated or polyunsaturated, and therefore less damaging to your arteries than saturated fat. As a matter of fact, tuna is high in heart-healthy omega-3 fatty acids, substances shown to lower the liver's production of triglycerides, a particularly bad type of blood fat.

SIDE DISHES

BAKED POTATO *with Cheese Sauce and Bacon*

Serving: 1 (about 10¾ oz.)

Calories	451
Fat	25.9 g
Saturated	10.1 g
Monounsaturated	9.7 g
Polyunsaturated	4.8 g
Calories from fat	52%
Cholesterol	30 mg
Sodium	972 mg
Protein	18.4 g
Carbohydrate	44.4 g
Dietary fiber	N/A

CHIEF NUTRIENTS

NUTRIENT	AMOUNT	% RDA
Vitamin C	28.7 mg	48
Calcium	308.0 mg	38
Vitamin B_6	0.8 mg	38
Iron	3.1 mg	31
Potassium	1,178.1 mg	31
Niacin	4.0 mg	21
Magnesium	68.8 mg	20

Caution: Dieters, beware: A spud topped with bacon and cheese has more calories than 2 slices of pepperoni pizza. But wait, there's more bad news: It also has *4 times* as much saturated fat, and loads of sodium. So, in addition to adding to your weight, a cheese and bacon–topped potato is unkind to your heart and blood pressure.

Eat with: Restraint. You may feel virtuous ordering a potato instead of a burger and fries, but the nutrients come at a high price.

Strengths: On the plus side, this hearty side dish serves up beneficial amounts of resistance-boosting vitamin C. You also get an impressive quantity of valuable minerals such as bone-strengthening calcium, and a rich dose of potassium. A very hearty amount of anemia-fighting iron is present.

A Better Idea: Order a plain baked potato, and top with cottage cheese or grated Parmesan from the salad bar.

BAKED POTATO *with Cheese Sauce and Broccoli*

Serving: 1 (about 12 oz.)

Calories	403
Fat	21.4 g
Saturated	8.5 g
Monounsaturated	7.7 g
Polyunsaturated	4.2 g
Calories from fat	48%
Cholesterol	20 mg
Sodium	485 mg
Protein	13.7 g
Carbohydrate	46.6 g
Dietary fiber	N/A

CHIEF NUTRIENTS

NUTRIENT	AMOUNT	% RDA
Vitamin C	48.5 mg	81
Calcium	335.6 mg	42
Vitamin B6	0.8 mg	39
Potassium	1,440.8 mg	38
Iron	3.3 mg	33
Folate	61.0 mg	31
Vitamin A	278.0 RE	28
Magnesium	78.0 mg	22

Caution: Although it may be a painless way to consume broccoli, this fast food special has almost 100 more calories than a baked potato topped with butter. It also has twice as much fat—something to keep in mind if you're counting calories or trying to nurture a healthier heart. Note, too, that the sodium content may make this off-limits to those on sodium-restricted diets.

Eat with: A salad with lots of greens, veggies, and beans.

Strengths: Although fiber values are not available, there is some fiber in broccoli, so this combination may contribute to digestive health. You also get an impressive array of vitamins and minerals—notably, whopping amounts of resistance-building vitamin C, shored up by generous quantities of iron, folate, and bone-strengthening calcium.

A Better Idea: Order a plain baked potato and top with low-fat dressing from the salad bar. (But avoid the salad bar section displaying macaroni, bacon bits, or other fatty fixings.)

BAKED POTATO *with Cheese Sauce and Chili*

Serving: 1 (about 10¾ oz.)

Calories	482
Fat	21.8 g
Saturated	13.0 g
Monounsaturated	6.8 g
Polyunsaturated	0.9 g
Calories from fat	41%
Cholesterol	32 mg
Sodium	699 mg
Protein	23.2 g
Carbohydrate	55.9 g
Dietary fiber	N/A

CHIEF NUTRIENTS

NUTRIENT	AMOUNT	% RDA
Iron	6.1 mg	61
Vitamin C	31.6 mg	53
Calcium	410.8 mg	51
Vitamin B6	1.0 mg	48
Potassium	1,572.1 mg	42
Magnesium	110.6 mg	32
Folate	51.4 mcg	26
Zinc	3.8 mg	25

Caution: A potato embellished with cheese sauce and chili has more fat than 6 slices of plain pizza—and most of the fat is saturated, which tends to raise blood cholesterol. You also get deluged with sodium.

Strengths: A baked potato topped with cheese and chili gives you a bonus of high-quality protein and iron, along with stupendous amounts of resistance-building vitamin C and bone-building calcium.

A Better Idea: Order a plain baked potato, then proceed at warp speed to the salad bar. By omitting the extras atop the potato, you'll spare yourself needless calories and fat. And with fresh crunchies from the salad bar, you'll get more fiber to help lower cholesterol and keep your GI tract humming along efficiently.

BAKED POTATO with Sour Cream and Chives

Serving: 1 (about 10¾ oz.)

Calories	393
Fat	22.3 g
Saturated	10.0 g
Monounsaturated	7.9 g
Polyunsaturated	3.3 g
Calories from fat	51%
Cholesterol	24 mg
Sodium	181 mg
Protein	6.7 g
Carbohydrate	50.0 g
Dietary fiber	N/A

CHIEF NUTRIENTS

NUTRIENT	AMOUNT	% RDA
Vitamin C	33.8 mg	56
Vitamin B$_6$	0.8 mg	40
Potassium	1,383.2 mg	37
Iron	3.1 mg	31
Vitamin A	277.8 RE	28
Magnesium	69.5 mg	20
Niacin	3.7 mg	20
Thiamine	0.3 mg	18
Folate	33.2 mcg	17
Calcium	105.7 mg	13
Vitamin B$_{12}$	0.2 mcg	11
Zinc	0.9 mg	6

Caution: The crème de la fat of potatoes, a potato topped with sour cream contains as much fat as *2* hot fudge sundaes. Nearly half the fat is saturated, and that concentration can push cholesterol levels up to the ceiling.

Eat with: A leafy salad, accompanied by a tall glass of juice. If that's your entire meal, you'll keep the fat content relatively reasonable.

Strengths: This is a terrific source of vitamin C and a very good source of iron—and it has a good supply of other resistance-boosting nutrients, such as folate and vitamin A, with a little help from zinc. You also get some hearty vitamin B$_6$ and a kick of magnesium.

A Better Idea: Order a plain baked potato and top it with low-fat ranch dressing at the salad bar: It has one-tenth as much fat as a baked potato with sour cream.

FRENCH FRIES

Serving: 20–25 1"–2" strips

Calories	235
Fat	12.2 g
Saturated	3.8 g
Monounsaturated	6.0 g
Polyunsaturated	1.9 g
Calories from fat	47%
Cholesterol	0
Sodium	124 mg
Protein	3.0 g
Carbohydrate	29.3 g
Dietary fiber	N/A

CHIEF NUTRIENTS

NUTRIENT	AMOUNT	% RDA
Potassium	541.1 mg	14
Folate	25.1 mcg	13
Iron	1.0 mg	10
Vitamin B$_6$	0.2 mg	10

Caution: Whether fries are cooked in vegetable or animal fat, nearly half the calories are derived from fat. Compare that to less than 1% for a baked potato! Moderately high levels of sodium are increased if you add catsup. So avoid french fries, or make them an occasional treat. Other fries to avoid are shoestring potatoes, thick steak fries, and microwave fries. However, you can cut up your own potatoes to make oven-baked french fries that will contain less than half the fat found in restaurant fries. *Allergy alert:* Some people with asthma are severely allergic to french fries treated with sulfites.

Strengths: They contain good amounts of potassium, folate, vitamin B$_6$, and iron.

Selection: Values shown are for french fries cooked in vegetable oil. Crisp, golden, lightly cooked fries are easier to digest than overcooked fries. If your fries arrive smoky-brown and sodden, send them back.

HUSH PUPPIES

Serving: 5 (about 2¾ oz.)

Calories	257
Fat	11.6 g
Saturated	2.7 g
Monounsaturated	7.8 g
Polyunsaturated	0.4 g
Calories from fat	41%
Cholesterol	135 mg
Sodium	965 mg
Protein	4.9 g
Carbohydrate	34.9 g
Dietary fiber	N/A

CHIEF NUTRIENTS

NUTRIENT	AMOUNT	% RDA
Iron	1.4 mg	14
Folate	21.1 mcg	11
Niacin	2.0 mg	11
Calcium	68.6 mg	9
Vitamin B$_{12}$	0.2 mcg	9
Potassium	188.0 mg	5
Vitamin B$_6$	0.1 mg	5

Caution: These Southern favorites are high in calories and sodium, with about as much fat as 1 Tbsp. of butter. The fat and calorie counts rival those of french fries, making this a poor choice for weight watchers, diabetics, or those on heart-healthy diets. *Allergy alert:* Contain corn, milk, and eggs, all common food allergens.

Strengths: Hush puppies offer a good amount of iron (presumably from cornmeal) and a little calcium, potassium, and B vitamins (from milk), but all these nutrients come at a price.

Curiosity: To keep hungry dogs from begging while dinner was being prepared, cooks used to toss scraps of the fried batter to their pets and admonish, "Hush, puppy!"

Origin: Small dumplings popularized during the Civil War.

Description: Hush puppies are made of cornmeal, milk, eggs, onions, baking powder, and salt. They're deep-fried and served piping hot.

Serve: With grilled or broiled catfish, without butter.

ONION RINGS, *Breaded and Fried*

Serving: 8–9 (about 3 oz.)

Calories	276
Fat	15.5 g
Saturated	7.0 g
Monounsaturated	6.7 g
Polyunsaturated	0.7 g
Calories from fat	51%
Cholesterol	14 mg
Sodium	430 mg
Protein	3.7 g
Carbohydrate	31.3 g
Dietary fiber	N/A

CHIEF NUTRIENTS

NUTRIENT	AMOUNT	% RDA
Calcium	73.0 mg	9
Iron	0.9 mg	9
Folate	11.6 mcg	6
Riboflavin	0.1 mg	6
Vitamin B$_{12}$	0.1 mcg	6
Thiamine	0.1 mg	5

Caution: Don't count this as one of your vegetables for the day: Onion rings have about as much fat and calories as french fries. All that deep-fat frying annihilates the nutritional merits of a raw onion. Liberal amounts of sodium further downgrade this respected member of the cancer-fighting allium genus. *Allergy alert:* The batter may contain corn, wheat, and milk, all common food allergens.

Eat with: Low-fat entrées like fish or chicken, broiled without butter, or a lean roast beef sandwich to keep the overall fat content of the meal at reasonable levels.

A Better Idea: If you can't resist onion rings, split the order with a friend. Or eat one or two and throw the rest out.

PIEROGIES, *Boiled*

Serving: 2 (about 2¾ oz.)

Calories	120
Fat	1.0 g
Saturated	N/A
Monounsaturated	N/A
Polyunsaturated	N/A
Calories from fat	8%
Cholesterol	4 mg
Sodium	340 mg
Protein	4.0 g
Carbohydrate	2.2 g
Dietary fiber	1.6 g

CHIEF NUTRIENTS

NUTRIENT	AMOUNT	% RDA
Thiamine	0.2 mg	12
Niacin	1.6 mg	8
Riboflavin	0.1 mg	8
Iron	0.7 mg	7

Caution: Pierogies are high in sodium. Calories and fat are reasonable, but there's a problem: People often serve pierogies with lots of butter or sour cream. *Drug interaction:* Because of the tyramine in the cheese, blood pressure may soar rapidly if you eat pierogies while taking MAO inhibitors. *Allergy alert:* Some people report that cheese triggers migraines, while others say the eggs in pierogies give them hives.

Strengths: If pierogies are boiled rather than fried, they're a low-fat, low-cholesterol treat containing a fairly reasonable number of calories. A good amount of thiamine and a bit of riboflavin help to give you energy, and they contain niacin and iron for healthier blood.

Description: A pierogie is a Polish pasta dumpling filled with mashed potatoes and cheese. It's often sautéed.

A Better Idea: Instead of spreading butter or sour cream on them, as is traditional, add some healthy salsa or nonfat yogurt. Make sure you boil your pierogies rather than fry them to keep fat and calories down.

Fats, Oils, and Salad Dressings
FATS

CHICKEN

Serving: 1 Tbsp.

Calories	115
Fat	12.8 g
Saturated	3.8 g
Monounsaturated	5.7 g
Polyunsaturated	2.6 g
Calories from fat	100%
Cholesterol	11 mg
Sodium	0
Protein	0
Carbohydrate	0
Dietary fiber	0

CHIEF NUTRIENTS

(None of the nutrients meet or exceed 5% of the RDA.)

Caution: This is 100% fat and cholesterol—so every glob of fat you trim from chicken spares your hips and arteries considerable damage.

Strength: Chicken fat is a less saturated fat than lard, but it's no better than shortening.

Curiosity: Chicken fat is used quite a bit in Jewish cooking, where kosher restrictions prohibit the use of meat and dairy products (such as butter) in the same dish or meal.

A Better Idea: Use diet margarine—it has about half as much fat, most of it unsaturated, and less than half the calories.

LARD

Serving: 1 Tbsp.

Calories	115
Fat	12.8 g
Saturated	5.0 g
Monounsaturated	5.8 g
Polyunsaturated	1.4 g
Calories from fat	100%
Cholesterol	12 mg
Sodium	0
Protein	0
Carbohydrate	0
Dietary fiber	0

CHIEF NUTRIENTS

(None of the nutrients meet or exceed 5% of the RDA.)

Caution: Lard makes the flakiest biscuits, pastry, and pie crusts, but it's one of the most highly saturated fats you can use. So it's murder on your arteries. And if you eat too much of it, you'll begin to tip the scales in the direction of infinity.

Eat with: The utmost restraint.

Description: Lard is simply purified pork fat. Unprocessed lard has a stronger flavor than lard that's been filtered, bleached, hydrogenated, and emulsified.

A Better Idea: Substitute soft or diet margarine for lard when making biscuits or pastry—you get a fraction of the saturated fat and half as many calories. If you must fry food, use corn, safflower, or other highly unsaturated oil.

SHORTENING

Serving: 1 Tbsp.

Calories	113
Fat	12.8 g
Saturated	3.2 g
Monounsaturated	5.7 g
Polyunsaturated	3.3 g
Calories from fat	100%
Cholesterol	0
Sodium	0
Protein	0
Carbohydrate	0
Dietary fiber	0

CHIEF NUTRIENTS

NUTRIENT	AMOUNT	% RDA
Vitamin E	1.0 mg αTE	10

Caution: Fairly high in saturated fat, shortening is a potential threat to the health of your heart and arteries. *Allergy alert:* May contain soy, a common food allergen.

Strengths: No cholesterol and a good amount of the antioxidant vitamin E. Shortening has a slightly higher percentage of monounsaturated fat than lard.

Description: To make shortening, a vegetable oil like soy, corn, or cottonseed is processed to make it solidify. During this processing, polyunsaturated fat is converted to saturated fat. Shortening resembles white lard. It helps to hold air in food and prevent proteins and carbohydrates from cooking into one solid indigestible mass. So it makes light, delicate cake batters, icing, biscuits, cookies, pie crusts, and rolls.

A Better Idea: Use margarine instead of shortening. Margarine has a higher percentage of monounsaturated fat and a lower percentage of saturated fat.

OILS

ALMOND

Serving: 1 Tbsp.

Calories	120
Fat	13.6 g
Saturated	1.1 g
Monounsaturated	9.5 g
Polyunsaturated	2.4 g
Calories from fat	100%
Cholesterol	0
Sodium	0
Protein	0
Carbohydrate	0
Dietary fiber	0

CHIEF NUTRIENTS

NUTRIENT	AMOUNT	% RDA
Vitamin E	5.3 mg αTE	53

Strengths: Although almond oil is high in fat, most of it is the monounsaturated form that can help raise levels of protective HDL cholesterol. It's an excellent source of vitamin E, an antioxidant that safeguards other nutrients in the body from harmful oxidative reactions. Like other vegetable oil, almond oil is cholesterol- and sodium-free.

Eat with: Fiber-rich foods, such as salads or lightly cooked vegetables.

Caution: Although this is a "healthy" oil, it is still 100% fat and should be used sparingly. *Allergy alert:* If you're allergic to either almonds or aspirin (almonds contain natural salicylate), you may react to the oil. Consult your allergist for advice.

Description: A light oil with a delicate almond flavor. French oil is often considered the best.

Serve: Brush a very light coating of the oil on hearty whole grain bread in place of butter or other highly saturated spreads.

AVOCADO

Serving: 1 Tbsp.

Calories	124
Fat	14.0 g
Saturated	1.6 g
Monounsaturated	9.9 g
Polyunsaturated	1.9 g
Calories from fat	100%
Cholesterol	0
Sodium	0
Protein	0
Carbohydrate	0
Dietary fiber	0

CHIEF NUTRIENTS

(None of the nutrients meet or exceed 5% of the RDA.)

Strengths: Like the avocado from which it's derived, this oil is high in fat. Fortunately, most of it is the heart-healthy monounsaturated form that can help control high cholesterol. In addition, there's no sodium or cholesterol.

Eat with: Foods rich in fiber, which oils lack.

Caution: Although this is a desirable oil, it is still a concentrated source of fat and calories—use it very cautiously. *Drug interaction:* Avocados contain tyramine, which may cause problems for those taking MAO-inhibitor antidepressants. Check with your doctor to see whether the oil is also problematic.

Description: A mild, virtually flavorless oil that goes well with all types of food. Most of the brands on the market come from California. Because the oil breaks down easily in high heat, it's not recommended for frying or baking.

Storage: To keep the oil from turning rancid, buy small quantities and refrigerate or store in a cool, dark place.

Serve: As a vinaigrette with herb vinegar; use on potato, rice, or vegetable salads.

CANOLA

Serving: 1 Tbsp.

Calories	124
Fat	14.0 g
Saturated	1.0 g
Monounsaturated	8.3 g
Polyunsaturated	4.1 g
Calories from fat	100%
Cholesterol	0
Sodium	0
Protein	0
Carbohydrate	0
Dietary fiber	0

CHIEF NUTRIENTS

(None of the nutrients meet or exceed 5% of the RDA.)

Strengths: Although high in fat, canola oil is lower in saturated fat than any other vegetable oil. Like olive oil, it contains a large proportion of monounsaturates; these fats apparently lower total cholesterol and also raise beneficial HDLs—the high-density lipoproteins that help rid your body of cholesterol. Canola oil is one of the few plant sources of heart-healthy omega-3 fatty acids.

Caution: Make sure that you use canola oil as a *replacement* for highly saturated fats, such as butter, rather than in addition to them. Your goal is to keep total fat intake low—under 30% of calories.

Curiosity: Also known as rapeseed, colza, or mustard oil, canola derives its name from the words *Canada,* where most rapeseed is grown, and *oil.*

Description: This oil is light-colored and flavorless, making it suitable for salad dressing and other uncooked dishes. In addition, it has a high smoking point, so it's good for all-purpose cooking.

COCONUT

Serving: 1 Tbsp.

Calories	120
Fat	13.6 g
Saturated	11.8 g
Monounsaturated	0.8 g
Polyunsaturated	0.2 g
Calories from fat	100%
Cholesterol	0
Sodium	0
Protein	0
Carbohydrate	0
Dietary fiber	0

CHIEF NUTRIENTS

(None of the nutrients meet or exceed 5% of the RDA.)

Caution: Very high in saturated fat, which is known to raise cholesterol levels and, in turn, can increase the risk of heart disease. Although not available for home cooking, coconut oil may be present in a wide variety of processed foods, including cereal, crackers, baked goods, candy, margarine, shortening, pressurized toppings, and coatings for ice cream bars. Read labels carefully to avoid it.

Description: Obtained from dried coconut meat, coconut oil is one of the tropical oils that help extend the shelf life of foods, making it a favorite of food manufacturers.

A Better Idea: Select processed products that say they contain no tropical oils. But be aware that hydrogenated oils, often used as substitutes, are also highly saturated and not the best alternative.

CORN ⇩

Serving: 1 Tbsp.

Calories	120
Fat	13.6 g
Saturated	1.7 g
Monounsaturated	3.3 g
Polyunsaturated	7.9 g
Calories from fat	100%
Cholesterol	0
Sodium	0
Protein	0
Carbohydrate	0
Dietary fiber	0

CHIEF NUTRIENTS

NUTRIENT	AMOUNT	% RDA
Vitamin E	1.9 mg αTE	19

Strengths: Although high in fat, corn oil is a rich source of polyunsaturated fatty acids; these tend to help reduce total blood cholesterol, but may lower levels of protective HDLs. This oil is a good source of vitamin E, which is essential for the normal functioning of the nervous system.

Eat with: Almost any vegetable that contains the vitamins and fiber oil lacks.

Caution: Quite high in fat and calories; use judiciously. *Allergy alert:* Corn is a powerful allergen, often causing hives, headaches and other reactions. Anyone allergic to corn should also avoid the oil, which may be hidden in foods such as margarine, salad dressing, and baked goods.

Curiosity: Corn oil is used in paint, varnish, soap, and linoleum in addition to food products.

Description: Odorless and tasteless; with its high smoking point, corn oil is suitable for cooking and baking. Unrefined corn oil from health food stores has a stronger corn flavor than supermarket oil.

COTTONSEED

Serving: 1 Tbsp.

Calories	120
Fat	13.6 g
Saturated	3.5 g
Monounsaturated	2.4 g
Polyunsaturated	7.0 g
Calories from fat	100%
Cholesterol	0
Sodium	0
Protein	0
Carbohydrate	0
Dietary fiber	0

CHIEF NUTRIENTS

NUTRIENT	AMOUNT	% RDA
Vitamin E	4.8 mg αTE	48

Strengths: Although cottonseed oil is high in fat, most of it is the unsaturated kind. There's also quite a bit of vitamin E, which acts as an antioxidant in the body. The E protects vitamins A and C, polyunsaturated fatty acids, and other substances from harmful oxidative reactions.

Caution: This oil's relatively high amount of saturated fat makes it less desirable than other common oils, such as canola, corn, and olive. *Allergy alert:* Cottonseed, which is present in many processed foods, is a potent allergen. The oil is less allergenic but may affect some people who are highly sensitive.

History: Cottonseed oil was the most important edible oil in the U.S. until the 1940s. Even today, it is our second most widely consumed vegetable oil.

Description: Neutral-tasting; commonly found in cooking oil, shortening, salad dressing, margarine, and baked goods.

A Better Idea: Other vegetable oils (with the exception of the undesirable tropical oils) are even lower in saturated fat.

GRAPESEED

Serving: 1 Tbsp.

Calories	120
Fat	13.6 g
Saturated	1.3 g
Monounsaturated	2.2 g
Polyunsaturated	9.5 g
Calories from fat	100%
Cholesterol	0
Sodium	0
Protein	0
Carbohydrate	0
Dietary fiber	0

CHIEF NUTRIENTS

NUTRIENT	AMOUNT	% RDA
Vitamin E	3.9 mg αTE	39

Strengths: Although high in fat, most of it is polyunsaturated, the form that can help reduce levels of total cholesterol. There's also a very good amount of vitamin E, which is essential for healthy nerves and may boost immunity to certain diseases.

Eat with: High-fiber foods.

Caution: Too high in fat for excessive consumption. Make sure that when you increase your intake of grapeseed oil you decrease other fats in your diet.

Description: This light-colored oil is extracted from grape seeds. The flavor can range from bland to lightly grapey. Most of the brands on the market come from France, Italy, or Switzerland, although some U.S. sources are being developed. With a subtle flavor and a high smoking point, grapeseed oil is suitable for sautéing.

Storage: Keep in the refrigerator or a cool place.

Serve: Try using grapeseed oil in muffins, grain casseroles, or stir-fries. This versatile oil is also a good dressing for vegetable or fruit salad.

HAZELNUT

Serving: 1 Tbsp.

Calories	120
Fat	13.6 g
Saturated	1.0 g
Monounsaturated	10.6 g
Polyunsaturated	1.4 g
Calories from fat	100%
Cholesterol	0
Sodium	0
Protein	0
Carbohydrate	0
Dietary fiber	0

CHIEF NUTRIENTS

NUTRIENT	AMOUNT	% RDA
Vitamin E	6.4 mg αTE	64

Strengths: Hazelnut oil is an excellent source of vitamin E, a powerful nutrient that helps prevent the breakdown of other vitamins. Vitamin E also helps to protect cell membranes from damage from a variety of toxicants, including heavy metals such as lead and mercury.

Description: A strong, deep-flavored oil with a rich, roasted-nut aroma.

Selection: Hazelnut oil is expensive and hard to find. It's usually imported from France. Look for it in gourmet shops.

Storage: As with other oils, hazelnut is best stored in a cool, dark place, such as a refrigerator.

Serve: In a vinaigrette, hazelnut oil adds an unexpected nutty flavor to greens such as escarole and endive. It can also be used in marinades and to add flavor to baked goods containing hazelnuts. It's often diluted with a lighter oil.

OLIVE

Serving: 1 Tbsp.

Calories	119
Fat	13.5 g
Saturated	1.8 g
Monounsaturated	10.0 g
Polyunsaturated	1.2 g
Calories from fat	100%
Cholesterol	0
Sodium	0
Protein	0
Carbohydrate	0
Dietary fiber	0

CHIEF NUTRIENTS

NUTRIENT	AMOUNT	% RDA
Vitamin E	1.6 mg αTE	16

Strengths: Very high in monounsaturated fat, which has been shown to reduce total cholesterol without lowering the HDLs that help keep arteries clear of plaque. In addition, olive oil may help lower high blood pressure and even glucose levels. There's a good amount of vitamin E to help strengthen immunity.

Caution: Olive oil is high in fat. Make sure you use it as a replacement for saturated fats in your diet, not as an addition to them.

Selection: Color can range from deep green to light amber; the flavor from bland to overpowering. Extra-virgin, which comes from the first pressing of the olives, is considered the finest. Lesser grades include superfine, semifine, and pure. ''Light'' olive oil is very mild tasting, so it's suitable for dishes where a strong olive taste is undesirable.

Serve: With pasta, bread, grains, vegetables, meat, poultry, seafood—practically anything that can be enhanced by the oil's distinctive flavor. Excellent in salad dressings.

PALM

Serving: 1 Tbsp.

Calories	120
Fat	13.6 g
Saturated	6.7 g
Monounsaturated	5.0 g
Polyunsaturated	1.3 g
Calories from fat	100%
Cholesterol	0
Sodium	0
Protein	0
Carbohydrate	0
Dietary fiber	0

CHIEF NUTRIENTS

(None of the nutrients meet or exceed 5% of the RDA.)

Caution: Along with palm, coconut, and palm kernel oils, plain palm oil is one of the most highly saturated vegetable oils. Its saturated fat is known to raise cholesterol and increase the risk of heart disease. Although there is a controversy over whether palm oil promotes atherosclerosis, far healthier oils are available. You would be wise to opt for them whenever possible. Read labels: All the tropical oils are used in all sorts of processed food, especially baked goods.

Curiosity: Palm oil is one of the world's most important soap-making oils.

Description: Palm oil is extracted from the fruit pulp of the oil palm. Thanks to some beta-carotene, it's reddish-orange in its crude state. After processing, it turns white. At room temperature palm oil is solid—which is why it's used in vegetable shortening and margarine.

PEANUT ⇩

Serving: 1 Tbsp.

Calories	119
Fat	13.5 g
Saturated	2.3 g
Monounsaturated	6.2 g
Polyunsaturated	4.3 g
Calories from fat	100%
Cholesterol	0
Sodium	0
Protein	0
Carbohydrate	0
Dietary fiber	0

CHIEF NUTRIENTS

NUTRIENT	AMOUNT	% RDA
Vitamin E	1.6 mg αTE	16

Strengths: With a high level of monounsaturated and polyunsaturated fat, peanut oil can help keep serum cholesterol down. It also has a good portion of vitamin E, which is essential for a healthy nervous system. Like other oils, the peanut variety is sodium- and cholesterol-free.

Eat with: Any food high in fiber and vitamins, such as fresh fruit and vegetables.

Caution: High in calories and total fat, both of which can contribute to obesity. Use peanut oil in moderation as a substitute for butter and other saturated fats.

Description: In general, American peanut oil is quite bland, whereas Chinese brands have a distinct peanut flavor. The oil has a high smoking point, making it good for all types of stove-top cooking.

Storage: Keep in a refrigerator or some other cool, dark place.

Serve: In tasty salad dressings. Combine peanut oil with fruit vinegar for use in fruit salads or with balsamic vinegar for green salads. The oil is also perfect for stir-fries and other oriental dishes.

SAFFLOWER

Serving: 1 Tbsp.

Calories	120
Fat	13.6 g
Saturated	1.2 g
Monounsaturated	1.7 g
Polyunsaturated	10.1 g
Calories from fat	100%
Cholesterol	0
Sodium	0
Protein	0
Carbohydrate	0
Dietary fiber	0

CHIEF NUTRIENTS

NUTRIENT	AMOUNT	% RDA
Vitamin E	4.6 mg αTE	46

Strengths: Of all the common types of oil, safflower has the highest amount of polyunsaturated fat, which is valued for its ability to help lower total cholesterol. It is a very good source of vitamin E, which helps prevent the polyunsaturates from undergoing destructive oxidative reactions in the body.

Caution: Too high in total fat for hearty use. For best health benefits, make sure you *replace* other more saturated fats in your diet with safflower oil rather than adding it to them.

History: Safflower is a thistlelike plant that was used in ancient times to make dye. Oil from its seeds lighted the lamps of Egypt's pharaohs.

Description: Bland and almost colorless, safflower oil has a high smoking point, making it suitable for cooking. Doesn't solidify when chilled, so it's good for salad dressing. Very often found in soft and liquid-oil margarine, commercial salad dressing, and diet ice cream.

Serve: On all types of salad ingredients, including chicken and tuna, as well as potatoes, greens, and other vegetables.

SESAME

Serving: 1 Tbsp.

Calories	120
Fat	13.6 g
Saturated	1.9 g
Monounsaturated	5.4 g
Polyunsaturated	5.7 g
Calories from fat	100%
Cholesterol	0
Sodium	0
Protein	0
Carbohydrate	0
Dietary fiber	0

CHIEF NUTRIENTS

(None of the nutrients meet or exceed 5% of the RDA.)

Strengths: Although high in total fat, most of it is unsaturated—the type of fat that can help lower serum cholesterol. Like other vegetable oil, sesame oil contains no dietary cholesterol or sodium.

Caution: Too high in fat and calories for unlimited use. Remember that even heart-healthy oil should be consumed in moderation.

Description: There are two basic types of sesame oil. One is very light in color and has a subtle sesame flavor; the other is much darker, with a stronger flavor and fragrance. Both types have a high smoking point, so they're suitable for sautéing. The light oil is good for salad dressings because its delicate flavor won't overwhelm other ingredients. The dark oil is best as a flavor accent; try it in oriental and Indian dishes. Both kinds are available at most health food stores.

Serve: With spinach and other salad greens, tofu, and oriental vegetables such as bok choy and snow peas.

SOYBEAN ⇩

Serving: 1 Tbsp.

Calories	120
Fat	13.6 g
Saturated	2.0 g
Monounsaturated	3.2 g
Polyunsaturated	7.9 g
Calories from fat	100%
Cholesterol	0
Sodium	0
Protein	0
Carbohydrate	0
Dietary fiber	0

CHIEF NUTRIENTS

NUTRIENT	AMOUNT	% RDA
Vitamin E	1.5 mg αTE	15

Strengths: High in polyunsaturated fat, which can help lower cholesterol. Unrefined soybean oil is also a good source of omega-3's, which may help protect against not only heart disease but also arthritis and cancer. As with many other vegetable oils, there's a good bit of vitamin E and no sodium or cholesterol.

Caution: Calories and total grams of fat are high.

Description: A light, yellowish oil. Its bland flavor makes it suitable for use in salad dressings. It has a high smoking point, so it's also good for cooking and baking (especially bran muffins, whole grain waffles, or bean salads). Cold-pressed soy oil, sold in health-food stores, is darker and has a nutty taste.

Selection: Soybean oil is the leading oil consumed in the U.S. In addition to being sold as an oil, it's a major component of margarine, shortening, and commercial salad dressing.

SUNFLOWER ☀⇩

Serving: 1 Tbsp.

Calories	120
Fat	13.6 g
Saturated	1.4 g
Monounsaturated	2.7 g
Polyunsaturated	8.9 g
Calories from fat	100%
Cholesterol	0
Sodium	0
Protein	0
Carbohydrate	0
Dietary fiber	0

CHIEF NUTRIENTS

NUTRIENT	AMOUNT	% RDA
Vitamin E	6.1 mg αTE	61

Strengths: With its high amount of polyunsaturated fat, sunflower oil may be a good choice to help lower high cholesterol levels. It's also an excellent source of vitamin E, which may enhance the body's immune response to certain diseases. Like other oils, it has no cholesterol and no sodium.

Eat with: Practically any food. Those highest in fiber and vitamins contribute nutrients that the oil lacks.

Caution: Like other oils, sunflower oil is too high in calories and total fat to eat with abandon.

Curiosity: Eighty percent of the vegetable oil produced in Russia comes from sunflowers.

Description: A mild, pale yellow oil with a faintly nutty flavor. Suitable for salad dressings, it's also a satisfactory cooking oil, even though it has a relatively low smoking point. A new type of sunflower oil, higher in monounsaturated fat, is being used in some commercially prepared salad dressings, frying and roasting applications, and for spray coatings to replace saturated fats such as coconut and palm kernel oils.

WALNUT ⇩

Serving: 1 Tbsp.

Calories	120
Fat	13.6 g
Saturated	1.2 g
Monounsaturated	3.1 g
Polyunsaturated	8.6 g
Calories from fat	100%
Cholesterol	0
Sodium	0
Protein	0
Carbohydrate	0
Dietary fiber	0

CHIEF NUTRIENTS

(None of the nutrients meet or exceed 5% of the RDA.)

Strength: Although this oil is high in total fat, most of it is the beneficial polyunsaturated type that can help lower high cholesterol.

Caution: Like other vegetable oils, walnut oil is too high in fat and calories for frequent consumption.

Description: A distinctively nutty flavor and fragrance. Many brands are imported from France.

Selection: Some walnut oils have a stronger flavor than others, so you may need to experiment to find the one you like best. The type of walnut oil found in health-food stores is much blander and far less expensive.

Storage: Keep in a cool, dark place. To store for more than 3 months, place in the refrigerator. Walnut oil may turn thick or cloudy under refrigeration; just warm it a bit before using.

Serve: On any salad.

A Better Idea: Because walnut oil is so strong-tasting—and so expensive—you might want to dilute it with blander oils, such as canola or corn, for salad dressings or sautéing.

SALAD DRESSINGS

BLUE CHEESE OR ROQUEFORT, *Low-Cal/Low-Fat*

Serving: 2 Tbsp.

Calories	6
Fat	0.3 g
Saturated	trace
Monounsaturated	N/A
Polyunsaturated	N/A
Calories from fat	54%
Cholesterol	0
Sodium	340 mg
Protein	0.4 g
Carbohydrate	0.4 g
Dietary fiber	trace

CHIEF NUTRIENTS

(None of the nutrients meet or exceed 5% of the RDA.)

Strengths: This convenient alternative to regular blue cheese dressing contains hardly any calories and a negligible amount of fat.

Caution: 340 mg of sodium in a mere 2 Tbsp. of dressing could be a problem if you have high blood pressure or you're on a low-salt diet. *Allergy alert:* Aged cheese (like blue-vein or Roquefort cheese) contains tyramine, a substance that may trigger migraine headaches in sensitive individuals.

Serve: With salad or crudités (cut-up raw vegetables).

A Better Idea: Choose a low-sodium or sodium-free non-cheese-type dressing. Some companies make single-serving low-calorie dressings that enable you to carry dressing with you when you dine out.

BLUE CHEESE OR ROQUEFORT, *Regular*

Serving: 2 Tbsp.

Calories	154
Fat	16.0 g
Saturated	3.0 g
Monounsaturated	3.8 g
Polyunsaturated	8.3 g
Calories from fat	93%
Cholesterol	5 mg
Sodium	335 mg
Protein	1.5 g
Carbohydrate	2.3 g
Dietary fiber	0

CHIEF NUTRIENTS

NUTRIENT	AMOUNT	% RDA
Vitamin E	1.8 mg αTE	18

Caution: This dressing has more calories than hot fudge sauce, with as much saturated fat and 3 times as much total fat. So it's not a wise choice for those who are trying to lose weight or follow a heart-healthy diet. Regular blue cheese or Roquefort dressing may also contain too much sodium for those on a low-sodium diet. *Allergy alert:* Read labels. May contain milk, soy, or other food allergens. May also trigger headaches or other reactions in those who are sensitive to mold or tyramine (a compound found in aged cheese).

Strengths: Blue cheese dressing supplies a handsome amount of vitamin E, a nutrient that protects other nutrients, such as vitamins A and C, from breaking down.

Serve: On a salad of dark leafy greens, like endive; with crunchy, high-fiber vegetables, like chopped broccoli, cabbage, carrots, and cauliflower.

A Better Idea: Look for low-calorie, reduced-sodium versions of blue cheese or Roquefort dressing—preferably with less than 3 g of fat per Tbsp.

FRENCH *(Vinaigrette)*

Serving: 2 Tbsp.

Calories	177
Fat	19.7 g
Saturated	3.5 g
Monounsaturated	5.8 g
Polyunsaturated	9.4 g
Calories from fat	100%
Cholesterol	0
Sodium	184 mg
Protein	trace
Carbohydrate	1.0 g
Dietary fiber	0

CHIEF NUTRIENTS

NUTRIENT	AMOUNT	% RDA
Vitamin E	2.2 mg αTE	22

Caution: Fairly high in fat, although most of it is unsaturated. *Allergy alert:* Vinegar may trigger hives or other reactions in those who are allergic to molds, yeasts, or other fungi.

Strengths: Thanks to the presence of vegetable oil, this basic vinaigrette contains a respectable amount of vitamin E, a nutrient that helps enhance resistance to certain diseases.

Description: Consists of 3 parts olive oil to 1 part wine vinegar, seasoned with salt, pepper, and prepared mustard.

Storage: Chill.

Preparation: Combine all ingredients in a Mason jar, cover tightly, and shake vigorously.

Serve: Over cooked artichokes, green salads, or cold meat. It's a nice alternative to more highly saturated toppings such as butter or hollandaise sauce.

A Better Idea: To reduce sodium and calories, omit the salt and increase the proportion of vinegar.

FRENCH, *Low-Cal/Low-Fat*   ⇩

Serving: 2 Tbsp.

Calories	44
Fat	1.9 g
Saturated	0.3 g
Monounsaturated	0.5 g
Polyunsaturated	1.1 g
Calories from fat	39%
Cholesterol	2 mg
Sodium	257 mg
Protein	0.1 g
Carbohydrate	7.1 g
Dietary fiber	0.1 g

CHIEF NUTRIENTS

(None of the nutrients meet or exceed 5% of the RDA.)

Strengths: Low-calorie French dressing contains about one-third as many calories and one-seventh as much fat as regular French dressing.

Caution: Bottled French dressing is fairly high in sodium, which may be off-limits if you're trying to keep your blood pressure within a healthy/normal range. *Allergy alert:* Read labels if you're sensitive to small amounts of monosodium glutamate (MSG), eggs, soy, citrus, or other ingredients that tend to trigger reactions in some people.

Description: A reduced-calorie version of the popular orange-colored tabletop salad dressing.

Serve: On a salad of endive or other dark leafy greens and lots of raw vegetables like chopped broccoli, cabbage, carrots, and cauliflower.

A Better Idea: If you're trying to control your blood pressure by monitoring your diet, shop around for a reduced-calorie dressing that's also low in sodium.

FRENCH, *Regular*

Serving: 2 Tbsp.

Calories	134
Fat	12.8 g
Saturated	3.0 g
Monounsaturated	2.5 g
Polyunsaturated	6.3 g
Calories from fat	86%
Cholesterol	18 mg
Sodium	427 mg
Protein	0.2 g
Carbohydrate	5.5 g
Dietary fiber	0

CHIEF NUTRIENTS

NUTRIENT	AMOUNT	% RDA
Vitamin E	1.6 mg αTE	16

Caution: The high sodium is a potential problem if you're prone to high blood pressure. Also, fat content and percentage of calories from fat exceed the maximum level recommended by the American Heart Association. *Allergy alert:* Read labels. Bottled French dressing may contain egg, soy, MSG, or other common food allergens. Also, those allergic to molds may not tolerate dressing with vinegar.

Strengths: French dressing offers a respectable amount of vitamin E, a nutrient that may enhance the immune response to certain diseases.

Description: A reddish-orange American dressing that's creamy and somewhat tart. It bears no resemblance to the classic French vinaigrette.

Serve: On a salad of dark leafy greens, such as spinach or romaine lettuce, heaped with crunchy, high-fiber vegetables.

A Better Idea: Look for low-calorie French dressing—you get a fraction of the fat and calories (and possibly less sodium).

ITALIAN, *Low-Cal*

Serving: 2 Tbsp.

Calories	32
Fat	2.9 g
Saturated	0.4 g
Monounsaturated	0.6 g
Polyunsaturated	1.8 g
Calories from fat	84%
Cholesterol	2 mg
Sodium	236 mg
Protein	trace
Carbohydrate	1.5 g
Dietary fiber	0.1 g

CHIEF NUTRIENTS

(None of the nutrients meet or exceed 5% of the RDA.)

Strength: Dieters' Italian dressing contains less than one-fourth the calories of regular Italian dressing.

Caution: Some low-calorie dressings are low in sodium, but this one is high. Read labels if you're trying to limit your sodium intake. *Allergy alert:* As with any bottled salad dressing, people with allergies should scrutinize labels for ingredients like eggs, soy, and monosodium glutemate (MSG), which may trigger adverse reactions.

Description: This is a reduced-calorie version of the popular bottled Italian dressing. Water, cellulose gel, and modified food starch are substituted for a large part of the oil to keep the fat content low.

Serve: With a heaping bowl of low-cal greens and crunchy, high-fiber beans and veggies. Skip the fat-sodden croutons, rich macaroni salad, and other fatty fixin's.

ITALIAN, *Regular*

Serving: 2 Tbsp.

Calories	137
Fat	14.2 g
Saturated	2.1 g
Monounsaturated	3.3 g
Polyunsaturated	8.2 g
Calories from fat	93%
Cholesterol	0
Sodium	231 mg
Protein	0.2 g
Carbohydrate	3.0 g
Dietary fiber	0

CHIEF NUTRIENTS

NUTRIENT	AMOUNT	% RDA
Vitamin E	1.5 mg αTE	15

Caution: Nearly all the calories in Italian dressing come from fat, although very little of it is saturated. The sodium content could pose problems if you've been advised to limit your daily intake. *Allergy alert:* Read labels. Bottled Italian dressing may contain small amounts of monosodium glutamate (MSG), soy, or other ingredients that tend to trigger reactions in sensitive people.

Strengths: Contains a good amount of vitamin E, a nutrient that helps to ensure that other nutrients become available to the body. As an antioxidant, vitamin E may also help protect against some of the effects of certain pollutants, like mercury.

Preparation: Wait until the last minute to dress salad greens. Otherwise, they'll wilt.

Serve: With endive, chicory, escarole, and other slightly bitter Mediterranean greens.

A Better Idea: Look for one of the many Italian dressings that are low in fat and sodium.

MAYONNAISE-TYPE, *Low-Cal*

Serving: 1 Tbsp.

Calories	22
Fat	2.0 g
Saturated	0.4 g
Monounsaturated	N/A
Polyunsaturated	N/A
Calories from fat	84%
Cholesterol	8 mg
Sodium	19 mg
Protein	0.2 g
Carbohydrate	0.8 g
Dietary fiber	0

CHIEF NUTRIENTS

(None of the nutrients meet or exceed 5% of the RDA.)

Strengths: Low-calorie mayo-type salad dressing contains half the fat of regular salad dressing and less than a quarter of the fat of mayonnaise. It also contains far less sodium.

Caution: Products may vary. Check labels if you're watching your sodium or fat intake. *Allergy alert:* Mayonnaise and other products containing vinegar may cause adverse reactions in those who are allergic to molds or yeasts.

Serve: If you're watching your intake of fat and calories, use low-calorie mayonnaise-type salad dressing instead of regular spread in potato salad, on sandwiches, and in other luncheon foods.

A Better Idea: If you're on a sodium-restricted diet, look for diet mayonnaise without added salt.

MAYONNAISE-TYPE, *Regular*

Serving: 1 Tbsp.

Calories	57
Fat	4.9 g
Saturated	0.7 g
Monounsaturated	1.3 g
Polyunsaturated	2.7 g
Calories from fat	77%
Cholesterol	4 mg
Sodium	104 mg
Protein	0.1 g
Carbohydrate	3.5 g
Dietary fiber	0

CHIEF NUTRIENTS

NUTRIENT	AMOUNT	% RDA
Vitamin E	0.6 mg αTE	6

Strengths: Contains less than one-third the calories and less than half the fat of mayonnaise. This salad dressing is slightly lower in calories and fat than some products labeled imitation mayonnaise.

Eat with: Tuna packed in water, turkey breast, and other low-fat fare to balance the overall fat content of the meal.

Caution: This dressing contains moderate amounts of cholesterol. And it's high in fat and calories, so don't go overboard. *Allergy alert:* Check the labels for eggs and other potential food allergens.

Description: Mayonnaise-type salad dressing contains less oil than mayonnaise, accounting for the lower calorie and fat count.

Serve: Mayonnaise-type salad dressing is a calorie-saving alternative when spread on sandwich bread. It can also be mixed in tuna or chicken salad, or used to make coleslaw and other luncheon dishes.

OIL AND VINEGAR ⬇

Serving: 2 Tbsp.

Calories	140
Fat	15.6 g
Saturated	2.8 g
Monounsaturated	4.6 g
Polyunsaturated	7.5 g
Calories from fat	100%
Cholesterol	0
Sodium	0.2 mg
Protein	0
Carbohydrate	0.8 g
Dietary fiber	0

CHIEF NUTRIENTS

NUTRIENT	AMOUNT	% RDA
Vitamin E	1.2 mg αTE	12

Caution: All the calories come from fat—and nearly all of those fat calories are converted to body fat. *Allergy alert:* Individuals who are allergic to molds may have to avoid vinegar. **Strengths:** Only a small percentage of the fat in oil and vinegar (vinaigrette) dressing is saturated. You also get a good amount of immunity-boosting vitamin E.

Curiosity: When serving vinaigrette with calf's head, French chefs sometimes like to thicken the sauce with pounded calves' brains.

Serve: With a few added ingredients, vinaigrette also makes an excellent topping for steamed asparagus, green beans, cauliflower, or other good-for-you vegetables.

Description: Usually made by adding three parts oil to one part vinegar.

A Better Idea: When making this dressing, be sure to use canola, olive, safflower, or another oil high in heart-healthy monounsaturated oils.

RUSSIAN, *Low-Cal* ⬇ 🗑

Serving: 2 Tbsp.

Calories	46
Fat	1.3 g
Saturated	0.2 g
Monounsaturated	0.3 g
Polyunsaturated	0.7 g
Calories from fat	25%
Cholesterol	2 mg
Sodium	283 mg
Protein	0.2 g
Carbohydrate	9.0 g
Dietary fiber	0

CHIEF NUTRIENTS

(None of the nutrients meet or exceed 5% of the RDA.)

Caution: This low-calorie form of Russian dressing gets a nod of approval for low fat and modest calories, but the sodium content in a couple of tablespoons is mean. *Allergy alert:* If you are allergic to eggs or other commonly used ingredients in prepared food, read the label carefully to determine whether a brand is safe for you.

Eat with: A hearty low-fat, high-fiber tossed salad. If you use a low-cal dressing instead of full-fat Russian dressing on a Reuben sandwich, you'll cut out 14 g of fat.

Strengths: Contains a minimum of fat and practically no cholesterol.

A Better Idea: Choose from one of the growing number of bottled dressings that are low in fat *and* sodium.

RUSSIAN, *Regular*

Serving: 2 Tbsp.

Calories	151
Fat	15.5 g
Saturated	2.2 g
Monounsaturated	3.6 g
Polyunsaturated	9.0 g
Calories from fat	93%
Cholesterol	6 mg
Sodium	266 mg
Protein	0.5 g
Carbohydrate	3.2 g
Dietary fiber	0

CHIEF NUTRIENTS

NUTRIENT	AMOUNT	% RDA
Vitamin E	1.8 mg αTE	18
Vitamin A	63.3 RE	6
Vitamin B$_{12}$	0.1 mcg	5

Caution: In Russian dressing, most of the calories come from fat, so use it sparingly. And the sodium content is so high that you'll probably want to avoid it if you're on a sodium-restricted diet. *Allergy alert:* May contain many of the allergens found in mayonnaise, such as eggs, vinegar, and monosodium glutamate.

Eat with: Low-fat, filling salad fixings like greens, chopped cabbage, sliced mushrooms, and chickpeas and other legumes.

Strengths: Like other dressings high in vegetable oil, Russian dressing offers a noticeable amount of vitamin E, which can help boost immunity. It also offers a little resistance-boosting vitamin A and a small amount of vitamin B$_{12}$.

Description: Russian dressing isn't Russian at all, but an American concoction consisting of mayonnaise mixed with chili sauce, pimentos, and other seasonings and ingredients.

A Better Idea: Scout around for low-fat and lower-sodium versions of this favored salad topping.

THOUSAND ISLAND, *Low-Cal* ⇩

Serving: 2 Tbsp.

Calories	49
Fat	3.3 g
Saturated	0.5 g
Monounsaturated	0.7 g
Polyunsaturated	1.9 g
Calories from fat	61%
Cholesterol	5 mg
Sodium	306 mg
Protein	0.2 g
Carbohydrate	5.0 g
Dietary fiber	0.4 g

CHIEF NUTRIENTS

(None of the nutrients meet or exceed 5% of the RDA.)

Strengths: This particular low-cal version of Thousand Island dressing supplies less than half the fat and calories of regular Thousand Island dressing.

Caution: With more than 300 mg of sodium per serving, this is one of the saltier dressings around. Thousand Island may be off-limits to many who have been advised to curb their sodium intake in order to reduce the risks of stroke and kidney disease. *Allergy alert:* If you are allergic to corn, eggs, soy, wheat, mold, or other common allergens, avoid salad dressings.

A Better Idea: With so many reduced-calorie dressings on the market, you should be able to find one that's even lower in fat, with far less sodium.

THOUSAND ISLAND, *Regular*

Serving: 2 Tbsp.

Calories	118
Fat	11.1 g
Saturated	1.9 g
Monounsaturated	2.6 g
Polyunsaturated	5.9 g
Calories from fat	85%
Cholesterol	8 mg
Sodium	218 mg
Protein	0.3 g
Carbohydrate	4.7 g
Dietary fiber	0.6 g

CHIEF NUTRIENTS

NUTRIENT	AMOUNT	% RDA
Vitamin E	1.2 mg αTE	12

Caution: This dressing contains a fair amount of fat and calories, with hardly a nutritional plus. If you're watching your waistline or your fat intake, go easy. *Allergy alert:* Like other bottled dressings, Thousand Island may contain vinegar, a potential problem for those allergic to mold. Read labels.

Strength: Slightly lower in fat than blue cheese or Roquefort dressing.

History: Said to be named for the Thousand Islands in the St. Lawrence Seaway, the dressing contains chunky ingredients that resemble thousands of tiny islands.

A Better Idea: You can make your own sodium-restricted Thousand Island dressing. Whisk together ½ tsp. lemon juice and ½ cup nonfat yogurt. Stir in 2 Tbsp. tomato paste, along with some finely chopped cucumber, minced garlic, shallots, and celery. Season with a pinch of dill.

Fish and Shellfish
FISH

ANCHOVIES, *Canned in Olive Oil*

Serving: 5 (about ¾ oz.)

Calories	42
Fat	1.9 g
Saturated	0.4 g
Monounsaturated	0.8 g
Polyunsaturated	0.5 g
Calories from fat	42%
Cholesterol	17 mg
Sodium	734 mg
Protein	5.8 g
Carbohydrate	0
Dietary fiber	0

CHIEF NUTRIENTS

NUTRIENT	AMOUNT	% RDA
Niacin	4.0 mg	21
Iron	0.9 mg	9
Vitamin B$_{12}$	0.2 mcg	9
Calcium	46.4 mg	6

Strengths: A good source of heart-healthy omega-3's and niacin, anchovies have fair amounts of vitamin B$_{12}$, calcium, and iron.

Caution: High in sodium and purines: People with high blood pressure or gout may want to avoid them. *Drug interaction:* Should not be eaten by those taking isoniazid, a drug prescribed for the prevention and treatment of tuberculosis. *Allergy alert:* Fish can occasionally cause an allergic reaction.

History: The anchovy was part of the bounty harvested from the Mediterranean Sea by the ancient Greeks. Although it is still fished off both Atlantic and Pacific coasts, pollution has caused it to vanish from the Mediterranean.

Description: The live fish is a 3- to 6-inch flash of silver with a green back. When anchovies are cooked, sliced, pickled, and canned, they usually change color.

BASS, *Sea, Mixed Species*

Serving: 3 oz., cooked, dry heat

Calories	105
Fat	2.2 g
Saturated	0.6 g
Monounsaturated	0.5 g
Polyunsaturated	0.8 g
Calories from fat	19%
Cholesterol	45 mg
Sodium	74 mg
Protein	20.1 g
Carbohydrate	0
Dietary fiber	0

CHIEF NUTRIENTS

NUTRIENT	AMOUNT	% RDA
Vitamin B$_6$	0.4 mg	20
Magnesium	45.1 mg	13
Vitamin B$_{12}$	0.3 mcg	13
Niacin	1.6 mg	9
Riboflavin	0.1 mg	8
Potassium	278.8 mg	7
Thiamine	0.1 mg	7

Strengths: A delicate flavor. Sea bass is a good, low-fat source of heart-healthy omega-3's, magnesium, protein, and vitamins B$_6$ and B$_{12}$.

Caution: *Allergy alert:* Fish can occasionally cause an allergic reaction. Sea bass caught in tropical waters may contain a deadly toxin that causes ciguatera poisoning. To minimize risk, do not eat the liver, intestines, eyes, brains, roe, or sex organs of this fish.

Description: Pure black, generally weighing from 1½ to 3 pounds. Sea bass may reach 5 pounds in the fall.

Preparation: Try steaming, poaching, chowders—anything but frying. This fish should be handled gently, however, since it bruises easily. If you need to scale sea bass, be careful: Its skin tears easily.

BASS, *Striped*

꙳◁〈 ⬇ 🏛

Serving: 3 oz., uncooked

Calories	82
Fat	2.0 g
Saturated	0.4 g
Monounsaturated	0.6 g
Polyunsaturated	0.7 g
Calories from fat	22%
Cholesterol	68 mg
Sodium	59 mg
Protein	15.1 g
Carbohydrate	0
Dietary fiber	0

CHIEF NUTRIENTS

NUTRIENT	AMOUNT	% RDA
Vitamin B_{12}	3.3 mcg	163
Vitamin B_6	0.3 mg	13
Magnesium	34.0 mg	10
Niacin	1.8 mg	9
Iron	0.7 mg	7
Potassium	217.6 mg	6
Thiamine	0.1 mg	6

Strengths: Called a "striper" on the Atlantic coast and a "rockfish" in the Chesapeake Bay area, the striped bass is a great low-fat source of blood-building vitamin B_{12}. It also has a nice shot of heart-healthy omega-3's and a good amount of vitamin B_6 to keep your immune system on the alert.

Caution: Stick to commercially caught fish taken at sea. Recreationally caught striped bass from freshwater ponds, lakes, and streams or from sheltered ocean bays and harbors may be contaminated with PCBs (polychlorinated biphenyls, chemicals that have been shown to cause cancer in laboratory animals). Pregnant and nursing women may want to check with their local health department before eating striped bass. Also, check before serving it to children.

Selection: Saltwater bass has a better flavor than freshwater bass. And the fresher the bass, the better the flavor. Bass should have bright, unsunken eyes, red-colored gillrakers, and a firm, silver body that will not hold fingerprints when it's squeezed.

Preparation: Striped bass is delicious poached, steamed, grilled, or broiled.

BLUEFISH

꙳◁〈

Serving: 3 oz., uncooked

Calories	105
Fat	3.6 g
Saturated	0.8 g
Monounsaturated	1.5 g
Polyunsaturated	0.9 g
Calories from fat	31%
Cholesterol	50 mg
Sodium	51 mg
Protein	17.0 g
Carbohydrate	0
Dietary fiber	0

CHIEF NUTRIENTS

NUTRIENT	AMOUNT	% RDA
Vitamin B_{12}	4.6 mcg	229
Niacin	5.1 mg	27
Vitamin B_6	0.3 mg	17
Vitamin A	101.2 RE	10
Magnesium	28.1 mg	8
Potassium	316.2 mg	8

Strengths: An excellent source of nerve-protecting vitamin B_{12}, bluefish has a good measure of niacin and a fair amount of omega-3's, special fats that seem to help protect against artery disease.

Caution: If not refrigerated immediately after being caught, bluefish can cause histamine poisoning. When these fish are over 20 inches long, they sometimes have excessive levels of PCBs, chemical pollutants that have been found to cause cancer in animals. *Allergy alert:* Fish can occasionally cause an allergic reaction.

Curiosity: Bluefish travel in savage schools that chop, mangle, and devour everything they can sink their razor-sharp teeth into. Found along the Atlantic coast from Maine to Mexico, they are particularly plentiful off Long Island.

History: For 17th-century English settlers on Nantucket, the bluefish was a tasty staple. Two centuries later, it kept many Depression-era families on Long Island and Manhattan from going hungry.

CARP

Serving: 3 oz., cooked, dry heat

Calories	138
Fat	6.1 g
Saturated	1.2 g
Monounsaturated	2.5 g
Polyunsaturated	1.6 g
Calories from fat	40%
Cholesterol	71 mg
Sodium	54 mg
Protein	19.4 g
Carbohydrate	0
Dietary fiber	0

CHIEF NUTRIENTS

NUTRIENT	AMOUNT	% RDA
Vitamin B_{12}	1.3 mcg	63
Iron	1.4 mg	14
Zinc	1.6 mg	11
Potassium	363.0 mg	10
Vitamin B_6	0.2 mg	10
Magnesium	32.3 mg	9
Niacin	1.8 mg	9
Thiamine	0.1 mg	8
Folate	14.7 mcg	7

Strengths: A good all-around body builder. Carp is an excellent source of vitamin B_{12}, which helps protect your nervous system. A good supply of iron helps build red blood cells, while zinc aids clotting and helps to heal wounds. This fish is a fair source of vitamin B_6, niacin, thiamine, folate, potassium, magnesium, calcium, and omega-3's.

Caution: *Allergy alert:* Fish can occasionally cause an allergic reaction.

Curiosity: Smaller varieties are sold as goldfish in pet stores.

History: Carp probably originated in China and was introduced to Europe during the Middle Ages. First imported to North America in 1876.

Selection: Carp frequently accumulate pollutants because they feed off the bottom of rivers and lakes. To avoid this problem, look for commercially caught fish, which are more likely to come from unpolluted areas.

CATFISH, *Channel, Breaded and Fried*

Serving: 3 oz.

Calories	195
Fat	11.3 g
Saturated	2.8 g
Monounsaturated	4.8 g
Polyunsaturated	2.8 g
Calories from fat	52%
Cholesterol	69 mg
Sodium	238 mg
Protein	15.4 g
Carbohydrate	6.8 g
Dietary fiber	N/A

CHIEF NUTRIENTS

NUTRIENT	AMOUNT	% RDA
Vitamin B_{12}	1.6 mcg	81
Iron	1.2 mg	12
Niacin	1.9 mg	10
Potassium	289.0 mg	8
Vitamin B_6	0.2 mg	8
Folate	14.0 mcg	7
Magnesium	23.0 mg	7
Riboflavin	0.1 mg	6

Strengths: Loaded with nerve-protecting vitamin B_{12}, catfish is a good source of protein and iron, as well as niacin—the metabolic broker that enables your body to convert calories into energy. It also has a fair amount of omega-3's to help protect against artery disease and a touch of folate, magnesium, and potassium.

Eat with: Fruits and vegetables that are high in vitamin C.

Caution: *Allergy alert:* Fish can occasionally cause an allergic reaction. Breading and frying adds fat and calories, so cook catfish by dry heat if you're on a weight-loss diet.

Curiosity: Some catfish go for a walk. They can survive out of water for many hours and are reported to travel from one pond or waterway to another.

Selection: Catfish are bottom feeders, which means that they frequently develop a "muddy" flavor. The delicately flavored channel catfish are often raised in fish farms to prevent this problem. Fish farm ponds are also less likely to harbor the pollutants that bottom feeders can accumulate.

CATFISH, *Channel, Uncooked*

Serving: 3 oz.

Calories	99
Fat	3.6 g
Saturated	0.8 g
Monounsaturated	1.4 g
Polyunsaturated	0.9 g
Calories from fat	33%
Cholesterol	49 mg
Sodium	54 mg
Protein	15.5 g
Carbohydrate	0
Dietary fiber	0

CHIEF NUTRIENTS

NUTRIENT	AMOUNT	% RDA
Vitamin B_{12}	1.9 mcg	94
Niacin	1.8 mg	10
Vitamin B_6	0.2 mg	9
Iron	0.8 mg	8
Potassium	296.7 mg	8
Folate	12.8 mcg	6
Magnesium	21.3 mg	6
Riboflavin	0.1 mg	5

Strengths: Catfish is fundamentally lean. High in protein, catfish that's grilled or broiled has about one-third the fat and calories of breaded, fried catfish—and less than one-fourth of the sodium. A super dose of vitamin B_{12} helps to keep nerves healthy. And catfish is a fair source of omega-3 fatty acids, special fats that seem to help protect against artery disease.

Caution: Contains purines, substances that should be eaten sparingly by anyone with gout. *Allergy alert:* While it's rare, fish can occasionally cause an allergic reaction.

Selection: Buy catfish raised on fish farms—they're less likely to harbor the pollutants that accumulate in fish that feed on the bottom of contaminated lakes or ponds.

Curiosity: Long, whiskerlike feelers hanging down from around its mouth give the catfish its name.

Preparation: Grill or broil.

CAVIAR, *Black or Red*

Serving: 1 Tbsp.

Calories	40
Fat	2.9 g
Saturated	0.7 g
Monounsaturated	0.7 g
Polyunsaturated	N/A
Calories from fat	64%
Cholesterol	94 mg
Sodium	240 mg
Protein	3.9 g
Carbohydrate	0.6 g
Dietary fiber	0

CHIEF NUTRIENTS

NUTRIENT	AMOUNT	% RDA
Vitamin B_{12}	3.2 mcg	160
Iron	1.9 mg	19
Magnesium	48.0 mg	14
Vitamin A	89.6 RE	9
Calcium	44.0 mg	6
Riboflavin	0.1 mg	6

Strengths: Caviar is packed with vitamin B_{12} to help build red blood cells and nerves (though it's unfortunately high in percent of calories from fat). It's also a good source of iron and magnesium.

Caution: Loaded with sodium. People with high blood pressure should avoid it. *Drug interaction:* The tyramine in caviar can combine with MAO-inhibitor antidepressants to cause dangerously high blood pressure. *Allergy alert:* Like other fish products, caviar can occasionally cause an allergic reaction.

Origin: The caviar that's considered highest quality comes from the beluga sturgeon that swim along the Iranian and Soviet shores of the Caspian Sea. These sturgeon can be 3,500 pounds and up to 100 years old.

Description: Caviar is made of raw fish ovaries and eggs that may be black, silver-gray, yellow-golden or deep red.

Selection: To buy caviar that is preserved with the least amount of salt, look for the word *malossol* on the package label.

Storage: Caviar is extremely perishable. Refrigerate, then serve in a bowl surrounded by ice.

COD, *Atlantic*   ⚡💧 ⬇ 🏛

Serving: 3 oz., cooked, dry heat

Calories	89
Fat	0.7 g
Saturated	trace
Monounsaturated	trace
Polyunsaturated	trace
Calories from fat	7%
Cholesterol	47 mg
Sodium	66 mg
Protein	19.4 g
Carbohydrate	0
Dietary fiber	0

CHIEF NUTRIENTS

NUTRIENT	AMOUNT	% RDA
Vitamin B_{12}	0.9 mcg	45
Vitamin B_6	0.2 mg	12
Niacin	2.1 mg	11
Magnesium	35.7 mg	10
Potassium	207.4 mg	6

Strengths: A dieter's delight, cod is low in fat and calories, high in protein, rich in vitamin B_{12}, and a good source of niacin, vitamin B_6, and magnesium. It has just a trace of omega-3's.

Caution: Do not eat raw: Cod may contain a parasite that is killed only through cooking. *Allergy alert:* Fish can occasionally cause an allergic reaction.

Curiosity: When the English adventurer Bartholomew Gosnold sailed south from Nova Scotia in 1602, he found so many cod off the American coast that he named the area's peninsula "Cape Cod." Today, despite the fact that each cod lays 4½ million eggs per year, giant trawlers and fish-processing factory ships have overfished the area.

Preparation: Cook by dry heat. Cod is also delicious prepared in a low-fat, tomato-based fish chowder.

CROAKER, *Atlantic, Breaded and Fried*

Serving: 3 oz.

Calories	188
Fat	10.8 g
Saturated	3.0 g
Monounsaturated	4.5 g
Polyunsaturated	2.5 g
Calories from fat	52%
Cholesterol	71 mg
Sodium	296 mg
Protein	15.5 g
Carbohydrate	6.4 g
Dietary fiber	N/A

CHIEF NUTRIENTS

NUTRIENT	AMOUNT	% RDA
Vitamin B_{12}	1.8 mcg	90
Niacin	3.7 mg	19
Vitamin B_6	0.2 mg	11
Magnesium	35.7 mg	10
Folate	15.4 mcg	8
Potassium	289.0 mg	8
Iron	0.7 mg	7
Riboflavin	0.1 mg	6
Thiamine	0.1 mg	5

Strengths: Croaker is an excellent source of vitamin B_{12}, with a good amount of protein, niacin, magnesium, and vitamin B_6. It also contains a fair amount of omega-3's, the special fats that seem to help protect against artery disease.

Caution: Croaker gains a great deal of fat, calories, and sodium in the breading-and-frying process. People with high blood pressure will probably want to avoid breaded, fried croaker since it is has a lot of sodium. *Allergy alert:* Some people occasionally experience an allergic reaction to fish.

Curiosity: Croakers—also called "drums"—contain a muscle that contracts against the fish's gas bladder to produce a drumming or croaking sound. The sound can be heard on land as croakers swim over shallow sandy bottoms near estuaries or in the surf along the Atlantic coast.

Selection: Croaker is frequently sold already breaded and ready to fry.

A Better Idea: Since omega-3's are destroyed by frying, try another method of cooking fresh croaker.

CROAKER, *Atlantic, Uncooked*

Serving: 3 oz.

Calories	88
Fat	2.7 g
Saturated	0.9 g
Monounsaturated	1.0 g
Polyunsaturated	0.4 g
Calories from fat	27%
Cholesterol	52 mg
Sodium	48 mg
Protein	15.1 g
Carbohydrate	0
Dietary fiber	0

CHIEF NUTRIENTS

NUTRIENT	AMOUNT	% RDA
Vitamin B$_{12}$	2.1 mcg	107
Niacin	3.6 mg	19
Vitamin B$_6$	0.3 mg	13
Magnesium	34.0 mg	10
Potassium	293.3 mg	8
Folate	12.8 mcg	6

Strengths: Croaker is moderately lean and is high in unsaturated fat. When prepared by dry heat instead of being breaded and fried, croaker has one-quarter the fat, less than half the calories, and a fraction of the sodium. And croaker is fundamentally loaded with vitamin B$_{12}$, and has small amounts of other valuable B vitamins, which combine to promote healthy blood and nerves.

Caution: Croaker should never be eaten raw: It may contain trematodes, parasites that are killed during cooking. Croaker contains purines, substances that should be eaten sparingly if you have gout. *Allergy alert:* Fish sometimes triggers allergic reactions, but this is rare.

Curiosity: Croakers have unusually large ear bones, which some American Indians wore as neck charms to ward off illness.

Description: Croaker is a light-fleshed fish. Its flavor is more pronounced than that of perch but not as strong as that of mackerel.

Preparation: Grill, bake, or broil on a rack.

EEL

Serving: 3 oz., cooked, dry heat

Calories	201
Fat	12.7 g
Saturated	2.6 g
Monounsaturated	7.8 g
Polyunsaturated	1.0 g
Calories from fat	57%
Cholesterol	137 mg
Sodium	55 mg
Protein	20.1 g
Carbohydrate	0
Dietary fiber	0

CHIEF NUTRIENTS

NUTRIENT	AMOUNT	% RDA
Vitamin B$_{12}$	2.5 mcg	123
Vitamin A	965.6 RE	97
Niacin	3.8 mg	20
Zinc	1.8 mg	12
Thiamine	0.2 mg	11
Potassium	296.7 mg	8
Folate	14.7 mcg	7
Magnesium	22.1 mg	6

Strengths: Eel is an excellent source of vitamin A and that great nerve protector, vitamin B$_{12}$. It also has plentiful niacin, thiamine, and zinc to keep your immune system on track.

Caution: *Allergy alert:* Fish can occasionally cause an allergic reaction.

Curiosity: Both American and European eels are born in the southwest part of the North Atlantic known as the Sargasso Sea. Immediately after birth, the American eels head toward North America and the European eels head toward Europe. The Americans arrive in about a year, while the Europeans—which have to travel almost 5,000 miles—arrive 3 years after they're born. When eels start their journey, their bodies are transparent, but as they approach the coast, they begin to take on coloration.

Preparation: When the Spanish prepare young eels, they rub the bottom of an earthenware pan with hot red pepper; then the eels are sautéed in olive oil and garlic.

FLOUNDER

Serving: 3 oz., cooked, dry heat

Calories	99
Fat	1.3 g
Saturated	0.3 g
Monounsaturated	0.3 g
Polyunsaturated	0.5 g
Calories from fat	12%
Cholesterol	58 mg
Sodium	89 mg
Protein	20.5 g
Carbohydrate	0
Dietary fiber	0

CHIEF NUTRIENTS

NUTRIENT	AMOUNT	% RDA
Vitamin B$_{12}$	2.1 mcg	107
Magnesium	49.3 mg	14
Niacin	1.9 mg	10
Vitamin B$_6$	0.2 mg	10
Potassium	292.4 mg	8
Riboflavin	0.1 mg	6

Strengths: An excellent source of body-building protein and vitamin B$_{12}$. Flounder is a good, low-fat source of magnesium, niacin, and vitamin B$_6$, and it has a fair amount of omega-3's.
Caution: *Allergy alert:* Fish can occasionally cause an allergic reaction.
Curiosity: These flatfish always swim with one side down. Although the fish are born with eyes on both sides of their head, one eye migrates to the upper side shortly after birth. The mouth scrunches down to the lower side—handy when you eat off the bottom.
Selection: All sole fish—with the exception of European Dover sole—are actually varieties of flounder. Fishmongers frequently stick the label *fillet of sole* on flounder and sell it for a higher price. To many palates, a fresh fillet of flounder is just as desirable as sole.
Preparation: Flounder's delicate flavor is enhanced by the simplest possible preparation: Poach or grill, with just a brushing of oil or a drizzle of lemon juice.

GROUPER

Serving: 3 oz., cooked, dry heat

Calories	100
Fat	1.1 g
Saturated	0.3 g
Monounsaturated	0.2 g
Polyunsaturated	0.3 g
Calories from fat	10%
Cholesterol	40 mg
Sodium	45 mg
Protein	21.1 g
Carbohydrate	0
Dietary fiber	0

CHIEF NUTRIENTS

NUTRIENT	AMOUNT	% RDA
Vitamin B$_{12}$	0.6 mcg	30
Vitamin B$_6$	0.3 mg	15
Potassium	403.8 mg	11
Iron	1.0 mg	10
Magnesium	31.5 mg	9

Strengths: Grouper is a great low-fat source of protein. It has decent amounts of potassium and omega-3's, and is rich in vitamins B$_6$ and B$_{12}$.
Caution: Avoid the yellowfin grouper and misty grouper when traveling in the Bahamas and West Indies. Both types are frequently implicated in ciguatera poisoning. *Allergy alert:* Fish can occasionally cause an allergic reaction.
Preparation: The skin has such a strong flavor that most cooks remove it. Grouper is delicious poached or added to chowders and bouillabaisse. Baking and broiling can dry it out if you're not careful. Try basting the fish with low-fat marinade. Chuck the roe: It's soggy and tasteless.

HADDOCK

Serving: 3 oz., cooked, dry heat

Calories	95
Fat	0.8 g
Saturated	trace
Monounsaturated	trace
Polyunsaturated	trace
Calories from fat	7%
Cholesterol	63 mg
Sodium	74 mg
Protein	20.6 g
Carbohydrate	0
Dietary fiber	0

CHIEF NUTRIENTS

NUTRIENT	AMOUNT	% RDA
Vitamin B_{12}	1.2 mcg	59
Niacin	3.9 mg	21
Vitamin B_6	0.3 mg	15
Iron	1.2 mg	12
Magnesium	42.5 mg	12
Potassium	339.2 mg	9
Folate	11.3 mcg	6

Strengths: Tasty haddock is an excellent low-fat source of protein and nerve-strengthening vitamin B_{12}. Also a good source of magnesium, niacin, iron, and vitamin B_6, with just a touch of omega-3's.

Caution: *Allergy alert:* Fish can occasionally cause an allergic reaction.

Curiosity: The male haddock is a romantic. He sweetly hums to the female as they mate.

Description: The haddock has a black shoulder blotch called ''St. Peter's thumbprint.'' Supposedly St. Peter caught the haddock, squeezed it, then tossed it back into the sea. (The only problem with this bit of apocrypha is that St. Peter worked the shores of Galilee—and haddock are caught mostly in the Atlantic.)

Selection: Buy it fresh. A soft-fleshed fish, haddock does not respond well to salt-drying—and of course the process increases sodium levels tremendously.

Preparation: Haddock is perfect for baking, poaching, sautéing, grilling, or even smoking.

HALIBUT

Serving: 3 oz., cooked, dry heat

Calories	119
Fat	2.5 g
Saturated	0.4 g
Monounsaturated	0.8 g
Polyunsaturated	0.8 g
Calories from fat	19%
Cholesterol	35 mg
Sodium	59 mg
Protein	22.7 g
Carbohydrate	0
Dietary fiber	0

CHIEF NUTRIENTS

NUTRIENT	AMOUNT	% RDA
Vitamin B_{12}	1.2 mcg	58
Niacin	6.1 mg	32
Magnesium	91.0 mg	26
Vitamin B_6	0.3 mg	17
Potassium	489.6 mg	13
Iron	0.9 mg	9
Calcium	51.0 mg	6
Folate	11.7 mcg	6

Strengths: King of the flatfishes, halibut is an excellent low-fat source of vitamin B_{12} and a very good source of niacin, both of which keep your body building red blood cells. It's also a good source of two blood pressure stabilizers, magnesium and potassium. A good chunk of vitamin B_6 helps to keep your immune system on the job, and a dash of omega-3's helps keep your heart happy.

Caution: May contain mercury. Pregnant and nursing women should avoid halibut. *Allergy alert:* Fish can occasionally cause an allergic reaction.

Curiosity: At birth, a halibut looks like a normal fish. But as it grows, the eyes usually migrate to the upper side.

Description: The largest of the flatfishes, halibut cruises the sea bottom lying on its side.

Preparation: Halibut tends to be dry, so poaching or sautéing is recommended. Sauté garlic and mushrooms in margarine, add the fish and a little white wine and light soy sauce. Simmer until fish flakes with a fork. Delicious!

HERRING, *Atlantic*

Serving: 3 oz., cooked, dry heat

Calories	173
Fat	9.9 g
Saturated	2.2 g
Monounsaturated	4.1 g
Polyunsaturated	2.3 g
Calories from fat	51%
Cholesterol	66 mg
Sodium	98 mg
Protein	19.6 g
Carbohydrate	0
Dietary fiber	0

CHIEF NUTRIENTS

NUTRIENT	AMOUNT	% RDA
Vitamin B_{12}	11.2 mcg	559
Niacin	3.5 mg	19
Riboflavin	0.3 mg	15
Vitamin B_6	0.3 mg	15
Iron	1.2 mg	12
Magnesium	34.9 mg	10
Potassium	356.2 mg	9
Calcium	62.9 mg	8
Thiamine	0.1 mg	7
Zinc	1.1 mg	7

Strengths: A well-balanced fish from a nutritional standpoint. Fresh or frozen Atlantic herring that has been cooked by dry heat is an excellent source of heart-healthy omega-3's and nerve-protecting vitamin B_{12}. It's also a good source of niacin, riboflavin, vitamin B_6, and iron, with a fair amount of thiamine, calcium, magnesium, potassium, and zinc.

Eat with: Foods that are high in vitamin C.

Caution: Do not eat raw: may contain parasites. Herring is loaded with purines; people with gout may want to avoid it. "Kippers" or "kippered" herring has three times more salt than fresh herring. People with high blood pressure may want to avoid fish prepared in this way. *Allergy alert:* Fish can occasionally cause an allergic reaction.

Selection: Most Atlantic and Pacific herring sold in the U.S. has been salt-cured or salt-dried, then pickled in vinegar. If you're watching your salt intake, buy fresh or fresh-frozen fish whenever you have the chance.

HERRING, *Kippered*

Serving: 1½ oz. (about 1 fillet)

Calories	87
Fat	5.0 g
Saturated	1.1 g
Monounsaturated	2.0 g
Polyunsaturated	1.2 g
Calories from fat	51%
Cholesterol	33 mg
Sodium	367 mg
Protein	9.8 g
Carbohydrate	0
Dietary fiber	0

CHIEF NUTRIENTS

NUTRIENT	AMOUNT	% RDA
Vitamin B_{12}	7.5 mcg	374
Niacin	1.8 mg	9
Vitamin B_6	0.2 mg	9
Riboflavin	0.3 mg	8
Iron	0.6 mg	6
Magnesium	18.4 mg	5
Potassium	178.8 mg	5

Caution: A Scandinavian favorite, kippered herring has more than 3 times as much salt as fresh herring. People with gout should avoid herring—it's high in purines. *Drug interaction:* Smoked herring contains tyramine, a substance that can cause severe hypertension in people taking MAO inhibitors.

Strengths: Kippered herring is loaded with vitamin B_{12}. And this form of herring is a nice source of omega-3's.

Curiosity: The term "red herring" refers to something that distracts attention from the real issue. It originated with the practice of dragging a heavily cured reddened herring across a trail to confuse hunting dogs.

Origin: The fish dish "kippers" was invented in Northumberland, England, in 1843.

Description: Usually sold canned, sometimes labeled as sardines, kippered herring is brined and smoked.

Selection: Look for fat, pale kippers, with no gamy smell.

A Better Idea: Look for fresh or frozen herring instead.

HERRING, *Pickled*

Serving: ¹/₂ oz.

Calories	39
Fat	2.7 g
Saturated	0.4 g
Monounsaturated	1.8 g
Polyunsaturated	0.3 g
Calories from fat	62%
Cholesterol	2 mg
Sodium	131 mg
Protein	2.1 g
Carbohydrate	1.5 g
Dietary fiber	0

CHIEF NUTRIENTS

NUTRIENT	AMOUNT	% RDA
Vitamin B_{12}	0.6 mcg	32

Caution: Like kippered herring, the pickled variety is a bit on the salty side and should be eaten sparingly by people with high blood pressure. Because it's high in purines, pickled herring should be avoided by those with gout. *Drug interaction:* Extraordinarily high in tyramine, a substance that can cause dangerously high increases in blood pressure in people taking MAO-inhibitor antidepressants. *Allergy alert:* Tyramine precipitates migraine headaches in some people.

Strengths: Pickled herring is rich in vitamin B_{12}, and it's a nice source of omega-3's.

Curiosity: Herring sneeze when fished out of the water.

Description: Pickled herring is marinated in vinegar and spices, then bottled in either a sour cream or wine sauce.

MACKEREL, *Atlantic*

Serving: 3 oz., cooked, dry heat

Calories	223
Fat	15.1 g
Saturated	3.6 g
Monounsaturated	6.0 g
Polyunsaturated	1.4 g
Calories from fat	61%
Cholesterol	64 mg
Sodium	71 mg
Protein	20.3 g
Carbohydrate	0
Dietary fiber	0

CHIEF NUTRIENTS

NUTRIENT	AMOUNT	% RDA
Vitamin B_{12}	16.2 mcg	808
Niacin	5.8 mg	31
Magnesium	82.5 mg	24
Riboflavin	0.4 mg	21
Vitamin B_6	0.4 mg	20
Iron	1.3 mg	13
Potassium	340.9 mg	9
Thiamine	0.1 mg	9
Zinc	0.8 mg	5

Strengths: One of the healthiest finfish you can eat, Atlantic mackerel is loaded with omega-3's to protect your heart, keep your arteries flexible, and lower your triglycerides. These fish oils can also lower your risk of breast cancer and boost your immune system. Mackerel in general is a super source of vitamin B_{12}, a very good source of niacin, and it has respectable amounts of magnesium, iron, riboflavin, and vitamin B_6.

Caution: If not refrigerated immediately after being caught, mackerel can cause histamine poisoning. Do not eat raw: may contain parasites. It's loaded with purines, so people who have gout may want to avoid it. *Drug interaction:* May intensify the effect of propranolol, a drug frequently used to treat high blood pressure.

Preparation: Best when bought absolutely fresh and cooked as soon as possible. Excellent broiled with a sprinkle of lemon juice or, as the French prefer, with a sour sauce made of gooseberries.

MACKEREL, *King*

Serving: 3 oz., uncooked

Calories	89
Fat	1.7 g
Saturated	0.3 g
Monounsaturated	0.7 g
Polyunsaturated	0.4 g
Calories from fat	17%
Cholesterol	45 mg
Sodium	134 mg
Protein	17.2 g
Carbohydrate	0
Dietary fiber	0

CHIEF NUTRIENTS

NUTRIENT	AMOUNT	% RDA
Vitamin B_{12}	13.3 mcg	663
Niacin	7.3 mg	38
Riboflavin	0.4 mg	24
Vitamin A	185.3 RE	19
Vitamin B_6	0.4 mg	19
Iron	1.5 mg	15
Potassium	369.8 mg	10
Magnesium	27.2 mg	8
Thiamine	0.1 mg	6

Strengths: Also called kingfish, king mackerel is much leaner and lower in calories than Atlantic mackerel. However, this particular species has just a fraction of the omega-3's of fattier species like Atlantic mackerel. King mackerel is a rich source of niacin and riboflavin and a phenomenal source of vitamin B_{12}, nutrients that help keep your nerves healthy. King mackerel is a bit richer in anemia-fighting iron than Atlantic mackerel.

Caution: Always cook mackerel to destroy parasites.

Curiosity: In French, *mackerel* is a slang term for a pimp.

Description: Related to tuna, king mackerel has white flesh. It's stronger in flavor than most other mackerels.

Storage: If you catch your own mackerel, ice it within 6 hours to prevent dangerous proliferation of poisonous bacteria.

Preparation: Some chefs suggest marinating king mackerel in lime juice for about an hour to offset the oily taste. Then bake, broil, poach, or steam the fish.

MACKEREL, *Spanish*

Serving: 3 oz., cooked, dry heat

Calories	134
Fat	5.4 g
Saturated	1.5 g
Monounsaturated	1.8 g
Polyunsaturated	1.5 g
Calories from fat	36%
Cholesterol	62 mg
Sodium	56 mg
Protein	20.1 g
Carbohydrate	0
Dietary fiber	0

CHIEF NUTRIENTS

NUTRIENT	AMOUNT	% RDA
Vitamin B_{12}	6.0 mcg	298
Niacin	4.3 mg	22
Vitamin B_6	0.4 mg	20
Potassium	470.9 mg	13
Riboflavin	0.2 mg	11
Magnesium	32.3 mg	9
Thiamine	0.1 mg	7
Iron	0.6 mg	6

Strengths: This mild-flavored cousin of the tuna fish is quite a bit leaner than Atlantic mackerel. It has the same amount of vitamin B_6 and a little more potassium, but less iron, niacin, and other nutrients. Spanish mackerel contains a fair amount of omega-3's, special substances in fish that lower the liver's production of triglycerides and reduce the tendency of blood to form clots.

Caution: Always cook mackerel to kill parasitic larvae.

Description: A beautifully colored blue fish dappled with olive spots, fading to a silvery underside, Spanish mackerel has a delicate flavor. Spanish mackerel are known for their spectacular leaps out of the water.

Selection: Available filleted, fresh, or frozen. Like all oily fish, Spanish mackerel is best when eaten absolutely fresh.

Preparation: Bake or broil. (Allow 10 minutes per inch of thickness at 450°F.)

MAHIMAHI

Serving: 3 oz., uncooked

Calories	72
Fat	0.6 g
Saturated	trace
Monounsaturated	trace
Polyunsaturated	trace
Calories from fat	8%
Cholesterol	62 mg
Sodium	75 mg
Protein	16.0 g
Carbohydrate	0
Dietary fiber	0

CHIEF NUTRIENTS

NUTRIENT	AMOUNT	% RDA
Niacin	5.2 mg	27
Vitamin B_{12}	0.5 mcg	26
Vitamin B_6	0.3 mg	17
Iron	1.0 mg	10
Potassium	353.6 mg	9
Magnesium	25.5 mg	7

Strengths: Mahimahi is a high-protein, low-fat source of vitamins B_6 and B_{12}. It is fair in iron, magnesium, and potassium, with just a tad of omega-3's.

Caution: *Allergy alert:* Some people experience an allergic reaction to fish.

Origin: Mahimahi is a sport fish that's usually caught in tropical and subtropical seas. Another name for mahimahi is dolphinfish. Despite the name, it's a true fish, unlike dolphins and porpoises, which are mammals.

Description: Brilliant green and golden colors begin to fade after the fish is caught. Mahimahi may weigh 2 to 5 pounds during the late spring and summer, 40 to 50 pounds in the winter.

Preparation: If you happen to catch mahimahi, it should be iced immediately to prevent scombroid poisoning. It tastes better if the tail is cut off and the fish is allowed to bleed before it is iced.

MONKFISH

Serving: 3 oz., uncooked

Calories	65
Fat	1.3 g
Saturated	0.3 g
Monounsaturated	0.2 g
Polyunsaturated	N/A
Calories from fat	18%
Cholesterol	21 mg
Sodium	15 mg
Protein	12.3 g
Carbohydrate	0
Dietary fiber	0

CHIEF NUTRIENTS

NUTRIENT	AMOUNT	% RDA
Vitamin B_{12}	0.8 mcg	39
Vitamin B_6	0.2 mg	10
Niacin	1.8 mg	9
Potassium	340.0 mg	9
Magnesium	17.9 mg	5

Strengths: Sometimes called "the poor man's lobster," sweet-tasting monkfish is a good source of protein. It's low in fat, low in cholesterol, low in sodium, and it's a good source of vitamin B_6 and rich in vitamin B_{12}.

Caution: *Allergy alert:* Fish can occasionally cause an allergic reaction.

Curiosity: A monkfish "fishes" for its own food by dangling a long, thin, wormlike spine in front of its mouth. It wiggles the spine to attract its prey, which includes other fish as well as seagulls and ducks. Sometimes called "angler fish."

Description: The monkfish is actually a flattened, bottom-dwelling shark. Some species grow up to 10 feet long, with pectoral fins that are modified to form "wings" to thrust the fish forward. The monkfish has been known to grab the feet of bathers. (Yes, it does have teeth!)

MULLET, *Striped*

Serving: 3 oz., cooked, dry heat

Calories	128
Fat	4.1 g
Saturated	1.2 g
Monounsaturated	1.2 g
Polyunsaturated	0.5 g
Calories from fat	29%
Cholesterol	54 mg
Sodium	60 mg
Protein	21.9 g
Carbohydrate	0
Dietary fiber	0

CHIEF NUTRIENTS

NUTRIENT	AMOUNT	% RDA
Niacin	5.4 mg	28
Vitamin B_6	0.4 mg	21
Iron	1.2 mg	12
Vitamin B_{12}	0.2 mcg	11
Potassium	389.3 mg	10
Magnesium	28.1 mg	8
Thiamine	0.1 mg	6
Riboflavin	0.1 mg	5

Strengths: Mullet is a good low-fat source of protein, niacin, iron, potassium, and vitamins B_6 and B_{12}.

Caution: Mullet that are caught near harbors taste awful and have guts full of oil. If you're buying fresh mullet, ask where it's caught. And if the first bite tastes foul, discard the rest. *Allergy alert:* Fish can occasionally cause an allergic reaction.

Curiosity: With over 100 species worldwide, mullet have numerous common names. In the U.S. the striped mullet (*Mugil cephalus*) and the white mullet (*Mugil curema*) are popular, particularly among the bayous of Florida, Alabama, Louisiana, and Texas. The red mullet—the mullet that people seem to like most—isn't really a mullet. It's a member of the goatfish family.

Selection: Fresh is the best. Frozen mullet is fine, but watch out for dark meat, which breaks down during freezing. Also, any mullet that comes from Canada is actually something else; unfortunately, the Canadian government allows suckers (freshwater fish) to be labeled "mullet."

OCEAN PERCH, *Atlantic*

Serving: 3 oz., cooked, dry heat

Calories	103
Fat	1.8 g
Saturated	0.3 g
Monounsaturated	0.7 g
Polyunsaturated	0.5 g
Calories from fat	16%
Cholesterol	46 mg
Sodium	82 mg
Protein	20.3 g
Carbohydrate	N/A
Dietary fiber	0

CHIEF NUTRIENTS

NUTRIENT	AMOUNT	% RDA
Vitamin B_{12}	1.0 mcg	49
Calcium	116.5 mg	15
Vitamin B_6	0.2 mg	12
Niacin	2.1 mg	11
Iron	1.0 mg	10
Magnesium	33.2 mg	10
Potassium	297.5 mg	8
Riboflavin	0.1 mg	7
Thiamine	0.1 mg	7

Strengths: Ocean perch is a very good low-fat source of protein and vitamin B_{12}, with good shots of niacin, vitamin B_6, calcium, and iron to build blood and bones and help strengthen your immune system. This fish also has a trace of omega-3's.

Caution: Do not eat raw; may contain parasites. *Allergy alert:* Fish can occasionally cause an allergic reaction.

Curiosity: The ocean perch comes from a strange family. Its African cousin, the "walking perch," not only walks on land but also climbs trees.

Selection: If it's ocean perch you really want, buy the *fresh* fish. Frozen fillets labeled "ocean perch" are frequently rosefish, redfish, red perch, or sea perch, not *ocean* perch.

OCTOPUS, *Common*

Serving: 3 oz., uncooked

Calories	70
Fat	0.9 g
Saturated	trace
Monounsaturated	trace
Polyunsaturated	trace
Calories from fat	11%
Cholesterol	41 mg
Sodium	196 mg
Protein	12.7 g
Carbohydrate	1.9 g
Dietary fiber	0

CHIEF NUTRIENTS

NUTRIENT	AMOUNT	% RDA
Vitamin B$_{12}$	17.0 mcg	850
Iron	4.5 mg	45
Vitamin B$_6$	0.3 mg	16
Zinc	1.4 mg	10
Niacin	1.8 mg	9
Potassium	297.5 mg	8
Folate	13.6 mcg	7
Magnesium	25.5 mg	7
Vitamin C	4.3 mg	7
Calcium	45.1 mg	6

Strengths: A phenomenal source of nerve-protecting vitamin B$_{12}$, octopus is rich in iron and has a good dose of vitamin B$_6$ to build red blood cells and beef up your immune system.

Caution: *Allergy alert:* Fish can occasionally cause an allergic reaction.

Curiosity: Octopi are shy, gentle creatures that roam the ocean floor and pile up stones to make lairs for themselves; when disturbed, they crawl in and pull some stones across the front to cover the entrance. Despite their magnified horror-film image, they are generally small, weighing about 3 pounds.

Preparation: If the octopus is alive, standard practice is to turn it inside out to kill it. Once the octopus has been killed, the chef prepares it for cooking by beating the flesh with a bottle or rolling pin.

A Better Idea: Buy predressed octopus, either fresh or frozen. It's available in many supermarkets and specialty fish markets. Only the 8 tentacles (minus suckers) and the body (minus eyes and beak) are edible.

ORANGE ROUGHY

Serving: 3 oz., uncooked

Calories	59
Fat	0.6 g
Saturated	trace
Monounsaturated	0.4 g
Polyunsaturated	trace
Calories from fat	9%
Cholesterol	17 mg
Sodium	54 mg
Protein	12.5 g
Carbohydrate	0
Dietary fiber	0

CHIEF NUTRIENTS

NUTRIENT	AMOUNT	% RDA
Vitamin B$_{12}$	1.7 mcg	85
Niacin	2.6 mg	13
Vitamin B$_6$	0.3 mg	13
Riboflavin	0.1 mg	8
Magnesium	25.5 mg	7
Potassium	255.0 mg	7
Thiamine	0.1 mg	6

Strengths: A great low-cholesterol source of body-building protein, niacin, and vitamin B$_6$, packed with vitamin B$_{12}$.

Caution: *Allergy alert:* Fish can occasionally cause an allergic reaction.

Curiosity: Called a "slimehead" by fisherfolk.

Description: Orange roughy is a mild-tasting fish usually imported frozen from New Zealand.

Preparation: Can be poached, baked, broiled, or fried.

PIKE, *Northern*

Serving: 3 oz., cooked, dry heat

Calories	96
Fat	0.8 g
Saturated	trace
Monounsaturated	trace
Polyunsaturated	trace
Calories from fat	7%
Cholesterol	43 mg
Sodium	42 mg
Protein	21.0 g
Carbohydrate	0
Dietary fiber	0

CHIEF NUTRIENTS

NUTRIENT	AMOUNT	% RDA
Vitamin B_{12}	2.0 mcg	98
Niacin	2.4 mg	13
Magnesium	34.0 mg	10
Calcium	62.1 mg	8
Potassium	281.4 mg	8
Folate	14.7 mcg	7
Iron	0.6 mg	6
Vitamin B_6	0.1 mg	6
Vitamin C	3.2 mg	5

Strengths: Pike is a super high-protein, low-fat, low-sodium source of vitamin B_{12} and a good source of niacin. It has a trace of omega-3's.

Caution: Do not eat the roe: It may be poisonous. *Allergy alert:* Fish can occasionally cause an allergic reaction.

Curiosity: Nicknamed the ''river wolf'' or the ''freshwater shark'' for its savage nature. The pike launches itself like a fast torpedo at any prey, swallows it whole, then rasps it to death with the more than 700 teeth covering its tongue and palate. Loved by sport fishermen.

Preparation: Some culinary experts feel that pike would be a more popular fish if only Americans would learn to cook it properly. Their suggestion? Prepare a bouillon by simmering celery, carrots, an onion studded with cloves, and a bouquet garni in water for 30 minutes. Wine, lemon juice, or vinegar may be added if you like. Add the pike and poach 25 minutes for a 3-pounder.

POLLACK

Serving: 3 oz., cooked, dry heat

Calories	96
Fat	1.0 g
Saturated	0.2 g
Monounsaturated	0.2 g
Polyunsaturated	0.4 g
Calories from fat	9%
Cholesterol	82 mg
Sodium	99 mg
Protein	20.0 g
Carbohydrate	0
Dietary fiber	0

CHIEF NUTRIENTS

NUTRIENT	AMOUNT	% RDA
Vitamin B_{12}	3.6 mcg	179
Magnesium	62.1 mg	18
Potassium	329.0 mg	9
Niacin	1.4 mg	7

Strengths: A great low-fat source of vitamin B_{12} that will keep your nerves well-protected. Pollack is also a good source of protein and magnesium, with just a dollop of omega-3's.

Caution: Do not eat raw; may contain parasites. *Allergy alert:* Fish can occasionally cause an allergic reaction.

Selection: Buy fresh, young pollack. Look for a translucent fillet that smells like fresh seawater. Pollack that looks white has been dipped in a preservative. Pollack is frequently mixed with crab and shrimp to create an affordable imitation of these shellfish called surimi.

Preparation: Pollack is best baked or broiled. To keep the inside moist and the outside crisp, first brush the fillets with oil and lemon juice, then follow with a dusting of seasoned flour or bread crumbs.

POMPANO, *Florida*

Serving: 3 oz., cooked, dry heat

Calories	179
Fat	10.3 g
Saturated	3.8 g
Monounsaturated	2.8 g
Polyunsaturated	0.4 g
Calories from fat	52%
Cholesterol	54 mg
Sodium	65 mg
Protein	20.1 g
Carbohydrate	0
Dietary fiber	0

CHIEF NUTRIENTS

NUTRIENT	AMOUNT	% RDA
Vitamin B$_{12}$	1.0 mcg	51
Thiamine	0.6 mg	39
Niacin	3.2 mg	17
Potassium	540.6 mg	14
Vitamin B$_6$	0.2 mg	10
Magnesium	26.4 mg	8
Riboflavin	0.1 mg	8
Folate	14.7 mcg	7
Iron	0.6 mg	6

Strengths: The Florida favorite, pompano is loaded with vitamin B$_{12}$ and a chunk of niacin to keep your red blood cells on their toes. It's also rich in thiamine and a good source of potassium, with just a smidgen of omega-3's.

Caution: *Allergy alert:* Fish can occasionally cause an allergic reaction.

Curiosity: Pompano can hurl themselves out of the water and leap across the surface like a skipping stone. Occasionally they've been known to jump—uninvited but deeply welcomed—into a nearby boat. Known among many cooks as America's finest fish.

Preparation: Perfect either grilled or broiled. In Tarpon Springs, Florida, where pompano is served Greek fisherman–style, the fillets are brushed with a little olive oil spiked with either lime or lemon juice.

ROE, *Mixed Species*

Serving: 3 oz., uncooked

Calories	119
Fat	5.5 g
Saturated	1.2 g
Monounsaturated	1.4 g
Polyunsaturated	2.3 g
Calories from fat	41%
Cholesterol	318 mg
Sodium	77 mg
Protein	19.0 g
Carbohydrate	1.3 g
Dietary fiber	0

CHIEF NUTRIENTS

NUTRIENT	AMOUNT	% RDA
Vitamin B$_{12}$	8.5 mcg	425
Riboflavin	0.6 mg	37
Folate	68.0 mcg	34
Vitamin C	13.6 mg	23
Thiamine	0.2 mg	13
Niacin	1.5 mg	8
Vitamin A	67.2 RE	7
Vitamin B$_6$	0.1 mg	7
Zinc	0.9 mg	6

Strengths: Roe is an excellent source of blood-building vitamin B$_{12}$, as well as a very good source of vitamin C, folate, and riboflavin. Roe also rivals some species of mackerel in omega-3's.

Caution: Loaded with cholesterol. *Allergy alert:* Fish can occasionally cause an allergic reaction.

Curiosity: In various cultures, men have believed that eating another animal's sperm will somehow increase the production of their own. Evidence is lacking, but the myth persists.

Description: The gonads of female and male fish. The female produces two saclike ovaries, through which the unripe eggs are visible. Male roe consists of two saclike testes that contain sperm. When male roe is pink or red, the testes are immature. When it is white—"white roe"—sperm is present.

Preparation: Wash the roe gently, being sure to eliminate every trace of blood. Care must be taken to not break the gall sac or the roe will taste bitter.

SABLEFISH

Serving: 3 oz., smoked

Calories	219
Fat	17.1 g
Saturated	3.6 g
Monounsaturated	9.0 g
Polyunsaturated	2.3 g
Calories from fat	71%
Cholesterol	54 mg
Sodium	627 mg
Protein	15.0 g
Carbohydrate	0
Dietary fiber	0

CHIEF NUTRIENTS

NUTRIENT	AMOUNT	% RDA
Vitamin B$_{12}$	1.7 mcg	85
Niacin	4.5 mg	24
Magnesium	62.9 mg	18
Vitamin B$_6$	0.3 mg	17
Iron	1.4 mg	14
Potassium	400.4 mg	11
Vitamin A	103.7 RE	10
Folate	16.8 mcg	8
Thiamine	0.1 mg	7
Riboflavin	0.1 mg	6
Calcium	42.5 mg	5

Strengths: Rich in nutrients, sablefish is an excellent source of blood-building vitamin B$_{12}$ and a good source of omega-3's, niacin, vitamin A, vitamin B$_6$, and iron. Magnesium and potassium help to keep your blood pressure down, and they improve the strength of your heart.

Eat with: Foods that are high in vitamin C.

Caution: *Drug interaction:* The tyramine in smoked sablefish can combine with MAO-inhibitor antidepressants to cause dangerously high blood pressure. *Allergy alert:* Smoked fish can occasionally set off an allergic reaction.

Selection: Sablefish is commonly marketed as "smoked black cod." It's not cod, nor does it taste good in cod recipes.

SALMON, *Chinook*

Serving: 3 oz., smoked

Calories	99
Fat	3.7 g
Saturated fat	0.8 g
Monounsaturated fat	1.7 g
Polyunsaturated fat	0.9 g
Calories from fat	33%
Cholesterol	20 mg
Sodium	666 mg
Protein	15.5 g
Carbohydrate	0
Dietary fiber	0

CHIEF NUTRIENTS

NUTRIENT	AMOUNT	% RDA
Vitamin B$_{12}$	2.8 mcg	139
Niacin	4.0 mg	21
Vitamin B$_6$	0.2 mg	12
Iron	0.7 mg	7
Riboflavin	0.1 mg	5

Strengths: The best of the Pacific salmon, Chinook is a great low-cholesterol source of blood-building vitamin B$_{12}$ and a good source of niacin and vitamin B$_6$, with a touch of heart-healthy omega-3's.

Caution: Loaded with sodium. People who have high blood pressure may want to avoid it. *Allergy alert:* Fish can occasionally cause an allergic reaction.

History: Artifacts from around the Columbia River in the northwestern U.S. indicate that salmon was popular as food and totem as early as 11,000 B.C. The lives of Chinook Indians were so intertwined with salmon that they believed their tribes would die out if it disappeared.

Preparation: Although the Northwest Indians cooked "planked salmon" by attaching the fish to driftwood and allowing it to cook in the embers of a fire, today Chinook salmon is used primarily to make lox, a salt-cured, cold-smoked fish that is traditionally served with bagels and cream cheese.

SALMON, *Chum, Canned*

Serving: 3 oz., drained, w/bones, w/out salt

Calories	120
Fat	4.7 g
Saturated	1.3 g
Monounsaturated	1.6 g
Polyunsaturated	1.3 g
Calories from fat	35%
Cholesterol	33 mg
Sodium	64 mg
Protein	18.0 g
Carbohydrate	0
Dietary fiber	0

CHIEF NUTRIENTS

NUTRIENT	AMOUNT	% RDA
Vitamin B_{12}	3.7 mcg	187
Niacin	6.0 mg	31
Calcium	211.7 mg	26
Vitamin B_6	0.3 mg	16
Folate	17.0 mcg	9
Riboflavin	0.1 mg	8
Magnesium	25.5 mg	7
Potassium	255.0 mg	7
Iron	0.6 mg	6

Strengths: Also called dog salmon, chum is one of the leanest, lightest-colored salmons. It has a generous supply of high-quality protein, and it's a good source of omega-3 fatty acids. (Omega-3's are the fish oils that appear to protect the heart and arteries by lowering triglycerides and reducing the tendency of blood to clot.) If you eat the small, soft bones, you'll also get bone-building calcium. Chum salmon is rich in niacin and contains whopping amounts of vitamin B_{12} to protect skin and nerves.

Curiosity: Cave paintings suggest that salmon were greatly esteemed by troglodytes.

Serve: In dishes such as casseroles, croquettes, or soufflés.

SALMON, *Coho*

Serving: 3 oz., cooked, moist heat

Calories	157
Fat	6.4 g
Saturated	1.2 g
Monounsaturated	2.2 g
Polyunsaturated	1.9 g
Calories from fat	37%
Cholesterol	42 mg
Sodium	50 mg
Protein	23.3 g
Carbohydrate	0
Dietary fiber	0

CHIEF NUTRIENTS

NUTRIENT	AMOUNT	% RDA
Vitamin B_{12}	3.1 mcg	153
Niacin	7.1 mg	38
Vitamin B_6	0.4 mg	20
Potassium	453.9 mg	12
Thiamine	0.2 mg	11
Riboflavin	0.2 mg	10
Magnesium	31.5 mg	9
Iron	0.8 mg	8

Strengths: Coho is an excellent source of blood-building vitamin B_{12}, with plenty of niacin, a metabolic cell-broker that allows your body to convert calories into energy. Coho is also a pretty good source of omega-3's, riboflavin, thiamine, vitamin B_6, and potassium.

Caution: Do not eat raw; may contain parasites. *Allergy alert:* Fish can occasionally cause an allergic reaction.

Origin: A good sport fish, coho salmon is taken mostly from rivers along the Pacific coast. Some fish have now migrated—with the help of man—to rivers in New Hampshire.

Preparation: To prepare planked salmon, Indian-style, clean the fish, stake it out on a plank, set the plank upright next to a bed of aromatic wood coals, then cook the fish by reflected heat. You can buy a Chinook cedar baking plank specifically for this purpose.

SALMON, Pink, Canned

Serving: 3 oz., w/bones & liquid, w/out salt

Calories	118
Fat	5.1 g
Saturated	1.3 g
Monounsaturated	1.5 g
Polyunsaturated	1.7 g
Calories from fat	39%
Cholesterol	47 mg
Sodium	64 mg
Protein	16.8 g
Carbohydrate	0
Dietary fiber	0

CHIEF NUTRIENTS

NUTRIENT	AMOUNT	% RDA
Vitamin B_{12}	3.7 mcg	187
Niacin	5.6 mg	29
Calcium	181.1 mg	23
Vitamin B_6	0.3 mg	13
Riboflavin	0.2 mg	9
Magnesium	28.9 mg	8
Folate	13.1 mcg	7
Iron	0.7 mg	7
Potassium	277.1 mg	7

Strengths: A super source of heart-healthy omega-3's and vitamin B_{12}. Canned pink salmon is also a good source of niacin, vitamin B_6, and bone-building calcium.

Caution: Pink salmon that is canned with salt has more than 7 times more sodium than salmon that is canned without salt, so read labels. *Allergy alert:* Fish can occasionally cause an allergic reaction.

Selection: The brighter the color of its meat and the deeper the color of its oil, the better the quality of the fish. Red or black skin indicates a sexually mature or late-run fish; pale flesh is present during the fasting period prior to spawning. These result in a loss of both nutrients and flavor.

Serve: The small flakes and delicate flavor of canned pink salmon make it a good choice for sandwiches, cooked dishes, and chowders.

SALMON, Sockeye, Canned

Serving: 3 oz., drained, w/bones, w/out salt

Calories	130
Fat	6.2 g
Saturated	1.4 g
Monounsaturated	2.4 g
Polyunsaturated	1.6 g
Calories from fat	43%
Cholesterol	37 mg
Sodium	64 mg
Protein	17.4 g
Carbohydrate	0
Dietary fiber	0

CHIEF NUTRIENTS

NUTRIENT	AMOUNT	% RDA
Calcium	203.2 mg	25
Niacin	4.7 mg	25
Vitamin B_6	0.3 mg	13
Vitamin B_{12}	0.3 mcg	13
Iron	0.9 mg	9
Potassium	320.5 mg	9
Riboflavin	0.2 mg	9
Magnesium	24.7 mg	7

Strengths: Canned sockeye is a good source of heart-healthy omega-3's, niacin, calcium, and vitamins B_6 and B_{12}.

Caution: Sockeye that is canned with salt has more than 7 times more sodium than sockeye that is canned without salt. *Allergy alert:* Fish can occasionally cause an allergic reaction.

Selection: Buy canned sockeye that was caught in the Pacific Ocean or in western rivers, where it goes to spawn. Sockeye from the Great Lakes should be avoided—it feeds on a regional fish, resulting in off-tasting, colorless meat.

Serve: Great for sandwiches and cooked dishes.

SALMON, *Sockeye, Fresh*

Serving: 3 oz., cooked, dry heat

Calories	184
Fat	9.3 g
Saturated	1.6 g
Monounsaturated	4.5 g
Polyunsaturated	1.3 g
Calories from fat	46%
Cholesterol	74 mg
Sodium	56 mg
Protein	23.2 g
Carbohydrate	0
Dietary fiber	0

CHIEF NUTRIENTS

NUTRIENT	AMOUNT	% RDA
Vitamin B$_{12}$	4.9 mcg	247
Niacin	5.7 mg	30
Thiamine	0.2 mg	12
Vitamin B$_6$	0.2 mg	10
Potassium	318.8 mg	9
Riboflavin	0.2 mg	9
Magnesium	26.4 mg	8
Vitamin A	53.6 RE	5

Strengths: An excellent source of nerve-protecting vitamin B$_{12}$. Fresh sockeye is also a good source of body-building niacin, thiamine, and omega-3's.

Caution: *Allergy alert:* Fish can occasionally cause an allergic reaction.

Selection: Buy fresh sockeye that has been caught in the Pacific Ocean or in western rivers, where it goes to spawn. Sockeye from the Great Lakes should be avoided—it feeds on a regional fish, resulting in off-tasting, colorless meat. Fresh sockeye is less nutritious than canned sockeye—partly because canned sockeye includes edible chunks of bone.

Preparation: Whole, fresh salmon that is poached in a fish stock with vegetables and a bouquet garni is delectable. So is fresh salmon that is baked in your oven on a Chinook cedar baking plank—the modern-day translation of an ancient Indian cooking technique that imbues salmon with mouth-watering aromas and flavors.

SARDINES, *Atlantic, Canned in Oil*

Serving: 2 (about 1 oz.), drained, w/bones

Calories	50
Fat	2.8 g
Saturated	0.4 g
Monounsaturated	0.9 g
Polyunsaturated	1.2 g
Calories from fat	50%
Cholesterol	34 mg
Sodium	121 mg
Protein	5.9 g
Carbohydrate	0
Dietary fiber	0

CHIEF NUTRIENTS

NUTRIENT	AMOUNT	% RDA
Vitamin B$_{12}$	2.2 mcg	108
Calcium	91.7 mg	11
Iron	0.7 mg	7
Niacin	1.3 mg	7

Strengths: Sardines contain an entire day's supply of vitamin B$_{12}$, with a good shot of calcium and a touch of omega-3's.

Caution: Half the calories come from fat, and those on low-salt diets will want to take note of the sodium level. If not refrigerated immediately after they're caught, sardines may cause histamine poisoning, particularly in those who are taking the drug isoniazid. *Allergy alert:* Fish can occasionally cause an allergic reaction. And sardines are loaded with purines: People who have gout may want to avoid them.

Curiosity: Sardines were first canned on the Mediterranean island of Sardinia.

Description: Sardines are actually immature Atlantic herring. The word *sardine* is simply a nickname that is used to refer to any tiny, soft-boned fish that is related to the herring.

Preparation: Canned sardines, which are just about the only kind available in the U.S., are generally mixed with other ingredients—curries, fruits, and cheeses, for example—to make salads, sandwiches, and canapés. To reduce the fat, salt, and calorie content, rinse sardines well before you eat them.

SHAD, *American*

Serving: 3 oz., uncooked

Calories	167
Fat	11.7 g
Saturated	3.8 g
Monounsaturated	4.1 g
Polyunsaturated	N/A
Calories from fat	63%
Cholesterol	64 mg
Sodium	43 mg
Protein	14.4 g
Carbohydrate	0
Dietary fiber	0

CHIEF NUTRIENTS

NUTRIENT	AMOUNT	% RDA
Niacin	7.1 mg	38
Vitamin B$_6$	0.3 mg	17
Riboflavin	0.2 mg	12
Potassium	326.4 mg	9
Thiamine	0.1 mg	9
Iron	0.8 mg	8
Magnesium	25.5 mg	7
Vitamin B$_{12}$	0.1 mcg	7
Folate	12.8 mcg	6

Strengths: A very good source of metabolism-enhancing niacin, with a good chunk of riboflavin and vitamin B$_6$ to boot.
Caution: High in fat. *Allergy alert:* Fish can occasionally cause an allergic reaction.
Origin: Shad come from Atlantic waters. They are caught in the spring when they swim upriver to spawn.
History: In 1732 a bunch of local boys from Philadelphia formed The Schuylkill Fishing Co., North America's first catch-it-and-cook-it club. They cooked their shad seasoned, wrapped with bacon, and staked to a white oak plank.
Selection: Be sexist. Many culinary experts prefer a female fillet because it's larger and fatter than the male. The male shad, although just as tasty as his better half, is so thin and small that there's not a lot to work with once you've removed the bones.

SHARK, *Mixed Species*

Serving: 3 oz., uncooked

Calories	111
Fat	3.8 g
Saturated	0.8 g
Monounsaturated	1.3 g
Polyunsaturated	1.0 g
Calories from fat	31%
Cholesterol	43 mg
Sodium	67 mg
Protein	17.8 g
Carbohydrate	0
Dietary fiber	0

CHIEF NUTRIENTS

NUTRIENT	AMOUNT	% RDA
Vitamin B$_{12}$	1.3 mcg	63
Vitamin B$_6$	0.3 mg	17
Niacin	2.5 mg	13
Magnesium	41.7 mg	12
Iron	0.7 mg	7
Vitamin A	59.5 RE	6

Strengths: Despite its reputation for blood-letting, shark is also a great source of blood-building—specifically vitamin B$_{12}$, a key nutrient that enhances red blood cells. Shark is also a good source of magnesium, niacin, vitamin B$_6$, and omega-3's.
Caution: Because shark is a large fish, heavy metals such as mercury can accumulate in its flesh. Pregnant or nursing women may want to avoid it. *Allergy alert:* Fish can occasionally cause an allergic reaction.
Curiosity: Shark's fin is reputed to be an aphrodisiac. More practically, its cartilage—reduced to a gelatinous goo—is used in Chinese cooking to thicken soup.
History: The spiny dogfish and the porbeagle—two types of shark—have historically been the fish part of British fish-and-chips.
Preparation: Because shark meat is flooded with urea, it must first be soaked in lemon juice or—if it will subsequently be fried—milk to remove the aroma and flavor of ammonia.

SMELT, *Rainbow*

Serving: 3 oz., cooked, dry heat

Calories	105
Fat	2.6 g
Saturated	0.5 g
Monounsaturated	0.7 g
Polyunsaturated	1.0 g
Calories from fat	23%
Cholesterol	77 mg
Sodium	66 mg
Protein	19.2 g
Carbohydrate	0
Dietary fiber	0

CHIEF NUTRIENTS

NUTRIENT	AMOUNT	% RDA
Vitamin B_{12}	3.4 mcg	169
Zinc	1.8 mg	12
Iron	1.0 mg	10
Magnesium	32.3 mg	9
Calcium	65.5 mg	8
Niacin	1.5 mg	8
Potassium	316.2 mg	8
Riboflavin	0.1 mg	7
Vitamin B_6	0.1 mg	7

Strengths: A great low-fat source of nerve-protecting vitamin B_{12}, smelts also have a fair amount of zinc to keep your immune system healthy, and a good amount of omega-3's to keep your arteries clean.

Caution: *Allergy alert:* Fish can occasionally cause an allergic reaction.

Curiosity: One species of smelt is so rich in oils that it was used by the Chinook Indians as a candle. Thus, it came to be known as a ''candlefish.''

Description: ''Smelt'' comes from the Anglo-Saxon word *smoelt,* which means smooth or shining—an appropriate name for this silvery little fish. Generally, smelts are so small that it takes 10 or 12 to make a pound.

Selection: Fresh smelts are best from September through May, although they may be hard to find. Since they're highly perishable, the fish are usually flash-frozen immediately after they've been caught.

Preparation: Culinary experts suggest that this little fish be dipped in egg, dredged in seasoned flour, and pan-fried for no more than 2½ minutes, then devoured in their entirety.

A Better Idea: Cut the smelts into bite-size chunks and bake them in a casserole.

SNAPPER, *Mixed Species*

Serving: 3 oz., cooked, dry heat

Calories	109
Fat	1.5 g
Saturated	0.3 g
Monounsaturated	0.3 g
Polyunsaturated	0.4 g
Calories from fat	12%
Cholesterol	40 mg
Sodium	49 mg
Protein	22.4 g
Carbohydrate	0
Dietary fiber	0

CHIEF NUTRIENTS

NUTRIENT	AMOUNT	% RDA
Vitamin B_{12}	3.0 mcg	149
Vitamin B_6	0.4 mg	20
Potassium	443.7 mg	12
Magnesium	31.5 mg	9

Strengths: A dieter's delight. Snapper is a super low-fat, low-salt source of both protein and vitamin B_{12}. It also has an extra kick of blood pressure–lowering potassium and immunity-enhancing vitamin B_6, with a touch of heart-boosting omega-3's.

Caution: *Allergy alert:* Fish can occasionally cause an allergic reaction. Large snapper caught in tropical waters may harbor a deadly toxin that causes ciguatera poisoning. To minimize your risk of encountering this problem, do not eat the liver, intestines, eyes, brains, roe, or sex organs of red snapper taken in tropical waters.

Selection: The snapper in American markets is largely from Atlantic waters.

SOLE

☀ ⬇ ▥

Serving: 3 oz., cooked, dry heat

Calories	99
Fat	1.3 g
Saturated	0.3 g
Monounsaturated	0.3 g
Polyunsaturated	0.5 g
Calories from fat	12%
Cholesterol	58 mg
Sodium	89 mg
Protein	20.5 g
Carbohydrate	0
Dietary fiber	0

CHIEF NUTRIENTS

NUTRIENT	AMOUNT	% RDA
Vitamin B_{12}	2.1 mcg	107
Magnesium	49.3 mg	14
Niacin	1.9 mg	10
Vitamin B_6	0.2 mg	10
Potassium	292.4 mg	8
Riboflavin	0.1 mg	6

Strengths: Every chef's favorite fish. Sole is an excellent low-fat source of body-building protein and vitamin B_{12}. It also has a good amount of magnesium to keep your blood pressure on an even keel and a dash of omega-3's to keep your arteries flexible.

Caution: *Allergy alert:* Fish can occasionally cause an allergic reaction.

Selection: Real sole is called Dover sole, imported English sole, or imported Dover sole in the U.S. and is available only as a frozen import from England or France. Anything fresh that's marked "sole" or "fillet of sole" is more likely to be the slightly less flavorful American flounder.

Preparation: "Fillet of sole" does not make the grade in a recipe that's been designed for Dover sole. The delicate flavor of real sole is brought out by the elaborate use of sauces, herbs, spices, fruits, and vegetables.

SQUID, *Mixed Species*

☀

Serving: 3 oz., fried

Calories	149
Fat	6.4 g
Saturated	1.6 g
Monounsaturated	2.3 g
Polyunsaturated	1.8 g
Calories from fat	38%
Cholesterol	221 mg
Sodium	260 mg
Protein	15.3 g
Carbohydrate	6.6 g
Dietary fiber	0

CHIEF NUTRIENTS

NUTRIENT	AMOUNT	% RDA
Vitamin B_{12}	1.0 mcg	52
Riboflavin	0.4 mg	23
Niacin	2.2 mg	12
Zinc	1.5 mg	10
Iron	0.9 mg	9
Magnesium	32.3 mg	9
Potassium	237.2 mg	6
Vitamin C	3.6 mg	6

Caution: Fried squid is loaded with cholesterol and sodium. People with high blood pressure or those who are on a cholesterol-lowering diet may want to avoid it. *Allergy alert:* Can occasionally cause an allergic reaction.

Strengths: Squid, or calamari as it is frequently known, has a sweet taste, is packed with vitamin B_{12}, and has a decent amount of niacin, riboflavin, and omega-3's.

Selection: The nonfrozen squid that you buy at the fish counter should be small and whole, with clear eyes and a fresh, day-at-the-beach scent.

Preparation: Do not eat raw. Squid may contain a particularly nasty parasite that is killed in cooking.

A Better Idea: For lower sodium and more omega-3's, try it baked or boiled. (Remember that deep-frying removes any heart-healthy omega-3's.)

STURGEON, *Mixed Species*

Serving: 3 oz., smoked

Calories	147
Fat	3.7 g
Saturated	0.9 g
Monounsaturated	2.0 g
Polyunsaturated	0.4 g
Calories from fat	23%
Cholesterol	68 mg
Sodium	628 mg
Protein	26.5 g
Carbohydrate	0
Dietary fiber	0

CHIEF NUTRIENTS

NUTRIENT	AMOUNT	% RDA
Vitamin B_{12}	2.5 mcg	124
Niacin	9.4 mg	50
Vitamin A	238.0 RE	24
Vitamin B_6	0.2 mg	12
Magnesium	40.0 mg	11
Folate	17.0 mcg	9
Potassium	322.2 mg	9
Iron	0.8 mg	8

Strengths: Although sturgeon is primarily famous for the caviar made from its eggs, its flesh is also a nutritious delicacy. It's an excellent low-fat source of nerve-protecting vitamin B_{12} and of niacin, and it's a good source of magnesium and vitamins A and B_6.

Caution: Because this is a smoked fish, sodium levels are high—so avoid it if you're on a salt-free diet. *Allergy alert:* Smoked fish can occasionally cause an allergic reaction.

Origin: Sturgeon generally comes from the Black and Caspian seas of the Middle East and the Soviet Union or from the Pacific Northwest and Southern Atlantic coasts of the U.S.

Selection: Look for fresh-smoked white sturgeon. Its quality is somewhat better than that of the green smoked sturgeon, which is sold primarily as a canned product. Also, fresh sturgeon is sometimes available.

Preparation: Fresh nonsmoked sturgeon works well in any swordfish recipe. It can be braised, grilled, broiled, sautéed, or baked.

SURIMI

Serving: 3 oz.

Calories	84
Fat	0.7 g
Saturated	trace
Monounsaturated	trace
Polyunsaturated	trace
Calories from fat	8%
Cholesterol	26 mg
Sodium	122 mg
Protein	12.9 g
Carbohydrate	5.8 g
Dietary fiber	0

CHIEF NUTRIENTS

NUTRIENT	AMOUNT	% RDA
Vitamin B_{12}	1.4 mcg	68
Magnesium	36.6 mg	10

Strengths: A very good source of blood-building vitamin B_{12} with respectable amounts of magnesium and protein, surimi is extremely low in fat.

Caution: The nutritional content of surimi varies from manufacturer to manufacturer—mostly because the fish contained in this preparation is rinsed several times during processing. Of particular concern is the fact that the sodium content is totally unpredictable. In 4 samples taken from 4 different manufacturers, the sodium content ranged from 143 to 1,085 mg per serving. *Allergy alert:* Fish can occasionally cause an allergic reaction. Surimi can sometimes contain shellfish.

Description: Surimi is a tasteless, formless paste of minced fish—usually Pacific pollack. A slew of substances is added to the mix in order to add flavor, keep it from turning into rubber, and hold it in a variety of shapes.

Selection: Read labels. Whether surimi is sold as a natural product such as shrimp or in a mix such as "seafood salad," the manufacturer is required to list the product ingredients.

SWORDFISH

Serving: 3 oz., cooked, dry heat

Calories	132
Fat	4.4 g
Saturated	1.2 g
Monounsaturated	1.7 g
Polyunsaturated	1.0 g
Calories from fat	30%
Cholesterol	43 mg
Sodium	98 mg
Protein	21.6 g
Carbohydrate	0
Dietary fiber	0

CHIEF NUTRIENTS

NUTRIENT	AMOUNT	% RDA
Vitamin B_{12}	1.7 mcg	86
Niacin	10.0 mg	53
Vitamin B_6	0.3 mg	16
Iron	0.9 mg	9
Magnesium	28.9 mg	8
Potassium	313.7 mg	8
Zinc	1.3 mg	8
Riboflavin	0.1 mg	6

Strengths: A decidedly rich source of blood-building vitamin B_{12} and niacin, swordfish is also a good, low-fat supplier of vitamin B_6. Its omega-3's also help keep your arteries healthy.
Caution: Usually contains mercury, so pregnant and nursing women may want to avoid it. For that matter, everyone should check with local health authorities before eating any *imported* swordfish. The Food and Drug Administration periodically bans imported swordfish because of dangerously high levels of mercury. *Allergy alert:* Fish can occasionally cause an allergic reaction.
Selection: Fresh swordfish from U.S. waters is available through the summer and fall months.
Preparation: Swordfish is tough enough to withstand sautéing, grilling, broiling, baking, or poaching. Many people like it marinated in herbs and olive oil, then grilled, basted with the marinade.

TILEFISH

Serving: 3 oz., cooked, dry heat

Calories	125
Fat	4.0 g
Saturated	0.7 g
Monounsaturated	1.1 g
Polyunsaturated	1.1 g
Calories from fat	29%
Cholesterol	54 mg
Sodium	50 mg
Protein	20.8 g
Carbohydrate	0
Dietary fiber	0

CHIEF NUTRIENTS

NUTRIENT	AMOUNT	% RDA
Vitamin B_{12}	2.1 mcg	107
Niacin	3.0 mg	16
Vitamin B_6	0.3 mg	13
Potassium	435.2 mg	12
Riboflavin	0.2 mg	9
Magnesium	28.1 mg	8
Thiamine	0.1 mg	8
Folate	14.7 mcg	7

Strengths: A yellow polka-dotted fish with iridescent purple and blue fins, tilefish is loaded with blood-building vitamin B_{12}. Tilefish is a good low-fat source of niacin, vitamin B_6, potassium, and omega-3's to keep your arteries young.
Caution: *Allergy alert:* Fish can occasionally cause an allergic reaction.
Selection: The Pacific tilefish—also known as "ocean whitefish"—sometimes develops a bitter flavor when it wanders out of its normal 1,000-foot depth and is caught in shallow waters. Deep-dwelling common tilefish of the same species—as well as the common blackline tilefish found at the edge of the Atlantic Gulfstream—tastes like the crustaceans it loves to eat.
Preparation: The rich taste of tilefish is enhanced by brushing the fillet with canola oil, then sprinkling it with ground white pepper and paprika. Bake until fully cooked, chill, cut into chunks, then toss with tomatoes, cucumbers, carrots, fresh basil, and salad greens in a large bowl. Sprinkle with a low-fat Italian dressing and serve. You'll never laugh at anything with yellow polka dots again.

TROUT, *Rainbow*

Serving: 3 oz., cooked, dry heat

Calories	128
Fat	3.7 g
Saturated	0.7 g
Monounsaturated	1.1 g
Polyunsaturated	1.3 g
Calories from fat	26%
Cholesterol	62 mg
Sodium	29 mg
Protein	22.4 g
Carbohydrate	0
Dietary fiber	0

CHIEF NUTRIENTS

NUTRIENT	AMOUNT	% RDA
Vitamin B_{12}	3.0 mcg	149
Niacin	5.9 mg	31
Iron	2.1 mg	21
Vitamin B_6	0.4 mg	20
Potassium	538.9 mg	14
Riboflavin	0.2 mg	11
Calcium	73.1 mg	9
Magnesium	33.2 mg	9

Strengths: Trout is a great low-fat source of vitamin B_{12}, with good chunks of body-building protein, niacin, iron, potassium, and vitamin B_6. It also has a very good amount of omega-3's. **Caution:** *Allergy alert:* Fish can occasionally cause an allergic reaction.

Description: Fresh trout has a delicate taste and firm texture.

Selection: Hook your own, preferably from a clear, swift-flowing stream. Trout raised on a fish farm or caught in a lake does not carry the full, rich flavor of free-swimming fish.

Preparation: Try planked trout. Dig a pit 6 inches deep, 3 feet by 3 feet, and fill it with charcoal. When the charcoal is white-hot, nail the trout to a water-soaked hardwood plank that is free of preservatives. Prop the plank in a vertical position at the edge of the pit, facing the charcoal. (The fish should be head-down on the plank.) Bake the trout for 20 to 25 minutes, or until the edges are lightly crisped.

TUNA, *Fresh*

Serving: 3 oz., cooked, dry heat

Calories	156
Fat	5.3 g
Saturated	1.4 g
Monounsaturated	1.8 g
Polyunsaturated	1.6 g
Calories from fat	31%
Cholesterol	42 mg
Sodium	43 mg
Protein	25.4 g
Carbohydrate	0
Dietary fiber	0

CHIEF NUTRIENTS

NUTRIENT	AMOUNT	% RDA
Vitamin B_{12}	9.3 mcg	463
Vitamin A	642.6 RE	64
Niacin	9.0 mg	47
Vitamin B_6	0.5 mg	23
Magnesium	54.4 mg	16
Thiamine	0.2 mg	16
Riboflavin	0.3 mg	15
Iron	1.1 mg	11
Potassium	274.6 mg	7

Strengths: One of the most health-giving fish you can buy, bluefin tuna is loaded with niacin and vitamins A and B_{12} to build both healthy red blood cells and an effective immune system. It's also a good source of riboflavin, thiamine, iron, magnesium, vitamin B_6, and omega-3's, special fats that seem to help protect against artery disease.

Caution: Do not eat raw. May contain parasites that are killed in cooking. If not refrigerated immediately after being caught, tuna can cause histamine poisoning. *Drug interaction:* People who are taking isoniazid (for tuberculosis) or levodopa (for Parkinson's disease) should not eat tuna. *Allergy alert:* Fish can occasionally cause an allergic reaction.

Curiosity: Pliny the Elder (A.D. 23–79) suggested that the ashes of tuna heads be mixed with salt and honey ''to keep away boils or pimples from the parts of which we are ashamed.''

TUNA, *Light Meat, Canned in Water*

Serving: 3 oz., w/out salt

Calories	111
Fat	0.4 g
Saturated	trace
Monounsaturated	trace
Polyunsaturated	trace
Calories from fat	3%
Cholesterol	15 mg
Sodium	43 mg
Protein	25.1 g
Carbohydrate	0
Dietary fiber	0

CHIEF NUTRIENTS

NUTRIENT	AMOUNT	% RDA
Vitamin B_{12}	1.9 mcg	94
Niacin	10.5 mg	55
Iron	2.7 mg	27
Vitamin B_6	0.3 mg	16
Magnesium	24.7 mg	7
Potassium	266.9 mg	7
Riboflavin	0.1 mg	6

Strengths: Pure health, light canned tuna is a super source of body-building protein, niacin, and vitamin B_{12}, while it's low in fat, cholesterol, and sodium. It's also a good source of vitamin B_6 and iron.

Eat with: Fruits and vegetables that are high in vitamin C. The presence of C will enhance the absorption of the iron in tuna.

Caution: Fresh tuna can cause histamine poisoning if it's not properly refrigerated. (Canned tuna, however, is precooked and safe to eat.) *Drug interaction:* People who are taking isoniazid or levodopa should not eat tuna. *Allergy alert:* Fish can occasionally cause an allergic reaction.

Curiosity: Ancient Mediterranean peoples erected wooden watchtowers to spot schools of tuna. Eventually, the villages near the watchtowers became the trading centers of Tunis, Tripoli, Palarmo, Messina, and Setubal.

Description: Canned tuna meat labeled ''light'' comes from the yellowfin tuna, skipjack, and bluefin. Meat labeled ''white'' comes from the albacore.

Storage: If you don't finish tuna from the can, transfer to a nonmetal container and refrigerate.

TUNA, *White, Canned in Water*

Serving: 3 oz., drained

Calories	116
Fat	2.1 g
Saturated	0.6 g
Monounsaturated	0.6 g
Polyunsaturated	0.8 g
Calories from fat	16%
Cholesterol	36 mg
Sodium	333 mg
Protein	22.7 g
Carbohydrate	0
Dietary fiber	0

CHIEF NUTRIENTS

NUTRIENT	AMOUNT	% RDA
Vitamin B_{12}	1.9 mcg	94
Niacin	4.9 mg	26
Vitamin B_6	0.4 mg	19
Magnesium	28.9 mg	8
Potassium	240.6 mg	6
Iron	0.5 mg	5

Strengths: With only 2 g of fat per serving, white meat tuna is leaner than chicken breast, making it an excellent staple for low-fat diets. White meat tuna is a fair source of heart-healthy omega-3's, while light meat tuna has very little. While the white meat type has half as much niacin and not nearly as much iron as the light, it has equally substantial amounts of vitamin B_{12} and a little more magnesium and vitamin B_6.

Caution: *Drug interaction:* Avoid tuna if you're taking levodopa (for Parkinson's disease) or isoniazid (for tuberculosis).

Description: Albacore tuna is the only fish sold as ''white meat'' tuna.

Serve: In pasta salad, antipasto, salade niçoise, or old favorites like tuna-noodle casserole. The tuna adds a substantial amount of high-quality protein.

A Better Idea: Look for reduced-salt versions of canned albacore. Or drain tuna and rinse well under running water to flush away the salt.

TURBOT, *European*

Serving: 3 oz., uncooked

Calories	81
Fat	2.5 g
Saturated	0.6 g
Monounsaturated	0.5 g
Polyunsaturated	N/A
Calories from fat	28%
Cholesterol	41 mg
Sodium	128 mg
Protein	13.6 g
Carbohydrate	0
Dietary fiber	0

CHIEF NUTRIENTS

NUTRIENT	AMOUNT	% RDA
Vitamin B_{12}	1.9 mcg	94
Magnesium	43.4 mg	12
Niacin	1.9 mg	10
Vitamin B_6	0.2 mg	9
Potassium	202.3 mg	5

Strengths: Appreciated for its delicate flavor, turbot is an excellent low-fat source of blood-building vitamin B_{12} and a good source of magnesium.

Caution: *Allergy alert:* Fish can occasionally cause an allergic reaction.

Origin: Found in Atlantic waters from Iceland to the Mediterranean.

Preparation: The subtle flavor of turbot is enhanced only by the simplest preparation. Many cooks say it tastes best when it is either steamed or poached and bathed in butter.

A Better Idea: Lightly brush with oil instead of loading on the butter.

WHITEFISH, *Mixed Species*

Serving: 3 oz., smoked

Calories	92
Fat	0.8 g
Saturated	trace
Monounsaturated	trace
Polyunsaturated	trace
Calories from fat	8%
Cholesterol	28 mg
Sodium	866 mg
Protein	19.9 g
Carbohydrate	0
Dietary fiber	0

CHIEF NUTRIENTS

NUTRIENT	AMOUNT	% RDA
Vitamin B_{12}	2.8 mcg	139
Vitamin B_6	0.3 mg	17
Niacin	2.0 mg	11
Potassium	359.6 mg	10
Magnesium	19.6 mg	6
Riboflavin	0.1 mg	5

Strengths: Whitefish is a great source of vitamin B_{12}, as well as a good source of niacin and vitamin B_6 to keep your immune system on its toes.

Caution: Smoked whitefish is loaded with sodium. People who have high blood pressure may want to avoid it. *Allergy alert:* Fish can occasionally cause an allergic reaction.

Description: A sweet, delicate flavor suggests the essence of ice-cold mountain streams and lakes.

Origin: Whitefish is found in deep cold waters from New England across Canada to the Northwest Territories.

Selection: Available frozen, smoked, and fresh all year.

Preparation: Fresh or frozen whitefish can be poached, baked, broiled, or grilled.

WHITING, *Mixed Species*

Serving: 3 oz., cooked, dry heat

Calories	98
Fat	1.4 g
Saturated	0.3 g
Monounsaturated	0.3 g
Polyunsaturated	0.8 g
Calories from fat	13%
Cholesterol	71 mg
Sodium	112 mg
Protein	20 g
Carbohydrate	0
Dietary fiber	0

CHIEF NUTRIENTS

NUTRIENT	AMOUNT	% RDA
Vitamin B_{12}	2.2 mcg	111
Potassium	368.9 mg	10
Vitamin B_6	0.2 mg	8
Calcium	52.7 mg	7
Magnesium	23.0 mg	7
Niacin	1.4 mg	7
Folate	12.8 mcg	6

Strengths: Whiting is a wonderful low-fat source of blood-building vitamin B_{12}, with a dash of omega-3's thrown in.
Caution: *Allergy alert:* Fish can occasionally cause an allergic reaction.
Description: Sometimes called a silver hake, whiting is actually a small gray and silver fish that is the hake's first cousin.
Preparation: Whiting can be poached, steamed, broiled, pan-fried, or baked. If you decide to broil it, sprinkle lemon juice, Romano cheese, and paprika over the fillet before cooking to enhance its very slight sweet flavor.

SHELLFISH

ABALONE

Serving: 3 oz., fried

Calories	161
Fat	5.8 g
Saturated	1.4 g
Monounsaturated	2.3 g
Polyunsaturated	1.4 g
Calories from fat	32%
Cholesterol	80 mg
Sodium	502 mg
Protein	16.7 g
Carbohydrate	9.4 g
Dietary fiber	0

CHIEF NUTRIENTS

NUTRIENT	AMOUNT	% RDA
Iron	3.2 mg	32
Vitamin B_{12}	0.6 mcg	30
Magnesium	47.6 mg	14
Thiamine	0.2 mg	13
Niacin	1.6 mg	9
Vitamin B_6	0.1 mg	7
Potassium	241.4 mg	6
Riboflavin	0.1 mg	6

Strengths: A pearl of the sea, abalone is loaded with protein to power your immune system and packed with nearly one-third the RDA of iron to help your body build red blood cells. It's also a good source of thiamine and vitamin B_{12}, with just a dash of omega-3's, special fats that seem to help protect against heart disease.
Eat with: Foods that are high in vitamin C.
Caution: Abalone is high in sodium. People with high blood pressure may want to avoid it. *Allergy alert:* Fish can occasionally cause an allergic reaction.
Description: Made up of a single shell and a tough, muscular foot, the abalone creeps across the ocean floor and clings to underwater rocks. Guess which part you eat.
Preparation: Fresh abalone must be pounded before cooking; it is generally sautéed briefly.
Serve: Abalone goes well in salads.

CLAMS, *Mixed Species, Breaded and Fried*

Serving: 20 small (about 6¾ oz.)

Calories	380
Fat	21.0 g
Saturated	5.0 g
Monounsaturated	8.5 g
Polyunsaturated	5.3 g
Calories from fat	50%
Cholesterol	115 mg
Sodium	684 mg
Protein	26.8 g
Carbohydrate	19.4 g
Dietary fiber	N/A

CHIEF NUTRIENTS

NUTRIENT	AMOUNT	% RDA
Vitamin B₁₂	75.7 mcg	3,786
Iron	26.2 mg	262
Vitamin C	18.8 mg	31
Riboflavin	0.5 mg	27
Niacin	3.9 mg	20
Zinc	2.7 mg	18
Folate	34.2 mcg	17
Vitamin A	169.2 RE	17
Potassium	612.9 mg	16
Calcium	118.4 mg	15

Caution: Batter-fried clams are popular seashore fare, but they have much more fat than steamed clams. They also have more than 6 times as much sodium, which makes them a poor choice if you're on a diet to protect your heart and guard against high blood pressure. Deep-frying almost triples the calorie count, and if you dip fried clams in catsup or cocktail sauce, you increase total sodium. Clams contain purines, substances that should be eaten sparingly if you have gout. *Allergy alert:* Clams sometimes trigger allergic reactions.

Selection: Breaded, fried clams are nearly as rich in vitamins and minerals as steamed clams, but the high fat and calorie counts cancel out any potential benefits.

A Better Idea: Steamed clams, served with clam broth and lemon wedges, would be a much wiser choice.

CLAMS, *Mixed Species, Canned*

Serving: 3 oz.

Calories	126
Fat	1.7 g
Saturated	0.2 g
Monounsaturated	0.2 g
Polyunsaturated	0.5 g
Calories from fat	12%
Cholesterol	57 mg
Sodium	95 mg
Protein	21.7 g
Carbohydrate	4.4 g
Dietary fiber	0

CHIEF NUTRIENTS

NUTRIENT	AMOUNT	% RDA
Vitamin B₁₂	84.1 mcg	4,203
Iron	23.8 mg	238
Vitamin C	18.8 mg	31
Riboflavin	0.4 mg	21
Niacin	2.9 mg	15
Vitamin A	145.4 RE	15
Zinc	2.3 mg	15
Potassium	533.8 mg	14

Strengths: Like steamed clams, a single serving of canned clams contains a 6-week supply of vitamin B₁₂, plus a 2-day supply of resistance-building iron. And they're even lower in calories, fat, and cholesterol than steamed clams. Clams have a touch of omega-3 fatty acids, special fats that seem to help protect against artery disease. (Some studies have indicated that clams can reduce blood levels of "bad" cholesterol and triglycerides.)

Eat with: To help enhance absorption of vitamin B₁₂, eat with calcium-rich foods such as sauces and dips made with skim milk or other nonfat dairy products.

Caution: Like most shellfish, clams are high in purines and should be eaten sparingly if you have gout. *Allergy alert:* Clams occasionally cause allergic reactions.

Preparation: Drain and rinse well to remove excess salt.

Serve: Canned clams are convenient to have on hand to make quick-and-easy chowder, dip, or clam sauce.

CLAMS, *Mixed Species, Steamed*

Serving: 20 small (about 3 oz.)

Calories	133
Fat	1.8 g
Saturated	0.2 g
Monounsaturated	0.2 g
Polyunsaturated	0.5 g
Calories from fat	12%
Cholesterol	60 mg
Sodium	101 mg
Protein	23.0 g
Carbohydrate	4.6 g
Dietary fiber	0

CHIEF NUTRIENTS

NUTRIENT	AMOUNT	% RDA
Vitamin B$_{12}$	89.0 mcg	4,450
Iron	25.2 mg	252
Vitamin C	19.9 mg	33
Riboflavin	0.4 mg	22
Niacin	3.0 mg	16
Zinc	2.5 mg	16
Potassium	565.2 mg	15
Vitamin A	153.9 RE	15
Folate	25.9 mcg	13
Calcium	82.8 mg	10

Strengths: One of nature's superfoods. A single serving of clams has a 44-day supply of vitamin B$_{12}$, plus a 2-day supply of iron. Clams are also a good, low-fat source of protein, calcium, potassium, zinc, folate, and vitamin A, and they are rich in vitamin C. Some studies indicate that clams can reduce the amount of "bad" cholesterol and triglycerides circulating in your bloodstream.

Caution: Buy commercially harvested clams. Since this shellfish is susceptible to red tide microorganisms, check with your local public health department before you dig up your own. *Allergy alert:* Clams occasionally cause allergic reactions.

Curiosity: The shells were used as money—*wampum*—by early American Indians.

Selection: Choose hard-shelled clams that are tightly closed. Soft-shelled clams should move when the neck is lightly touched.

Preparation: Boil water, drop in clams, and steam for 4 to 6 minutes after water returns to a boil. Avoid eating raw.

CRAB, *Alaskan King*

Serving: 3 oz., steamed

Calories	82
Fat	1.3 g
Saturated	0.1 g
Monounsaturated	0.2 g
Polyunsaturated	0.5 g
Calories from fat	14%
Cholesterol	45 mg
Sodium	911 mg
Protein	16.5 g
Carbohydrate	0
Dietary fiber	0

CHIEF NUTRIENTS

NUTRIENT	AMOUNT	% RDA
Vitamin B$_{12}$	9.8 mcg	489
Zinc	6.5 mg	43
Folate	43.4 mcg	22
Magnesium	53.6 mg	15
Vitamin C	6.5 mg	11
Vitamin B$_6$	0.2 mg	8
Iron	0.7 mg	7

Strengths: Alaskan king crab is very high in vitamin B$_{12}$ and is low in fat. It has a generous amount of zinc, with a good chunk of magnesium, folate, and vitamin C on the side. There's a fair amount of omega-3's to keep your heart running smoothly. Some studies indicate that crab can reduce the amount of "bad" cholesterol and triglycerides circulating in your bloodstream.

Caution: Crab has a fair amount of sodium and purines. People with high blood pressure or gout may want to avoid it. *Allergy alert:* Shellfish can occasionally cause an allergic reaction.

Preparation: Frozen king crab has been precooked. Just let it thaw, then steam or boil until it's heated through.

CRAB, *Blue*

Serving: 3 oz., cooked, moist heat

Calories	87
Fat	1.5 g
Saturated	0.2 g
Monounsaturated	0.2 g
Polyunsaturated	0.6 g
Calories from fat	16%
Cholesterol	85 mg
Sodium	237 mg
Protein	17.2 g
Carbohydrate	0
Dietary fiber	0

CHIEF NUTRIENTS

NUTRIENT	AMOUNT	% RDA
Vitamin B$_{12}$	6.2 mcg	311
Zinc	3.6 mg	24
Folate	43.2 mcg	22
Niacin	2.8 mg	15
Calcium	88.4 mg	11
Iron	0.8 mg	8
Magnesium	28.1 mg	8
Vitamin B$_6$	0.2 mg	8
Potassium	275.4 mg	7

Strengths: An Atlantic jewel. Like almost every shellfish, blue crab is an excellent source of vitamin B$_{12}$. In addition, it contains a trace of omega-3's and a respectable amount of body-building protein, zinc, calcium, folate, and niacin. It's also low in fat. Some studies indicate that crab can reduce the amount of "bad" cholesterol and triglycerides circulating in your bloodstream.

Caution: Blue crab is high in sodium and purines. People with high blood pressure or gout may want to avoid it. *Allergy alert:* Eating crab or inhaling steam from the water in which it is boiled can occasionally cause an allergic reaction.

Origin: Blue crab roams the Atlantic shoreline from Nova Scotia to Uruguay, with heavy populations in the Chesapeake Bay and the Gulf of Mexico.

Description: Blue crabs have an oval, dark-blue-green shell and blue claws.

CRAB, *Soft-Shell*

Serving: 1 (about 4½ oz.), fried

Calories	334
Fat	17.9 g
Saturated	4.4 g
Monounsaturated	7.7 g
Polyunsaturated	4.9 g
Calories from fat	48%
Cholesterol	45 mg
Sodium	1,118 mg
Protein	11.0 g
Carbohydrate	31.2 g
Dietary fiber	N/A

CHIEF NUTRIENTS

NUTRIENT	AMOUNT	% RDA
Vitamin B$_{12}$	4.5 mcg	224
Iron	1.8 mg	18
Folate	20.0 mcg	10
Niacin	1.8 mg	9
Vitamin B$_6$	0.2 mg	8
Calcium	55.0 mg	7
Magnesium	25.0 mg	7
Thiamine	0.1 mg	7

Strengths: Soft-shell crab is an excellent source of vitamin B$_{12}$, with good shots of iron and folate to keep your immune system on the alert.

Caution: Fried soft-shell crab is high in purines and very high in sodium. People with high blood pressure or gout may want to avoid it. *Allergy alert:* Shellfish occasionally causes an allergic reaction.

Description: Soft-shell crab is actually a stage, not a species. There's a short, vulnerable period of time after the blue crab has shed one shell but before it has grown another. Crabs caught during that stage are "soft-shells."

Preparation: Less than half the recipe is actually crab meat. The rest is bread crumbs and butter.

A Better Idea: A 3-oz. steamed blue crab has almost double the protein, one-third the calories, less than half the fat, and only one-fifth the sodium contained in a fried soft-shell crab. It also has a trace of omega-3's, which would be lost in the frying process.

CRAYFISH

Serving: 3 oz., steamed

Calories	97
Fat	1.2 g
Saturated	0.2 g
Monounsaturated	0.3 g
Polyunsaturated	0.3 g
Calories from fat	11%
Cholesterol	151 mg
Sodium	58 mg
Protein	20.3 g
Carbohydrate	0
Dietary fiber	0

CHIEF NUTRIENTS

NUTRIENT	AMOUNT	% RDA
Vitamin B_{12}	2.9 mcg	147
Iron	2.7 mg	27
Niacin	2.5 mg	13
Thiamine	0.2 mg	10
Zinc	1.4 mg	10
Magnesium	26.4 mg	8
Potassium	298.4 mg	8
Vitamin B_6	0.2 mg	8

Strengths: A low-fat gift from the Louisiana bayous. Crayfish—also known as "crawfish," "crawdaddys," and "mudbugs"—have a ton of protein and vitamin B_{12}, with a good amount of iron, niacin, and thiamine to help your body build red blood cells and use carbohydrates efficiently. They also have a touch of omega-3's.

Caution: *Allergy alert:* Eating crayfish or inhaling the steam from water in which they are boiled can occasionally cause an allergic reaction.

Curiosity: Although crayfish are found in freshwater creeks, marshes, bayous, and waterways around the world, 90 percent come from Louisiana. Eighty percent are also eaten there. A typical event is the "crawfish boil," a Louisiana social gathering similar to a clambake.

Description: A cousin of the lobster, crayfish found in the U.S. are 1 to 5 inches long.

Preparation: Crayfish are supposed to be cooked while still alive.

LOBSTER

Serving: 3 oz., cooked, moist heat

Calories	83
Fat	0.5 g
Saturated	trace
Monounsaturated	trace
Polyunsaturated	trace
Calories from fat	5%
Cholesterol	61 mg
Sodium	323 mg
Protein	17.4 g
Carbohydrate	1.1 g
Dietary fiber	0

CHIEF NUTRIENTS

NUTRIENT	AMOUNT	% RDA
Vitamin B_{12}	2.6 mcg	132
Zinc	2.5 mg	17
Magnesium	29.8 mg	9
Potassium	299.2 mg	8
Calcium	51.9 mg	6

Strengths: A bona fide treasure from the sea, lobster is an excellent low-fat source of vitamin B_{12}, with a decent amount of zinc and a trace of omega-3's.

Caution: Since lobster is high in sodium and has a fair amount of purines, people with either high blood pressure or gout may want to select another seafood. *Allergy alert:* Lobster and other shellfish are potent allergens.

Selection: Maine lobster has sweeter, softer meat than its southern cousin, the spiny—or "rock"—lobster. Make sure these crustaceans are alive when you buy them, since bacteria can build up quickly as soon as they die. If the tail curls underneath when you lift the lobster, it's a good indication that you're holding a live one.

Preparation: Unless you happen to have a bucket of seawater handy, wrap lobster in a wet cloth and keep it refrigerated on a bed of ice until you're ready to cook it. Lobster should be cooked the same day you buy it.

MUSSELS, *Blue*

Serving: 3 oz., cooked, moist heat

Calories	146
Fat	3.8 g
Saturated	0.7 g
Monounsaturated	0.9 g
Polyunsaturated	1.0 g
Calories from fat	23%
Cholesterol	48 mg
Sodium	314 mg
Protein	20.2 g
Carbohydrate	6.3 g
Dietary fiber	0

CHIEF NUTRIENTS

NUTRIENT	AMOUNT	% RDA
Vitamin B$_{12}$	20.4 mcg	1,020
Iron	5.7 mg	57
Folate	64.3 mcg	32
Riboflavin	0.4 mg	21
Vitamin C	11.6 mg	19
Thiamine	0.3 mg	17
Zinc	2.3 mg	15
Niacin	2.6 mg	13
Magnesium	31.5 mg	9

Strengths: Want to live long and prosper? Eat mussels! Loaded with iron and vitamin B$_{12}$, they are also a good source of protein, zinc, vitamin C, thiamine, riboflavin, and niacin, and they are rich in folate. Low in fat with a moderate amount of cholesterol, they have a touch of omega-3's. Studies indicate that mussels may lower triglycerides and one of the "bad" kinds of cholesterol in your body.

Caution: Since mussels are susceptible to red tide microorganisms, check with your local public health department before you harvest your own. Contain a lot of sodium; people with high blood pressure may want to avoid them. *Allergy alert:* Fish can occasionally cause an allergic reaction.

Selection: Buy small, fresh, commercially harvested mussels with tightly closed shells. Avoid those that feel heavy—they're full of sand. And reject those that feel light and loose when you shake them—they're dead.

Preparation: Mussels should be thoroughly scrubbed with a wire brush before cooking. Discard any mussel that doesn't open during cooking.

OYSTERS, *Eastern, Breaded and Fried*

Serving: 6 medium (about 3 oz.)

Calories	173
Fat	11.1 g
Saturated	2.8 g
Monounsaturated	4.1 g
Polyunsaturated	2.8 g
Calories from fat	57%
Cholesterol	71 mg
Sodium	367 mg
Protein	7.7 g
Carbohydrate	10.2 g
Dietary fiber	0

CHIEF NUTRIENTS

NUTRIENT	AMOUNT	% RDA
Vitamin B$_{12}$	13.8 mcg	688
Zinc	76.7 mg	511
Iron	6.1 mg	61
Magnesium	51.0 mg	15
Riboflavin	0.2 mg	11
Thiamine	0.1 mg	9
Niacin	1.5 mg	8
Vitamin A	79.2 RE	8
Calcium	54.6 mg	7

Caution: Batter-fried oysters have 3 times as many calories and over 5 times as much fat as steamed oysters. The batter accounts for a lot of sodium—and deep-frying leaches out significant amounts of heart-healthy omega-3's. So if you're trying to avoid weight gain, heart disease, or high blood pressure, fried oysters are not a prudent choice.

Strengths: Oysters are high in immunity-boosting minerals like iron and zinc: Concentrations are high, no matter how the oysters are cooked.

A Better Idea: Order your oysters steamed, and flavor tartly with a spritz of lemon juice.

OYSTERS, *Eastern, Steamed*

Serving: 6 medium (1½ oz.)

Calories	57
Fat	2.1 g
Saturated	0.6 g
Monounsaturated	0.3 g
Polyunsaturated	0.8 g
Calories from fat	33%
Cholesterol	44 mg
Sodium	177 mg
Protein	5.9 g
Carbohydrate	3.3 g
Dietary fiber	0

CHIEF NUTRIENTS

NUTRIENT	AMOUNT	% RDA
Vitamin B$_{12}$	14.7 mcg	735
Zinc	76.3 mg	509
Iron	5.0 mg	50
Magnesium	40.0 mg	11
Riboflavin	0.1 mg	8
Thiamine	0.1 mg	8
Niacin	1.0 mg	5
Vitamin C	2.5 mg	4
Vitamin A	23.0 RE	2

Strengths: Oysters that are steamed or otherwise cooked by moist heat have many strengths: They are loaded with iron and zinc—which do good things for your immune system. A bonus of vitamin B$_{12}$ ensures healthy nerves. Also a good source of magnesium, with a fair amount of omega-3's. Studies indicate that an oyster-rich diet may lower cholesterol and triglyceride levels.

Caution: Buy commercially harvested oysters. Since oysters are susceptible to red tide microorganisms, check with your local public health department before you harvest your own. *Allergy alert:* Fish occasionally cause allergic reactions.

Curiosity: The old salt's tale that you shouldn't eat oysters during months without an ''r'' in the name—May, June, July, and August—may be more than just a rumor. When oysters spawn during the summer months, they produce an excessive amount of glycogen, which gives them a spoiled-seafood taste. **Preparation:** Cook oysters until the edges curl, but don't overcook—it makes them tough. Deep-fat frying destroys omega-3's and almost doubles the fat content.

OYSTERS, *Eastern, Uncooked*

Serving: 6 medium (about 3 oz.)

Calories	58
Fat	2.1 g
Saturated	0.5 g
Monounsaturated	0.2 g
Polyunsaturated	0.6 g
Calories from fat	32%
Cholesterol	46 mg
Sodium	94 mg
Protein	5.9 g
Carbohydrate	3.3 g
Dietary fiber	0

CHIEF NUTRIENTS

NUTRIENT	AMOUNT	% RDA
Vitamin B$_{12}$	16.1 mcg	804
Zinc	76.4 mg	509
Iron	5.6 mg	56
Magnesium	45.4 mg	13
Thiamine	0.1 mg	9
Riboflavin	0.1 mg	8
Vitamin A	75.6 RE	8
Vitamin C	4.2 mg	7

Caution: Eating raw oysters is a bad idea. In a single day, the average oyster pumps about 25 gallons of water through its body. If the water's polluted—which is often the case—the oyster will suck up significant quantities of potentially harmful organisms. Oysters are high in purines and should be eaten sparingly by anyone with gout. *Allergy alert:* Like other bivalves, oysters sometimes cause allergic reactions.

Strengths: Oysters are a fair source of omega-3 fatty acids, which are purported to lower triglycerides, a particularly bad type of blood fat. They may qualify as weight-loss aids if eaten prudently: They have moderate calories, and just over 30 percent of calories are from fat.

Selection: Oysters are at their best during fall and winter.

A Better Idea: Make a reduced-fat oyster stew with skim milk, onions, potatoes, and oyster broth. Or bake oysters on the half-shell, topped with chopped spinach or watercress and bread crumbs and seasoned with Tabasco (sort of a stripped-down version of oysters Rockefeller).

SCALLOPS, *Mixed Species, Breaded and Fried*

Serving: 2 large (about 1 oz.)

Calories	67
Fat	3.4 g
Saturated	0.8 g
Monounsaturated	1.4 g
Polyunsaturated	0.9 g
Calories from fat	46%
Cholesterol	19 mg
Sodium	144 mg
Protein	5.6 g
Carbohydrate	3.1 g
Dietary fiber	N/A

CHIEF NUTRIENTS

NUTRIENT	AMOUNT	% RDA
Vitamin B$_{12}$	0.4 mcg	21
Magnesium	18.3 mg	5

Strengths: Breaded and fried scallops have lots of taste and a good chunk of vitamin B$_{12}$.

Caution: Loaded with purines. People who have gout may want to avoid scallops. *Allergy alert:* Fish can occasionally cause an allergic reaction.

Selection: There are basically two groups of scallops used for breading and frying—bay scallops, which are available fresh only during the fall, and sea scallops, generally available fresh from mid-fall to mid-spring. The latter are a little less tender, a little more chewy. When you're buying fresh scallops, select the ones that range from pale beige to creamy pink. Scallops that look pure white have been soaked in water to increase their weight, and the soaking steals some of the flavor.

Preparation: Cook them hot and fast. Scallops will lose both moisture and flavor if overcooked.

A Better Idea: Skewer and grill with green peppers and cherry tomatoes, or poach in chicken broth.

SCALLOPS, *Mixed Species, Uncooked*

Serving: 3 oz.

Calories	75
Fat	0.7 g
Saturated	trace
Monounsaturated	trace
Polyunsaturated	trace
Calories from fat	8%
Cholesterol	28 mg
Sodium	137 mg
Protein	14.3 g
Carbohydrate	2.0 g
Dietary fiber	0

CHIEF NUTRIENTS

NUTRIENT	AMOUNT	% RDA
Vitamin B$_{12}$	1.3 mcg	65
Magnesium	47.6 mg	14
Folate	13.6 mcg	7
Potassium	273.7 mg	7
Vitamin B$_6$	0.1 mg	7
Niacin	1.0 mg	5
Zinc	0.8 mg	5

Strengths: Relatively low in cholesterol, scallops have some omega-3 fatty acids as long as they aren't fried. They're also extremely high in vitamin B$_{12}$. There's a good amount of magnesium and some potassium, minerals that can help keep blood pressure at healthy levels. Reasonably low in calories, scallops have a negligible amount of fat, so they're fine for people on weight-loss diets.

Eat with: A baked potato or other high-potassium food.

Caution: People who have gout should avoid eating scallops: They're loaded with purines. *Allergy alert:* Fish can occasionally cause an allergic reaction.

Selection: Look for the smaller, more succulent bay scallops. The larger, more widely available sea scallops are slightly chewy (but still good).

Storage: Refrigerate scallops immediately after purchase, and use within a day or two.

Preparation: To keep fat and calorie content low, scallops should be broiled, poached, or added to seafood chowder. Avoid breaded, fried scallops—they are high in fat and calories.

SHRIMP, *Mixed Species, Breaded and Fried*

Serving: 3 oz.

Calories	206
Fat	10.4 g
Saturated	1.8 g
Monounsaturated	3.2 g
Polyunsaturated	4.3 g
Calories from fat	46%
Cholesterol	151 mg
Sodium	292 mg
Protein	18.2 g
Carbohydrate	9.8 g
Dietary fiber	N/A

CHIEF NUTRIENTS

NUTRIENT	AMOUNT	% RDA
Vitamin B$_{12}$	1.6 mcg	80
Niacin	2.6 mg	14
Iron	1.1 mg	11
Magnesium	34.0 mg	10
Zinc	1.2 mg	8
Calcium	57.0 mg	7
Riboflavin	0.1 mg	7
Thiamine	0.1 mg	7
Potassium	191.3 mg	5

Caution: Fried shrimp is high in fat, high in calories, and high in sodium. People with elevated blood pressure or a history of heart disease may wish to avoid it. *Allergy alert:* Eating shrimp can occasionally cause an allergic reaction.

Eat with: Fruits and vegetables that contain vitamins A and C.

Strengths: Fried shrimp is an excellent source of blood-building vitamin B$_{12}$. It also has a decent amount of iron and niacin.

A Better Idea: Steamed shrimp. It has less fat, less sodium, and fewer calories. It also has a fair amount of omega-3's. Fried shrimp lacks the omega-3's because they're destroyed in frying.

SHRIMP, *Mixed Species, Steamed*

Serving: 3 oz.

Calories	84
Fat	0.9 g
Saturated	trace
Monounsaturated	trace
Polyunsaturated	trace
Calories from fat	10%
Cholesterol	166 mg
Sodium	190 mg
Protein	17.8 g
Carbohydrate	0
Dietary fiber	0

CHIEF NUTRIENTS

NUTRIENT	AMOUNT	% RDA
Vitamin B$_{12}$	1.3 mcg	63
Iron	2.6 mg	26
Niacin	2.2 mg	12
Zinc	1.3 mg	9
Magnesium	28.9 mg	8
Vitamin A	56.1 RE	6
Vitamin B$_6$	0.1 mg	6

Strengths: America's favorite shellfish, shrimp is a great low-fat source of protein and nerve-protecting vitamin B$_{12}$. It also has a decent amount of iron and niacin, with just a kiss of heart-healthy omega-3's.

Eat with: Fruits and vegetables that contain vitamin C.

Caution: *Allergy alert:* Eating shrimp or inhaling the steam from water in which it is boiled can occasionally cause an allergic reaction.

Selection: Shrimp is generally marketed by size, ranging from "colossal" (10 or fewer per pound) to "small" (36 to 45 per pound). The largest are usually called prawns in the U.S., although the name "prawns" is actually borrowed from another species. Pick shrimp that smell of the sea with no hint of ammonia. Since shrimp spoils rapidly, most of the shrimp you buy will have been previously frozen or cooked in seawater.

Preparation: Remove the dark intestinal vein of larger shrimp since it contains an unpleasant grit. Smaller shrimp do not need to be deveined. Simmer until the moment they change color and become opaque. Do not overcook.

WHELKS

Serving: 3 oz., cooked, moist heat

Calories	234
Fat	0.7 g
Saturated	trace
Monounsaturated	trace
Polyunsaturated	trace
Calories from fat	3%
Cholesterol	111 mg
Sodium	350 mg
Protein	40.5 g
Carbohydrate	13.2 g
Dietary fiber	0

CHIEF NUTRIENTS

NUTRIENT	AMOUNT	% RDA
Vitamin B_{12}	15.4 mcg	771
Iron	8.6 mg	86
Magnesium	146.2 mg	42
Vitamin B_6	0.6 mg	28
Zinc	2.8 mg	18
Potassium	589.9 mg	16
Calcium	96.1 mg	12
Riboflavin	0.2 mg	11
Vitamin C	5.8 mg	10
Niacin	1.7 mg	9

Strengths: Packed with nutrients, whelks are a super source of iron, magnesium, and vitamin B_{12}, as well as a good source of protein, riboflavin, vitamin B_6, calcium, potassium, and zinc. They're also one of the lowest-fat marine foods you can eat.

Caution: Contain a tad too much sodium. People who have high blood pressure may want to avoid them. *Allergy alert:* Fish can occasionally cause an allergic reaction.

Description: A snail with a large, beautiful shell and a tough, tasty foot.

Selection: Fresh whelks are available in the spring and fall, although cooked whelks—canned in vinegar—are available all year round. If you have difficulty finding whelks, check with a market that specializes in Chinese or Italian food.

Preparation: Walking the ocean floor makes whelks tough. So tenderize your catch by pounding, then keep them tender by cooking as little as possible.

Fruit

ACEROLA CHERRIES

Serving: 3 (about ½ oz.)

Calories	5
Fat	trace
Saturated	N/A
Monounsaturated	N/A
Polyunsaturated	N/A
Calories from fat	6%
Cholesterol	0
Sodium	1.0 mg
Protein	0.1 g
Carbohydrate	1.1 g
Dietary fiber	0.2 g

CHIEF NUTRIENTS

NUTRIENT	AMOUNT	% RDA
Vitamin C	241.6 mg	403

Strengths: Considered the richest natural source of vitamin C, acerola cherries are valuable for collagen formation, wound healing, infection fighting, and possible prevention of certain types of cancer. In addition, women who take birth control pills or anyone on aspirin or tetracycline therapy may benefit from the extra C. The fruit has virtually no calories, fat, or sodium.

Eat with: Iron-rich foods like meats to enhance iron absorption.

Description: This white, cherrylike fruit is also known as Barbados cherry or West Indian cherry. Because it has a very tart flavor, it usually needs sweetening of some kind. In the U.S., it's generally sold in dried form. It has long been used to make vitamin C supplements.

Preparation: Crush the dried fruit with a rolling pin and use like raisins in baked goods and marinades. (Add a little extra sweetener to compensate for the tartness; apple juice concentrate is good.) Make into tea, preserves, or fruit butters.

APPLE

Serving: 1 (about 5 oz.), w/skin

Calories	81
Fat	0.5 g
Saturated	trace
Monounsaturated	trace
Polyunsaturated	trace
Calories from fat	6%
Cholesterol	0
Sodium	1 mg
Protein	0.3 g
Carbohydrate	21.1 g
Dietary fiber	3.0 g

CHIEF NUTRIENTS

NUTRIENT	AMOUNT	% RDA
Vitamin C	7.9 mg	13
Vitamin E	0.8 mg αTE	8

Strengths: Modest in calories and low in fat and sodium; cholesterol-free. Lots of insoluble fiber (to help prevent constipation and protect against colon cancer) and pectin (a fiber component that fights cholesterol). The apple is a good source of vitamin C, which helps heal wounds. It also has plenty of boron, a mineral that may boost alertness. (Boron also appears to curb the calcium losses that may lead to osteoporosis.)

Eat with: Foods rich in iron (vitamin C aids iron absorption); low-fiber foods, such as meats and poultry; cheddar cheese to help counteract certain fruit sugars that may lead to cavities.

Caution: Apples may cause intestinal gas, especially if eaten in large quantities. *Allergy alert:* Some people develop an allergic rash from handling raw apples. Others experience oral itching from fresh apples; slicing or grating the fruit and exposing the flesh to air for a few minutes tends to break down the offending substance. Those allergic to aspirin may react to the natural salicylate in apples.

APPLES, *Dried*

Serving: 10 rings (about 2 oz.)

Calories	156
Fat	0.2 g
Saturated	trace
Monounsaturated	trace
Polyunsaturated	trace
Calories from fat	1%
Cholesterol	0
Sodium	56 mg
Protein	0.6 g
Carbohydrate	42.2 g
Dietary fiber	5.6 g

CHIEF NUTRIENTS

NUTRIENT	AMOUNT	% RDA
Iron	0.9 mg	9
Potassium	288.0 mg	8
Riboflavin	0.1 mg	6

Strengths: Apples have barely any fat, only moderate calories (if you don't eat too many), and no cholesterol. Fiber content varies; apples dried with their skin contain more fiber than those that are peeled before drying. Iron, potassium, and riboflavin become more concentrated in the drying process, but vitamin C is lost.

Caution: Dried fruit often leaves a sticky residue on teeth, possibly leading to cavities; brush after eating. *Allergy alert:* May contain sulfites, which help preserve some vitamins but can cause asthmatic wheezing, flushing, and tingling sensations in sensitive people. Those allergic to aspirin may react to the natural salicylate in apples.

Selection: Look for dried apples that have good color and are free from defects and insect damage. Squeeze the package: Quality dried fruit is pliable.

Storage: Store in a cool, dry place.

Preparation: You can dry your own apples without sulfites or other preservatives by using a dehydrator.

APRICOTS

Serving: 3 (about 4 oz.)

Calories	51
Fat	0.4 g
Saturated	trace
Monounsaturated	trace
Polyunsaturated	trace
Calories from fat	7%
Cholesterol	0
Sodium	1 mg
Protein	1.5 g
Carbohydrate	11.8 g
Dietary fiber	2.0 g

CHIEF NUTRIENTS

NUTRIENT	AMOUNT	% RDA
Vitamin A	276.7 RE	28
Vitamin C	10.6 mg	18
Potassium	313.8 mg	8
Iron	0.6 mg	6

Strengths: Low in calories and fat; no cholesterol—perfect for weight control and heart health. The combination of low sodium and a nice amount of potassium may help prevent high blood pressure and stroke. Apricots are rich in fiber and beta-carotene (the plant form of vitamin A), both of which help prevent certain types of cancer. And they contain a good amount of vitamin C, which may help their iron be better absorbed.

Caution: Apricots contain some oxalic acid, which should be restricted by those with calcium-oxalate stones. Eaten in large amounts, apricots may cause intestinal gas. *Allergy alert:* Fresh apricots can cause an itchy mouth, especially in those who are also sensitive to peaches, plums, or cherries. Those allergic to aspirin may react to the natural salicylate in apricots.

Origin: Apricots originated in China. Alexander the Great is reported to have introduced them to Greece.

Selection: Avoid greenish apricots, which will never ripen well at home.

APRICOTS, Dried

Serving: 10 halves

Calories	83
Fat	0.2 g
Saturated	trace
Monounsaturated	trace
Polyunsaturated	trace
Calories from fat	2%
Cholesterol	0
Sodium	4 mg
Protein	1.3 g
Carbohydrate	21.6 g
Dietary fiber	2.7 g

CHIEF NUTRIENTS

NUTRIENT	AMOUNT	% RDA
Vitamin A	253.4 RE	25
Iron	1.7 mg	17
Potassium	482.3 mg	13
Niacin	1.1 mg	6

Strengths: Barely any fat and not too many calories (if you don't munch them absentmindedly); no cholesterol. Dried apricots are high in fiber, and they have a nice kick of vitamin A for healthy skin, good night vision, and strong immunity. A tasty source of iron.

Caution: Dried fruit often leaves a sticky residue on teeth that could lead to cavities; brush after eating. *Allergy alert:* Dried apricots may contain sulfites, which help preserve some vitamins but destroy thiamine and can cause asthmatic wheezing, flushing, and tingling sensations in sensitive people. Those allergic to aspirin may react to the natural salicylate in apricots.

Selection: Look for dried apricots that have good color and are free from defects and insect damage. Squeeze the package: Quality dried fruit is pliable.

Storage: Store in a cool, dry place.

Preparation: You can dry your own apricots without sulfites or other preservatives by using a dehydrator.

AVOCADO

Serving: ¹/₂ (about 3 oz.)

Calories	162
Fat	15.4 g
Saturated	2.5 g
Monounsaturated	9.7 g
Polyunsaturated	2.0 g
Calories from fat	86%
Cholesterol	0
Sodium	10 mg
Protein	2.0 g
Carbohydrate	7.4 g
Dietary fiber	2.5 g

CHIEF NUTRIENTS

NUTRIENT	AMOUNT	% RDA
Folate	62.2 mcg	31
Potassium	602.0 mg	16
Vitamin B₆	0.3 mg	14
Vitamin C	7.9 mg	13
Magnesium	39.2 mg	11
Iron	1.0 mg	10
Niacin	1.9 mg	10
Riboflavin	0.1 mg	7
Thiamine	0.1 mg	7
Vitamin A	61.3 RE	6

Strengths: Very high in monounsaturated fat, the type that's been shown to help lower cholesterol. Although rich and buttery in taste and texture, the avocado contains no cholesterol. Another benefit: It's a very good source of folate, which is important for the formation of hemoglobin. And women who take birth control pills may benefit from extra folate. Ounce for ounce, an avocado contains more potassium than a banana, which is considered a good source. Avocado eaters get plentiful amounts of other nutrients as well, including vitamin B₆, vitamin C, magnesium, and iron.

Caution: Because the avocado *is* quite high in fat, weight watchers should take note. *Drug interaction:* Large quantities may cause problems for those taking MAO-inhibitor antidepressants.

Curiosity: Avocados were first called alligator pears, in homage to both their shape and rough skin.

Selection: Hass avocados have thick, pebbly skin that turns from green to purple-black when ripe. Green-skinned varieties have a thin, smoother skin. Both can be tested for ripeness: When they yield to gentle pressure, they're ready to eat. To speed ripening, store in a paper bag at room temperature.

BANANA

Serving: 1 (about 4 oz.), w/out skin

Calories	105
Fat	0.6 g
Saturated	trace
Monounsaturated	trace
Polyunsaturated	trace
Calories from fat	5%
Cholesterol	0
Sodium	1 mg
Protein	1.2 g
Carbohydrate	26.7 g
Dietary fiber	1.8 g

CHIEF NUTRIENTS

NUTRIENT	AMOUNT	% RDA
Vitamin B_6	0.7 mg	33
Vitamin C	10.4 mg	17
Potassium	451.4 mg	12
Folate	21.8 mcg	11
Magnesium	33.1 mg	9
Riboflavin	0.1 mg	6

Strengths: High in vitamin B_6, which helps fight infection and is essential for the synthesis of heme, the iron-containing part of hemoglobin. The banana is a good source of vitamin C, needed to help fight infections. The combination of a good amount of potassium and low sodium may help protect against high blood pressure. As an easily digestible source of fiber, the banana is valuable in preventing both diarrhea and constipation. It has a creamy texture often associated with fatty foods, yet is virtually fat-free.

Caution: *Drug interaction:* People taking levodopa should not eat too many bananas, because the vitamin B_6 may interfere with the drug. Large quantities may also cause problems for those taking MAO-inhibitor antidepressants. *Allergy alert:* People allergic to ragweed may also be sensitive to bananas.

Preparation: Wash bananas before peeling because bacteria on the peel can be transferred to the fruit.

BLACKBERRIES

Serving: 1 cup

Calories	75
Fat	0.6 g
Saturated	N/A
Monounsaturated	N/A
Polyunsaturated	N/A
Calories from fat	7%
Cholesterol	0
Sodium	0
Protein	1.0 g
Carbohydrate	18.4 g
Dietary fiber	7.2 g

CHIEF NUTRIENTS

NUTRIENT	AMOUNT	% RDA
Vitamin C	30.2 mg	50
Folate	49.0 mcg	24
Vitamin E	0.9 mg αTE	9
Iron	0.8 mg	8
Magnesium	28.8 mg	8
Potassium	282.2 mg	8
Calcium	46.1 mg	6

Strengths: Super-high in fiber (thanks to their multitude of edible seeds), blackberries contain substantial vitamin C: One cup supplies half the daily requirement. Surprisingly, they're also a fair source of calcium and iron. Like most fresh fruit, blackberries have a modest amount of calories, and they're low in fat and sodium. They're also cholesterol-free. Pregnant women can benefit from the folate in blackberries, since folate helps proper cell and tissue growth in babies. And blackberries have vitamin E, which is essential for healthy nerves.

Eat with: Meat and poultry. (The vitamin C in blackberries enhances absorption of meat's iron.)

Caution: These berries contain some oxalic acid, which should be restricted by those with calcium-oxalate stones. *Allergy alert:* Those allergic to aspirin may react to the salicylates in blackberries.

Selection: Look for plump, firm, dark-colored berries with no hulls. Attached hulls are a sign of immature berries, which may be suited for jams but will be excessively tart for out-of-hand eating.

BLUEBERRIES

Serving: 1 cup

Calories	81
Fat	0.6 g
Saturated	N/A
Monounsaturated	N/A
Polyunsaturated	N/A
Calories from fat	6%
Cholesterol	0
Sodium	9 mg
Protein	1.0 g
Carbohydrate	20.5 g
Dietary fiber	3.3 g

CHIEF NUTRIENTS

NUTRIENT	AMOUNT	% RDA
Vitamin C	18.9 mg	31

Strengths: Moderate in calories and low in fat and sodium. No cholesterol. Blueberries have an excellent helping of fiber, and they contain lots of vitamin C to help heal wounds, fight infections, and enhance absorption of iron from other foods. Women on birth control pills may also benefit from extra C.
Eat with: Meat, poultry, and other foods low in fiber but high in iron.
Caution: Heat destroys vitamin C, so blueberries are best eaten raw. Frozen berries have a bit more folate than fresh but have suffered considerable vitamin C losses. *Allergy alert:* Berries in general are a common trigger of food allergy. More specifically, those allergic to aspirin may react to the natural salicylate in blueberries.
Selection: Look for plump, firm, indigo-blue berries with a silvery frost. Do not wash until ready to use.
Serve: Perfect sprinkled on all kinds of cereals. A nice low-cal snack all by themselves.

BREADFRUIT

Serving: ¼ (about 3 oz.)

Calories	99
Fat	0.2 g
Saturated	N/A
Monounsaturated	N/A
Polyunsaturated	N/A
Calories from fat	2%
Cholesterol	0
Sodium	2 mg
Protein	1.0 g
Carbohydrate	26.0 g
Dietary fiber	4.2 g

CHIEF NUTRIENTS

NUTRIENT	AMOUNT	% RDA
Vitamin C	27.8 mg	46
Potassium	470.4 mg	13
Magnesium	24.0 mg	7
Thiamine	0.1 mg	7
Iron	0.5 mg	5

Strengths: Rich in vitamin C if eaten raw. Very low in fat, with no cholesterol and a good amount of fiber. Breadfruit has barely any sodium, and it is a good source of potassium—a beneficial combination for controlling high blood pressure. An unexpected source of iron.
History: When the notorious mutiny on the *Bounty* took place, the ship was on its way to the West Indies with breadfruit tree cuttings.
Description: Large, round fruit 8 to 10 inches in diameter, with bumpy green skin. Bland-tasting cream-colored center. Generally picked when green, the fruit needs to be eaten before it becomes overripe—which can happen quickly since it doesn't store well. Ripe fruit has a brown, speckled peel and smooth flesh with an avocado-like texture. Cooking brings out a bread-like aroma, which explains the fruit's name.
Preparation: Cut and peel, discarding the spongy core. It's often cooked like a potato—baked, mashed, boiled, or steamed.

CANTALOUPE

Serving: 1 cup cubes

Calories	56
Fat	0.5 g
Saturated	N/A
Monounsaturated	N/A
Polyunsaturated	N/A
Calories from fat	7%
Cholesterol	0
Sodium	14 mg
Protein	1.4 g
Carbohydrate	13.4 g
Dietary fiber	1.3 g

CHIEF NUTRIENTS

NUTRIENT	AMOUNT	% RDA
Vitamin C	67.5 mg	113
Vitamin A	515.2 RE	52
Folate	27.2 mcg	14
Potassium	494.4 mg	13
Vitamin B$_6$	0.2 mg	9
Magnesium	17.6 mg	5
Niacin	0.9 mg	5

Strengths: Hard to beat as an excellent source of vitamins A and C—which help promote wound healing and enhance immunity. Cantaloupe contains few calories, virtually no fat, and little sodium. A good amount of folate also helps enhance immunity, while potassium helps protect against high blood pressure. Cantaloupe also has some B vitamins to promote healthy skin and nerves. A great source of beta-carotene for possible cancer protection. And it contains a nice amount of fiber.

Caution: *Allergy alert:* Some people who react to ragweed or grass pollen may develop an itchy throat or mouth after eating cantaloupe.

History: Named after a little town called Cantalupa near Rome. (True cantaloupes grow only in Europe and Asia; what we buy are "muskmelons.")

Selection: Cantaloupes should have a slightly golden undercolor, with a netting pattern that stands out prominently and covers the melon. A ripe cantaloupe will have a distinctive, sweet aroma.

CARAMBOLA

Serving: 1 (about 4 oz.)

Calories	42
Fat	0.4 g
Saturated	N/A
Monounsaturated	N/A
Polyunsaturated	N/A
Calories from fat	9%
Cholesterol	0
Sodium	3 mg
Protein	0.7 g
Carbohydrate	9.9 g
Dietary fiber	2.2 g

CHIEF NUTRIENTS

NUTRIENT	AMOUNT	% RDA
Vitamin C	26.9 mg	45
Potassium	207.0 mg	6
Vitamin A	62.2 RE	6

Strengths: Plenty of vitamin C to help fight infection, heal wounds, and possibly prevent certain types of stomach and esophageal cancers. Low in calories and fat, with barely any sodium. Carambola has a nice touch of vitamin A for healthy skin and better night vision.

Caution: Contains some oxalic acid, which should be restricted by those with calcium-oxalate stones.

Description: A five-ribbed fruit about 3 to 5 inches long. The star-shaped cross section accounts for carambola's nickname, "starfruit." When ripe, fruit turns deep yellow. Flavor varies from tart and citrusy to sweet.

Selection: In general, the broader the ribs, the sweeter the fruit. Choose fruit with bright, blemish-free skin.

Storage: Keep at room temperature until it yields slightly to touch, then refrigerate. May keep up to 2 weeks when refrigerated.

Serve: Raw in salads or as garnish for poultry, seafood, pork, and other meats. Goes well with avocado. Can be sautéed or cooked into compotes, chutneys, or stir-fries.

Preparation: Discard any seeds before serving.

CASABA

Serving: 1 cup cubes

Calories	45
Fat	0.2 g
Saturated	N/A
Monounsaturated	N/A
Polyunsaturated	N/A
Calories from fat	3%
Cholesterol	0
Sodium	20 mg
Protein	1.5 g
Carbohydrate	10.5 g
Dietary fiber	1.4 g

CHIEF NUTRIENTS

NUTRIENT	AMOUNT	% RDA
Vitamin C	27.2 mg	45
Potassium	357.0 mg	10
Iron	0.7 mg	7
Thiamine	0.1 mg	7

Strengths: Low in calories, fat, and sodium; no cholesterol. Casaba is a surprising source of iron, with lots of vitamin C to aid in its absorption. With considerably more potassium than sodium, the melon is a good selection if you have sodium-sensitive high blood pressure. Like other melons, the casaba is a source of fiber.

Caution: *Allergy alert:* Melons are among the foods that often cause allergic reactions, especially for people sensitive to ragweed, grass pollen, or tomato.

History: Named for Kasaba, Turkey.

Description: A golden-yellow, globe-shaped muskmelon, slightly pointed at the stem end, with lengthwise ridges. The flesh is thick, white, juicy, and sweet.

Selection: Look for a slightly soft stem end. If the melon has a fruity aroma and uniform color, it's a good choice. The casaba should feel heavy for its size.

Serve: With any meal, or as a snack. It's a delicious way to add vitamin C. Great with low-fat cheeses or popcorn-rice cakes.

CHAYOTE

Serving: ½ cup, chopped, boiled

Calories	19
Fat	0.4 g
Saturated	N/A
Monounsaturated	N/A
Polyunsaturated	N/A
Calories from fat	18%
Cholesterol	0
Sodium	1 mg
Protein	0.5 g
Carbohydrate	4.1 g
Dietary fiber	2.4 g

CHIEF NUTRIENTS

NUTRIENT	AMOUNT	% RDA
Vitamin C	6.4 mg	11
Folate	14.5 mcg	7

Strengths: Very low in calories; excellent for dieters. Good fiber, no cholesterol, and just a trace of sodium. Chayote has a chunk of vitamin C and some folate, both of which are important for enhanced immunity.

Curiosity: West Indians call chayote *christophine,* in homage to Christopher Columbus.

History: Chayote was once cultivated by the ancient Aztecs and Mayans.

Description: Also known as mirliton. Gourdlike in shape, it's usually about the size of a very large pear. Beneath the furrowed green skin is white, bland-tasting flesh and a large seed.

Preparation: The seed is edible. The skin becomes tough when it's cooked, so peel the chayote before or after cooking.

Serve: With meat, poultry, or seafood entrées. Popular in Cajun and Creole recipes; the mild, squashlike flavor marries well with spices.

CHERIMOYA

Serving: ½ (about 10 oz.)

Calories	257
Fat	1.1 g
Saturated	N/A
Monounsaturated	N/A
Polyunsaturated	N/A
Calories from fat	4%
Cholesterol	0
Sodium	N/A
Protein	3.6 g
Carbohydrate	65.6 g
Dietary fiber	8.7 g

CHIEF NUTRIENTS

NUTRIENT	AMOUNT	% RDA
Vitamin C	24.6 mg	41
Niacin	3.6 mg	19
Riboflavin	0.3 mg	18
Thiamine	0.3 mg	18
Iron	1.4 mg	14
Calcium	62.9 mg	8

Strengths: Cherimoya is very high in vitamin C (women who take oral contraceptives may benefit from the extra C). In addition, this fruit is a good source of several B vitamins that help form red blood cells and boost energy. It has a surprising—and significant—amount of iron and a bit of calcium. Cherimoya is low in fat, and it's super-high in fiber.
Caution: Rather high in calories for a fruit.
Curiosity: Also known as custard apple.
Origin: The name comes from the ancient Incan. Cherimoya originated in the South American highlands and may be the earliest recorded New World fruit.
Description: Resembles a plump pine cone. Has the texture of an avocado and a taste that mingles the flavors of honey, banana, and pineapple.
Selection: Select a cherimoya that is pale green and fairly evenly colored. Let the fruit ripen at room temperature until it is as soft as a thin-skinned avocado.
Preparation: Can stand alone as a dessert. Cut fruit lengthwise, remove the seeds, and scoop out the white, creamy flesh.

CHERRIES, *Sour*

Serving: 1 cup, w/out pits

Calories	78
Fat	0.5 g
Saturated	trace
Monounsaturated	trace
Polyunsaturated	trace
Calories from fat	5%
Cholesterol	0
Sodium	5 mg
Protein	1.6 g
Carbohydrate	18.9 g
Dietary fiber	1.9 g

CHIEF NUTRIENTS

NUTRIENT	AMOUNT	% RDA
Vitamin C	15.5 mg	26
Vitamin A	198.4 RE	20
Potassium	268.2 mg	7
Folate	11.6 mcg	6

Strengths: Sour cherries are moderate in calories if you don't add too much sweetener to counterbalance their tart flavor. They're also low in fat and sodium, with no cholesterol! A nice source of dietary fiber for both digestive health and control of high blood cholesterol. The sour varieties have much more vitamin A than sweet cherries, giving an extra dose of immune power and possible protection from some forms of cancer. A good amount of vitamin C and some folate help those taking birth control pills or aspirin (both of which deplete these vitamins).
Caution: Cooking sour cherries, which tends to make them a little more palatable, destroys some vitamin C. Cherries contain some oxalic acid, which should be restricted by those with calcium-oxalate stones. *Allergy alert:* May cause itchy mouth or throat in some people. Those allergic to apples, apricots, peaches, plums, or pears may be more likely to react to cherries.
Curiosity: An old-time remedy for gout.
Serve: With Cornish game hens or as part of a rice stuffing for winter squash.

CHERRIES, *Sweet* ⬇ 🗄

Serving: 1 cup, w/out pits

Calories	104
Fat	1.4 g
Saturated	0.3 g
Monounsaturated	0.4 g
Polyunsaturated	0.4 g
Calories from fat	12%
Cholesterol	0
Sodium	3 mg
Protein	1.7 g
Carbohydrate	24.0 g
Dietary fiber	1.6 g

CHIEF NUTRIENTS

NUTRIENT	AMOUNT	% RDA
Vitamin C	10.2 mg	17
Potassium	324.8 mg	9
Iron	0.6 mg	6
Riboflavin	0.1 mg	5

Strengths: A moderate amount of calories with little fat; no cholesterol. Sweet cherries have a good helping of fiber, some of which can help lower cholesterol. They have a good amount of vitamin C to enhance absorption of the nonheme iron that is present. With some potassium and a small amount of sodium, sweet cherries can help prevent stroke and sodium-induced high blood pressure.

Caution: Cherries contain some oxalic acid, which should be restricted by those with calcium-oxalate stones. *Allergy alert:* Some people may get an itchy mouth and throat from eating cherries, especially if they're allergic to apples, apricots, peaches, plums, or pears.

Selection: Look for cherries that are firm, with glossy skin and pliable stems.

Serve: With poultry, rice, pork, and duck dishes. They make a sweet topping for pancakes or waffles.

CRANBERRIES 🔄 🧪 ⬇ 🗄

Serving: ½ cup

Calories	23
Fat	0.1 g
Saturated	N/A
Monounsaturated	N/A
Polyunsaturated	N/A
Calories from fat	2%
Cholesterol	0
Sodium	trace
Protein	0.2 g
Carbohydrate	6.0 g
Dietary fiber	2.0 g

CHIEF NUTRIENTS

NUTRIENT	AMOUNT	% RDA
Vitamin C	6.4 mg	11

Strengths: Very low in fat, sodium, and calories (if unsweetened); no cholesterol. Cranberries contain a good amount of vitamin C to help enhance immunity and iron absorption, as well as aid calcium absorption. A decent source of dietary fiber.

Caution: Dieters, note: These berries are very tart all by themselves, so they are often accompanied by a lot of sweetener. Cranberries contain some oxalic acid, which should be restricted by those with calcium-oxalate stones.

Curiosity: Alternate names include bounceberries (good-quality berries bounce) and craneberries (the pale pink flowers resemble crane heads).

History: The early American Indians used cranberries as a poultice on wounds. (They have an astringent effect that contracts tissues and stops bleeding.)

Preparation: Although cranberries are most often cooked (which diminishes their vitamin C content), you can make uncooked cranberry sauce. Coarsely grind the berries along with a sweet fruit like apples, oranges, or dried apricots. If needed, add a touch of sweetener.

CURRANTS, Black

Serving: 1 cup

Calories	71
Fat	0.5 g
Saturated	trace
Monounsaturated	trace
Polyunsaturated	trace
Calories from fat	6%
Cholesterol	0
Sodium	2 mg
Protein	1.6 g
Carbohydrate	17.2 g
Dietary fiber	4.4 g

CHIEF NUTRIENTS

NUTRIENT	AMOUNT	% RDA
Vitamin C	202.7 mg	338
Iron	1.7 mg	17
Vitamin E	1.1 mg αTE	11
Potassium	360.6 mg	10
Calcium	61.6 mg	8
Magnesium	26.9 mg	8

Strengths: Black currants are just overflowing with vitamin C. (In fact, a cup of black currants has almost 3 times as much vitamin C as an orange). This vitamin boosts cancer protection, strengthens immunity, helps heal wounds, and enhances absorption of the currants' iron. And currants are also a good source of vitamin E, which helps protect the vitamin C from destructive oxidation.

Caution: Currants contain some oxalic acid, which should be restricted by those with calcium-oxalate stones. *Allergy alert:* Those allergic to aspirin may react to the natural salicylate in currants.

History: European black currants were widely grown throughout the northern U.S. until the end of the 1800s. After an outbreak of white pine blister rust was traced to black currants, their culture was prohibited in many pine-growing areas. Today they're mostly found growing wild or cultivated in small gardens.

Preparation: Black currants are commonly used for preserves. (They're also in some liqueurs, such as cassis.)

CURRANTS, Red or White

Serving: 1 cup

Calories	63
Fat	0.2 g
Saturated	trace
Monounsaturated	trace
Polyunsaturated	trace
Calories from fat	3%
Cholesterol	0
Sodium	1 mg
Protein	1.6 g
Carbohydrate	15.5 g
Dietary fiber	4.8 g

CHIEF NUTRIENTS

NUTRIENT	AMOUNT	% RDA
Vitamin C	45.9 mg	77
Iron	1.1 mg	11
Potassium	308.0 mg	8

Strengths: Low in fat, calories, and sodium, red and white currants contain no cholesterol. High in fiber, they may help improve digestive health and lower cholesterol levels. An excellent source of vitamin C, which helps in the absorption of the iron in these currants. (But they have about one-quarter the vitamin C of black currants.) A nice source of potassium, which helps keep blood pressure on an even keel and may protect against stroke.

Caution: Currants contain some oxalic acid, which should be restricted by those with calcium-oxalate stones. *Allergy alert:* Those allergic to aspirin may react to the natural salicylate in currants.

Curiosity: Red and white varieties of currants are among the top 2 dozen fruit crops in the world. Most of the harvest goes into processed jams and jellies.

Description: White currants are a variety of red currants without pigmentation. White ones are generally smaller and sweeter.

CURRANTS, *Zante, Dried*

Serving: ½ cup

Calories	204
Fat	0.2 g
Saturated	trace
Monounsaturated	trace
Polyunsaturated	trace
Calories from fat	0.1%
Cholesterol	0
Sodium	6 mg
Protein	2.9 g
Carbohydrate	53.3 g
Dietary fiber	4.9 g

CHIEF NUTRIENTS

NUTRIENT	AMOUNT	% RDA
Iron	2.3 mg	23
Potassium	642.2 mg	17
Vitamin B$_6$	0.2 mg	11
Calcium	61.9 mg	8
Magnesium	29.5 mg	8
Thiamine	0.1 mg	8
Niacin	1.2 mg	6
Riboflavin	0.1 mg	6
Vitamin C	3.4 mg	6

Strengths: A good amount of iron, along with some vitamin C, which helps iron absorption. A variety of B vitamins for healthy skin, muscles, and nervous system. Very low in sodium but high in potassium for possible protection against strokes. High fiber, a mere trace of fat, and no cholesterol. Zante currants are an unexpected source of calcium for strong bones and teeth. Their magnesium helps promote resistance to tooth decay.

Caution: Dried fruit often leaves a sticky residue on teeth, possibly leading to cavities; brush after eating. *Allergy alert:* Dried currants may contain sulfites, which help preserve some vitamins but destroy thiamine and can cause asthmatic wheezing, flushing, and tingling sensations in sensitive people.

Description: Despite their name, Zante currants are unrelated to fresh black, red, or white currants. They're actually tiny purple Zante grapes that have been dried.

Serve: In fruit salads, stuffings, and rice casseroles. Also used in baked goods.

DATES

Serving: 5 (about 1½ oz.)

Calories	114
Fat	0.2 g
Saturated	N/A
Monounsaturated	N/A
Polyunsaturated	N/A
Calories from fat	1%
Cholesterol	0
Sodium	1 mg
Protein	0.8 g
Carbohydrate	30.5 g
Dietary fiber	3.5 g

CHIEF NUTRIENTS

NUTRIENT	AMOUNT	% RDA
Potassium	270.6 mg	7
Iron	0.5 mg	5
Niacin	0.9 mg	5

Strengths: Very low in fat and sodium, dates are a nice source of potassium, essential for the normal functioning of nerves and muscles. They contain a bit of iron, and also a healthy amount of fiber for good digestive health and cholesterol control. Dates are among the few fruits that can be successfully dried and stored without sulfites.

Eat with: Other fruit (such as apples and oranges) that contains vitamin C: They'll help enhance the absorption of dates' iron. Dates also go well with cheddar cheese, which can help counteract certain fruit sugars that can lead to cavities.

Caution: Dried fruit often leaves a sticky residue on teeth, possibly leading to cavities; brush after eating.

Curiosity: One of the few fruits that doesn't contain *any* vitamin C.

History: According to Muslim legend, the date palm was made from dust left over from the creation of Adam.

FIGS

Serving: 3 (about 5 oz.)

Calories	111
Fat	0.5 g
Saturated	trace
Monounsaturated	trace
Polyunsaturated	trace
Calories from fat	4%
Cholesterol	0
Sodium	2 mg
Protein	1.1 g
Carbohydrate	28.8 g
Dietary fiber	5.0 g

CHIEF NUTRIENTS

NUTRIENT	AMOUNT	% RDA
Potassium	348.0 mg	9
Vitamin B_6	0.2 mg	9
Calcium	52.5 mg	7
Magnesium	25.5 mg	7
Iron	0.6 mg	6
Thiamine	0.1 mg	6
Riboflavin	0.1 mg	5

Strengths: Plump, ripe figs are a real taste treat for those who are only familiar with dried figs. Moderate in calories, with minimal fat and sodium, they have no cholesterol. Figs are a nice source of both calcium and iron. Some potassium and several B vitamins help ensure healthy skin, muscles, and nerves. Figs are rich in fiber, thanks to their multitude of tiny seeds.
Caution: Figs contain some oxalic acid, which should be restricted by those with calcium-oxalate stones.
Curiosity: Early Olympic athletes wore figs—as a sign of honor and good health—*and* ate them.
Selection: Choose plump, soft (but not mushy) figs; skin color can vary from yellowish to deep purple.
Store: In the refrigerator and eat within a few days.
Serve: With other fruit rich in vitamin C to help absorption of figs' nonheme iron. They're wonderful with smoked turkey, low-fat cheese, or omelets.

FIGS, *Dried*

Serving: 3 (about 2 oz.)

Calories	143
Fat	0.7 g
Saturated	trace
Monounsaturated	trace
Polyunsaturated	trace
Calories from fat	4%
Cholesterol	0
Sodium	6 mg
Protein	1.7 g
Carbohydrate	36.7 g
Dietary fiber	5.2 g

CHIEF NUTRIENTS

NUTRIENT	AMOUNT	% RDA
Iron	1.3 mg	13
Potassium	399.0 mg	11
Calcium	80.7 mg	10
Magnesium	33.0 mg	10
Vitamin B_6	0.1 mg	6

Strengths: Like all dried fruit, figs are nutrient-dense. A good source of both calcium and iron, plus plentiful potassium and little sodium for better blood pressure and possible protection against strokes. A nice amount of vitamin B_6 helps those taking oral contraceptives, phenobarbital, or certain corticosteroids (all of which tend to deplete the B_6). A super dose of fiber aids in controlling high cholesterol and digestive problems.
Eat with: Cheddar cheese; it may help neutralize certain fruit sugars that could lead to tooth decay.
Caution: Figs contain some oxalic acid, which should be restricted by those with calcium-oxalate stones. *Allergy alert:* May contain sulfites, which can cause asthmatic wheezing, flushing, and tingling sensations in sensitive people.
History: Figs were so highly valued by the Greeks that their export was forbidden.
Description: Colors range from light tan to deep purple, and consistency varies from moist to dry.

GRAPEFRUIT, *Pink or Red*

Serving: ¹/₂ (about 4 oz.)

Calories	37
Fat	0.1 g
Saturated	trace
Monounsaturated	trace
Polyunsaturated	trace
Calories from fat	3%
Cholesterol	0
Sodium	0
Protein	0.7 g
Carbohydrate	9.5 g
Dietary fiber	0.7 g

CHIEF NUTRIENTS

NUTRIENT	AMOUNT	% RDA
Vitamin C	46.9 mg	78
Folate	15.0 mcg	8

Strengths: Low-calorie, low-fat, sodium- and cholesterol-free, red or pink grapefruit is somewhat higher in vitamin C than the white variety. It gets its color from beta-carotene, which turns to vitamin A in the body and may help prevent some forms of cancer. Although grapefruit isn't a terribly rich source of beta-carotene, every bit helps.

Eat with: Cooked carrots or sweet potatoes for extra vitamin A.

Caution: May cause intestinal gas, especially if eaten in large quantities. *Allergy alert:* Allergy to citrus fruit may trigger bedwetting in some children. Citrus fruit may also cause allergic headaches.

Curiosity: The name grapefruit may come from a description of a 19th-century variety. It was said that they grew "in trusses [clusters] like grapes."

Selection: The redder the flesh, the sweeter it is. Look for heavy, thin-skinned grapefruit. Don't worry too much about minor defects such as scars, scratches, and discoloration.

Serve: Grapefruit is excellent in all types of salads, including chicken, tuna, shrimp, Waldorf, chef's, and spinach.

GRAPEFRUIT, *White*

Serving: ¹/₂ (about 4 oz.)

Calories	39
Fat	0.1 g
Saturated	trace
Monounsaturated	trace
Polyunsaturated	trace
Calories from fat	3%
Cholesterol	0
Sodium	0
Protein	0.8 g
Carbohydrate	9.9 g
Dietary fiber	0.7 g

CHIEF NUTRIENTS

NUTRIENT	AMOUNT	% RDA
Vitamin C	39.3 mg	65
Folate	11.8 mcg	6

Strengths: Very low in fat and calories, with no sodium or cholesterol. White grapefruit is a nice source of fiber to help dieters get their day off to a good start. (The type of insoluble fiber found in grapefruit and other fruit can also help lower cholesterol.) Grapefruit contains some folate and lots of vitamin C to help enhance immunity.

Eat with: Calcium-rich foods such as low-fat cheese (the vitamin C enhances calcium absorption). If you eat grapefruit with iron-rich foods, such as meats, your body will better utilize their iron.

Caution: May cause intestinal gas, especially if eaten in large quantities. *Allergy alert:* Allergy to citrus fruit may trigger bedwetting in some children. Citrus fruit may also cause allergic headaches.

Selection: Thin-skinned grapefruit that feels heavy for its size will generally contain more juice than a puffy, spongy, or coarse-skinned grapefruit.

GRAPES, *American* ⬇ 🏛

Serving: 1 cup

Calories	58
Fat	0.3 g
Saturated	trace
Monounsaturated	trace
Polyunsaturated	trace
Calories from fat	5%
Cholesterol	0
Sodium	2 mg
Protein	0.6 g
Carbohydrate	15.8 g
Dietary fiber	0.6 g

CHIEF NUTRIENTS

NUTRIENT	AMOUNT	% RDA
Vitamin C	3.7 mg	6
Potassium	175.7 mg	5
Thiamine	0.1 mg	5

Strengths: American grapes are low in calories, fat, and sodium, and they're cholesterol-free. They have some vitamin C, but only about one-fifth as much as their European cousins. The amount of fiber depends upon the variety, but of course you get more fiber if you actually eat the skin and seeds (if any). Grapes contain a substance called ellagic acid, which may kill certain cancer-causing compounds in the body.

Caution: Grapes (especially Concord) contain some oxalic acid, which should be restricted by those with calcium-oxalate stones. *Allergy alert:* Those allergic to aspirin may react to the natural salicylate in grapes.

History: It's believed that grapes were domesticated as early as 5000 B.C. in western Asia. When blight affected European grapes in the late 1800s, the wine industry was rescued by grafting healthy vines onto disease-resistant American rootstock.

Description: American grapes have skins that slip off easily, such as are found on the black-purple Concord, purple-red Catawba, and pink Delaware. Although they can be eaten fresh, these varieties are often processed into jellies, jams, and juice.

GRAPES, *European* 🄯 ☀ 🗲 ⬇ 🏛

Serving: 1 cup

Calories	114
Fat	0.9 g
Saturated	trace
Monounsaturated	trace
Polyunsaturated	trace
Calories from fat	7%
Cholesterol	0
Sodium	3 mg
Protein	1.1 g
Carbohydrate	28.4 g
Dietary fiber	1.1 g

CHIEF NUTRIENTS

NUTRIENT	AMOUNT	% RDA
Vitamin C	17.3 mg	29
Thiamine	0.2 mg	10
Vitamin B$_6$	0.2 mg	9
Potassium	296.0 mg	8
Riboflavin	0.1 mg	5

Strengths: Moderately low in calories, with barely any fat, little sodium, and no cholesterol. European grapes are a good source of vitamin C. They have nice amounts of some B vitamins for healthy skin, nerves, and cardiovascular system. Also a decent amount of fiber for good bowel health. Grapes contain a substance called ellagic acid, which may kill certain cancer-causing compounds in the body.

Caution: Pediatricians caution that whole grapes are among the foods most likely to cause choking in small children; cut grapes in halves or quarters before serving. Grapes contain some oxalic acid, which should be restricted by those with calcium-oxalate stones. *Allergy alert:* Those allergic to aspirin may react to the natural salicylate in grapes.

Description: European grapes have skin that adheres tightly to the fruit, while American grapes are the slip-skin varieties. The European kind are the grapes we usually eat out-of-hand. Varieties include the green Thompson seedless and the red Tokay and emperor.

GUAVA

Serving: 1 (about 3 oz.)

Calories	46
Fat	0.5 g
Saturated	trace
Monounsaturated	trace
Polyunsaturated	trace
Calories from fat	11%
Cholesterol	0
Sodium	3 mg
Protein	0.7 g
Carbohydrate	10.7 g
Dietary fiber	4.9 g

CHIEF NUTRIENTS

NUTRIENT	AMOUNT	% RDA
Vitamin C	165.2 mg	275
Potassium	255.6 mg	7
Vitamin A	71.1 RE	7
Vitamin B$_6$	0.1 mg	7
Niacin	1.1 mg	6

Strengths: Guava has a tremendous amount of vitamin C— almost three times the RDA. That's of special interest to women taking birth control pills (which seem to increase the need for vitamins C and B$_6$). Low in calories, fat, and sodium, the guava is also a very good source of fiber. (The National Cancer Institute recommends 20 to 35 g of fiber per day.)

Eat with: Any iron-rich food such as meat.

Caution: Note that canned guava juice may contain a fair amount of sugar as well as artificial color.

Description: The guava comes in many varieties ranging from round to pear-shaped and from 1 inch to 4 inches in diameter. The skin can vary from pale yellow to yellow-green. The taste can be either sweet or sour, with overtones of strawberry, pineapple, or banana.

Selection: Choose a smooth-skinned, unblemished guava with a rich fragrance. The rind will yield to gentle pressure when the guava is ripe.

Serve: Try adding chopped guava to the braising liquid, then puree to make a thick fruity sauce for duck, pork, or chicken.

HONEYDEW

Serving: 1 cup cubes

Calories	60
Fat	0.2 g
Saturated	N/A
Monounsaturated	N/A
Polyunsaturated	N/A
Calories from fat	3%
Cholesterol	0
Sodium	17 mg
Protein	0.8 g
Carbohydrate	15.6 g
Dietary fiber	1.4 g

CHIEF NUTRIENTS

NUTRIENT	AMOUNT	% RDA
Vitamin C	42.0 mg	70
Potassium	460.7 mg	12

Strengths: Low in calories, with only a trace of fat and some fiber, honeydew is perfect for those watching their weight. It's fairly low in sodium and has a good amount of potassium to balance it out and help prevent high blood pressure in sodium-sensitive people. And honeydew melon is a terrific source of vitamin C. Some of the newer, orange-fleshed varieties contain some vitamin A for a measure of cancer protection and stronger immunity.

Caution: Melons are among the foods that often cause allergic reactions.

Description: Although we're most used to seeing honeydews with pale green flesh, newer orange, pink, and gold varieties are becoming popular and increasingly available in supermarkets.

Selection: Aroma is not a sure indicator of ripeness due to differences in varieties; instead, test the blossom end. If it's slightly soft, the melon is ready to eat.

KIWIFRUIT

Serving: 1 (about 2½ oz.)

Calories	46
Fat	0.3 g
Saturated	N/A
Monounsaturated	N/A
Polyunsaturated	N/A
Calories from fat	6%
Cholesterol	0
Sodium	4 mg
Protein	0.8 g
Carbohydrate	11.3 g
Dietary fiber	2.6 g

CHIEF NUTRIENTS

NUTRIENT	AMOUNT	% RDA
Vitamin C	74.5 mg	124
Magnesium	22.8 mg	7
Potassium	252.3 mg	7

Strengths: The kiwi has moderate calories but no fat or sodium to speak of. This fruit is an excellent source of vitamin C for protection against certain cancers, enhanced immunity, and wound-healing power. Kiwi also contains some magnesium, which (like potassium) is associated with healthier blood pressure. A good amount of fiber helps keep cholesterol levels under control.

Eat with: Low-fiber foods, such as poultry and fish.

Caution: *Allergy alert:* Some people experience itching of the throat and mouth after eating kiwis. Others experience nasal and eye symptoms as well.

Curiosity: Like the papaya, the kiwi contains a meat-tenderizing element that prevents gelatin from setting.

Origin: The kiwi originally came from China; its alternate name is Chinese gooseberry.

Storage: Unripe kiwi keeps in the refrigerator for 4 to 6 weeks. To hasten ripening, place in a plastic bag with a ripe apple, pear, or banana, all of which give off ethylene gas that speeds ripening.

Serve: With chicken, turkey, or tuna salads.

KUMQUATS

Serving: 5 (about 3½ oz.)

Calories	60
Fat	0.1 g
Saturated	N/A
Monounsaturated	N/A
Polyunsaturated	N/A
Calories from fat	1%
Cholesterol	0
Sodium	6 mg
Protein	0.9 g
Carbohydrate	16.4 g
Dietary fiber	3.7 g

CHIEF NUTRIENTS

NUTRIENT	AMOUNT	% RDA
Vitamin C	35.6 mg	59
Riboflavin	0.1 mg	6
Calcium	41.9 mg	5
Potassium	185.7 mg	5
Thiamine	0.1 mg	5

Strengths: Low in calories, kumquats are a surprising source of calcium. They have some B vitamins for healthy skin, nerves, and muscles, and they're an excellent source of vitamin C. Kumquats provide a rich amount of fiber when you eat them skin and all.

Caution: *Allergy alert:* Some people are allergic to citrus peel, which is very similar in composition to kumquat peel. If you've noted such allergies, avoid kumquat peel. Also, citrus fruit in general sometimes triggers allergic headaches.

Curiosity: The kumquat is more winter-hardy than most other citrus, so it has been crossed with limes and oranges to produce new cold-resistant hybrids such as limequats and orangequats.

Description: Resembles a tiny orange. Unlike other citrus, kumquats have a rind that is deliciously sweet, though the pulp is puckery-sour.

Preparation: Use raw or blanch briefly; slice thinly.

Serve: With meat, fish, and fruit or vegetable salads. Can substitute for orange in duck à l'orange.

LEMON

Serving: ¹/₂ (about 1 oz.), w/out skin

Calories	8
Fat	0.1 g
Saturated	trace
Monounsaturated	trace
Polyunsaturated	trace
Calories from fat	9%
Cholesterol	0
Sodium	1 mg
Protein	0.3 g
Carbohydrate	2.7 g
Dietary fiber	0.6 g

CHIEF NUTRIENTS

NUTRIENT	AMOUNT	% RDA
Vitamin C	15.4 mg	26

Strengths: The lemon is practically calorie-free, and it has only tiny amounts of fat and sodium. It's a good source of vitamin C, which helps boost immunity and heal wounds.

Caution: The acid in lemons may erode tooth enamel; don't make a habit of eating them raw. In large enough quantities, oxalic acid in the peel and juice may cause problems for those with calcium-oxalate stones. *Allergy alert:* Lemons may produce contact dermatitis in sensitized individuals because of oil found in the peel.

Selection: Choose deep-yellow lemons that are heavy for their size, with fine-textured skin.

Serve: Thinly sliced with fish, poultry, or meat dishes (such as Moroccan lamb and couscous).

LIME

Serving: ¹/₂ (about 1 oz.), w/out skin

Calories	10
Fat	0.1 g
Saturated	trace
Monounsaturated	trace
Polyunsaturated	trace
Calories from fat	6%
Cholesterol	0
Sodium	1 mg
Protein	0.2 g
Carbohydrate	3.5 g
Dietary fiber	0.7 g

CHIEF NUTRIENTS

NUTRIENT	AMOUNT	% RDA
Vitamin C	9.7 mg	16

Strengths: Practically calorie-free, with minimal fat and sodium. The lime is a good source of vitamin C, which can help boost immunity and heal wounds.

Caution: May cause intestinal gas, especially if eaten in large quantities. The acid in limes may erode tooth enamel, so don't make a habit of eating them raw. *Allergy alert:* Limes may produce contact dermatitis in sensitized individuals because of oil found in the peel. Like other citrus fruit, the lime may cause allergic headaches in some people and possibly cause bedwetting in children.

Curiosity: Because limes are a good source of vitamin C, they were fed to British sailors long ago as a way to prevent scurvy. That led to the sailors being called "limeys."

Selection: The variety known as Key lime is rounder and smaller. Some are grown in this country, but most of those sold are dark green Tahitian or Persian limes. Look for bright, heavy, smooth-skinned specimens.

Serve: With fish and other types of seafood (the juice makes a nice marinade). A squeezed lime section adds a little vitamin C to all sorts of drinks.

LITCHIS

Serving: 10 (about 3½ oz.)

Calories	63
Fat	0.4 g
Saturated	N/A
Monounsaturated	N/A
Polyunsaturated	N/A
Calories from fat	6%
Cholesterol	0
Sodium	1 mg
Protein	0.8 mg
Carbohydrate	15.9 g
Dietary fiber	0.5 g

CHIEF NUTRIENTS

NUTRIENT	AMOUNT	% RDA
Vitamin C	68.6 mg	114

Strengths: Low in calories, fat, and sodium; no cholesterol. Litchis provide more than a full day's supply of vitamin C for immune power and healthy skin and teeth.

Eat with: Iron-containing foods: The vitamin C in litchis helps the iron be absorbed better.

Caution: Although they are sometimes called litchi *nuts*, the ''nut'' (central seed) itself should not be consumed.

Description: Litchis are pearl-like fruit the size of a walnut. They have a soft, juicy texture and taste like green grapes. The inner fruit is covered with a thin red shell that can be peeled off easily.

Selection: Look for litchis that have a bright rosy shell and feel heavy for their size. Canned litchis generally come in heavy syrup, which will add calories (but not diminish flavor).

Serve: Try adding litchis to sweet-and-sour stir-fries and other oriental dishes.

LITCHIS, *Dried*

Serving: 1 oz.

Calories	78
Fat	0.3 g
Saturated	N/A
Monounsaturated	N/A
Polyunsaturated	N/A
Calories from fat	4%
Cholesterol	0
Sodium	1 mg
Protein	1.1 g
Carbohydrate	19.8 g
Dietary fiber	0.6 g

CHIEF NUTRIENTS

NUTRIENT	AMOUNT	% RDA
Vitamin C	51.2 mg	85
Riboflavin	0.2 mg	9
Potassium	310.8 mg	8
Iron	0.5 mg	5
Niacin	0.9 mg	5

Strengths: Low in fat and sodium, but for a dried fruit litchis are surprisingly packed with vitamin C. The C gives you an extra measure of immunity in general, along with some protection against certain types of cancer. A fair amount of potassium paired with a small amount of sodium helps guard against high blood pressure in some people. A bit of iron helps in the fight against anemia. Added benefit: Vitamin C will enhance the iron's absorption.

Caution: High in calories. Dried fruit often leaves a sticky residue on teeth, possibly leading to cavities; brush after eating. *Allergy alert:* Like other dried fruit, litchis may contain sulfites, which can cause asthmatic wheezing, flushing, and tingling sensations in sensitive people.

Description: Dried litchis, which resemble nuts, have a chewy texture and smoky flavor. Most brands probably still contain the central seed, which is not edible, so chew with caution (or take out the seed first) if you're snacking on them.

LONGANS

Serving: 20 (about 2 oz.)

Calories	38
Fat	0.1 g
Saturated	N/A
Monounsaturated	N/A
Polyunsaturated	N/A
Calories from fat	2%
Cholesterol	0
Sodium	0
Protein	0.8 g
Carbohydrate	9.7 g
Dietary fiber	0.7 g

CHIEF NUTRIENTS

NUTRIENT	AMOUNT	% RDA
Vitamin C	53.8 mg	90
Riboflavin	0.1 mg	5

Strengths: Low in fat, with no sodium or cholesterol and some fiber, longans are also an excellent source of vitamin C. That's of special interest to those on oral contraceptives, aspirin, tetracycline, or other drugs that deplete the vitamin. Surprisingly, they also retain their vitamin C during drying.

Curiosity: Also known as dragon's eyes, no doubt for the beady appearance of these small (about 1 inch in diameter), hard-shelled, brown fruits.

Description: Sweet grapelike pulp surrounds a central seed. The hard shell is easily cracked.

Selection: In recipes, they can often be substituted for litchis. Dried longans are sometimes available and can be used like other dried fruit. (Dried longans contain about 5 times the nutrient levels of the fresh fruit.)

Serve: With poultry, fish (longans make good sauces), or citrus fruit.

LOQUATS

Serving: 10 (about 3½ oz.)

Calories	47
Fat	0.2 g
Saturated	trace
Monounsaturated	trace
Polyunsaturated	trace
Calories from fat	4%
Cholesterol	0
Sodium	1 mg
Protein	0.4 g
Carbohydrate	12.1 g
Dietary fiber	1.7 g

CHIEF NUTRIENTS

NUTRIENT	AMOUNT	% RDA
Vitamin A	153.0 RE	15
Potassium	266.0 mg	7

Strengths: Low calories, sodium, and fat; no cholesterol. With a good amount of vitamin A, loquats contribute to better vision, healthy skin, and possibly cancer protection.

Curiosity: Loquats contain virtually no vitamin C—unusual for a fruit.

Description: Pear-shaped fruit with thin skin that ranges from pale yellow to deep orange. The flesh—creamy to orange in color—is juicy and tender, reminiscent of plums or cherries. Loquats taste somewhat like a mixture of apple and apricot.

Selection: Fresh loquats bruise easily, so they're not often found fresh outside their native growing areas (Florida and California in the U.S.). Choose big loquats that are tender and sweet-scented.

Preparation: Loquats should be peeled and the large seeds removed before serving.

Serve: With chicken and shrimp.

MAMEY

Serving: ¼ (about 7½ oz.)

Calories	108
Fat	1.1 g
Saturated	N/A
Monounsaturated	N/A
Polyunsaturated	N/A
Calories from fat	9%
Cholesterol	0
Sodium	32 mg
Protein	1.1 g
Carbohydrate	26.4 g
Dietary fiber	6.3 g

CHIEF NUTRIENTS

NUTRIENT	AMOUNT	% RDA
Vitamin C	29.6 mg	49
Iron	1.5 mg	15
Riboflavin	0.1 mg	5

Strengths: The mamey has a creamy, avocado-like texture but with fewer calories and much less fat than avocados. There's lots of fiber for digestive health and vitamin C for stronger immunity and possible prevention of certain types of cancer. The mamey also has some riboflavin, which is essential for the metabolism of other B vitamins. The mamey is a good source of iron (which the vitamin C renders more easily absorbed by the body).

Description: Also known as mammee apple or mammy-apple. Shaped like a football 3 to 6 inches in diameter with coarse brown skin; the orange-pink flesh tastes like pumpkin custard with a hint of toasted almond. You can tell whether it's ripe by pressing gently—the flesh should yield.

Preparation: Put some mamey in a blender for a quick breakfast. Can be stewed or made into jam (but the C content will doubtless decline).

Serve: With meats and poultry. Always appropriate in fruit salads.

MANGO

Serving: ½ (about 3½ oz.)

Calories	67
Fat	0.3 g
Saturated	trace
Monounsaturated	trace
Polyunsaturated	trace
Calories from fat	4%
Cholesterol	0
Sodium	2 mg
Protein	0.5 g
Carbohydrate	17.6 g
Dietary fiber	2.1 g

CHIEF NUTRIENTS

NUTRIENT	AMOUNT	% RDA
Vitamin C	28.7 mg	48
Vitamin A	402.6 RE	40
Vitamin E	1.2 mg αTE	12
Vitamin B_6	0.1 mg	7

Strengths: Nice and low in calories and fat, the mango is a very good source of fiber for bowel health and cholesterol control. It has rich amounts of vitamins A and C for possible protection against certain cancers. Its vitamin E content enhances the immune response to certain diseases. And it has vitamin B_6, which is essential for the synthesis of heme (the iron-containing part of hemoglobin).

Caution: Some people suffer stomach complaints if they drink milk or alcohol shortly after eating mangoes. *Allergy alert:* Mangoes belong to the same family as poison ivy: The skin of the fruit (as well as parts of the tree itself) may produce rashes in susceptible people.

Description: A flattish oval fruit with yellow-green or reddish skin, the mango is ripe when it's very fragrant and yields to gentle pressure. There's a large, clinging pit that can only be removed by cutting it away from the flesh.

Preparation: The mango makes an easy, high-fiber sauce for pancakes and waffles; just puree the peeled flesh.

Serve: With all sorts of poultry and fish.

NECTARINE

Serving: 1 (5 oz.)

Calories	67
Fat	0.6 g
Saturated	N/A
Monounsaturated	N/A
Polyunsaturated	N/A
Calories from fat	9%
Cholesterol	0
Sodium	0
Protein	1.3 g
Carbohydrate	16.0 g
Dietary fiber	2.2 g

CHIEF NUTRIENTS

NUTRIENT	AMOUNT	% RDA
Vitamin C	7.3 mg	12
Vitamin A	100.6 RE	10
Potassium	288.3 mg	8
Niacin	1.4 mg	7

Strengths: The nectarine has good amounts of vitamins A and C for protection against certain types of cancer; the C also enhances immunity to some diseases. It's nice and low in calories, with no sodium or cholesterol and barely a trace of fat. The nectarine provides a very good helping of fiber in a delicious package, along with some niacin for healthy skin and nervous system.

Caution: *Allergy alert:* Those allergic to aspirin may react to the natural salicylate in nectarines.

Curiosity: Named after the Olympic god Nektar.

Description: Contrary to rumors, the nectarine is not a cross between a peach and a plum. It is a smooth-skinned variety of peach.

Serve: With poultry or pork (to make fine kabobs, thread slices onto skewers with turkey, chicken, or pork; add onions and zucchini, then baste with a soy-based sauce). Also delicious on a platter with pasta salads. Or add to muffins to contribute extra fiber.

ORANGE

Serving: 1 (about 4½ oz.)

Calories	62
Fat	0.2 g
Saturated	trace
Monounsaturated	trace
Polyunsaturated	trace
Calories from fat	2%
Cholesterol	0
Sodium	0
Protein	1.2 g
Carbohydrate	15.4 g
Dietary fiber	3.1 g

CHIEF NUTRIENTS

NUTRIENT	AMOUNT	% RDA
Vitamin C	69.7 mg	116
Folate	39.7 mcg	20
Calcium	52.4 mg	7
Thiamine	0.1 mg	7
Potassium	237.1 mg	6

Strengths: An awesome source of vitamin C, each orange has more than one day's RDA of that valuable nutrient. Plus an orange is low in calories and has no sodium or cholesterol and no more than a trace of fat. The orange has a very nice amount of fiber to help lower cholesterol. A good amount of the B vitamin folate boosts healthy blood and strengthens immunity. And an orange is a surprising source of calcium.

Eat with: Any vegetable! Eating an orange with your meal can boost your body's absorption of iron from plant foods by as much as 400%.

Caution: Oranges contain some oxalic acid, which should be restricted by those with calcium-oxalate stones. May also cause intestinal gas. *Allergy alert:* Some people have allergic reactions to oranges (especially to the peels), sometimes causing contact dermatitis or headaches. Those allergic to grass pollen or aspirin may also react to oranges.

PAPAYA

Serving: ½ (about 5½ oz.)

Calories	59
Fat	0.2 g
Saturated	trace
Monounsaturated	trace
Polyunsaturated	trace
Calories from fat	3%
Cholesterol	0
Sodium	5 mg
Protein	0.9 g
Carbohydrate	14.9 g
Dietary fiber	2.6 g

CHIEF NUTRIENTS

NUTRIENT	AMOUNT	% RDA
Vitamin C	93.9 mg	157
Folate	57.8 mcg	29
Potassium	390.6 mg	10

Strengths: Low in fat, calories, and sodium. Considered a good source of fiber. Papaya has plenty of folate and contains an extraordinary amount of vitamin C. With all that C, it helps build stronger immunity and healthy skin and may offer some protection against certain types of cancer. The beneficial sodium/potassium ratio is good for those with sodium-induced high blood pressure. Papayas also contain beta-carotene (the deep color is a clue) to help prevent cancer.

Caution: Some people develop hives or other allergic reactions from handling or eating papayas.

Curiosity: No relation to the wild American pawpaws, even though papayas often go by that name.

Selection: Look for fruit that's mostly yellow. Let it stand in a dark place at room temperature until it turns golden and yields slightly to pressure.

Serve: With chicken and seafood salads.

PASSION FRUIT

Serving: 5 (3½ oz.)

Calories	97
Fat	0.7 g
Saturated	N/A
Monounsaturated	N/A
Polyunsaturated	N/A
Calories from fat	7%
Cholesterol	0
Sodium	28 mg
Protein	2.2 g
Carbohydrate	23.4 g
Dietary fiber	1.7 g

CHIEF NUTRIENTS

NUTRIENT	AMOUNT	% RDA
Vitamin C	30.0 mg	50
Iron	1.6 mg	16
Potassium	348.0 mg	9
Magnesium	29.0 mg	8
Niacin	1.5 mg	8
Riboflavin	0.1 mg	8
Vitamin A	70.0 RE	7

Strengths: Purple passion fruit have plenty of nutrients with only moderate calories. They contain no fat or cholesterol. Lots of vitamin C and some vitamin A means better immunity and possible cancer protection. A surprising source of iron for healthy blood (and the vitamin C helps enhance the absorption of the iron). Although passion fruit are higher in sodium than most fruit, there's a nice amount of potassium to help offset the sodium—and potassium also helps prevent stroke. With some B vitamins and magnesium, passion fruit contribute to healthy skin, nerves, and muscles. The small edible seeds boost fiber in your diet.

Curiosity: The name doesn't refer to sensual passion but rather to the flowers, which were said to symbolize the instruments of Christ's passion and crucifixion.

Description: Each round purple fruit is approximately the size of an egg. The skin turns from smooth to dusty and lumpy as the fruit ripens. Filled with edible seeds surrounded by chartreuse-tinted flesh, passion fruit have a sour-sweet taste.

Serve: Make glorious tropical drinks and interesting marinades that complement poultry, seafood, and meat.

PEACH   ⇩ ▥

Serving: 1

Calories	37
Fat	0.1 g
Saturated	trace
Monounsaturated	trace
Polyunsaturated	trace
Calories from fat	2%
Cholesterol	0
Sodium	0
Protein	0.6 g
Carbohydrate	9.7 g
Dietary fiber	1.4 g

CHIEF NUTRIENTS

NUTRIENT	AMOUNT	% RDA
Vitamin C	5.7 mg	10

Strengths: Low in calories, fat, and sodium; cholesterol-free. A good amount of dietary fiber to help keep cholesterol levels down and promote digestive health. Peaches are a good source of vitamin C for healthy skin, wound healing, and infection fighting.

Eat with: Seafood, poultry, desserts, and rice salads. Peaches add extra fiber plus vitamin C to both hot and cold breakfast cereals.

Caution: Some people experience contact dermatitis (rash or hives) after handling peaches. *Allergy alert:* Those allergic to aspirin or otherwise sensitive to apricots, plums, and cherries may react to peaches; they may also prompt itching of the throat and mouth in some people.

History: Native to China, peaches came to Europe (and later America) through Persia, which accounts for their former name, Persian apples.

PEACHES, *Dried*   ⊘ ✳ ◉ ✳ ⇩ ▥

Serving: 5 halves (about 2 oz.)

Calories	155
Fat	0.5 g
Saturated	trace
Monounsaturated	trace
Polyunsaturated	trace
Calories from fat	3%
Cholesterol	0
Sodium	5 mg
Protein	2.4 g
Carbohydrate	39.9 g
Dietary fiber	5.3 g

CHIEF NUTRIENTS

NUTRIENT	AMOUNT	% RDA
Iron	2.6 mg	26
Potassium	647.4 mg	17
Niacin	2.9 mg	15
Vitamin A	140.4 RE	14
Magnesium	27.3 mg	8
Riboflavin	0.2 mg	8
Vitamin C	3.1 mg	5

Strengths: When they're dried, peaches become a concentrated source of vitamins and minerals. Count on plenty of iron for healthy blood, plus a bit of vitamin C to help the iron be absorbed more easily. And a very favorable sodium/potassium ratio may help ward off stroke and may benefit those with sodium-induced high blood pressure. An excellent source of fiber for both intestinal health and control of elevated cholesterol.

Eat with: Meats and poultry, complementing their flavors and adding both nutrients and fiber they lack.

Caution: Dried fruit often leaves a sticky residue on teeth, possibly leading to cavities; brush after eating. Also, drying raises the calorie concentration. *Allergy alert:* Dried peaches may contain sulfites, which help preserve some vitamins but destroy thiamine and can cause asthmatic wheezing, flushing, and tingling sensations in sensitive people. In addition, those allergic to aspirin may react to the natural salicylate in peaches.

PEAR

Serving: 1 (6 oz.)

Calories	98
Fat	0.7 g
Saturated	trace
Monounsaturated	trace
Polyunsaturated	trace
Calories from fat	6%
Cholesterol	0
Sodium	1 mg
Protein	0.7 g
Carbohydrate	25.1 g
Dietary fiber	4.3 g

CHIEF NUTRIENTS

NUTRIENT	AMOUNT	% RDA
Vitamin C	6.6 mg	11
Folate	12.1 mcg	6
Potassium	207.5 mg	6

Strengths: Low in calories, fat, and sodium, a fresh pear has no cholesterol. It's a very good source of fiber, which can lower cholesterol. Vitamin C and folate help build immunity to infections and may play a role in preventing some cancers. The pear has a good sodium/potassium ratio to help prevent strokes and control high blood pressure in sodium-sensitive individuals.

Caution: Pears contain some oxalic acid, which should be restricted by those with calcium-oxalate stones. *Allergy alert:* Those allergic to birch pollen may react to pears (as well as cherries and apples).

Curiosity: Pears are among the few fruits that are picked while still green. They ripen well off the tree.

Preparation: Pears are easily shredded if you want to include them in muffins. Can be cooked into an all-fruit conserve to spread (instead of butter) on toast and pancakes.

Serve: With a little peanut butter or cheese (pears are a perfect snack!). Good on all cereals.

PEARS, *Dried*

Serving: 5 halves (3 oz.)

Calories	229
Fat	0.6 g
Saturated	trace
Monounsaturated	trace
Polyunsaturated	trace
Calories from fat	2%
Cholesterol	0
Sodium	5 mg
Protein	1.6 g
Carbohydrate	61.0 g
Dietary fiber	11.5 g

CHIEF NUTRIENTS

NUTRIENT	AMOUNT	% RDA
Iron	1.8 mg	18
Potassium	466.4 mg	12
Vitamin C	6.1 mg	10
Magnesium	28.9 mg	8
Riboflavin	0.1 mg	7
Niacin	1.2 mg	6

Strengths: Dried pears are a great source of fiber and a concentrated source of vitamins and minerals (but also calories). Their B vitamins promote healthy skin and nerves, and they have a good amount of iron (plus vitamin C to help with its absorption). Magnesium and potassium (with little sodium) help prevent high blood pressure in some people.

Caution: Dried fruit often leaves a sticky residue on teeth, possibly leading to cavities; brush after eating. *Allergy alert:* Dried pears may contain sulfites. Though helping to preserve some vitamins, sulfites also destroy thiamine and can cause asthmatic wheezing, flushing, and tingling sensations in sensitive people.

Serve: With fresh fruit in a compote. Perfect when braised along with meats.

A Better Idea: Dry your own pears by using a dehydrator.

PERSIMMON

Serving: 1 (6 oz.)

Calories	118
Fat	0.3 g
Saturated	N/A
Monounsaturated	N/A
Polyunsaturated	N/A
Calories from fat	2%
Cholesterol	0
Sodium	2 mg
Protein	1.0 g
Carbohydrate	31.2 g
Dietary fiber	2.9 g

CHIEF NUTRIENTS

NUTRIENT	AMOUNT	% RDA
Vitamin A	364.6 RE	36
Vitamin C	12.6 mg	21
Potassium	270.5 mg	7
Folate	12.6 mcg	6

Strengths: Only a moderate amount of calories, with practically no fat or sodium. The persimmon contains a hefty dose of fiber. It has lots of vitamin A plus a good amount of vitamin C for stronger immunity and an extra dose of anticancer power. With a nice supply of potassium, it may help blood pressure stay on an even keel. A good choice for those on oral contraceptives, who tend to need extra C and folate.

Caution: Nothing's more unpleasant than an underripe Hachiya or native persimmon: Tannic acid in the fruit gives it a bitter, harsh taste that'll literally make your mouth pucker.

Description: There are two main types: The large (3-inch-diameter) acorn-shaped Hachiya, which should be quite soft before eating, and the smaller tomato-shaped Fuyu, which may be eaten while hard. Small native persimmons, which grow wild, are hard to find, but they're richer in some nutrients. Like the Hachiyas, they must eaten when soft.

PINEAPPLE

Serving: 1 cup cubes

Calories	76
Fat	0.7 g
Saturated	trace
Monounsaturated	trace
Polyunsaturated	trace
Calories from fat	8%
Cholesterol	0
Sodium	2 mg
Protein	0.6 g
Carbohydrate	19.2 g
Dietary fiber	1.9 g

CHIEF NUTRIENTS

NUTRIENT	AMOUNT	% RDA
Vitamin C	23.9 mg	40
Thiamine	0.1 mg	9
Folate	16.4 mg	8
Vitamin B$_6$	0.1 mcg	7
Iron	0.6 mg	6
Magnesium	21.7 mg	6

Strengths: Low in fat, calories, and sodium, pineapple is a decent source of dietary fiber, which may help control high cholesterol. Pineapple also has nice amounts of various vitamins and minerals that assist in boosting immunity (vitamin C, folate, iron) and breaking down food for energy (folate, thiamine, vitamin B$_6$).

Caution: *Allergy alert:* The juice of a pineapple contains bromelin, an enzyme that can cause dermatitis in some people.

Curiosity: Named for its resemblance to a pine cone, the pineapple has long been a symbol of hospitality.

Selection: The fruit is ripe when the scales have brown tips. The characteristic ''pineapple smell'' is also a sign of ripeness. Some people test pineapple by pulling out leaves, but this is not a reliable indicator of maturity.

Serve: With meat, poultry, seafood. Makes a good tenderizer for meats—use in marinades. (Fresh pineapple must be cooked for use in gelatin-based dishes.)

PLANTAINS

Serving: ½ cup slices, cooked

Calories	89
Fat	0.1 g
Saturated	N/A
Monounsaturated	N/A
Polyunsaturated	N/A
Calories from fat	1%
Cholesterol	0
Sodium	4 mg
Protein	0.6 g
Carbohydrate	24.0 g
Dietary fiber	1.8 g

CHIEF NUTRIENTS

NUTRIENT	AMOUNT	% RDA
Vitamin C	8.4 mg	14
Folate	20.0 mcg	10
Potassium	358.1 mg	10
Vitamin B_6	0.2 mg	9
Magnesium	24.6 mg	7
Vitamin A	70.1 RE	7

Strengths: Very low in fat and sodium, with no cholesterol. Like their close relatives, bananas, plantains are a decent source of fiber. They have a nice complement of nutrients, including vitamins A and C and folate for immune power. Their magnesium and potassium (coupled with low sodium) are good for fighting high blood pressure in sodium-sensitive individuals. The quantity of vitamins B_6 and C in plantains should be of special interest to those taking birth control pills, which tend to deplete those vitamins.

Curiosity: The plantain is a fruit, but is most often regarded as a starchy vegetable to be used in side dishes. Although a close relative of the banana, it cannot be eaten raw.

Selection: Choose plantains with green to yellowish or black skin; the yellow ones are riper and sweeter than the green, and the black ones are fully ripe and sweetest.

Preparation: Peel by cutting the skin off. Can be baked, boiled, roasted, sautéed, or steamed.

Serve: With any meat, fish, or poultry. Plantains are an excellent substitute for french fries if you cut them into thick fries and sauté in a tiny amount of oil.

PLUMS

Serving: 2 (about 4¾ oz.)

Calories	73
Fat	0.8 g
Saturated	trace
Monounsaturated	trace
Polyunsaturated	trace
Calories from fat	10%
Cholesterol	0
Sodium	0
Protein	1.0 g
Carbohydrate	17.2 g
Dietary fiber	2.0 g

CHIEF NUTRIENTS

NUTRIENT	AMOUNT	% RDA
Vitamin C	12.5 mg	21
Riboflavin	0.1 mg	7
Potassium	227.0 mg	6
Vitamin B_6	0.1 mg	5

Strengths: Low in fat; no sodium or cholesterol. Like most fruit, plums are a good source of vitamin C, the vitamin that helps heal wounds, fight infections, and form collagen, an integral part of connective tissue. A fair amount of potassium may help counteract any sodium in the diet and possibly prevent stroke. A nice amount of fiber may aid in preventing certain types of cancer. (Plums also contain another cancer-fighting substance: ellagic acid.)

Caution: Contain some oxalic acid, which should be restricted by those with calcium-oxalate stones. *Allergy alert:* May cause itching of the throat and mouth. May trigger problems in those who are also allergic to apricots, cherries, peaches, or aspirin.

Selection: Literally dozens of varieties exist. Choose plums that yield slightly to palm pressure. Very firm plums may be ripened by enclosing them in a paper bag for a few days.

Serve: With meat (plums make excellent compotes).

POMEGRANATE

Serving: ½ (about 3 oz.)

Calories	52
Fat	0.2 g
Saturated	N/A
Monounsaturated	N/A
Polyunsaturated	N/A
Calories from fat	4%
Cholesterol	0
Sodium	2 mg
Protein	0.7 g
Carbohydrate	13.2 g
Dietary fiber	2.8 g

CHIEF NUTRIENTS

NUTRIENT	AMOUNT	% RDA
Vitamin C	4.7 mg	8
Potassium	199.4 mg	5

Strengths: Very low in fat and sodium. An excellent source of fiber (if you eat the seeds) for good digestive health and some help lowering high cholesterol levels. Pomegranates have some vitamin C for immune power, along with potassium for proper nerve and muscle function.
Caution: Pomegranate juice stains, which is why it has long been used as a dye.
Curiosity: Pomegranates are a symbol of fertility in folklore throughout the world.
Description: About the size of an orange, with a red, leathery skin. Inside, clusters of seeds burst with sweet juice.
Selection: Look for richly colored fruit with no signs of deterioration.
Preparation: Some people suggest opening the fruit in a bowl of water. Slit the skin from top to bottom in several places and carefully break apart the fruit. The seeds and the surrounding pulp are edible, but discard the light membrane inside the skin.
Serve: With almost anything. Makes a beautiful edible garnish for salads, soups, meat sauces, poultry, or desserts.

PRICKLY PEAR

Serving: 1 (about 3½ oz.)

Calories	42
Fat	0.5 g
Saturated	N/A
Monounsaturated	N/A
Polyunsaturated	N/A
Calories from fat	11%
Cholesterol	0
Sodium	5 mg
Protein	0.8 g
Carbohydrate	9.9 g
Dietary fiber	3.7 g

CHIEF NUTRIENTS

NUTRIENT	AMOUNT	% RDA
Magnesium	87.6 mg	25
Vitamin C	14.4 mg	24
Calcium	57.7 mg	7
Potassium	226.6 mg	6

Strengths: Low in calories, fat, and sodium. Good amounts of vitamin C and magnesium, which help form strong teeth. Prickly pear is also a surprising source of calcium. The calcium, magnesium, and potassium are all associated with healthy blood pressure. Some varieties have lots of small edible seeds, which contribute to the nice share of insoluble fiber.
Caution: Handle with care or wear gloves: Although the prickly pear's sharp spines are removed at harvest, you'll still encounter tiny barbed hairs that can get embedded in the skin. It's said that the prickly pear has a strong laxative effect, but if you just eat 1 or 2 you shouldn't have a problem.
Curiosity: The juice of this cactus can be used as an emergency water source in the desert.
Description: Not at all related to the tree-grown pear, prickly pear is actually the fruit of the prickly pear cactus (*opuntia*).
Preparation: Must be peeled to be eaten. Using tongs to hold the fruit, slice off the ends, then slit the skin lengthwise. Peel back the skin to remove. Slice the pulp and eat as is.

PRUNES

Serving: 5 (about 1½ oz.)

Calories	100
Fat	0.2 g
Saturated	trace
Monounsaturated	trace
Polyunsaturated	trace
Calories from fat	2%
Cholesterol	0
Sodium	2 mg
Protein	1.1 g
Carbohydrate	26.3 g
Dietary fiber	3.0 g

CHIEF NUTRIENTS

NUTRIENT	AMOUNT	% RDA
Iron	1.0 mg	10
Potassium	312.9 mg	8
Vitamin A	83.6 RE	8
Vitamin B₆	0.1 mg	6
Magnesium	18.9 mg	5

Strengths: Prunes are a super source of fiber, which is why they have always been valued for their laxative effect. (Part of the fiber is the soluble form that helps control cholesterol.) This fruit is a good source of iron, and it has some beta-carotene to help prevent cancer. Prunes are among the few dried fruits that can be successfully produced without sulfites. Their magnesium and potassium (coupled with low sodium) may help maintain normal blood pressure in some individuals.

Eat with: Foods rich in vitamin C to promote better absorption of the iron.

Caution: Dried fruit often leaves a sticky residue on teeth, possibly leading to cavities; brush after eating. *Allergy alert:* Those allergic to aspirin may react to the natural salicylate in prunes.

Curiosity: Only certain types of plums (those that can be dried without removing the pit) are used to make prunes.

Serve: Try adding prunes to a citrus-based compote. Or bake up some prune muffins that you can serve with strawberries and a juice high in vitamin C.

PUMMELOS

Serving: 1 cup sections

Calories	72
Fat	0.1 g
Saturated	N/A
Monounsaturated	N/A
Polyunsaturated	N/A
Calories from fat	1%
Cholesterol	0
Sodium	2 mg
Protein	1.4 g
Carbohydrate	18.3 g
Dietary fiber	N/A

CHIEF NUTRIENTS

NUTRIENT	AMOUNT	% RDA
Vitamin C	115.9 mg	193
Potassium	410.4 mg	11

Strengths: Pummelos are low in calories, fat, and sodium (with a good amount of potassium to benefit those with sodium-sensitive high blood pressure). An awesome source of vitamin C with its myriad benefits (ranging from infection fighting to possible cancer prevention). Although exact fiber figures aren't available, you might expect pummelos to offer a little fiber—like other citrus fruit.

Eat with: Fruit and green salads. Like grapefruit, pummelos are an excellent way to add vitamin C and a little fiber to meat dishes.

Caution: *Allergy alert:* May trigger the usual allergies associated with citrus.

Description: Also called pomelo and pommelo. The largest citrus fruit, they look like giant grapefruit (and may in fact be its ancestor). The flesh may be pale yellow to pink. One distinct advantage: They don't have grapefruit's bitterness.

Serve: Pummelos are good with cottage cheese and crisp raw vegetables.

QUINCE

Serving: 1 (about 3 oz.)

Calories	52
Fat	0.1 g
Saturated	trace
Monounsaturated	trace
Polyunsaturated	trace
Calories from fat	2%
Cholesterol	0
Sodium	4 mg
Protein	0.4 g
Carbohydrate	14.1 g
Dietary fiber	1.7 g

CHIEF NUTRIENTS

NUTRIENT	AMOUNT	% RDA
Vitamin C	13.8 mg	23
Iron	0.6 mg	6
Potassium	181.2 mg	5

Strengths: Low in fat and sodium; no cholesterol. The quince has long been valued as an abundant source of pectin (a type of fiber that can help control cholesterol). Surprisingly, it's a source of iron, and it also has a good amount of vitamin C to help increase its absorption. (Quinces are generally served cooked, so some of the C will be lost.)

History: Ancient Romans used quince flowers and fruit for everything from perfume to honey. The fruit was a symbol of love. (Some hold that quince is the forbidden fruit referred to in Genesis.)

Description: Looks like a golden apple, but you can expect to see rusty blotches on the skin. The flesh is yellow, hard, and sour but turns soft, fragrant, and red when cooked. Its flavor is a mix of apple, pear, and tropical fruit.

Preparation: Cook a quince as you would an apple or a pear.

Serve: Add to fruit compotes or combine with other fruit for high-fiber preserves. Or try it as an accompaniment to pork or poultry.

RAISINS, *Golden, Seedless*

Serving: 1/2 cup

Calories	219
Fat	0.3 g
Saturated	trace
Monounsaturated	trace
Polyunsaturated	trace
Calories from fat	1%
Cholesterol	0
Sodium	9 mg
Protein	2.5 g
Carbohydrate	57.7 g
Dietary fiber	3.9 g

CHIEF NUTRIENTS

NUTRIENT	AMOUNT	% RDA
Potassium	540.8 mg	14
Iron	1.3 mg	13
Vitamin B$_6$	0.2 mg	12
Riboflavin	0.1 mg	8
Magnesium	25.4 mg	7

Strengths: Low in fat, with no cholesterol. High in potassium but low in sodium to benefit those with sodium-related high blood pressure. A good source of iron, golden raisins have lots of fiber, a good portion of which is soluble, which can help lower cholesterol. Some B vitamins for energy.

Eat with: Foods high in vitamin C for better iron absorption.

Caution: Pediatricians warn that children under 4 could easily choke on raisins. Dried fruit often leaves a sticky residue on teeth; brush after eating. *Allergy alert:* Golden raisins contain sulfites (to maintain color), which can cause reactions in some people. Those allergic to aspirin may react to the natural salicylate in raisins.

History: Cave drawings indicate that our prehistoric ancestors dined on raisins.

RAISINS, *Seedless*

Serving: ½ cup

Calories	218
Fat	0.3 g
Saturated	trace
Monounsaturated	trace
Polyunsaturated	trace
Calories from fat	1%
Cholesterol	0
Sodium	9 mg
Protein	2.3 g
Carbohydrate	57.4 g
Dietary fiber	3.9 g

CHIEF NUTRIENTS

NUTRIENT	AMOUNT	% RDA
Iron	1.5 mg	15
Potassium	544.5 mg	15
Vitamin B$_6$	0.2 mg	9
Thiamine	0.1 mg	8
Magnesium	23.9 mg	7

Strengths: Low in fat, with no cholesterol, raisins are high in potassium but low in sodium to benefit those with sodium-related high blood pressure. A good source of iron. Lots of fiber, a good portion of which is the soluble form that can help lower cholesterol. Dark raisins are among the few fruits that can be dried and stored without sulfuring.

Eat with: Foods high in vitamin C to help the iron be absorbed better. Raisins are good in rice and poultry dishes, and they add natural sweetness to baked goods, chutneys, and stuffings.

Caution: Pediatricians warn that children under 4 could easily choke on raisins. Dried fruit often leaves a sticky residue on teeth, possibly leading to cavities; brush after eating. *Allergy alert:* Those allergic to aspirin may react to the natural salicylate in raisins.

Curiosity: Only 4 varieties of grapes are suitable for drying into raisins.

RASPBERRIES

Serving: 1 cup

Calories	60
Fat	0.7 g
Saturated	trace
Monounsaturated	trace
Polyunsaturated	trace
Calories from fat	10%
Cholesterol	0
Sodium	0
Protein	1.1 g
Carbohydrate	14.2 g
Dietary fiber	6.0 g

CHIEF NUTRIENTS

NUTRIENT	AMOUNT	% RDA
Vitamin C	30.7 mg	51
Folate	32.0 mcg	16
Iron	0.7 mg	7
Magnesium	22.1 mg	6
Niacin	1.1 mg	6
Riboflavin	0.1 mg	6
Potassium	187.0 mg	5

Strengths: Low in fat and calories, raspberries have no cholesterol or sodium. And, thanks to the seeds, they have lots of dietary fiber. Many of their nutrients boost immunity, including folate, iron, and vitamin C. (And the tremendous amount of C boosts absorption of the iron.) They have a nice amount of B vitamins for energy, with some magnesium and potassium for good measure.

Eat with: Everything! These delicious little gems perk up every meal, contributing the fiber and vitamin C that meat, seafood, and low-fat cheese lack.

Caution: Contain some oxalic acid, which should be restricted by those with calcium-oxalate stones. *Drug interaction:* May cause problems for those taking MAO-inhibitor antidepressants. *Allergy alert:* Berries often cause allergic reactions. And raspberries contain natural salicylate, which may affect those sensitive to aspirin.

Curiosity: Black raspberries tend to have more calories than red ones.

SAPOTE

Serving: ½ (about 4 oz.)

Calories	151
Fat	0.7 g
Saturated	N/A
Monounsaturated	N/A
Polyunsaturated	N/A
Calories from fat	4%
Cholesterol	0
Sodium	11 mg
Protein	2.4 g
Carbohydrate	38.0 g
Dietary fiber	4.4 g

CHIEF NUTRIENTS

NUTRIENT	AMOUNT	% RDA
Vitamin C	22.5 mg	38
Iron	1.1 mg	11
Niacin	2.0 mg	11
Magnesium	33.8 mg	10
Potassium	387.0 mg	10
Calcium	43.9 mg	5
Vitamin A	46.1 RE	5

Strengths: Little sodium and no cholesterol. Plenty of immune-power nutrients (vitamins A and C and iron). The A and C may also help prevent certain types of cancer. The sapote is an excellent source of fiber, and it has a good amount of potassium and magnesium, which may help control blood pressure. Although their silky, creamy texture suggests high fat, they're actually quite low in all kinds.

Description: The most common variety is the white sapote, which has smooth green skin and ivory-yellow flesh. Its slightly tart taste is reminiscent of apricot and banana. The small seeds in the center are inedible. Eat as you would a ripe persimmon.

Selection: Choose plump specimens, free of blemishes. Let ripen at room temperature until soft as a ready-to-eat plum.

Preparation: The flesh darkens on exposure to air; treat with lemon juice if not using immediately. Puree into a fluffy sauce for pancakes and waffles.

Serve: With muffins and quick breads (as an instant low-fat spread). One sapote makes a fine dessert by itself.

STRAWBERRIES

Serving: 1 cup

Calories	45
Fat	0.6 g
Saturated	trace
Monounsaturated	trace
Polyunsaturated	trace
Calories from fat	11%
Cholesterol	0
Sodium	1 mg
Protein	0.9 g
Carbohydrate	10.5 g
Dietary fiber	3.9 g

CHIEF NUTRIENTS

NUTRIENT	AMOUNT	% RDA
Vitamin C	84.5 mg	141
Folate	26.4 mcg	13
Potassium	247.3 mg	7
Iron	0.6 mg	6
Riboflavin	0.1 mg	6

Strengths: Strawberries are super-low in calories, fat, and sodium—and they overflow with vitamin C for enhanced immunity and healthy skin and teeth. Those on oral contraceptives, aspirin, or tetracycline therapy—all of which deplete C—can really benefit. Strawberries have an excellent amount of fiber, and they contain a substance called ellagic acid that seems capable of fighting certain cancer-causing agents. (Note that frozen berries are similar in nutrients to fresh but may contain added sugar; check the label.)

Caution: Strawberries contain some oxalic acid, which should be restricted by those with calcium-oxalate stones. *Allergy alert:* May trigger various allergies. May also cause trouble for those allergic to aspirin.

Serve: With all sorts of salads, meats, poultry. Good any time of the day, and especially for snacks. Eating strawberries fresh preserves their vitamin C content.

TANGERINE

Serving: 1 (3 oz.)

Calories	37
Fat	0.2 g
Saturated	trace
Monounsaturated	trace
Polyunsaturated	trace
Calories from fat	4%
Cholesterol	0
Sodium	1 mg
Protein	0.5 g
Carbohydrate	9.4 g
Dietary fiber	N/A

CHIEF NUTRIENTS

NUTRIENT	AMOUNT	% RDA
Vitamin C	25.9 mg	43
Folate	17.1 mcg	9
Vitamin A	77.3 RE	8
Thiamine	0.1 mg	6

Strengths: Wonderfully low in calories, fat, and sodium, the tangerine is a very good source of vitamin C for healthy skin, immune power, and wound-healing ability. It's higher in vitamin A (and beta-carotene) than other citrus fruit, so it gives an extra measure of protection from certain types of cancer. Although exact figures are not available, tangerines do contain fiber (especially pectin, which is credited with lowering cholesterol).

Eat with: All sorts of meats, poultry, and seafood (foods that are low in vitamin C). Or spinach salad, so the nonheme iron in spinach will be absorbed better.

Caution: Excess consumption could cause intestinal gas. *Allergy alert:* Citrus fruit in general is a common allergen, so tangerines may provoke symptoms—including bed-wetting in children and allergic headaches. Oil in the peel may cause allergic reactions. (Tangerine juice may contain enough peel to cause problems, too.)

WATERMELON

Serving: 1 cup cubes

Calories	51
Fat	0.7 g
Saturated	N/A
Monounsaturated	N/A
Polyunsaturated	N/A
Calories from fat	12%
Cholesterol	0
Sodium	3 mg
Protein	1.0 g
Carbohydrate	11.5 g
Dietary fiber	0.6 g

CHIEF NUTRIENTS

NUTRIENT	AMOUNT	% RDA
Vitamin C	15.4 mg	26
Vitamin B_6	0.2 mg	12
Thiamine	0.1 mg	9
Vitamin A	59.2 RE	6
Magnesium	17.6 mg	5
Potassium	185.6 mg	5

Strengths: Watermelon is low in calories, fat, and sodium, and it has no cholesterol. It has some fiber (every bit helps). Nice amounts of magnesium and potassium (with little sodium) may help protect against stroke and sodium-induced high blood pressure. There's a good amount of vitamin C, along with some vitamin A and beta-carotene, which may protect against certain types of cancer. The vitamin B_6 helps in the production of hemoglobin. Women on oral contraceptives can benefit from extra B_6 and C.

Eat with: Any food that's not high in vitamin C.

Caution: *Allergy alert:* May cause itching of the mouth and throat, especially in people also allergic to ragweed, grass pollen, tomatoes, or other melons.

Curiosity: Some watermelons can weigh as much as 45 pounds.

Serve: With chicken, rice, or seafood salads. Always a welcome snack. Refreshing for summer breakfasts—delicious with whole grain muffins.

Grains, Hot Cereals, and Pasta
GRAINS

AMARANTH SEEDS

Serving: ¼ cup, raw

Calories	183
Fat	3.2 g
Saturated	0.8 g
Monounsaturated	0.7 g
Polyunsaturated	1.4 g
Calories from fat	16%
Cholesterol	0
Sodium	10 mg
Protein	7.1 g
Carbohydrate	32.4 g
Dietary fiber	7.5 g

CHIEF NUTRIENTS

NUTRIENT	AMOUNT	% RDA
Iron	3.7 mg	37
Magnesium	130.3 mg	37
Folate	24.0 mcg	12
Zinc	1.6 mg	10
Calcium	75.0 mg	9
Riboflavin	0.1 mg	6
Vitamin B_6	0.1 mg	6

Strengths: Amaranth seeds have a moderate supply of protein, and they're a very good source of vitamins and minerals. One serving provides more than one-third of the RDA of iron and magnesium, good amounts of folate and zinc, and some bone-strengthening calcium. The fair amounts of riboflavin and vitamin B_6 supplied by amaranth seeds are important for energy metabolism.

Curiosity: Aztecs in pre-Columbian America pounded amaranth with honey and human blood into dough that they baked in the form of snakes, birds, and other deities.

Description: Has a delicious toasted-sesame flavor—but old amaranth smells like linseed oil.

Selection: Found in most health food stores.

Preparation: Grind into flour and combine with 50% wheat flour for a yeast dough, or try popping it like popcorn in a hot wok or heavy skillet. (Pop only 1 Tbsp. of seeds at a time, and use a small pastry brush to keep the seeds moving.)

Serve: As a hot cereal, using 2 parts liquid to 1 part grain.

BARLEY, *Pearled*

Serving: ½ cup, cooked

Calories	97
Fat	0.4 g
Saturated	trace
Monounsaturated	trace
Polyunsaturated	trace
Calories from fat	3%
Cholesterol	0
Sodium	2 mg
Protein	1.8 g
Carbohydrate	22.3 g
Dietary fiber	4.4 g

CHIEF NUTRIENTS

NUTRIENT	AMOUNT	% RDA
Iron	1.1 mg	11
Niacin	1.6 mg	9
Folate	12.6 mcg	6

Strengths: Pearled barley is high in cholesterol-lowering fiber and low in fat. A serving offers 11% of the RDA of iron and fair amounts of folate and niacin, all important for healthy blood.

Caution: Add a little oil if you're cooking barley in a pressure cooker; otherwise, starch may bubble up and clog the cooker's safety valve.

Curiosity: Barley water, a famous and soothing drink for invalids, is made by boiling barley in water, then flavoring the strained liquid with lemon.

History: Dates back to the Stone Age. A recipe for barley wine was found engraved on a Babylonian brick.

Description: Pearled barley has had the hull and bran removed and has been steamed and polished.

Serve: With lentils, split peas, cabbage, carrots, onions, mushrooms, and beef—the flavor complements all these foods.

BRAN, *Corn*

Serving: 2 Tbsp., raw

Calories	21
Fat	0.1 g
Saturated	trace
Monounsaturated	trace
Polyunsaturated	trace
Calories from fat	4%
Cholesterol	0
Sodium	1 mg
Protein	0.8 g
Carbohydrate	8.0 g
Dietary fiber	7.9 g

CHIEF NUTRIENTS

(None of the nutrients meet or exceed 5% of the RDA.)

Strengths: With more fiber than oat, wheat, or rice bran, corn bran offers a whopping amount of cancer-protective fiber. It is possibly as effective as oat bran in lowering blood levels of cholesterol.

Description: Corn bran consists of the fibrous outer cover of the corn kernel that has been separated from the more nutritious germ and endosperm. The bran is coarse, tan and white, and smells like cornmeal.

Serve: Can be substituted for cornmeal in muffin, biscuit, bread, and pancake recipes.

Preparation: To enhance the taste, toast unprocessed corn bran in a medium-hot heavy skillet, stirring until the mixture takes on an amber hue. Grind to desired texture in a blender or food processor. Mix with sugar and a bit of butter to use as a topping for muffins and cakes.

BRAN, *Oat*

Serving: 2 Tbsp., raw

Calories	29
Fat	0.8 g
Saturated	trace
Monounsaturated	trace
Polyunsaturated	trace
Calories from fat	26%
Cholesterol	0
Sodium	trace
Protein	2.0 g
Carbohydrate	7.7 g
Dietary fiber	1.8 g

CHIEF NUTRIENTS

NUTRIENT	AMOUNT	% RDA
Thiamine	0.1 mg	9
Magnesium	27.2 mg	8
Iron	0.6 mg	6

Strengths: Oat bran has plenty of soluble fiber, with significantly more than the same amount of oatmeal. It's a nice source of thiamine and iron, which help boost energy levels, and of magnesium, for converting vitamin D into its usable form.

Description: Made of the ground-up outer layers of the oat kernel, which are rich in soluble fiber.

Selection: Plain oat bran cooked as a hot cereal generally has more soluble fiber than commercial oat bran products such as crackers, bread, and ready-to-eat cereal.

Preparation: To enhance the flavor of unprocessed bran, toast it in a medium-hot skillet, stirring until the mixture takes on an amber hue. Can be substituted for a portion of wheat bran in some recipes to produce moister breads, pancakes, and cookies.

Serve: As a hot cereal with fruit and nuts. Add to soup or stew as a thickener, and use instead of bread crumbs for toppings or to coat meat or fish.

BRAN, *Rice*

Serving: 2 Tbsp., raw

Calories	33
Fat	2.2 g
Saturated	0.4 g
Monounsaturated	0.8 g
Polyunsaturated	0.8 g
Calories from fat	59%
Cholesterol	0
Sodium	1 mg
Protein	1.4 g
Carbohydrate	5.2 g
Dietary fiber	2.3 g

CHIEF NUTRIENTS

NUTRIENT	AMOUNT	% RDA
Magnesium	82.0 mg	23
Vitamin B$_6$	0.4 mg	21
Iron	2.0 mg	19
Niacin	3.6 mg	19
Thiamine	0.3 mg	19

Strengths: Rice bran has knock-out nutritional credentials: It provides a high percentage of the RDA of blood pressure–modulating magnesium and it has healthy amounts of niacin, thiamine, and vitamin B$_6$—B vitamins that are essential for healthy nerves and proper energy prodcuction. It's also a very good source of fiber. One study found that rice bran lowered cholesterol levels as much as oat bran.

Caution: High in fat! Almost 60% of rice bran's calories are from fat.

Description: Rice bran is the outer shell of a grain of rice. Medium brown, the texture of fine cornmeal, with a sweet, nutty taste.

Storage: Keep rice bran in an airtight, moisture-free container in the refrigerator.

Preparation: Can replace up to half of the flour in quickbread and muffin recipes. Use instead of bread crumbs or oats, or add to meat loaf and casseroles.

BRAN, *Wheat*

Serving: 2 Tbsp., raw

Calories	15
Fat	0.3 g
Saturated	trace
Monounsaturated	trace
Polyunsaturated	trace
Calories from fat	18%
Cholesterol	0
Sodium	0.1 mg
Protein	1.1 g
Carbohydrate	4.5 g
Dietary fiber	3.0 g

CHIEF NUTRIENTS

NUTRIENT	AMOUNT	% RDA
Magnesium	42.8 mg	12
Iron	0.7 mg	7

Strengths: A great source of fiber, wheat bran also provides a good deal of magnesium, which helps regulate blood pressure, and it has a bit of blood-building iron.

Description: Wheat bran is the nutritious, fibrous outer covering of wheat grain. Unprocessed bran can be flaky or powdery-fine, depending on how it's milled. Color varies from yellowish to dark reddish-brown.

Preparation: To enhance the taste of unprocessed bran, toast it in a medium-hot skillet, stirring, until the mixture takes on an amber color. You can grind bran in a blender to a finer texture.

Serve: In hot cereal, muffins, or bread; or use instead of bread crumbs to coat meats or fish before baking. Or sprinkle on top of a casserole.

BUCKWHEAT GROATS

Serving: 1/2 cup, cooked

Calories	91
Fat	0.6 g
Saturated	trace
Monounsaturated	trace
Polyunsaturated	trace
Calories from fat	6%
Cholesterol	0
Sodium	4 mg
Protein	3.4 g
Carbohydrate	19.7 g
Dietary fiber	N/A

CHIEF NUTRIENTS

NUTRIENT	AMOUNT	% RDA
Magnesium	50.5 mg	14
Iron	0.8 mg	8
Folate	13.9 mcg	7

Strengths: A great source of low-fat, low-sodium fiber. Buckwheat groats have a good supply of magnesium, a mineral that's important for proper muscle function and stable blood pressure. In addition, groats are a fair source of blood-building iron and folate.

Curiosity: Despite its name, buckwheat is not related to wheat, nor is it a grain; it is a tiny triangular seed. The plant itself has heart-shaped leaves and pink or white flowers, which are rich in nectar (buckwheat honey is delicious!).

Origin: Native to central Asia. Famous Russian *blini* are made with buckwheat flour. So are Japanese soba noodles.

Description: Groats are hulled, crushed (rather than ground) buckwheat kernels. Roasted buckwheat (kasha) has a strong, nutty taste that lends a nice flavor to pork, beef, and lamb. Unroasted buckwheat has a light, delicate flavor that goes well with veal or fish.

Serve: As a side dish or in meat casseroles instead of rice. Buckwheat flour makes delicious, hearty pancakes.

BULGUR WHEAT

Serving: 1/2 cup, cooked

Calories	76
Fat	0.2 g
Saturated	trace
Monounsaturated	trace
Polyunsaturated	trace
Calories from fat	3%
Cholesterol	0
Sodium	5 mg
Protein	2.8 g
Carbohydrate	16.9 g
Dietary fiber	4.1 g

CHIEF NUTRIENTS

NUTRIENT	AMOUNT	% RDA
Iron	0.9 mg	9
Folate	16.4 mcg	8
Magnesium	29.0 mg	8

Strengths: Bulgur is a great low-fat, low-sodium source of cancer-protective fiber, and it has a fair amount of iron and folate for healthy blood. Its magnesium assists muscle and nerve functions; this mineral is also associated with healthy blood pressure.

Caution: *Allergy alert:* People who are are allergic to wheat should avoid bulgur.

Description: Has a tender, chewy texture. Made from steamed, dried, cracked wheat kernels. Can be cooked, boiled, or simply soaked before use.

Selection: Comes in coarse, medium, and fine grinds. Use medium grind for salad, fine grind for bread and desserts.

Storage: If possible, store in an airtight jar in a cool, dark spot.

Preparation: Tabbouleh, a Middle Eastern salad, is made with bulgur, parsley, mint, olive oil, and lemon juice.

Serve: As a pilaf, or instead of rice in meat and vegetable dishes.

CORNMEAL, *Whole Grain, White or Yellow*

Serving: ¼ cup, raw

Calories	109
Fat	1.1 g
Saturated	0.2 g
Monounsaturated	0.3 g
Polyunsaturated	0.5 g
Calories from fat	9%
Cholesterol	0
Sodium	11 mg
Protein	2.4 g
Carbohydrate	23.1 g
Dietary fiber	3.3 g

CHIEF NUTRIENTS

NUTRIENT	AMOUNT	% RDA
Magnesium	38.1 mg	11
Iron	1.0 mg	10
Thiamine	0.1 mg	8
Niacin	1.1 mg	6

Strengths: Low in fat and sodium and high in fiber, cornmeal contains good amounts of iron, important for healthy blood, and magnesium, which helps regulate blood pressure and plays a role in nerve and muscle function. Cornmeal also offers a fair share of thiamine and niacin, vitamins that enhance energy metabolism.

Origin: Corn is native to the Americas.

History: Colonists used corn as money and even traded corn for marriage licenses. It was so vital to the Hopi Indians that a special ear was dedicated to each newborn child as its "corn mother."

Description: Dried corn kernels ground to a coarse texture.

Storage: Because of the oil content of the germ, whole grain cornmeal turns rancid more quickly than other meals. Store in an airtight container in a cool, dry place—or preferably the refrigerator or freezer.

Serve: In cornbread, muffins, and pancakes; or in hot cereal. Adding another flour to a cornmeal recipe produces a softer, lighter product. Add corn bran to cornmeal for even more fiber.

CORN GRITS, *White or Yellow*

Serving: ½ cup, cooked

Calories	73
Fat	0.2 g
Saturated	trace
Monounsaturated	trace
Polyunsaturated	trace
Calories from fat	3%
Cholesterol	0
Sodium	0
Protein	1.7 g
Carbohydrate	15.7 g
Dietary fiber	1.9 g

CHIEF NUTRIENTS

NUTRIENT	AMOUNT	% RDA
Iron	0.8 mg	8
Thiamine	0.1 mg	8
Niacin	1.0 mg	5

Strengths: A good fiber food, low in fat and sodium, corn grits contain fair amounts of iron, thiamine, and niacin. These nutrients are essential for healthy blood and high energy levels.

Description: Starchy, coarsely ground, hulled, dried corn kernels that may or may not include the bran and germ.

Selection: Stone-ground grits include the heart, or germ, of the corn and are usually more nutritious than mass-produced grits.

Storage: Grits can be stored up to 5 months in an airtight jar or plastic bag in the refrigerator. Frozen grits will keep for about a year.

Preparation: Boil and eat as a cereal, or as a dessert with fruit or preserves. Fry grits fat-free in a nonstick skillet and flavor with fruit butter rather than syrup or gravy.

Serve: With yams, cooked greens, and lean baked country-style ham or turkey sausage. "Grillades" (braised meats) and grits is a Louisiana specialty.

COUSCOUS

Serving: ½ cup, cooked

Calories	101
Fat	0.1 g
Saturated	trace
Monounsaturated	trace
Polyunsaturated	trace
Calories from fat	1%
Cholesterol	0
Sodium	5 mg
Protein	3.4 g
Carbohydrate	20.9 g
Dietary fiber	0.9 g

CHIEF NUTRIENTS

NUTRIENT	AMOUNT	% RDA
Folate	13.5 mcg	7

Strengths: A low-fat, low-sodium wheat product, couscous has a fair amount of folate, a B vitamin important for the growth and regeneration of red blood cells and for the formation of hemoglobin.

Curiosity: The word *couscous* may actually be onomatopoeic—a verbal approximation of the sound of steam hissing its way through a mass of grain.

Origin: A staple of North African countries—Morocco, Algeria, and Tunisia.

Description: Couscous is coarse-ground hard wheat endosperm, the starchy center of the grain.

Selection: Instant couscous is available in many supermarkets and is much easier to prepare than the traditional variety.

Serve: With milk, as a hot cereal; with dressing, as a salad; sweetened and mixed with fruits for dessert. Or as a substitute for rice.

HOMINY, *White or Yellow, Canned*

Serving: ½ cup, raw

Calories	58
Fat	0.7 g
Saturated	trace
Monounsaturated	trace
Polyunsaturated	trace
Calories from fat	11%
Cholesterol	0
Sodium	168 mg
Protein	1.2 g
Carbohydrate	11.4 g
Dietary fiber	2.0 g

CHIEF NUTRIENTS

NUTRIENT	AMOUNT	% RDA
Zinc	0.8 mg	6
Iron	0.5 mg	5

Strengths: Hominy has no cholesterol, is low in calories, and gets only 11% of those calories from fat. It has fair amounts of iron and zinc, nutrients that help build healthy blood and a strong immune system.

Description: Hominy is dried white or yellow corn kernels from which the hull and germ have been removed. It's sold canned, ready-to-eat, or dried.

Serve: With traditional southern fare, such as yams, cooked greens, and lean ham or turkey sausage.

MILLET

Serving: 1/2 cup, cooked

Calories	143
Fat	1.2 g
Saturated	0.2 g
Monounsaturated	0.2 g
Polyunsaturated	0.6 g
Calories from fat	8%
Cholesterol	0
Sodium	2 mg
Protein	4.2 g
Carbohydrate	28.4 g
Dietary fiber	1.8 g

CHIEF NUTRIENTS

NUTRIENT	AMOUNT	% RDA
Magnesium	52.8 mg	15
Folate	22.8 mcg	11
Thiamine	0.1 mg	9
Iron	0.8 mg	8
Niacin	1.6 mg	8
Vitamin B$_6$	0.1 mg	7
Zinc	1.1 mg	7
Riboflavin	0.1 mg	6

Strengths: An all-around nutritious grain, millet has a good amount of magnesium, which helps maintain normal blood pressure. It also offers a good share of folate, important for the body's production of red blood cells. Its niacin, riboflavin, thiamine, vitamin B$_6$, and iron are essential for healthy blood and energy metabolism. And the zinc helps the body perform many functions, including healing wounds and fighting infection. Millet generally has more protein than wheat, rice, or corn.

History: Extremely drought-resistant, millet has been a staple for millions who live in dry areas of India, Africa, China, and Russia.

Description: Small, buff-colored, round kernels that resemble sand. The bland flavor blends well with seasonings.

Storage: Stores well—up to 6 months in a tightly closed container. Stale millet smells bad.

Preparation: Cook as a grain.

Serve: In vegetable soup: Puffed millet adds a creamy, low-fat texture. It also adds a light touch to muffins.

QUINOA

Serving: 1/4 cup, raw

Calories	159
Fat	2.5 g
Saturated	0.3 g
Monounsaturated	0.7 g
Polyunsaturated	1.0 g
Calories from fat	14%
Cholesterol	0
Sodium	9 mg
Protein	5.6 g
Carbohydrate	29.3 g
Dietary fiber	N/A

CHIEF NUTRIENTS

NUTRIENT	AMOUNT	% RDA
Iron	3.9 mg	39
Magnesium	89.3 mg	26
Folate	20.8 mcg	10
Riboflavin	0.2 mg	10
Zinc	1.4 mg	9
Potassium	314.5 mg	8
Niacin	1.3 mg	7

Strengths: One of the best vegetable sources of protein, quinoa (*KI-no-a*) is similar to whole dried milk in amino acid content. It's higher than true grains in the amino acids lysine, methionine, and cysteine. It's also higher in cholesterol-lowering fiber and blood-building iron. Quinoa also contains a respectable amount of magnesium, which helps regulate blood pressure.

Curiosity: Quinoa (meaning ''mother grain'') was the sustaining food of the Inca Indians.

Origin: The grain was first found growing at high altitudes in Peru and Bolivia.

Description: Quinoa seeds look like round sesame seeds. When cooked, they have a fluffy texture and a mild flavor.

Selection: Usually all quinoa is grown organically.

Storage: Quinoa spoils easily. Use within a month of purchase.

Preparation: Rinse quinoa well to remove the bitter-tasting coating that may cling to the grains. Cook like rice.

Serve: In casseroles, cold salads, or desserts.

RICE, *Brown*

Serving: 1/2 cup, cooked

Calories	110
Fat	0.8 g
Saturated	trace
Monounsaturated	trace
Polyunsaturated	trace
Calories from fat	7%
Cholesterol	0
Sodium	1 mg
Protein	2.3 g
Carbohydrate	23.0 g
Dietary fiber	1.7 g

CHIEF NUTRIENTS

NUTRIENT	AMOUNT	% RDA
Magnesium	43.1 mg	12
Vitamin B$_6$	0.2 mg	8
Niacin	1.0 mg	7
Thiamine	0.1 mg	7
Iron	0.5 mg	5

Strengths: Brown rice is a good source of magnesium, which helps maintain proper blood pressure. It also contains some niacin, thiamine, and vitamin B$_6$, vitamins essential for energy metabolism and the production of red blood cells. A fair source of blood-building iron, this rice is low in fat and has no cholesterol.

Curiosity: Rice is an essential staple: In some languages, the word *eat* means "to eat rice."

Description: Brown rice is the entire grain with only the inedible outer husk removed. It has a light tan color, nutlike flavor, and chewy texture. Compared to most other grains, it's low in fiber.

Selection: Comes in long-, medium-, and short-grain varieties. Long-grain cooks up dry and fluffy; short-grain is moist and sticky, which makes it easy to handle with chopsticks. Medium-grain is fairly fluffy right after it's cooked but begins to clump as it cools.

RICE, *White, Enriched*

Serving: 1/2 cup, cooked

Calories	133
Fat	0.2 g
Saturated	trace
Monounsaturated	trace
Polyunsaturated	trace
Calories from fat	1%
Cholesterol	0
Sodium	0
Protein	2.4 g
Carbohydrate	29.2 g
Dietary fiber	0.2 g

CHIEF NUTRIENTS

NUTRIENT	AMOUNT	% RDA
Iron	1.5 mg	15
Thiamine	0.2 mg	11
Niacin	1.9 mg	10

Strengths: Because it has been enriched with some of the nutrients removed during processing, white rice contains good amounts of iron, thiamine, and niacin. These nutrients are essential for the body's production of red blood cells and for energy metabolism. White rice is low in fat and has no cholesterol.

Curiosity: Even though brown rice is more nutritious, most of the rice-eating world prefers white. Even Confucius liked his rice white.

Description: The inedible husk, fiber-rich bran, and nutritious germ are removed, leaving only the starchy part of the grain.

Selection: Long-grain white rice cooks up drier and fluffier than medium- or short-grain. Enriched rice has been sprayed with a solution of nutrients.

Preparation: No two types of rice cook identically, so experiment.

RICE, *Wild*   ⬇ 🏛

Serving: ½ cup, cooked

Calories	83
Fat	0.3 g
Saturated	trace
Monounsaturated	trace
Polyunsaturated	trace
Calories from fat	3%
Cholesterol	0
Sodium	2 mg
Protein	3.3 g
Carbohydrate	17.5 g
Dietary fiber	0.6 g

CHIEF NUTRIENTS

NUTRIENT	AMOUNT	% RDA
Folate	21.3 mcg	11
Magnesium	26.2 mg	8
Zinc	1.1 mg	7
Niacin	1.1 mg	6
Vitamin B$_6$	0.1 mg	6

Strengths: One serving of wild rice provides about 11% of the RDA for folate, important for the growth and regeneration of red blood cells. It's a nice source of zinc, essential for proper wound healing and immune function. It also has magnesium, which helps maintain proper blood pressure, plus niacin and vitamin B$_6$, nutrients that can improve energy metabolism.

Curiosity: Wild rice, no cousin to regular rice, is the seed of a wild grass native to Minnesota's lake shores.

Description: Purplish-black husks, with a nutty flavor and chewy texture.

Preparation: Clean wild rice thoroughly before cooking. Soak it in cold water for a few minutes, then pour off any debris that floats to the surface.

Serve: With turkey, pheasant, quail, oysters, and mushrooms— all foods that complement wild rice's flavor. Wild rice is expensive, but you can extend it by combining it with brown rice, white rice, or bulgur. It's ideal for pilaf, stuffing, and salad.

RYE   🔄 ⏱ ⬇

Serving: ¼ cup, raw

Calories	141
Fat	1.1 g
Saturated	0.1 g
Monounsaturated	0.1 g
Polyunsaturated	0.5 g
Calories from fat	7%
Cholesterol	0
Sodium	3 mg
Protein	6.2 g
Carbohydrate	29.3 g
Dietary fiber	N/A

CHIEF NUTRIENTS

NUTRIENT	AMOUNT	% RDA
Magnesium	50.8 mg	15
Folate	25.2 mcg	13
Iron	1.1 mg	11
Niacin	1.8 mg	10
Zinc	1.6 mg	10
Thiamine	0.1 mg	9
Riboflavin	0.1 mg	6
Vitamin B$_6$	0.1 mg	6
Vitamin E	0.5 mg αTE	5

Strengths: Compared with other grains, rye offers an impressive array of vitamins and minerals. Although fiber values for rye are not available from the USDA, an equivalent amount of rye flour offers a healthy 6.1 g of cholesterol-lowering fiber. It has good amounts of folate and iron, both important for healthy blood and proper immune function. A good source of magnesium, which helps maintain proper blood pressure, rye has immunity-boosting zinc and a supply of other nutrients that are essential for many body functions.

Caution: Avoid rye flour that smells musty. Although store-bought rye is rarely contaminated, one type of mold, ergot, causes an array of bizarre symptoms, including hallucinations. (LSD was originally derived from ergot.)

Curiosity: Some dieters swear by rye crisp crackers because they tend to swell in the stomach, giving the sensation of fullness. Rye crisps are made by mixing rye meal with snow or powdered ice. (The expansion of the ice-cold foam makes the dough rise when it's placed in the oven.)

Description: Rye flour, a dirty-gray color, is usually combined with other lighter flours for baking. Rye bread has a rich, sour taste, in part because it's made by the sourdough method.

TRITICALE

Serving: $^1/_2$ *cup, raw*

Calories	161
Fat	1.0 g
Saturated	0.2 g
Monounsaturated	0.1 g
Polyunsaturated	0.4 g
Calories from fat	6%
Cholesterol	0
Sodium	2 mg
Protein	6.3 g
Carbohydrate	34.6 g
Dietary fiber	8.7 g

CHIEF NUTRIENTS

NUTRIENT	AMOUNT	% RDA
Folate	35.0 mcg	18
Magnesium	62.4 mg	18
Thiamine	0.2 mg	13
Iron	1.2 mg	12
Zinc	1.6 mg	11

Strengths: Triticale (*triti-KAY-lee*) is a good source of folate, thiamine, and iron, nutrients that help build healthy blood and maintain good energy levels. A good amount of magnesium enhances many body functions, including muscle and nerve activity, and its zinc is essential for wound healing. Triticale contains more protein and less gluten than wheat and is an excellent source of cancer-protective fiber.

Caution: *Allergy alert:* This grain does contain gluten and should be avoided by those with gluten sensitivity.

Origin: A cross between rye and wheat, this hybrid cereal crop was developed in Scotland in 1875.

Description: Has a nutty-sweet flavor.

Selection: Available in several forms—berries, flakes, and flour. Can be found in health-food stores.

Preparation: Cook berries for 40 to 50 minutes and serve as a breakfast cereal.

WHEAT GERM

Serving: $^1/_4$ *cup, toasted*

Calories	108
Fat	3.0 g
Saturated	0.5 g
Monounsaturated	0.4 g
Polyunsaturated	1.9 g
Calories from fat	25%
Cholesterol	0
Sodium	1 mg
Protein	8.3 g
Carbohydrate	14.1 g
Dietary fiber	3.7 g

CHIEF NUTRIENTS

NUTRIENT	AMOUNT	% RDA
Folate	100.0 mcg	50
Vitamin E	4.0 mg αTE	40
Zinc	4.7 mg	32
Thiamine	0.5 mg	31
Iron	2.6 mg	26
Magnesium	90.9 mg	26
Riboflavin	0.2 mg	14
Vitamin B$_6$	0.3 mg	14
Niacin	1.6 mg	8
Potassium	269.0 mg	7

Strengths: Deserves its "health food" reputation. Wheat germ offers 50% of the RDA for folate, a B vitamin important for the production of healthy red blood cells. It also has a top-notch dose of vitamin E, a nutrient with some powerful protective properties. Wheat germ also has very good amounts of thiamine and zinc.

Caution: *Allergy alert:* Some people are allergic to wheat; they are likely to react to wheat germ as well.

Description: Wheat germ is the vitamin- and mineral-rich embryo of the wheat berry that is removed when flour is refined from whole wheat to white. It has a crumblike texture and a nutty taste. It's available in both raw and toasted forms.

Storage: Because of its high oil content, wheat germ often spoils quickly. Once a container has been opened, it should be stored in the refrigerator. It may be kept chilled for up to 6 months.

Serve: Sprinkled on frozen yogurt. Or add to casserole, meat loaf, bread, muffins, and pancakes.

HOT CEREALS

CREAM OF RICE   ⬇ 🏛

Serving: ³/₄ cup, cooked

Calories	95
Fat	0.2 g
Saturated	N/A
Monounsaturated	N/A
Polyunsaturated	N/A
Calories from fat	2%
Cholesterol	0
Sodium	0
Protein	1.7 g
Carbohydrate	21.1 g
Dietary fiber	N/A

CHIEF NUTRIENTS

(None of the nutrients meet or exceed 5% of the RDA.)

Strengths: Cream of rice is extremely low in fat and has no cholesterol or sodium if it's prepared without adding salt, butter, or margarine.

Caution: This cereal has little to offer in the way of nutrients. A serving has less than 5% of the RDA of any vitamin or mineral.

Description: Consists of finely ground white rice, from which the fibrous bran and nutrient-rich germ have been removed. When cooked, it is creamy-white and bland, with a smooth texture.

Serve: Besides being served as a hot cereal, cream of rice can be used to thicken soups or added to baked goods.

A Better Idea: If you eat cream of rice regularly, you may want to add some corn or wheat bran. Or choose a whole grain cereal that's naturally higher in fiber and nutrients.

CREAM OF WHEAT, *Quick-Cooking*   ☀ ✳ ⬇ 🏛

Serving: ³/₄ cup, cooked

Calories	97
Fat	0.4 g
Saturated	N/A
Monounsaturated	N/A
Polyunsaturated	N/A
Calories from fat	3%
Cholesterol	0
Sodium	104 mg
Protein	2.7 g
Carbohydrate	20.0 g
Dietary fiber	1.0 g

CHIEF NUTRIENTS

NUTRIENT	AMOUNT	% RDA
Iron	7.7 mg	77
Thiamine	0.2 mg	12
Niacin	1.1 mg	6

Strengths: Cream of wheat is very low in fat and has no cholesterol. Because it is fortified with iron, a serving provides a whopping amount of this blood-building nutrient. It also offers a good amount of thiamine and some niacin, B-complex vitamins that are needed for energy, and it has a small amount of fiber.

Caution: Quick-cooking cream of wheat does have a noticeable amount of sodium, which comes from disodium phosphate, an ingredient added to make the cereal cook faster.

Origin: Cream of wheat was an instant hit in 1893. It was first marketed in New York City by an enterprising Grand Forks, North Dakota, flour mill employee who packaged leftovers from the wheat milling process, called "middlings."

Description: Contains wheat farina, the starchy part of the wheat kernel, and wheat germ, which offers some nutrients.

A Better Idea: Add a tablespoon of wheat bran, corn bran, or oat bran to boost the fiber content.

FARINA, *Enriched*

Serving: ¾ cup, cooked

Calories	88
Fat	0.2 g
Saturated	trace
Monounsaturated	trace
Polyunsaturated	trace
Calories from fat	2%
Cholesterol	0
Sodium	0
Protein	2.4 g
Carbohydrate	18.6 g
Dietary fiber	2.5 g

CHIEF NUTRIENTS

NUTRIENT	AMOUNT	% RDA
Thiamine	0.1 mg	10
Iron	0.9 mg	9
Niacin	1.0 mg	5
Riboflavin	0.1 mg	5

Strengths: Farina is low in fat, contains no cholesterol, and is a good source of fiber. Because it's enriched, farina provides nice amounts of thiamine, niacin, and riboflavin, nutrients essential for converting food into energy. It also contains some iron for healthy blood.

Caution: If cooked with the ⅛ tsp. of salt suggested on the package, a serving of farina has 265 mg of sodium.

Description: Farina is a creamy-white, soothingly bland cereal made from the starchy part of soft wheat. It is similar to cream of wheat. (Cream of wheat, however, contains added wheat germ. Farina does not.)

Serve: In soups, as a thickener.

A Better Idea: Add corn bran, wheat bran, wheat germ, or fresh fruit to this cereal to boost its fiber content.

OATMEAL

Serving: ¾ cup, cooked

Calories	109
Fat	1.8 g
Saturated	0.3 g
Monounsaturated	0.6 g
Polyunsaturated	0.7 g
Calories from fat	15%
Cholesterol	0
Sodium	2 mg
Protein	4.6 g
Carbohydrate	18.9 g
Dietary fiber	3.9 g

CHIEF NUTRIENTS

NUTRIENT	AMOUNT	% RDA
Thiamine	0.2 mg	13
Iron	1.2 mg	12
Magnesium	42.0 mg	12
Zinc	0.9 mg	6

Strengths: Cooked oats are high in fiber, which has been shown to help reduce cholesterol. Oats provide respectable amounts of thiamine and magnesium for healthy nerves and muscles, as well as iron for oxygen-rich blood. The zinc is essential for wound-healing and proper immune function. A filling breakfast, oatmeal sticks to your ribs!

Curiosity: Legend has it that Attila the Hun fed his barbarian troops only oat soups, which reputedly made them fierce as tigers.

History: Once considered the poor man's food, oats are the Horatio Alger of grains, having progressed from the status of a disdained weed to that of a health food.

Description: Old-fashioned oatmeal (also called rolled oats) is oat grain that has been hulled, then steamed and rolled flat. Quick-cooking oatmeal is made by cutting oats before steaming, then rolling them into thinner flakes. Instant oatmeal is precooked and dried (to make it very soft) before it's rolled.

Serve: In fruit crisp, vegetable soup, biscuits, muffins, cake, and granola, or as a hot cereal.

RALSTON

Serving: ¾ cup, cooked

Calories	101
Fat	0.6 g
Saturated	N/A
Monounsaturated	N/A
Polyunsaturated	N/A
Calories from fat	5%
Cholesterol	0
Sodium	0
Protein	4.2 g
Carbohydrate	21.3 g
Dietary fiber	6.0 g

CHIEF NUTRIENTS

NUTRIENT	AMOUNT	% RDA
Iron	1.2 mg	12
Magnesium	43.7 mg	12
Thiamine	0.2 mg	10
Niacin	1.5 mg	8
Riboflavin	0.1 mg	8
Folate	13.3 mcg	7
Zinc	1.1 mg	7

Strengths: Ralston is a standout among high-fiber cereals, with 6 g per serving. It's low in fat and has no cholesterol. Because it's a whole grain cereal, it also offers good amounts of magnesium and iron, both essential for many body functions, along with folate, niacin, riboflavin, and thiamine, B-complex vitamins that aid in energy production. It also contains a bit of zinc for healthy skin.

Description: A hearty, stick-to-the-ribs cereal made with milled wheat and added wheat bran.

Storage: If you won't be using this whole grain cereal within a few weeks, store in an airtight container in the refrigerator.

Preparation: Cook with water. Generally, a dash of salt is added, along with toppings of butter, milk, and sugar.

A Better Idea: Use a nonsugar sweetener, skip the butter, and serve with skim milk. Note that you can also add Ralston cereal to casserole, soup, meat loaf, and baked goods if you want to add good amounts of fiber.

WHEATENA

Serving: ¾ cup, cooked

Calories	102
Fat	0.9 g
Saturated	N/A
Monounsaturated	N/A
Polyunsaturated	N/A
Calories from fat	8%
Cholesterol	0
Sodium	0
Protein	3.6 g
Carbohydrate	21.5 g
Dietary fiber	4.0 g

CHIEF NUTRIENTS

NUTRIENT	AMOUNT	% RDA
Iron	1.0 mg	10
Magnesium	36.4 mg	10
Zinc	1.3 mg	8
Folate	12.7 mcg	6
Niacin	1.0 mg	5

Strengths: Wheatena provides 4 g of fiber per serving, so it's an excellent source of fiber—second only to Ralston among prepared hot cereals. It also offers a good amount of blood-building iron, along with some folate and niacin, B-complex vitamins that help produce energy. This whole grain product has a good amount of magnesium, which helps regulate blood pressure, and a bit of zinc, involved in immune functions. It's low in fat and has no cholesterol or sodium if cooked without added salt.

History: Wheatena was first marketed in 1880 in New York City.

Description: A golden-brown cereal made from toasted crushed wheat kernels, which include wheat bran and wheat germ. Toasting gives the grain a pleasant, nutty flavor.

Storage: Store in an airtight container in a cool, dry place.

Serve: As a hot cereal or a topping for casseroles. May also be added to muffins or meat loaf.

PASTA

MACARONI, *Enriched*

Serving: 1 cup, cooked

Calories	197
Fat	0.9 g
Saturated	trace
Monounsaturated	trace
Polyunsaturated	trace
Calories from fat	4%
Cholesterol	0
Sodium	1 mg
Protein	6.7 g
Carbohydrate	39.7 g
Dietary fiber	2.2 g

CHIEF NUTRIENTS

NUTRIENT	AMOUNT	% RDA
Iron	2.0 mg	20
Thiamine	0.3 mg	19
Niacin	2.3 mg	12
Riboflavin	0.1 mg	8
Magnesium	25.2 mg	7

Strengths: Plain, cooked macaroni is low in fat and sodium and has no cholesterol. Because this white-flour product is enriched, it contains good amounts of thiamine, niacin, and iron, all important for healthy blood, and fair portions of riboflavin and magnesium. And it offers some fiber in each serving.

Caution: *Allergy alert:* Avoid this if you are allergic to wheat.

Curiosity: In 18th-century England, the word *macaroni* meant a fop or dandy—someone who assumed foreign airs, such as eating that dang-fangled Italian pasta. The song ''Yankee Doodle Dandy'' was meant to poke fun at these types.

Description: Macaroni usually refers to tube-shaped pasta made without eggs. Among the best-known shapes are elbows (short, curved tubes), ditalini (tiny, very short tubes), mostaccioli (2-inch tubes cut diagonally), penne (large straight tubes), and ziti (long, thin tubes). Macaroni forms also include shells, twists, and ribbons.

MACARONI, *Vegetable, Enriched*

Serving: 1 cup, cooked

Calories	172
Fat	0.2 g
Saturated	trace
Monounsaturated	trace
Polyunsaturated	trace
Calories from fat	1%
Cholesterol	0
Sodium	8 mg
Protein	6.1 g
Carbohydrate	35.7 g
Dietary fiber	N/A

CHIEF NUTRIENTS

NUTRIENT	AMOUNT	% RDA
Thiamine	0.2 mg	10
Niacin	1.4 mg	8
Iron	0.7 mg	7
Magnesium	25.5 mg	7

Strengths: Plain, cooked vegetable macaroni is extremely low in fat and sodium, and has no cholesterol. Because it's enriched, it also contains a good amount of thiamine, which helps convert carbohydrates into energy; and fair portions of niacin, blood-building iron, and magnesium.

Caution: *Allergy alert:* Since it's mostly wheat flour, vegetable macaroni should be avoided by those with wheat allergies. (If you're allergic to wheat but not corn, look for corn pasta products.)

Description: Vegetable macaroni contains small amounts of vegetables—spinach, tomato, artichoke, red beet, parsley, carrot—usually as a concentrated paste or powder. These small amounts give color to the pasta but generally do not impart much taste.

Selection: Color, more than taste, may determine your selection of vegetable macaroni. Red sauce goes well with green spinach noodles; pesto with red tomato noodles. Carbonara sauce confers confetti-like dots of color to noodles.

Serve: With anything you'd eat regular pasta with.

MACARONI, *Whole Wheat*

Serving: 1 cup, cooked

Calories	174
Fat	0.8 g
Saturated	trace
Monounsaturated	trace
Polyunsaturated	trace
Calories from fat	4%
Cholesterol	0
Sodium	4 mg
Protein	7.5 g
Carbohydrate	37.2 g
Dietary fiber	4.2 g

CHIEF NUTRIENTS

NUTRIENT	AMOUNT	% RDA
Iron	1.5 mg	15
Magnesium	42.0 mg	12
Thiamine	0.2 mg	10
Zinc	1.1 mg	8
Vitamin B_6	0.1 mg	6
Niacin	1.0 mg	5

Strengths: A good source of blood-building iron, whole wheat macaroni also contains a good amount of magnesium, which helps regulate blood pressure. And it has a decent amount of thiamine, which aids in energy production. The macaroni contains some niacin, vitamin B_6, and zinc, and it's low in fat and sodium. Also a better source of fiber than regular macaroni.

Caution: *Allergy alert:* Like other wheat products, whole wheat macaroni contains gluten. You'll want to avoid this pasta if you're allergic to wheat products.

Description: Light-brown noodles with a slightly nutty taste. Federal regulations require whole wheat noodles to be made with 100% whole wheat or whole durum wheat flour. They have a coarser, chewier texture than white-flour noodles.

Storage: Kitchen critters love to munch on whole wheat noodles, so store the pasta in airtight containers.

Preparation: Cook al dente—until the noodles are slightly chewy but have no starchy taste. Overcooked whole wheat noodles start to break up in the cooking water. Drain, then rinse in cold water to stop the cooking process.

Serve: In cheese and vegetable casseroles and hearty meals.

NOODLES, *Chinese Cellophane, Dehydrated*

Serving: 1 cup, raw

Calories	491
Fat	0.1 g
Saturated	trace
Monounsaturated	trace
Polyunsaturated	trace
Calories from fat	0.1%
Cholesterol	0
Sodium	14 mg
Protein	0.2 g
Carbohydrate	120.5 g
Dietary fiber	0.1 g

CHIEF NUTRIENTS

NUTRIENT	AMOUNT	% RDA
Iron	3.0 mg	30
Thiamine	0.2 mg	14

Strengths: Each serving is rich in iron and contains a good amount of thiamine, both important for high energy levels. Chinese cellophane noodles are low in fat and sodium, with no cholesterol.

Caution: These noodles are very high in calories, mostly from carbohydrates. So choose another kind of pasta if you're on a weight-loss diet.

Description: Also called bean thread or pea-starch noodles, these translucent strings are made from mung bean starch. Rather tasteless, they have a gelatinous texture when cooked.

Selection: Cellophane noodles can be found in some supermarkets and in oriental grocery stores.

Preparation: Presoak in hot water before adding to most dishes, except soups. They can also be deep-fried; when prepared this way, they blow up like a white bird's nest.

Serve: Good in one-dish Chinese meals; their bland taste is the perfect foil for hot, spicy foods.

NOODLES, *Egg, Enriched*

Serving: 1 cup, cooked

Calories	213
Fat	2.4 g
Saturated	0.5 g
Monounsaturated	0.7 g
Polyunsaturated	0.7 g
Calories from fat	10%
Cholesterol	53 mg
Sodium	11 mg
Protein	7.6 g
Carbohydrate	39.7 g
Dietary fiber	3.5 g

CHIEF NUTRIENTS

NUTRIENT	AMOUNT	% RDA
Iron	2.5 mg	25
Thiamine	0.3 mg	20
Niacin	2.4 mg	13
Magnesium	30.4 mg	9
Riboflavin	0.1 mg	8
Vitamin B$_{12}$	0.1 mcg	7
Zinc	1.0 mg	7
Folate	11.2 mcg	6

Strengths: Because enriched egg noodles are made with nutrient-supplemented flour, they provide a good amount of iron, essential for healthy blood. They also offer respectable servings of thiamine and niacin, nutrients that help produce energy, and some magnesium, zinc, riboflavin, and folate. Egg noodles are low in fat, and they're a good source of fiber, which may help prevent colon cancer. Because they contain egg, these noodles have about 7½ g of protein per serving.

Caution: Unlike most other pastas, egg noodles do contain cholesterol, but it's not especially high. A 1-cup serving has about one-fourth as much cholesterol as an egg. *Allergy alert:* Avoid if you're allergic to eggs or flour.

Selection: Choose thin noodles for light sauces; thicker noodles hold up better in stews.

Preparation: Cook in plenty of water but do not overcook. Drain and rinse; you may want to toss the noodles with a small amount of oil to prevent them from sticking together.

NOODLES, *Japanese Soba*

Serving: 1 cup, cooked

Calories	113
Fat	0.1 g
Saturated	trace
Monounsaturated	trace
Polyunsaturated	trace
Calories from fat	1%
Cholesterol	0
Sodium	68 mg
Protein	5.8 g
Carbohydrate	24.4 g
Dietary fiber	N/A

CHIEF NUTRIENTS

NUTRIENT	AMOUNT	% RDA
Thiamine	0.1 mg	7
Iron	0.5 mg	6

Strengths: These low-fat noodles have a fair amount of iron and thiamine, both important for healthy energy levels.

Caution: *Allergy alert:* Since these noodles contain some wheat flour, avoid them if you are allergic to wheat.

Curiosity: In ancient Japan, soba noodles were supposedly eaten by warriors before battle.

Description: Flat, linguini-type noodles made with buckwheat flour and white wheat flour. The ratio of buckwheat to wheat determines the taste and color of the noodle. Chasoba, a form of soba noodle, is made with buckwheat and green tea.

Serve: In wintertime, soba noodles are traditionally served hot, in a rich broth that has been used to cook bits of meat and vegetables. In summertime, they are traditionally eaten cold, with a dipping sauce that may include soy sauce, rice vinegar, sake or sherry, ginger, and walnuts.

NOODLES, *Japanese Somen*

Serving: 1 cup, cooked

Calories	231
Fat	0.3 g
Saturated	trace
Monounsaturated	trace
Polyunsaturated	trace
Calories from fat	1%
Cholesterol	0
Sodium	283 mg
Protein	7.0 g
Carbohydrate	48.5 g
Dietary fiber	N/A

CHIEF NUTRIENTS

NUTRIENT	AMOUNT	% RDA
Iron	0.9 mg	9

Strengths: Japanese somen noodles contain a fair amount of iron, which is important for healthy blood.

Curiosity: In Japan, sucking noodles loudly into your mouth is considered a proper display of gusto, not beastly bad manners.

Caution: *Allergy alert:* These noodles contain wheat flour, a common allergen.

Description: Somen are very fine white wheat flour noodles. Cha somen also have green tea powder; tomago somen are bright yellow noodles made with egg yolk; ume somen are an attractive pink, from plum and shiso oil in the dough.

Serve: Traditionally, somen noodles are served cold, along with mounds of grated ginger, thinly sliced green onion, and a dipping sauce. They may also be used to garnish clear broths.

NOODLES, *Spinach, Enriched*

Serving: 1 cup, cooked

Calories	211
Fat	2.5 g
Saturated	0.6 g
Monounsaturated	0.8 g
Polyunsaturated	0.6 g
Calories from fat	11%
Cholesterol	53 mg
Sodium	19 mg
Protein	8.1 g
Carbohydrate	38.8 g
Dietary fiber	N/A

CHIEF NUTRIENTS

NUTRIENT	AMOUNT	% RDA
Thiamine	0.4 mg	26
Folate	34.0 mcg	17
Iron	1.7 mg	17
Niacin	2.4 mg	12
Riboflavin	0.2 mg	12
Magnesium	38.4 mg	11
Vitamin B_{12}	0.2 mcg	11
Vitamin B_6	0.2 mg	9
Zinc	1.0 mg	7

Strengths: Because they are made with enriched white flour, spinach egg noodles are a good source of iron and B-complex vitamins—folate, niacin, riboflavin, thiamine, and vitamin B_{12}—important for red blood cell production and better energy. They also provide a good share of magnesium, which helps regulate blood pressure, and they offer some vitamin B_6 and zinc, which are both involved in immune functions.

Caution: *Allergy alert:* These noodles contain eggs and wheat.

Description: Flat green noodles made with white flour, eggs, spinach powder or paste, and water. Spinach egg noodles contain very little spinach; nutritionally, they are not much different from white egg noodles.

Preparation: Cook as you would any egg noodle, using plenty of water. Do not overcook.

Serve: Mix with regular egg noodles for a colorful dish called ''hay and straw.'' They're delectable in soups, cheese casseroles, and vegetable dishes.

PASTA, *Fresh*

⟳ 🕐 ⬇ 🗑

Serving: 1 cup, cooked

Calories	183
Fat	1.5 g
Saturated	0.2 g
Monounsaturated	0.2 g
Polyunsaturated	0.6 g
Calories from fat	7%
Cholesterol	46 mg
Sodium	8 mg
Protein	7.2 g
Carbohydrate	34.9 g
Dietary fiber	2.2 g

CHIEF NUTRIENTS

NUTRIENT	AMOUNT	% RDA
Thiamine	0.3 mg	20
Iron	1.6 mg	16
Riboflavin	0.2 mg	12
Vitamin B_{12}	0.2 mcg	10
Magnesium	25.2 mg	7
Niacin	1.4 mg	7

Strengths: Fresh pasta offers good amounts of thiamine and iron, both important for healthy energy levels. And it contains fair to good amounts of blood-boosting niacin, riboflavin, and vitamin B_{12}. Fresh pasta also has some magnesium, which helps maintain proper blood pressure.

Caution: *Allergy alert:* Fresh-made pasta is more likely to contain egg than dried, packaged pasta, so be wary if you're allergic. People who are allergic to wheat should also take note that pasta is made with wheat flour.

Description: Fresh pasta usually comes in flat strips, similar to long noodles.

Selection: Buy fresh refrigerated noodles, or make fresh pasta dough using flour and water.

Storage: Refrigerated, fresh pasta keeps for about a week; frozen, it keeps 2 to 3 months; and dried, for about 2 weeks. On fresh pasta sold in sealed packages, sell-by dates are marked on the package.

PASTA, *Spinach, Fresh*

⬇ 🗑

Serving: 1 cup, cooked

Calories	182
Fat	1.3 g
Saturated	0.3 g
Monounsaturated	0.4 g
Polyunsaturated	0.3 g
Calories from fat	7%
Cholesterol	46 mg
Sodium	8 mg
Protein	7.1 g
Carbohydrate	35.1 g
Dietary fiber	N/A

CHIEF NUTRIENTS

NUTRIENT	AMOUNT	% RDA
Thiamine	0.3 mg	17
Iron	1.6 mg	16
Folate	25.2 mcg	13
Riboflavin	0.2 mg	11
Magnesium	34.0 mg	10
Vitamin B_{12}	0.2 mcg	10
Vitamin B_6	0.2 mg	8
Niacin	1.4 mg	7

Strengths: Fresh spinach pasta offers good amounts of thiamine and iron, both important for healthy energy levels. And it has minor to decent amounts of blood-boosting folate, niacin, riboflavin, and vitamin B_{12}. Fresh spinach pasta also has some magnesium, which helps maintain proper blood pressure.

Caution: *Allergy alert:* Those allergic to eggs should note that fresh-made pasta is more likely to contain egg than dried, packaged pasta. Also contains wheat, a common allergen.

Description: Flat, thin, green noodles made with white wheat flour, eggs, spinach powder or puree, and water.

Storage: May be refrigerated for up to a week or frozen for 2 to 3 months. An unopened package of fresh pasta may be kept until the date marked on the package.

Preparation: Fresh spinach pasta cooks much faster than dried pasta—in as little as 2 minutes. Watch it from the minute you put it in the pot, and drain quickly once it's done.

Serve: With cottage cheese and chive-tomato sauce.

SPAGHETTI, *Enriched*

Serving: 1 cup, cooked

Calories	197
Fat	0.9 g
Saturated	trace
Monounsaturated	trace
Polyunsaturated	trace
Calories from fat	4%
Cholesterol	0
Sodium	1 mg
Protein	6.7 g
Carbohydrate	39.7 g
Dietary fiber	2.2 g

CHIEF NUTRIENTS

NUTRIENT	AMOUNT	% RDA
Iron	2.0 mg	20
Thiamine	0.3 mg	19
Niacin	2.3 mg	12
Riboflavin	0.1 mg	8
Magnesium	25.2 mg	7

Strengths: Because it is made with enriched white flour, spaghetti contains economical amounts of thiamine and iron, both important for high energy levels. Spaghetti also offers a good portion of the day's RDA of niacin and some riboflavin, B-complex vitamins involved in the formation of red blood cells. And it has some magnesium, a mineral involved in muscle and nerve function.

Curiosity: Spaghetti comes from the Italian word meaning "little strings."

History: Italians were probably not the only ones to invent pasta. China, India, and the Middle East all had their versions of spaghetti by the year 1200.

Description: Long, thin strands are made from semolina and water.

Preparation: Cook in plenty of water so the strands don't stick together.

SPAGHETTI, *Spinach*

Serving: 1 cup, cooked

Calories	182
Fat	0.9 g
Saturated	trace
Monounsaturated	trace
Polyunsaturated	trace
Calories from fat	4%
Cholesterol	0
Sodium	20 mg
Protein	6.4 g
Carbohydrate	36.6 g
Dietary fiber	N/A

CHIEF NUTRIENTS

NUTRIENT	AMOUNT	% RDA
Magnesium	86.8 mg	25
Iron	1.5 mg	15
Niacin	2.1 mg	11
Zinc	1.5 mg	10
Thiamine	0.1 mg	9
Folate	16.8 mcg	8
Riboflavin	0.1 mg	8
Vitamin B_6	0.1 mg	7
Calcium	42.0 mg	5

Strengths: A serving offers one-fourth of the day's requirement of magnesium, helpful in maintaining proper blood pressure; also has a good share of blood-building iron. Spinach spaghetti also includes fair portions of folate, niacin, riboflavin, thiamine, and vitamin B_6—vitamins involved in many metabolic functions. A good amount of zinc and some calcium are also present.

Caution: *Allergy alert:* May contain eggs, and always contains wheat—both are common allergens.

Description: The long, thin, green strands of spinach spaghetti are made from semolina, water, spinach powder or paste, and sometimes eggs.

Preparation: Cook in plenty of water until just tender. Drain and rinse in cold water as soon as the spaghetti is cooked.

SPAGHETTI, *Whole Wheat*

Serving: 1 cup, cooked

Calories	174
Fat	0.8 g
Saturated	trace
Monounsaturated	trace
Polyunsaturated	trace
Calories from fat	4%
Cholesterol	0
Sodium	4 mg
Protein	7.5 g
Carbohydrate	37.2 g
Dietary fiber	5.4 g

CHIEF NUTRIENTS

NUTRIENT	AMOUNT	% RDA
Iron	1.5 mg	15
Magnesium	42.0 mg	12
Thiamine	0.2 mg	10
Zinc	1.1 mg	8
Vitamin B$_6$	0.1 mg	6
Niacin	1.0 mg	5

Strengths: Whole wheat spaghetti offers 15% of the RDA of iron, a mineral that is important for healthy blood and better immunity. It also has good amounts of magnesium, which helps regulate blood pressure, and thiamine, a B-complex vitamin involved in energy production. Some niacin, vitamin B$_6$, and zinc are also present. Because it includes wheat bran, whole wheat spaghetti contains more fiber than the white flour variety.

Caution: *Allergy alert:* People allergic to wheat should be cautious about trying this whole wheat product.

Curiosity: Legend has it that northern Italians would toss a strand of cooking spaghetti at a wall. If it stuck, the spaghetti was considered done. Unfortunately, this technique inevitably leads to starchy, overcooked pasta.

Description: Long, thin, light brown strands that are usually round and solid; it's made from whole wheat flour and water.

Storage: Bugs love to munch on whole wheat products. Store in an airtight, insectproof container, and tuck in a few bay leaves to repel insects.

Preparation: Cook in plenty of water. Check often: overcooking disintegrates the noodles.

Serve: With vegetable sauce.

Gravies and Sauces
GRAVIES

AU JUS, *Canned*

Serving: 1/4 cup

Calories	10
Fat	trace
Saturated	trace
Monounsaturated	trace
Polyunsaturated	trace
Calories from fat	11%
Cholesterol	0.3 mg
Sodium	30 mg
Protein	0.7 g
Carbohydrate	1.5 g
Dietary fiber	N/A

CHIEF NUTRIENTS

(None of the nutrients meet or exceed 5% of the RDA.)

Strengths: Au jus gravy is lower in fat, sodium, and calories than beef gravy, but it lacks significant amounts of vitamins or minerals.

Description: Natural juice, usually from beef.

Serve: With top round or other lean cuts of meat.

AU JUS, *Dehydrated*

Serving: 1/4 cup, made w/water

Calories	8
Fat	0.3 g
Saturated	trace
Monounsaturated	trace
Polyunsaturated	trace
Calories from fat	37%
Cholesterol	0.6 mg
Sodium	241 mg
Protein	0.3 g
Carbohydrate	1.0 g
Dietary fiber	N/A

CHIEF NUTRIENTS

(None of the nutrients meet or exceed 5% of the RDA.)

Caution: Dehydrated au jus mix contains 8 times as much sodium as canned au jus. That amount could nudge blood pressure to unhealthy levels in salt-sensitive people. Also, au jus is high in purines and should be avoided by those with gout. *Allergy alert:* Dehydrated gravy mixes tend to be high in monosodium glutamate (MSG), a flavor enhancer that triggers migraine headaches in some people.

Strengths: Au jus is low in calories and has very little fat.

Description: Dehydrated au jus gravy is a commercial product made from beef juices and other ingredients.

A Better Idea: To serve meat au jus, skim all fat from the stock, then boil the stock down to half its original volume. You'll get a rich, concentrated taste with absolutely no fat or added salt. (If desired, season with herbs and maybe a little prepared mustard.)

BEEF, *Canned*

Serving: ¼ cup

Calories	31
Fat	1.4 g
Saturated	0.7 g
Monounsaturated	0.6 g
Polyunsaturated	0.1 g
Calories from fat	40%
Cholesterol	2 mg
Sodium	326 mg
Protein	2.2 g
Carbohydrate	2.8 g
Dietary fiber	0.2 g

CHIEF NUTRIENTS

(None of the nutrients meet or exceed 5% of the RDA.)

Caution: Sodium content is too high to justify if you're on a blood pressure–lowering diet, and the fat content won't recommend it to dieters. Gravy is high in purines and should be eaten sparingly by people with gout.

Eat with: Lean meat like top round and low-fat, high-carbohydrate grains, beans, and potatoes.

Strengths: None. You'd have to drink an entire cup of beef gravy to get measurable amounts of any vitamins, minerals, or protein.

Description: Gravy is a flour-based sauce made from pan drippings and meat juices.

BROWN, *Dehydrated*

Serving: ¼ cup, made w/water

Calories	19
Fat	0.4 g
Saturated	trace
Monounsaturated	trace
Polyunsaturated	trace
Calories from fat	21%
Cholesterol	0.7 mg
Sodium	269 mg
Protein	0.6 g
Carbohydrate	3.3 g
Dietary fiber	0.3 g

CHIEF NUTRIENTS

(None of the nutrients meet or exceed 5% of the RDA.)

Caution: Salty embellishments, like this one, can send blood pressure into overdrive in salt-sensitive people. Also, gravy is high in purines and should be avoided by those with gout. *Allergy alert:* Read labels. Gravy made from a mix may contain wheat or milk, two common food allergens. Also, dehydrated gravy mixes tend to be high in monosodium glutamate (MSG), a flavor enhancer that tends to trigger migraine headaches.

Strengths: Low in calories and very flavorful, brown gravy can add a nice touch to lean meat if not poured on heavily.

Description: Brown gravy made from a mix is a commercial version of stovetop-cooked meat gravy.

Serve: With well-trimmed small cuts of meat, skinless poultry, or mashed potatoes.

A Better Idea: You can make a flavorful, reduced-sodium gravy with degreased pan juices, skim milk, browned flour, and seasonings. Although recipes vary, you can always skip the salt.

CHICKEN, *Canned*

Serving: ¼ cup

Calories	47
Fat	3.4 g
Saturated	0.8 g
Monounsaturated	1.5 g
Polyunsaturated	0.9 g
Calories from fat	65%
Cholesterol	1 mg
Sodium	344 mg
Protein	1.2 g
Carbohydrate	3.2 g
Dietary fiber	0.2 g

CHIEF NUTRIENTS

NUTRIENT	AMOUNT	% RDA
Vitamin A	66.2 RE	7

Caution: Chicken gravy provides a high percentage of calories from fat—more than double the amount recommended by the American Heart Association. It's high in sodium, so keep portions modest if you're trying to restrict your sodium intake. High in purines, gravy should be eaten sparingly by people with gout. *Allergy alert:* Chicken is a common food allergen.

Eat with: Lean meat like chicken breast and low-fat, high-carbohydrate grains, beans, potatoes, fruit, and vegetables—the mainstays of a healthy diet.

Strength: Contains a little bit of vitamin A, which, in large quantities, helps to boost resistance to infection, protect against cancer, keep skin healthy, promote wound healing, build bones, and sharpen night vision.

Description: A flour-based sauce made from pan drippings and meat juices.

CHICKEN, *Dehydrated*

Serving: ¼ cup, made w/water

Calories	21
Fat	0.5 g
Saturated	trace
Monounsaturated	trace
Polyunsaturated	trace
Calories from fat	21%
Cholesterol	0.7 mg
Sodium	283 mg
Protein	0.7 g
Carbohydrate	3.6 g
Dietary fiber	0.3 g

CHIEF NUTRIENTS

(None of the nutrients meet or exceed 5% of the RDA.)

Caution: This popular embellishment to stuffed chicken is high in sodium. People who have been advised to limit intake of salty foods should use it sparingly. Also, chicken gravy is loaded with purines and should be avoided by those with gout. *Allergy alert:* May contain wheat and milk, which can trigger allergies. Like other dehydrated gravy mixes, the chicken variety tends to be high in monosodium glutamate (MSG), a flavor enhancer that may induce migraine headaches in sensitive people.

Strengths: Low in fat and calories.

Description: A chicken-flavored gravy mix.

Serve: With lean chicken breast, with vegetables on the side.

A Better Idea: To reduce sodium, make your own chicken gravy with degreased pan drippings, flour, and stock. Omit the salt.

MUSHROOM, *Canned*

Serving: ¼ cup

Calories	30
Fat	1.6 g
Saturated	0.2 g
Monounsaturated	0.7 g
Polyunsaturated	0.6 g
Calories from fat	49%
Cholesterol	0
Sodium	340 mg
Protein	0.8 g
Carbohydrate	3.3 g
Dietary fiber	0.2 g

CHIEF NUTRIENTS

(None of the nutrients meet or exceed 5% of the RDA.)

Caution: Salty stuff—probably off-limits for those on a sodium-restricted diet to control blood pressure. Mushrooms are moderately high in purines and should be eaten sparingly by those with gout. And dipping into the gravy boat too often can swamp a meal with calories. *Allergy alert:* Read labels; mushroom gravy may contain wheat, milk, or other troublesome ingredients.

A Better Idea: To put a cap on sodium levels, make your own mushroom gravy with fresh sliced mushrooms, fat-free stock, flour, and thyme or other herbs—and omit the salt.

ONION, *Dehydrated*

Serving: ¼ cup, made w/water

Calories	20
Fat	0.2 g
Saturated	trace
Monounsaturated	trace
Polyunsaturated	trace
Calories from fat	8%
Cholesterol	0
Sodium	253 mg
Protein	0.5 g
Carbohydrate	4.1 g
Dietary fiber	N/A

CHIEF NUTRIENTS

(None of the nutrients meet or exceed 5% of the RDA.)

Caution: Watch out for the sodium in this savory sauce. *Allergy alert:* Like other dehydrated gravy mixes, this product may be high in monosodium glutamate (MSG), an ingredient that tends to trigger migraine headaches in some people. May also include wheat or other common food allergens.

Strengths: Lower in calories and fat than many homemade gravies.

Description: A packaged, onion-flavored gravy mix.

Serve: With lean beef, fish, poultry, or vegetables.

A Better Idea: You can make your own onion gravy that's lower in sodium than the dehydrated mix.

PORK, *Dehydrated*

⬇ 🗑

Serving: ¼ cup, made w/water

Calories	19
Fat	0.5 g
Saturated	trace
Monounsaturated	trace
Polyunsaturated	trace
Calories from fat	22%
Cholesterol	0.7 mg
Sodium	309 mg
Protein	0.5 g
Carbohydrate	3.4 g
Dietary fiber	0.3 g

CHIEF NUTRIENTS

(None of the nutrients meet or exceed 5% of the RDA.)

Caution: High in sodium; may tend to crank up blood pressure in those who are salt-sensitive. *Allergy alert:* Like other dehydrated meat sauces, pork gravy mix may be high in monosodium glutamate (MSG), a flavor enhancer that tends to trigger migraine headaches in some people. Also may contain wheat, milk, or other common allergens, so read labels if you have food allergies.

Strengths: Low in calories and fat.

Serve: Sparingly, with very lean cuts of pork, like tenderloin.

A Better Idea: You can make homemade pork gravy that's sodium-free, using skim milk or fat-free stock, diet margarine, and no salt.

TURKEY, *Canned*

Serving: ¼ cup

Calories	30
Fat	1.3 g
Saturated	0.4 g
Monounsaturated	0.5 g
Polyunsaturated	0.3 g
Calories from fat	37%
Cholesterol	1 mg
Sodium	344 mg
Protein	1.6 g
Carbohydrate	3.0 g
Dietary fiber	0.2 g

CHIEF NUTRIENTS

(None of the nutrients meet or exceed 5% of the RDA.)

Caution: Too high in sodium for those on a diet designed to lower blood pressure. Gravy is high in purines and should be eaten sparingly by people with gout.

Strengths: Moderate amounts of fat and calories.

Serve: With turkey breast and low-fat, high-carbohydrate grains, beans, or potatoes. Serving size should be limited—only enough to add flavor.

TURKEY, *Dehydrated*

Serving: ¼ cup, made w/water

Calories	22
Fat	0.5 g
Saturated	trace
Monounsaturated	trace
Polyunsaturated	trace
Calories from fat	20%
Cholesterol	0.7 mg
Sodium	374 mg
Protein	0.7 g
Carbohydrate	3.8 g
Dietary fiber	0.3 g

CHIEF NUTRIENTS

(None of the nutrients meet or exceed 5% of the RDA.)

Caution: One of the saltiest gravies you can buy, this item is probably off-limits to anyone on a sodium-restricted diet to control high blood pressure. Turkey gravy is also high in purines and should be avoided by those with gout. *Allergy alert:* Read labels; may contain milk, wheat, or other common food allergens. Dehydrated gravy mixes are often high in monosodium glutamate (MSG), a flavor enhancer that tends to trigger migraine headaches.

Strength: Low in calories.

Description: A commercial version of made-from-scratch turkey gravy.

Serve: With skinless turkey and roasted potatoes.

A Better Idea: You can make a reduced-sodium turkey gravy with low fat content if you use pan drippings, degreased turkey stock, and flour—but no salt.

SAUCES

BARBECUE, *Ready-to-Serve*

Serving: ¼ cup

Calories	47
Fat	1.1 g
Saturated	0.2 g
Monounsaturated	0.5 g
Polyunsaturated	0.4 g
Calories from fat	22%
Cholesterol	0
Sodium	509 mg
Protein	1.1 g
Carbohydrate	8.0 g
Dietary fiber	N/A

CHIEF NUTRIENTS

NUTRIENT	AMOUNT	% RDA
Vitamin C	4.4 mg	7
Iron	0.6 mg	6
Vitamin A	54.4 RE	5

Caution: Commercial concoctions tend to be high in sodium—around 500 mg in a ¼-cup serving. For people on a sodium-restricted diet, scout around for a good homemade recipe, and omit the salt.

Eat with: Baked potatoes, leafy green salads, and other mainstays of a healthy diet.

Strengths: A little vitamin C, with some iron and vitamin A. But you really can't count on barbecue sauce for much in the way of nutrients.

Curiosity: The word barbecue comes from the Mexican-Spanish *barbacua,* the grilling of meat over a wood fire.

Description: Basic ingredients include catsup, onions, vinegar, Worcestershire sauce, brown sugar, and other seasonings, but the variations are endless.

Serve: With lean meat, like center-cut pork chops, minus all visible fat. Or brush lightly on skinless chicken breasts.

BÉARNAISE, *Dehydrated*

Serving: ¼ cup, made w/milk & butter

Calories	175
Fat	17.1 g
Saturated	10.5 g
Monounsaturated	5.0 g
Polyunsaturated	0.8 g
Calories from fat	88%
Cholesterol	47 mg
Sodium	316 mg
Protein	2.1 g
Carbohydrate	4.4 g
Dietary fiber	N/A

CHIEF NUTRIENTS

NUTRIENT	AMOUNT	% RDA
Vitamin A	189.2 RE	19
Calcium	57.3 mg	7
Vitamin B$_{12}$	0.1 mcg	6

Caution: With 17 g of fat per serving, this rich sauce is a Triple Bypass Special. The sodium level runs high, too, so it's not for anyone on a salt-free diet. *Allergy alert:* Milk is a common food allergen. Also, many dehydrated sauces contain significant amounts of monosodium glutamate (MSG), which commonly triggers headaches in sensitive individuals.

Eat with: Restraint. Or with a lean cut of meat to help limit the total fat content of the meal.

Strengths: This sauce contributes a good amount of resistance-building vitamin A, chiefly from egg yolk. And you get a little calcium and vitamin B$_{12}$. But the saturated fat content cancels out any potential benefits you might get from the other nutrients.

Curiosity: The name is in homage to Henry IV, a popular French king who was born in Béarn, France.

Description: A classic French sauce made with butter, egg yolks, vinegar, wine, tarragon, shallots, and sometimes lemon juice. Pronounced *bair-NAYZ.*

Serve: Drizzle sparingly on poultry, meat, fish, eggs, or vegetables.

CURRY, *Dehydrated*

Serving: ¼ cup, made w/milk

Calories	67
Fat	3.7 g
Saturated	1.5 g
Monounsaturated	1.3 g
Polyunsaturated	0.7 g
Calories from fat	49%
Cholesterol	9 mg
Sodium	319 mg
Protein	2.7 g
Carbohydrate	6.4 g
Dietary fiber	N/A

CHIEF NUTRIENTS

NUTRIENT	AMOUNT	% RDA
Calcium	121.2 mg	15
Vitamin B$_{12}$	0.3 mcg	14
Riboflavin	0.1 mg	8

Caution: High in sodium. This sauce is off-limits for many on sodium-restricted diets. While actual fat and calorie contents are moderate, the percentage of calories from fat exceeds the American Heart Association's recommendation of 30% or less. *Allergy alert:* Milk is a common food allergen. Dehydrated sauces often contain significant amounts of monosodium glutamate (MSG), which often triggers headaches in sensitive individuals.

Strengths: When prepared with milk, powdered curry sauce provides a good amount of useful bone-building calcium and vitamin B$_{12}$, along with a little riboflavin.

Description: Curry is not one spice but a blend of turmeric and up to 16 other spices.

A Better Idea: For a less salty curry, make your own sauce. Recipes vary, but they rarely (if ever) call for salt, since curry is so spicy to begin with.

HOLLANDAISE, *Dehydrated*

Serving: ¼ cup, made w/milk & butter

Calories	176
Fat	17.1 g
Saturated	10.5 g
Monounsaturated	5.0 g
Polyunsaturated	0.7 g
Calories from fat	87%
Cholesterol	47 mg
Sodium	283 mg
Protein	2.1 g
Carbohydrate	4.5 g
Dietary fiber	0.2 g

CHIEF NUTRIENTS

NUTRIENT	AMOUNT	% RDA
Vitamin A	173.9 RE	17
Calcium	59.9 mg	7
Vitamin B$_{12}$	0.1 mcg	6

Caution: This rich, velvety sauce is high in calories and saturated fat, posing a double threat to the heart and arteries. It's even riskier when served over eggs, as in eggs Benedict. And the high sodium level could aggravate blood pressure problems in salt-sensitive individuals. *Allergy alert:* Depending on the ingredients, hollandaise sauce may contain milk or eggs, two common food allergens. Dehydrated sauce mix may also contain monosodium glutamate (MSG), which often triggers headaches.

Strengths: Hollandaise supplies a good amount of vitamin A. Unfortunately, the nutrient is contained in egg yolk, which is high in cholesterol. Milk contributes some bone-building calcium and a little vitamin B$_{12}$, but not a lot.

A Better Idea: For a lighter-than-air alternative to butter-laden Hollandaise, fold together ⅓ cup nonfat yogurt, 3 Tbsp. thawed egg substitute, 1 Tbsp. lemon juice, and a pinch of cayenne.

HOT PEPPER

Serving: 1 tsp.

Calories	0
Fat	0
Saturated	0
Monounsaturated	0
Polyunsaturated	0
Calories from fat	0
Cholesterol	N/A
Sodium	22 mg
Protein	0.1 g
Carbohydrate	0.1 g
Dietary fiber	0.1 g

CHIEF NUTRIENTS

(None of the nutrients meet or exceed 5% of the RDA.)

Strengths: With its fiery flavor, hot pepper sauce can enhance many dishes. It has no fat and zero calories.

Caution: Has sodium—though just a dash doesn't contain much.

Origin: *The* hot pepper sauce, McIlhenny's Tabasco Sauce, is actually a food trademark. The sauce was created by members of the McIlhenny family of Louisiana. The seeds of the famed red peppers were imported from Tabasco, Mexico, in 1848, and 20 years later the family bottled the first sauce. The family-owned business now produces over 200,000 bottles of Tabasco sauce daily.

Description: Peppers are crushed, salted, and fermented for 3 years. After this blend is mixed with vinegar, it is bottled.

Serve: As a pungent addition to soups, stews, and casseroles. Many people like hot pepper sauce on raw clams and oysters, or added to pasta sauce.

MARINARA, *Canned*

Serving: ½ *cup*

Calories	85
Fat	4.2 g
Saturated	0.6 g
Monounsaturated	2.1 g
Polyunsaturated	1.2 g
Calories from fat	44%
Cholesterol	0
Sodium	786 mg
Protein	2.0 g
Carbohydrate	12.7 g
Dietary fiber	1.9 g

CHIEF NUTRIENTS

NUTRIENT	AMOUNT	% RDA
Vitamin C	16.0 mg	27
Vitamin B$_6$	0.3 mg	16
Potassium	530.0 mg	14
Vitamin A	120.0 RE	12
Iron	1.0 mg	10
Niacin	2.0 mg	10
Magnesium	30.0 mg	9
Folate	16.9 mcg	8

Strengths: Marinara is a good source of resistance-building vitamins A and C, with significant amounts of B vitamins, especially B$_6$. This sauce also has a decent level of fiber. Virtually all the fat is unsaturated.

Caution: Canned marinara may be high in sodium, so make your own sauce if you're cutting back on sodium. *Allergy alert:* Tomatoes are a common food allergen and may trigger hives, headaches, mouth itching, or other reactions in sensitive people. People who are allergic to aspirin may need to avoid tomatoes: They contain salicylate, a basic component of aspirin.

Curiosity: The name means ''mariner's style,'' perhaps referring to the sailors' custom of plunking the fresh catch of the day into a steaming stewpot of tomatoes, onions, and seasonings for an impromptu meal.

Description: A quick-cooking sauce that's composed of tomatoes and seasonings, including garlic, oregano, parsley, olive oil, and sometimes wine.

Serve: With pasta, seafood, or meat.

SOY, *Regular*

Serving: 1 Tbsp.

Calories	7
Fat	trace
Saturated	trace
Monounsaturated	trace
Polyunsaturated	trace
Calories from fat	1%
Cholesterol	0
Sodium	1,024 mg
Protein	0.4 g
Carbohydrate	1.4 g
Dietary fiber	0

CHIEF NUTRIENTS

(None of the nutrients meet or exceed 5% of the RDA.)

Caution: Soy is very high in sodium: Use sparingly. *Drug interaction:* Fermented foods like soy sauce contain tyramine, an amino acid derivative that can raise blood pressure to dangerously high levels in anyone taking MAO inhibitors. *Allergy alert:* Made from soybeans, a common food allergen. May also contain other allergens like corn, wheat, or gluten—so read the label. Also, large amounts of soy sauce may trigger headaches in people sensitive to fermented foods.

Strengths: Hardly any calories, free of cholesterol, with only a trace of fat.

Description: Traditionally, a dark, salty liquid made by fermenting boiled soybeans and roasted wheat or barley. A chemically fermented soy sauce prepared from hydrolyzed soy protein, salt, caramel coloring, corn syrup, and other additives is common in the U.S.

Selection: Japanese soy sauce (shoyu) is lighter, sweeter, and less salty than Chinese soy sauce.

Serve: As a condiment in oriental dishes.

A Better Idea: Choose low-sodium varieties.

SOY, *Shoyu*

Serving: 1 Tbsp.

Calories	10
Fat	trace
Saturated	trace
Monounsaturated	trace
Polyunsaturated	trace
Calories from fat	1%
Cholesterol	0
Sodium	1,029 mg
Protein	0.9 g
Carbohydrate	1.5 g
Dietary fiber	0

CHIEF NUTRIENTS

(None of the nutrients meet or exceed 5% of the RDA.)

Caution: Very high in sodium. Definitely to be avoided by those on a sodium-restricted diet. *Drug interaction:* Like other fermented foods, shoyu soy sauce contains tyramine, an amino acid derivative that can raise blood pressure to dangerously high levels in anyone taking MAO inhibitors. *Allergy alert:* Made from soybeans and wheat, common food allergens. Large amounts of soy sauce can trigger headaches in people sensitive to other fermented foods, like aged cheese.

Strengths: Moderate in calories, with hardly any fat and no cholesterol.

Origin: Soy sauce originated in China some 2,500 years ago and was introduced to Japan in the 7th century by Buddhist priests.

Description: A fermented Japanese soy sauce made from soybeans, wheat, salt, and *Aspergillus soyae* mold. Tastes like a salty meat extract.

Serve: As a condiment in oriental dishes.

A Better Idea: There are many types of soy sauce. Read labels, choose one lower in sodium, and use sparingly.

SOY, *Tamari*

Serving: 1 Tbsp.

Calories	11
Fat	trace
Saturated	trace
Monounsaturated	trace
Polyunsaturated	trace
Calories from fat	2%
Cholesterol	0
Sodium	1,006 mg
Protein	1.9 g
Carbohydrate	1.0 g
Dietary fiber	0

CHIEF NUTRIENTS

(None of the nutrients meet or exceed 5% of the RDA.)

Caution: Very high in sodium. *Drug interaction:* Fermented foods like tamari contain tyramine, an amino acid derivative that can raise blood pressure to dangerous levels in anyone taking MAO inhibitors. *Allergy alert:* Made from soybeans, a common food allergen. May also contain some wheat. Large amounts of tamari may trigger headaches in people sensitive to fermented foods.

Strengths: Moderate calories and free of cholesterol, with almost no fat.

Description: Like shoyu or regular soy sauce, tamari is a dark sauce made from soybeans, but it contains less wheat and is usually stronger in flavor.

A Better Idea: If you're restricting sodium intake, look for low-sodium soy sauce.

SPAGHETTI, Canned

Serving: ¹/₂ cup

Calories	136
Fat	5.9 g
Saturated	0.9 g
Monounsaturated	3.0 g
Polyunsaturated	1.6 g
Calories from fat	39%
Cholesterol	0
Sodium	618 mg
Protein	2.3 g
Carbohydrate	19.8 g
Dietary fiber	4.2 g

CHIEF NUTRIENTS

NUTRIENT	AMOUNT	% RDA
Vitamin C	14.0 mg	23
Vitamin B$_6$	0.4 mg	22
Vitamin A	153.1 RE	15
Folate	26.9 mcg	13
Potassium	478.1 mg	13
Niacin	1.9 mg	10
Magnesium	29.9 mg	9
Iron	0.8 mg	8

Strengths: Spaghetti sauce is a decent source of B vitamins, especially vitamin B$_6$. It has significant amounts of vitamins A and C to boost resistance, and a hearty helping of fiber. Virtually all the fat is unsaturated.

Eat with: A generous handful of sautéed green peppers and mushrooms to boost the fiber. To increase iron and protein, add lean ground meat.

Caution: Those on a sodium-restricted diet should look for low-sodium sauce or make their own. *Allergy alert:* Tomatoes are a common food allergen and may trigger hives or other reactions in sensitive people. Also, people who are allergic to salicylate in aspirin should avoid tomatoes; they contain the same substance.

Description: Spaghetti sauce can be any thick tomato sauce that is simmered with onions, garlic, basil, olive oil, and other seasonings.

SWEET-AND-SOUR, Dehydrated

Serving: ¹/₄ cup, made w/water & vinegar

Calories	74
Fat	trace
Saturated	0
Monounsaturated	trace
Polyunsaturated	trace
Calories from fat	0.2%
Cholesterol	0
Sodium	195 mg
Protein	0.2 g
Carbohydrate	18.2 g
Dietary fiber	0.2 g

CHIEF NUTRIENTS

(None of the nutrients meet or exceed 5% of the RDA.)

Caution: This oriental favorite contains a fair amount of sugar and sodium, but very little else. *Allergy alert:* Read labels; dehydrated sauce mixes often contain significant amounts of monosodium glutamate (MSG) or other ingredients that trigger reactions in sensitive individuals.

Strengths: This is a tasty accompaniment to Chinese-style food. Eaten in careful moderation, it gives food a nice flavor with a moderate number of calories, no cholesterol, and hardly any fat.

Serve: With diced chicken breast, lean pork strips, lean meatballs, or chopped vegetables.

A Better Idea: To control the sodium level, make your own sweet-and-sour sauce. Most recipes call for honey, pineapple juice or sugar, lemon juice or vinegar, and ginger or other seasonings.

TARTAR

Serving: 1 Tbsp.

Calories	74
Fat	8.1 g
Saturated	1.5 g
Monounsaturated	N/A
Polyunsaturated	N/A
Calories from fat	98%
Cholesterol	7 mg
Sodium	99 mg
Protein	0.2 g
Carbohydrate	0.6 g
Dietary fiber	N/A

CHIEF NUTRIENTS

(None of the nutrients meet or exceed 5% of the RDA.)

Caution: Heavy on mayo, tartar sauce gets 98% of its calories from fat. *Allergy alert:* The ingredients in tartar sauce can cause a number of allergic reactions in people sensitive to eggs. This sauce also can lead to giant hives around the mouth, caused by contact with the allergenic substance. Check labels if you're allergic to soybeans or corn—the mayonnaise in tartar sauce may contain either or both.

Strengths: Has tiny amounts of calcium, potassium, iron, protein, and vitamin C—but not enough to provide much nutrient value.

Description: A sauce that includes mayonnaise, capers, olives, parsley, and chopped pickles.

Origin: The French first created what they call "tartare" sauce. Their version—containing minced shallots and French Dijon mustard—is more pungent than what Americans traditionally consume.

A Better Idea: Make your own low-fat mayonnaise for homemade tartar sauce, using yogurt as a base.

TERIYAKI, *Dehydrated*

Serving: ¼ cup, made w/water

Calories	33
Fat	trace
Saturated	trace
Monounsaturated	trace
Polyunsaturated	trace
Calories from fat	6%
Cholesterol	0
Sodium	1,198 mg
Protein	1.0 g
Carbohydrate	6.9 g
Dietary fiber	1.3 g

CHIEF NUTRIENTS

NUTRIENT	AMOUNT	% RDA
Iron	0.7 mg	7
Magnesium	21.2 mg	6

Caution: Contains slightly less sodium than the Bonneville Salt Flats—too much for those who have been warned to steer clear of the white stuff. *Drug interaction:* Teriyaki sauce usually contains soy sauce, a potential problem for those taking MAO-inhibitor drugs. *Allergy alert:* Read labels for soy, MSG, or other ingredients that may cause problems.

Strengths: Teriyaki marinade contains a little iron and magnesium, but varying amounts of these nutrients remain after food has been grilled.

Curiosity: *Teri-* means glossy and *-yaki* means grilled. The sugar in this sauce gives a glazed surface to food grilled with the sauce.

Origin: Teriyaki sauce has long been a staple of Japanese cuisine.

Serve: As a marinade for chicken, beef, and seafood.

A Better Idea: To reduce sodium content, make your own teriyaki sauce. Most recipes call for ginger root, garlic, lemon juice, sugar, soy sauce, and sometimes small amounts of sherry or other cooking wine. Use low-sodium soy sauce.

TERIYAKI, *Ready-to-Serve*

Serving ¼ cup

Calories	60
Fat	0
Saturated	0
Monounsaturated	0
Polyunsaturated	0
Calories from fat	0
Cholesterol	0
Sodium	2,760 mg
Protein	4.3 g
Carbohydrate	11.5 g
Dietary fiber	0.1 g

CHIEF NUTRIENTS

NUTRIENT	AMOUNT	% RDA
Magnesium	43.9 mg	13
Iron	1.2 mg	12
Folate	14.4 mcg	7

Caution: Contains more sodium than a teaspoon of salt—enough to aggravate high blood pressure in some people. *Drug interaction:* May cause potential problems for those taking MAO inhibitors. *Allergy alert:* Read labels for MSG and other potentially troublesome ingredients.

Strengths: Contains good amounts of anemia-fighting iron and magnesium, with a little folate for good measure. But varying amounts of those nutrients remain after a food has been marinated and cooked in this sauce.

Storage: The sauce can be stored in a closed container in the refrigerator for up to 2 weeks.

Serve: As a marinade for chicken, beef, and seafood.

A Better Idea: To tame sodium levels, make your own teriyaki sauce using light soy sauce.

TOMATO, *Canned*

Serving: ½ cup

Calories	37
Fat	0.2 g
Saturated	trace
Monounsaturated	trace
Polyunsaturated	trace
Calories from fat	5%
Cholesterol	0
Sodium	738 mg
Protein	1.6 g
Carbohydrate	8.8 g
Dietary fiber	1.8 g

CHIEF NUTRIENTS

NUTRIENT	AMOUNT	% RDA
Vitamin C	16.0 mg	27
Potassium	452.6 mg	12
Vitamin A	119.6 RE	12
Vitamin B$_6$	0.2 mg	10
Iron	0.9 mg	9
Magnesium	23.2 mg	7
Niacin	1.4 mg	7
Folate	11.5 mcg	6
Thiamine	0.1 mg	5

Strengths: Tomato sauce contains a little fiber, which we all need for good digestive health. It also contains some B vitamins, especially B$_6$, as well as a respectable amount of vitamin C to help boost iron absorption. That extra C is a big plus for those prone to iron-deficiency anemia. With less fat than spaghetti sauce, tomato sauce is a health-enhancer on many counts.

Caution: *Allergy alert:* Tomatoes are a common food allergen and may trigger hives, headaches, mouth itching, or other reactions in sensitive people. Also, many allergists advise people who are allergic to aspirin to avoid tomatoes, which, like aspirin, contain salicylates.

Description: Pure tomato sauce may be less highly seasoned than spaghetti sauce. Some brands may contain flour, which is used as a thickener—so read the labels to find out what's in the brand you're buying.

A Better Idea: If you're on a sodium-restricted diet, look for low-sodium varieties.

TOMATO CHILI, *Low Sodium*

Serving: 1 Tbsp.

Calories	16
Fat	0.1 g
Saturated	0
Monounsaturated	N/A
Polyunsaturated	N/A
Calories from fat	3%
Cholesterol	0
Sodium	3 mg
Protein	0.4 g
Carbohydrate	3.7 g
Dietary fiber	0.9 g

CHIEF NUTRIENTS

(None of the nutrients meet or exceed 5% of the RDA.)

Caution: Tomatoes may cause bed-wetting in young children. Spices and condiments are associated with allergic tension-fatigue syndrome. *Allergy alert:* Those allergic to mold may have an allergic reaction to the vinegar in the sauce. Those who get hives and allergic headaches from tomato products should also beware.

Strengths: Low-sodium tomato chili sauce provides a high-powered dose of spicy flavor with very little salt.

Description: This sauce—which can be hot or mild—is a blend of tomato puree, chili pepper pulp (or powder), spices, onions, green peppers, vinegar, and sugar.

Origin: Chilis are native to Central and South America and were once enjoyed by the Incas of Peru and ancient Aztecs of Mexico.

History: Chilies were brought to India in 1611; that country is now the largest producer.

A Better Idea: Make your own tomato chili sauce using fresh tomatoes.

WHITE, *Dehydrated*

Serving: ¼ cup, made w/milk

Calories	60
Fat	3.4 g
Saturated	1.6 g
Monounsaturated	1.2 g
Polyunsaturated	0.4 g
Calories from fat	50%
Cholesterol	9 mg
Sodium	199 mg
Protein	2.6 g
Carbohydrate	5.4 g
Dietary fiber	0.2 g

CHIEF NUTRIENTS

NUTRIENT	AMOUNT	% RDA
Magnesium	66.0 mg	19
Calcium	106.2 mg	13
Vitamin B$_{12}$	0.3 mcg	13
Riboflavin	0.1 mg	7

Caution: The sodium content of this basic sauce is quite high, even in a ¼-cup portion. But the calorie count isn't too bad—provided you keep portions modest. *Allergy alert:* Contains milk and wheat, common food allergens. Also, dehydrated sauce mixes tend to contain a significant amount of monosodium glutamate (MSG), which sometimes triggers headaches.

Strengths: White sauce has appreciable amounts of magnesium, calcium, and B vitamins, such as B$_{12}$ and riboflavin, all of which are nutrients your body needs for healthy bones, nerves, and skin.

Description: A smooth white sauce made from butter, flour, and either milk or chicken or fish stock.

A Better Idea: To bring sodium content down to a manageable level, make your own white sauce with melted margarine, flour, and scalded skim milk or lean chicken stock. Skip the salt.

WORCESTERSHIRE

⬇ 🏛

Serving: 1 tsp.

Calories	4
Fat	0
Saturated	0
Monounsaturated	0
Polyunsaturated	0
Calories from fat	0
Cholesterol	N/A
Sodium	55 mg
Protein	0.1 g
Carbohydrate	0.9 g
Dietary fiber	0

CHIEF NUTRIENTS

NUTRIENT	AMOUNT	% RDA
Vitamin C	9.0 mg	15

Strength: This sauce has a good supply of vitamin C.

Caution: Even the low-sodium version of Worcestershire sauce packs some salt in each teaspoon.

Origin: In the 1830s, Lea and Perrins, the two owners of a pharmacy in Worcester, England, made a sauce for a former governor general using a recipe he brought back from India. He tasted the mixture but didn't like it. Lea and Perrins put the leftover sauce aside. When they came across it a few years later, they found the sauce much improved by the aging process. Bottled for sale, Lea & Perrins ''original'' Worcestershire sauce became an instant success.

Description: A blend of 20 natural ingredients, including onions, vinegar, molasses, corn sweeteners, tamarind, salt, garlic, shallots, cloves, chili peppers, and anchovies or sardines.

Serve: As a flavor enhancer on lean meats, in soups, or in salad dressings.

Meat and Poultry
BEEF

BEEF LIVER

Serving: 3 oz., braised

Calories	137
Fat	4.2 g
Saturated	1.6 g
Monounsaturated	0.6 g
Polyunsaturated	0.8 g
Calories from fat	27%
Cholesterol	331 mg
Sodium	60 mg
Protein	20.7 g
Carbohydrate	3.0 g
Dietary fiber	0

CHIEF NUTRIENTS

NUTRIENT	AMOUNT	% RDA
Vitamin B_{12}	60.4 mcg	3,018
Vitamin A	9,011.7 RE	901
Riboflavin	3.5 mg	205
Folate	184.5 mcg	92
Iron	5.8 mg	58
Niacin	9.1 mg	48
Vitamin B_6	0.8 mg	39
Zinc	5.2 mg	34
Vitamin C	19.6 mg	33

Strengths: On the plus side, beef liver is unsurpassed as a source of vitamins A, B_{12}, and riboflavin. It's also rich in zinc and iron.

Eat with: Steamed onions. Since beef liver is completely devoid of fiber, the onions help to offset some of the potentially harmful effects of fat and cholesterol.

Caution: Liver is a major depot for cholesterol, a fatlike substance that can clog your arteries, so it's a poor choice for those who have been advised to lower their intake of cholesterol. Another problem: The liver filters and collects some of the chemicals, medicines, and hormones that are fed to beef cattle. The older the animal, the greater the accumulation of unwanted substances, so many people choose calves' liver over beef liver. *Drug interaction:* People taking levodopa (an anti-Parkinson's drug) are often advised to avoid liver and other foods high in vitamin B_6. Also, people who have gout should avoid liver—it's high in purines, substances that may trigger an attack.

BLADE ROAST

Serving: 3 oz., lean only, braised

Calories	213
Fat	11.1 g
Saturated	4.3 g
Monounsaturated	4.8 g
Polyunsaturated	0.4 g
Calories from fat	47%
Cholesterol	90 mg
Sodium	60 mg
Protein	26.4 g
Carbohydrate	0
Dietary fiber	0

CHIEF NUTRIENTS

NUTRIENT	AMOUNT	% RDA
Vitamin B_{12}	2.1 mcg	105
Zinc	8.7 mg	58
Iron	3.1 mg	31
Riboflavin	0.2 mg	14
Vitamin B_6	0.3 mg	13

Strengths: An excellent source of zinc, which promotes proper blood clotting, helps white blood cells fight off infection-causing invaders, and speeds wound healing. Blade roast is rich in heme iron, the form best absorbed by the body. Like other cuts of meat, it's chock-full of vitamin B_{12}, which is essential for the release of folate, another B vitamin, which is present in many vegetables. Its riboflavin is required for B vitamin metabolism, and the vitamin B_{12} helps fight infection.

Eat with: Potatoes, carrots, peas, and other high-carbohydrate, low-fat, low-cholesterol foods that have fiber.

Caution: Blade roast is higher in calories, fat, and cholesterol than pot roast—and nearly half the fat is saturated.

Preparation: If you're trying to cut back on saturated fat and lower your cholesterol, limit portions by cubing meat for use in stew, stir-fries, or shish kabobs.

BOTTOM ROUND ROAST

Serving: 3 oz., lean only, braised

Calories	178
Fat	7.0 g
Saturated	2.4 g
Monounsaturated	3.1 g
Polyunsaturated	0.3 g
Calories from fat	35%
Cholesterol	82 mg
Sodium	43 mg
Protein	26.9 g
Carbohydrate	0
Dietary fiber	0

CHIEF NUTRIENTS

NUTRIENT	AMOUNT	% RDA
Vitamin B_{12}	2.1 mcg	105
Zinc	4.7 mg	31
Iron	2.9 mg	29
Niacin	3.5 mg	18
Vitamin B_6	0.3 mg	16
Riboflavin	0.2 mg	13
Potassium	261.8 mg	7
Magnesium	21.3 mg	6

Strengths: One serving of bottom round supplies nearly half the daily requirement for protein. It's an excellent source of vitamin B_{12} to help prevent skin and nerve abnormalities. Rich in zinc, which promotes proper blood clotting, helps white blood cells fight off infection-causing invaders, and speeds wound healing. Like other meats, bottom round is valuable as a source of heme iron, the form best absorbed by the body. Some magnesium helps promote resistance to tooth decay, and its potassium is a catalyst in protein metabolism.

Eat with: Baked potatoes, baked beans, pasta salad, and other high-carbohydrate foods to offset the fat and calories.

Caution: About the same amount of iron, zinc, and B vitamins as top round, but with slightly more fat, calories, and calories from fat. Bottom round is not the best choice if you're trying to cut back on saturated fat and lower your cholesterol.

BRISKET

Serving: 3 oz., lean only, braised

Calories	206
Fat	10.9 g
Saturated	3.9 g
Monounsaturated	5.0 g
Polyunsaturated	0.3 g
Calories from fat	48%
Cholesterol	79 mg
Sodium	60 mg
Protein	25.3 g
Carbohydrate	0
Dietary fiber	0

CHIEF NUTRIENTS

NUTRIENT	AMOUNT	% RDA
Vitamin B_{12}	2.2 mcg	111
Zinc	5.9 mg	39
Iron	2.4 mg	24
Niacin	3.2 mg	17
Vitamin B_6	0.3 mg	13
Riboflavin	0.2 mg	11
Potassium	242.3 mg	7
Magnesium	19.6 mg	6

Strengths: Brisket is a treasure trove of zinc, which promotes proper blood clotting, helps white blood cells fight infection-causing invaders, and speeds wound healing. It also has decent amounts of heme iron, the form best absorbed by the body. The dose of niacin is essential for red blood cells, and the riboflavin is required for B vitamin metabolism. In addition to infection-fighting vitamin B_6, beef also contributes some magnesium, which helps promote resistance to tooth decay, and its potassium is a catalyst in protein metabolism.

Eat with: High-carbohydrate foods with fiber, like potatoes, beans, and vegetables, to offset the high fat and calorie count.

Caution: Rich in nutrients but also high in fat, cholesterol, and calories. Consumption of too much meat and other sources of saturated fat has been linked to a higher risk of colon and breast cancer. If the brisket is cured in salt brine to make corned beef brisket, the high sodium content is a potential problem.

EYE ROUND ROAST

Serving: 3 oz., lean only, roasted

Calories	143
Fat	4.2 g
Saturated	1.5 g
Monounsaturated	1.8 g
Polyunsaturated	0.1 g
Calories from fat	26%
Cholesterol	59 mg
Sodium	53 mg
Protein	24.6 g
Carbohydrate	0
Dietary fiber	0

CHIEF NUTRIENTS

NUTRIENT	AMOUNT	% RDA
Vitamin B$_{12}$	1.8 mcg	92
Zinc	4.0 mg	27
Iron	1.7 mg	17
Niacin	3.2 mg	17
Vitamin B$_6$	0.3 mg	16
Potassium	335.8 mg	9
Riboflavin	0.1 mg	8
Magnesium	23.0 mg	7

Strengths: Supplies nearly half the daily requirement for protein. Lower in fat and calories than blade roast, bottom round, and brisket, eye round is a good choice for those who are trying to improve their dietary supply of iron without consuming too much fat. Its good supply of zinc promotes proper blood clotting, helps white blood cells fight off infection-causing invaders, and speeds wound healing. With fairly decent amounts of other resistance-boosting B vitamins, eye round also contributes more potassium than the average cut of beef.

Eat with: Foods like potatoes, beans, and other vegetables that will add carbohydrates and fiber to your diet. These foods will also contribute magnesium and potassium to help control blood pressure.

Caution: Consumption of too much meat and other sources of saturated fat has been linked to a higher risk of colon and breast cancer.

FILET MIGNON

Serving: 3 oz., lean only, broiled

Calories	179
Fat	8.5 g
Saturated	3.2 g
Monounsaturated	3.2 g
Polyunsaturated	0.3 g
Calories from fat	43%
Cholesterol	71 mg
Sodium	54 mg
Protein	24.0 g
Carbohydrate	0
Dietary fiber	0

CHIEF NUTRIENTS

NUTRIENT	AMOUNT	% RDA
Vitamin B$_{12}$	2.2 mcg	109
Zinc	4.8 mg	32
Iron	3.0 mg	30
Vitamin B$_6$	0.4 mg	19
Niacin	3.3 mg	18
Riboflavin	0.3 mg	15
Potassium	356.2 mg	10
Magnesium	25.5 mg	7
Thiamine	0.1 mg	7

Strengths: Even staunch vegetarians sometimes salivate at the prospect of a tender filet mignon. What you get for your indulgence is well over a day's supply of vitamin B$_{12}$, along with nearly one-third of the day's supply of iron, ample amounts of zinc, and nice amounts of thiamine, magnesium, and potassium.

Caution: High in fat, calories, and cholesterol. That could spell trouble for people on fat-restricted diets. The fat content makes this a less-than-ideal source of protein for diabetics, who fare better on lean meat and complex carbohydrates. Although its potassium and magnesium might otherwise help lower blood pressure, this benefit may be offset by the saturated fat content, which is high.

A Better Idea: If you're going to treat yourself to filet mignon, order surf and turf. Chances are you'll get a smaller medallion of beef (and consequently less fat), along with a lean tail of lobster.

FLANK STEAK

Serving: 3 oz., lean only, broiled

Calories	176
Fat	8.6 g
Saturated	3.7 g
Monounsaturated	3.5 g
Polyunsaturated	0.3 g
Calories from fat	44%
Cholesterol	57 mg
Sodium	71 mg
Protein	23.0 g
Carbohydrate	0
Dietary fiber	0

CHIEF NUTRIENTS

NUTRIENT	AMOUNT	% RDA
Vitamin B_{12}	2.8 mcg	138
Zinc	4.1 mg	27
Niacin	4.3 mg	23
Iron	2.2 mg	22
Vitamin B_6	0.3 mg	15
Potassium	351.9 mg	9
Riboflavin	0.2 mg	9
Magnesium	20.4 mg	6
Thiamine	0.1 mg	6

Strengths: The high zinc promotes proper blood clotting, helps white blood cells fight off infection-causing invaders, and speeds wound healing. Steak is also valuable as a source of heme iron, the form best absorbed by the body. Steak eaters also benefit from niacin, too little of which can trigger confusion and irritability. Steak contributes a fair amount of thiamine and contributes a nice amount of potassium. However, flank steak has a little less protein than leaner cuts of beef like pot roast.

Eat with: High-carbohydrate foods with fiber, like potatoes, beans, and vegetables, to offset the fat and calorie count.

Caution: Flank steak's nutritional attributes come at a price: Nearly half the calories are from fat, and it's moderately high in cholesterol. So you'll probably want to avoid a steak entrée if you've been advised to watch your fat intake.

GROUND BEEF, *Extra-Lean*

Serving: 3 oz., broiled

Calories	218
Fat	13.9 g
Saturated	5.5 g
Monounsaturated	6.1 g
Polyunsaturated	0.5 g
Calories from fat	57%
Cholesterol	71 mg
Sodium	60 mg
Protein	21.6 g
Carbohydrate	0
Dietary fiber	0

CHIEF NUTRIENTS

NUTRIENT	AMOUNT	% RDA
Vitamin B_{12}	1.8 mcg	92
Zinc	4.6 mg	31
Niacin	4.2 mg	22
Iron	2.0 mg	20
Riboflavin	0.2 mg	14
Vitamin B_6	0.2 mg	12
Potassium	266.1 mg	7
Magnesium	17.9 mg	5

Caution: Consuming too much meat and other sources of saturated fat has been linked to a higher risk of colon and breast cancer. So choose the leanest blend of ground beef—ground round—instead of ground chuck.

Eat with: Whole wheat bread, peppers, and whole grain rice. Other high-carbohydrate foods with fiber, like potatoes, beans, and vegetables, can also help offset the fat and calorie count. These foods help protect against the cancer associated with diets high in animal fat.

Strengths: You get ample amounts of zinc and vitamin B_{12}, along with other plentiful B vitamins and some potassium.

Selection: Most stores label the lean-to-fat ratio of ground beef. Look for ground meat that, by weight, is 10% fat or less. Don't confuse percentage of fat by weight with percentage of calories from fat, which is much higher.

A Better Idea: For the leanest possible ground beef, buy top or eye round and grind it yourself. And instead of broiling, try microwaving: It zaps more fat from ground beef.

PORTERHOUSE STEAK

Serving: 3 oz., lean only, broiled

Calories	185
Fat	9.2 g
Saturated	3.7 g
Monounsaturated	3.7 g
Polyunsaturated	0.3 g
Calories from fat	45%
Cholesterol	68 mg
Sodium	56 mg
Protein	23.9 g
Carbohydrate	0
Dietary fiber	0

CHIEF NUTRIENTS

NUTRIENT	AMOUNT	% RDA
Vitamin B_{12}	1.9 mcg	97
Zinc	4.6 mg	31
Iron	2.6 mg	26
Niacin	3.9 mg	21
Vitamin B_6	0.3 mg	17
Riboflavin	0.2 mg	12
Potassium	346.0 mg	9
Magnesium	24.7 mg	7
Thiamine	0.1 mg	6

Strengths: A very good source of infection-fighting zinc, porterhouse also contains generous amounts of heme iron, the form best absorbed by the body. This cut has respectable amounts of B vitamins (with an excellent portion of B_{12}), and it also helps replenish a little bit of thiamine, which may be depleted when you drink coffee or tea.

Eat with: Broccoli and other members of the cruciferous family, which contain substances that may slow the spread of digestive system cancers. High-fiber vegetables like corn can also offer some protection against cancer associated with fat intake.

Caution: At more than 9 g of fat per serving, porterhouse is best eaten in moderation, especially if you're trying to cut back on saturated fat and lower your cholesterol. Consumption of too much meat and other sources of saturated fat has been linked to a higher risk of colon and breast cancer.

Selection: Restaurants tend to serve steak in king-size portions only. Do your arteries a favor: Eat only half, and enjoy the baked potato instead.

POT ROAST, *Arm*

Serving: 3 oz., lean only, braised

Calories	184
Fat	7.1 g
Saturated	2.6 g
Monounsaturated	3.0 g
Polyunsaturated	0.3 g
Calories from fat	35%
Cholesterol	86 mg
Sodium	56 mg
Protein	28.1 g
Carbohydrate	0
Dietary fiber	0

CHIEF NUTRIENTS

NUTRIENT	AMOUNT	% RDA
Vitamin B_{12}	2.9 mcg	145
Zinc	7.4 mg	49
Iron	3.2 mg	32
Niacin	3.2 mg	17
Riboflavin	0.3 mg	15
Vitamin B_6	0.3 mg	14
Potassium	245.7 mg	7
Magnesium	20.4 mg	6

Strengths: Supplies nearly half the daily requirement for protein. Arm pot roast is a superb source of vitamin B_{12} to guard against anemia and nerve degeneration. It's also a good source of niacin, essential for red blood cells, and of riboflavin, which is required for B vitamin metabolism. Pot roast is rich in heme iron, the form best absorbed by the body, and it supplies generous amounts of zinc—which promotes proper blood clotting, helps white blood cells fight off infection-wielding invaders, and speeds wound healing. Also contributes some blood pressure–lowering magnesium and potassium.

Eat with: Lots of fiber-rich, low-fat, high-carbohydrate foods like potatoes or whole grain rice.

Caution: The nutritional goodies in pot roast come at a price: Each serving has moderate amounts of fat and cholesterol.

RIB EYE STEAK

Serving: 3 oz., lean only, broiled

Calories	191
Fat	10.0 g
Saturated	4.0 g
Monounsaturated	4.2 g
Polyunsaturated	0.3 g
Calories from fat	47%
Cholesterol	68 mg
Sodium	59 mg
Protein	23.8 g
Carbohydrate	0
Dietary fiber	0

CHIEF NUTRIENTS

NUTRIENT	AMOUNT	% RDA
Vitamin B_{12}	2.8 mcg	141
Zinc	5.9 mg	40
Iron	2.1 mg	22
Niacin	4.1 mg	22
Vitamin B_6	0.3 mg	17
Riboflavin	0.2 mg	11
Potassium	334.9 mg	9
Magnesium	23.0 mg	7
Thiamine	0.1 mg	6

Strengths: Rib eye is a good source of protein, iron, zinc, and B vitamins—and a super source of vitamin B_{12}.

Eat with: Vegetables and fruits associated with a lower risk of rectal cancer. Look for vitamin C–rich foods—sweet potatoes, red peppers, cantaloupe, and strawberries—and foods high in carotenoids, such as carrots and spinach. High-fiber foods, such as beans and grains, may offer some additional protection.

Caution: Nutritionally, this cut is in the same league as porterhouse steak. (It's a close neighbor—just three cuts away on the steer.) High in minerals and B vitamins, it's also high in fat and calories, with twice as much fat as tip round or top round steak. Evidence suggests that increased consumption of meats like hamburger, steak, and cold cuts seems to boost the risk of rectal cancer, especially among men. As with porterhouse steak and other popular restaurant meats, try to limit portions to about 3 oz., roughly the size of an audiocassette.

SHANK CROSS CUTS

Serving: 3 oz., lean only, simmered

Calories	171
Fat	5.4 g
Saturated	2.0 g
Monounsaturated	2.4 g
Polyunsaturated	0.2 g
Calories from fat	29%
Cholesterol	66 mg
Sodium	54 mg
Protein	28.6 g
Carbohydrate	0
Dietary fiber	0

CHIEF NUTRIENTS

NUTRIENT	AMOUNT	% RDA
Vitamin B_{12}	3.2 mcg	161
Zinc	8.9 mg	60
Iron	3.3 mg	33
Niacin	5.0 mg	26
Vitamin B_6	0.3 mg	16
Riboflavin	0.2 mg	11
Potassium	380.0 mg	10
Thiamine	0.1 mg	8
Magnesium	25.5 mg	7

Strengths: Hard to beat as a source of vitamin B_{12}. People taking cimetidine (a peptic ulcer drug) or colestipol (a lipid-lowering drug) can use the extra B_{12}, since these drugs deplete the vitamin. Shank cross cuts are also fairly low in fat, and they're an excellent source of zinc for wound healing. They're a superb source of heme iron, the form best absorbed by the body, and they also have fair amounts of potassium and magnesium.

Eat with: Other magnesium- and potassium-rich vegetables, like onions, beans, and greens, to maximize the blood pressure–lowering potential.

Caution: Although lower in saturated fat than many other cuts of beef, shank cross cuts should still be eaten in moderation, especially if you're cutting back on saturated fat and cholesterol. Consumption of too much meat has been linked to a higher risk of colon and breast cancer.

Serve: Use as soup meat.

SHORT RIBS

Serving: 3 oz., lean only, braised

Calories	251
Fat	15.4 g
Saturated	6.6 g
Monounsaturated	6.8 g
Polyunsaturated	0.5 g
Calories from fat	55%
Cholesterol	79 mg
Sodium	49 mg
Protein	26.2 g
Carbohydrate	0
Dietary fiber	0

CHIEF NUTRIENTS

NUTRIENT	AMOUNT	% RDA
Vitamin B$_{12}$	2.9 mcg	147
Zinc	6.6 mg	44
Iron	2.9 mg	29
Niacin	2.7 mg	14
Vitamin B$_6$	0.2 mg	12
Riboflavin	0.2 mg	10
Potassium	266.1 mg	7
Magnesium	18.7 mg	5

Caution: The granddaddy of fatty meat, with more than half the calories derived from fat. People who have been advised to lose weight or lower their intake of saturated fat, or both, should eat short ribs very rarely, if at all. Diabetics need to steer clear of this high-fat source of protein, especially if the ribs are slathered with sugary barbecue sauce.

Strengths: Short ribs contain very respectable amounts of niacin, which is essential for red blood cells. In addition, they have the riboflavin required for B vitamin metabolism, and plentiful infection-fighting vitamin B$_6$. Ribs are an awesome source of vitamin B$_{12}$, with generous amounts of heme iron, the form best absorbed by the body. Rich in zinc, too, which is important for healthy skin.

A Better Idea: Marinated eye steak, top round, or chicken breast. Any one of these is lower in fat and calories, and therefore a much better choice.

SIRLOIN STEAK, *Wedge Bone*

Serving: 3 oz., lean only, broiled

Calories	166
Fat	6.1 g
Saturated	2.4 g
Monounsaturated	2.6 g
Polyunsaturated	0.2 g
Calories from fat	33%
Cholesterol	76 mg
Sodium	56 mg
Protein	25.8 g
Carbohydrate	0
Dietary fiber	0

CHIEF NUTRIENTS

NUTRIENT	AMOUNT	% RDA
Vitamin B$_{12}$	2.4 mcg	121
Zinc	5.5 mg	37
Iron	2.9 mg	29
Niacin	3.6 mg	19
Vitamin B$_6$	0.4 mg	19
Riboflavin	0.3 mg	15
Potassium	342.6 mg	9
Magnesium	27.2 mg	8
Thiamine	0.1 mg	7

Strengths: One of the better cuts of beef, this sirloin's measure of fat and calories is about the same as tip round's. It's a princely source of vitamin B$_{12}$, which helps stave off anemia and nerve problems. Appreciable amounts of niacin help build red blood cells, while its riboflavin is required for B vitamin metabolism. Infection-fighting vitamin B$_6$ is also present. Rich in infection-fighting zinc, this sirloin also has ample amounts of heme iron, the form best absorbed by the body.

Eat with: High-carbohydrate foods with fiber, like potatoes, beans, and vegetables. Also, foods high in potassium and magnesium, like beans or greens, may help your body make the most of potassium and magnesium, which help lower blood pressure. Beef will also enhance absorption of iron from vegetables such as spinach.

Caution: To keep fat and cholesterol in check, be sure to limit yourself to a 3-oz. serving.

T-BONE STEAK

Serving: 3 oz., lean only, broiled

Calories	182
Fat	8.8 g
Saturated	3.5 g
Monounsaturated	3.5 g
Polyunsaturated	0.3 g
Calories from fat	44%
Cholesterol	68 mg
Sodium	56 mg
Protein	23.9 g
Carbohydrate	0
Dietary fiber	0

CHIEF NUTRIENTS

NUTRIENT	AMOUNT	% RDA
Vitamin B$_{12}$	1.9 mcg	97
Zinc	4.6 mg	31
Iron	2.6 mg	26
Niacin	3.9 mg	21
Vitamin B$_6$	0.3 mg	17
Riboflavin	0.2 mg	12
Potassium	346.0 mg	9
Magnesium	24.7 mg	7

Strengths: On a par with other steaks, T-bone has lots of protein, minerals, and B vitamins, offset by a fair amount of fat. Its heme iron is the form best absorbed by the body. A rich source of zinc, critical for speedy wound healing and resistance, and a storehouse of vitamin B$_{12}$. The fat content is too high for people who've been told to lose weight to help control blood pressure.

Eat with: Foods that have fiber and are high in carbohydrates, like potatoes and beans, to offset the fat content.

Caution: Consumption of too much meat and other sources of saturated fat has been linked to a higher risk of heart disease and certain kinds of cancer. If you are watching your weight, trying to control your blood sugar, or keeping a watchful eye on your cholesterol, consider leaner cuts, such as top round.

TIP ROUND STEAK

Serving: 3 oz., lean only, roasted

Calories	157
Fat	5.9 g
Saturated	2.1 g
Monounsaturated	2.3 g
Polyunsaturated	0.2 g
Calories from fat	34%
Cholesterol	69 mg
Sodium	55 mg
Protein	24.4 g
Carbohydrate	0
Dietary fiber	0

CHIEF NUTRIENTS

NUTRIENT	AMOUNT	% RDA
Vitamin B$_{12}$	2.5 mcg	123
Zinc	6.0 mg	40
Iron	2.5 mg	25
Niacin	3.2 mg	17
Vitamin B$_6$	0.3 mg	17
Riboflavin	0.2 mg	14
Potassium	328.1 mg	9
Magnesium	23.0 mg	7

Strengths: With only 34% of the calories from fat, this is one of the best cuts of beef for people watching their intake of fat, calories, and cholesterol—provided you limit yourself to a 3-oz. portion, which is about the size of a deck of cards. What you get is well over a third of a day's supply of protein, superb amounts of vitamin B$_{12}$, and decent amounts of other B vitamins. Tip round is also a very good source of zinc, which is needed during pregnancy, and quite a decent source of heme iron, the form best absorbed by the body.

Eat with: Foods rich in potassium and magnesium, like beans or greens, to maximize the potential blood pressure–lowering effect of those minerals. Will also enhance nonheme forms of iron in broccoli, grains, or beans.

Caution: Consumption of too much meat and other sources of saturated fat has been linked to a higher risk of heart disease and certain kinds of cancer.

TOP LOIN STEAK
☀☀ ⚒

Serving: 3 oz., lean only, broiled

Calories	176
Fat	8.0 g
Saturated	3.1 g
Monounsaturated	3.2 g
Polyunsaturated	0.3 g
Calories from fat	41%
Cholesterol	65 mg
Sodium	58 mg
Protein	24.3 g
Carbohydrate	0
Dietary fiber	0

CHIEF NUTRIENTS

NUTRIENT	AMOUNT	% RDA
Vitamin B_{12}	1.7 mcg	85
Zinc	4.4 mg	30
Niacin	4.5 mg	24
Iron	2.1 mg	21
Vitamin B_6	0.4 mg	18
Riboflavin	0.2 mg	10
Potassium	336.6 mg	9
Magnesium	23.0 mg	7

Strengths: A very good supply of zinc, which promotes proper blood clotting, helps white blood cells fight off infection-causing invaders, and speeds wound healing. Top loin steak also contributes appreciable amounts of vitamin B_6 and niacin, with an awesome amount of vitamin B_{12}.

Eat with: High-carbohydrate foods with fiber, like potatoes, green beans, and vegetables, to offset the potentially harmful effects of saturated fat.

Caution: Overconsumption of meat and other animal foods has been linked to higher risk of breast and colorectal cancer. Special caution is advised in eating fatty meat if you have a family history of cancer. To avoid saturated fat and lower your cholesterol, try to limit yourself to 3 oz. when you do eat steak; otherwise, you might consume even more fat and calories than what's shown here.

Preparation: Broil rather than pan-fry to allow the fat to drip away from the steak.

TOP ROUND STEAK
☀☀ ⚒ ⬇ 🏛

Serving: 3 oz., lean only, broiled

Calories	153
Fat	4.2 g
Saturated	1.4 g
Monounsaturated	1.6 g
Polyunsaturated	0.2 g
Calories from fat	25%
Cholesterol	71 mg
Sodium	52 mg
Protein	26.9 g
Carbohydrate	0
Dietary fiber	0

CHIEF NUTRIENTS

NUTRIENT	AMOUNT	% RDA
Vitamin B_{12}	2.1 mcg	106
Zinc	4.7 mg	32
Niacin	5.1 mg	27
Iron	2.5 mg	25
Vitamin B_6	0.5 mg	24
Riboflavin	0.2 mg	14
Potassium	375.7 mg	10
Magnesium	26.4 mg	8

Strengths: One of the leanest cuts of beef you can buy, top round is leaner than most cuts of veal, with less saturated fat than fried chicken. With less than *one-third* as much fat as extra-lean ground beef, this cut is a sensible alternative to grilled hamburgers. Supplying almost half the daily quota for protein in a 3-oz. serving, top round gets high marks for hard-to-get B vitamins and minerals. It also supplies a bit of folate.

Eat with: Since meat does not have a bit of fiber, corn on the cob, salad, and baked potatoes nicely complement a small entrée of top round.

Caution: To keep fat and calories at a sane level, each serving should be quite small.

TRIPE, *Pickled*   ⬇ 🏛

Serving: 3 oz.

Calories	53
Fat	1.1 g
Saturated	0.4 g
Monounsaturated	N/A
Polyunsaturated	N/A
Calories from fat	19%
Cholesterol	58 mg
Sodium	39 mg
Protein	10.0 g
Carbohydrate	0
Dietary fiber	0

CHIEF NUTRIENTS

NUTRIENT	AMOUNT	% RDA
Iron	1.4 mg	14
Calcium	108.0 mg	13
Niacin	1.4 mg	7
Riboflavin	0.1 mg	7

Caution: *Drug interaction:* People taking MAO inhibitors should avoid pickled, dried, or aged meats because they contain a substance called tyramine. The interaction could raise blood pressure considerably. *Allergy alert:* Pickled meats of all kinds may cause reactions in people who are allergic to molds.

Strengths: Moderately low in calories, pickled tripe is a source of some protein to feed your muscles. It also supplies good amounts of calcium for sturdy bones and iron for healthy blood.

Description: Meat from the lining of beef stomach.

GAME

BEAR   ⚡ 🐟

Serving: 3 oz., simmered

Calories	220
Fat	11.4 g
Saturated	N/A
Monounsaturated	N/A
Polyunsaturated	N/A
Calories from fat	47%
Cholesterol	N/A
Sodium	N/A
Protein	27.6 g
Carbohydrate	0
Dietary fiber	0

CHIEF NUTRIENTS

NUTRIENT	AMOUNT	% RDA
Iron	9.1 mg	91
Riboflavin	0.7 mg	41
Thiamine	0.1 mg	6

Strengths: Bear meat is a very good source of protein and an excellent source of heme iron, the form most readily absorbed by the body. It's also rich in riboflavin, which aids in the formation of red blood cells.

Caution: Cook thoroughly to avoid trichinosis, a parasitic infection. Since bear is moderate in fat and high in calories, keep portions modest if you're dieting or following a low-fat diet.

Curiosity: Bear's paw, wrapped in clean mud and baked in the oven, is a delicacy in north China.

Description: Looks like beef and has the consistency of pork. Opinions differ about the flavor.

Preparation: Remove all fat. Older game is best if marinated and braised, or used in stew. Bear meat may also be roasted like pork loin. Just be sure, if you're trying it out, that the meat is very well done, with no pink.

BISON

Serving: 3 oz., roasted

Calories	122
Fat	2.1 g
Saturated	0.8 g
Monounsaturated	0.8 g
Polyunsaturated	0.2 g
Calories from fat	15%
Cholesterol	70 mg
Sodium	49 mg
Protein	24.2 g
Carbohydrate	0
Dietary fiber	0

CHIEF NUTRIENTS

NUTRIENT	AMOUNT	% RDA
Iron	2.9 mg	29
Zinc	2.3 mg	21
Potassium	306.9 mg	8
Magnesium	22.1 mg	6

Strengths: Bison is a very good source of protein, needed to maintain muscle mass, and a good source of heme iron, the form of this mineral most readily absorbed by the body. Significant amounts of zinc help fight infections and speed wound healing. Bison is actually preferable to beef—lower in fat, cholesterol, and calories.

Description: Now raised on ranches like cattle, bison meat is surprisingly tender, with a taste like lean beef.

Selection: Buffalo and bison are the same—the meat is sometimes found in restaurants and specialty markets.

Preparation: Like beef.

BOAR, *Wild*

Serving: 3 oz., roasted

Calories	136
Fat	3.7 g
Saturated	1.1 g
Monounsaturated	1.5 g
Polyunsaturated	0.5 g
Calories from fat	25%
Cholesterol	N/A
Sodium	N/A
Protein	24.1 g
Carbohydrate	0
Dietary fiber	0

CHIEF NUTRIENTS

NUTRIENT	AMOUNT	% RDA
Niacin	3.6 mg	19
Thiamine	0.3 mg	17
Riboflavin	0.1 mg	7

Strengths: A very good source of protein, wild boar is low in fat and moderately low in calories. This unusual meat also supplies decent amounts of niacin and thiamine, with a little riboflavin.

Caution: Cook thoroughly to avoid trichinosis, a parasitic infection.

Curiosity: Boars may live as long as 30 years.

Description: Boar is dark, dry, and tough, with a distinctly wild taste. The older the boar, the tougher the meat—although marinating helps. The most palatable meat comes from animals that are less than a year old. Edible cuts include the leg, saddle, loin, shoulder, and back. Also called wild pig.

Preparation: Best if marinated and roasted.

CARIBOU

Serving: 3 oz., roasted

Calories	142
Fat	3.8 g
Saturated	1.5 g
Monounsaturated	1.1 g
Polyunsaturated	0.5 g
Calories from fat	24%
Cholesterol	93 mg
Sodium	51 mg
Protein	25.3 g
Carbohydrate	0
Dietary fiber	0

CHIEF NUTRIENTS

NUTRIENT	AMOUNT	% RDA
Vitamin B$_{12}$	5.6 mcg	282
Iron	5.2 mg	52
Riboflavin	0.8 mg	45
Zinc	4.5 mg	30
Niacin	4.9 mg	26
Thiamine	0.2 mg	14
Vitamin B$_6$	0.3 mg	14
Magnesium	23.0 mg	7
Potassium	263.5 mg	7

Strengths: If you happen to find caribou meat, you're onto a great source of protein, iron, zinc, and B vitamins. This game contains some potassium and magnesium; it's low in fat and moderately low in calories.

Eat with: Beans, grains, vegetables, and other plant foods. The heme iron in caribou will help enhance absorption of nonheme iron from these foods.

Description: Wild caribou are the same as domesticated reindeer. The meat is finer-grained than that of moose.

Preparation: Best if marinated and braised, cooked like pot roast, or used in stew.

Serve: With strong-flavored vegetables such as red cabbage, turnips, or beets.

DEER *(Venison)*

Serving: 3 oz., roasted

Calories	134
Fat	2.7 g
Saturated	1.1 g
Monounsaturated	0.8 g
Polyunsaturated	0.5 g
Calories from fat	18%
Cholesterol	95 mg
Sodium	46 mg
Protein	25.7 g
Carbohydrate	0
Dietary fiber	0

CHIEF NUTRIENTS

NUTRIENT	AMOUNT	% RDA
Iron	3.8 mg	38
Niacin	5.7 mg	30
Riboflavin	0.5 mg	30
Zinc	2.3 mg	16
Thiamine	0.2 mg	10
Potassium	284.8 mg	8
Magnesium	20.4 mg	6

Strengths: Lower in fat and calories than skinless chicken breast, venison is a good choice in game meat for those watching their weight or restricting intake of saturated fat. And it's a very good source of the protein needed to maintain muscle mass while dieting. Rich in heme iron, the form most readily absorbed by the body. Venison has generous amounts of B vitamins and plentiful immunity-boosting zinc, with a little potassium and magnesium.

Caution: Wild deer must be field-dressed and chilled promptly to prevent spoilage. Because of the danger of spoiled meat, it's illegal as well as inadvisable to purchase freshly killed deer from a private hunter.

Description: Flavor and tenderness vary, depending on the animal's diet and age, the time of year it was killed, and the way the meat was handled after the kill.

Preparation: Marinate and cook like beef. Broil steaks and chops, roast the legs, and pot roast less tender cuts.

ELK

Serving: 3 oz., roasted

Calories	124
Fat	1.6 g
Saturated	0.6 g
Monounsaturated	0.4 g
Polyunsaturated	0.3 g
Calories from fat	12%
Cholesterol	62 mg
Sodium	52 mg
Protein	25.7 g
Carbohydrate	0
Dietary fiber	0

CHIEF NUTRIENTS

NUTRIENT	AMOUNT	% RDA
Iron	3.1 mg	31
Zinc	2.7 mg	18
Potassium	278.8 mg	7
Magnesium	20.4 mg	6

Strengths: Lower in fat and calories than skinless chicken breast, elk is a perfectly acceptable—if somewhat exotic—meat to try, even for those watching their weight or restricting saturated fat intake. A very good source of protein, the game meat has a rich supply of heme iron, the form most readily absorbed by the body. It contains a significant amount of zinc, which boosts resistance to infection.

Eat with: Beans, whole grains, and other foods high in non-heme iron, to boost absorption of heme iron.

Caution: The game must be field-dressed, hung, and chilled quickly to prevent spoilage.

Description: Avid hunters use words like ''superb'' to describe elk meat, but like most wild game, elk is an acquired taste.

Preparation: Best if marinated and braised, pot-roasted, or used in stew. Can also be broiled.

MOOSE

Serving: 3 oz., roasted

Calories	114
Fat	0.8 g
Saturated	trace
Monounsaturated	trace
Polyunsaturated	trace
Calories from fat	7%
Cholesterol	66 mg
Sodium	59 mg
Protein	24.9 g
Carbohydrate	0
Dietary fiber	0

CHIEF NUTRIENTS

NUTRIENT	AMOUNT	% RDA
Iron	3.6 mg	36
Niacin	4.5 mg	24
Zinc	3.1 mg	21
Riboflavin	0.3 mg	17
Potassium	283.9 mg	8
Vitamin C	4.3 mg	7
Magnesium	20.4 mg	6

Strengths: An excellent source of protein with practically no fat, one serving of moose meat can be an excellent substitute for beef—though availability is another matter. Its supply of B vitamins is a benefit, since deficits of B vitamins may trigger skin and nerve problems. Moose also contains a little vitamin C. It's a very good source of heme iron, the form most readily absorbed by the body.

Caution: Freshly killed moose must be dressed, hung, and chilled quickly to prevent spoilage.

Description: A member of the deer family, moose is better-tasting than beef—at least to some palates. Flavor and tenderness depend on the animal's diet and age, the time of year it was killed, and how it was dressed and transported following the kill.

Preparation: Best if marinated and braised, roasted, or used in stew like deer meat.

PHEASANT

Serving: 3 oz., meat only, uncooked

Calories	113
Fat	3.1 g
Saturated	1.1 g
Monounsaturated	1.0 g
Polyunsaturated	0.5 g
Calories from fat	25%
Cholesterol	56 mg
Sodium	32 mg
Protein	20.0 g
Carbohydrate	0
Dietary fiber	0

CHIEF NUTRIENTS

NUTRIENT	AMOUNT	% RDA
Vitamin B_{12}	0.7 mcg	36
Vitamin B_6	0.6 mg	31
Niacin	5.8 mg	30
Iron	1.0 mg	10
Vitamin C	5.1 mg	9
Riboflavin	0.1 mg	7
Potassium	222.7 mg	6

Strengths: A phenomenally good source of protein, just 3 oz. of pheasant supplies approximately one-third of a day's quota. The meat is also rich in niacin and other B vitamins, including B_6 and B_{12}, which somewhat affect learning and memory. This game bird supplies a nice amount of heme iron, the form most readily absorbed by the body, and some vitamin C, not usually found in animal foods. It also contains a bit of potassium.

Caution: Moderately high in calories, so keep portions modest and remove skin.

Curiosity: Derived from the Greek word *phasianos,* a bird of the Phasis River in the Caucasus.

Selection: May be purchased frozen.

Preparation: Usually roasted like chicken. Baste while cooking to keep the meat moist.

Serve: With other ''wild'' foods, such as watercress, chestnuts, and wild rice.

QUAIL

Serving: 1, meat only, uncooked

Calories	123
Fat	4.2 g
Saturated	1.2 g
Monounsaturated	1.2 g
Polyunsaturated	1.0 g
Calories from fat	30%
Cholesterol	64 mg
Sodium	47 mg
Protein	20.0 g
Carbohydrate	0
Dietary fiber	0

CHIEF NUTRIENTS

NUTRIENT	AMOUNT	% RDA
Iron	4.2 mg	42
Niacin	7.5 mg	40
Vitamin B_6	0.5 mg	25
Vitamin B_{12}	0.4 mcg	22
Thiamine	0.3 mg	17
Zinc	2.5 mg	17
Riboflavin	0.3 mg	15
Vitamin C	6.6 mg	11
Magnesium	23.0 mg	7

Strengths: This small game bird is a very good source of heme iron and has respectable amounts of many B vitamins, especially niacin and vitamin B_6. A fine source of protein, with only moderate amounts of fat. It's a good choice for people limiting their intake of saturated fat. Quail also supplies nice amounts of valuable minerals, including zinc, magnesium, and a good amount of vitamin C—which is not usually found in animal foods.

Curiosity: The true quail is a migratory European game bird. In America the quail is an unrelated bird, sometimes called a bobwhite or partridge.

Description: The meat is white and delicately flavored.

Selection: Quail is available fresh or frozen from specialty butchers.

Preparation: Roast or broil young quail; cook older birds with slow, moist heat; braise or use in stew.

Serve: With wild rice.

RABBIT, *Domestic*

Serving: 3 oz., roasted

Calories	131
Fat	5.4 g
Saturated	1.6 g
Monounsaturated	1.5 g
Polyunsaturated	1.0 g
Calories from fat	37%
Cholesterol	54 mg
Sodium	32 mg
Protein	19.4 g
Carbohydrate	0
Dietary fiber	0

CHIEF NUTRIENTS

NUTRIENT	AMOUNT	% RDA
Vitamin B_{12}	5.5 mcg	277
Niacin	5.6 mg	30
Vitamin B_6	0.3 mg	16
Iron	1.5 mg	15
Zinc	1.5 mg	10
Riboflavin	0.1 mg	8

Strengths: Rabbit meat is extraordinarily rich in vitamin B_{12}, which is needed to prevent anemia and nerve problems. It also supplies respectable amounts of other B vitamins, as well as heme iron, which is the form most readily absorbed by the body.

Caution: Contains a moderate amount of fat, so keep portions modest.

Curiosity: The Aztecs worshipped a rabbit god.

Description: Virtually all white meat. Domestic, ranch-raised rabbit is plumper and less strongly flavored than wild rabbit.

Selection: Young rabbit is more tender than mature rabbit. You can select rabbit meat fresh or frozen, whole or cut up.

Preparation: Young rabbit may be roasted, broiled, or grilled like chicken. Older (or wild) rabbit must be cooked with slow, moist heat. It's often braised or used as stew meat.

SQUAB *(Pigeon)*

Serving: 1, meat only, uncooked

Calories	239
Fat	12.6 g
Saturated	3.3 g
Monounsaturated	4.5 g
Polyunsaturated	2.3 g
Calories from fat	48%
Cholesterol	151 mg
Sodium	86 mg
Protein	29.0 g
Carbohydrate	0
Dietary fiber	0

CHIEF NUTRIENTS

NUTRIENT	AMOUNT	% RDA
Iron	7.6 mg	76
Niacin	11.5 mg	61
Vitamin B_6	0.9 mg	45
Vitamin B_{12}	0.8 mcg	40
Thiamine	0.5 mg	32
Zinc	4.5 mg	30
Riboflavin	0.5 mg	28
Vitamin C	12.0 mg	20
Magnesium	42.0 mg	12
Potassium	398.2 mg	11

Strengths: An excellent source of protein and niacin, squab is also rich in other B vitamins and is loaded with iron. It has good portions of magnesium and potassium. A high amount of zinc can speed wound healing and fight infections. Surprisingly, squab also has generous amounts of vitamin C, not usually found in animal foods. Almost half the fat is monounsaturated, making squab less harmful to your arteries than meat having more highly saturated forms of fat.

Caution: The percentage of calories from fat is far above the 30% recommended by the American Heart Association. Squab is high in calories, so only eat a modest portion if you're watching your weight.

Curiosity: A dietary mainstay during the Depression.

Description: Squab sold at the poultry counter is a domesticated, farm-raised pigeon, and the meat is extremely tender, dark, and delicately flavored.

Selection: Available frozen. Also served in restaurants.

Preparation: Cook like chicken.

LAMB

ARM ROAST

Serving: 3 oz., lean only, roasted

Calories	163
Fat	7.9 g
Saturated	3.1 g
Monounsaturated	3.2 g
Polyunsaturated	0.7 g
Calories from fat	43%
Cholesterol	73 mg
Sodium	57 mg
Protein	21.6 g
Carbohydrate	0
Dietary fiber	0

CHIEF NUTRIENTS

NUTRIENT	AMOUNT	% RDA
Vitamin B$_{12}$	2.2 mcg	111
Zinc	4.5 mg	30
Niacin	5.4 mg	28
Iron	1.9 mg	19
Riboflavin	0.2 mg	14
Folate	21.3 mcg	11

Strengths: With a little less fat than porterhouse steak, lean lamb arm has roughly the same nutritional profile. Loads of vitamin B$_{12}$, which helps ward off anemia and nerve problems; an abundance of zinc. This mineral promotes proper blood clotting, helps white blood cells fight off infection-causing invaders, and is necessary for healing wounds. Like beef, lamb is a good source of heme iron, so it's a good choice for women and others who run a higher-than-average risk of iron deficiency.

Eat with: Potatoes, grains, and vegetables: In lamb curry, stew, and other dishes they add fiber to the meal.

Caution: Moderately high in fat, lamb is not the best choice for people trying to limit artery-clogging saturated fat and cholesterol. Consumption of high amounts of animal fat has been linked to increased risk of certain kinds of cancer. Contains purines—should be eaten sparingly by people with gout.

BLADE ROAST

Serving: 3 oz., lean only, roasted

Calories	178
Fat	9.8 g
Saturated	3.7 g
Monounsaturated	4.0 g
Polyunsaturated	0.9 g
Calories from fat	50%
Cholesterol	74 mg
Sodium	58 mg
Protein	20.9 g
Carbohydrate	0
Dietary fiber	0

CHIEF NUTRIENTS

NUTRIENT	AMOUNT	% RDA
Vitamin B$_{12}$	2.3 mcg	117
Zinc	5.5 mg	37
Niacin	4.7 mg	25
Iron	1.8 mg	18
Riboflavin	0.2 mg	12
Folate	21.3 mcg	11
Vitamin B$_6$	0.1 mg	7
Magnesium	21.3 mg	6
Potassium	219.3 mg	6
Thiamine	0.1 mg	5

Strengths: A terrific source of vitamin B$_{12}$ to help prevent anemia and nerve problems, lamb is also a very good source of zinc. This mineral promotes proper blood clotting, helps white blood cells fight off infectious invaders, and is essential for wound healing. Lamb is a good source of heme iron, the form best absorbed by the body, so it's a good choice for women and others who run a higher-than-average risk of iron deficiency. It's a useful alternative to beef, pork, and chicken for people with allergies to those meats.

Eat with: High-fiber, high-carbohydrate foods like sweet potatoes and beans, and vegetables such as broccoli and fresh red cabbage.

Caution: Moderately high in fat, lamb is not the best choice for people trying to limit artery-clogging saturated fat and cholesterol. Consumption of high amounts of animal fat has been linked to increased risk of certain kinds of cancer. Diabetics, who need to eat less fat and more complex carbohydrates, should choose leaner cuts of meat. Contains purines—should be eaten sparingly by people with gout.

FORESHANK

Serving: 3 oz., lean only, braised

Calories	159
Fat	5.1 g
Saturated	1.8 g
Monounsaturated	2.2 g
Polyunsaturated	0.3 g
Calories from fat	29%
Cholesterol	88 mg
Sodium	63 mg
Protein	26.4 g
Carbohydrate	0
Dietary fiber	0

CHIEF NUTRIENTS

NUTRIENT	AMOUNT	% RDA
Vitamin B$_{12}$	1.9 mcg	96
Zinc	7.4 mg	49
Niacin	14.3 mg	23
Iron	1.9 mg	19
Riboflavin	0.2 mg	9
Folate	16.2 mcg	8
Magnesium	19.6 mg	6
Potassium	227.0 mg	6

Strengths: Lower in fat than other cuts of lamb, foreshank is a great source of protein, and it's a fine alternative choice for people who have beef allergies. People watching their fat and calorie intake, or those at risk of iron-deficiency anemia, may also choose this cut of meat. Rich in infection-fighting zinc, lamb also has some potassium and magnesium.

Caution: Lamb foreshank is a better choice than ground lamb or chops, but people trying to restrict their intake of artery-clogging saturated fat and cholesterol will want to eat modest portions (about 3 oz.). Contains purines—should be eaten sparingly by people with gout.

Selection: Nutritionally, New Zealand lamb is about equal to domestic lamb, with slightly more B vitamins but less zinc. Fat content is virtually the same, but the New Zealand cut is lower in sodium.

GROUND LAMB

Serving: 3 oz., broiled

Calories	241
Fat	16.7 g
Saturated	6.9 g
Monounsaturated	7.1 g
Polyunsaturated	1.2 g
Calories from fat	62%
Cholesterol	82 mg
Sodium	69 mg
Protein	21.0 g
Carbohydrate	0
Dietary fiber	0

CHIEF NUTRIENTS

NUTRIENT	AMOUNT	% RDA
Vitamin B$_{12}$	2.2 mcg	111
Niacin	5.7 mg	30
Zinc	4.0 mg	26
Iron	1.5 mg	15
Riboflavin	0.2 mg	12
Folate	16.2 mcg	8
Potassium	288.2 mg	8
Magnesium	20.4 mg	6
Thiamine	0.1 mg	6

Caution: One of the fattiest meats going, ground lamb is loaded with artery-clogging saturated fat and cholesterol. Consumption of high amounts of animal fat has been linked to increased risk of certain kinds of cancer. Since it contains purines, lamb should be eaten sparingly by people with gout.

Strengths: A fabulous source of vitamin B$_{12}$, which helps prevent anemia and nerve problems, ground lamb is also rich in niacin, with nice amounts of other essential B vitamins. It provides a good amount of heme iron, plus a respectable amount of infection-fighting zinc. The potential blood pressure–lowering benefits of potassium and magnesium may be offset by the fat content; people who have been told to lose weight to help control blood pressure should think thrice before eating ground lamb.

A Better Idea: Buying leg of lamb or lamb shank and grinding it yourself gives you leaner patties.

LEG

Serving: 3 oz., lean only, roasted

Calories	162
Fat	6.6 g
Saturated	2.4 g
Monounsaturated	2.9 g
Polyunsaturated	0.4 g
Calories from fat	36%
Cholesterol	76 mg
Sodium	58 mg
Protein	24.1 g
Carbohydrate	0
Dietary fiber	0

CHIEF NUTRIENTS

NUTRIENT	AMOUNT	% RDA
Vitamin B_{12}	2.2 mcg	112
Niacin	5.4 mg	28
Zinc	4.2 mg	28
Iron	1.8 mg	18
Riboflavin	0.3 mg	15
Folate	19.6 mcg	10
Potassium	287.3 mg	8
Vitamin B_6	0.1 mg	7
Magnesium	22.1 mg	6

Strengths: Leg of lamb is a useful alternative to beef, pork, and chicken for people with allergies to those meats. It's a good source of heme iron, the form best absorbed by the body—therefore, a good choice for women and others who run a higher-than-average risk of iron deficiency. Its zinc promotes proper blood clotting, helps white blood cells fight off infection-causing invaders, and helps wound healing. An excellent source of vitamin B_{12}, lamb also has fair amounts of other B vitamins, including the niacin essential for red blood cells.

Eat with: Balancing lamb with high-fiber complex carbohydrates slims the meal and boosts the nutritional benefits.

Caution: This lamb has moderate levels of fat, so keep portions modest (3 oz. or less). Contains purines—should be eaten sparingly by people with gout.

Selection: New Zealand and domestic lamb are about equal nutritionally. For those on low-sodium diets, the New Zealand cut is preferable.

Preparation: Cube and stir-fry with fresh vegetables.

Serve: With baked sweet potatoes.

LIVER

Serving: 3 oz., braised

Calories	187
Fat	7.5 g
Saturated	2.9 g
Monounsaturated	1.6 g
Polyunsaturated	1.1 g
Calories from fat	36%
Cholesterol	426 mg
Sodium	48 mg
Protein	26.0 g
Carbohydrate	2.2 g
Dietary fiber	0

CHIEF NUTRIENTS

NUTRIENT	AMOUNT	% RDA
Vitamin B_{12}	65.0 mcg	3,252
Vitamin A	6,366.5 RE	637
Riboflavin	3.4 mg	202
Iron	7.0 mg	70
Niacin	10.3 mg	54
Zinc	6.7 mg	45
Folate	62.1 mcg	31

Strengths: Lamb liver contributes super-awesome amounts of immunity-boosting vitamins like A and an even more impressive amount of vitamin B_{12}. It's an excellent source of heme iron, the form best absorbed by the body, so it's a valuable meal for those who run a higher-than-average risk of iron deficiency.

Caution: As with beef and chicken liver, the nutritional attributes come at a price. Lamb liver has moderate amounts of fat, and its artery-clogging cholesterol far exceeds the upper limit of 300 mg per day recommended by the American Heart Association. As the depot for chemicals, medicines, and hormones consumed by the animal, the liver accumulates potentially harmful substances. Contains purines—should be eaten sparingly by people with gout.

LOIN ROAST

Serving: 3 oz., lean only, roasted

Calories	172
Fat	8.3 g
Saturated	3.2 g
Monounsaturated	3.4 g
Polyunsaturated	0.7 g
Calories from fat	44%
Cholesterol	74 mg
Sodium	56 mg
Protein	22.6 g
Carbohydrate	0
Dietary fiber	0

CHIEF NUTRIENTS

NUTRIENT	AMOUNT	% RDA
Vitamin B_{12}	1.8 mcg	92
Niacin	5.8 mg	31
Zinc	3.5 mg	23
Iron	2.1 mg	21
Riboflavin	0.2 mg	14
Folate	21.3 mcg	11
Magnesium	23.0 mg	7
Potassium	227.0 mg	6

Strengths: Lamb loin roast or chops will supply almost an entire day's supply of vitamin B_{12}. And loin also gets high marks for niacin. Appreciable amounts of iron and zinc make this a fairly decent choice for people who risk deficiencies of those hard-to-get minerals. Potassium and magnesium have the potential to lower blood pressure, but it comes at a price: The moderate fat content is *not* a benefit for people who've been told to lose weight to help control blood pressure.

Eat with: Broccoli, cauliflower, or other vitamin C–rich foods to boost the iron absorption. Balancing lamb loin with high-fiber complex carbohydrates like sweet potatoes or black-eyed peas boosts the nutritional benefits of the meal.

Caution: Lamb loin is not the best choice for people trying to restrict their intake of artery-clogging saturated fat and cholesterol. Advice: Eat very limited portions. Contains purines—should be eaten sparingly by people with gout.

RIB ROAST

Serving: 3 oz., lean only, roasted

Calories	197
Fat	11.3 g
Saturated	4.1 g
Monounsaturated	5.0 g
Polyunsaturated	0.8 g
Calories from fat	52%
Cholesterol	75 mg
Sodium	69 mg
Protein	22.2 g
Carbohydrate	0
Dietary fiber	0

CHIEF NUTRIENTS

NUTRIENT	AMOUNT	% RDA
Vitamin B_{12}	1.8 mcg	92
Niacin	5.2 mg	28
Zinc	3.8 mg	25
Iron	1.5 mg	15
Riboflavin	0.2 mg	12
Folate	18.7 mcg	9
Potassium	267.8 mg	7
Vitamin B_6	0.1 mg	7
Magnesium	19.6 mg	6

Strengths: One serving of a rib roast of lamb—also called a crown roast or rack of lamb—boasts nearly an entire day's supply of vitamin B_{12}, along with plentiful zinc and niacin. Respectable amounts of other B vitamins, as well as iron, are also present. Potassium and magnesium may help lower blood pressure, but the fat content is too high, especially for people who've been told to lose weight to help control blood pressure.

Eat with: High-fiber, high-carbohydrate foods like sweet potatoes, brussels sprouts, or fruit compote.

Caution: One of the fattier cuts of meat sold. People trying to restrict their intake of artery-clogging saturated fat and cholesterol should limit their consumption of lamb rib roast and keep portions modest. Contains purines—should be eaten sparingly by people with gout. Also, consumption of high amounts of animal fat has been linked to increased risk of certain kinds of cancer.

SHANK

Serving: 3 oz., lean only, roasted

Calories	153
Fat	5.7 g
Saturated	2.0 g
Monounsaturated	2.5 g
Polyunsaturated	0.4 g
Calories from fat	33%
Cholesterol	74 mg
Sodium	56 mg
Protein	23.9 g
Carbohydrate	0
Dietary fiber	0

CHIEF NUTRIENTS

NUTRIENT	AMOUNT	% RDA
Vitamin B_{12}	2.3 mcg	115
Niacin	5.4 mg	29
Zinc	4.3 mg	28
Iron	1.8 mg	18
Riboflavin	0.2 mg	14
Folate	20.4 mcg	10
Potassium	290.7 mg	8
Vitamin B_6	0.1 mg	7
Magnesium	22.1 mg	6

Strengths: On a par with other cuts of lamb, lean shank is hard to beat as a source of vitamin B_{12}, which can help ward off anemia and nerve problems. High in resistance-boosting zinc and iron, lamb is also a good source of heme iron. And it's a good choice for women and others who run a higher-than-average risk of iron deficiency.

Eat with: High-fiber, high-carbohydrate foods like sweet potatoes, beans, green peas, and corn on the cob.

Caution: Consumption of high amounts of animal fat has been linked to increased risk of certain kinds of cancer, so keep to the modest 3-oz. portion—about the size of a deck of cards. Contains purines—should be eaten sparingly by people with gout.

SHOULDER

Serving: 3 oz., lean only, roasted

Calories	173
Fat	9.2 g
Saturated	3.5 g
Monounsaturated	3.7 g
Polyunsaturated	0.8 g
Calories from fat	47%
Cholesterol	74 mg
Sodium	58 mg
Protein	21.2 g
Carbohydrate	0
Dietary fiber	0

CHIEF NUTRIENTS

NUTRIENT	AMOUNT	% RDA
Vitamin B_{12}	2.3 mcg	115
Zinc	5.1 mg	34
Niacin	4.9 mg	26
Iron	1.8 mg	18
Riboflavin	0.2 mg	13
Folate	21.3 mcg	11
Vitamin B_6	0.1 mg	7
Magnesium	21.3 mg	6
Potassium	225.3 mg	6

Strengths: Comparable to other moderately high-fat cuts of lamb, lamb shoulder is an excellent source of vitamin B_{12} and provides nice amounts of other B vitamins as well. As a rich source of immunity-boosting zinc, it can help out anyone plagued with colds or other infections. There are potential blood pressure–lowering benefits of potassium and magnesium, but these pluses may be offset by the fat content, especially if you're trying to lose weight to help control blood pressure. People who are allergic to beef, pork, or chicken can rely on lamb as a source of protein.

Eat with: High-fiber, high-carbohydrate foods like sweet potatoes, beans, and vegetables.

Caution: Not the best choice for people trying to restrict their intake of artery-clogging saturated fat and cholesterol. Also, consumption of high amounts of animal fat has been linked to increased risk of certain kinds of cancer. Keep portions modest. Contains purines—should be eaten sparingly by people with gout.

Serve: In stew, chili, or cassoulet.

SIRLOIN

Serving: 3 oz., lean only, roasted

Calories	173
Fat	7.8 g
Saturated	2.8 g
Monounsaturated	3.4 g
Polyunsaturated	0.5 g
Calories from fat	40%
Cholesterol	78 mg
Sodium	60 mg
Protein	24.1 g
Carbohydrate	0
Dietary fiber	0

CHIEF NUTRIENTS

NUTRIENT	AMOUNT	% RDA
Vitamin B$_{12}$	2.2 mcg	110
Niacin	5.3 mg	28
Zinc	4.1 mg	27
Iron	1.9 mg	19
Riboflavin	0.3 mg	15
Folate	17.9 mcg	9
Potassium	283.1 mg	8
Thiamine	0.1 mg	7
Magnesium	21.3 mg	6

Strengths: Has nearly one-third of a day's supply of protein, with a significant amount of zinc and appreciable amounts of B vitamins. The vitamin B$_{12}$ exceeds 100% of the RDA, helping to prevent anemia and nerve problems. Like beef, lamb is a good source of heme iron, the form best absorbed by the body, so it's a good choice for women and others who run a higher-than-average risk of iron deficiency. Although potassium and magnesium may help lower blood pressure, the fat content is too high for people who are trying to lose weight.

Eat with: High-fiber, high-carbohydrate foods like sweet potatoes, beans, and vegetables.

Caution: Not the best choice for people trying to restrict their intake of artery-clogging saturated fat and cholesterol. Also, consumption of high amounts of animal fat has been linked to increased risk of certain kinds of cancer. Keep portions modest (3 oz. or less). Contains purines—should be eaten sparingly by people with gout.

STEW/KABOB MEAT

Serving: 3 oz., lean only, broiled

Calories	158
Fat	6.2 g
Saturated	2.2 g
Monounsaturated	2.5 g
Polyunsaturated	0.6 g
Calories from fat	35%
Cholesterol	77 mg
Sodium	65 mg
Protein	23.9 g
Carbohydrate	0
Dietary fiber	0

CHIEF NUTRIENTS

NUTRIENT	AMOUNT	% RDA
Vitamin B$_{12}$	2.6 mcg	129
Zinc	4.9 mg	33
Niacin	5.6 mg	30
Iron	2.0 mg	20
Riboflavin	0.3 mg	15
Folate	19.6 mcg	10
Magnesium	26.4 mg	8
Potassium	284.8 mg	8

Strengths: The beauty of lamb stew or kabobs is that you get all of the nutritional benefits—plentiful zinc, iron, and B vitamins—but you consume smaller portions (and therefore less fat and calories) because you're making the meat with lots of vegetables. Like beef, lamb is a good source of heme iron, the form best absorbed by the body, so it's a good choice for anyone who runs a higher-than-average risk of iron deficiency. Lamb in general is an exceptionally rich source of zinc, which promotes proper blood clotting, helps white blood cells fight off infection-causing invaders, and is essential for wound healing. Very lean lamb kabobs served with lots of vegetables can be a useful source of protein.

Eat with: High-carbohydrate foods like potatoes, mushrooms, peppers. Since lamb also boosts the absorption of nonheme iron from beans and whole grains, serve with rice on the side.

Caution: Contains purines—should be eaten sparingly by people with gout.

LUNCH MEAT AND SAUSAGE

BEEF, *Cured*

Serving: 7 thin slices (about 1 oz.)

Calories	50
Fat	1.1 g
Saturated	0.5 g
Monounsaturated	0.5 g
Polyunsaturated	0.1 g
Calories from fat	20%
Cholesterol	12 mg
Sodium	408 mg
Protein	8.0 g
Carbohydrate	1.6 g
Dietary fiber	0

CHIEF NUTRIENTS

NUTRIENT	AMOUNT	% RDA
Vitamin B_{12}	0.7 mcg	37
Iron	0.8 mg	8
Niacin	1.5 mg	8
Zinc	1.0 mg	8
Vitamin C	4.1 mg	7
Vitamin B_6	0.1 mg	5

Caution: Cured beef is loaded with sodium, which is certainly a potential problem for those prone to high blood pressure. Cured beef usually contains nitrites, preservatives that prevent botulism but are suspected of contributing to cancer. *Drug interaction:* Smoked foods have tyramine, an amino acid derivative that can raise blood pressure to dangerously high levels in anyone taking MAO inhibitors. *Allergy alert:* May trigger headaches or other complaints in those sensitive to nitrites.

Strengths: A decent source of protein, and it's low in fat. Cured beef also has nice amounts of easy-to-absorb heme iron, zinc, and vitamin C, and a generous amount of vitamin B_{12}.

A Better Idea: Plain sliced beef—not smoked or cured—is a better choice for those watching sodium intake.

BOLOGNA, *Beef*

Serving: 2 slices (about 2 oz.)

Calories	76
Fat	16.2 g
Saturated	6.9 g
Monounsaturated	7.8 g
Polyunsaturated	0.6 g
Calories from fat	83%
Cholesterol	32 mg
Sodium	556 mg
Protein	6.9 g
Carbohydrate	0.4 g
Dietary fiber	0

CHIEF NUTRIENTS

NUTRIENT	AMOUNT	% RDA
Vitamin B_{12}	0.8 mcg	40
Vitamin C	12.0 mg	20
Iron	0.9 mg	9
Zinc	1.2 mg	8
Niacin	1.4 mg	7

Strengths: Beef bologna is very high in vitamin B_{12}, with appreciable amounts of vitamin C derived primarily from sodium ascorbate.

Caution: Bologna is high in fat, and it's also quite high in sodium, a potential problem if you're prone to high blood pressure. The nitrites added to prevent botulism are suspected of contributing to cancer. *Allergy alert:* May contain corn, soy, milk, wheat, or other common food allergens. May trigger headaches or other complaints in those sensitive to nitrites.

Serve: With whole wheat bread and mustard. If you're using spreads or cheese on a bologna sandwich, choose low-fat varieties to minimize fat content.

BOLOGNA, *Lebanon*

Serving: 2 slices (about 2 oz.)

Calories	120
Fat	7.5 g
Saturated	3.3 g
Monounsaturated	3.4 g
Polyunsaturated	0.3 g
Calories from fat	56%
Cholesterol	40 mg
Sodium	758 mg
Protein	11.0 g
Carbohydrate	1.5 g
Dietary fiber	0

CHIEF NUTRIENTS

NUTRIENT	AMOUNT	% RDA
Vitamin B_{12}	1.4 mcg	72
Vitamin C	12.4 mg	21
Zinc	2.3 mg	15
Iron	1.4 mg	14
Niacin	2.5 mg	13
Riboflavin	0.1 mg	7
Vitamin B_6	0.1 mg	7

Caution: Lebanon bologna contributes an extra dose of calories from fat—far above the 30% maximum recommended by the American Heart Association. It's also quite high in sodium, which is a potential problem for those prone to high blood pressure. Usually contains nitrites, preservatives that are suspected of contributing to cancer. *Drug interaction:* Contains tyramine, an amino acid derivative that can raise blood pressure to dangerously high levels in anyone taking MAO inhibitors. *Allergy alert:* May trigger headaches or other complaints in those sensitive to nitrites or smoked foods.

Strengths: A whopping amount of vitamin B_{12}. As a bonus, you also get a surprisingly decent amount of vitamin C, derived primarily from sodium ascorbate, added to some (but not all) cured meats. One of the leaner lunch meats available, making this a fair choice for those watching fat and calorie intake.

Origin: Lebanon, Pennsylvania.

BOLOGNA, *Pork*

Serving: 2 slices (about 2 oz.)

Calories	140
Fat	11.3 g
Saturated	3.9 g
Monounsaturated	5.5 g
Polyunsaturated	1.2 g
Calories from fat	72%
Cholesterol	34 mg
Sodium	671 mg
Protein	8.7 g
Carbohydrate	0.4 g
Dietary fiber	0

CHIEF NUTRIENTS

NUTRIENT	AMOUNT	% RDA
Vitamin C	20.0 mg	33
Vitamin B_{12}	0.5 mcg	26
Thiamine	0.3 mg	20
Niacin	2.2 mg	12
Vitamin B_6	0.2 mg	8
Zinc	1.2 mg	8

Caution: The percentage of calories from fat in pork bologna is way over the 30% recommended by the American Heart Association. It's also quite high in sodium, definitely a problem for those prone to high blood pressure. Usually contains nitrites as well; these preservatives prevent botulism, but they are also suspected of contributing to cancer. *Drug interaction:* Smoked foods contain tyramine, an amino acid derivative that can raise blood pressure to dangerously high levels in anyone taking MAO inhibitors. *Allergy alert:* May contain corn, soy, milk, wheat, or other common food allergens. May trigger headaches or other complaints in those sensitive to nitrites.

Strengths: The respectable amounts of various B vitamins in pork bologna are essential for healthy skin and nerves. As a bonus, you also get an impressive amount of vitamin C, derived primarily from sodium ascorbate. (This form of C is added to many cured meats.)

BOLOGNA, *Turkey*

Serving: 2 slices (about 2 oz.)

Calories	113
Fat	8.6 g
Saturated	2.9 g
Monounsaturated	2.7 g
Polyunsaturated	2.4 g
Calories from fat	69%
Cholesterol	56 mg
Sodium	498 mg
Protein	7.8 g
Carbohydrate	0.6 g
Dietary fiber	0

CHIEF NUTRIENTS

NUTRIENT	AMOUNT	% RDA
Niacin	2.0 mg	11
Iron	0.1 mg	9
Vitamin B$_{12}$	0.2 mcg	8
Zinc	1.0 mg	7
Calcium	47.6 mg	6
Vitamin B$_6$	0.1 mg	6
Riboflavin	0.1 mg	5

Caution: If you eat turkey bologna, you'll be getting more than two-thirds of your calories from fat—far above the level recommended by the American Heart Association. And the processed meat is loaded with sodium—so limit yourself to very few slices. Bologna usually contains nitrites, preservatives suspected of contributing to cancer. *Drug interaction:* Contains tyramine, an amino acid derivative that can raise blood pressure to dangerously high levels in anyone taking MAO inhibitors. *Allergy alert:* May contain corn, soy, milk, wheat, or other common food allergens. May trigger headaches or other complaints in those sensitive to nitrites.

Eat with: Whole wheat bread and mustard or other low-fat spreads to keep a lid on fat content.

Strengths: Turkey bologna has a smattering of various B vitamins and minerals, with a fair amount of protein. To its credit, it contains less fat than beef bologna. However, it's not as lean as turkey ham or smoked turkey breast.

BRATWURST, *Fresh*

Serving: 1 link (about 3 oz.)

Calories	256
Fat	22.0 g
Saturated	7.9 g
Monounsaturated	10.4 g
Polyunsaturated	2.3 g
Calories from fat	77%
Cholesterol	51 mg
Sodium	474 mg
Protein	12.0 g
Carbohydrate	1.8 g
Dietary fiber	0

CHIEF NUTRIENTS

NUTRIENT	AMOUNT	% RDA
Vitamin B$_{12}$	0.8 mcg	41
Thiamine	0.4 mg	29
Niacin	2.7 mg	14
Zinc	2.0 mg	13
Iron	1.1 mg	11
Riboflavin	0.2 mg	9
Vitamin B$_6$	0.2 mg	9
Potassium	180.0 mg	5

Caution: Bratwurst is loaded with fat and calories, is high in sodium, and has a moderate amount of cholesterol. *Drug interaction:* Watch out for *smoked* bratwurst if you happen to be taking MAO inhibitors; the smoked meat contains tyramine, an amino acid derivative that can raise blood pressure to dangerously high levels if you're on this medication. *Allergy alert:* Smoked foods may trigger headaches or other complaints in some people.

Strengths: Bratwurst has significant amounts of B vitamins, especially vitamin B$_{12}$ and thiamine. It also contains a chunk of easy-to-absorb iron, with a good amount of zinc. This meat has a fair amount of protein, although it's less than half the amount supplied by lean beef or chicken.

Description: A seasoned sausage made from chopped pork or veal.

Preparation: Fresh bratwurst must be well grilled or sautéed before eating.

BRAUNSCHWEIGER

Serving: 2 oz.

Calories	204
Fat	18.2 g
Saturated	6.2 g
Monounsaturated	8.5 g
Polyunsaturated	2.1 g
Calories from fat	80%
Cholesterol	88 mg
Sodium	648 mg
Protein	7.7 g
Carbohydrate	1.8 g
Dietary fiber	0

CHIEF NUTRIENTS

NUTRIENT	AMOUNT	% RDA
Vitamin B_{12}	11.4 mcg	570
Vitamin A	2,392.7 RE	239
Iron	5.3 mg	53
Riboflavin	0.9 mg	51
Niacin	4.7 mg	25
Folate	24.9 mcg	13
Zinc	1.6 mg	11
Thiamine	0.1 mg	9
Vitamin B_6	0.2 mg	9
Vitamin C	5.4 mg	9

Strengths: Made from liver, this sausage shares some of liver's nutritional strengths: It's a treasure trove of valuable nutrients like vitamins A and B_{12}, with a hefty amount of iron and appreciable amounts of vitamin C and zinc.

Caution: To get all those lusty nutrients, you're consuming a meat that's loaded with fat and calories and high in sodium. *Drug interaction:* Like other smoked foods, braunschweiger contains tyramine, an amino acid derivative that can raise blood pressure to dangerously high levels in anyone taking MAO inhibitors. *Allergy alert:* Contains milk and eggs, common food allergens. And smoked foods may trigger allergic reactions in some people.

Origin: Brunswick Province (Braunschweig), Germany.

Description: A kind of liverwurst that's cured and often smoked, braunschweiger is also called liver sausage.

CHICKEN ROLL, *Light Meat*

Serving: 2 slices (about 2 oz.)

Calories	90
Fat	4.2 g
Saturated	1.2 g
Monounsaturated	1.7 g
Polyunsaturated	0.9 g
Calories from fat	42%
Cholesterol	28 mg
Sodium	331 mg
Protein	11.1 g
Carbohydrate	1.4 g
Dietary fiber	0

CHIEF NUTRIENTS

NUTRIENT	AMOUNT	% RDA
Niacin	3.0 mg	16
Iron	0.6 mg	6
Vitamin B_6	0.1 mg	6

Caution: This processed meat is high in sodium. And the percentage of calories from fat of chicken roll is 42%, making it twice as fatty as chicken breast.

Strengths: Chicken roll has a good amount of niacin, with some vitamin B_6. This is an asset because deficits of B vitamins can lead to skin and nerve problems. It also has a fair amount of protein, with a little anemia-fighting iron.

Description: Contains chopped and reconstituted white or dark chicken meat (or both), plus gelatin, sugar, and other fillers and flavorings.

A Better Idea: Why not eat fresh chicken instead of chicken roll? The same amount of sliced chicken breast has a bit more protein (14 g) and half as much fat (2 g) as chicken roll. And sliced fresh chicken breast has almost 3 times as much niacin and vitamin B_6 as chicken roll, plus a fair amount of vitamin B_{12}, zinc, and other nourishing vitamins and minerals.

CHORIZO, *Dried*

Serving: 1 link (about 2 oz.)

Calories	273
Fat	23.0 g
Saturated	8.6 g
Monounsaturated	11.0 g
Polyunsaturated	2.0 g
Calories from fat	76%
Cholesterol	53 mg
Sodium	741 mg
Protein	14.5 g
Carbohydrate	1.1 g
Dietary fiber	0

CHIEF NUTRIENTS

NUTRIENT	AMOUNT	% RDA
Vitamin B_{12}	1.2 mcg	60
Thiamine	0.4 mg	25
Niacin	3.1 mg	16
Vitamin B_6	0.3 mg	16
Zinc	2.1 mg	14
Riboflavin	0.2 mg	11
Iron	1.0 mg	10
Potassium	238.8 mg	6

Caution: Chorizo is very, very high in fat and high in calories and sodium, with moderate amounts of cholesterol. So eat sparingly. *Drug interaction:* Smoked sausages like chorizo contain tyramine, an amino acid derivative that can raise blood pressure to dangerously high levels in anyone taking MAO inhibitors. *Allergy alert:* Smoked foods may trigger allergic reactions in some people.

Eat with: Low-fat skinless white meat chicken, seafood, rice, and peas (as in paella), to keep fat intake to a minimum.

Strengths: Gets high marks for B vitamins (especially B_{12} and thiamine), along with a passing grade for zinc and other valuable minerals. Chorizo has a fair amount of protein, too.

Description: A spicy Spanish sausage of beef and pork seasoned with hot pepper, garlic, salt, pepper, and other seasonings. Dried chorizo has usually been smoked in processing.

CORNED BEEF

Serving: 2 slices (about 2 oz.)

Calories	142
Fat	8.5 g
Saturated	3.5 g
Monounsaturated	3.4 g
Polyunsaturated	0.4 g
Calories from fat	54%
Cholesterol	49 mg
Sodium	570 mg
Protein	15.4 g
Carbohydrate	0
Dietary fiber	0

CHIEF NUTRIENTS

NUTRIENT	AMOUNT	% RDA
Vitamin B_{12}	0.9 mcg	46
Zinc	2.0 mg	14
Iron	1.2 mg	12
Niacin	1.4 mg	7

Strengths: For a deli meat, corned beef is one of the leaner choices. It's rich in vitamin B_{12}, with respectable portions of iron and zinc, and a bit of niacin. And it has a good amount of protein, too.

Caution: Watch out for the sodium level—it's a little steep. Corned beef also contains cholesterol. And avoid the other corned beef classic, a grilled Reuben sandwich that contains an overdose of fat from oil, Swiss cheese, and Russian dressing. *Drug interaction:* Since it's a cured meat, corned beef has tyramine, an amino acid derivative that can raise blood pressure to dangerously high levels in anyone taking MAO inhibitors. *Allergy alert:* May trigger reactions in people allergic to molds.

Description: Chuck, round, or brisket of beef cured in brine.

Serve: On rye bread, of course. (But go easy on the mustard.)

FRANKFURTER, *Beef*

Serving: 1 (about 1½ oz.)

Calories	142
Fat	12.8 g
Saturated	5.4 g
Monounsaturated	6.1 g
Polyunsaturated	0.6 g
Calories from fat	81%
Cholesterol	27 mg
Sodium	462 mg
Protein	5.4 g
Carbohydrate	0.8 g
Dietary fiber	0

CHIEF NUTRIENTS

NUTRIENT	AMOUNT	% RDA
Vitamin B_{12}	0.7 mcg	35
Vitamin C	10.8 mg	18
Zinc	1.0 mg	7
Iron	0.6 mg	6
Niacin	1.1 mg	6

Caution: Beef franks are high in sodium, and nearly all the calories come from fat. Frankfurters usually contain nitrites, preservatives that are suspected of contributing to cancer. *Drug interaction:* May contain tyramine, an amino acid derivative that can raise blood pressure to dangerously high levels in anyone taking MAO inhibitors. *Allergy alert:* May trigger headaches or other reactions in people sensitive to tyramine.

Strengths: Beef franks have a good amount of Vitamin C, and they're rich in vitamin B_{12}, which contributes to healthy skin and nerves. They also have a fair amount of iron and immunity-boosting zinc.

Description: A cooked, smoked, cured, and seasoned sausage made of beef (labeled all-beef) or beef and pork with fillers. Also called a frank, hot dog, weenie, wiener, dog, or red hot.

A Better Idea: People allergic to soy, corn, milk, wheat or other common food allergens should look for all-beef franks, which are less likely to contain fillers.

FRANKFURTER, *Chicken*

Serving: 1 (about 1½ oz.)

Calories	116
Fat	8.8 g
Saturated	2.5 g
Monounsaturated	3.8 g
Polyunsaturated	1.8 g
Calories from fat	68%
Cholesterol	45 mg
Sodium	617 mg
Protein	5.8 g
Carbohydrate	3.1 g
Dietary fiber	0

CHIEF NUTRIENTS

NUTRIENT	AMOUNT	% RDA
Iron	0.9 mg	9
Niacin	1.4 mg	7
Vitamin B_6	0.1 mg	7
Vitamin B_{12}	0.1 mcg	6
Calcium	42.8 mg	5

Strengths: Chicken franks are a bit lower in fat than beef franks, and they're higher in protein. They have a smattering of B vitamins, and a decent amount of iron.

Caution: High in sodium. Although chicken franks have less fat than beef franks, there's still a fair amount. They usually contain nitrites, preservatives suspected of contributing to cancer. *Drug interaction:* May contain tyramine, an amino acid derivative that can raise blood pressure to dangerously high levels in anyone taking MAO inhibitors. *Allergy alert:* May trigger headaches or other reactions in people sensitive to tyramine. Fillers may include soy, corn, milk, wheat, or other common food allergens.

Description: A cooked, smoked, cured, and seasoned sausage made of chicken.

FRANKFURTER, *Turkey*

Serving: 1 (about 1½ oz.)

Calories	102
Fat	8.0 g
Saturated	2.7 g
Monounsaturated	2.5 g
Polyunsaturated	2.3 g
Calories from fat	71%
Cholesterol	48 mg
Sodium	642 mg
Protein	6.4 g
Carbohydrate	0.7 g
Dietary fiber	0

CHIEF NUTRIENTS

NUTRIENT	AMOUNT	% RDA
Niacin	1.9 mg	10
Zinc	1.4 mg	9
Iron	0.8 mg	8
Vitamin B$_{12}$	0.1 mcg	7
Calcium	47.7 mg	6
Vitamin B$_6$	0.1 mg	5

Strengths: A bit lower in fat than beef franks, turkey varieties are high in protein. They have a smattering of B vitamins and some minerals.

Caution: High in sodium, these franks have a steep percentage of calories from fat. Usually contain nitrites, preservatives that are suspected of contributing to cancer. *Drug interaction:* May contain tyramine, an amino acid derivative that can raise blood pressure to dangerously high levels in anyone taking MAO inhibitors. *Allergy alert:* May trigger headaches or other reactions in people sensitive to tyramine. May contain soy, corn, milk, wheat, or other common food allergens as fillers.

Description: A cooked, smoked, cured, and seasoned sausage made of light and dark turkey meat.

HAM

Serving: 2 slices (about 2 oz.)

Calories	103
Fat	6.0 g
Saturated	1.9 g
Monounsaturated	2.8 g
Polyunsaturated	0.7 g
Calories from fat	52%
Cholesterol	32 mg
Sodium	747 mg
Protein	10.0 g
Carbohydrate	1.8 g
Dietary fiber	0

CHIEF NUTRIENTS

NUTRIENT	AMOUNT	% RDA
Thiamine	0.5 mg	33
Vitamin C	15.8 mg	27
Vitamin B$_{12}$	0.5 mcg	24
Niacin	3.0 mg	16
Vitamin B$_6$	0.2 mg	10
Riboflavin	0.1 mg	8
Zinc	1.2 mg	8
Iron	0.6 mg	6
Potassium	188.2 mg	5

Strengths: One of the leaner sandwich meats, sliced ham is a satisfactory choice for people watching their intake of fat and calories, as long as the portions are modest. Ham is rich in thiamine and supplies respectable amounts of other B vitamins for skin and nerve health. As a bonus, you also get an impressive amount of vitamin C, derived primarily from sodium ascorbate, an additive in some cured meats.

Caution: High in sodium, ham usually contains nitrites, suspected of contributing to cancer. *Drug interaction:* May contain tyramine, an amino acid derivative that can raise blood pressure to dangerously high levels in anyone taking MAO inhibitors. *Allergy alert:* May trigger headaches or other complaints in some people.

HEAD CHEESE

Serving: 2 slices (about 2 oz.)

Calories	120
Fat	9.0 g
Saturated	2.8 g
Monounsaturated	4.6 g
Polyunsaturated	0.9 g
Calories from fat	67%
Cholesterol	45.9 mg
Sodium	713 mg
Protein	9.1 g
Carbohydrate	0.2 g
Dietary fiber	0

CHIEF NUTRIENTS

NUTRIENT	AMOUNT	% RDA
Vitamin B_{12}	0.6 mcg	30
Vitamin C	12.5 mg	21
Iron	0.7 mg	7
Riboflavin	0.1 mg	6
Vitamin B_6	0.1 mg	6

Caution: Head cheese is high in sodium and fat, with moderate amounts of cholesterol. *Drug interaction:* May contain tyramine, an amino acid derivative that can raise blood pressure to dangerously high levels in anyone taking MAO inhibitors. *Allergy alert:* May trigger complaints in people sensitive to molds.

Strengths: Rich in vitamin B_{12}, with a significant amount of vitamin C, head cheese also has a bit of anemia-fighting iron and a fair amount of protein.

Description: A gelatinous, molded, cured sausage made by grinding up meat from a hog's head with the heart, tongue, and sometimes feet. The American Meat Institute calls head cheese ''attractive and colorful.''

Serve: Head cheese is usually eaten cold, as an appetizer. Thinly sliced, it's often served with Dijon mustard and pickles.

HONEY LOAF

Serving: 2 slices (about 2 oz.)

Calories	73
Fat	2.5 g
Saturated	0.8 g
Monounsaturated	1.1 g
Polyunsaturated	0.3 g
Calories from fat	32%
Cholesterol	19 mg
Sodium	748 mg
Protein	8.9 g
Carbohydrate	3.0 g
Dietary fiber	0

CHIEF NUTRIENTS

NUTRIENT	AMOUNT	% RDA
Vitamin B_{12}	0.6 mcg	31
Vitamin C	11.9 mg	20
Thiamine	0.3 mg	18
Vitamin B_6	0.2 mg	10
Niacin	1.8 mg	9
Zinc	1.4 mg	9
Iron	0.8 mg	8
Riboflavin	0.1 mg	8
Potassium	194.5 mg	5

Strengths: With moderate levels of fat, cholesterol, and calories, this is one of the better choices among lunch meats for dieters watching their fat intake.

Eat with: Foods containing potassium, like cubed cantaloupe or sliced tomato, to help counteract the effects of sodium on blood pressure.

Caution: The sodium content is sky-high. *Drug interaction:* May contain tyramine, an amino acid derivative that can raise blood pressure to dangerous levels if you're taking MAO inhibitors. *Allergy alert:* May contain corn, soy, milk or other common food allergens, which are used as fillers or binders. May trigger headaches or other reactions in people sensitive to tyramine.

Description: Honey loaf is almost like bologna, but it is seasoned with honey, spices, and sometimes pickles, pimentos, or both. It may or may not be cured, and it sometimes contains fillers or binders.

KIELBASA, *Smoked*

Serving: 2 slices (about 2 oz.)

Calories	176
Fat	15.4 g
Saturated	5.6 g
Monounsaturated	7.3 g
Polyunsaturated	1.7 g
Calories from fat	79%
Cholesterol	38 mg
Sodium	610 mg
Protein	7.5 g
Carbohydrate	1.2 g
Dietary fiber	0

CHIEF NUTRIENTS

NUTRIENT	AMOUNT	% RDA
Vitamin B_{12}	0.9 mcg	46
Vitamin C	11.9 mg	20
Niacin	1.6 mg	9
Thiamine	0.1 mg	8
Zinc	1.1 mg	8
Riboflavin	0.1 mg	7
Vitamin B_6	0.1 mg	5

Caution: High in fat, sodium, and calories, and moderately high in cholesterol, kielbasa usually contains nitrites, preservatives suspected of contributing to cancer. *Drug interaction:* Contains tyramine, an amino acid derivative that can raise blood pressure to dangerously high levels in anyone taking MAO inhibitors. *Allergy alert:* May trigger headaches or other reactions in people sensitive to molds or nitrites.

Strengths: Rich in vitamin B_{12}, kielbasa also has fair amounts of other B vitamins. Also contains a decent amount of vitamin C, derived primarily from sodium ascorbate, an additive used in some cured meats.

Curiosity: *Kielbasa* is Polish for sausage.

Description: A highly seasoned sausage of pork and beef. The most common kind is smoked.

Preparation: Usually precooked, kielbasa tastes best heated.

KNOCKWURST, *Smoked*

Serving: 1 link (about 2½ oz.)

Calories	209
Fat	18.9 g
Saturated	6.9 g
Monounsaturated	8.7 g
Polyunsaturated	2.0 g
Calories from fat	81%
Cholesterol	39 mg
Sodium	687 mg
Protein	8.1 g
Carbohydrate	1.2 g
Dietary fiber	0

CHIEF NUTRIENTS

NUTRIENT	AMOUNT	% RDA
Vitamin B_{12}	0.8 mcg	40
Vitamin C	18.4 mg	31
Thiamine	0.2 mg	15
Niacin	1.9 mg	10
Zinc	1.1 mg	8
Iron	0.6 mg	6
Riboflavin	0.1 mg	6
Vitamin B_6	0.1 mg	6

Caution: One of the fattiest sausage meats you can eat, smoked knockwurst is also high in sodium. So reserve it for special occasions only—like Oktoberfest. *Drug interaction:* Smoked foods contain tyramine, an amino acid derivative that can raise blood pressure to dangerously high levels in anyone taking MAO inhibitors. *Allergy alert:* May trigger headaches or other complaints in people sensitive to molds or nitrites. Read labels: Knockwurst may contain nonfat dry milk, soy flour, and other fillers, all common food allergens.

Strengths: With some useful amounts of B vitamins, knockwurst is especially rich in vitamin B_{12}. As a bonus, you get a beneficial amount of vitamin C from sodium ascorbate, which is added to this cured meat.

Description: Resembling a plump, garlic-flavored frankfurter, knockwurst is made of smoked pork, beef, or both.

Preparation: Boil, grill, or steam before serving.

LIVERWURST, *Fresh*

Serving: 3 slices (about 2 oz.)

Calories	185
Fat	16.2 g
Saturated	6.0 g
Monounsaturated	7.6 g
Polyunsaturated	1.5 g
Calories from fat	79%
Cholesterol	90 mg
Sodium	488 mg
Protein	8.0 g
Carbohydrate	1.3 g
Dietary fiber	0

CHIEF NUTRIENTS

NUTRIENT	AMOUNT	% RDA
Vitamin A	4,706.1 RE	471
Vitamin B$_{12}$	7.6 mcg	382
Iron	3.6 mg	36
Riboflavin	0.6 mg	34
Niacin	1.4 mg	13
Thiamine	0.2 mg	11
Folate	17.0 mcg	9
Zinc	1.3 mg	9
Vitamin B$_6$	0.1 mg	5

Strengths: Has copious amounts of resistance-building vitamin A and blood-building vitamin B$_{12}$. Rich in anemia-fighting iron, with decent amounts of zinc and other B vitamins.
Eat with: Restraint.
Caution: Liverwurst is loaded with fat and calories, and it's high in cholesterol and sodium. It usually contains nitrites, preservatives that prevent botulism but are suspected of contributing to cancer. *Drug interaction:* Contains tyramine, an amino acid derivative that can raise blood pressure to dangerously high levels in anyone taking MAO inhibitors.
Description: A well-seasoned, ready-to-eat sausage made from pork liver and pork or other meat. Liverwurst that's prepared for crackers is smooth and spreadable, while the kind made for sandwiches is firm enough to slice.

MORTADELLA

Serving: 4 slices (about 2 oz.)

Calories	187
Fat	15.2 g
Saturated	5.7 g
Monounsaturated	6.8 g
Polyunsaturated	1.9 g
Calories from fat	74%
Cholesterol	34 mg
Sodium	748 mg
Protein	9.8 g
Carbohydrate	1.8 g
Dietary fiber	0

CHIEF NUTRIENTS

NUTRIENT	AMOUNT	% RDA
Vitamin B$_{12}$	0.9 mcg	44
Vitamin C	15.6 mg	26
Iron	0.8 mg	9
Niacin	1.6 mg	9
Zinc	1.3 mg	9

Caution: High in sodium and calories and visibly high in fat, mortadella has moderately high levels of cholesterol. It usually contains nitrites, preservatives suspected of contributing to cancer. *Drug interaction:* Smoked foods contain tyramine, an amino acid derivative that can raise blood pressure to dangerously high levels if you're taking MAO inhibitors. *Allergy alert:* May trigger headaches or other reactions in people sensitive to tyramine or molds.
Eat with: Restraint.
Strengths: Rich in vitamin B$_{12}$, with respectable amounts of immunity-bolstering vitamin C, and a touch of other nutrients such as iron and zinc.
Origin: Bologna, Italy.
Description: An Italian-style smoked sausage of cured pork and beef, with large flecks of white fat. Mortadella is seasoned with garlic and anise.

OLIVE LOAF

Serving: 2 slices (about 2 oz.)

Calories	133
Fat	9.4 g
Saturated	3.3 g
Monounsaturated	4.5 g
Polyunsaturated	1.1 g
Calories from fat	63%
Cholesterol	22 mg
Sodium	841 mg
Protein	6.7 g
Carbohydrate	5.2 g
Dietary fiber	0

CHIEF NUTRIENTS

NUTRIENT	AMOUNT	% RDA
Vitamin B_{12}	0.7 mcg	36
Thiamine	0.2 mg	11
Riboflavin	0.2 mg	9
Calcium	61.8 mg	8
Vitamin C	5.0 mg	8
Vitamin B_6	0.1 mg	7
Niacin	1.0 mg	5
Zinc	0.8 mg	5

Caution: High in sodium and fat, and moderately high in calories and cholesterol. *Drug interaction:* May contain tyramine, an amino acid derivative that can raise blood pressure to dangerously high levels in anyone taking MAO inhibitors. *Allergy alert:* May contain corn, soy, milk, and other common food allergens, which are used as filler or binders. May trigger headaches or other reactions in people sensitive to tyramine.

Eat with: Foods or beverages high in potassium, like sliced bananas or fruit juice, to help counteract some of the effects of sodium on blood pressure.

Strengths: Olive loaf is rich in vitamin B_{12}, with a smattering of other useful B vitamins for healthy skin and nerves. It has a fair amount of resistance-building vitamin C, too.

Description: A blend of pork and beef containing whole, stuffed olives.

PASTRAMI, *Beef*

Serving: 2 slices (about 2 oz.)

Calories	198
Fat	16.6 g
Saturated	5.9 g
Monounsaturated	8.2 g
Polyunsaturated	0.6 g
Calories from fat	75%
Cholesterol	53 mg
Sodium	696 mg
Protein	9.8 g
Carbohydrate	1.7 g
Dietary fiber	0

CHIEF NUTRIENTS

NUTRIENT	AMOUNT	% RDA
Vitamin B_{12}	1.0 mcg	50
Zinc	2.4 mg	16
Niacin	2.9 mg	15
Iron	1.1 mg	11
Riboflavin	0.1 mg	6

Caution: High in calories, fat, and sodium, and moderately high in cholesterol, beef pastrami may contain nitrites, preservatives suspected of contributing to cancer. *Drug interaction:* Smoked, cured meats contain tyramine, an amino acid derivative that can raise blood pressure to dangerously high levels in anyone taking MAO inhibitors. *Allergy alert:* May trigger headaches or other reactions in people sensitive to tyramine or molds.

Strengths: Supplies a super amount of vitamin B_{12}, with respectable amounts of protein, niacin, and minerals (iron, zinc). This combination of B_{12} and iron helps fight anemia.

Curiosity: "Pastrami" comes from the Rumanian word *pastra,* which means to preserve.

Description: Smoked and cured brisket or round of beef seasoned with garlic, black pepper, coriander, and other pungent spices.

PASTRAMI, *Turkey*

Serving: 2 slices (about 2 oz.)

Calories	80
Fat	3.5 g
Saturated	1.0 g
Monounsaturated	1.2 g
Polyunsaturated	0.9 g
Calories from fat	40%
Cholesterol	31 mg
Sodium	593 mg
Protein	10.4 g
Carbohydrate	0.9 g
Dietary fiber	0

CHIEF NUTRIENTS

NUTRIENT	AMOUNT	% RDA
Niacin	2.0 mg	11
Iron	0.9 mg	9
Riboflavin	0.1 mg	8
Vitamin B_6	0.2 mg	8
Zinc	1.2 mg	8
Vitamin B_{12}	0.1 mcg	7

Strengths: Turkey pastrami is considerably lower in fat, calories, and cholesterol than beef pastrami. And it's also a better choice than turkey bologna or salami. With nice amounts of various B vitamins, it contributes to healthy skin and nerves. The anemia-fighting iron and zinc are accompanied by a respectable amount of protein.

Eat with: Foods or beverages containing potassium, like bananas or fruit juice, to offset some of the effects of sodium on blood pressure.

Caution: Watch out for the sodium—it's high. Usually contains nitrites, preservatives suspected of contributing to cancer. *Drug interaction:* Smoked, cured foods contain tyramine, an amino acid derivative that can raise blood pressure to dangerously high levels in anyone taking MAO inhibitors. *Allergy alert:* May trigger headaches or other reactions in people sensitive to tyramine or molds.

Description: A cured, smoked, highly seasoned cold cut made from turkey instead of beef.

PEPPERONI

Serving: 10 slices (about 2 oz.)

Calories	273
Fat	24.2 g
Saturated	8.9 g
Monounsaturated	11.6 g
Polyunsaturated	2.4 g
Calories from fat	80%
Cholesterol	44 mg
Sodium	1,122 mg
Protein	11.5 g
Carbohydrate	1.6 g
Dietary fiber	0

CHIEF NUTRIENTS

NUTRIENT	AMOUNT	% RDA
Vitamin B_{12}	1.4 mcg	70
Niacin	2.7 mg	14
Thiamine	0.2 mg	13
Zinc	1.4 mg	9
Iron	0.8 mg	8
Riboflavin	0.1 mg	6
Potassium	190.9 mg	5
Vitamin B_6	0.1 mg	5

Caution: Pepperoni has an exorbitant amount of fat, calories, and sodium. A 10-slice serving—about what you'd get on a couple pieces of pepperoni pizza—has more than 200 calories from fat. This processed meat usually contains nitrites, suspected of contributing to cancer. *Drug interaction:* Contains tyramine, an amino acid derivative that can raise blood pressure to dangerously high levels in anyone taking MAO inhibitors. *Allergy alert:* May trigger headaches or other reactions in people sensitive to tyramine or molds.

Strengths: Extremely high in vitamin B_{12}, pepperoni also has a smattering of useful vitamin B_6. It contains some anemia-fighting iron, zinc, and muscle-building protein.

Description: A slender, firm, fermented Italian salami made of pork and beef, highly seasoned with black and red pepper.

PICKLE AND PIMIENTO LOAF

Serving: 2 slices (about 2 oz.)

Calories	149
Fat	12.0 g
Saturated	4.5 g
Monounsaturated	5.4 g
Polyunsaturated	1.5 g
Calories from fat	73%
Cholesterol	21 mg
Sodium	788 mg
Protein	6.5 g
Carbohydrate	3.3 g
Dietary fiber	0

CHIEF NUTRIENTS

NUTRIENT	AMOUNT	% RDA
Vitamin B_{12}	0.7 mcg	34
Vitamin C	7.4 mg	12
Thiamine	0.2 mg	11
Riboflavin	0.1 mg	8
Calcium	53.9 mg	7
Iron	0.6 mg	6
Niacin	1.2 mg	6
Vitamin B_6	0.1 mg	6
Potassium	192.8 mg	5
Zinc	0.8 mg	5

Caution: Not only is this lunch meat loaf high in sodium, fat, and calories, it's also moderately high in cholesterol. *Drug interaction:* May contain tyramine, an amino acid derivative that can raise blood pressure to dangerously high levels in anyone taking MAO inhibitors. *Allergy alert:* The fillers or binders may contain corn, soy, milk, or other common food allergens. May trigger headaches or other reactions in people sensitive to tyramine.

Eat with: Foods and beverages containing potassium, like sliced bananas or a tall glass of tomato juice, to counteract some of the effects of sodium on blood pressure.

Strengths: Rich in vitamin B_{12}, pickle and pimiento loaf has a smattering of other useful B vitamins for healthy skin and nerves. A fair amount of resistance-building vitamin C is also present.

Description: A blend of pork and beef, with sweet pickles and pimientos added. It's not always cured and may or may not contain fillers or binders.

SALAMI, *Beef*

Serving: 2 slices (about 2 oz.)

Calories	148
Fat	11.7 g
Saturated	5.1 g
Monounsaturated	5.4 g
Polyunsaturated	0.6 g
Calories from fat	71%
Cholesterol	36 mg
Sodium	668 mg
Protein	8.5 g
Carbohydrate	1.6 g
Dietary fiber	0

CHIEF NUTRIENTS

NUTRIENT	AMOUNT	% RDA
Vitamin B_{12}	1.7 mcg	50
Zinc	1.2 mg	18
Vitamin C	9.8 mg	16
Iron	1.2 mg	12
Niacin	1.8 mg	10
Riboflavin	0.1 mg	6
Vitamin B_6	0.1 mg	5

Caution: Beef salami is dramatically high in sodium. And the percentage of calories from fat is way above the 30% recommended by the American Heart Association. As in other smoked meats, nitrites are usually used as preservatives; while they prevent botulism, they are also suspected of contributing to cancer. *Drug interaction:* Smoked foods contain tyramine, an amino acid derivative that can raise blood pressure to dangerously high levels in anyone taking MAO inhibitors. *Allergy alert:* May trigger headaches or other reactions in people sensitive to tyramine or molds.

Strengths: Like many pork products, beef salami is also loaded with vitamin B_{12}, with small amounts of other B vitamins. You also get appreciable iron and zinc. The surprise bonus is a decent amount of vitamin C, derived primarily from sodium ascorbate, an additive used in some cured meats.

Description: A cured, smoked lunch meat.

SALAMI, *Pork*

Serving: 3 slices (about 2 oz.)

Calories	230
Fat	19.1 g
Saturated	6.7 g
Monounsaturated	9.0 g
Polyunsaturated	2.1 g
Calories from fat	75%
Cholesterol	45 mg
Sodium	1,277 mg
Protein	12.8 g
Carbohydrate	0.9 g
Dietary fiber	0

CHIEF NUTRIENTS

NUTRIENT	AMOUNT	% RDA
Vitamin B_{12}	1.6 mcg	79
Thiamine	0.5 mg	35
Niacin	3.2 mg	17
Vitamin B_6	0.3 mg	16
Zinc	2.4 mg	16
Riboflavin	0.2 mg	11
Iron	0.7 mg	7
Potassium	213.6 mg	6

Caution: The sodium content of this salami soars right through the roof. And it's higher in fat and calories than beef salami. *Drug interaction:* Salami and other smoked or fermented sausages contain tyramine, an amino acid derivative that can raise blood pressure to dangerously high levels in anyone taking MAO inhibitors. *Allergy alert:* May trigger headaches or other reactions in people sensitive to smoked meats or molds.

Strengths: Rich in vitamin B_{12} and thiamine, pork salami has respectable amounts of other B vitamins that contribute to healthy skin and nerves. A healthy amount of zinc is present, with small amounts of other valuable minerals.

Description: A fermented sausage. May or may not be smoked.

SALAMI, *Turkey*

Serving: 2 slices (about 2 oz.)

Calories	111
Fat	7.8 g
Saturated	2.3 g
Monounsaturated	2.6 g
Polyunsaturated	2.0 g
Calories from fat	63%
Cholesterol	46 mg
Sodium	569 mg
Protein	9.3 g
Carbohydrate	0.3 g
Dietary fiber	0

CHIEF NUTRIENTS

NUTRIENT	AMOUNT	% RDA
Niacin	2.0 mg	11
Iron	0.9 mg	9
Vitamin B_6	0.1 mg	7
Zinc	1.0 mg	7
Riboflavin	0.1 mg	6
Vitamin B_{12}	0.1 mcg	6

Strengths: Lower in fat and calories than pork or beef salami, turkey salami has a smattering of B vitamins, with a fair amount of anemia-fighting iron. It also has some zinc and a modest amount of protein.

Eat with: Foods containing potassium, like potato salad or green peppers, to help counteract the effects of sodium on blood pressure.

Caution: Turkey salami is high in sodium, though it contains less than the beef and pork varieties. Fat content, which is always steep, varies from product to product. So read the labels! *Drug interaction:* Smoked foods contain tyramine, an amino acid derivative that can raise blood pressure to dangerously high levels in anyone taking MAO inhibitors. *Allergy alert:* May trigger headaches or other reactions in people sensitive to tyramine or molds.

Description: A smoked lunch meat. Looks and tastes like beef or pork salami.

SAUSAGE, *Beef, Smoked*

Serving: 1 link (about 1½ oz.)

Calories	134
Fat	11.6 g
Saturated	4.9 g
Monounsaturated	5.6 g
Polyunsaturated	0.5 g
Calories from fat	78%
Cholesterol	29 mg
Sodium	486 mg
Protein	6.1 g
Carbohydrate	1.0 g
Dietary fiber	0

CHIEF NUTRIENTS

NUTRIENT	AMOUNT	% RDA
Vitamin B$_{12}$	0.8 mcg	40
Vitamin C	5.2 mg	9
Iron	0.8 mg	8
Zinc	1.2 mg	8
Niacin	1.4 mg	7

Caution: Like most bacon and sausage, smoked beef sausage has way over the 30% of calories from fat recommended by the American Heart Association. It's also quite high in sodium, which presents a potential problem for those prone to high blood pressure. And the sausage usually contains nitrites, preservatives that are suspected of contributing to cancer. *Drug interaction:* Smoked foods have tyramine, an amino acid derivative that can raise blood pressure to dangerously high levels in anyone taking MAO inhibitors. *Allergy alert:* May trigger headaches or other complaints in those sensitive to nitrites. Also, sausage may contain corn and soy, other common food allergens.

Eat with: Egg substitute and other low-fat foods to offset the high fat content.

Strengths: Rich in vitamin B$_{12}$ (for healthy nerves), smoked beef sausage offers some easy-to-absorb iron and fair amounts of resistance-boosting vitamin C and zinc.

Preparation: Braise, pan-fry, roast, or bake.

SAUSAGE, *Italian, Fresh*

Serving: 1 link (about 2½ oz.)

Calories	216
Fat	17.2 g
Saturated	6.1 g
Monounsaturated	8.0 g
Polyunsaturated	2.2 g
Calories from fat	72%
Cholesterol	52 mg
Sodium	618 mg
Protein	13.4 g
Carbohydrate	1.0 g
Dietary fiber	0

CHIEF NUTRIENTS

NUTRIENT	AMOUNT	% RDA
Vitamin B$_{12}$	0.9 mcg	44
Thiamine	0.4 mg	28
Niacin	2.8 mg	15
Vitamin B$_6$	0.2 mg	11
Zinc	1.6 mg	11
Iron	1.0 mg	10
Riboflavin	0.2 mg	9
Potassium	203.7 mg	5

Strengths: Italian sausage has a respectable dose of various B vitamins, and is rich in vitamin B$_{12}$. It also has decent amounts of zinc and other essential minerals, and roughly twice as much protein as kielbasa or Vienna sausage. This is a fresh sausage; if you must take MAO inhibitors, this is a good choice since it isn't cured or smoked, and therefore contains no tyramine.

Eat with: Steamed peppers and onions to counterbalance the lack of fiber and the high fat content.

Caution: The calories from fat are very high—over 40% above the level that the American Heart Association considers maximum. Like most sausage, it has lots of sodium and cholesterol as well.

Selection: Highly seasoned with garlic and other ingredients. "Hot" sausage is made with hot red peppers; "sweet" sausage is not.

Preparation: As with all fresh pork products, cook thoroughly in little or no added fat; drain well on paper towels to wick away excess grease.

SAUSAGE, Pork, Fresh

Serving: 4 links (about 2 oz.)

Calories	192
Fat	16.2 g
Saturated	5.6 g
Monounsaturated	7.2 g
Polyunsaturated	2.0 g
Calories from fat	76%
Cholesterol	43 mg
Sodium	673 mg
Protein	10.2 g
Carbohydrate	0.5 g
Dietary fiber	0

CHIEF NUTRIENTS

NUTRIENT	AMOUNT	% RDA
Vitamin B_{12}	0.9 mcg	44
Thiamine	0.4 mg	27
Niacin	2.4 mg	12
Zinc	0.7 mg	9
Vitamin B_6	0.2 mg	8
Riboflavin	0.1 mg	7
Iron	0.6 mg	6
Potassium	187.7 mg	5

Strengths: Like other pork products, fresh sausage is rich in vitamin B_{12}, with handsome amounts of niacin and thiamine, and a smattering of other B vitamins (for healthy nerves and skin). It also boasts some zinc, which can have some immunity-boosting benefits. Since fresh sausage is made from meat that has not been smoked or cured, it may not contain nitrites, so it may be acceptable for people who are sensitive to these and other preservatives. Check labels, however; some brands may include nitrites or nitrates as preservatives.

Caution: Pork sausage is high in fat, calories, and sodium, with moderate amounts of cholesterol—and the calories from fat should send a warning signal to dieters.

Curiosity: The longest sausage in the world was 9.89 miles long. (That's more than twice as long as the longest banana split.)

Description: Ground pork and pork fat, seasoned with sage, pepper, or other ingredients. Also available in patties. Fresh pork sausage may also be labeled "country-style."

Preparation: Must be thoroughly cooked.

SAUSAGE, Pork, Smoked

Serving: 1 link (about 2½ oz.)

Calories	265
Fat	21.6 g
Saturated	7.7 g
Monounsaturated	10.0 g
Polyunsaturated	2.6 g
Calories from fat	74%
Cholesterol	46 mg
Sodium	1,020 mg
Protein	15.1 g
Carbohydrate	1.4 g
Dietary fiber	0

CHIEF NUTRIENTS

NUTRIENT	AMOUNT	% RDA
Vitamin B_{12}	1.1 mcg	56
Thiamine	0.5 mg	32
Niacin	3.1 mg	16
Zinc	1.9 mg	13
Vitamin B_6	0.2 mg	12
Riboflavin	0.2 mg	10
Iron	0.8 mg	8
Potassium	228.5 mg	6

Caution: Pork sausage is sky-high in fat, sodium, and calories. In addition, it may contain nitrites, preservatives that prevent botulism but are suspected of contributing to cancer. *Drug interaction:* Smoked foods contain tyramine, an amino acid derivative that can raise blood pressure to dangerously high levels in anyone taking MAO inhibitors. *Allergy alert:* May trigger headaches or other reactions in people sensitive to tyramine or molds.

Strengths: Like other pork products, smoked sausage is rich in thiamine and vitamin B_{12}. Handsome amounts of niacin, vitamin B_6, and riboflavin help promote healthy nerves and skin. This sausage also boasts significant levels of zinc.

Description: Mildly cured and smoked pork sausage.

Preparation: Cook thoroughly.

SAUSAGE, *Vienna*

Serving: 3½ (about 2 oz.)

Calories	158
Fat	14.2 g
Saturated	5.2 g
Monounsaturated	7.1 g
Polyunsaturated	1.0 g
Calories from fat	81%
Cholesterol	29 mg
Sodium	538 mg
Protein	5.8 g
Carbohydrate	1.2 g
Dietary fiber	0

CHIEF NUTRIENTS

NUTRIENT	AMOUNT	% RDA
Vitamin B_{12}	0.6 mcg	29
Zinc	0.9 mg	6
Iron	0.5 mg	5

Caution: One of the fattiest sausages you can buy—so think twice before overindulging, especially if you're on a fat-restricted diet. That high sodium level is definitely a red flag for those restricting sodium intake. *Drug interaction:* Smoked foods contain tyramine, an amino acid derivative that can raise blood pressure to dangerously high levels in anyone taking MAO inhibitors. *Allergy alert:* May trigger headaches or other reactions in people sensitive to tyramine or molds.

Strengths: Like most pork products, Vienna sausage contributes handsome amounts of vitamin B_{12}, which is essential for keeping nerves, skin, and blood in tip-top shape.

Description: A cooked, smoked sausage made of beef and pork, similar to frankfurters, Vienna sausage usually comes canned.

SCRAPPLE

Serving: 2 oz.

Calories	120
Fat	7.6 g
Saturated	2.8 g
Monounsaturated	N/A
Polyunsaturated	N/A
Calories from fat	57%
Cholesterol	25 mg
Sodium	536 mg
Protein	4.9 g
Carbohydrate	8.2 g
Dietary fiber	N/A

CHIEF NUTRIENTS

NUTRIENT	AMOUNT	% RDA
Iron	0.7 mg	7
Thiamine	0.1 mg	7
Niacin	1.0 mg	5

Caution: With 57% of its calories coming from fat, scrapple is also chock-full of sodium. Scrapple is made with pork, which contains purines, substances that can aggravate gout. *Allergy alert:* The pork in scrapple is a meat that causes allergic reactions in some people.

Strengths: Scrapple has some nutrients, but only in modest supply. Along with fair amounts of niacin and iron, it has thiamine, a B vitamin that helps convert carbohydrates to energy.

Curiosity: The Pennsylvania Dutch also call scrapple "poor-do" because it often is made from leftovers.

History: The dish was a Pennsylvania Dutch favorite as early as 1817.

Description: A grayish-brown Pennsylvania Dutch dish made from boiling scraps of pork, organ meats, cornmeal, sage, and other spices in a loaf pan. The mixture is cooled, then sliced and fried before serving.

A Better Idea: If you'd like an accompaniment to morning French toast or pancakes, try apple slices instead of scrapple.

TURKEY BREAST

Serving: 2 slices (about 1 1/2 oz.)

Calories	47
Fat	0.7 g
Saturated	trace
Monounsaturated	trace
Polyunsaturated	trace
Calories from fat	13%
Cholesterol	17 mg
Sodium	608 mg
Protein	9.6 g
Carbohydrate	0
Dietary fiber	0

CHIEF NUTRIENTS

NUTRIENT	AMOUNT	% RDA
Vitamin B_{12}	0.9 mcg	43
Niacin	3.8 mg	19
Vitamin B_6	0.2 mg	8

Strengths: Sliced turkey breast earns a blue ribbon in the cold cut competition, taking top honors for minimal fat content and low calorie count. This lunch meat contains only a tiny fraction of the saturated fat found in bologna, pastrami, and other cold cuts.

Caution: Despite its relative merits, turkey breast is still high in sodium. Look for low-salt versions at the deli counter or supermarket. *Allergy alert:* If you're allergic to corn, soy, wheat, nitrites, or other food allergens, watch out for wrongly labeled turkey "breast" that has been processed and filled with additives or modified food starch.

TURKEY HAM

Serving: 2 slices (about 2 oz.)

Calories	73
Fat	2.8 g
Saturated	1.0 g
Monounsaturated	0.7 g
Polyunsaturated	0.9 g
Calories from fat	36%
Cholesterol	32 mg
Sodium	565 mg
Protein	10.7 g
Carbohydrate	0.2 g
Dietary fiber	0

CHIEF NUTRIENTS

NUTRIENT	AMOUNT	% RDA
Iron	1.6 mg	16
Niacin	2.0 mg	11
Zinc	1.7 mg	11
Riboflavin	0.1 mg	8
Vitamin B_6	0.1 mg	7
Vitamin B_{12}	0.1 mcg	7

Strengths: This is a very respectable source of heme iron, the form most readily absorbed by the body. And the calories from fat aren't too bad, compared with other lunch meats. Turkey ham is a good source of zinc, which improves skin health and helps heal wounds. Zinc also benefits the immune system and sharpens the senses of taste and smell. Turkey ham contributes fair amounts of B vitamins, like vitamin B_6, specifically needed by women who take birth control pills.

Caution: It has the classic drawback of cured meat—high sodium. *Drug interaction:* May contain tyramine, an amino acid derivative that can raise blood pressure to dangerously high levels in anyone taking MAO inhibitors. *Allergy alert:* May trigger headaches or other reactions in people sensitive to ingredients in cured meats.

Description: Cured turkey thigh meat.

TURKEY ROLL, *Light Meat*

Serving: 2 slices (about 2 oz.)

Calories	83
Fat	4.1 g
Saturated	1.2 g
Monounsaturated	1.4 g
Polyunsaturated	1.0 g
Calories from fat	44%
Cholesterol	24 mg
Sodium	277 mg
Protein	10.6 g
Carbohydrate	0.3 g
Dietary fiber	0

CHIEF NUTRIENTS

NUTRIENT	AMOUNT	% RDA
Niacin	4.0 mg	21
Vitamin B_6	0.2 mg	9
Riboflavin	0.1 mg	8
Iron	0.7 mg	7
Vitamin B_{12}	0.1 mcg	7
Zinc	0.9 mg	6

Caution: If you're trying to lower sodium intake, steer clear of lunch meat rolls. The percentage of calories from fat for turkey roll is 44%, making it more than 3 times as fatty as sliced turkey breast, and it's well above the 30% of calories from fat recommended by the American Heart Association.

Strengths: With good amounts of niacin, turkey roll also has some vitamins B_6, B_{12}, and riboflavin. All three are worthwhile, because deficits of B vitamins can lead to skin and nerve problems. A fair amount of protein, with a little zinc and anemia-fighting iron.

Description: Contains chopped and reconstituted turkey meat plus gelatin, sugar, and other fillers and flavorings.

A Better Idea: Why not eat sliced turkey breast instead: The same serving has less than 1 g of fat. In addition, a 2-oz. serving of fresh turkey breast has almost 10 times as much vitamin B_{12}.

PORK

BACON

Serving: 3 medium slices (about ¾ oz.)

Calories	109
Fat	9.4 g
Saturated	3.3 g
Monounsaturated	4.5 g
Polyunsaturated	1.1 g
Calories from fat	77%
Cholesterol	16 mg
Sodium	303 mg
Protein	5.8 g
Carbohydrate	0.1 g
Dietary fiber	N/A

CHIEF NUTRIENTS

NUTRIENT	AMOUNT	% RDA
Vitamin B_{12}	0.3 mcg	17
Vitamin C	6.4 mg	11
Thiamine	0.1 mg	9
Niacin	1.4 mg	7

Caution: Bacon has way over the 30% of calories from fat recommended by the American Heart Association. It's also high in sodium, which is a potential problem for those prone to high blood pressure. And it usually contains nitrites to prevent botulism; these preservatives are suspected of contributing to cancer. *Allergy alert:* May trigger headaches or other complaints in people sensitive to nitrites.

Eat with: Egg substitute and other low-fat versions of breakfast favorites to offset the high fat level.

Strengths: Bacon has some useful amounts of B vitamins, which help prevent skin and nerve problems. You also get a surprisingly decent amount of vitamin C, derived primarily from sodium ascorbate.

Preparation: Broil, pan-fry, or roast.

A Better Idea: To reduce fat content, microwave bacon and drain well on paper towels.

BACON, *Canadian*

Serving: 2 medium slices (about 1½ oz.)

Calories	86
Fat	3.9 g
Saturated	1.3 g
Monounsaturated	1.9 g
Polyunsaturated	0.4 g
Calories from fat	41%
Cholesterol	27 mg
Sodium	719 mg
Protein	11.3 g
Carbohydrate	0.6 g
Dietary fiber	0

CHIEF NUTRIENTS

NUTRIENT	AMOUNT	% RDA
Thiamine	0.4 mg	25
Vitamin B$_{12}$	0.4 mcg	18
Niacin	3.2 mg	17
Vitamin C	10.0 mg	17
Vitamin B$_6$	0.2 mg	11
Potassium	181.4 mg	5
Riboflavin	0.1 mg	5
Zinc	0.8 mg	5

Strengths: For bacon lovers watching their diet, Canadian bacon is a good alternative to strip bacon. It's lower in fat and calories; the percentage of calories from fat is also lower in Canadian bacon. It's also higher in protein, while offering respectable amounts of B vitamins. (Deficits of B vitamins can lead to skin and nerve problems.) As a bonus, you also get a surprisingly decent amount of vitamin C.

Caution: Like other bacon, this kind is high in sodium, which is a potential problem for those prone to high blood pressure. It usually contains nitrites, preservatives that are suspected of contributing to cancer. *Drug interaction:* May contain tyramine, an amino acid derivative that can raise blood pressure to dangerously high levels in anyone taking MAO inhibitors. *Allergy alert:* May trigger headaches or other complaints in those sensitive to nitrites.

Description: Cut from the long loin muscle along the pig's back.

Preparation: Grill.

BLADE LOIN ROAST

Serving: 3 oz., lean only, roasted

Calories	237
Fat	16.4 g
Saturated	5.7 g
Monounsaturated	7.4 g
Polyunsaturated	2.0 g
Calories from fat	62%
Cholesterol	76 mg
Sodium	58 mg
Protein	21.0 g
Carbohydrate	0
Dietary fiber	0

CHIEF NUTRIENTS

NUTRIENT	AMOUNT	% RDA
Vitamin B$_{12}$	0.7 mcg	34
Thiamine	0.5 mg	33
Niacin	4.0 mg	21
Zinc	3.1 mg	20
Vitamin B$_6$	0.4 mg	19
Riboflavin	0.3 mg	17
Iron	1.1 mg	11
Potassium	294.1 mg	8

Caution: Blade loin of pork is one of the fattiest cuts of meat you can buy; for a heart-healthy diet, avoid it completely. Lean ham or fresh pork leg may be a better choice. If you have gout, eat pork sparingly, as it contains purines. *Drug interaction:* People taking levodopa are often advised against eating excessive amounts of vitamin B$_6$–rich foods such as pork. *Allergy alert:* May trigger migraine headaches in susceptible people.

Strengths: Pork in general is a very good source of B vitamins—especially thiamine and B$_{12}$—and a good source of zinc, iron, and high-quality protein.

Selection: Look for cuts with little marbling.

Preparation: Although trichinosis is rare, there is some danger. Be sure to cook pork to an internal temperature of at least 160°F—"medium," with only traces of pink. Freezing meat at 5°F for 20 days will also kill trichinae, the hazardous organisms.

A Better Idea: If you want to eat pork, cuts of tenderloin, lean ham, and fresh pork leg are better choices.

CENTER LOIN

Serving: 3 oz., lean only, roasted

Calories	204
Fat	11.1 g
Saturated	3.8 g
Monounsaturated	5.0 g
Polyunsaturated	1.4 g
Calories from fat	49%
Cholesterol	77 mg
Sodium	59 mg
Protein	24.2 g
Carbohydrate	0
Dietary fiber	0

CHIEF NUTRIENTS

NUTRIENT	AMOUNT	% RDA
Thiamine	0.8 mg	51
Vitamin B_{12}	0.5 mcg	26
Niacin	4.6 mg	24
Vitamin B_6	0.4 mg	19
Riboflavin	0.2 mg	13
Zinc	1.9 mg	13
Iron	0.9 mg	9
Potassium	307.7 mg	8

Strengths: An excellent source of thiamine, with generous amounts of other B vitamins, pork is also a good source of infection-fighting zinc and a fair source of other minerals.

Caution: Contains purines—should be eaten sparingly by people who have gout. *Drug interaction:* People taking levodopa are often advised not to eat excessive amounts of vitamin B_6–rich foods such as pork. *Allergy alert:* May trigger migraine headaches in susceptible people.

Selection: Look for cuts with little marbling. Commonly called pork chops.

Preparation: Because of the slight risk of trichinosis, be sure to cook pork to an internal temperature of at least 160°F—"medium," with only traces of pink. Freezing meat at 5°F for 20 days will also kill trichinae, the hazardous organisms.

CENTER RIB

Serving: 3 oz., lean only, roasted

Calories	208
Fat	11.7 g
Saturated	4.1 g
Monounsaturated	5.3 g
Polyunsaturated	1.4 g
Calories from fat	51%
Cholesterol	67 mg
Sodium	39 mg
Protein	24.0 g
Carbohydrate	0
Dietary fiber	0

CHIEF NUTRIENTS

NUTRIENT	AMOUNT	% RDA
Thiamine	0.5 mg	36
Niacin	4.6 mg	24
Vitamin B_{12}	0.5 mcg	24
Vitamin B_6	0.3 mg	17
Riboflavin	0.3 mg	16
Zinc	1.9 mg	13
Potassium	359.6 mg	10
Iron	0.9 mg	9

Caution: This meat is too high in fat for people who are watching their diet for health reasons. Contains purines—should be eaten sparingly by people who have gout. *Drug interaction:* Those taking levodopa are often advised not to eat excessive amounts of vitamin B_6–rich foods such as pork. *Allergy alert:* May trigger migraine headaches in susceptible people.

Strengths: The center rib cut is a rich source of thiamine, with generous amounts of other B vitamins. A good source of infection-fighting zinc and other minerals, it contains a smaller percentage of saturated fat than beef.

Selection: Look for cuts with little marbling.

Preparation: Because of the slight risk of trichinosis, be sure to cook pork to an internal temperature of at least 160°F—"medium," with only traces of pink. Freezing meat at 5°F for 20 days will also kill trichinae, the hazardous organisms.

A Better Idea: Tenderloin, lean ham, and fresh pork leg are better choices than a center cut of pork.

HAM, *Cured*

Serving: 3 oz., boneless, roasted

Calories	140
Fat	6.5 g
Saturated	2.2 g
Monounsaturated	3.2 g
Polyunsaturated	0.9 g
Calories from fat	42%
Cholesterol	48 mg
Sodium	1,177 mg
Protein	18.7 g
Carbohydrate	0.4 g
Dietary fiber	0

CHIEF NUTRIENTS

NUTRIENT	AMOUNT	% RDA
Thiamine	0.6 mg	42
Vitamin C	18.7 mg	31
Vitamin B_{12}	0.6 mcg	29
Niacin	4.5 mg	24
Vitamin B_6	0.3 mg	15
Zinc	2.2 mg	15
Riboflavin	0.2 mg	14
Iron	1.2 mg	12
Potassium	307.7 mg	8

Caution: The sodium content is sky-high. Cured ham usually contains nitrites, preservatives that prevent botulism but are suspected of contributing to cancer. *Drug interaction*: In processed hams you'll get tyramine, an amino acid derivative that can raise blood pressure to dangerously high levels if you're taking MAO inhibitors. *Allergy alert:* May trigger headaches or other complaints in people sensitive to nitrites or mold.

Strengths: Relatively low in fat and calories, as pork products go, roasted cured ham has significant amounts of various important B vitamins. It also has an impressive amount of vitamin C from sodium ascorbate, an additive.

Preparation: Read the label. Hams that are labeled "uncooked" or "partially cooked" must be cooked prior to serving. Others, labeled "heat and serve" or "ready to eat," may be served cold or slightly warmed.

A Better Idea: To avoid nitrites and sky-high sodium, look for fresh, unprocessed ham.

HAM, *Fresh (Leg)*

Serving: 3 oz., lean only, roasted

Calories	187
Fat	9.4 g
Saturated	3.2 g
Monounsaturated	4.2 g
Polyunsaturated	1.1 g
Calories from fat	45%
Cholesterol	80 mg
Sodium	54 mg
Protein	24.1 g
Carbohydrate	0
Dietary fiber	0

CHIEF NUTRIENTS

NUTRIENT	AMOUNT	% RDA
Thiamine	0.6 mg	39
Vitamin B_{12}	0.6 mcg	31
Niacin	4.2 mg	22
Vitamin B_6	0.4 mg	19
Zinc	2.8 mg	19
Riboflavin	0.3 mg	18
Iron	1.0 mg	10
Potassium	317.1 mg	9

Strengths: Somewhat lower in fat and calories than center-cut pork, fresh ham is a very good source of protein and of thiamine and vitamin B_{12}. It's also a good source of infection-fighting zinc and other minerals.

Caution: To minimize salt intake, avoid cured hams. Contains purines—should be eaten sparingly by people who have gout. *Drug interaction:* People taking levodopa are often advised not to eat excessive amounts of vitamin B_6–rich foods such as pork. *Allergy alert:* May trigger migraine headaches in susceptible people.

Selection: Look for cuts with little marbling.

Preparation: Trim all fat. Because of the slight risk of trichinosis, be sure to cook pork to an internal temperature of at least 160°F—"medium," with only traces of pink. Freezing meat at 5°F for 20 days will also kill trichinae, the hazardous organisms.

LIVER

Serving: 3 oz., braised

Calories	140
Fat	3.7 g
Saturated	1.2 g
Monounsaturated	0.5 g
Polyunsaturated	0.9 g
Calories from fat	24%
Cholesterol	302 mg
Sodium	42 mg
Protein	22.1 g
Carbohydrate	3.2 g
Dietary fiber	0

CHIEF NUTRIENTS

NUTRIENT	AMOUNT	% RDA
Vitamin B_{12}	15.9 mg	794
Vitamin A	4,589.2 RE	459
Iron	15.2 mg	152
Riboflavin	1.9 mg	110
Folate	138.6 mcg	69
Niacin	7.2 mg	38
Zinc	5.7 mg	38
Vitamin C	20.1 mg	33
Vitamin B_6	0.5 mg	24

Strengths: A highly concentrated source of vitamin A, vitamin B_{12}, iron, riboflavin, and folate, pork liver also has rich supplies of other important nutrients.

Caution: High in cholesterol—302 mg. (The American Heart Association recommends less than 300 mg *per day*.) Pork liver is a depot for hormones, medicines, and other unwanted chemicals in the animal's system. Since it contains purines, this liver should be eaten sparingly by people who have gout. *Drug interaction:* People taking levodopa are often advised not to eat excessive amounts of vitamin B_6–rich foods such as this. *Allergy alert:* Pork may trigger migraine headaches in susceptible people.

PICNIC SHOULDER ARM

Serving: 3 oz., lean only, roasted

Calories	194
Fat	10.7 g
Saturated	3.7 g
Monounsaturated	4.8 g
Polyunsaturated	1.3 g
Calories from fat	50%
Cholesterol	81 mg
Sodium	68 mg
Protein	22.7 g
Carbohydrate	0
Dietary fiber	0

CHIEF NUTRIENTS

NUTRIENT	AMOUNT	% RDA
Thiamine	0.5 mg	33
Vitamin B_{12}	0.7 mcg	33
Zinc	3.5 mg	23
Niacin	3.7 mg	19
Riboflavin	0.3 mg	18
Vitamin B_6	0.4 mg	18
Iron	1.2 mg	12

Strengths: A very good source of protein, thiamine, and vitamin B_{12}, with respectable amounts of other B vitamins. This cut of pork has significant amounts of immunity-boosting zinc and heme iron.

Caution: To limit the fat in your diet, serve only small portions. If you have gout, eat pork sparingly, as it contains purines. *Drug interaction:* People taking levodopa are often advised not to eat excessive amounts of vitamin B_6–rich foods such as pork. *Allergy alert:* May trigger migraine headaches in susceptible people.

Selection: Look for cuts with little marbling.

Preparation: Because of the slight risk of trichinosis, be sure to cook pork to an internal temperature of at least 160°F—''medium,'' with only traces of pink. Freezing meat at 5°F for 20 days will also kill trichinae, the hazardous organisms.

PIG'S FEET, *Pickled*

Serving: 3 oz.

Calories	173
Fat	13.7 g
Saturated	4.7 g
Monounsaturated	6.5 g
Polyunsaturated	1.5 g
Calories from fat	72%
Cholesterol	78 mg
Sodium	785 mg
Protein	11.5 g
Carbohydrate	trace
Dietary fiber	0

CHIEF NUTRIENTS

NUTRIENT	AMOUNT	% RDA
Vitamin B_{12}	0.5 mcg	27
Vitamin B_6	0.3 mg	17
Zinc	1.0 mg	7
Potassium	199.9 mg	5

Caution: High in fat and sodium, pig's feet are moderately high in cholesterol and calories as well. This pickled product usually contains nitrites, preservatives suspected of contributing to cancer. *Allergy alert:* May trigger headaches or other reactions in people sensitive to tyramine or molds.

Strengths: Like many pork products, pig's feet are high in vitamin B_{12}, with a good amount of vitamin B_6. You also get some potassium and zinc.

Curiosity: Known as trotters in Britain and *pieds de porc* in France.

Description: The flavorful feet and ankles of a pig are pickled in a brine that typically has salt, sugar, and sodium—sometimes with potassium nitrite or nitrate.

SHOULDER BLADE

Serving: 3 oz., lean only, roasted

Calories	218
Fat	14.3 g
Saturated	4.9 g
Monounsaturated	6.4 g
Polyunsaturated	1.7 g
Calories from fat	59%
Cholesterol	83 mg
Sodium	62 mg
Protein	20.7 g
Carbohydrate	0
Dietary fiber	0

CHIEF NUTRIENTS

NUTRIENT	AMOUNT	% RDA
Vitamin B_{12}	0.8 mcg	41
Thiamine	0.5 mg	33
Zinc	3.7 mg	25
Niacin	3.7 mg	19
Riboflavin	0.3 mg	18
Iron	1.4 mg	14
Vitamin B_6	0.3 mg	13
Potassium	300.1 mg	8

Caution: Shoulder blade is too high in fat for any heart-healthy diet. If you have gout, you should eat pork sparingly, as it contains purines. *Drug interaction:* People taking levodopa are often advised not to eat excessive amounts of vitamin B_6–rich foods such as pork. *Allergy alert:* May trigger migraine headaches in susceptible people.

Eat with: Chunky applesauce, sweet potatoes, or other high-fiber, low-fat fruits and vegetables to partially offset high fat content.

Strengths: Rich in thiamine and vitamin B_{12}, with commendable amounts of other B vitamins, pork is a good source of infection-fighting zinc and other minerals. This cut has a fair amount of potassium, although its potential benefit—lowering blood pressure—may be canceled by the high fat content.

Preparation: Because of the slight risk of trichinosis, be sure to cook pork to an internal temperature of at least 160°F—''medium,'' with only traces of pink. Freezing meat at 5°F for 20 days will also kill trichinae, the hazardous organisms.

SIRLOIN

Serving: 3 oz., lean only, roasted

Calories	201
Fat	11.2 g
Saturated	3.9 g
Monounsaturated	5.0 g
Polyunsaturated	1.4 g
Calories from fat	50%
Cholesterol	77 mg
Sodium	53 mg
Protein	23.4 g
Carbohydrate	0
Dietary fiber	0

CHIEF NUTRIENTS

NUTRIENT	AMOUNT	% RDA
Thiamine	0.7 mg	45
Vitamin B_{12}	0.7 mcg	33
Niacin	4.7 mg	25
Vitamin B_6	0.4 mg	18
Riboflavin	0.3 mg	17
Zinc	2.1 mg	14
Iron	0.9 mg	9
Potassium	314.5 mg	8

Caution: Avoid this high-fat meat if you're on a heart-healthy diet. If you have gout, eat pork sparingly, as it contains purines. *Drug interaction:* People taking levodopa are often advised not to eat excessive amounts of vitamin B_6–rich foods such as pork. *Allergy alert:* Pork may trigger migraine headaches in susceptible people.

Strengths: A rich source of thiamine—but avoid drinking large amounts of tea or taking antacids, since they can quickly deplete this nutrient. Pork has decent amounts of other critical B vitamins and contributes some iron and zinc. Although magnesium and potassium are present, they probably have little effect in lowering high blood pressure.

Preparation: Because of the slight risk of trichinosis, be sure to cook pork to an internal temperature of at least 160°F—"medium," with only traces of pink. Freezing meat at 5°F for 20 days will also kill trichinae, the hazardous organisms.

A Better Idea: Tenderloin, lean ham, or fresh pork leg.

SPARERIBS

Serving: 3 oz., lean only, braised

Calories	337
Fat	25.8 g
Saturated	10.0 g
Monounsaturated	12.0 g
Polyunsaturated	3.0 g
Calories from fat	69%
Cholesterol	103 mg
Sodium	79 mg
Protein	24.7 g
Carbohydrate	0
Dietary fiber	0

CHIEF NUTRIENTS

NUTRIENT	AMOUNT	% RDA
Vitamin B_{12}	0.9 mcg	46
Zinc	3.9 mg	26
Niacin	4.7 mg	25
Thiamine	0.4 mg	23
Riboflavin	0.3 mg	19
Iron	1.6 mg	16
Vitamin B_6	0.3 mg	15
Potassium	272.0 mg	7

Caution: Too high in fat to justify on a heart-healthy diet. If you have gout, eat pork sparingly, as it contains purines. *Drug interaction:* People taking levodopa are often advised not to eat excessive amounts of vitamin B_6–rich foods such as pork. *Allergy alert:* May trigger migraine headaches in susceptible people.

Strengths: Ample amounts of zinc and iron. Spareribs are rich in vitamin B_{12}, with a good supply of other B vitamins, especially niacin and thiamine. (Deficits of B vitamins can show up as various skin and nerve problems.)

Preparation: Because of the slight risk of trichinosis, be sure to cook pork to an internal temperature of at least 160°F—"medium," with only traces of pink. Freezing meat at 5°F for 20 days will also kill trichinae, the hazardous organisms.

A Better Idea: Tenderloin, lean ham, or fresh pork leg.

TENDERLOIN

Serving: 3 oz., lean only, roasted

Calories	141
Fat	4.1 g
Saturated	1.4 g
Monounsaturated	1.8 g
Polyunsaturated	0.5 g
Calories from fat	26%
Cholesterol	79 mg
Sodium	57 mg
Protein	25.0 g
Carbohydrate	0
Dietary fiber	0

CHIEF NUTRIENTS

NUTRIENT	AMOUNT	% RDA
Thiamine	0.8 mg	53
Vitamin B_{12}	0.5 mcg	24
Niacin	4.0 mg	21
Riboflavin	0.3 mg	19
Vitamin B_6	0.4 mg	18
Zinc	2.6 mg	17
Iron	1.3 mg	13
Potassium	457.3 mg	12
Magnesium	21.3 mg	6

Strengths: A good source of protein, which is necessary to maintain muscle mass while losing weight. The nutritional profile of tenderloin is similar to other cuts of pork, with the added bonus of a lower-than-average fat content. It's good insurance against iron deficiency and also a superb source of thiamine, with high marks for other B vitamins.

Caution: As with any meat, keep portions modest to minimize fat intake. Contains purines—should be eaten sparingly by people who have gout. *Drug interaction:* People taking levodopa are often advised not to eat excessive amounts of vitamin B_6–rich foods such as pork. *Allergy alert:* May trigger migraine headaches in susceptible people.

Selection: Look for cuts with little marbling.

Preparation: Because of the slight risk of trichinosis, be sure to cook pork to an internal temperature of at least 160°F—''medium,'' with only traces of pink. Freezing meat at 5°F for 20 days will also kill trichinae, the hazardous organisms.

TOP LOIN

Serving: 3 oz., lean only, roasted

Calories	209
Fat	11.7 g
Saturated	4.1 g
Monounsaturated	5.3 g
Polyunsaturated	1.4 g
Calories from fat	51%
Cholesterol	67 mg
Sodium	39 mg
Protein	24.0 g
Carbohydrate	0
Dietary fiber	0

CHIEF NUTRIENTS

NUTRIENT	AMOUNT	% RDA
Thiamine	0.5 mg	36
Niacin	4.6 mg	24
Vitamin B_{12}	0.5 mcg	24
Vitamin B_6	0.3 mg	17
Riboflavin	0.3 mg	16
Zinc	1.9 mg	13
Potassium	359.6 mg	10
Iron	0.9 mg	9

Caution: Too high in fat to justify on a heart-healthy diet. If you have gout, eat pork sparingly, as it contains purines. *Drug interaction:* People taking levodopa are often advised not to eat excessive amounts of vitamin B_6–rich foods such as pork. *Allergy alert:* May trigger migraine headaches in susceptible people.

Strengths: Rich in protein and thiamine, top loin is a decent source of other B vitamins. The additional vitamin B_6 is especially beneficial for women who take birth control pills. Also a good source of infection-fighting zinc and other minerals.

Selection: Look for cuts with little marbling. Also called pork chops.

Preparation: Because of the slight risk of trichinosis, be sure to cook pork to an internal temperature of at least 160°F—''medium,'' with only traces of pink. Freezing meat at 5°F for 20 days will also kill trichinae, the hazardous organisms.

A Better Idea: Tenderloin, lean ham, or fresh pork leg.

POULTRY

CAPON

Serving: 3 oz., meat only, roasted

Calories	195
Fat	9.9 g
Saturated	2.8 g
Monounsaturated	4.0 g
Polyunsaturated	2.1 g
Calories from fat	46%
Cholesterol	73 mg
Sodium	42 mg
Protein	24.6 g
Carbohydrate	0
Dietary fiber	0

CHIEF NUTRIENTS

NUTRIENT	AMOUNT	% RDA
Niacin	7.6 mg	40
Vitamin B_6	0.4 mg	18
Vitamin B_{12}	0.3 mcg	14
Iron	1.3 mg	13
Zinc	1.5 mg	10
Riboflavin	0.1 mg	8
Magnesium	20.4 mg	6
Potassium	216.8 mg	6

Strengths: Much of the fat in capon is unsaturated. It's a good source of protein, needed to prevent loss of muscle mass when dieting. Significant amounts of B vitamins help prevent various skin and nerve problems. Capon is also a decent source of heme iron, the form best absorbed by the body.

Caution: Be sure to cook chicken thoroughly: If contaminated with salmonella bacteria, undercooked chicken can cause food poisoning. High in purines; should be eaten sparingly by people with gout. *Allergy alert:* Chicken is a food allergen that affects some people.

Description: A capon is a rooster that was castrated when young and fed a fattening diet.

A Better Idea: Remove skin before or after cooking—it contains most of the fat. Microwaving does a good job of liquefying poultry fat so it drains off.

CHICKEN *Back, Broiler/Fryer*

Serving: 1 (about 3 oz.), meat only

Calories	191
Fat	10.5 g
Saturated	2.9 g
Monounsaturated	3.9 g
Polyunsaturated	2.4 g
Calories from fat	50%
Cholesterol	72 mg
Sodium	77 mg
Protein	22.6 g
Carbohydrate	0
Dietary fiber	0

CHIEF NUTRIENTS

NUTRIENT	AMOUNT	% RDA
Niacin	5.7 mg	30
Vitamin B_6	0.3 mg	14
Zinc	2.1 mg	14
Vitamin B_{12}	0.2 mcg	12
Iron	1.1 mg	11
Riboflavin	0.2 mg	11
Magnesium	17.6 mg	5
Potassium	189.6 mg	5

Strengths: A rich source of niacin, chicken has good amounts of other B vitamins to help prevent various skin and nerve problems. It also offers a chunk of infection-fighting zinc, along with heme iron.

Caution: Be sure to cook thoroughly: If contaminated with salmonella bacteria, undercooked chicken can cause food poisoning. High in purines; should be eaten sparingly by people with gout. *Allergy alert:* Chicken is a food allergen that affects some people.

Selection: Broilers and fryers are lower in fat than roasters.

Preparation: If the skin is on, be sure to remove it before or after roasting—it contains most of the fat.

A Better Idea: Microwaving does a good job of liquefying poultry fat so it drains off.

CHICKEN *Breast, Broiler/Fryer*

Serving: ½ (about 3 oz.), meat only

Calories	142
Fat	3.1 g
Saturated	0.9 g
Monounsaturated	1.1 g
Polyunsaturated	0.6 g
Calories from fat	19%
Cholesterol	73 mg
Sodium	64 mg
Protein	26.7 g
Carbohydrate	0
Dietary fiber	0

CHIEF NUTRIENTS

NUTRIENT	AMOUNT	% RDA
Niacin	11.8 mg	62
Vitamin B_6	0.5 mg	26
Vitamin B_{12}	0.3 mcg	15
Iron	0.9 mg	9
Magnesium	24.9 mg	7
Potassium	220.2 mg	6
Riboflavin	0.1 mg	6
Zinc	0.9 mg	6

Strengths: Chicken breast is a good source of protein needed to prevent loss of muscle mass when dieting. The percentage of calories from fat is 19%, a figure considered acceptable by the American Heart Association. A good source of B vitamins, which may have a positive influence on memory and learning.

Caution: Be sure to cook thoroughly: If contaminated with salmonella bacteria, undercooked chicken can cause food poisoning. High in purines; should be eaten sparingly by people with gout. *Allergy alert:* Chicken is a food allergen that affects some people.

Preparation: Remove skin before or after roasting—it contains most of the fat.

A Better Idea: Microwaving does a good job of liquefying poultry fat so it drains off.

CHICKEN *Drumsticks, Broiler/Fryer*

Serving: 2 (about 3 oz.), meat only

Calories	151
Fat	5.0 g
Saturated	1.3 g
Monounsaturated	1.6 g
Polyunsaturated	1.2 g
Calories from fat	30%
Cholesterol	82 mg
Sodium	84 mg
Protein	24.9 g
Carbohydrate	0
Dietary fiber	0

CHIEF NUTRIENTS

NUTRIENT	AMOUNT	% RDA
Niacin	5.3 mg	28
Zinc	2.8 mg	19
Vitamin B_6	0.3 mg	17
Vitamin B_{12}	0.3 mcg	15
Riboflavin	0.2 mg	12
Iron	1.1 mg	11

Strengths: The protein in chicken drumsticks helps to prevent loss of muscle mass when dieting. The percentage of calories from fat, 30%, is about the maximum considered acceptable by the American Heart Association. Since chicken drumsticks are a decent source of niacin, it can help prevent some skin and nerve problems. It also has a respectable amount of readily absorbable iron.

Caution: Be sure to cook chicken thoroughly: If contaminated with salmonella bacteria, undercooked chicken can cause food poisoning. High in purines; should be eaten sparingly by people with gout. *Allergy alert:* Chicken is a food allergen that affects some people.

Preparation: Remove skin before or after roasting—it contains most of the fat.

A Better Idea: Microwaving does a good job of liquefying poultry fat so it drains off.

CHICKEN *Giblets, Broiler/Fryer*

Serving: 3 oz, simmered

Calories	133
Fat	4.1 g
Saturated	1.3 g
Monounsaturated	1.0 g
Polyunsaturated	0.9 g
Calories from fat	27%
Cholesterol	334 mg
Sodium	49 mg
Protein	22.0 g
Carbohydrate	0.8 g
Dietary fiber	0

CHIEF NUTRIENTS

NUTRIENT	AMOUNT	% RDA
Vitamin B$_{12}$	8.6 mcg	431
Vitamin A	1,894.7 RE	189
Folate	319.6 mcg	160
Iron	5.5 mg	55
Riboflavin	0.8 mg	48
Zinc	3.9 mg	26
Niacin	3.5 mg	18
Vitamin B$_6$	0.3 mg	14
Vitamin C	6.8 mg	11
Thiamine	0.1 mg	5

Strengths: Deficits of B vitamins may trigger various skin and nerve problems, and in some instances may impair certain mental processes. Being high in many B vitamins, chicken can help prevent some of these adverse effects. The giblets are also an excellent source of heme iron, the form best absorbed by the body, so they help to insure against iron-deficiency anemia.
Caution: The high cholesterol exceeds the daily maximum of 300 mg recommended by the American Heart Association. Giblets are also high in purines and should be eaten sparingly by people with gout. *Drug interaction:* If giblets include the liver, you will ingest tyramine, which may trigger dangerously high increases in blood pressure in people taking monoamine oxidase (MAO) inhibitors. *Allergy alert:* Chicken is a food allergen that affects some people.
Description: The heart, liver, and gizzard of a chicken.
Serve: In soup and gravy.

CHICKEN *Gizzards, Broiler/Fryer*

Serving: 3 oz., simmered

Calories	130
Fat	3.1 g
Saturated	0.9 g
Monounsaturated	0.8 g
Polyunsaturated	0.9 g
Calories from fat	22%
Cholesterol	165 mg
Sodium	57 mg
Protein	23.1 g
Carbohydrate	1.0 g
Dietary fiber	0

CHIEF NUTRIENTS

NUTRIENT	AMOUNT	% RDA
Vitamin B$_{12}$	1.7 mcg	82
Iron	3.5 mg	35
Zinc	3.7 mg	25
Folate	45.1 mcg	23
Niacin	3.4 mg	18
Riboflavin	0.2 mg	12

Strengths: A great source of vitamin B$_{12}$, chicken gizzards also have appreciable amounts of folate, niacin, and riboflavin which help prevent anemia and various skin and nerve problems. Its rich supply of heme iron is useful for those prone to iron-deficiency anemia. Gizzards are also a good source of zinc.
Eat with: Whole grains, dried beans, and other plant sources of iron to boost absorption of nonheme iron.
Caution: High in cholesterol. The American Heart Association recommends limiting total cholesterol intake to less than 300 mg per day—so 3 oz. of gizzards has more than half a day's supply. Chicken is also high in purines and should be eaten sparingly by people with gout. *Allergy alert:* Chicken is a food allergen that affects some people.

CHICKEN *Leg, Broiler/Fryer*

Serving: 1 (about 3 oz.), meat only

Calories	181
Fat	8.0 g
Saturated	2.2 g
Monounsaturated	2.9 g
Polyunsaturated	1.8 g
Calories from fat	40%
Cholesterol	89 mg
Sodium	86 mg
Protein	25.7 g
Carbohydrate	0
Dietary fiber	0

CHIEF NUTRIENTS

NUTRIENT	AMOUNT	% RDA
Niacin	6.0 mg	32
Vitamin B$_6$	0.4 mg	18
Zinc	2.7 mg	18
Vitamin B$_{12}$	0.3 mcg	15
Riboflavin	0.2 mg	13
Iron	1.2 mg	12
Magnesium	22.8 mg	7

Strengths: A whole chicken leg—thigh and drumstick—is a good source of protein needed to prevent loss of muscle mass when dieting. Chicken legs are rich in niacin, with satisfactory amounts of other B vitamins. (A deficit of B vitamins may affect mental skills and trigger various skin and nerve problems.) It's also a good source of heme iron, the form best absorbed by the body. Plentiful zinc helps fight off infections and is necessary for wound healing.

Caution: Be sure to cook chicken thoroughly: If contaminated with salmonella bacteria, undercooked chicken can cause food poisoning. High in purines; chicken legs should be eaten sparingly by people with gout. *Allergy alert:* Chicken is a food allergen that affects some people.

Preparation: Remove skin before or after roasting—it contains most of the fat.

A Better Idea: Microwaving does a good job of liquefying poultry fat so it drains off.

CHICKEN *Liver, Broiler/Fryer*

Serving: 3 oz., simmered

Calories	133
Fat	4.6 g
Saturated	1.6 g
Monounsaturated	1.1 g
Polyunsaturated	0.7 g
Calories from fat	31%
Cholesterol	536 mg
Sodium	43 mg
Protein	20.7 g
Carbohydrate	0.8 g
Dietary fiber	0

CHIEF NUTRIENTS

NUTRIENT	AMOUNT	% RDA
Vitamin B$_{12}$	16.5 mcg	824
Vitamin A	4,176.1 RE	418
Folate	654.5 mcg	327
Riboflavin	1.5 mg	87
Iron	7.2 mg	72
Vitamin B$_6$	0.5 mg	25
Zinc	3.7 mg	25
Vitamin C	13.4 mg	22
Niacin	3.8 mg	20

Strengths: Incredibly high in resistance-building vitamin A, chicken liver has phenomenal amounts of vitamin B$_{12}$ and generous amounts of other B vitamins. The exceptionally high concentration of folate should be noted by women who are pregnant or taking oral contraceptives: They may run low on this B vitamin. Chicken liver is also a superb source of heme iron, helping to insure against iron-deficiency anemia. A good supply of zinc is accompanied by high vitamin C.

Caution: There's so much cholesterol that just a single 3-oz. serving puts you way over the daily maximum of 300 mg recommended by the American Heart Association. Another drawback is that liver tends to accumulate medicines, hormones, and other unwanted chemicals fed to chickens. *Drug interaction:* Contains tyramine, which may trigger dangerously high increases in blood pressure in people taking MAO inhibitors.

CHICKEN Thigh, Broiler/Fryer

Serving: 1 (about 2 oz.), meat only

Calories	109
Fat	5.7 g
Saturated	1.6 g
Monounsaturated	2.2 g
Polyunsaturated	1.3 g
Calories from fat	47%
Cholesterol	49 mg
Sodium	46 mg
Protein	13.5 g
Carbohydrate	0
Dietary fiber	0

CHIEF NUTRIENTS

NUTRIENT	AMOUNT	% RDA
Niacin	3.4 mg	18
Vitamin B$_6$	0.2 mg	9
Zinc	1.3 mg	9
Vitamin B$_{12}$	0.2 mcg	8
Iron	0.7 mg	7
Riboflavin	0.1 mg	7

Strengths: Much of the fat in chicken thigh is unsaturated. The thigh is a decent source of niacin, with fair amounts of other B vitamins to help prevent skin and nerve problems. Chicken also supplies some heme iron, the form best absorbed by the body, and a little zinc to help people on the mend.

Caution: Be sure to cook chicken thoroughly: If contaminated with salmonella bacteria, undercooked chicken can cause food poisoning. Chicken is high in purines and should be eaten sparingly by people with gout. *Allergy alert:* Chicken is a food allergen that affects some people.

Preparation: Remove skin before or after roasting—it contains most of the fat.

A Better Idea: Microwaving does a good job of liquefying poultry fat so it drains off.

CHICKEN Wings, Broiler/Fryer

Serving: 4 (about 3 oz.), meat only

Calories	171
Fat	6.8 g
Saturated	1.9 g
Monounsaturated	2.2 g
Polyunsaturated	1.4 g
Calories from fat	36%
Cholesterol	71 mg
Sodium	77 mg
Protein	25.6 g
Carbohydrate	0
Dietary fiber	0

CHIEF NUTRIENTS

NUTRIENT	AMOUNT	% RDA
Niacin	6.2 mg	32
Vitamin B$_6$	0.5 mg	24

Strengths: Chicken wings are very rich in niacin, and they have a good amount of vitamin B$_6$.

Caution: Fat content almost quadruples if chicken wings are served with the skin. Be sure to cook thoroughly: If contaminated with salmonella bacteria, undercooked chicken can cause food poisoning. Chicken is high in purines and should be eaten sparingly by people with gout. *Allergy alert:* Chicken is a food allergen that affects some people.

Preparation: Remove skin before or after roasting—it contains most of the fat.

A Better Idea: If you're accustomed to chicken wings cooked Buffalo-style, try them plain, without the heavy sauce, for reduced fat.

DUCK

Serving: 3 oz., meat only, roasted

Calories	171
Fat	9.5 g
Saturated	3.6 g
Monounsaturated	3.2 g
Polyunsaturated	1.2 g
Calories from fat	50%
Cholesterol	76 mg
Sodium	55 mg
Protein	20.0 g
Carbohydrate	0
Dietary fiber	0

CHIEF NUTRIENTS

NUTRIENT	AMOUNT	% RDA
Riboflavin	0.4 mg	24
Iron	2.3 mg	23
Niacin	4.3 mg	23
Vitamin B_{12}	0.3 mcg	17
Thiamine	0.2 mg	15
Zinc	2.2 mg	15
Vitamin B_6	0.2 mg	11
Potassium	214.2 mg	6

Strengths: Roast duck is a valuable source of heme iron, the form most easily absorbed by the body. It also has a high profile in B vitamins, especially the niacin that is essential for healthy red blood cells.

Eat with: Vegetable sources of iron like beans or fruit to boost absorption of their mineral content. Serving low-fat side dishes, such as potatoes, complements duck without adding much fat.

Caution: Duck is fattier than chicken or turkey. Be sure to discard the skin before eating, especially if you're on a heart-healthy diet. Also, duck is high in purines and should be eaten sparingly by people with gout.

Selection: Some types of duck are lower in fat than others. Muscovy is a lean, meaty, tasty breed often available in farmers' markets.

GOOSE

Serving: 3 oz., meat only, roasted

Calories	202
Fat	10.8 g
Saturated	3.9 g
Monounsaturated	3.7 g
Polyunsaturated	1.3 g
Calories from fat	48%
Cholesterol	82 mg
Sodium	65 mg
Protein	24.6 g
Carbohydrate	0
Dietary fiber	0

CHIEF NUTRIENTS

NUTRIENT	AMOUNT	% RDA
Iron	2.4 mg	24
Vitamin B_{12}	0.4 mcg	21
Vitamin B_6	0.4 mg	20
Riboflavin	0.3 mg	19
Niacin	3.5 mg	18
Zinc	2.7 mg	18
Potassium	329.8 mg	9
Magnesium	21.3 mg	6
Folate	10.2 mcg	5

Strengths: Like pork and some cuts of chicken, goose is a good source of immunity-boosting zinc and essential B vitamins. It supplies generous amounts of protein and plentiful heme iron, the form best absorbed by the body. Magnesium and potassium may help lower blood pressure, but the high fat content is a real drawback if you're trying to lose weight.

Eat with: Vegetable sources of iron like beans or fruit to boost their absorption of this nonheme mineral. Low-fat side dishes such as potatoes or apricot and apple stuffing help to complement this high-fat poultry.

Caution: Goose tends to be higher in fat and calories than turkey, so it's fortunate that many people consider roast goose a holiday treat rather than everyday fare. If you're trying to lose weight or avoid dietary fat, keep portions to a modest 3 oz.— or less.

TURKEY, *Dark Meat*

Serving: 3 oz., meat only, roasted

Calories	159
Fat	6.1 g
Saturated	2.1 g
Monounsaturated	1.4 g
Polyunsaturated	1.9 g
Calories from fat	35%
Cholesterol	72 mg
Sodium	67 mg
Protein	24.3 g
Carbohydrate	0
Dietary fiber	0

CHIEF NUTRIENTS

NUTRIENT	AMOUNT	% RDA
Zinc	3.8 mg	25
Iron	2.0 mg	20
Niacin	3.1 mg	16
Vitamin B_{12}	0.3 mcg	16
Vitamin B_6	0.3 mg	15
Riboflavin	0.2 mg	13
Potassium	246.5 mg	7
Magnesium	20.4 mg	6

Strengths: An excellent source of protein necessary for maintaining muscle mass when dieting, turkey has respectable amounts of B vitamins. This is a benefit, since lack of B vitamins can trigger various skin and nerve problems. Dark meat turkey is a valuable source of heme iron, the form best absorbed by the body, so it's good insurance against iron-deficiency anemia. With a good amount of immunity-boosting zinc, the poultry has some magnesium and potassium.

Caution: Avoid self-basting turkeys, which have been injected with butter or oil: These are fattier than any other kind. Turkey is high in purines and should be eaten sparingly by people with gout.

Preparation: Do not partially roast a stuffed turkey one day and complete roasting the next. Interrupted cooking may raise bacteria concentrations to dangerous levels. Also, refrigerate turkey as soon as possible after cooking—within 2 hours. Remove the stuffing before refrigerating.

TURKEY *Gizzards*

Serving: 3 oz., simmered

Calories	139
Fat	3.3 g
Saturated	0.9 g
Monounsaturated	0.6 g
Polyunsaturated	0.9 g
Calories from fat	21%
Cholesterol	197 mg
Sodium	46 mg
Protein	25.0 g
Carbohydrate	0.5 g
Dietary fiber	0

CHIEF NUTRIENTS

NUTRIENT	AMOUNT	% RDA
Vitamin B_{12}	1.6 mcg	81
Iron	4.6 mg	46
Zinc	3.5 mg	24
Folate	44.2 mcg	22
Riboflavin	0.3 mg	16
Niacin	2.6 mg	14
Vitamin B_6	0.1 mg	5

Strengths: An excellent source of protein and vitamin B_{12}, turkey gizzards are also rich in immunity-boosting iron, with respectable amounts of zinc and B vitamins. It's a good source of folate. Pregnant women and those on oral contraceptives may benefit from the extra folate.

Eat with: Restraint.

Caution: High in cholesterol. The American Heart Association recommends limiting total daily intake of cholesterol to less than 300 mg per day: 1 small serving of turkey gizzard has two-thirds of that. Turkey is high in purines and should be eaten sparingly by people with gout.

TURKEY *Liver*

Serving: 3 oz., simmered

Calories	144
Fat	5.1 g
Saturated	1.6 g
Monounsaturated	1.3 g
Polyunsaturated	0.9 g
Calories from fat	32%
Cholesterol	532 mg
Sodium	54 mg
Protein	20.4 g
Carbohydrate	2.9 g
Dietary fiber	0

CHIEF NUTRIENTS

NUTRIENT	AMOUNT	% RDA
Vitamin B_{12}	40.4 mcg	2,019
Vitamin A	3,179.9 RE	318
Folate	566.1 mcg	283
Riboflavin	1.2 mg	71
Iron	6.6 mg	66
Niacin	5.1 mg	27
Vitamin B_6	0.4 mg	22
Zinc	2.6 mg	18

Strengths: As a phenomenally good source of vitamin B_{12} and a super source of heme iron, turkey liver is ideal for people who are prone to anemia. It's also a terrific source of resistance-building vitamin A, folate, and riboflavin, and it has a good supply of infection-fighting zinc.

Eat with: Restraint.

Caution: This food is extraordinarily high in cholesterol. The American Heart Association recommends less than 300 mg of cholesterol per day; a 3-oz. helping of turkey liver puts you over 500 mg. Turkey is high in purines and should be eaten sparingly by people with gout. Liver in general is a depot for hormones, medicines, and other unwanted chemicals that are filtered from the animal's body.

TURKEY, *White Meat*

Serving: 3 oz., meat only, roasted

Calories	133
Fat	2.7 g
Saturated	0.9 g
Monounsaturated	0.5 g
Polyunsaturated	0.7 g
Calories from fat	18%
Cholesterol	59 mg
Sodium	54 mg
Protein	25.4 g
Carbohydrate	0
Dietary fiber	0

CHIEF NUTRIENTS

NUTRIENT	AMOUNT	% RDA
Niacin	5.8 mg	31
Vitamin B_6	0.5 mg	23
Vitamin B_{12}	0.3 mcg	16
Iron	1.2 mg	11
Zinc	1.7 mg	12
Magnesium	23.8 mg	7
Potassium	259.3 mg	7

Strengths: The white meat in turkey breast and wings has less fat than the dark meat in drumsticks and thighs. In general, turkey is a superb source of protein, which is needed to maintain muscle mass when dieting. This poultry gets high marks for B vitamins and iron and other minerals.

Caution: Turkey is high in purines and should be eaten sparingly by people with gout. To avoid unnecessary fat, avoid self-basting turkeys—they have been injected with butter or oil.

Preparation: Do not partially roast a stuffed turkey one day and complete roasting the next: Interrupted cooking may raise bacteria growth to dangerous levels. Also, refrigerate turkey as soon as possible after cooking—within 2 hours. Remove the stuffing before refrigerating.

Serve: On whole grain bread for sandwiches. Sliced extra thin and piled high with lettuce and tomatoes, or cubed into stew or stir-fries, a little turkey can go a long way. Add turkey meat to a chef's salad that has lots of high-fiber, low-fat vegetables. Top with nonfat dressing.

VEAL

ARM ROAST

Serving: 3 oz., lean only, roasted

Calories	139
Fat	4.9 g
Saturated	2.0 g
Monounsaturated	1.8 g
Polyunsaturated	0.4 g
Calories from fat	32%
Cholesterol	93 mg
Sodium	77 mg
Protein	22.2 g
Carbohydrate	0
Dietary fiber	0

CHIEF NUTRIENTS

NUTRIENT	AMOUNT	% RDA
Vitamin B$_{12}$	1.3 mcg	67
Niacin	7.0 mg	37
Zinc	3.7 mg	24
Riboflavin	0.3 mg	16
Vitamin B$_6$	0.3 mg	13
Iron	1.0 mg	10
Potassium	302.6 mg	8

Strengths: A very nice profile of hard-to-get minerals and B vitamins—and less fat than most cuts of beef. For those on a low-fat diet, an occasional serving of veal arm is an alternative to poultry. It has about the same amount of cholesterol as a chicken leg. Veal contains a fair amount of folate, but it has less iron than an equivalent serving of beef.

Eat with: Generous servings of high-fiber vegetables like brussels sprouts, beans, or grains. Since veal, like other animal foods, has no fiber, the high-fiber vegetables are an important complement.

Caution: Since veal arm exceeds the amount of saturated fat recommended by the American Heart Assocaition, avoid frequent consumption. A 3-oz. portion is about the size of a deck of cards.

BLADE ROAST

Serving: 3 oz., lean only, roasted

Calories	145
Fat	5.9 g
Saturated	2.2 g
Monounsaturated	2.1 g
Polyunsaturated	0.5 g
Calories from fat	36%
Cholesterol	101 mg
Sodium	87 mg
Protein	21.8 g
Carbohydrate	0
Dietary fiber	0

CHIEF NUTRIENTS

NUTRIENT	AMOUNT	% RDA
Vitamin B$_{12}$	1.8 mcg	88
Zinc	4.9 mg	32
Niacin	5.0 mg	26
Riboflavin	0.3 mg	18
Vitamin B$_6$	0.2 mg	10
Iron	0.9 mg	9
Potassium	263.5 mg	7
Magnesium	20.4 mg	6

Strengths: This cut has a little more fat and calories than veal arm, but not much. Veal blade contains good amounts of niacin for red blood cells, riboflavin for B vitamin metabolism, and vitamin B$_6$ for infection fighting. It provides nearly an entire day's quota of vitamin B$_{12}$. As for minerals, it's rich in zinc and has modest amounts of iron, potassium, and magnesium to help keep blood healthy and free-flowing. This is a good choice for people prone to canker sores, as studies have found that about 15 percent of people who suffer mouth sores are deficient in iron, folate, or other B vitamins.

Eat with: High-fiber complex carbohydrates like sweet potatoes or black-eyed peas to boost the nutritional benefits of the meal.

Caution: Saturated fat exceeds the amount recommended by the American Heart Association. A caveat for cholesterol watchers: Veal blade has a little more cholesterol than beef, poultry, or other cuts of veal.

Selection: Often labeled veal shoulder.

GROUND VEAL

Serving: 3 oz., broiled

Calories	146
Fat	6.4 g
Saturated	2.6 g
Monounsaturated	2.4 g
Polyunsaturated	0.5 g
Calories from fat	40%
Cholesterol	88 mg
Sodium	71 mg
Protein	20.7 g
Carbohydrate	0
Dietary fiber	0

CHIEF NUTRIENTS

NUTRIENT	AMOUNT	% RDA
Vitamin B_{12}	1.1 mcg	54
Niacin	6.8 mg	36
Zinc	3.3 mg	22
Vitamin B_6	0.3 mg	17
Riboflavin	0.2 mg	14
Iron	0.8 mg	8
Potassium	286.5 mg	8
Magnesium	20.4 mg	6

Strengths: Leaner than the leanest of ground beef, ground veal is an excellent source of protein. The nutritional profile is handsome: Veal is a good source of vitamin B_6, a very good source of niacin, and a superb source of vitamin B_{12}. Although it has good amounts of zinc, ground veal has far less iron than ground beef.

Eat with: High-fiber foods like kidney beans, lentils, corn on the cob, cooked beets, asparagus, mixed grain bread, or pumpernickel pita pockets: They counterbalance the lack of fiber and carbohydrates in meat.

Caution: Ground veal is high in saturated fat—not recommended for those on low-fat diets, even though it's preferable to ground beef.

Serve: In hamburgers, meatballs, meat loaf, stuffed peppers, or chili.

LEG

Serving: 3 oz., lean only, roasted

Calories	128
Fat	2.9 g
Saturated	1.0 g
Monounsaturated	1.0 g
Polyunsaturated	0.3 g
Calories from fat	20%
Cholesterol	88 mg
Sodium	58 mg
Protein	23.9 g
Carbohydrate	0
Dietary fiber	0

CHIEF NUTRIENTS

NUTRIENT	AMOUNT	% RDA
Vitamin B_{12}	1.0 mcg	50
Niacin	8.6 mg	45
Zinc	2.6 mg	17
Riboflavin	0.3 mg	16
Vitamin B_6	0.3 mg	13
Potassium	334.1 mg	9
Iron	0.8 mg	8
Folate	13.6 mcg	7

Strengths: The saturated fat is lower than in any other cut of veal, and cholesterol levels are about average, making veal leg a nice change of pace for dinner. Total fat and calories are even lower than roast chicken breast, and the level of vitamin B_{12} is 3 times as high. As an added bonus, veal leg contains some folate, making it a good choice for people who tend to develop anemia.

Eat with: High-fiber, low-fat side dishes like broccoli, carrots, and baked apples. A calcium-rich side dish or dessert (such as yogurt) can boost the magnesium and potassium in veal to help lower blood pressure.

Caution: Even though veal is one of the leanest meats, it's best to limit a portion to 3 oz.

LIVER

Serving: 3 oz., braised

Calories	140
Fat	5.9 g
Saturated	2.2 g
Monounsaturated	1.3 g
Polyunsaturated	0.9 g
Calories from fat	38%
Cholesterol	477 mg
Sodium	45 mg
Protein	18.4 g
Carbohydrate	2.3 g
Dietary fiber	0

CHIEF NUTRIENTS

NUTRIENT	AMOUNT	% RDA
Vitamin B_{12}	31.0 mcg	1,552
Vitamin A	6,841.7 RE	684
Folate	645.2 mcg	323
Riboflavin	1.7 mg	97
Zinc	8.1 mg	54
Vitamin C	26.4 mg	44
Niacin	7.2 mg	38
Iron	2.2 mg	22
Vitamin B_6	0.4 mg	21

Caution: Veal liver is a prime repository for cholesterol, a fatlike substance that can clog arteries. In addition, it's high in saturated fat—above the 10% level recommended by the American Heart Association. If you have hypercholesterolemia—a tendency toward super-high blood cholesterol that may be genetic—this liver is not a prudent choice.

Strengths: As with beef liver, veal liver supplies super levels of vitamin B_{12}—two weeks' worth in a single serving. Added to that is almost a week's supply of vitamin A and more folate than you'll find in just about any other food. Both of these nutrients help boost immunity. Vitamin A turns on the body's T-lymphocyte cells, and folate is required to make special white blood cells called macrophages, which help fight infection. Veal liver is also a superb source of riboflavin, which is required for B vitamin metabolism.

LOIN

Serving: 3 oz., lean only, roasted

Calories	149
Fat	5.9 g
Saturated	2.2 g
Monounsaturated	2.1 g
Polyunsaturated	0.5 g
Calories from fat	36%
Cholesterol	90 mg
Sodium	82 mg
Protein	22.4 g
Carbohydrate	0
Dietary fiber	0

CHIEF NUTRIENTS

NUTRIENT	AMOUNT	% RDA
Vitamin B_{12}	1.1 mcg	56
Niacin	8.0 mg	42
Zinc	2.8 mg	18
Vitamin B_6	0.3 mg	16
Riboflavin	0.3 mg	15
Potassium	289.0 mg	8
Folate	13.6 mcg	7

Strengths: Veal is an outstanding source of protein, which is needed to maintain muscle mass while losing weight. With a superlative supply of vitamin B_{12}, it's also a very good source of niacin, and has good supplies of riboflavin and vitamin B_6. A serving of veal can benefit women on birth control pills, as oral contraceptives may increase requirements for B_6. It contributes significant amounts of zinc, and also contains fair amounts of potassium. A very good source of protein, this meat is better than most if you're counting calories—but saturated fat is high.

Eat with: Chunky applesauce, beets, corn muffins, or other high-fiber fare. Veal, like other animal foods, has no fiber.

Caution: Veal, since it comes from young animals, is lean but is *not* fat-free, and it does contribute some cholesterol. Limit portions to 3 oz. or less.

RIB ROAST

Serving: 3 oz., lean only, roasted

Calories	150
Fat	6.3 g
Saturated	1.8 g
Monounsaturated	2.3 g
Polyunsaturated	0.6 g
Calories from fat	38%
Cholesterol	98 mg
Sodium	82 mg
Protein	21.9 g
Carbohydrate	0
Dietary fiber	0

CHIEF NUTRIENTS

NUTRIENT	AMOUNT	% RDA
Vitamin B$_{12}$	1.3 mcg	67
Niacin	6.4 mg	34
Zinc	3.8 mg	25
Riboflavin	0.3 mg	15
Vitamin B$_6$	0.2 mg	12
Iron	0.8 mg	8
Potassium	264.4 mg	7
Folate	11.9 mcg	6

Strengths: With a whopping amount of vitamin B$_{12}$, this cut of veal is a good choice for people who tend to develop anemia or other symptoms of B$_{12}$ deficiency. Veal rib is also rich in niacin, with significant amounts of other B vitamins. As a source of iron, it's fair—supplying about the same as a chicken breast.

Eat with: Beans, rice, potatoes, carrots, and other starchy foods containing fiber. These will also supply some carbohydrates lacking in meat.

Caution: Veal rib roast has a bit more fat and calories than veal loin but less than chuck pot roast, blade steak, brisket, ribs, and other fatty cuts of beef. Just remember to keep portion size down to 3 oz.—the size of a deck of cards.

SIRLOIN

Serving: 3 oz., lean only, roasted

Calories	143
Fat	5.3 g
Saturated	2.1 g
Monounsaturated	1.9 g
Polyunsaturated	0.4 g
Calories from fat	33%
Cholesterol	88 mg
Sodium	72 mg
Protein	22.4 g
Carbohydrate	0
Dietary fiber	0

CHIEF NUTRIENTS

NUTRIENT	AMOUNT	% RDA
Vitamin B$_{12}$	1.3 mcg	64
Niacin	7.9 mg	42
Zinc	3.0 mg	20
Riboflavin	0.3 mg	18
Vitamin B$_6$	0.3 mg	15
Iron	0.8 mg	8
Potassium	310.3 mg	8
Folate	13.6 mcg	7
Magnesium	23.0 mg	7

Strengths: The nutrient profile for veal sirloin is similar to that of veal rib, making it a good choice for those who want infection-fighting, energy-boosting B vitamins. Veal sirloin is particularly rich in niacin (skin problems, mental confusion, diarrhea, and irritability can result from niacin deficiency). It's also a significant source of zinc, which is needed to maintain normal taste and smell.

Eat with: Peas, sweet potatoes, and other low-fat, high-fiber sources of carbohydrates.

Caution: Like other lean cuts of veal, the sirloin contains fat and cholesterol. With that in mind, anyone on a cholesterol-lowering regimen should only eat a modest portion.

OTHER MEAT

FROGS' LEGS

⬇ 🏛

Serving: 3 oz., uncooked

Calories	62
Fat	0.3 g
Saturated	0
Monounsaturated	trace
Polyunsaturated	trace
Calories from fat	4%
Cholesterol	43 mg
Sodium	49 mg
Protein	13.9 g
Carbohydrate	0
Dietary fiber	0

CHIEF NUTRIENTS

NUTRIENT	AMOUNT	% RDA
Iron	1.3 mg	13
Riboflavin	0.2 mg	13
Thiamine	0.1 mg	8
Potassium	242.3 mg	6
Niacin	1.0 mg	5

Strengths: Frogs' legs provide a good amount of blood-building iron, along with some niacin, riboflavin, and thiamine—B vitamins important for the production of energy. They are extremely lean, getting about 4% of their calories from fat.
Curiosity: In rare cases, men can get priapism—a prolonged, painful erection—from eating frogs' legs. Priapism occurs when "Spanish fly" *(cantharidin)* is present; some frogs gorge on beetles containing this substance.
Description: The meat is tender and lightly sweet.
Selection: Frogs' legs are usually sold in pairs, fresh or frozen. Fresh, plump, pink legs are much preferred.
Storage: May be stored, loosely wrapped, in the refrigerator for up to 2 days.
Preparation: Dust lightly with seasoned flour, then quickly sauté in butter or olive oil. Overcooking toughens frogs' legs.

GOAT

☀ ⚡ ⬇ 🏛

Serving: 3 oz., roasted

Calories	122
Fat	2.6 g
Saturated	0.8 g
Monounsaturated	1.2 g
Polyunsaturated	0.2 g
Calories from fat	19%
Cholesterol	64 mg
Sodium	73 mg
Protein	23.0 g
Carbohydrate	0
Dietary fiber	0

CHIEF NUTRIENTS

NUTRIENT	AMOUNT	% RDA
Vitamin B_{12}	1.0 mcg	51
Iron	3.2 mg	32
Riboflavin	0.5 mg	31
Zinc	4.5 mg	30
Niacin	3.4 mg	18
Potassium	344.3 mg	9
Thiamine	0.1 mg	5

Strengths: Goat meat is rich in B vitamins, especially riboflavin and vitamin B_{12}. A good source of protein, it's slightly lower in fat and calories than roast chicken breast. Since it's a very good source of heme iron, the form most readily absorbed by the body, goat is good insurance against iron-deficiency anemia. The meat is rich in zinc and has a little potassium.
Eat with: Beans, grains, and other foods high in nonheme iron, to boost total absorption of iron.
Curiosity: Goat is often served as a main dish in arab countries.
Description: Though closely related to sheep, goats are less discerning in their feeding habits. Old goat tastes, well, goaty—but young kid resembles lamb and is more agreeable.

410

Nuts and Seeds

ALMONDS, Dried, Unblanched

Serving: 1 oz.

Calories	167
Fat	14.8 g
Saturated	1.4 g
Monounsaturated	9.6 g
Polyunsaturated	3.1 g
Calories from fat	80%
Cholesterol	0
Sodium	3 mg
Protein	5.7 g
Carbohydrate	5.8 g
Dietary fiber	1.9 g

CHIEF NUTRIENTS

NUTRIENT	AMOUNT	% RDA
Magnesium	84.1 mg	24
Riboflavin	0.2 mg	13
Iron	1.0 mg	10
Calcium	75.5 mg	9
Folate	16.7 mcg	8
Potassium	207.9 mg	6
Zinc	0.8 mg	6

Strengths: A good source of fiber, almonds are low in sodium and have no cholesterol. Although they're high in fat, most of it is monounsaturated, which can help control serum cholesterol. There are fair amounts of the nutrients folate and zinc. Almonds are a good source of magnesium, which helps regulate blood pressure, and also of iron, to help prevent anemia. They also supply fair amounts of potassium and calcium.

Eat with: Foods rich in vitamin C to enhance iron absorption. (Try sprinkling chopped toasted almonds over spinach salad or steamed broccoli.)

Caution: Almonds are too high in fat and calories for absent-minded snacking, and they contain oxalic acid, which should be restricted by those with calcium-oxalate stones. *Allergy alert:* Nuts are highly allergenic, often causing hives, headaches, and other reactions. Those allergic to aspirin may react to natural salicylates found in almonds.

Preparation: Toasting or dry-roasting almonds does not significantly change their nutrient content. But salting them raises sodium to 222 mg per oz.

BRAZIL NUTS, Dried, Unblanched

Serving: 1 oz.

Calories	186
Fat	18.8 g
Saturated	4.6 g
Monounsaturated	6.5 g
Polyunsaturated	6.8 g
Calories from fat	91%
Cholesterol	0
Sodium	1 mg
Protein	4.1 g
Carbohydrate	3.6 g
Dietary fiber	1.6 g

CHIEF NUTRIENTS

NUTRIENT	AMOUNT	% RDA
Thiamine	0.3 mg	19
Magnesium	63.9 mg	18
Iron	1.0 mg	10
Zinc	1.3 mg	9
Calcium	50.0 mg	6

Strengths: Low in sodium, with no cholesterol, Brazil nuts have a nice amount of fiber. They contain a substance called ellagic acid that seems capable of eliminating certain cancer-causing agents from the body. Although the nuts are high in fat, most of it is monounsaturated and polyunsaturated (both of which can help control cholesterol). They are also a nice source of iron and calcium.

Eat with: Oranges and other foods high in vitamin C to enhance iron absorption. (Sprinkle chopped nuts on a fruit cup or over lightly sautéed red peppers.)

Caution: Too high in fat and calories for absent-minded eating. Certain components of the nuts may inhibit iron absorption in people on no-meat, low-C diets. *Allergy alert:* Nuts are highly allergenic, often causing hives, headaches, and other reactions.

Storage: Shelled Brazil nuts should be kept refrigerated; their high oil content makes them susceptible to rancidity if they are stored in a warm place.

Preparation: To make shelling easier, boil the nuts for 3 minutes, drain, and cool. Use a nutcracker.

CASHEW NUTS, *Dry-Roasted*

Serving: 1 oz.

Calories	163
Fat	13.2 g
Saturated	2.6 g
Monounsaturated	7.8 g
Polyunsaturated	2.2 g
Calories from fat	73%
Cholesterol	0
Sodium	5 mg
Protein	4.4 g
Carbohydrate	9.3 g
Dietary fiber	0.9 g

CHIEF NUTRIENTS

NUTRIENT	AMOUNT	% RDA
Magnesium	73.8 mg	21
Iron	1.7 mg	17
Zinc	1.6 mg	11
Folate	219.7 mcg	10

Strengths: As long as cashews are unsalted, they're low in sodium. They have no cholesterol—and even though they're high in fat, most of it is monounsaturated, which can help control serum cholesterol. Their immunity-enhancing nutrients include iron, zinc, and folate. And they have plentiful magnesium, which is important for the normal functioning of nerves and muscles. Cashews also contain some fiber to aid digestion.

Eat with: Foods high in vitamin C for better iron absorption. (Sprinkle some chopped cashews over steamed green beans or add them to a fruity chicken salad.)

Caution: These nuts are too high in fat and calories for absent-minded snacking. And they contain oxalic acid, which should be restricted by those with calcium-oxalate stones. *Allergy alert:* Nuts are highly allergenic, often causing hives, headaches, and other reactions.

Curiosity: Cashews are related to poison ivy. In fact, their shells contain a caustic oil that can blister the hands. It must be removed by heat or solvents before the nuts can be extracted.

CHESTNUTS, *European, Roasted*

Serving: 1 oz.

Calories	70
Fat	0.6 g
Saturated	trace
Monounsaturated	trace
Polyunsaturated	trace
Calories from fat	8%
Cholesterol	0
Sodium	1 mg
Protein	0.9 g
Carbohydrate	15.0 g
Dietary fiber	3.7 g

CHIEF NUTRIENTS

NUTRIENT	AMOUNT	% RDA
Vitamin C	7.4 mg	12
Folate	19.9 mcg	10
Vitamin B$_6$	0.1 mg	7

Strengths: Chestnuts have barely any fat, which is unusual for a nut. Calories and sodium are low, and there's no cholesterol. Combined with an excellent amount of fiber, chestnuts also yield a surprising amount of vitamin C, which helps heal wounds, fight infections, and form collagen. Their B vitamins help promote healthy skin and nerves.

Curiosity: It is said that Mount Olympus, home of the gods, was covered with chestnut trees.

Selection: The nutrient figures given here are for European chestnuts—the ones most often sold around the holidays. Chinese and Japanese chestnuts, the types generally grown in the U.S., are slightly lower in fat and calories, with similar nutrient values.

Preparation: Chestnuts are difficult to peel unless they've been cooked. But always pierce the shells before boiling or roasting so the nuts don't explode.

Serve: With anything. Chestnuts are good in stuffings, pilafs, vegetable side dishes, and soups. They're also excellent snacks all by themselves and make a delicious low-fat topping for ice milk.

COCONUT, *Raw*

Serving: 1 oz., grated

Calories	99
Fat	9.4 g
Saturated	8.3 g
Monounsaturated	0.4 g
Polyunsaturated	0.1 g
Calories from fat	85%
Cholesterol	0
Sodium	6 mg
Protein	0.9 g
Carbohydrate	4.3 g
Dietary fiber	2.5 g

CHIEF NUTRIENTS

NUTRIENT	AMOUNT	% RDA
Iron	0.7 mg	7

Caution: Raw coconut is too high in fat for frequent consumption; almost all the fat is the unhealthy saturated form that can raise cholesterol levels.

Eat with: Restraint.

Strengths: Low in sodium, with no cholesterol. Contains quite a bit of fiber to help prevent constipation, plus some iron that builds strong blood. But there are far healthier sources of these nutrients.

Curiosity: The word *coconut* is derived from the Portuguese word for goblin or monkey and reflects the fact that markings on the nut resemble a face.

Preparation: To open a fresh coconut, first pierce the indented "eyes" with an ice pick and drain out the liquid. Bake the coconut at 350°F for 30 minutes. Then tap with a hammer until the shell opens.

A Better Idea: For an occasional treat, chew on a piece of fresh coconut meat. A 1 × 1 × ½-inch piece—which will keep your jaws busy for a *long* time—has a respectable 1 g of fiber but only 40 calories and 3.8 g of fat.

COCONUT, *Sweetened, Flaked*

Serving: 1 oz.

Calories	126
Fat	9.0 g
Saturated	8.0 g
Monounsaturated	0.4 g
Polyunsaturated	0.1 g
Calories from fat	64%
Cholesterol	0
Sodium	6 mg
Protein	1.0 g
Carbohydrate	11.7 g
Dietary fiber	4.7 g

CHIEF NUTRIENTS

NUTRIENT	AMOUNT	% RDA
Iron	0.5 mg	5

Caution: Coconut is high in fat, and almost all the fat is in the unhealthy saturated form that can raise serum cholesterol levels. Contains added sugar, which some people might want to avoid.

Eat with: Restraint.

Strengths: Low in sodium, with no cholesterol. Contains a lot of fiber for healthy digestion and a nice amount of iron that helps prevent anemia. But, all things considered, there are far better sources of these nutrients.

Curiosity: Although its meat is very different, the coconut is anatomically similar to an almond. The coconut is an important food for about one-third of the world's population. And various parts of the tree are used for such diverse products as ropes, nets, fuel, baskets, mats, bowls, and wood for building.

Origin: Unknown. Coconuts have been used for so long and are so widespread in tropical countries that no one is sure where they originated.

COCONUT CREAM, *Canned*

Serving: ¼ cup

Calories	142
Fat	13.1 g
Saturated	11.6 g
Monounsaturated	0.6 g
Polyunsaturated	0.1 g
Calories from fat	83%
Cholesterol	0
Sodium	37 mg
Protein	2.0 g
Carbohydrate	6.2 g
Dietary fiber	N/A

CHIEF NUTRIENTS

NUTRIENT	AMOUNT	% RDA
Folate	10.5 mcg	5

Caution: Sweetened coconut cream is very high in fat, almost all of which is saturated and can contribute to high cholesterol, hardening of the arteries, and heart disease. If you buy this canned product, you'll consume a hefty amount of calories, considering the serving size.

Strength: The only health benefit is provided by a little bit of folate.

Description: Coconut cream is made by simmering 4 parts shredded coconut with 1 part water or milk until foamy. It's then strained through cheesecloth to extract as much of the liquid as possible. It may be sweetened for use in desserts and tropical drinks. A similar product, coconut milk, is made in the same way but with a larger quantity of water.

Selection: Store-bought coconut cream is not the same as the watery liquid inside raw coconuts, which is very low in fat and calories and can be drunk straight from the coconut.

FILBERTS, *Dried, Unblanched*

Serving: 1 oz.

Calories	179
Fat	17.8 g
Saturated	1.3 g
Monounsaturated	13.9 g
Polyunsaturated	1.7 g
Calories from fat	89%
Cholesterol	0
Sodium	1 mg
Protein	3.7 g
Carbohydrate	4.4 g
Dietary fiber	1.8 g

CHIEF NUTRIENTS

NUTRIENT	AMOUNT	% RDA
Magnesium	80.9 mg	23
Folate	20.4 mcg	10
Iron	0.9 mg	9
Thiamine	0.1 mg	9
Vitamin B_6	0.2 mg	9
Calcium	53.4 mg	7

Strengths: Filberts are a pretty good source of fiber, are low in sodium, and have no cholesterol. The fat they contain is largely monounsaturated, which can help control serum cholesterol. They have good amounts of folate and iron, which boost immunity, while magnesium and some calcium help regulate blood pressure. Some B vitamins contribute to healthy skin, nerves, and muscles.

Eat with: Foods rich in vitamin C to enhance absorption of the iron. (Sprinkle chopped filberts over mashed sweet potatoes or cooked kale.)

Caution: Too high in fat and calories for everyday use. Certain components of nuts may inhibit iron absorption if you're on a no-meat, low-C diet. *Allergy alert:* Nuts in general are highly allergenic, often causing hives, headaches, and other reactions. Filberts may cause itching of the throat and mouth, especially if you're allergic to birch pollen.

Curiosity: Filberts are thought to be named for Saint Philibert, whose feast day occurs at the time of the nut harvest in Europe. Their alternate name is hazelnuts (they grow on hazel trees).

HICKORY NUTS, *Dried*

Serving: 1 oz.

Calories	187
Fat	18.3 g
Saturated	2.0 g
Monounsaturated	9.3 g
Polyunsaturated	6.2 g
Calories from fat	88%
Cholesterol	0
Sodium	0.3 mg
Protein	3.6 g
Carbohydrate	5.2 g
Dietary fiber	1.8 g

CHIEF NUTRIENTS

NUTRIENT	AMOUNT	% RDA
Thiamine	0.3 mg	17
Magnesium	49.1 mg	14
Zinc	1.2 mg	8
Folate	11.4 mcg	6
Iron	0.6 mg	6

Strengths: Fiber-rich hickory nuts are low in sodium and contain no cholesterol. Although they're high in fat, most of it is in the beneficial forms—monounsaturated and polyunsaturated—that can help control serum cholesterol. The nuts have immunity-strengthening nutrients, including zinc, folate, and iron. They're also a good source of thiamine and magnesium for a healthy nervous system.

Eat with: Foods rich in vitamin C to enhance absorption of the iron. (For a nice breakfast topping, sprinkle chopped hickory nuts and toasted oats on peaches, blueberries, and raspberries mixed with yogurt.)

Caution: Too high in fat and calories for absent-minded snacking. *Allergy alert:* Nuts in general are highly allergenic, often causing hives, headaches, and other reactions.

Preparation: Hickory nuts have a very hard shell that's best cracked with a hammer. If gathering wild nuts, allow them to dry for several weeks before you attempt to crack them open.

Serve: As substitutes for pecans, which are actually a type of hickory nut.

MACADAMIA NUTS, *Dried*

Serving: 1 oz.

Calories	199
Fat	20.9 g
Saturated	3.1 g
Monounsaturated	16.5 g
Polyunsaturated	0.4 g
Calories from fat	95%
Cholesterol	0
Sodium	1 mg
Protein	2.4 g
Carbohydrate	3.9 g
Dietary fiber	N/A

CHIEF NUTRIENTS

NUTRIENT	AMOUNT	% RDA
Magnesium	32.9 mg	9
Iron	0.7 mg	7
Thiamine	0.1 mg	7

Strengths: Unsalted macadamia nuts are low in sodium and they have no cholesterol. Although the nuts are high in fat, virtually all of it is monounsaturated, which can help regulate serum cholesterol. Some magnesium helps control blood pressure, and thiamine contributes to healthy nerves and muscles. They also have some iron.

Eat with: Foods rich in vitamin C, such as chunks of pineapple, papaya, mango, cantaloupe, or honeydew.

Caution: Too high in fat and calories for frequent snacking. *Allergy alert:* May cause hives, rashes, headaches, or other reactions.

Curiosity: Macadamia trees were once prized only for their ornamental foliage.

Origin: Queensland, Australia, which accounts for macadamias' alternate names: Queensland nut and Australian nut.

Preparation: Roasting the nuts in oil does not significantly change calorie or fat content. However, the addition of salt raises sodium to 74 mg per oz.

MIXED NUTS, *Dry-Roasted*

Serving: 1 oz.

Calories	169
Fat	14.6 g
Saturated	2.0 g
Monounsaturated	8.9 g
Polyunsaturated	3.1 g
Calories from fat	78%
Cholesterol	0
Sodium	3 mg
Protein	4.9 g
Carbohydrate	7.2 g
Dietary fiber	2.6 g

CHIEF NUTRIENTS

NUTRIENT	AMOUNT	% RDA
Magnesium	63.9 mg	18
Iron	1.1 mg	11
Folate	14.3 mcg	7
Niacin	1.3 mg	7
Zinc	1.1 mg	7

Strengths: Unsalted mixed nuts are low in sodium with no cholesterol. They're a good source of fiber, although the exact amount depends on which nuts are used. Most of their fat is monounsaturated, which can help control serum cholesterol. And they have nice amounts of the immunity-strengthening nutrients iron, folate, and zinc. Also a good source of magnesium for healthy blood pressure.

Eat with: Orange juice, grapefruit juice, or other drinks high in vitamin C to enhance absorption of the iron.

Caution: Too high in fat and calories for regular snacking. Almonds and cashews contain oxalic acid, which should be restricted by those with calcium-oxalate stones. *Allergy alert:* Nuts in general are highly allergenic, often triggering hives and other reactions. Those allergic to aspirin may react to the natural salicylate in almonds.

Description: The nutrients shown are for a mixture of cashews, almonds, peanuts, filberts, and pecans.

Selection: Avoid salted nuts: Their sodium is about 190 mg per oz.

PEANUTS, *Dry-Roasted*

Serving: 1 oz.

Calories	164
Fat	13.9 g
Saturated	1.9 g
Monounsaturated	6.9 g
Polyunsaturated	4.4 g
Calories from fat	76%
Cholesterol	0
Sodium	2 mg
Protein	6.6 g
Carbohydrate	6.0 g
Dietary fiber	2.2 g

CHIEF NUTRIENTS

NUTRIENT	AMOUNT	% RDA
Folate	40.7 mcg	20
Niacin	3.8 mg	20
Magnesium	49.3 mg	14
Thiamine	0.1 mg	8
Iron	0.6 mg	6
Zinc	0.9 mg	6
Potassium	184.2 mg	5

Strengths: A good source of fiber, with a valuable amount of protein, peanuts are also high in folate. They're a good source of other B vitamins, especially niacin. Those B vitamins are valuable in helping to avoid skin or nerve problems. Peanuts boast more magnesium than most other legumes.

Eat with: Generous servings of vitamin C–rich foods and beverages to enhance iron absorption.

Caution: Peanuts are very high in fat, which can up the risk of heart disease and certain types of cancer. And they're extremely high in weight-boosting calories. Hypertensives who need to restrict sodium intake should avoid salted peanuts. Also, peanuts contain phytates and other substances that inhibit absorption of iron from plant foods unless vitamin C is consumed with the meal. *Allergy alert:* Peanuts are a potent and common food allergen.

Curiosity: Called goobers in the southern U.S.

Description: Peanuts are really legumes, not nuts.

PECANS, *Dried*

Serving: 1 oz.

Calories	189
Fat	19.2 g
Saturated	1.5 g
Monounsaturated	12.0 g
Polyunsaturated	4.7 g
Calories from fat	91%
Cholesterol	0
Sodium	0.3 mg
Protein	2.2 g
Carbohydrate	5.2 g
Dietary fiber	1.9 g

CHIEF NUTRIENTS

NUTRIENT	AMOUNT	% RDA
Thiamine	0.2 mg	16
Magnesium	36.4 mg	10
Zinc	1.6 mg	10
Folate	11.1 mcg	6
Iron	0.6 mg	6

Strengths: A good source of fiber, pecans have a minimal amount of sodium and no cholesterol. Their fat is largely the monounsaturated variety, which can help control serum cholesterol. Pecans also contain nice amounts of the immunity-boosting nutrients zinc, folate, and iron. Their thiamine and magnesium contribute to healthy nerves and muscles.

Eat with: Foods rich in vitamin C to enhance absorption of the nonheme iron. (Pecans are delicious in berry desserts.)

Caution: These nuts are too high in fat and calories for everyday use. *Allergy alert:* Nuts are highly allergenic, often causing hives, headaches, and other reactions.

Selection: Neither oil-roasting nor dry-roasting significantly raises levels of calories or fat. However, salted nuts have sodium levels near 222 mg per oz.

Storage: Like other nuts with a high fat content, pecans must be stored properly to prevent rancidity. Refrigerating shelled pecans in an airtight container keeps them fresh for at least 3 months. They can be frozen for up to 6 months.

PINE NUTS, *Pignolia, Dried*

Serving: 1 oz.

Calories	146
Fat	14.4 g
Saturated	2.2 g
Monounsaturated	5.4 g
Polyunsaturated	6.1 g
Calories from fat	89%
Cholesterol	0
Sodium	1 mg
Protein	6.8 g
Carbohydrate	4.0 g
Dietary fiber	4.1 g

CHIEF NUTRIENTS

NUTRIENT	AMOUNT	% RDA
Iron	2.6 mg	26
Magnesium	66.2 mg	19
Thiamine	0.2 mg	15
Folate	16.3 mcg	8
Zinc	1.2 mg	8
Niacin	1.0 mg	5

Strengths: Pignolia nuts are high in fiber. A fine source of iron, they can help prevent anemia, help in the production of collagen, and help boost resistance to infections. In addition, their folate and zinc can strengthen immunity. Although these nuts are high in fat, most of it is in the beneficial monounsaturated and polyunsaturated forms that can help control high cholesterol. Having a good amount of magnesium and a very low sodium content, pignolias may help prevent high blood pressure. The B vitamins thiamine and niacin are good for steady nerves and healthy skin and muscles.

Eat with: Tomatoes and other foods high in vitamin C to help increase absorption of the iron. (Sprinkle over a salad of garden-fresh tomatoes, fresh basil, and part-skim mozzarella; use a reduced-fat vinaigrette.)

Caution: Pignolias are too high in calories and fat for indiscriminate snacking. *Allergy alert:* May cause anaphylaxis or other allergic reactions.

Description: A type of pine nut, these ivory-colored, torpedo-shaped nuts are generally imported from Europe. They have a light, delicate flavor.

PINE NUTS, *Piñon, Dried*

Serving: 1 oz.

Calories	161
Fat	17.3 g
Saturated	2.7 g
Monounsaturated	6.5 g
Polyunsaturated	7.3 g
Calories from fat	97%
Cholesterol	0
Sodium	20 mg
Protein	3.3 g
Carbohydrate	5.5 g
Dietary fiber	3.0 g

CHIEF NUTRIENTS

NUTRIENT	AMOUNT	% RDA
Thiamine	0.4 mg	23
Magnesium	66.5 mg	19
Iron	0.9 mg	9
Folate	16.4 mcg	8
Zinc	1.2 mg	8
Niacin	1.2 mg	7
Potassium	178.4 mg	5

Strengths: Piñon nuts have a substantial amount of fiber, no cholesterol, and not much sodium. Along with some immunity-boosting nutrients—iron, folate, and zinc—are the blood pressure regulators magnesium and potassium. These nuts are high in monounsaturated and polyunsaturated fat, which can help control high cholesterol and prevent heart disease.
Caution: Too high in calories and fat for heavy consumption. *Allergy alert:* May cause anaphylaxis or other allergic reactions.
Description: Similar to pignolias. Most piñons come from trees that grow wild in the Southwest. The nuts, formed inside large pine cones, have to be extracted by hand—a labor-intensive process.
Storage: Due to their high fat content, these nuts can turn rancid quickly. Always store them in airtight containers in the refrigerator or freezer.
Serve: With turkey or chicken salad, pilaf, lamb with couscous, or in a pesto sauce on pasta.

PISTACHIO NUTS, *Dried*

Serving: 1 oz.

Calories	164
Fat	13.7 g
Saturated	1.7 g
Monounsaturated	9.3 g
Polyunsaturated	2.1 g
Calories from fat	75%
Cholesterol	0
Sodium	2 mg
Protein	5.8 g
Carbohydrate	7.1 g
Dietary fiber	3.1 g

CHIEF NUTRIENTS

NUTRIENT	AMOUNT	% RDA
Iron	1.9 mg	19
Thiamine	0.2 mg	15
Magnesium	44.9 mg	13
Folate	16.5 mcg	8
Potassium	310.4 mg	8

Strengths: Pistachios are quite high in fiber, with barely any sodium and no cholesterol. A good amount of iron contributes to healthy blood and enhances resistance to infections. Most of the fat is monounsaturated, which can help prevent high serum cholesterol. Pistachios also have a nice amount of magnesium and some potassium.
Eat with: Foods rich in vitamin C to enhance absorption of the iron. (Try sprinkling the nuts over lightly sautéed cabbage or a pilaf made from rice and lots of red peppers.)
Caution: Too high in fat and calories to eat by the bagful. *Allergy alert:* May cause hives, anaphylaxis, or other allergic reactions.
Selection: If unshelled, select ones that are partly opened; closed shells indicate unripe nuts.
Preparation: Dry-roasting the nuts does not significantly raise calories or fat, but it does cut iron by half. Salting raises sodium to a hefty 222 mg per oz. Nuts with red shells have been dyed.

PUMPKIN SEEDS, *Dried*

Serving: 1 oz., hulled

Calories	154
Fat	13.0 g
Saturated	2.5 g
Monounsaturated	4.1 g
Polyunsaturated	5.9 g
Calories from fat	76%
Cholesterol	0
Sodium	5 mg
Protein	7.0 g
Carbohydrate	5.1 g
Dietary fiber	3.9 g

CHIEF NUTRIENTS

NUTRIENT	AMOUNT	% RDA
Iron	4.3 mg	43
Magnesium	151.9 mg	43
Zinc	2.1 mg	14
Folate	16.3 mcg	8
Potassium	229.2 mg	6
Riboflavin	0.1 mg	5

Strengths: A rich source of iron, which (along with zinc and folate) can bolster immunity. Their high magnesium content, along with some potassium, may help prevent high blood pressure. Pumpkin seeds have no cholesterol and very little sodium (as long as the seeds are unsalted), as well as substantial fiber. Although the seeds are high in fat, most of it is unsaturated, which may help lower high serum cholesterol.

Eat with: Foods high in beta-carotene and vitamin C, such as oranges, cantaloupes, papayas, and apricots. Iron helps convert the beta-carotene in these fruits into vitamin A for possible protection against some cancers. And vitamin C helps the iron be absorbed better.

Caution: These seeds are too high in calories and fat for steady snacking. But keep in mind that 1 oz. is about 140 kernels—a hefty portion.

Selection: Pumpkin seeds (also called pepitas), are green when hulled. You can buy them raw or roasted, salted or unsalted, and with or without hulls.

SESAME SEEDS, *Dried*

Serving: 1 Tbsp., hulled

Calories	47
Fat	4.4 g
Saturated	0.6 g
Monounsaturated	1.7 g
Polyunsaturated	1.9 g
Calories from fat	84%
Cholesterol	0
Sodium	3 mg
Protein	2.1 g
Carbohydrate	0.8 g
Dietary fiber	0.8 g

CHIEF NUTRIENTS

NUTRIENT	AMOUNT	% RDA
Magnesium	27.8 mg	8
Iron	0.6 mg	6
Zinc	0.8 mg	5

Strengths: These seeds are low in sodium, and they have no cholesterol. Although a high percentage of the calories comes from fat, most of it is unsaturated. Just a small measure of sesame seeds can perk up the flavors of other foods, so you don't get much fat in one serving. Magnesium helps steady nerves, and a little iron helps prevent anemia. The zinc in sesame seeds helps strengthen immunity to infections.

Eat with: All sorts of fruits and vegetables. For example, sprinkle toasted sesame seeds over compotes, fruit cups, steamed green beans, or baked potatoes.

Caution: High in fat. Foods made almost entirely of sesame seeds, such as halvah and other confections, should be eaten in moderation.

Storage: Keep these seeds in the refrigerator or freezer to prevent them from turning rancid.

Preparation: Toasting the seeds gives them a particularly rich, nutty flavor—and does not significantly alter their nutritional value.

SUNFLOWER SEEDS, Dried

Serving: 1 oz.

Calories	162
Fat	14.1 g
Saturated	1.5 g
Monounsaturated	2.7 g
Polyunsaturated	9.3 g
Calories from fat	78%
Cholesterol	0
Sodium	0.9 mg
Protein	6.5 g
Carbohydrate	5.3 g
Dietary fiber	1.9 g

CHIEF NUTRIENTS

NUTRIENT	AMOUNT	% RDA
Thiamine	0.7 mg	43
Folate	64.6 mcg	32
Magnesium	100.5 mg	29
Iron	1.9 mg	19
Vitamin B$_6$	0.2 mg	11
Zinc	1.4 mg	10
Niacin	1.3 mg	7
Potassium	195.7 mg	5

Strengths: These super seeds are packed with nutrients. They have lots of thiamine and magnesium, along with a good amount of vitamin B$_6$ for healthy nerves and muscles. Also included are nutrients that help boost immunity—plenty of folate and good amounts of iron and zinc. The magnesium, along with some potassium and very little sodium (if the seeds are unsalted), helps keep blood pressure levels on an even keel. Although the seeds are high in fat, most of it is unsaturated and may help lower cholesterol levels.

Eat with: Foods rich in vitamin C and beta-carotene, such as orange juice or apricots. Iron helps convert the carotene to cancer-fighting vitamin A, and C renders the iron more absorbable.

Caution: Too high in fat for daily snacking. *Allergy alert:* Like nuts, seeds can be powerful allergens.

Curiosity: Sunflower seeds are the sole American native to become a significant world crop. Although they were cultivated here long before the Pilgrims arrived, Russia is now the largest producer.

WALNUTS, Black, Dried

Serving: 1 oz.

Calories	172
Fat	16.1 g
Saturated	1.0 g
Monounsaturated	3.6 g
Polyunsaturated	10.4 g
Calories from fat	84%
Cholesterol	0
Sodium	0.3 mg
Protein	6.9 g
Carbohydrate	3.4 g
Dietary fiber	1.4 g

CHIEF NUTRIENTS

NUTRIENT	AMOUNT	% RDA
Magnesium	57.4 mg	16
Folate	18.6 mcg	9
Iron	0.9 mg	9
Vitamin B$_6$	0.2 mg	8
Zinc	1.0 mg	6

Strengths: Black walnuts are a nice source of fiber, with very little sodium and no cholesterol. Although high in fat, it is mostly polyunsaturated, which can help lower total cholesterol. They contain some heart-healthy omega-3's, and nice amounts of immunity-boosting nutrients (folate, iron, zinc). A good source of magnesium and a bit of vitamin B$_6$ for steady nerves.

Eat with: Foods rich in vitamin C to enhance absorption of the nonheme iron. (Add black walnuts to a winter fruit cup featuring oranges, grapefruit, and pomegranates.)

Caution: Too high in fat and calories for everyday use. Certain components of the nuts may inhibit iron absorption if you're on a no-meat, low-C diet. *Allergy alert:* Nuts are highly allergenic, often causing hives and other reactions.

Curiosity: These nuts are so hard to crack that people sometimes spread them on the ground and run over them with a car.

Origin: North American native Indians were said to have eaten black walnuts 3,000 years ago.

WALNUTS, *English, Dried* ⇩

Serving: 1 oz.

Calories	182
Fat	17.6 g
Saturated	1.6 g
Monounsaturated	4.0 g
Polyunsaturated	11.0 g
Calories from fat	87%
Cholesterol	0
Sodium	3 mg
Protein	4.1 g
Carbohydrate	5.2 g
Dietary fiber	1.4 g

CHIEF NUTRIENTS

NUTRIENT	AMOUNT	% RDA
Magnesium	48.0 mg	14
Folate	18.7 mcg	9
Vitamin B$_6$	0.2 mg	8
Iron	0.7 mg	7
Thiamine	0.1 mg	7
Zinc	0.8 mg	5

Strengths: A nice amount of fiber, with little sodium and no cholesterol. Although high in fat, most of it is polyunsaturated, which can help lower serum cholesterol. English walnuts contain some omega-3's, which also have cholesterol-lowering properties. They have nice amounts of immunity nutrients, including folate, iron, and zinc. Some B vitamins and a good amount of magnesium help to steady nerves and promote healthy skin and muscles.

Eat with: Foods rich in vitamin C. (Try sprinkling chopped walnuts over steamed brussels sprouts, cooked turnip or collard greens, or baked sweet potatoes.)

Caution: Too high in fat and calories for heavy consumption. Certain components of the nuts might inhibit iron absorption if you're on a no-meat, low-C diet. *Allergy alert:* Nuts are highly allergenic, often causing hives, headaches, and other reactions. May be especially troublesome for people allergic to birch pollen.

Origin: Persia, which is why these nuts are also called Persian walnuts.

Prepared Foods and Mixed Dishes
BEAN DISHES

BAKED BEANS, *Homemade*

Serving: ¹/₂ cup

Calories	190
Fat	6.5 g
Saturated	2.5 g
Monounsaturated	2.7 g
Polyunsaturated	1.0 g
Calories from fat	31%
Cholesterol	6 mg
Sodium	532 mg
Protein	7.0 g
Carbohydrate	27.0 g
Dietary fiber	9.8 g

CHIEF NUTRIENTS

NUTRIENT	AMOUNT	% RDA
Folate	61.0 mcg	31
Iron	2.5 mg	25
Magnesium	54.2 mg	16
Potassium	451.1 mg	12
Thiamine	0.2 mg	11
Calcium	76.9 mg	10
Vitamin B$_6$	0.1 mg	6
Zinc	0.9 mg	6

Strengths: Packed with fiber, beans are digested slowly, so one serving can suppress your appetite for hours. They're an excellent source of complex carbohydrates, good for diabetics. Their healthy doses of anemia-fighting iron, nerve-soothing B vitamins, and blood pressure–lowering minerals may be especially needed by dieters.

Eat with: Vitamin C–rich foods to boost iron absorption. Rice, grains, or pasta to form complete protein.

Caution: If you're on a salt- or fat-restricted diet, omit the salt pork or bacon. Beans are moderately high in purines and should be eaten sparingly by anyone with gout. Also beans give some people gas. *Allergy alert:* Legumes (including beans) sometimes trigger migraine headaches in sensitive individuals.

Description: The dish called baked beans is usually made from navy beans with salt pork or bacon, molasses, brown sugar, onions, and other seasonings.

CHILI, *Canned, with Beans*

Serving: ¹/₂ cup

Calories	143
Fat	7.0 g
Saturated	3.0 g
Monounsaturated	3.0 g
Polyunsaturated	0.5 g
Calories from fat	44%
Cholesterol	22 mg
Sodium	668 mg
Protein	7.3 g
Carbohydrate	15.2 g
Dietary fiber	4.6 g

CHIEF NUTRIENTS

NUTRIENT	AMOUNT	% RDA
Iron	4.4 mg	44
Zinc	2.6 mg	17
Magnesium	57.6 mg	16
Folate	29.1 mcg	15
Potassium	467.2 mg	13
Vitamin B$_6$	0.2 mg	9
Calcium	602.0 mg	8
Riboflavin	0.1 mg	8

Strengths: One bowl of hearty chili is a good source of folate, zinc, and fiber, especially if made with red kidney beans. The dish is rich in available iron, due to the iron-enhancing presence of ground beef. Chili with beans is a good choice for women susceptible to iron deficiency due to heavy menstrual periods, frequent dieting, and low vitamin C intake. It has significant amounts of potassium and magnesium, important for good blood pressure control and stroke prevention.

Eat with: Chili peppers (to maximize iron absorption from beans). Rice (to maximize the vegetable protein in beans).

Caution: The canned product is very high in sodium. Beans are moderately high in purines and should be eaten sparingly by anyone with gout. And they give some people gas. *Allergy alert:* Beans sometimes trigger migraine headaches in sensitive individuals.

A Better Idea: For a lower-sodium version, make your own chili and beans and omit the salt.

REFRIED BEANS, *Canned*

Serving: ½ cup

Calories	135
Fat	1.4 g
Saturated	0.5 g
Monounsaturated	0.6 g
Polyunsaturated	0.2 g
Calories from fat	9%
Cholesterol	0
Sodium	534 mg
Protein	7.9 g
Carbohydrate	23.3 g
Dietary fiber	6.7 g

CHIEF NUTRIENTS

NUTRIENT	AMOUNT	% RDA
Folate	105.2 mcg	52
Iron	2.2 mg	22
Magnesium	49.1 mg	14
Potassium	495.2 mg	13
Vitamin C	7.6 mg	13
Zinc	1.7 mg	12
Calcium	58.0 mg	7

Strengths: A fairly decent source of protein and carbohydrates, refried beans are high in fiber as well. A ½-cup serving supplies more than one-half of the day's folate requirements—making it useful to pregnant women, who need adequate amounts of this B vitamin for healthy babies. The vitamin C content helps to boost absorption of nonheme iron, the plant form of iron that needs C to be well absorbed.

Eat with: Lean meat containing heme iron to enhance the absorption of nonheme iron. Also, eat with grains to form complete proteins.

Caution: High in sodium—a potential problem for hypertensives. Read labels: Some brands are fried in lard, a highly saturated and undesirable form of fat. The cautions that apply to pinto and kidney beans also hold for refried beans.

Description: Refried beans consist of cooked, mashed pinto or red kidney beans that have been seasoned with garlic, salt, and pepper. (The name is a misinterpretation of the Mexican term *frijoles refritos,* which means "well-fried beans."

Serve: As a filling in tortillas or as a side dish.

BEEF DISHES

BEEF POTPIE, *Homemade*

Serving: 1 piece (about 7½ oz.), baked

Calories	517
Fat	30.5 g
Saturated	8.4 g
Monounsaturated	N/A
Polyunsaturated	N/A
Calories from fat	53%
Cholesterol	44 mg
Sodium	596 mg
Protein	21.2 g
Carbohydrate	39.5 g
Dietary fiber	N/A

CHIEF NUTRIENTS

NUTRIENT	AMOUNT	% RDA
Iron	3.8 mg	38
Vitamin A	344.4 RE	34
Niacin	4.8 mg	25
Thiamine	0.3 mg	19
Riboflavin	0.3 mg	17
Vitamin C	6.3 mg	11
Potassium	333.9 mg	9

Caution: About 53% of the calories in beef potpie are derived from fat. It is also very high in sodium and has moderate levels of cholesterol. Beef contains purines, substances that may aggravate gout.

Strengths: A very good source of iron, a nutrient your body needs to make healthy tissue. This meat and vegetable dish also delivers respectable amounts of niacin, a B vitamin essential to the formation of red blood cells, as well as riboflavin and thiamine. A good amount of vitamin C and a bit of potassium contribute to beef potpie's nutritional appeal. It's a rich source of vitamin A, which may reduce the risk of certain cancers.

Curiosity: The first commercially produced frozen potpie hit grocery freezers in 1951.

Description: Cooked chunks of beef, vegetables, and gravy baked in a pie shell.

A Better Idea: If making your own potpie, start with lean ground beef and cook it in a microwave.

BEEF STROGANOFF

Serving: 6 oz. frozen entrée

Calories	185
Fat	9.1 g
Saturated	N/A
Monounsaturated	N/A
Polyunsaturated	N/A
Calories from fat	44%
Cholesterol	44 mg
Sodium	350 mg
Protein	20.5 g
Carbohydrate	5.3 g
Dietary fiber	N/A

CHIEF NUTRIENTS

NUTRIENT	AMOUNT	% RDA
Iron	3.2 mg	32
Riboflavin	0.2 mg	13
Potassium	277.0 mg	7

Caution: A large percentage of the calories in beef stroganoff are from fat, and it's also high in sodium. Beef contains purines, which may cause problems if you're prone to gout. *Allergy alert:* People allergic to mold may show a reaction to mushrooms in stroganoff.

Strengths: High in protein for maintaining muscle mass, beef stroganoff also is a very good source of iron, which the body needs to fight off fatigue and to convert beta-carotene into vitamin A. The riboflavin in beef stroganoff plays a role in the formation of red blood cells in bone marrow.

Origin: Said to be named after Count Paul Stroganov, a 19th-century Russian diplomat.

Description: Thin strips of beef tenderloin are sautéed with mushrooms and onions, combined with a sour cream sauce, and usually served with rice pilaf.

A Better Idea: Instead of buying the frozen entrée, make your own stroganoff—and replace the high-fat sour cream with low-fat yogurt or light sour cream.

CORNED BEEF HASH, *Canned*

Serving: 1 cup

Calories	398
Fat	24.9 g
Saturated	11.9 g
Monounsaturated	N/A
Polyunsaturated	N/A
Calories from fat	56%
Cholesterol	73 mg
Sodium	1,188 mg
Protein	19.4 g
Carbohydrate	23.5 g
Dietary fiber	1.1 g

CHIEF NUTRIENTS

NUTRIENT	AMOUNT	% RDA
Iron	4.4 mg	44
Niacin	4.6 mg	24
Potassium	440.0 mg	12
Riboflavin	0.2 mg	12

Caution: About 56% of this food's calories are from fat, and nearly half of that is the saturated type that has been linked to increased blood cholesterol levels. Each serving of corned beef hash is packed with more than 1,100 mg of sodium. *Drug interaction:* People taking MAO inhibitors should avoid eating corned beef because the meat contains tyramine, a substance that can raise blood pressure when it interacts with MAOs. *Allergy alert:* A pickled meat, corned beef may cause allergic reactions in some people. The beef contains purines that may aggravate gout.

Strengths: Perhaps the ultimate meal for a meat and potatoes lover, corned beef hash is a rich source of iron, a nutrient that helps your body fight off infections and combat fatigue and headaches. It has a respectable amount of niacin, a B vitamin that contributes to the formation of red blood cells.

Curiosity: Hash once was associated with cheap restaurants called "hash houses" or "hasheries."

Description: The main ingredient is chopped corned beef; it is combined with an assortment of vegetables, usually including potatoes.

CREAMED CHIPPED BEEF, *Homemade*

Serving: 1 cup

Calories	377
Fat	25.2 g
Saturated	13.7 g
Monounsaturated	N/A
Polyunsaturated	N/A
Calories from fat	60%
Cholesterol	98 mg
Sodium	1,754 mg
Protein	20.1 g
Carbohydrate	17.4 g
Dietary fiber	N/A

CHIEF NUTRIENTS

NUTRIENT	AMOUNT	% RDA
Calcium	257.3 mg	32
Riboflavin	0.5 mg	28
Iron	2.0 mg	20
Vitamin A	176.4 RE	18
Potassium	374.9 mg	10
Thiamine	0.2 mg	10
Niacin	1.5 mg	8

Caution: This home-on-the-range classic gets 60% of its calories from fat and has a whopping 1,754 mg of sodium per serving. More than half the fat is the saturated type that can contribute to increased blood cholesterol. *Drug interaction:* People taking MAO inhibitors should avoid chipped beef because the meat contains tyramine, a substance that may cause dangerous blood pressure increases if it interacts with MAOs. *Allergy alert:* Chipped beef contains purines, substances that may aggravate gout. Some people are generally allergic to meat and its by-products.

Strengths: This entrée contains a generous dose of calcium, a nutrient that helps your blood clot and assists in maintaining good blood pressure. Its riboflavin and thiamine help your body convert carbohydrates and other foods into energy. Strong in iron, it also has a nutritionally significant amount of potassium and a little niacin. And it has a good amount of vitamin A, which helps promote wound healing.

MEAT LOAF

Serving: 3 oz.

Calories	170
Fat	11.2 g
Saturated	5.1 g
Monounsaturated	N/A
Polyunsaturated	N/A
Calories from fat	59%
Cholesterol	55 mg
Sodium	1,105 mg
Protein	13.5 g
Carbohydrate	2.8 g
Dietary fiber	N/A

CHIEF NUTRIENTS

NUTRIENT	AMOUNT	% RDA
Iron	1.5 mg	15
Niacin	2.1 mg	11
Riboflavin	0.2 mg	11
Thiamine	0.1 mg	7
Potassium	195.5 mg	5

Caution: Depending on the types of meat used, this entrée can be more than 50% fat. Almost half of the fat is saturated, the kind that can lead to increased blood cholesterol levels. It also contains moderate amounts of cholesterol and more than 1,100 mg of sodium. *Allergy alert:* Meat loaf is commonly made with beef, veal, pork, tomatoes, and eggs, foods that may cause allergic reactions in some people. Meats used in this dish also contain purines, substances that may aggravate gout.

Strengths: Surprisingly, a 3-oz. serving of meat loaf only has 170 calories. Not surprisingly, given the amount of meat, it has a hefty amount of protein. It also has a good amount of iron, needed for healthy red blood cells, plus decent amounts of niacin and riboflavin, and some potassium and thiamine.

A Better Idea: By using lean meats, you'll reduce the fat content. To cut down on sodium intake, avoid seasoning with salt.

SALISBURY STEAK

Serving: 5 oz. frozen entrée

Calories	230
Fat	18.0 g
Saturated	N/A
Monounsaturated	N/A
Polyunsaturated	N/A
Calories from fat	70%
Cholesterol	35 mg
Sodium	766 mg
Protein	10.0 g
Carbohydrate	7.0 g
Dietary fiber	0

CHIEF NUTRIENTS

NUTRIENT	AMOUNT	% RDA
Iron	1.9 mg	19
Riboflavin	0.1 mg	7
Calcium	49.0 mg	6
Niacin	1.1 mg	6
Thiamine	0.1 mg	6

Caution: Salisbury steak contains an alarming percentage of calories from fat, and it's very high in sodium.

Strengths: Offers valuable blood-building iron, along with some thiamine, riboflavin, and niacin—B-complex vitamins that help maintain healthy nerves. The frozen entrée also contributes a bit of calcium, important for strong bones.

Curiosity: Named after the 19th-century English doctor J. H. Salisbury, who urged his patients to eat beef as a cure for just about every ailment.

Description: A pan-fried ground beef patty flavored with onions and seasonings, usually served with a mushroom gravy made from pan drippings.

Serve: With mashed potatoes and green beans for an all-American dinner.

A Better Idea: Make your own streamlined version of Salisbury steak: Use lean ground beef or turkey, broil instead of frying, and serve with a low-fat gravy and water-sautéed onions.

SWEDISH MEATBALLS

Serving: 6 oz. frozen entrée, w/sauce

Calories	220
Fat	13.2 g
Saturated	N/A
Monounsaturated	N/A
Polyunsaturated	N/A
Calories from fat	54%
Cholesterol	42 mg
Sodium	528 mg
Protein	12.0 g
Carbohydrate	13.2 g
Dietary fiber	N/A

CHIEF NUTRIENTS

NUTRIENT	AMOUNT	% RDA
Niacin	7.2 mg	38
Iron	1.2 mg	12
Riboflavin	0.2 mg	11
Thiamine	0.2 mg	11
Potassium	294.0 mg	8
Vitamin C	3.0 mg	5

Caution: Swedish meatballs are high in calories, fat, and sodium. And the beef contains uric acid, so be cautious if you're susceptible to gout. *Allergy alert:* Many people are allergic to the wheat and eggs found in Swedish meatball recipes.

Strengths: In addition to some protein, Swedish meatballs supply the B vitamins niacin and thiamine, which help transform food into energy.

Description: Small, seasoned meatballs are sautéed in butter or pan drippings, then served either in a gravy made with milk or cream or in beef consommé.

A Better Idea: Cook the meatballs in the microwave to reduce the fat content. Use skim milk or low-fat yogurt if your recipe calls for milk or cream.

BREAKFAST DISHES

FRENCH TOAST, *Homemade*

Serving: 1 slice (about 2 oz.)

Calories	153
Fat	6.7 g
Saturated	N/A
Monounsaturated	N/A
Polyunsaturated	N/A
Calories from fat	40%
Cholesterol	N/A
Sodium	257 mg
Protein	5.7 g
Carbohydrate	17.2 g
Dietary fiber	0.1 g

CHIEF NUTRIENTS

NUTRIENT	AMOUNT	% RDA
Vitamin B$_{12}$	0.3 mcg	15
Iron	1.3 mg	13
Riboflavin	0.2 mg	10
Calcium	72.2 mg	9
Thiamine	0.1 mg	8
Niacin	1.0 mg	5

Caution: More than one-third of the calories in French toast come from fat, and there's enough sodium to warrant some caution if you're on a low-salt diet. *Allergy alert:* The eggs, milk, and wheat used to prepare French toast are common allergens and may prompt a variety of reactions in sensitive people.

Strengths: French toast supplies carbohydrates the body needs for energy and contains a good amount of vitamin B$_{12}$, which helps guard the nervous system. Its iron boosts the overall health of blood cells, and some calcium may contribute to stronger bones.

A Better Idea: Use whole wheat bread, skim milk, and egg substitute. Top the French toast with a little apple juice or pureed fruit instead of syrup.

HAM AND CHEESE OMELET

Serving: 1 (about 4 oz.), made w/2 eggs

Calories	255
Fat	17.7 g
Saturated	N/A
Monounsaturated	N/A
Polyunsaturated	N/A
Calories from fat	62%
Cholesterol	446 mg
Sodium	399 mg
Protein	17.1 g
Carbohydrate	6.7 g
Dietary fiber	0

CHIEF NUTRIENTS

NUTRIENT	AMOUNT	% RDA
Vitamin B$_{12}$	1.2 mcg	59
Riboflavin	0.6 mg	32
Iron	2.9 mg	29
Folate	41.0 mcg	21
Calcium	116.0 mg	15
Zinc	2.2 mg	14
Vitamin B$_6$	0.2 mg	10
Vitamin C	6.0 mg	10
Thiamine	0.1 mg	9
Magnesium	19.0 mg	5
Potassium	189.0 mg	5

Caution: This brunch-time special is a high-fat way to start the day. More than half the calories in a ham and cheese omelet come from fat, and it also is very high in sodium and cholesterol. *Allergy alert:* Eggs and dairy products are common allergens.

Strengths: Protein from an egg is the easiest kind of animal protein to absorb, and an omelet contains a considerable amount. A bountiful supply of riboflavin helps bone marrow form red blood cells and aids in the metabolism of other B vitamins. An omelet also is a significant source of iron, a deficiency of which could cause fatigue or headaches, and zinc, which is important to metabolizing protein and maintaining healthy skin.

Curiosity: It is estimated that 270 billion eggs are eaten worldwide throughout the year.

A Better Idea: Reduce fat and sodium by eliminating ham and cheese from the omelet; replace with some green peppers and onions. Lower fat and cholesterol by using egg substitute.

JOHNNYCAKE *(Homemade Cornbread)* ⇩

Serving: 1 (about 2½ oz.)

Calories	195
Fat	3.8 g
Saturated	0.4 g
Monounsaturated	N/A
Polyunsaturated	N/A
Calories from fat	18%
Cholesterol	63 mg
Sodium	504 mg
Protein	6.4 g
Carbohydrate	33.2 g
Dietary fiber	N/A

CHIEF NUTRIENTS

NUTRIENT	AMOUNT	% RDA
Iron	1.3 mg	13
Riboflavin	0.2 mg	13
Calcium	81.0 mg	10
Thiamine	0.1 mg	10
Vitamin A	68.0 RE	7
Niacin	1.1 mg	6

Strengths: Low in fat, a johnnycake is loaded with carbohydrates, and it's a good source of riboflavin, which the body needs to convert those carbohydrates into energy. Its iron and calcium help contribute to healthy blood and stronger bones. A hearty amount of thiamine helps with energy production, and niacin assists in the formation of red blood cells.

Caution: Johnnycakes are high in sodium, so you'll probably want something else for breakfast if you're on a low-salt diet. *Allergy alert:* Cornmeal and milk are common allergens.

Curiosity: In Rhode Island, purists insist that the only real johnnycake is one made with an obsolete strain of low-yield Indian corn.

Origin: The name of this corn cake has been disputed for centuries. It may have come from the Indian word *joniken,* for flat cornmeal cake; from *Shawnee cake,* after an Indian tribe; or from the words *journey cake.*

Preparation: Some make johnnycakes with cornmeal, salt, and cold milk; others replace the milk with boiling water, which makes a smaller, thicker cake.

PANCAKE, *Buckwheat, from Mix*

Serving: 1 (about 1 oz.), made w/egg & milk

Calories	54
Fat	2.5 g
Saturated	0.8 g
Monounsaturated	N/A
Polyunsaturated	N/A
Calories from fat	41%
Cholesterol	18 mg
Sodium	125 mg
Protein	1.8 g
Carbohydrate	6.4 g
Dietary fiber	N/A

CHIEF NUTRIENTS

NUTRIENT	AMOUNT	% RDA
Calcium	59.4 mg	7

Caution: A buckwheat pancake made from a mix and using milk offers a bit of bone-building calcium. But even a plain buckwheat pancake gets 41% of its calories from fat.

Eat with: Low-fat toppings such as fruit butter, cottage cheese, sliced fruit, or yogurt to avoid going overboard on calories.

Curiosity: In early America, pancakes were sometimes called hoe cakes, because they were cooked on a flat hoe blade.

Description: A buckwheat pancake is made partially from wheat flour—but the buckwheat adds a delicious nutty flavor. Recipes for homemade pancakes usually specify a batter that is no more than half buckwheat—and mixes may contain even less buckwheat flour.

Preparation: If you're not serving buckwheat cakes straight from the griddle, lay them on a cloth-covered cookie sheet. Don't stack pancakes until you're ready to serve them; they'll get soggy.

A Better Idea: Mix your own buckwheat cakes, using a healthy portion of nutritious buckwheat flour; cook in a nonstick skillet.

PANCAKE, *Plain or Buttermilk, from Mix*

Serving: 1 (about 1 oz.), made w/egg & milk

Calories	61
Fat	2.0 g
Saturated	0.7 g
Monounsaturated	N/A
Polyunsaturated	N/A
Calories from fat	29%
Cholesterol	20 mg
Sodium	152 mg
Protein	1.9 g
Carbohydrate	8.8 g
Dietary fiber	0.4 g

CHIEF NUTRIENTS

NUTRIENT	AMOUNT	% RDA
Calcium	58.1 mg	7

Caution: Nutritional lightweights, both plain and buttermilk pancakes are mostly carbohydrates, with some protein. Even without a butter topping, these pancakes get almost 30% of their calories from fat.

Strength: Because it is made with milk, a pancake does offer some calcium.

Curiosity: The largest pancake ever made was 28 ft. in diameter and 2 in. deep. It contained 810 lbs. of buckwheat flour.

History: Pancakes were probably first made on hot rocks and are still sometimes made on a soapstone griddle.

Description: A thin batter of flour, milk, shortening, and eggs, leavened with baking powder.

Serve: With low-fat fruit butter and cottage cheese, or sliced fruit and yogurt drizzled with maple syrup.

A Better Idea: Mix up your own pancakes using whole wheat flour and less fat than the mix calls for.

WAFFLE, *Frozen, Enriched*

Serving: 1 (about 1¼ oz.)

Calories	86
Fat	2.1 g
Saturated	0.5 g
Monounsaturated	N/A
Polyunsaturated	N/A
Calories from fat	22%
Cholesterol	43 mg
Sodium	219 mg
Protein	2.4 g
Carbohydrate	14.3 g
Dietary fiber	0.8 g

CHIEF NUTRIENTS

NUTRIENT	AMOUNT	% RDA
Iron	0.8 mg	8
Calcium	41.5 mg	5

Strengths: Because it is made with enriched wheat flour, a frozen waffle contains a bit of bone-building calcium and some iron, essential for healthy blood. A frozen waffle is lower in fat than many homemade ones. With only 22% of calories from fat, it's kinder to your waistline than a buckwheat pancake made from a packaged mix.

Caution: Surprisingly high in sodium; each waffle has over 200 mg.

Description: Waffles are the equivalent of a crisp, light bread with a honeycombed surface. They are made with a batter of flour, milk, eggs, baking powder, and oil or melted shortening. The frozen, packaged variety may be browned in a toaster or under a broiler before serving.

Serve: With sliced fresh fruit and yogurt, or seafood or chicken served in a light, low-fat white sauce.

WAFFLE, *Homemade, Enriched*

Serving: 1 (about 1¾ oz.)

Calories	140
Fat	4.9 g
Saturated	1.6 g
Monounsaturated	N/A
Polyunsaturated	N/A
Calories from fat	32%
Cholesterol	63 mg
Sodium	238 mg
Protein	4.7 g
Carbohydrate	18.8 g
Dietary fiber	N/A

CHIEF NUTRIENTS

NUTRIENT	AMOUNT	% RDA
Iron	1.2 mg	12
Riboflavin	0.2 mg	9
Thiamine	0.1 mg	8
Calcium	56.5 mg	7
Niacin	1.0 mg	5

Strengths: Enriched wheat flour gives your homemade waffle a good amount of blood-building iron. It also offers some calcium, important for strong bones. The riboflavin and niacin are both B-complex vitamins that aid energy metabolism.

Caution: Perhaps because most recipes call for butter, homemade waffles contain almost twice as much saturated fat as frozen waffles.

Curiosity: In Baltimore, Maryland, kidney stew on waffles is a traditional Sunday specialty.

Description: Food values given here are for a waffle made with flour, milk, eggs, baking powder, salt, sugar, and oil or melted shortening. It is cooked in a waffle iron.

A Better Idea: Make your own lower-fat, high-fiber waffles using part wheat bran and less fat than the recipe call for. Cook in a nonstick waffle iron.

CHEESE DISHES

CHEESE BLINTZES

Serving: 8 oz. frozen entrée

Calories	432
Fat	25.6 g
Saturated	N/A
Monounsaturated	N/A
Polyunsaturated	N/A
Calories from fat	53%
Cholesterol	436 mg
Sodium	246 mg
Protein	19.2 g
Carbohydrate	31.2 g
Dietary fiber	N/A

CHIEF NUTRIENTS

NUTRIENT	AMOUNT	% RDA
Riboflavin	2.0 mg	118
Iron	4.8 mg	48
Calcium	336.0 mg	42
Vitamin A	313.6 RE	31
Thiamine	0.3 mg	22
Niacin	3.9 mg	21
Potassium	312.0 mg	8

Caution: High in sodium, fat, and cholesterol: They ought to be avoided if you're on a low-salt diet, trying to lose weight, or watching your cholesterol levels. *Allergy alert:* Not recommended for anyone allergic to tyramine. People who are lactose intolerant may have difficulty coping with the whey that's part of the cottage cheese in blintzes. And it's a good idea to avoid blintzes if you're sensitive to eggs.

Strengths: Each cheese blintze frozen entrée contains nearly half the RDA of iron, which contributes to healthy blood—as well as almost one-third the RDA of vitamin A, which helps boost immunity. And blintzes have a phenomenal amount of energy-producing riboflavin—well over a full day's supply. Along with a very generous amount of calcium for stronger bones, a cheese blintze also supplies nearly 20% of the RDA of protein.

Origin: A favorite side dish in Jewish kitchens, blintzes were served on Shevuoth.

Description: Superthin rolled-up pancakes, blintzes contain fillings such as cottage or ricotta cheese, fruit, or meat mixtures.

CHEESE FONDUE

Serving: ¼ cup

Calories	151
Fat	10.4 g
Saturated	5.1 g
Monounsaturated	N/A
Polyunsaturated	N/A
Calories from fat	62%
Cholesterol	128 mg
Sodium	309 mg
Protein	8.4 g
Carbohydrate	5.7 g
Dietary fiber	0

CHIEF NUTRIENTS

NUTRIENT	AMOUNT	% RDA
Calcium	180.7 mg	23
Vitamin A	176.0 RE	18
Riboflavin	0.2 mg	11
Iron	0.7 mg	7

Caution: About 62% of cheese fondue's calories are derived from fat. With more than 300 mg of sodium per serving, it should send up warning flares to anyone on a low-salt diet.

Strengths: Cheese fondue is a good source of calcium, a nutrient that is necessary for strong bones and teeth and has a good amount of vitamin A, which helps the body's immune reactions. It's high in protein and low in carbohydrates. Fondue also contains a decent amount of riboflavin as well as some iron. *Allergy alert:* Some people are allergic to cheese and may develop hives and other symptoms if they eat it.

Description: The cheeses commonly used in fondue are Swiss and Gruyère. Bite-size pieces of meat, fruit, or bread are impaled on skewers and dipped into the fondue pot.

History: Fondue was a Swiss peasant dish that was introduced to the U.S. about 40 years ago. The word is a form of a French verb meaning "to melt."

A Better Idea: Try using low-fat, low-sodium cheese. Some recipes call for white wine or kirsch; but you may omit the alcohol if you're concerned about calories.

CHEESE SOUFFLÉ, *Homemade*

Serving: 1 cup

Calories	207
Fat	16.2 g
Saturated	8.2 g
Monounsaturated	N/A
Polyunsaturated	N/A
Calories from fat	71%
Cholesterol	176 mg
Sodium	346 mg
Protein	9.4 g
Carbohydrate	5.9 g
Dietary fiber	N/A

CHIEF NUTRIENTS

NUTRIENT	AMOUNT	% RDA
Calcium	191.0 mg	24
Vitamin A	152.0 RE	15
Riboflavin	0.2 mg	13
Iron	1.0 mg	10

Caution: More than 70% of this entrée's calories are derived from fat. A 1-cup serving contains 346 mg of sodium and a high-level dose of cholesterol. *Allergy alert:* The eggs, cheese, and milk in soufflé may cause allergic reactions in some people.

Strengths: A serving has 24% of the RDA of calcium, a nutrient that may help maintain healthy blood pressure. And it has a good amount of vitamin A, which is good for vision. You'll also get a moderate amount of protein and some carbohydrates as well. Soufflé is a good source of iron and riboflavin, a B vitamin that helps your body convert food into energy.

Description: A basic soufflé mixture is similar to pancake batter, but stiffly beaten egg whites are added. Then the mixture is either baked, chilled, or frozen.

A Better Idea: Use low-fat cheeses such as "light" versions of cheddar, Swiss, and mozzarella. Use egg substitute in place of the egg yolks.

MACARONI AND CHEESE, *Homemade*

Serving: 1 cup, baked

Calories	430
Fat	22.2 g
Saturated	8.9 g
Monounsaturated	N/A
Polyunsaturated	N/A
Calories from fat	46%
Cholesterol	42 mg
Sodium	1,086 mg
Protein	16.8 g
Carbohydrate	40.2 g
Dietary fiber	N/A

CHIEF NUTRIENTS

NUTRIENT	AMOUNT	% RDA
Calcium	362.0 mg	45
Riboflavin	0.4 mg	24
Iron	1.8 mg	18
Vitamin A	172.0 RE	17
Thiamine	0.2 mg	13
Niacin	1.8 mg	9
Potassium	240.0 mg	6

Caution: This American favorite contains massive amounts of sodium, and 46% of its calories are from fat. It also has a moderate amount of cholesterol.

Strengths: A rich source of calcium, a nutrient needed for healthy bones and teeth. Calcium also helps prevent osteoporosis, a gradual loss of bone mass. Baked macaroni and cheese provides iron and has good levels of riboflavin and thiamine, two B vitamins that help your body convert carbohydrates and other foods into energy. It has a good amount of immunity-boosting vitamin A. *Allergy alert:* Cheese may cause allergic reactions, including hives, in some people.

History: According to legend, an Italian sovereign gave the noodle its name when he ate it and exclaimed, "Ma caroni," meaning "how very dear."

QUICHE LORRAINE

Serving: 1 slice (about 6¼ oz.)

Calories	600
Fat	48.0 g
Saturated	23.2 g
Monounsaturated	17.8 g
Polyunsaturated	4.1 g
Calories from fat	72%
Cholesterol	285 mg
Sodium	653 mg
Protein	13.0 g
Carbohydrate	29.0 g
Dietary fiber	N/A

CHIEF NUTRIENTS

NUTRIENT	AMOUNT	% RDA
Vitamin A	454.0 RE	45
Calcium	211.0 mg	26
Riboflavin	0.3 mg	19
Iron	1.0 mg	10
Potassium	283.0 mg	8
Thiamine	0.1 mg	7

Caution: If you want to give your arteries a hefty infusion of fat, sodium, and cholesterol, plus hundreds of calories in a few small bites, this is the food for you. *Drug interaction:* The bacon and cheese in quiche Lorraine could be a problem if you're taking an MAO inhibitor. The interaction between MAOs and naturally occurring tyramine can lead to high blood pressure, headaches, and nausea. *Allergy alert:* Contains wheat, eggs, cheese, and other dairy products, ingredients that are common food allergens.

Strengths: This dish is rich in vitamin A for keen vision and contains a good amount of calcium and iron for healthy blood.

History: The quiche got its name from the Alsace-Lorraine region of northeastern France.

Description: A savory pie filled with eggs, cream, onions, bacon, and sometimes Gruyère cheese.

A Better Idea: You can make a satisfying, low-fat quiche with skim milk, reduced-fat cheese, and egg substitute, plus onions and herbs, baked in an oil-based pie crust.

WELSH RAREBIT

Serving: 1 cup

Calories	415
Fat	31.6 g
Saturated	17.3 g
Monounsaturated	N/A
Polyunsaturated	N/A
Calories from fat	68%
Cholesterol	100 mg
Sodium	770 mg
Protein	18.8 g
Carbohydrate	14.6 g
Dietary fiber	N/A

CHIEF NUTRIENTS

NUTRIENT	AMOUNT	% RDA
Calcium	582.3 mg	73
Riboflavin	0.5 mg	31
Potassium	320.2 mg	9
Iron	0.7 mg	7
Thiamine	0.1 mg	6

Caution: Because of the high cheese and butter content, Welsh rarebit is loaded with calories and saturated fat. Salt watchers take note: Rarebit has a lofty level of sodium. *Allergy alert:* Welsh rarebit includes flour and cheese, and often beer—all of which are common allergens.

Strengths: One serving supplies almost three-quarters of the daily requirement of calcium, an important nutrient in combating osteoporosis. A good dose of riboflavin contributes to the formation of red blood in bone marrow and plays a role in the body's energy production.

Description: The quintessential teatime food in Great Britain, Welsh rarebit is made of melted cheddar cheese, usually mixed with beer and served on toast. It's also called Welsh *rabbit,* although there's no meat used.

A Better Idea: Use low-fat cheddar cheese and substitute skim milk for beer.

CHICKEN/TURKEY DISHES

CHICKEN À LA KING, *Homemade*

Serving: 1 cup

Calories	468
Fat	34.3 g
Saturated	12.7 g
Monounsaturated	N/A
Polyunsaturated	N/A
Calories from fat	66%
Cholesterol	186 mg
Sodium	760 mg
Protein	27.4 g
Carbohydrate	12.3 g
Dietary fiber	N/A

CHIEF NUTRIENTS

NUTRIENT	AMOUNT	% RDA
Niacin	5.4 mg	28
Iron	2.5 mg	25
Riboflavin	0.4 mg	25
Vitamin A	225.4 RE	23
Vitamin C	12.3 mg	20
Calcium	127.4 mg	16
Potassium	404.3 mg	11
Thiamine	0.1 mg	7

Caution: More à la king than chicken, this meal is very high in sodium and saturated fat—and a waist-bulging two-thirds of its calories come from fat. Since poultry also is high in purines like uric acid, people with gout may want to avoid chicken à la king. *Allergy alert:* Chicken can cause allergic reactions in some people; also, those allergic to mold may have a reaction to the mushrooms in chicken à la king.

Strengths: A great source of protein, chicken also contains good amounts of niacin and riboflavin, two B vitamins that help the body generate energy from food. This dish is a good source of blood-building iron, and it contains plentiful vitamin C to help in the absorption of iron.

Description: Diced chicken served with mushrooms, peppers, and pimentos in a thick, creamy sauce.

A Better Idea: To make a lower-calorie à la king, substitute low-fat yogurt or low-fat cottage cheese for sour cream. Broil the chicken and steam the vegetables instead of sautéing them, and remove the chicken skin before eating.

CHICKEN FRICASSEE, *Homemade*

Serving: 1 cup

Calories	386
Fat	22.3 g
Saturated	7.2 g
Moncunsaturated	N/A
Polyunsaturated	N/A
Calories from fat	52%
Cholesterol	96 mg
Sodium	370 mg
Protein	36.7 g
Carbohydrate	7.7 g
Dietary fiber	N/A

CHIEF NUTRIENTS

NUTRIENT	AMOUNT	% RDA
Niacin	5.8 mg	30
Iron	2.2 mg	22
Riboflavin	0.2 mg	10
Potassium	336.0 mg	9

Caution: Fricassee is a fatty fowl feast with moderate cholesterol and high sodium; more than half its calories come from fat. Gout sufferers may wisely choose to avoid it, since poultry is generally high in purines like uric acid. *Allergy alert:* Chicken may cause allergic reactions in some people.

Strengths: An excellent source of protein for the muscles, chicken fricassee also is high in niacin, a B vitamin required for the body to change food into energy and for the development of red blood cells. Its good amount of iron contributes to healthy blood. And there is a decent amount of the B vitamin riboflavin, as well as some potassium.

Description: A fricassee is similar to a stew: Chopped-up meat is simmered or sautéed, then cooked in a sauce.

A Better Idea: To reduce calories, cook the chicken in stock or water rather than sautéing it.

CHICKEN KIEV

Serving: 6¹/₂ oz. frozen entrée

Calories	240
Fat	9.5 g
Saturated	N/A
Monounsaturated	N/A
Polyunsaturated	N/A
Calories from fat	36%
Cholesterol	126 mg
Sodium	442 mg
Protein	23.6 g
Carbohydrate	14.9 g
Dietary fiber	N/A

CHIEF NUTRIENTS

NUTRIENT	AMOUNT	% RDA
Vitamin C	11.0 mg	18
Potassium	451.0 mg	12
Riboflavin	0.1 mg	8

Caution: Chicken Kiev contains more than 30% of calories from fat, the maximum recommended by the American Heart Association. So keep portions modest. It's also high in sodium. And since chicken is high in purines, it should be eaten sparingly by anyone with gout. *Allergy alert:* May contain eggs and wheat, two common food allergens.

Eat with: Cantaloupe, a baked potato, or other rich sources of potassium to help offset the potentially harmful effect of sodium.

Strengths: High in protein, chicken Kiev also offers good amounts of vitamin C and potassium, and also a bit of riboflavin.

Description: True chicken Kiev is made from seasoned, batter-fried chicken breast. It literally oozes fat from the butter rolled up inside the breast.

A Better Idea: Roast or broil boneless chicken breast with lemon and herbs—it's just as quick as cooking a frozen entrée, with less fat and sodium.

CHICKEN POTPIE, *Homemade*

Serving: 1 piece (about 8⅛ oz.), baked

Calories	545
Fat	31.3 g
Saturated	10.9 g
Monounsaturated	N/A
Polyunsaturated	N/A
Calories from fat	52%
Cholesterol	72 mg
Sodium	594 mg
Protein	23.4 g
Carbohydrate	42.5 g
Dietary fiber	N/A

CHIEF NUTRIENTS

NUTRIENT	AMOUNT	% RDA
Vitamin A	617.1 RE	62
Iron	3.0 mg	30
Niacin	4.9 mg	26
Thiamine	0.3 mg	21
Riboflavin	0.3 mg	19
Calcium	69.6 mg	9
Potassium	343.4 mg	9
Vitamin C	4.6 mg	8

Caution: Nutritionally, chicken potpie seems to have a little bit of everything in it. Unfortunately, not all those things are good for you. Nearly 300 of its calories are from fat, and it has a high level of sodium. Commercially made potpie has even more sodium and is made with chemical additives. Potpie has a moderate amount of cholesterol. The poultry in the pie contains high amounts of purines, substances that may aggravate gout. *Allergy alert:* Chicken may cause allergic reactions, such as skin rashes, in some people.

Strengths: Chicken potpie has an excellent quantity of vitamin A, which is essential for good vision and healthy skin. It also has good amounts of iron and niacin: Both nutrients contribute to the formation of healthy red blood cells, and iron is necessary to your body's ability to fight off infections. In addition, good amounts of two B vitamins, thiamine and riboflavin, help convert food into energy. This entrée also provides your body with some vitamin C, calcium, and potassium.

TURKEY POTPIE, *Homemade*

Serving: 1 piece (about 8⅛ oz.), baked

Calories	550
Fat	31.3 g
Saturated	10.5 g
Monounsaturated	N/A
Polyunsaturated	N/A
Calories from fat	51%
Cholesterol	72 mg
Sodium	633 mg
Protein	24.1 g
Carbohydrate	42.9 g
Dietary fiber	N/A

CHIEF NUTRIENTS

NUTRIENT	AMOUNT	% RDA
Vitamin A	617.1 RE	62
Niacin	6.5 mg	34
Iron	3.3 mg	33
Riboflavin	0.4 mg	21
Thiamine	0.3 mg	21
Potassium	459.4 mg	12
Calcium	62.6 mg	8
Vitamin C	4.6 mg	8

Caution: Like other potpies, a version made with turkey has some drawbacks. About one-third of the calories from fat are saturated, a type that may increase blood cholesterol levels. Turkey potpie also has a high level of sodium. *Allergy alert:* Turkey may cause allergic reactions, such as skin rashes, in some people. The meat also contains purines, substances that may aggravate gout.

Strengths: There are many good reasons to gobble up all the turkey you can. It's extremely high in vitamin A, which is a great immunity builder. Potpie also has generous amounts of iron and niacin, two nutrients that play a vital role in keeping red blood cells functioning properly. It also has significant amounts of thiamine and riboflavin, two B vitamins that help your body break down food. And its calcium provides some benefits for bones and teeth.

Selection: Commercially made potpies contain a lot of sodium, and they have chemical additives. So you're better off with a homemade version—or, even better, cooked turkey breast with fresh-cooked vegetables on the side.

FISH/SEAFOOD DISHES

CRAB CAKE

Serving: 1 (about 2 oz.), fried

Calories	93
Fat	4.5 g
Saturated	0.9 g
Monounsaturated	1.7 g
Polyunsaturated	1.4 g
Calories from fat	44%
Cholesterol	90 mg
Sodium	198 mg
Protein	12.1 g
Carbohydrate	0.3 g
Dietary fiber	N/A

CHIEF NUTRIENTS

NUTRIENT	AMOUNT	% RDA
Vitamin B_{12}	3.6 mcg	178
Zinc	2.5 mg	16
Folate	24.9 mcg	12
Niacin	1.7 mg	9
Calcium	63.0 mg	8
Iron	0.7 mg	7
Magnesium	19.8 mg	6
Potassium	194.4 mg	5

Strengths: A great source of blood-building vitamin B_{12}, with a good bit of folate and zinc to pep up your immune system. Some studies indicate that crab can reduce the amount of "bad" cholesterol and triglycerides circulating in your bloodstream.

Caution: Crab cakes are high in sodium and purines. People with high blood pressure or gout may want to avoid them. *Allergy alert:* Crab can occasionally cause an allergic reaction.

A Better Idea: Steamed crab. Although steamed blue crab has about the same amount of sodium as fried crab cakes, it also has half the fat, less cholesterol, and more folate, vitamin B_{12}, niacin, and thiamine. And steamed crab has more omega-3's, since the cooking process involved in frying crab cakes destroys these heart-healthy fatty acids.

CRAB IMPERIAL

Serving: 1 cup

Calories	323
Fat	16.7 g
Saturated	8.8 g
Monounsaturated	N/A
Polyunsaturated	N/A
Calories from fat	47%
Cholesterol	308 mg
Sodium	1,602 mg
Protein	32.1 g
Carbohydrate	8.6 g
Dietary fiber	N/A

CHIEF NUTRIENTS

NUTRIENT	AMOUNT	% RDA
Iron	2.0 mg	20
Vitamin C	11.0 mg	18
Calcium	132.0 mg	17
Riboflavin	0.3 mg	15
Niacin	2.4 mg	13
Thiamine	0.1 mg	9
Potassium	288.2 mg	8

Caution: Crab imperial is loaded with cholesterol and sodium. It is also high in purines. People with high blood pressure, heart disease, or gout may want to avoid it. *Allergy alert:* Eating crab can occasionally cause an allergic reaction.

Strengths: Crab imperial contributes a good amount of iron for healthy blood and stronger immunity. Its vitamin C assists in iron absorption, and is also important in healing wounds. In addition, crab imperial has a fair amount of calcium for stronger teeth and bones, as well as the B vitamins riboflavin and niacin to assist in energy production.

Description: Crab imperial is generally made with crabmeat, bread crumbs, mayonnaise, spices, and cheddar or Parmesan cheese.

A Better Idea: Steamed blue crab. It has more protein, half the cholesterol, less than half the fat, and only a smidgen of the sodium that crab imperial contains.

GEFILTE FISH

Serving: 1 piece, cooked, in broth

Calories	35
Fat	0.7 g
Saturated	trace
Monounsaturated	trace
Polyunsaturated	trace
Calories from fat	19%
Cholesterol	13 mg
Sodium	220 mg
Protein	3.8 g
Carbohydrate	3.1 g
Dietary fiber	0

CHIEF NUTRIENTS

NUTRIENT	AMOUNT	% RDA
Vitamin B$_{12}$	0.4 mcg	18
Iron	1.0 mg	10

Strengths: A traditional, low-fat dish, gefilte fish is a good source of iron and vitamin B$_{12}$. It has just a dash of omega-3's.
Caution: *Allergy alert:* Fish occasionally causes an allergic reaction.
History: Gefilte fish is a fish dumpling that is usually made of carp, walleye, whitefish, or all three. It originated in Poland, Hungary, and Czechoslovakia, then spread around the world. A traditional Jewish dish, it's generally made for holidays and the Sabbath.
Preparation: The fish is ground, mixed with eggs, matzo meal, and seasonings, then formed into small balls and simmered in a fish or vegetable stock.

LOBSTER NEWBURG

Serving: 1 cup

Calories	485
Fat	26.5 g
Saturated	15.0 g
Monounsaturated	N/A
Polyunsaturated	N/A
Calories from fat	49%
Cholesterol	455 mg
Sodium	572 mg
Protein	46.3 g
Carbohydrate	12.8 g
Dietary fiber	N/A

CHIEF NUTRIENTS

NUTRIENT	AMOUNT	% RDA
Calcium	217.5 mg	27
Iron	2.3 mg	23
Niacin	3.8 mg	20
Riboflavin	0.3 mg	16
Thiamine	0.2 mg	12
Potassium	427.5 mg	11

Caution: Lobster Newburg contains almost a 2-day quota of cholesterol in a single serving. It is also extremely high in calories, with nearly half of them from fat. And it can load you up with sodium and purines. People who have heart disease, high cholesterol levels, high blood pressure, or gout may wish to avoid it. *Allergy alert:* Eating lobster or inhaling steam from the water in which it is boiled can occasionally cause an allergic reaction.
Strengths: Lobster Newburg is a good source of protein and bone-building calcium, and it adds to your daily requirement of niacin, riboflavin, thiamine, and iron for strong blood.
Origin: A West Indies sea captain named Ben Wenberg brought the recipe to Delmonico's restaurant in New York around 1876. The dish was called "lobster à la Wenberg" until Wenberg and the restaurant's owner allegedly had an argument. In a vengeful pique, Delmonico reversed the first 3 letters of Wenberg's name and renamed the dish "lobster à la Newberg," later changed to lobster Newburg.
Description: Rich. It's loaded with lobster meat, sherry, egg yolks, cream, and cayenne pepper.
A Better Idea: Steamed Maine lobster.

LOBSTER THERMIDOR

Serving: 5½ oz.

Calories	405
Fat	26.6 g
Saturated	N/A
Monounsaturated	N/A
Polyunsaturated	N/A
Calories from fat	59%
Cholesterol	N/A
Sodium	N/A
Protein	28.5 g
Carbohydrate	14.8 g
Dietary fiber	0.5 g

CHIEF NUTRIENTS

NUTRIENT	AMOUNT	% RDA
Calcium	290.0 mg	36
Riboflavin	0.5 mg	30
Niacin	4.8 mg	25
Iron	1.9 mg	19
Thiamine	0.2 mg	10

Caution: Although the USDA doesn't provide exact data, this rich concoction probably has a fair amount of cholesterol. Lobster is high in purines and should be eaten sparingly by anyone with gout. *Allergy alert:* Like other shellfish, lobster is a common allergen.

Strength: Because there are copious portions of cream and milk in the recipe, lobster Thermidor contributes a very good amount of calcium. It's also rich in riboflavin, with decent amounts of iron and other B vitamins. If you're on a heart-healthy diet, the high fat content may cancel out any nutritional benefits.

Description: Made with boiled lobster meat, butter, shallots or onions, white wine, tomato puree, mushrooms, heavy cream or white sauce, Parmesan cheese, and seasonings.

A Better Idea: Plain boiled or broiled lobster, spritzed with fresh-squeezed lemon, is a much better bet: You get about one-fifth the calories and a tiny amount of fat per serving.

SEAFOOD CREOLE

Serving: 6 oz. frozen entrée

Calories	137
Fat	3.8 g
Saturated	N/A
Monounsaturated	N/A
Polyunsaturated	N/A
Calories from fat	25%
Cholesterol	21 mg
Sodium	372 mg
Protein	10.4 g
Carbohydrate	15.4 g
Dietary fiber	N/A

CHIEF NUTRIENTS

NUTRIENT	AMOUNT	% RDA
Vitamin C	38.0 mg	63
Calcium	223.0 mg	28
Vitamin A	244.0 RE	24
Iron	1.6 mg	16
Potassium	537.0 mg	14
Niacin	1.8 mg	9
Thiamine	0.1 mg	7
Riboflavin	0.1 mg	6

Caution: This spicy dish contains more than 370 mg of sodium; individuals on salt-restricted diets should be wary of it. *Allergy alert:* Seafood creole is made with scallops, fish, and garlic, ingredients that may cause allergic reactions such as headaches and hives. In addition, scallops contain purines, substances that may aggravate gout.

Strengths: Since creole sauce is made with tomato sauce and chopped vegetables, it's a phenomenal source of vitamin C, a nutrient that enhances your body's ability to combat infections and heal wounds. It also has a good level of vitamin A, which is essential for healthy skin and good vision. Its high iron boosts your immune system. Seafood creole provides respectable amounts of calcium and potassium, and it has B vitamins that help convert food into energy.

History: The origin of creole cooking—a mixture of French, African, and Spanish influences—dates back to 18th-century Louisiana.

ITALIAN DISHES

BEEF RAVIOLI

Serving: 5 oz. frozen entrée

Calories	185
Fat	5.7 g
Saturated	N/A
Monounsaturated	N/A
Polyunsaturated	N/A
Calories from fat	28%
Cholesterol	59 mg
Sodium	446 mg
Protein	9.2 g
Carbohydrate	24.4 g
Dietary fiber	N/A

CHIEF NUTRIENTS

NUTRIENT	AMOUNT	% RDA
Vitamin A	179.0 RE	18
Iron	1.2 mg	12
Calcium	83.0 mg	10
Vitamin C	6.0 mg	10
Potassium	348.0 mg	9
Riboflavin	0.2 mg	9
Niacin	1.2 mg	6
Thiamine	0.1 mg	5

Strengths: Beef ravioli has almost as much protein as manicotti but less than half as much fat. From the frozen entrée you get an array of vitamins and minerals, including good bone-building calcium and resistance-boosting iron and vitamins A and C.

Caution: The sodium content is a bit steep, especially if you're trying to limit your salt intake. Beef is high in purines and should be eaten sparingly if you have gout. *Allergy alert:* If you're allergic to wheat or eggs, ravioli could be a problem for you.

Description: Ravioli is Italy's answer to Polish pierogies and Chinese wontons. This Italian specialty consists of little pillows of dough filled with ground beef, cheese, or vegetables, usually boiled and served with tomato sauce.

Serve: With a tossed green salad and Italian bread.

CHICKEN CACCIATORE

Serving: 10 oz. frozen entrée

Calories	260
Fat	5.0 g
Saturated	N/A
Monounsaturated	N/A
Polyunsaturated	N/A
Calories from fat	17%
Cholesterol	40 mg
Sodium	510 mg
Protein	17.0 g
Carbohydrate	35.0 g
Dietary fiber	N/A

CHIEF NUTRIENTS

NUTRIENT	AMOUNT	% RDA
Iron	3.0 mg	30
Vitamin A	135.6 RE	14
Thiamine	0.2 mg	13
Niacin	2.3 mg	12
Riboflavin	0.2 mg	12
Calcium	42.0 mg	5
Potassium	191.0 mg	5

Strengths: With only 5 g of fat and less than 300 calories per serving, this Italian-American classic makes a fine entrée. You get lots of easy-to-absorb, anemia-fighting heme iron, along with a fair amount of protein, good amounts of B vitamins and resistance-boosting vitamin A, and a little potassium and calcium.

Caution: With over 510 mg of sodium, frozen chicken cacciatore may be a problem if you're on a sodium-restricted diet. Chicken is high in purines and should be eaten sparingly by those who have gout.

Description: *Cacciatore* refers to food prepared ''hunter style,'' with mushrooms, onions, tomatoes, and herbs.

A Better Idea: You can easily make a low-sodium version of chicken cacciatore. Cut up a frying chicken, remove the skin, and braise with mushrooms, onions, tomatoes, garlic, basil, and thyme—but no added salt.

EGGPLANT PARMIGIANA

Serving: 6½ oz. frozen entrée

Calories	223
Fat	12.0 g
Saturated	N/A
Monounsaturated	N/A
Polyunsaturated	N/A
Calories from fat	48%
Cholesterol	5 mg
Sodium	982 mg
Protein	10.8 g
Carbohydrate	17.9 g
Dietary fiber	N/A

CHIEF NUTRIENTS

NUTRIENT	AMOUNT	% RDA
Vitamin C	19.0 mg	32
Vitamin A	287.2 RE	29
Calcium	218.0 mg	27
Iron	2.2 mg	22
Potassium	462.0 mg	12
Riboflavin	0.2 mg	11
Thiamine	0.2 mg	10
Niacin	1.7 mg	9

Caution: High in fat and sodium, this dish may be off-limits if you are on a heart-healthy, stroke-prevention diet.

Strengths: Eggplant parmigiana is a good source of calcium (for stronger teeth and bones) and blood-building iron. A good amount of potassium aids in energy production. It's rich in vitamins C and A, which help heal wounds and keep immunity up to par.

Description: Eggplant parmigiana is made by dipping sliced eggplant in egg, milk, and bread crumbs; the slices are fried and baked under a layer of tomato sauce and grated Parmesan cheese—hence, the name parmigiana.

A Better Idea: You can make very low-fat eggplant parmigiana by baking, steaming, or microwaving sliced eggplant, then layering it with reduced-fat mozzarella and meatless tomato sauce. Bake until the cheese is melted.

SAUSAGE AND PEPPERS

Serving: 6 oz. frozen entrée

Calories	327
Fat	29.4 g
Saturated	N/A
Monounsaturated	N/A
Polyunsaturated	N/A
Calories from fat	81%
Cholesterol	59 mg
Sodium	665 mg
Protein	7.2 g
Carbohydrate	9.6 g
Dietary fiber	N/A

CHIEF NUTRIENTS

NUTRIENT	AMOUNT	% RDA
Vitamin C	361.0 mg	602
Thiamine	0.3 mg	19
Vitamin A	186.8 RE	19
Iron	1.8 mg	18
Niacin	1.9 mg	10
Riboflavin	0.1 mg	8

Caution: More than 80% of the calories in this dish are derived from fat. It also has 665 mg of sodium per serving, a much-too-big number for anyone on a low-salt diet. *Allergy alert:* Pork, the main ingredient of Italian sausage, wins the "least congenial" award: It causes more allergic reactions than any other meat.

Strengths: Italian sausage and peppers is a stupendous source of vitamin C, containing more than six times the recommended daily allowance of the nutrient that helps your body heal wounds and combat infections. It also has good amounts of vitamin A and niacin, along with iron for healthy blood cells.

Curiosity: Christopher Columbus brought sweet and hot peppers back to Europe from the New World 500 years ago.

History: A Mediterranean staple for more than 3,000 years, sausage is one of the oldest known processed foods.

LASAGNA

Serving: 6 oz. frozen entrée

Calories	244
Fat	10.8 g
Saturated	N/A
Monounsaturated	N/A
Polyunsaturated	N/A
Calories from fat	40%
Cholesterol	21 mg
Sodium	816 mg
Protein	15.6 g
Carbohydrate	21.0 g
Dietary fiber	N/A

CHIEF NUTRIENTS

NUTRIENT	AMOUNT	% RDA
Thiamine	0.5 mg	35
Vitamin C	19.0 mg	32
Iron	2.8 mg	28
Vitamin A	261.6 RE	26
Niacin	4.4 mg	23
Calcium	167.0 mg	21
Riboflavin	0.3 mg	18
Potassium	271.0 mg	7

Strengths: Nutritionally, lasagna is multidimensionally good: It's a good source of iron for healthy blood, calcium for stronger bones, vitamin A for stronger resistance to infections, and niacin for energy and healthy blood. It's rich in vitamin C and thiamine, *and* it's relatively low in cholesterol, with moderate amounts of fat and calories.

Caution: The frozen version of this standard Italian-American favorite is very high in sodium, posing a potential problem if you're trying to eat less sodium. *Allergy alert:* If you're allergic to milk, wheat, or eggs, lasagna could be a problem for you.

Curiosity: The plural of lasagna is lasagne.

Description: Broad, flat noodles, layered with mozzarella and other cheese, topped with tomato sauce, and baked until bubbly.

A Better Idea: If you can't find a reduced-sodium version of lasagna, make your own with a minimum amount of added salt. And in the homemade version you can use reduced-fat cheese and meatless sauce to keep the fat content from getting out of hand.

LINGUINI with White Clam Sauce, Canned

Serving: 11 oz.

Calories	391
Fat	10.5 g
Saturated	N/A
Monounsaturated	N/A
Polyunsaturated	N/A
Calories from fat	24%
Cholesterol	N/A
Sodium	753 mg
Protein	17.8 g
Carbohydrate	56.4 g
Dietary fiber	N/A

CHIEF NUTRIENTS

NUTRIENT	AMOUNT	% RDA
Iron	6.2 mg	62
Thiamine	0.6 mg	40
Niacin	2.5 mg	13
Riboflavin	0.2 mg	12
Calcium	45.0 mg	6
Vitamin C	3.0 mg	5

Caution: This Italian dish contains a tremendous amount of sodium per serving. *Allergy alert:* Linguini noodles are made with flour, water or milk, and sometimes eggs, ingredients that may cause allergic reactions in some people. Clams, clam juice, and milk may be used in the sauce and also may cause adverse reactions.

Strengths: Linguini with white clam sauce is a tremendous source of iron, a nutrient needed to maintain a strong immune system. This entrée also is rich in thiamine and has a respectable amount of riboflavin, B vitamins that help your body convert food into energy. Its niacin is vital to the production of healthy red blood cells. Calcium and vitamin C are also present.

Curiosity: In Italian, *linguini* means "little tongues."

A Better Idea: Since white sauce is laden with whole milk and butter, do your arteries a favor and prepare a light tomato sauce with fresh clams.

MANICOTTI

Serving: 5 oz. frozen entrée

Calories	229
Fat	13.5 g
Saturated	N/A
Monounsaturated	N/A
Polyunsaturated	N/A
Calories from fat	53%
Cholesterol	56 mg
Sodium	421 mg
Protein	13.0 g
Carbohydrate	12.5 g
Dietary fiber	N/A

CHIEF NUTRIENTS

NUTRIENT	AMOUNT	% RDA
Potassium	225.0 mg	6

Caution: When made with whole-milk ricotta, manicotti tends to be high in fat. Sodium content may be a problem if you're on a diet to control high blood pressure and avoid a stroke or kidney disease. *Allergy alert:* Manicotti may trigger a reaction if you're allergic to wheat, eggs, or milk and other dairy products.

Strengths: This Italian favorite contributes a fair amount of protein and a little potassium to the diet.

Description: These are tubular noodles about 4 inches long and an inch or so in diameter, stuffed with ricotta cheese and topped with tomato sauce.

Serve: With a green salad.

A Better Idea: You can reduce fat and sodium if you make your own manicotti with low-fat ricotta, egg substitute, tomato sauce, and no added salt.

PASTA PRIMAVERA

Serving: 6 oz. frozen entrée

Calories	305
Fat	14.2 g
Saturated	N/A
Monounsaturated	N/A
Polyunsaturated	N/A
Calories from fat	42%
Cholesterol	45 mg
Sodium	1,410 mg
Protein	11.6 g
Carbohydrate	32.3 g
Dietary fiber	N/A

CHIEF NUTRIENTS

NUTRIENT	AMOUNT	% RDA
Vitamin A	317.6 RE	32
Calcium	251.0 mg	31
Vitamin C	10.0 mg	17
Iron	1.4 mg	14
Thiamine	0.2 mg	13
Riboflavin	0.2 mg	9
Niacin	1.5 mg	8

Caution: This frozen food is extremely high in sodium. Since it's also high in calories and moderately high in fat, proceed with caution if you're watching your weight or guarding against high blood pressure.

Strengths: Because it's made with vitamin A–enriched milk (or cream) and with vegetables, pasta primavera is a very good source of vitamin A, which may help protect against cancer. The milk or cream provides a good share of bone-strengthening calcium. Pasta primavera also is high in vitamin C, important for a healthy immune system. Iron, thiamine, riboflavin, and niacin contribute to healthy blood and energy production.

Curiosity: *Primavera* in Italian means "springtime," but there is no traditional dish in Italian cuisine by this name.

Description: Noodles with quickly cooked vegetables, usually mixed in a cream sauce. May include zucchini, broccoli, snow peas, peppers, and sliced mushrooms, along with pine nuts and Parmesan cheese.

A Better Idea: Make (and freeze) your own skinny version of pasta primavera using low-fat milk.

SPAGHETTI, *Homemade*

Serving: 1 cup

Calories	260
Fat	8.8 g
Saturated	2.0 g
Monounsaturated	N/A
Polyunsaturated	N/A
Calories from fat	30%
Cholesterol	8 mg
Sodium	955 mg
Protein	8.8 g
Carbohydrate	37.0 g
Dietary fiber	N/A

CHIEF NUTRIENTS

NUTRIENT	AMOUNT	% RDA
Iron	2.3 mg	23
Vitamin C	12.5 mg	21
Thiamine	0.3 mg	17
Niacin	2.3 mg	12
Potassium	407.5 mg	11
Riboflavin	0.2 mg	11
Calcium	80.0 mg	10

Strengths: Very low in cholesterol, spaghetti offers an excellent supply of energy-giving carbohydrates. When tomato sauce is added, you have a good source of vitamin C for a better-functioning immune system. The C has another benefit: It helps your body absorb the iron also found in significant amounts in spaghetti.

Caution: This pasta dish is high in sodium, and homemade spaghetti with cheese derives almost one-third of its calories from fat. *Allergy alert:* Wheat, eggs, cheese, and tomatoes all can produce allergic reactions in sensitive people.

Curiosity: Italians were eating pasta long before Marco Polo returned from China and supposedly introduced the food to Europe. There now are some 200 distinct kinds of pasta differentiated by shape, size, and texture.

Description: Spaghetti is just one of many forms of pasta. *Pasta* means "dough" in Italy; *spaghetti* translates into "strings." It is served with tomato sauce and cheese.

SPAGHETTI AND MEATBALLS, *Homemade*

Serving: 1 cup

Calories	332
Fat	11.7 g
Saturated	3.3 g
Monounsaturated	N/A
Polyunsaturated	N/A
Calories from fat	32%
Cholesterol	74 mg
Sodium	1,009 mg
Protein	18.6 g
Carbohydrate	38.7 g
Dietary fiber	N/A

CHIEF NUTRIENTS

NUTRIENT	AMOUNT	% RDA
Iron	3.7 mg	37
Vitamin C	22.3 mg	37
Niacin	4.0 mg	21
Potassium	664.6 mg	18
Riboflavin	0.3 mg	18
Thiamine	0.3 mg	17
Calcium	124.0 mg	16

Strengths: Meatballs significantly increase the protein in this spaghetti dish, and they also boost the iron content. Homemade spaghetti and meatballs is served with tomato sauce, which adds a rich amount of vitamin C. A moderate amount of niacin helps the body form red blood cells—and this entrée also has a good amount of potassium.

Caution: There's more fat, sodium, and cholesterol in spaghetti with meatballs than in spaghetti that's made with a nonmeat sauce. And beef is high in such purines as uric acid, so people who suffer from gout should be wary. *Allergy alert:* Ingredients include wheat, eggs and tomatoes, which all can produce allergic reactions in some people.

Curiosity: It is a sign of spaghetti-eating expertise and sophistication to be able to twirl the pasta around the fork directly from the plate without the aid of a spoon.

A Better Idea: To get all the nutritional benefits of beef but reduce the fat, start with extra-lean beef and microwave the meatballs. (Beef that is microwaved retains less fat in the cooking process.) When shopping, look for ground round, which is leaner than ground sirloin and ground chuck.

TURKEY TETRAZZINI

Serving: 6 oz. frozen entrée

Calories	228
Fat	9.0 g
Saturated	N/A
Monounsaturated	N/A
Polyunsaturated	N/A
Calories from fat	36%
Cholesterol	23 mg
Sodium	66 mg
Protein	20.8 g
Carbohydrate	15.9 g
Dietary fiber	N/A

CHIEF NUTRIENTS

NUTRIENT	AMOUNT	% RDA
Calcium	250.0 mg	31
Niacin	3.5 mg	18
Riboflavin	0.2 mg	10
Iron	0.7 mg	7
Potassium	196.0 mg	5

Strengths: This dish is lower in sodium than most other frozen entrées. It's a very good source of calcium, a mineral that can help protect against high blood pressure and build strong bones when eaten in sufficient quantities. You also get appreciable niacin and riboflavin, B vitamins that help release energy within your cells.

Caution: This frozen entrée has moderate amounts of fat, so keep portions modest if you're watching your weight or limiting fat intake. Turkey is high in purines and should be eaten sparingly by anyone with gout.

Origin: Turkey tetrazzini was named after the Italian opera star Luisa Tetrazzini, who toured the U.S. in the early 1900s.

Description: First cousin to tuna noodle casserole, this baked dish consists of turkey, spaghetti, mushrooms, and a cream sauce of some sort that is flavored with grated Parmesan cheese. Almonds are optional.

VEAL MARSALA

Serving: 6 oz. frozen entrée

Calories	164
Fat	9.1 g
Saturated	N/A
Monounsaturated	N/A
Polyunsaturated	N/A
Calories from fat	50%
Cholesterol	52 mg
Sodium	366 mg
Protein	13.0 g
Carbohydrate	7.7 g
Dietary fiber	N/A

CHIEF NUTRIENTS

NUTRIENT	AMOUNT	% RDA
Niacin	4.5 mg	24
Iron	2.2 mg	22
Riboflavin	0.2 mg	10
Potassium	288.0 mg	8
Thiamine	0.1 mg	7

Caution: This dish is moderate in fat and cholesterol and high in sodium. *Allergy alert:* Veal contains purines, substances that can aggravate gout.

Strengths: Veal marsala contains significant levels of niacin and iron. Both nutrients are needed for healthy red blood cells. Iron also enhances your body's ability to combat infections. In addition, the presence of the B vitamin riboflavin helps convert food into energy.

Description: This dish is made with veal and marsala wine, a white wine made from Sicilian grapes.

A Better Idea: Instead of sampling the dish as a frozen entrée, make your own and you will be able to control the sodium content.

VEAL PARMIGIANA

Serving: 5 oz. frozen entrée

Calories	279
Fat	17.9 g
Saturated	N/A
Monounsaturated	N/A
Polyunsaturated	N/A
Calories from fat	58%
Cholesterol	67 mg
Sodium	858 mg
Protein	17.0 g
Carbohydrate	11.8 g
Dietary fiber	N/A

CHIEF NUTRIENTS

NUTRIENT	AMOUNT	% RDA
Calcium	133.0 mg	17
Iron	1.7 mg	17
Niacin	2.0 mg	11
Potassium	429.0 mg	11
Riboflavin	0.2 mg	9
Thiamine	0.1 mg	7

Caution: In a 5-oz. serving, about 160 calories come from fat. Veal parmigiana also contains more than 850 mg of sodium, so take a detour around this frozen food if you're on a low-salt diet. Veal contains purines, substances that may aggravate gout.
Strengths: The entrée has a good amount of iron, a nutrient that in combination with B vitamins may prevent canker sores. It also contains a respectable amount of the calcium needed for strong bones and teeth. In addition to being a good source of niacin, it's equally strong in potassium, a nutrient muscles and nerves need to function properly.
Curiosity: Authentic Parmesan cheese comes from the Parma region of Italy and is stamped "Parmigiano-Reggiano."
Description: This Italian dish is made with veal that's heavily covered with bread crumbs and Parmesan cheese. (Many people object to eating veal because it comes from calves between 1 and 3 months old.)
A Better Idea: Make your own veal parmesan without the heavy breading.

ORIENTAL DISHES

BEEF TERIYAKI

Serving: 6 oz. frozen entrée

Calories	154
Fat	4.4 g
Saturated	N/A
Monounsaturated	N/A
Polyunsaturated	N/A
Calories from fat	26%
Cholesterol	18 mg
Sodium	816 mg
Protein	12.3 g
Carbohydrate	16.1 g
Dietary fiber	N/A

CHIEF NUTRIENTS

NUTRIENT	AMOUNT	% RDA
Vitamin C	20.0 mg	33
Vitamin A	258.0 RE	26
Iron	2.2 mg	22
Potassium	382.0 mg	10
Niacin	1.7 mg	9
Riboflavin	0.2 mg	9
Calcium	46.0 mg	6

Strengths: Beef teriyaki is low in fat, moderate in calories, and very high in vitamin C, which may help protect against cancer. It's a good source of blood-building iron, and of vitamin A, essential for a healthy immune system and good skin and eyes. Beef teriyaki also offers some riboflavin and niacin, B-complex vitamins that help produce energy. Potassium and calcium are an added bonus: These minerals help regulate blood pressure.
Caution: Beef teriyaki is high in sodium, mostly from its soy sauce marinade.
Description: Beef strips are marinated in soy sauce, sugar, sherry, ginger, and garlic, then grilled. The entrée often includes rice and vegetables, such as broccoli, peas, and green onions.
Serve: With brown rice and some lightly steamed snow peas.
A Better Idea: It's easy to make your own beef teriyaki using low-sodium soy sauce and lots of fresh vegetables.

CHICKEN CHOW MEIN, *Homemade*

Serving: 1 cup, w/out noodles

Calories	255
Fat	10.0 g
Saturated	2.4 g
Monounsaturated	N/A
Polyunsaturated	N/A
Calories from fat	35%
Cholesterol	78 mg
Sodium	718 mg
Protein	31.0 g
Carbohydrate	10.0 g
Dietary fiber	N/A

CHIEF NUTRIENTS

NUTRIENT	AMOUNT	% RDA
Iron	2.5 mg	25
Niacin	4.3 mg	22
Vitamin C	10.0 mg	17
Riboflavin	0.2 mg	14
Potassium	472.5 mg	13
Calcium	57.5 mg	7
Vitamin A	55.0 RE	6

Strengths: The chicken in chow mein is very high in protein. A serving fulfills one-quarter of the body's daily need for iron—a long-term deficiency of which may cause fatigue and irritability. Niacin helps the body transform food into energy, and vitamin C assists in fighting infections.

Caution: Very high in sodium, with moderate fat and calories, chicken chow mein also has a moderate amount of cholesterol. Poultry is high in purines, so be wary if you're prone to gout. *Allergy alert:* Chicken causes allergic reactions in some people.

Curiosity: The word chow mein is derived from the Mandarin Chinese term *ch'ao mien,* which translates into "fried noodles."

Description: A hybrid culinary creation of the Chinese-Americans, chicken chow mein contains stewed vegetables and bean sprouts as well as chicken, and is often served over fried noodles.

CHOP SUEY, *Homemade*

Serving: 1 cup

Calories	300
Fat	17.0 g
Saturated	8.5 g
Monounsaturated	N/A
Polyunsaturated	N/A
Calories from fat	51%
Cholesterol	100 mg
Sodium	1,053 mg
Protein	26.0 g
Carbohydrate	12.8 g
Dietary fiber	N/A

CHIEF NUTRIENTS

NUTRIENT	AMOUNT	% RDA
Vitamin C	32.5 mg	54
Iron	4.8 mg	48
Niacin	5.0 mg	26
Riboflavin	0.4 mg	22
Thiamine	0.3 mg	19
Vitamin A	120.0 RE	12
Potassium	425.0 mg	11
Calcium	60.0 mg	8

Caution: In chop suey, a moderate amount of cholesterol is accompanied by very high amounts of saturated fat and sodium—and more than half the calories come from fat. If the chop suey contains chicken, it may aggravate gout in some people. *Allergy alert:* May include pork, poultry, and beans, which are known to cause allergic reactions in some people. And if you're allergic to molds, it's advisable to be cautious when eating meals that contain mushrooms.

Strengths: Protein-rich chop suey provides a kick of iron, which is an essential part of the hemoglobin in red blood cells. The meal also contains more than half of the RDA for vitamin C, which helps fight off infections, heal wounds, and improve the body's ability to absorb iron. Its important B vitamins—niacin, riboflavin, and thiamine—help boost energy, and niacin and riboflavin also aid in the formation of red blood cells.

Curiosity: Chop suey is from the Mandarin word *tsa sui,* meaning chopped-up odds and ends.

Description: Usually contains pork or chicken, bamboo shoots, bean sprouts, and water chestnuts, all served over rice.

GREEN-PEPPER STEAK

Serving: 10 oz. frozen entrée

Calories	310
Fat	5.0 g
Saturated	N/A
Monounsaturated	N/A
Polyunsaturated	N/A
Calories from fat	15%
Cholesterol	50 mg
Sodium	1,125 mg
Protein	21.0 g
Carbohydrate	44.0 g
Dietary fiber	N/A

CHIEF NUTRIENTS

NUTRIENT	AMOUNT	% RDA
Iron	3.3 mg	33
Niacin	2.7 mg	14
Thiamine	0.2 mg	14
Riboflavin	0.2 mg	11
Potassium	221.0 mg	6
Vitamin A	61.4 RE	6

Caution: Green-pepper steak contains more than 1,100 mg of sodium, an ultra-excessive level for people on salt-restricted diets. Studies have linked high-calorie, high-fat diets to increased incidence of rectal cancer. Steak is one of those meats that seems to boost the risk. *Allergy alert:* Beef can cause allergic reactions in some people. In fact, a few individuals are so sensitive that the fumes from cooking beef will cause itching or respiratory difficulties.

Strengths: You'll get a very good boost of iron, a nutrient that's essential for healthy red blood cells and for production of collagen. Green-pepper steak also has beneficial levels of thiamine, riboflavin, and niacin, three B vitamins that help your body convert food into energy. Nutritionally, potassium and vitamin A also are present in significant amounts.

Description: Made with round steak, soy sauce, beef bouillon, garlic, ginger, and green peppers.

A Better Idea: Prepare this dish with top round or round tip, which are leaner than some other cuts of beef.

SWEET-AND-SOUR PORK

Serving: 6 oz. frozen entrée

Calories	222
Fat	9.7 g
Saturated	N/A
Monounsaturated	N/A
Polyunsaturated	N/A
Calories from fat	39%
Cholesterol	38 mg
Sodium	801 mg
Protein	12.6 g
Carbohydrate	21.0 g
Dietary fiber	N/A

CHIEF NUTRIENTS

NUTRIENT	AMOUNT	% RDA
Vitamin C	44.0 mg	73
Niacin	2.6 mg	14
Iron	1.3 mg	13
Vitamin A	132.0 RE	13
Potassium	300.0 mg	8
Thiamine	0.1 mg	6

Caution: Sweet-and-sour pork gets more than one-third of its calories from fat and contains a moderate amount of cholesterol. It is also high in sodium. The purines in the meat may aggravate gout in people who are susceptible. *Allergy alert:* Pork may cause a number of allergic reactions, including hives, headaches and, it is speculated in certain instances, epilepsy.

Strengths: Vitamin C, which is essential for fighting infections and healing wounds, is packed into sweet-and-sour pork. The pork contributes a good boost of protein to help build energy, and there's an economical supply of iron for healthier blood. There's a good amount of vitamin A and some potassium for healthier nerves and muscles.

Description: Lean pork loin or shoulder is browned in hot vegetable oil, then simmered with pineapple juice, chicken stock, and other ingredients. The sweet in the recipe usually comes from sugar, the sour from vinegar. Other ingredients may be added as well: pineapple chunks, carrots, and green pepper or onion slices.

POTATO/RICE DISHES

HASH BROWN POTATOES

Serving: ½ cup

Calories	119
Fat	10.9 g
Saturated	4.2 g
Monounsaturated	4.9 g
Polyunsaturated	1.3 g
Calories from fat	82%
Cholesterol	N/A
Sodium	19 mg
Protein	1.9 g
Carbohydrate	5.8 g
Dietary fiber	1.6 g

CHIEF NUTRIENTS

NUTRIENT	AMOUNT	% RDA
Vitamin B$_6$	0.2 mg	11
Niacin	1.6 mg	8
Potassium	250.4 mg	7
Vitamin C	4.5 mg	7
Iron	0.6 mg	6

Caution: Contain almost as much fat as a glazed doughnut, custard éclair, or slice of cheesecake. Even microwave hash browns are soaked in fat. If you're watching your fat intake, hash browns should be eaten only occasionally. *Allergy alert:* Some people with asthma are severely allergic to hash browns treated with sulfites.

Eat with: Scrambled egg substitute (or omelets) and fat-free pita bread or unbuttered toast to offset the high fat content of this popular breakfast food.

Strength: Like french fries, hash browns contain some useful vitamins and minerals.

Description: Chopped potatoes fried in butter, oil, or fat.

A Better Idea: A baked potato! It contains 1% fat, in contrast to the 82% in hashed browns.

MASHED POTATOES

Serving: ½ cup

Calories	81
Fat	0.6 g
Saturated	trace
Monounsaturated	trace
Polyunsaturated	trace
Calories from fat	7%
Cholesterol	2 mg
Sodium	318 mg
Protein	2.0 g
Carbohydrate	18.4 g
Dietary fiber	2.1 g

CHIEF NUTRIENTS

NUTRIENT	AMOUNT	% RDA
Vitamin B$_6$	0.2 mg	12
Vitamin C	7.0 mg	12
Potassium	314.0 mg	8
Niacin	1.2 mg	6
Thiamine	0.1 mg	6
Magnesium	18.9 mg	5

Strengths: Potatoes aren't fattening: In fact, mashed spuds are just the kind of low-fat complex carbohydrates that can control your appetite and promote weight loss—as long as you don't drown them in butter and gravy. Though they're not quite as nutrient-rich as baked potatoes, they're close. And decent amounts of nerve-soothing B vitamins help to allay skin problems.

Caution: Choose a recipe for mashed potatoes with care, especially if you're on a low-salt diet. The recipe used by the USDA is high in sodium (318 mg per serving), but this amount can be easily reduced if you use little or no salt in cooking. (A baked potato has only 16 mg of sodium.)

Preparation: Substitute skim milk or low-fat yogurt for whole milk. To avoid added fat, use butter-flavor granules or diet margarine instead of regular margarine.

O'BRIEN POTATOES

Serving: ½ cup

Calories	79
Fat	1.2 g
Saturated	0.8 g
Monounsaturated	0.3 g
Polyunsaturated	0.1 g
Calories from fat	14%
Cholesterol	4 mg
Sodium	210 mg
Protein	2.3 g
Carbohydrate	15.0 g
Dietary fiber	N/A

CHIEF NUTRIENTS

NUTRIENT	AMOUNT	% RDA
Vitamin C	16.2 mg	27
Vitamin B$_6$	0.2 mg	11
Potassium	258.0 mg	7
Vitamin A	55.3 RE	6
Niacin	1.0 mg	5
Thiamine	0.1 mg	5

Strengths: The appreciable offering of vitamin C in potatoes is needed for stronger immunity and healthy bones, teeth, skin, and blood vessels. A respectable amount of vitamin B$_6$ is present, with small portions of other nerve-soothing B vitamins. The vitamin A in potatoes is believed to protect against cancer. O'Brien potatoes have less fat and fewer calories than hash brown potatoes.

Caution: Sodium content is a bit on the high side. *Allergy alert:* If you're allergic to milk, leave it out of your O'Brien recipe.

Description: No one knows who O'Brien was, but he liked his potatoes diced and fried with lots of chopped onions and sweet green peppers.

Preparation: Most recipes call for butter or bacon fat, whole milk, and salt. But if you're cutting back on fat and sodium, use diet margarine, skim milk, and black pepper, minced parsley, or paprika.

POTATO PANCAKE

Serving: 1 (about 2⅔ oz.)

Calories	234
Fat	12.6 g
Saturated	3.4 g
Monounsaturated	5.4 g
Polyunsaturated	2.5 g
Calories from fat	48%
Cholesterol	93 mg
Sodium	388 mg
Protein	4.6 g
Carbohydrate	26.4 g
Dietary fiber	1.5 g

CHIEF NUTRIENTS

NUTRIENT	AMOUNT	% RDA
Vitamin B$_6$	0.3 mg	15
Potassium	538.1 mg	14
Iron	1.2 mg	12
Folate	21.5 mcg	11
Niacin	1.6 mg	8
Magnesium	24.3 mg	7
Thiamine	0.1 mg	7
Riboflavin	0.1 mg	6

Strengths: Like other potato dishes, it offers decent amounts of various B vitamins. The need for folate and vitamin B$_6$ is generally higher in women who are pregnant or taking oral contraceptives. One potato pancake also supplies a helpful amount of anemia-fighting iron and some potentially useful potassium. Potato pancakes also have fiber for digestive health.

Caution: *Allergy alert:* Eggs and wheat, often used in potato pancakes, are common food allergens. Weight watchers should beware: These pancakes are high in fat, sodium, and calories.

Description: A fried patty containing egg, onion, flour, butter or margarine, and salt, as well as potatoes.

Preparation: To cut down on sodium, cholesterol, and saturated fat, omit salt and use skim milk, diet margarine, and egg substitute.

POTATOES AU GRATIN

Serving: ½ cup

Calories	161
Fat	9.2 g
Saturated	5.8 g
Monounsaturated	2.6 g
Polyunsaturated	0.3 g
Calories from fat	52%
Cholesterol	28 mg
Sodium	528 mg
Protein	6.2 g
Carbohydrate	13.8 g
Dietary fiber	2.2 g

CHIEF NUTRIENTS

NUTRIENT	AMOUNT	% RDA
Vitamin C	12.1 mg	20
Calcium	145.2 mg	18
Potassium	483.1 mg	13
Vitamin B₆	0.2 mg	11
Iron	0.8 mg	8
Riboflavin	0.1 mg	8
Magnesium	24.4 mg	7
Niacin	1.2 mg	6

Strengths: Along with nice amounts of various B vitamins, potatoes au gratin have a significant amount of bone-building calcium—thanks to the milk and cheese. And the potatoes contribute a good amount of resistance-building vitamin C. Potatoes au gratin have a good amount of fiber, which helps to improve digestive health.

Caution: Au gratin–style potatoes, if prepared with a high-salt recipe, are very high in sodium. The fat and calories in this dish are to be avoided if you're trying to lose weight or cut your risk of heart disease or stroke. *Allergy alert:* Usually contain milk, a common food allergen.

Description: *Au gratin* is a French phrase describing any food that's topped with bread crumbs, butter, and grated cheese like Gruyère or Parmesan. Browned in the oven to form a crisp, golden-brown crust.

Preparation: By omitting salt and using skim milk, diet margarine, and low-fat cheese instead of full-fat recipe ingredients, you can cut the sodium, cholesterol, and saturated fat content.

SCALLOPED POTATOES

Serving: ½ cup

Calories	105
Fat	4.5 g
Saturated	2.8 g
Monounsaturated	1.3 g
Polyunsaturated	0.2 g
Calories from fat	39%
Cholesterol	15 mg
Sodium	409 mg
Protein	3.5 g
Carbohydrate	13.2 g
Dietary fiber	2.3 g

CHIEF NUTRIENTS

NUTRIENT	AMOUNT	% RDA
Vitamin C	12.9 mg	22
Potassium	461.2 mg	12
Vitamin B₆	0.2 mg	11
Calcium	69.5 mg	9
Iron	0.7 mg	7
Magnesium	23.2 mg	7
Niacin	1.3 mg	7

Strengths: Scalloped potatoes have fewer calories, a little less sodium, and half as much fat as potatoes au gratin. Along with a respectable amount of resistance-building vitamin C are an array of B vitamins that tend to improve skin and nerve health. A decent amount of potassium and a bit of magnesium may help protect against high blood pressure and stroke if the dish is made without salt. There's a good amount of fiber for improved digestion.

Caution: High in sodium. *Allergy alert:* Contains milk and wheat flour, two common allergens.

Curiosity: According to food lore, potatoes in this style were originally served in scallop shells.

Description: Sliced potatoes, layered with cream sauce (or milk), flour, and butter, and baked until brown and bubbly.

A Better Idea: By omitting salt and using skim milk and diet margarine instead of whole milk and butter, you can cut the sodium, cholesterol, and saturated fat content.

SPANISH RICE, *Homemade*

Serving: ½ cup

Calories	107
Fat	2.1 g
Saturated	0
Monounsaturated	N/A
Polyunsaturated	N/A
Calories from fat	18%
Cholesterol	0
Sodium	387 mg
Protein	2.2 g
Carbohydrate	20.3 g
Dietary fiber	N/A

CHIEF NUTRIENTS

NUTRIENT	AMOUNT	% RDA
Vitamin C	18.4 mg	31
Potassium	283.0 mg	8
Iron	0.7 mg	7
Niacin	0.9 mg	5

Strengths: Moderate in calories, with zero cholesterol, Spanish rice is a low-fat source of carbohydrates. A Spanish rice dish is about one-half tomatoes, which are a potent source of vitamin C. This vitamin strengthens the immune system and may help to lower the risk of contracting some forms of cancer.

Caution: The sodium in Spanish rice is high, so try another dish if you're on a low-salt diet. Tomatoes may cause bed-wetting in children. *Allergy alert:* Some people are allergic to tomatoes and may suffer such reactions as headaches and hives.

Description: The basic dish consists of rice cooked with to-matoes, peppers, onion, and celery, often with some garlic. Al-most any kind of meat can be added to the recipe—traditionally pork, chicken, seafood, or beef.

A Better Idea: Use brown instead of white rice. Besides being more easily digested, brown rice is higher in nutrients, espe-cially thiamine and protein.

SALADS AND SANDWICHES

COLESLAW

Serving: ½ cup

Calories	41
Fat	1.6 g
Saturated	0.2 g
Monounsaturated	0.4 g
Polyunsaturated	0.8 g
Calories from fat	34%
Cholesterol	5 mg
Sodium	14 mg
Protein	0.8 g
Carbohydrate	7.5 g
Dietary fiber	N/A

CHIEF NUTRIENTS

NUTRIENT	AMOUNT	% RDA
Vitamin C	19.6 mg	33
Folate	16.0 mcg	8

Strengths: A very good source of vitamin C. (Aspirin, tetra-cycline, and oral contraceptives may deplete vitamin C.) Cole-slaw also contains a bit of folate, which may be depleted by aspirin, certain antacids, diuretics, and a number of other drugs. Since cabbage is the primary ingredient, you may help boost your resistance to cancer if you regularly order this salad.

Eat with: Grains or rice. The vitamin C in coleslaw will aid in absorption of plant iron.

Caution: *Allergy alert:* Some cases of cabbage allergy—with facial and throat swelling—have been reported. However, these cases are rare.

Description: Made of cabbage, celery, green pepper, onion, and seasoned dressing.

Selection: Look for coleslaw prepared with low-fat dressing, or make your own with light mayo and nonfat yogurt. (A high-fat dressing could cancel out health benefits.)

Origin: From the Dutch *kool sla,* or cabbage salad.

LOBSTER SALAD

Serving: ½ cup

Calories	286
Fat	16.6 g
Saturated	2.6 g
Monounsaturated	N/A
Polyunsaturated	N/A
Calories from fat	52%
Cholesterol	120 mg
Sodium	322 mg
Protein	26.3 g
Carbohydrate	6.0 g
Dietary fiber	N/A

CHIEF NUTRIENTS

NUTRIENT	AMOUNT	% RDA
Vitamin C	46.8 mg	78
Iron	2.3 mg	23
Niacin	3.9 mg	21
Potassium	686.4 mg	18
Thiamine	0.2 mg	15
Calcium	93.6 mg	12
Riboflavin	0.2 mg	12

Strengths: Lobster salad is a superb source of vitamin C, with noticeable shots of body-building niacin, riboflavin, thiamine, calcium, iron, and potassium.

Caution: Loaded with fat, sodium, calories, and high in purines. People who have high blood pressure or gout may wish to avoid it. *Allergy alert:* Eating lobster or inhaling steam from the water in which it is cooked can occasionally cause an allergic reaction.

Description: Everybody has their own recipe. Most seem to include lobster meat, mayonnaise, and diced celery or watercress as a foundation.

A Better Idea: When making lobster salad, use one of the new fat-free, cholesterol-free brands of mayonnaise.

PASTA SALAD

Serving: 3½ oz.

Calories	111
Fat	5.3 g
Saturated	N/A
Monounsaturated	N/A
Polyunsaturated	N/A
Calories from fat	43%
Cholesterol	N/A
Sodium	334 mg
Protein	2.4 g
Carbohydrate	13.4 g
Dietary fiber	N/A

CHIEF NUTRIENTS

NUTRIENT	AMOUNT	% RDA
Iron	1.6 mg	16

Caution: Pasta salad is high in fat and sodium. *Allergy alert:* The ingredients in pasta—eggs and wheat flour—are common allergens and could cause such reactions as headaches, migraines, and rashes.

Eat with: A variety of vegetables—tomatoes, peppers, eggplant, zucchini, and carrots—to boost the nutritional content.

Strength: A pasta salad contains some iron for healthy blood.

Description: If you eat just the pasta, it's low in fat and quite healthy; but if you add mayonnaise and oil-laden dressing to the noodles, you'll produce a salad that's a little too crammed with calories.

A Better Idea: Avoid cream- and mayonnaise-based dressings; instead, use a light low-fat yogurt dressing.

POTATO SALAD

Serving: ½ cup

Calories	179
Fat	10.3 g
Saturated	1.8 g
Monounsaturated	3.1 g
Polyunsaturated	4.6 g
Calories from fat	52%
Cholesterol	85 mg
Sodium	661 mg
Protein	3.4 g
Carbohydrate	14.0 g
Dietary fiber	N/A

CHIEF NUTRIENTS

NUTRIENT	AMOUNT	% RDA
Vitamin C	12.5 mg	21
Potassium	317.5 mg	9
Vitamin B$_6$	0.2 mg	9
Iron	0.8 mg	8
Thiamine	0.1 mg	7
Niacin	1.1 mg	6
Magnesium	18.8 mg	5

Caution: If made with full-fat ingredients, potato salad has nearly the same amount of fat and calories as a fudge brownie, so use restraint if you're on a diet. And if you follow a standard recipe, this salad is very high in sodium. *Allergy alert:* If you're allergic to eggs, beware: Many restaurants and take-out shops put chopped egg in their potato salad.

Eat with: Low-fat, high-fiber side dishes like corn on the cob, or low-fat sandwiches such as turkey breast to help counterbalance the fat content.

Strengths: A very respectable source of vitamin C, potato salad has some nerve-soothing B vitamins, a little anemia-fighting iron, and some potentially useful potassium.

Description: A hot or cold salad made with cubed potatoes, mayonnaise or salad dressing, and various other ingredients and seasonings.

Serve: With oven-baked, skinless chicken.

A Better Idea: By omitting salt and using low-fat, low-cholesterol mayonnaise or salad dressing, you can slash the fat, cholesterol, sodium, and calorie count of this tasty side dish.

THREE-BEAN SALAD

Serving: ½ cup

Calories	90
Fat	0.3 g
Saturated	trace
Monounsaturated	trace
Polyunsaturated	trace
Calories from fat	3%
Cholesterol	0
Sodium	920 mg
Protein	2.2 g
Carbohydrate	19.8 g
Dietary fiber	N/A

CHIEF NUTRIENTS

NUTRIENT	AMOUNT	% RDA
Iron	1.0 mg	10
Calcium	44.0 mg	6
Potassium	188.0 mg	5

Strengths: Beans are a wonderful source of low-fat, no-cholesterol complex carbohydrates for energy and stamina. They also have dietary fiber, although the exact amount in this salad is not available. Three-bean salad has a fair amount of iron to help build healthier blood and provides some calcium for stronger bones and teeth.

Eat with: Whole grains that supply protein.

Caution: A lot of sodium is added to this prepared version. Green beans are high in oxalic acid, which should be kept in mind if you have a history of calcium oxalate kidney stones. Beans also contain purines, which may cause gout in some people. *Allergy alert:* Beans may cause such allergic reactions as hives and headaches.

Description: The traditional three-bean salad is made with kidney beans, green beans, and either wax or lima beans.

A Better Idea: Skip the canned preparations and make the salad yourself. By controlling the ingredients in the homemade salad, you can reduce the sodium in this dish.

TUNA SALAD

Serving: 3 oz.

Calories	159
Fat	7.9 g
Saturated	1.3 g
Monounsaturated	2.5 g
Polyunsaturated	3.5 g
Calories from fat	45%
Cholesterol	11 mg
Sodium	342 mg
Protein	13.6 g
Carbohydrate	8.0 g
Dietary fiber	N/A

CHIEF NUTRIENTS

NUTRIENT	AMOUNT	% RDA
Vitamin B$_{12}$	1.0 mcg	51
Niacin	5.7 mg	30
Iron	0.9 mg	9

Caution: Tuna salad can be loaded with sodium, so people with high blood pressure may want to avoid it. *Drug interaction:* People who are taking isoniazid (for tuberculosis) or levodopa (for Parkinson's disease) should not eat tuna. *Allergy alert:* Fish can occasionally cause an allergic reaction.

Eat with: Fruit and vegetables that are high in vitamin C to enhance iron absorption.

Strengths: Tuna is a good source of vitamin B$_{12}$ and niacin.

Description: Prepared tuna salad includes salt, mayonnaise, celery, relish, and onion. The tuna in this salad is the traditional canned variety, loaded with salt and oil.

A Better Idea: Make your tuna salad with canned tuna that is water-packed without added salt. Drain the tuna and add the traditional chopped celery, but use a low-cal substitute in place of high-fat, high-cholesterol regular mayonnaise.

BACON, LETTUCE, AND TOMATO SANDWICH

Serving: 1 (about 4¾ oz.)

Calories	352
Fat	22.1 g
Saturated	4.9 g
Monounsaturated	6.3 g
Polyunsaturated	8.8 g
Calories from fat	56%
Cholesterol	26 mg
Sodium	640 mg
Protein	10.8 g
Carbohydrate	27.6 g
Dietary fiber	N/A

CHIEF NUTRIENTS

NUTRIENT	AMOUNT	% RDA
Thiamine	0.4 mg	24
Vitamin C	13.0 mg	22
Iron	2.0 mg	20
Vitamin B$_{12}$	0.4 mcg	19
Niacin	3.3 mg	17
Riboflavin	0.2 mg	12
Vitamin B$_6$	0.2 mg	12
Folate	17.8 mcg	9
Calcium	52.1 mg	7
Potassium	223.2 mg	6
Zinc	0.8 mg	5

Caution: High in fat and sodium, the BLT should be crossed off the lunch menu if you're on a low-fat or low-salt diet. Calories are high, and well over half the calories in this sandwich come from fat. *Drug interaction:* Anyone taking levodopa should limit consumption of bacon because it contains vitamin B$_6$. Some bacon also contains tyramine, so avoid it if you're taking MAO inhibitors. *Allergy alert:* Reactions such as hives or anaphylaxis may be caused by the tomatoes in BLTs.

Strengths: One BLT provides a good supply of the B vitamins thiamine, niacin, and riboflavin for more energy. It is also an economical source of iron and vitamin B$_{12}$ for healthy blood. There's a kick of vitamin C and some zinc for enhanced immunity and healthy skin, as well as potassium to help nerves and muscles function smoothly.

Description: A sandwich with bacon, iceberg lettuce, tomato slices, mayonnaise, and white bread.

A Better Idea: For a healthier BLT, try a single slice of turkey bacon or subtitute with one-quarter of an avocado; use a low-calorie mayonnaise dressing instead of regular mayonnaise, and whole grain instead of white bread.

PEANUT BUTTER AND JELLY SANDWICH

Serving: 1 (about 3½ oz.)

Calories	372
Fat	17.6 g
Saturated	3.4 g
Monounsaturated	7.6 g
Polyunsaturated	4.6 g
Calories from fat	43%
Cholesterol	2 mg
Sodium	410 mg
Protein	12.3 g
Carbohydrate	44.6 g
Dietary fiber	N/A

CHIEF NUTRIENTS

NUTRIENT	AMOUNT	% RDA
Niacin	5.9 mg	31
Iron	2.2 mg	22
Thiamine	0.2 mg	16
Magnesium	50.2 mg	14
Folate	25.0 mcg	13
Riboflavin	0.2 mg	9
Calcium	56.7 mg	7
Potassium	270.5 mg	7
Vitamin B_6	0.1 mg	6
Zinc	0.8 mg	5

Caution: High in sodium, a peanut butter and jelly sandwich should not be eaten by people on a low-salt diet. It's not recommended if you're trying to lose weight, either, since well over one-third of its calories come from fat. To prevent choking, do not feed chunky-style peanut butter to kids under four years of age. *Allergy alert:* If you're allergic to glutens, ask your doctor about safe breads to eat. Those allergic to salicylates or sodium benzoate may have a reaction to the jelly in a PB&J. Also, some people are simply allergic to peanuts.

Strengths: This favorite quick-lunch sandwich contains slightly more than one-fifth of the daily RDA of iron and an even larger dose of niacin for healthy blood. An economical supply of protein helps to grow and maintain tissues, and there's plentiful folate to help the regeneration of red blood cells. A peanut butter and jelly sandwich also gives you thiamine, magnesium, vitamin B_6, and riboflavin to assist in converting the food we eat into energy. A nice amount of calcium contributes to stronger bones, and you also get some zinc for healthy skin.

A Better Idea: To lower calories and salt, use low-sugar jelly and unsalted peanut butter.

OTHER DISHES

APPLESAUCE, *Unsweetened*

Serving: 1 cup

Calories	105
Fat	0.1 g
Saturated	trace
Monounsaturated	trace
Polyunsaturated	trace
Calories from fat	1%
Cholesterol	0
Sodium	5 mg
Protein	0.4 g
Carbohydrate	27.6 g
Dietary fiber	3.7 g

CHIEF NUTRIENTS

NUTRIENT	AMOUNT	% RDA
Potassium	183.0 mg	5

Strengths: Low in calories, fat, and sodium; cholesterol-free. Applesauce has a favorable sodium/potassium ratio, which may help prevent strokes. It's a good source of fiber—thought to help lower serum cholesterol and reduce the risk of colon cancer. High in complex carbohydrates, which may lower the risk of diabetes. Some canned brands have added vitamin C.

Caution: Overconsumption may cause intestinal gas. *Allergy alert:* Those allergic to aspirin may react to the natural salicylate in apples.

Preparation: If you make your own applesauce, puree the skin with the apples (use a blender or food processor) for the most fiber. Applesauce freezes well; make lots in season.

Serve: With pancakes and waffles (it makes an excellent low-fat, low-cal topping). Applesauce is a good side dish.

BREAD STUFFING *from Mix*

Serving: 1 cup, moist

Calories	416
Fat	25.6 g
Saturated	13.1 g
Monounsaturated	N/A
Polyunsaturated	N/A
Calories from fat	55%
Cholesterol	132 mg
Sodium	1,008 mg
Protein	8.8 g
Carbohydrate	39.4 g
Dietary fiber	N/A

CHIEF NUTRIENTS

NUTRIENT	AMOUNT	% RDA
Iron	2.0 mg	20
Vitamin A	168.0 RE	17
Riboflavin	0.2 mg	11
Calcium	80.0 mg	10
Niacin	1.6 mg	8
Thiamine	0.1 mg	7

Caution: Fat comprises 55% of bread stuffing's calories: More than half of that fat is saturated and may increase blood cholesterol levels. A 1-cup serving has a moderate amount of calories but is loaded with enough sodium to melt an ice cube. Bread stuffing is a source of purines, which may aggravate gout. *Allergy alert:* Some people may be allergic to the wheat, eggs, and seasonings used in this food.

Strengths: Provides 20% of the RDA of iron, a nutrient that boosts the body's immune system. It has a good amount of vitamin A, which is essential for healthy skin. Bread stuffing also has a good amount of calcium, a mineral that strengthens bones and teeth. Its B vitamins—niacin, riboflavin, and thiamine—help convert carbohydrates into energy.

Curiosity: In the late 19th century, prim Victorians thought "dressing" was a more appropriate word than "stuffing."

A Better Idea: Dry, crumbly bread stuffing is slightly more nutritious than moist stuffing made with eggs.

CORN FRITTER, *Yellow or White*

Serving: 1 (1¼ oz.)

Calories	132
Fat	7.5 g
Saturated	2.0 g
Monounsaturated	N/A
Polyunsaturated	N/A
Calories from fat	51%
Cholesterol	31 mg
Sodium	167 mg
Protein	2.7 g
Carbohydrate	13.9 g
Dietary fiber	N/A

CHIEF NUTRIENTS

NUTRIENT	AMOUNT	% RDA
Iron	0.6 mg	6
Riboflavin	0.1 mg	5
Thiamine	0.1 mg	5

Caution: Because it is generally deep-fried, more than 50% of the calories in a corn fritter come from fat. And it has a moderate amount of sodium. *Allergy alert:* Fritter batter contains egg, which may produce an allergic reaction in some people. And wheat flour is also a common allergen.

Strengths: In each fritter there's a bit of iron, riboflavin, and thiamine, all essential to healthy blood and high energy levels.

Origin: Corn fritters are a southern specialty, often served as a side dish or as a main course for breakfast.

Description: Corn fritters are made from a batter of corn, egg, flour, baking powder, salt, and usually milk. They're often deep-fried in hot fat.

A Better Idea: To reduce fat, make fritters with buttermilk and fry them like pancakes in a nonstick pan.

CRANBERRY SAUCE, *Canned* ⬇ 🏛

Serving: ½ cup

Calories	209
Fat	0.2 g
Saturated	N/A
Monounsaturated	N/A
Polyunsaturated	N/A
Calories from fat	1%
Cholesterol	0
Sodium	40 mg
Protein	0.3 g
Carbohydrate	53.9 g
Dietary fiber	1.2 g

CHIEF NUTRIENTS

NUTRIENT	AMOUNT	% RDA
Vitamin C	2.8 mg	5

Caution: Very high in calories due to the amount of sweetener needed to make the tart berries palatable. Cranberries contain some oxalic acid, which should be restricted by those with calcium-oxalate stones. A half-cup of cranberry sauce has more than 40 times as much sodium as an equal measure of raw berries: This is certainly a concern if you have sodium-sensitive high blood pressure.

Eat with: Turkey and other meat (the vitamin C in the sauce aids absorption of the iron in meat).

Strengths: Very low in fat, has no cholesterol, and contains some fiber.

Serve: Raw or cooked.

A Better Idea: You can prepare your own cranberry sauce using a high proportion of other fresh and dried fruits (try apples, apricots, pears, or oranges) for natural sweetness. Further enhance the flavor by adding vanilla, cinnamon, or other spices.

FRUIT COCKTAIL, *Canned* 🔄 ⏱ ⬇ 🏛

Serving: 1 cup, w/juice

Calories	114
Fat	trace
Saturated	trace
Monounsaturated	trace
Polyunsaturated	trace
Calories from fat	0.2%
Cholesterol	0
Sodium	10 mg
Protein	1.1 g
Carbohydrate	29.4 g
Dietary fiber	2.7 g

CHIEF NUTRIENTS

NUTRIENT	AMOUNT	% RDA
Vitamin C	6.7 mg	11
Vitamin A	76.9 RE	8
Vitamin B$_6$	0.1 mg	7
Potassium	235.6 mg	6
Iron	0.5 mg	5
Niacin	1.0 mg	5

Strengths: Fruit cocktail combines all the benefits of the individual fruits used. It's a good source of dietary fiber, with moderate calories, low sodium, and just a trace of fat. A good amount of vitamin C helps ensure that the fruit's iron is better absorbed. Some vitamin A (and beta-carotene) promotes healthy skin and good vision.

Caution: *Allergy alert:* Those who react to any of the individual fruits used should, of course, be wary.

Description: Government standards of identity demand that fruit cocktail always contain peaches, pears, seedless grapes, pineapple, and colored cherries.

History: Fruit cocktail is a California invention that dates back to the 1930s. Originally it was concocted from excess amounts of certain fruits that weren't quite perfect enough to be canned whole.

Serve: With cottage cheese. Makes a nice topping for pancakes and waffles, especially in winter.

POI

⬇ 🏛

Serving: ½ cup

Calories	134
Fat	0.2 g
Saturated	trace
Monounsaturated	trace
Polyunsaturated	trace
Calories from fat	1%
Cholesterol	0
Sodium	14 mg
Protein	0.5 g
Carbohydrate	32.7 g
Dietary fiber	0.5 g

CHIEF NUTRIENTS

NUTRIENT	AMOUNT	% RDA
Vitamin B$_6$	0.3 mg	17
Folate	25.7 mcg	13
Iron	1.1 mg	11
Thiamine	0.2 mg	11
Magnesium	28.8 mg	8
Vitamin C	4.8 mg	8
Niacin	1.3 mg	7
Potassium	219.6 mg	6

Strengths: Although further studies are needed, some experts think that those with a gluten allergy may find poi to be an excellent substitute for wheat. It's a good source of vitamin B$_6$, which helps boost energy and fight infection. The folate and iron help build healthy blood. Thiamine, magnesium, vitamin C, niacin, and potassium are also present.

Caution: Sour, acid-tasting poi is an acquired taste. Those without an appreciation for exotic food find it unpalatable.

Description: A staple in Hawaii, this pasty food is derived from the root of the taro plant. The root is pounded to make a smooth paste, which is generally eaten with the fingers.

Selection: Commercially available in Hawaii, poi can also be obtained in some specialty food stores elsewhere in the U.S.

SAUERKRAUT, *Canned*

🔄 ☀ 🛈 🎏 ⬇ 🏛

Serving: ½ cup, w/liquid

Calories	22
Fat	0.2 g
Saturated	trace
Monounsaturated	trace
Polyunsaturated	trace
Calories from fat	7%
Cholesterol	0
Sodium	780 mg
Protein	1.1 g
Carbohydrate	5.1 g
Dietary fiber	3.0 g

CHIEF NUTRIENTS

NUTRIENT	AMOUNT	% RDA
Vitamin C	17.4 mg	29
Iron	1.7 mg	17
Folate	28.0 mcg	14
Vitamin B$_6$	0.2 mg	8
Potassium	200.6 mg	5

Strengths: This salted, fermented form of cabbage qualifies as a member of the cancer-fighting cruciferous family. Containing an ample amount of fiber and a significant measure of vitamin C, it's also a decent source of iron. (As a bonus, the iron in sauerkraut is readily absorbed.) Sauerkraut has a good supply of folate, which is a critical nutrient for pregnant women. It also offers some vitamin B$_6$. While sauerkraut contains a bit of potassium, it's too little to offset the high sodium levels.

Eat with: Restraint! Sauerkraut is very high in sodium. If you're eating with pork, make sure you have very lean cuts of meat to reduce fat consumption.

Caution: Sauerkraut sometimes has a laxative effect. *Drug interaction:* People who are taking MAO-inhibitor drugs should avoid sauerkraut. It contains tyramine, which, when combined with an MAO drug, could raise blood pressure to dangerous levels.

Curiosity: Used to help prevent scurvy among Dutch sailors during long voyages of trading and exploration.

SPINACH SOUFFLÉ

Serving: 1 cup

Calories	219
Fat	18.4 g
Saturated	7.2 g
Monounsaturated	6.8 g
Polyunsaturated	3.1 g
Calories from fat	75%
Cholesterol	184 mg
Sodium	763 mg
Protein	11.0 g
Carbohydrate	2.8 g
Dietary fiber	N/A

CHIEF NUTRIENTS

NUTRIENT	AMOUNT	% RDA
Vitamin B_{12}	1.4 mcg	68
Vitamin A	674.6 RE	67
Folate	61.9 mcg	31
Calcium	229.8 mg	29
Riboflavin	0.3 mg	18
Iron	1.4 mg	14
Magnesium	38.1 mg	11
Zinc	1.3 mg	9
Thiamine	0.1 mg	6
Vitamin B_6	0.1 mg	6
Potassium	201.3 mg	5

Caution: Three-fourths of the calories in this dish come from fat, and it has significant amounts of sodium. Spinach contains purines, substances that may aggravate gout. *Allergy alert:* Dairy products such as milk and eggs are used in soufflés and may cause allergic reactions in some people.

Strengths: Spinach soufflé provides 67% of the recommended daily allowance of vitamin A, a nutrient that is important for good vision and healthy skin. This dish also provides 68% of the RDA of B_{12}, a vitamin essential for the production of healthy red blood cells. It has a very good level of folate, another nutrient that is important in the production of cells. Along with significant amounts of calcium and magnesium are a good amount of riboflavin and some thiamine, two B vitamins that help your body convert food into energy.

A Better Idea: Substitute low-fat milk in the recipe. Or go the *very* health-conscious route and eat plain spinach.

STEWED TOMATOES

Serving: ½ cup

Calories	40
Fat	1.4 g
Saturated	0.3 g
Monounsaturated	0.5 g
Polyunsaturated	0.5 g
Calories from fat	31%
Cholesterol	0
Sodium	230 mg
Protein	1.0 g
Carbohydrate	6.6 g
Dietary fiber	0.7 g

CHIEF NUTRIENTS

NUTRIENT	AMOUNT	% RDA
Vitamin C	9.2 mg	15
Iron	0.5 mg	5

Strengths: Stewed tomatoes are a good source of vitamin C. Since some kinds of physical stress increase the need for C, you might choose this dish before a long trek or heavy exertion. Vitamin C also enhances absorption of the iron in tomatoes.

Caution: Commercial brands may contain a high level of sodium and should be eaten with restraint if you're trying to limit dietary sodium. *Allergy alert:* Tomatoes are a common food allergen and may trigger hives, headaches, mouth itching, or other reactions in sensitive people. Also, if you are allergic to aspirin, you will probably want to avoid tomatoes, which contain salicylate, a basic component of aspirin.

Description: Stewed tomatoes may be made with onions, peppers, celery, or other ingredients.

A Better Idea: To cut down on sodium, stew your own tomatoes instead of buying them in cans.

STUFFED CABBAGE

Serving: 7¼ oz. frozen entrée

Calories	172
Fat	6.1 g
Saturated	N/A
Monounsaturated	N/A
Polyunsaturated	N/A
Calories from fat	32%
Cholesterol	43 mg
Sodium	479 mg
Protein	10.9 g
Carbohydrate	18.3 g
Dietary fiber	N/A

CHIEF NUTRIENTS

NUTRIENT	AMOUNT	% RDA
Vitamin C	39.0 mg	65
Iron	2.6 mg	26
Vitamin A	222.6 RE	22
Niacin	2.6 mg	14
Potassium	447.0 mg	12
Riboflavin	0.2 mg	9
Thiamine	0.1 mg	7
Calcium	43.0 mg	5

Strengths: Cabbage and tomatoes are an excellent source of vitamin C and contain an appreciable amount of vitamin A, both of which may help fight off certain cancers. The meal supplies a good amount of iron, which is necessary for healthy blood, and also a hearty amount of protein for maintenance of muscle mass. Riboflavin and niacin strengthen red blood cells, and some useful potassium helps to metabolize protein and carbohydrates.

Caution: High in sodium, stuffed cabbage should be avoided if you're on a low-salt diet. Because of the meat, it exceeds the 30% calories-from-fat ratio recommended by the American Heart Association. Beef is high in purines, such as uric acid, which tend to aggravate gout symptoms. *Allergy alert:* Beef can be a food allergen.

Description: Cabbage stuffed with rice, beef, and tomato sauce.

Preparation: Avoid overcooking cabbage; the longer cabbage is cooked, the more unpleasant its smell becomes. Also, overcooking destroys a lot of its vitamins.

STUFFED GREEN PEPPER

Serving: 1 (about 6½ oz.)

Calories	315
Fat	10.2 g
Saturated	4.8 g
Monounsaturated	N/A
Polyunsaturated	N/A
Calories from fat	29%
Cholesterol	70 mg
Sodium	581 mg
Protein	24.1 g
Carbohydrate	31.1 g
Dietary fiber	N/A

CHIEF NUTRIENTS

NUTRIENT	AMOUNT	% RDA
Vitamin C	74.0 mg	123
Iron	3.9 mg	39
Niacin	4.6 mg	24
Riboflavin	0.3 mg	18
Potassium	477.3 mg	13
Thiamine	0.2 mg	11
Calcium	77.7 mg	10

Strengths: One stuffed pepper provides more than a day's supply of vitamin C. High in protein and carbohydrates, it also is a very good source of iron, which is needed to help the body convert beta-carotene into usable vitamin A.

Caution: The meat filling contributes the fat and cholesterol to this dish. Because beef contains such purines as uric acid, it should be avoided or eaten in moderation by people who suffer from gout.

Description: Peppers stuffed with beef and bread crumbs.

A Better Idea: To reduce fat and cholesterol, prepare a stuffing that excludes beef. Try mixing rice with lentils for a filling that's high in vitamins and carbohydrates and provides a complete vegetable protein.

SUCCOTASH

Serving: ½ cup

Calories	110
Fat	0.8 g
Saturated	trace
Monounsaturated	trace
Polyunsaturated	trace
Calories from fat	6%
Cholesterol	0
Sodium	16 mg
Protein	4.9 g
Carbohydrate	23.4 g
Dietary fiber	5.2 g

CHIEF NUTRIENTS

NUTRIENT	AMOUNT	% RDA
Folate	31.5 mcg	16
Iron	1.5 mg	15
Magnesium	50.9 mg	15
Vitamin C	7.9 mg	13
Thiamine	0.2 mg	11
Potassium	393.6 mg	10
Niacin	1.3 mg	7
Vitamin B$_6$	0.1 mg	6
Riboflavin	0.1 mg	5

Strengths: Despite Sylvester the Cat's sibilant sigh and lisping lament, there is no suffering in succotash. Fat and sodium are negligible, cholesterol nonexistent. The bean and grain combination forms a complete vegetable protein, and succotash provides some dietary fiber. There's a decent amount of iron, although iron from lima beans is absorbed less efficiently than that from a meat source. Good amounts of potassium and magnesium help to maintain sharp nerve impulses.

Caution: *Allergy alert:* People who are allergic to corn may develop rashes, hives, or headaches.

Origin: The word succotash dates back to the 1700s and is derived from the Narraganset Indian term *misickquatash,* which refers to boiled whole kernels of corn.

Description: Usually made of lima beans and corn kernels, the succotash mix may include chopped red and green sweet peppers. Finely shredded green beans are sometimes substituted for lima beans.

Snacks

CHIPS, *Corn*

Serving: 30 small (about 1 oz.)

Calories	155
Fat	9.1 g
Saturated	1.5 g
Monounsaturated	3.4 g
Polyunsaturated	4.3 g
Calories from fat	53%
Cholesterol	0
Sodium	164 mg
Protein	1.7 g
Carbohydrate	16.9 g
Dietary fiber	1.6 g

CHIEF NUTRIENTS

NUTRIENT	AMOUNT	% RDA
Magnesium	21.9 mg	6
Calcium	37.1 mg	5

Caution: High in sodium and fat. *Allergy alert*: Anyone allergic to corn will want to avoid these chips. Some people with corn allergies have developed canker sores from eating corn chips; in other cases, people who are allergic get migraines and other headaches.

Strengths: Corn chips have no cholesterol and are a good source of carbohydrates. There's some calcium for healthy bones, and a fair amount of magnesium might help lower your blood pressure; unfortunately, all the salt sends blood pressure in the other direction.

History: In 1952, Elmer Doolin met a man who had come up with a recipe to make chips out of corn instead of potatoes. Elmer purchased the recipe for what became Fritos corn chips for a mere $100.

A Better Idea: Have a cup of unbuttered, unsalted popcorn instead—and do your blood pressure a favor.

CHIPS, *Potato*

Serving: 10 (about ¾ oz.)

Calories	105
Fat	7.1 g
Saturated	1.8 g
Monounsaturated	1.3 g
Polyunsaturated	3.6 g
Calories from fat	61%
Cholesterol	0
Sodium	94 mg
Protein	1.3 g
Carbohydrate	10.4 g
Dietary fiber	1.0 g

CHIEF NUTRIENTS

NUTRIENT	AMOUNT	% RDA
Vitamin C	8.3 mg	14
Potassium	259.6 mg	7

Strengths: Unbelievably, chips have some nutrition—a surprisingly useful amount of vitamin C, with a little potassium and a bit of protein.

Eat with: Dips like salsa or seasoned low-fat yogurt that supply a little vitamin C or calcium.

Caution: Potato chips are concentrated sources of calories and salt—and they're one of the most addictive foods around. Keep in mind that the fat, sodium, and calorie counts given here are for 10 chips. If you eat half a bag or a bowlful of chips, you can do some real damage to your waistline. And salted chips hurt your blood pressure if you're sodium sensitive. *Allergy alert:* Chips sometimes trigger a rash, dermatitis, or canker sores around the mouth in sensitive individuals.

Curiosity: In May, 1969, Australian university student Paul G. Tully ate 30 2-oz. bags of potato chips in 24 minutes, 33.6 seconds—without a drink. Talk about a snack attack!

CHIPS, *Tortilla*

Serving: 10 (about 1 oz.)

Calories	150
Fat	8.0 g
Saturated	N/A
Monounsaturated	N/A
Polyunsaturated	N/A
Calories from fat	48%
Cholesterol	N/A
Sodium	155 mg
Protein	2.0 g
Carbohydrate	18.0 g
Dietary fiber	1.8 g

CHIEF NUTRIENTS

(None of the nutrients meet or exceed 5% of the RDA.)

Caution: Nearly 50% of the calories come from fat. This snack is high in calories and moderately high in sodium. *Allergy alert:* Tortilla chips are commonly made with corn and corn oil, substances that may cause allergic reactions in some people.

Strengths: The chips contain almost 2 g of dietary fiber per serving, which can help reduce the risk of hemorrhoids and may relieve constipation. The food also supplies a moderate amount of carbohydrates.

Curiosity: Tortilla is the Spanish-American diminutive of the Spanish *torta,* which means ''round cake.''

History: The tortilla originated in Mexico. The first English mention in print of ''tortilla'' was in 1699.

A Better Idea: Unsalted and ''light'' versions are available in many grocery stores.

CRACKERS, *Animal*

Serving: 5 (½ oz.)

Calories	56
Fat	1.2 g
Saturated	0.3 g
Monounsaturated	N/A
Polyunsaturated	N/A
Calories from fat	20%
Cholesterol	0
Sodium	39 mg
Protein	0.9 g
Carbohydrate	10.4 g
Dietary fiber	trace

CHIEF NUTRIENTS

(None of the nutrients meet or exceed 5% of the RDA.)

Strengths: Compared with most other cookies and crackers, animal crackers are healthfully low in fat, calories, and sodium. Because they are made with enriched wheat flour, they offer a bit of blood-boosting iron.

Caution: *Allergy alert:* Contain wheat and/or gluten; may also contain whey, corn syrup, or cottonseed oil, all common allergens.

Curiosity: Poet-philosopher Christopher Morley reflects in his poem ''Animal Crackers'':

Animal crackers, and cocoa to drink,
That is the finest of suppers, I think;
When I'm grown up and can have what I please
I think I shall always insist upon these.

Origin: First nationally marketed in 1902 under the trade name Barnum's Animals.

Description: Slightly sweet wheat flour cookies cut into animal shapes. Nabisco makes the familiar circus wagon box. A newer brand, Small World Animal Grahams, donates part of its profits to groups working to protect animals.

CRACKERS, *Butter-Flavor*

Serving: 4 (½ oz.)

Calories	64
Fat	2.5 g
Saturated	0.8 g
Monounsaturated	N/A
Polyunsaturated	N/A
Calories from fat	35%
Cholesterol	0
Sodium	153 mg
Protein	1.0 g
Carbohydrate	9.4 g
Dietary fiber	0.2 g

CHIEF NUTRIENTS

NUTRIENT	AMOUNT	% RDA
Iron	0.5 mg	5

Caution: Butter-flavor crackers are moderately high in fat and sodium, with little to offer in the way of nutrition.

Strength: These crackers do contain a bit of blood-building iron.

Description: A golden-brown, light, flaky-crisp cracker with a distinct buttery taste. Sometimes called ''Ritz'' crackers, though that's a trade name.

Serve: With low-fat toppings, such as a cottage cheese– or yogurt-based dip and cucumber slices. Also good with a sprinkling of crisp, spicy radish seed sprouts.

A Better Idea: Look for low-salt butter crackers, which contain about half as much sodium as regular butter crackers. Or, for less fat and salt, substitute rye crisps or rice cakes.

CRACKERS, *Cheese*

Serving: 4 round (½ oz.)

Calories	67
Fat	3.0 g
Saturated	1.2 g
Monounsaturated	N/A
Polyunsaturated	N/A
Calories from fat	40%
Cholesterol	4 mg
Sodium	145 mg
Protein	1.6 g
Carbohydrate	8.5 g
Dietary fiber	0.1 g

CHIEF NUTRIENTS

NUTRIENT	AMOUNT	% RDA
Calcium	47.0 mg	6
Iron	0.5 mg	5

Caution: Cheese crackers are a high-calorie snack, with about 40% of calories from fat. They also contain an indecent amount of sodium—some 145 mg in 4 crackers. (And the feat of eating *only* 4 takes real restraint.) This type of cracker contains what nutritionists call ''hidden'' fat and salt: Most people don't realize how much they consume when they're idly munching away.

Strengths: Cheese crackers have a bit of calcium, along with some iron.

Description: Orangish crackers made with flour, water, and cheese and usually flavored with paprika. Cheez-Its is a popular brand.

Storage: Stored properly, cheese crackers will stay crisp for weeks. Keep in an airtight container. To absorb moisture from the package and keep the crackers crisp, enclose a teaball with a few grains of rice. Soggy crackers can also be restored to crispness by drying in a low-temperature oven for a few minutes.

A Better Idea: For less fat and sodium, stick with low-sodium saltines, rye crisps, or rice cakes.

CRACKERS, *Cheese and Peanut Butter*

Serving: 2 (½ oz.)

Calories	69
Fat	3.4 g
Saturated	0.9 g
Monounsaturated	N/A
Polyunsaturated	N/A
Calories from fat	44%
Cholesterol	2 mg
Sodium	139 mg
Protein	2.1 g
Carbohydrate	7.9 g
Dietary fiber	0.2 g

CHIEF NUTRIENTS

NUTRIENT	AMOUNT	% RDA
Thiamine	0.2 mg	11
Iron	0.5 mg	5

Caution: Cheese and peanut butter crackers are fairly high in fat, with 44% of calories from fat. They also contain more salt than you want if you're on a low-sodium diet. If you down them without thinking, this kind of snack food can conceivably sabotage well-intended efforts to cut calories, fat, and salt.

Eat with: A glass of skim milk to provide some calcium.

Strengths: Cheese and peanut butter crackers offer a bit of iron and a decent amount of thiamine, both of which are essential to help maintain healthy blood and achieve good energy levels. The crackers also provide some protein.

Description: Prepackaged cheese cracker sandwiches filled with peanut butter.

A Better Idea: As a snack, dry-popped popcorn has much less fat, a little more protein, and virtually no sodium.

CRACKERS, *Graham*

Serving: 1 (½ oz.)

Calories	55
Fat	1.3 g
Saturated	0.3 g
Monounsaturated	N/A
Polyunsaturated	N/A
Calories from fat	22%
Cholesterol	0
Sodium	95 mg
Protein	1.1 g
Carbohydrate	10.4 g
Dietary fiber	0.5 g

CHIEF NUTRIENTS

NUTRIENT	AMOUNT	% RDA
Iron	0.5 mg	5
Riboflavin	0.1 mg	5

Strengths: Like animal crackers, graham crackers are a great alternative to most cookies and other crackers. They are low in fat, salt, and calories and contain no cholesterol. Because they are made with enriched white flour and whole wheat flour, they provide a bit of iron and riboflavin, both essential for healthy blood and energy production.

Origin: The crackers were invented in 1829 by Sylvester Graham, an evangelical vegetarian who strongly promoted un-sifted, coarsely ground whole wheat flour. Eventually, the flour was named after him.

Description: In addition to the flour, graham crackers are made with sweeteners—usually honey, corn syrup, or molasses.

Serve: As a snack, with skim milk, apple butter, sliced bananas, low-calorie instant hot chocolate, or a dab of peanut butter. Also good with cottage cheese and apple butter. May be crushed to make graham-cracker crust for pies, especially cheesecake.

CRACKERS, *Saltine*

Serving: 5 (¹/₂ oz.)

Calories	61
Fat	1.7 g
Saturated	0.4 g
Monounsaturated	N/A
Polyunsaturated	N/A
Calories from fat	25%
Cholesterol	0
Sodium	156 mg
Protein	1.3 g
Carbohydrate	10.2 g
Dietary fiber	0.4 g

CHIEF NUTRIENTS

NUTRIENT	AMOUNT	% RDA
Iron	0.7 mg	7

Strengths: Saltine crackers are low in fat and calories and contain no cholesterol. Because they are made with enriched wheat flour, they offer a bit of blood-building iron.
Caution: Regular saltines are moderately high in sodium.
Curiosity: More than 35 billion of Nabisco's Premium Saltines are made each year—and that's just one brand of saltines!
Description: A thin, square, white cracker usually dusted with coarse salt. Made with white flour and water with yeast as a leavening agent.
Selection: Low-sodium saltines containing about one-third less salt are now available.
Storage: Stored in an airtight container, saltines will stay crisp for weeks.
Serve: With soup, chili, salads.

CRACKERS, *Soda*

Serving: 5 (¹/₂ oz.)

Calories	62
Fat	1.9 g
Saturated	0.4 g
Monounsaturated	N/A
Polyunsaturated	N/A
Calories from fat	27%
Cholesterol	0
Sodium	156 mg
Protein	1.3 g
Carbohydrate	10.0 g
Dietary fiber	0.3 g

CHIEF NUTRIENTS

NUTRIENT	AMOUNT	% RDA
Iron	0.6 mg	6

Strengths: Made with enriched wheat flour, soda crackers offer a smattering of blood-building iron. These crackers are fairly low in fat.
Caution: Soda crackers contain a moderate amount of sodium.
Curiosity: Crackers got their name from the ''cracking'' sound they make when broken.
Description: Soda crackers are plump, biscuity-looking saltines leavened with baking soda. Uneeda Biscuits is a popular brand.
Storage: Soda crackers are sorry food if they get soggy. Store in an airtight container, and if they get stale, crisp them up in a low-temperature oven.
Serve: As canapés at teas or with soup.

CRACKERS, *Wheat*

Serving: 7 thin (½ oz.)

Calories	61
Fat	1.8 g
Saturated	0.9 g
Monounsaturated	0.9 g
Polyunsaturated	0.7 g
Calories from fat	26%
Cholesterol	0
Sodium	121 mg
Protein	1.8 g
Carbohydrate	8.8 g
Dietary fiber	0.8 g

CHIEF NUTRIENTS

NUTRIENT	AMOUNT	% RDA
Iron	0.5 mg	5

Strengths: Wheat crackers are in the low-fat category range—if you eat a modest number. They contain no cholesterol.
Caution: Moderately high in sodium, wheat crackers are much higher in fat than Ry-Krisp or melba toast.
Description: A thin, square, light-brown, salted wheat flour cracker.
Storage: Like other light crackers, they need to be stored in an airtight tin. Once the package is unsealed, you might want to put in a teaball containing rice grains to absorb moisture.
Serve: With soup, salad, dip, spicy spread, or low-fat cheese.
A Better Idea: Ounce for ounce, Ry-Krisp is a far superior choice, with more fiber and nutrients and less fat and calories than wheat crackers. Low-sodium wheat crackers, with only half the sodium of regular wheat crackers, are also available.

GRANOLA BAR

Serving: 1 (about ¾ oz.)

Calories	109
Fat	4.2 g
Saturated	N/A
Monounsaturated	N/A
Polyunsaturated	N/A
Calories from fat	35%
Cholesterol	N/A
Sodium	67 mg
Protein	2.4 g
Carbohydrate	16.0 g
Dietary fiber	1.1 g

CHIEF NUTRIENTS

NUTRIENT	AMOUNT	% RDA
Iron	0.8 mg	8

Strengths: A granola bar has some protein, which is essential for growth, and it also has a fair amount of dietary fiber and some iron. *Allergy alert:* Granola often is made with oats, nuts, wheat, and cornmeal, substances that may cause allergic reactions in some people.
Description: This snack or breakfast food consists of toasted grains, nuts, and dried fruits bound together with honey and oils.
History: The first packaged breakfast cereal in America was called Granula and was created by a New Yorker named Dr. James Calik Jackson. In the 1890s, Dr. John Kellogg, one of the founders of the modern cereal industry, created a packaged cereal called Granola. Later the name was changed to avoid confusing it with the earlier product. In the 1960s, a surge in interest in natural grains led to granola as we know it today.

MELBA TOAST ⇩

Serving: 3 pieces (¹/₂ oz.)

Calories	60
Fat	trace
Saturated	trace
Monounsaturated	trace
Polyunsaturated	trace
Calories from fat	0
Cholesterol	0
Sodium	132 mg
Protein	3.0 g
Carbohydrate	12.0 g
Dietary fiber	0.9 g

CHIEF NUTRIENTS

(None of the nutrients meet or exceed 5% of the RDA.)

Strengths: Melba toast contains only a trace of fat and no cholesterol. It provides mostly carbohydrates and a bit of protein.
Caution: Low in nutrients, Melba toast has a good bit of sodium, with 132 mg per ½-oz. serving.
Origin: This toast was created by the famous French chef Auguste Escoffier for Australian opera singer Dame Nellie Melba. (For the same singer the chef also created Melba sauce, a fresh raspberry and red currant jelly creation that is spooned over peaches.)
Description: A very thin, crisp, dry, bland, cracker-like toast, commonly sold in packaged form.
Serve: With soup or salad. Also good with cottage cheese or a thin layer of cheese or meat spread.
A Better Idea: The same amount of Ry-Krisp is a clear winner over Melba toast: It has more nutrients and fiber and less sodium and calories.

POPCORN, *Air-Popped, Unsalted*

Serving: 1 cup

Calories	23
Fat	0.3 g
Saturated	trace
Monounsaturated	N/A
Polyunsaturated	N/A
Calories from fat	12%
Cholesterol	0
Sodium	0.2 mg
Protein	0.8 g
Carbohydrate	4.6 g
Dietary fiber	0.9 g

CHIEF NUTRIENTS

(None of the nutrients meet or exceed 5% of the RDA.)

Caution: If you're giving popcorn to children under 4, doctors recommend that you chop it up to prevent choking. *Allergy alert*: People sensitive to corn must avoid popcorn. Read microwave popcorn labels before you buy: Most are loaded with salt and fat.
Strengths: Plain popcorn is low in saturated fat and calories. It's a good source of fiber, too.
Description: Corn kernels are heated until they burst into white puffs of starchy goodness. The burst kernels are up to 30 times greater in size after they pop.
History: An Indian contribution to the first Thanksgiving was called popped corn—shortened by 1820 to popcorn.
Selection: Air-popped popcorn is generally lower in fat and calories than the kinds that are cooked in oil. Many ''gourmet'' popcorns are available.
Preparation: For the healthiest popcorn, use a hot-air popper and omit butter and salt.

POTATO STICKS

Serving: 1 oz.

Calories	148
Fat	9.8 g
Saturated	2.5 g
Monounsaturated	N/A
Polyunsaturated	5.0 g
Calories from fat	60%
Cholesterol	0
Sodium	71 mg
Protein	2.0 g
Carbohydrate	15.2 g
Dietary fiber	1.0 g

CHIEF NUTRIENTS

NUTRIENT	AMOUNT	% RDA
Potassium	351.0 mg	9
Iron	0.7 mg	7
Niacin	1.4 mg	7
Folate	11.0 mcg	6
Magnesium	18.0 mg	5

Caution: In potato sticks, well over half the calories come from fat—so choose another side dish if you're on a low-fat diet.

Strengths: Moderate in sodium, potato sticks have about half the amount found in the same quantity of potato chips. Because they are usually fried in vegetable fat, they contain no cholesterol. Small amounts of potassium and magnesium are present, and these minerals help regulate blood pressure. The iron, niacin, and folate in potato sticks are all nutrients that contribute to healthy blood.

Description: Potato sticks (also called shoestring potatoes) are julienned strips that look like wooden matchsticks. Like potato chips, they are fried in oil that may include palm oil and partially hydrogenated soybean or cottonseed oil, all of which contain some saturated fats. Potato sticks are usually vacuum-packed in cans to keep them fresh and crisp.

A Better Idea: "Light" potato chips or tortilla chips offer similar crunch and flavor with one-quarter to one-third the fat. Flavored rice cakes come in lower still.

PRETZELS, *Dutch-Type, Salted*

Serving: 2 large (about 1 oz.)

Calories	125
Fat	1.4 g
Saturated	0.3 g
Monounsaturated	N/A
Polyunsaturated	N/A
Calories from fat	10%
Cholesterol	0
Sodium	538 mg
Protein	3.1 g
Carbohydrate	24.3 g
Dietary fiber	0.9 g

CHIEF NUTRIENTS

(None of the nutrients meet or exceed 5% of the RDA.)

Caution: Salted pretzels are extremely high in sodium, and they're also relatively high in calories. And pretzels may cause flatulence in some people. *Allergy alert:* If you're sensitive to gluten, avoid pretzels made with wheat.

Strengths: Contain some fiber—which may afford protection against colon cancer. Pretzels are also high in carbohydrates for added energy. For a snack, they're very low in fat.

History: The Romans had pretzels; the Latin word for them is *brachiatus*—"having branches like arms." When Germans borrowed the term, this salty snack was named a *brezel*. The snack was probably brought to America by Dutch settlers.

Description: Food values given here are for large, knot-shaped, salted Dutch pretzels. Unsalted pretzels still have sodium—but less than 15% of the sodium found in salted varieties.

A Better Idea: Some unsalted pretzels made with oat bran are available from mail-order companies.

RICE CAKES

⬇ 🏛

Serving: 2 (about ½ oz.)

Calories	70
Fat	0.4 g
Saturated	N/A
Monounsaturated	N/A
Polyunsaturated	N/A
Calories from fat	5%
Cholesterol	0
Sodium	20 mg
Protein	0.8 g
Carbohydrate	16.0 g
Dietary fiber	0.8 g

CHIEF NUTRIENTS

(None of the nutrients meet or exceed 5% of the RDA.)

Strengths: Rice cakes are a great source of low-fat, low-calorie carbohydrates needed for energy. The sodium is quite low—and you can find no-salt rice cakes if you're on a really low-sodium diet. Some fiber helps contribute to digestive health.

Selection: Great to munch on, rice cakes are a healthy cut above other snack foods. But watch out for brands with cheese, teriyaki, or barbecue flavoring. Buy cakes made with brown rice that includes the germ and bran.

RY-KRISP

🔄 ⏱ ⬇ 🏛

Serving: ¼ large square (½ oz.)

Calories	40
Fat	0.2 g
Saturated	N/A
Monounsaturated	N/A
Polyunsaturated	N/A
Calories from fat	5%
Cholesterol	0
Sodium	112 mg
Protein	1.5 g
Carbohydrate	13.0 g
Dietary fiber	2.5 g

CHIEF NUTRIENTS

NUTRIENT	AMOUNT	% RDA
Magnesium	34.0 mg	10
Zinc	0.8 mg	5

Strengths: Ry-Krisp contains only a trace of fat and no cholesterol. It's low in calories and has a moderate amount of sodium. Each ½-oz. serving is impressively high in the kind of fiber that is particularly beneficial to weight watchers. (The fiber swells in the stomach, creating the sensation of fullness and slowing the digestion of sugars—an action that moderates blood-sugar levels.) This whole grain product offers a good amount of magnesium, which helps normalize blood pressure. And it has some zinc, a nutrient important for healthy skin and immune function.

Description: A dry, light wafer made of whole rye flour and corn bran and flavored with caraway seeds and salt.

Storage: Kept in an airtight tin, Ry-Krisp lasts a long time.

Serve: With soup, salad, and low-fat cheese. Ry-Krisp goes well with cottage cheese.

Soups and Stews
SOUPS

BEEF BROTH OR BOUILLON, *Cube* ⬇ 🗄

Serving: 1, in 6 fl. oz. water

Calories	5
Fat	0.1 g
Saturated	trace
Monounsaturated	trace
Polyunsaturated	N/A
Calories from fat	23%
Cholesterol	0
Sodium	869 mg
Protein	0.6 g
Carbohydrate	0.6 g
Dietary fiber	0

CHIEF NUTRIENTS

(None of the nutrients meet or exceed 5% of the RDA.)

Caution: A beef bouillon cube, which is dissolved to make beef broth, consists of compressed, flavor-concentrated dehydrated beef, and it's quite high in sodium. Avoid beef broth if you're on a low-salt diet to control your blood pressure. Because of the beef content, the broth contains some purines and should be avoided by anyone with gout. *Allergy alert:* Some bouillon cubes may contain wheat or gluten, common food allergens. May have some monosodium glutamate (MSG), which is contained in hydrolyzed vegetable protein; MSG is a common cause of headaches in sensitive individuals.

Eat with: Low-sodium, potassium-rich foods such as potatoes.

Strength: Low in calories.

Serve: Use as a base for soups and sauces.

A Better Idea: To make your own low-sodium beef broth, simmer beef (shin bones, shank, or flank) in water with onions, carrots, parsnips, celery, thyme, bay leaf, parsley, and cloves. Strain out the solids, reserving the liquid. Allow the broth to cool, then refrigerate. Skim fat from surface before serving.

BEEF BROTH OR BOUILLON, *Powder* 🗄

Serving: 1 packet, in 6 fl. oz. water

Calories	15
Fat	0.5 g
Saturated	trace
Monounsaturated	trace
Polyunsaturated	trace
Calories from fat	33%
Cholesterol	0
Sodium	1,021 mg
Protein	1.0 g
Carbohydrate	1.4 g
Dietary fiber	0

CHIEF NUTRIENTS

(None of the nutrients meet or exceed 5% of the RDA.)

Caution: Even higher in sodium than beef broth from cubes, the powder-made broth is sure to be off-limits if you're on a low-sodium diet. Beef broth contains purines and should be eaten sparingly by anyone with gout. *Allergy alert:* Some bouillon powder may contain wheat or gluten, common food allergens. May have some monosodium glutamate (MSG), which is contained in hydrolyzed vegetable protein; MSG may trigger migraines in sensitive individuals.

Eat with: Low-sodium, potassium-rich foods such as fruit salad.

Strength: Low in calories.

Description: Granules of dehydrated beef stock dissolved in hot water.

A Better Idea: Make a low-sodium beef broth by simmering beef in water with a variety of vegetables.

BEEF BROTH OR BOUILLON, *Ready-to-Serve*

Serving: 1 cup

Calories	17
Fat	0.5 g
Saturated	trace
Monounsaturated	trace
Polyunsaturated	trace
Calories from fat	28%
Cholesterol	0
Sodium	782 mg
Protein	2.7 g
Carbohydrate	0.1 g
Dietary fiber	0

CHIEF NUTRIENTS

NUTRIENT	AMOUNT	% RDA
Niacin	1.9 mg	10
Vitamin B$_{12}$	0.2 mcg	9

Caution: Although slightly lower in sodium than broth from powder or cubes, even the ready-to-serve kind of beef broth is not recommended if you're on a low-salt diet to control blood pressure. Beef broth contains purines and should be eaten sparingly by anyone with gout.

Eat with: Low-sodium, potassium-rich foods such as fresh vegetables.

Strengths: Canned broth has some niacin for production of red blood cells and vitamin B$_{12}$ for healthy nerves. It also has the benefit of being low in calories.

Serve: Use as a base for soups and sauces.

A Better Idea: You can make your own low-sodium beef broth with beef simmered with vegetables.

BEEF MUSHROOM, *Condensed*

Serving: 1 cup, made w/water

Calories	73
Fat	3.0 g
Saturated	1.5 g
Monounsaturated	1.2 g
Polyunsaturated	0.1 g
Calories from fat	37%
Cholesterol	7 mg
Sodium	941 mg
Protein	5.8 g
Carbohydrate	6.3 g
Dietary fiber	N/A

CHIEF NUTRIENTS

NUTRIENT	AMOUNT	% RDA
Vitamin B$_{12}$	0.2 mcg	10
Zinc	1.5 mg	10
Iron	0.9 mg	9
Vitamin C	4.6 mg	8
Niacin	1.0 mg	5

Strengths: Along with decent amounts of vitamin C and B vitamins, beef mushroom soup also offers anemia-fighting iron and immunity-boosting zinc. In general, soup can help suppress your appetite, so weight watchers may want to start off a meal with a cup of soup.

Eat with: Vitamin C–rich foods to boost iron absorption.

Caution: Although its fat is low, this soup is quite high in sodium—a potential problem if you're on a low-salt diet to control your blood pressure. Beef soup has purines and should be eaten sparingly by anyone with gout. *Allergy alert:* Read labels. May contain wheat or other common food allergens.

A Better Idea: Look for reduced-sodium or low-sodium beef mushroom soups. You get the same nutrients with just a fraction of the sodium.

BEEF NOODLE, *Condensed*

Serving: 1 cup, made w/water

Calories	83
Fat	3.1 g
Saturated	1.2 g
Monounsaturated	1.2 g
Polyunsaturated	0.5 g
Calories from fat	33%
Cholesterol	5 mg
Sodium	952 mg
Protein	4.8 g
Carbohydrate	9.0 g
Dietary fiber	N/A

CHIEF NUTRIENTS

NUTRIENT	AMOUNT	% RDA
Iron	1.1 mg	11
Vitamin B$_{12}$	0.2 mcg	10
Zinc	1.5 mg	10
Niacin	1.1 mg	6
Vitamin A	63.4 RE	6

Strengths: Decent amounts of anemia-fighting iron and immunity-boosting zinc are provided in beef noodle soup, along with vitamin B$_{12}$ for healthy nerves. This soup also offers some niacin to aid in red blood cell production and a bit of vitamin A to help protect against cancer. It has a little protein and a moderate number of calories. Eating soup can help suppress your appetite for more fatty food, preventing weight gain. **Eat with:** Low-sodium, potassium-rich foods such as broccoli. **Caution:** This is not an advisable choice if you're trying to control your blood pressure. And the beef is high in purines; eat sparingly if you have gout. *Allergy alert*: Read labels. May contain wheat, eggs, gluten or other food allergens. **A Better Idea:** Look for cans of condensed, low-sodium beef noodle soup: You get the same nutrients with just a fraction of the sodium. Better yet, make your own using a no-salt recipe.

BEEF NOODLE, *Dehydrated*

Serving: 1 cup, made w/water

Calories	40
Fat	0.8 g
Saturated	trace
Monounsaturated	trace
Polyunsaturated	trace
Calories from fat	18%
Cholesterol	3 mg
Sodium	1,042 mg
Protein	2.2 g
Carbohydrate	6.0 g
Dietary fiber	0.8 g

CHIEF NUTRIENTS

NUTRIENT	AMOUNT	% RDA
Thiamine	0.1 mg	8

Strengths: This soup has some useful thiamine to help convert carbohydrates into energy, and it's low in fat, calories, and cholesterol. Eating soup as a first course can help suppress your appetite, preventing weight gain. **Eat with:** Cottage cheese and diced pears, or other high-calcium, potassium-rich foods. **Caution:** Sky-high in sodium, this dehydrated soup is off-limits if you're on a low-salt diet to control blood pressure. Beef soup contains purines and should be eaten sparingly by anyone with gout. *Allergy alert:* Read labels. May contain wheat, eggs, or other common food allergens.

BLACK BEAN, *Condensed*

Serving: 1 cup, made w/water

Calories	116
Fat	1.5 g
Saturated	0.4 g
Monounsaturated	0.5 g
Polyunsaturated	0.4 g
Calories from fat	12%
Cholesterol	0
Sodium	1,198 mg
Protein	5.6 g
Carbohydrate	19.8 g
Dietary fiber	N/A

CHIEF NUTRIENTS

NUTRIENT	AMOUNT	% RDA
Iron	2.2 mg	22
Folate	24.7 mcg	12
Magnesium	42.0 mg	12
Zinc	1.4 mg	9
Potassium	274.2 mg	7
Calcium	44.5 mg	6
Thiamine	0.1 mg	5

Strengths: Low in fat, with no cholesterol and a nice bit of protein. Thanks to the beans, this soup provides some fiber, although exact amounts are not available. It also offers a handsome amount of anemia-fighting iron and a good amount of folate, essential for healthy red blood cells. The potassium, magnesium, and calcium can help lower blood pressure levels. There's also weight-loss potential: Eating soup that contains beans can help suppress your appetite.

Eat with: Whole grain crackers or rolls.

Caution: Brimming with sodium, black bean soup is certainly off-limits if you're trying to control your salt consumption.

Origin: This hearty soup is a favorite in the Caribbean and Central and South America.

A Better Idea: Look for low-sodium versions. You get the same nutrients, with just a fraction of the salt. Or make your own from black beans (also called turtle beans), vegetable stock, onions, celery, and garlic. Season with cayenne pepper.

CHEESE, *Condensed*

Serving: 1 cup, made w/water

Calories	156
Fat	10.5 g
Saturated	6.7 g
Monounsaturated	3.0 g
Polyunsaturated	0.3 g
Calories from fat	61%
Cholesterol	30 mg
Sodium	958 mg
Protein	5.4 g
Carbohydrate	10.5 g
Dietary fiber	N/A

CHIEF NUTRIENTS

NUTRIENT	AMOUNT	% RDA
Calcium	140.8 mg	18
Vitamin A	108.7 RE	11
Riboflavin	0.1 mg	8
Iron	0.7 mg	7

Caution: With more than 10 g of fat per cup (most of it saturated), cheese soup has twice the percentage of calories from fat considered acceptable by the American Heart Association. So cheese soup is best avoided if you're trying to lose weight or cut your risk of heart attack. And those on low-salt diets will want to avoid it because of the steep sodium level.

Strengths: A cup of cheese soup offers a good amount of bone-building calcium. There's a decent amount of vitamin A to help improve vision and boost immunity. Its riboflavin aids in the formation of red blood cells, and there's some iron to help synthesize collagen for healthier skin and bones.

A Better Idea: For an equally hearty and satisfying meal with a fraction of the fat, serve tomato soup. Preferably, select a reduced-sodium variety.

CHICKEN BROTH OR BOUILLON, *Condensed* ⬇ 🏛

Serving: 1 cup, made w/water

Calories	39
Fat	1.4 g
Saturated	0.4 g
Monounsaturated	0.6 g
Polyunsaturated	0.3 g
Calories from fat	32%
Cholesterol	0
Sodium	776 mg
Protein	4.9 g
Carbohydrate	1.0 g
Dietary fiber	0

CHIEF NUTRIENTS

NUTRIENT	AMOUNT	% RDA
Niacin	3.4 mg	18
Vitamin B$_{12}$	0.2 mcg	12
Iron	0.5 mg	5
Potassium	209.8 mg	5

Caution: The sodium content is steep, so condensed soup is not recommended if you're trying to control your blood pressure. Chicken and chicken broth contain purines and should be eaten sparingly by anyone with gout.

Strengths: Low in calories, this broth has no cholesterol. Its bit of iron helps bolster your defenses against disease, and some B vitamins assist in building healthier skin and nerves.

A Better Idea: If you're watching your sodium intake, shop for low-sodium broth. Or you can make your own by simmering a cut-up stewing chicken in water with onions, celery, carrots, parsley, sage, and thyme. Let cool, strain, and refrigerate, then skim fat from the surface.

CHICKEN BROTH OR BOUILLON, *Cube* ⬇ 🏛

Serving: 1, in 6 fl. oz. water

Calories	9
Fat	0.2 g
Saturated	trace
Monounsaturated	trace
Polyunsaturated	trace
Calories from fat	22%
Cholesterol	0
Sodium	593 mg
Protein	0.7 g
Carbohydrate	1.1 g
Dietary fiber	0

CHIEF NUTRIENTS

(None of the nutrients meet or exceed 5% of the RDA.)

Caution: If you're on a low-salt diet to avoid stroke, high blood pressure, or kidney disease, you'll want to stay away from this high-sodium chicken broth, which is made from bouillon cubes. Since chicken contains purines, this broth should be eaten sparingly by anyone with gout. *Allergy alert*: Some bouillon cubes may contain wheat or gluten, common food allergens. May have some monosodium glutamate (MSG), which is contained in hydrolyzed vegetable protein: MSG is a common cause of headaches in sensitive individuals.

Strengths: Practically no fat, calories, or cholesterol.

Description: A concentrated, dehydrated extract of chicken broth, dissolved in hot liquid.

Serve: Use as a base for soups and sauces.

A Better Idea: You can make low-sodium chicken stock by simmering a cut-up chicken in water with cut-up vegetables along with parsley, sage, and thyme. (The broth can be stored frozen in sealed 8-oz. tubs.)

CHICKEN BROTH OR BOUILLON, *Powder*

Serving: 1 packet, in 6 fl. oz. water

Calories	17
Fat	0.8 g
Saturated	trace
Monounsaturated	trace
Polyunsaturated	trace
Calories from fat	45%
Cholesterol	0
Sodium	1,113 mg
Protein	1.0 g
Carbohydrate	1.1 g
Dietary fiber	0

CHIEF NUTRIENTS

(None of the nutrients meet or exceed 5% of the RDA.)

Caution: The sodium is too high if you're following a salt-free diet to avoid stroke, high blood pressure, or kidney disease. Chicken and chicken broth have purines, so they should be eaten sparingly by anyone with gout. *Allergy alert:* May contain wheat or gluten, common food allergens. May have some monosodium glutamate (MSG), which is contained in hydrolyzed vegetable protein: MSG is a common cause of headaches in sensitive individuals.

Strength: Virtually no cholesterol.

Description: A concentrated, dehydrated extract of chicken broth, dissolved in hot liquid.

Serve: Use as a base for soups and sauces.

A Better Idea: Make your own chicken stock in large quantities and keep it in the freezer until you need it for cooking.

CHICKEN GUMBO, *Condensed*

Serving: 1 cup, made w/water

Calories	56
Fat	1.4 g
Saturated	0.3 g
Monounsaturated	0.7 g
Polyunsaturated	0.3 g
Calories from fat	23%
Cholesterol	5 mg
Sodium	954 mg
Protein	2.7 g
Carbohydrate	8.4 g
Dietary fiber	N/A

CHIEF NUTRIENTS

NUTRIENT	AMOUNT	% RDA
Iron	0.9 mg	9
Vitamin C	4.9 mg	8

Strengths: Low in fat, cholesterol, and calories.

Caution: The high sodium is a drawback if you're on a low-salt diet to control blood pressure. Chicken contains purines and should be eaten sparingly by anyone with gout.

Origin: This Creole dish comes straight from old New Orleans.

Description: Made with okra (a member of the hibiscus genus) or filé (a powder derived from the tender leaves of the sassafras tree and used to thicken soups or stews). Other ingredients in most chicken gumbo recipes include celery, onions, green peppers, and a variety of seasonings in addition to the poultry.

Serve: With cornbread—for a classic Creole meal.

A Better Idea: Look for the reduced-sodium version of gumbo soup. Or make your own gumbo by simmering celery, green peppers, onions, tomatoes, and okra in homemade chicken broth—and hold the salt. Add diced cooked chicken before serving.

CHICKEN MUSHROOM, *Condensed*

Serving: 1 cup, made w/water

Calories	132
Fat	9.2 g
Saturated	2.4 g
Monounsaturated	4.0 g
Polyunsaturated	2.3 g
Calories from fat	63%
Cholesterol	10 mg
Sodium	942 mg
Protein	4.4 g
Carbohydrate	9.3 g
Dietary fiber	N/A

CHIEF NUTRIENTS

NUTRIENT	AMOUNT	% RDA
Vitamin A	112.2 RE	11
Iron	0.9 mg	9
Niacin	1.6 mg	9
Riboflavin	0.1 mg	7
Zinc	1.0 mg	7

Strengths: A smattering of useful vitamins and minerals, including vitamin A, helps protect against cancer. Chicken mushroom soup has iron to fight anemia, zinc to boost immunity, and B vitamins for healthy skin and nerves. Some protein is also present.

Eat with: A hefty chef's salad of diced turkey breast, chunks of low-fat Swiss cheese, garbanzos, and other low-sodium, potassium-rich foods.

Caution: The soup has moderate amounts of fat and calories, and it's quite high in sodium—which raises blood pressure in some people. Chicken and mushrooms have purines and should be eaten sparingly by anyone with gout. *Allergy alert:* Read labels. May contain milk, wheat, or other common food allergens.

A Better Idea: Look for low-sodium versions. You get the same nutrients with just a fraction of the sodium.

CHICKEN NOODLE, *Condensed*

Serving: 1 cup, made w/water

Calories	75
Fat	2.5 g
Saturated	0.7 g
Monounsaturated	1.1 g
Polyunsaturated	0.5 g
Calories from fat	30%
Cholesterol	7 mg
Sodium	1,106 mg
Protein	4.1 g
Carbohydrate	9.4 g
Dietary fiber	N/A

CHIEF NUTRIENTS

NUTRIENT	AMOUNT	% RDA
Iron	0.8 mg	8
Niacin	1.4 mg	7
Vitamin A	72.3 RE	7
Vitamin B_{12}	0.1 mcg	7

Strengths: With low cholesterol and a moderate amount of calories, chicken noodle soup fills you up without contributing much fat to your diet. Also, this is "mom's special" for a head cold or stuffy nose: Chicken noodle soup does provide some relief.

Eat with: Whole grain bread and a hearty salad to make a filling, fiber-rich meal.

Caution: Beware the sodium: This soup just has too much if you're trying to limit salt intake to control your blood pressure. *Allergy alert*: May contain eggs, wheat, monosodium glutamate, or other allergy triggers.

Curiosity: The noodles in condensed chicken noodle soup appear in an almost infinite number of shapes, thanks to the ingenuity of food processors. Look for flat and curly noodles as well as stars, Os, and alphabet letters.

A Better Idea: Select low- or reduced-sodium chicken noodle soup. You get the same nutrients, with just a fraction of the sodium. Or make your own soup by adding diced chicken and noodles to chicken stock and simmering until done.

CHICKEN NOODLE, *Dehydrated*

Serving: 1 cup, made w/water

Calories	53
Fat	1.2 g
Saturated	0.3 g
Monounsaturated	0.5 g
Polyunsaturated	0.4 g
Calories from fat	20%
Cholesterol	3 mg
Sodium	1284 mg
Protein	3.0 g
Carbohydrate	7.0 g
Dietary fiber	0.8 g

CHIEF NUTRIENTS

NUTRIENT	AMOUNT	% RDA
Iron	0.5 mg	5

Caution: One of the saltier soups you can slurp, so steer clear if you've been advised to reduce your sodium intake to control high blood pressure. *Allergy alert:* Read labels. May contain eggs, wheat, or other common allergy triggers. Dehydrated soups contain up to 10% monosodium glutamate (MSG), which is contained in hydrolyzed vegetable protein; MSG is a common cause of headaches in sensitive individuals.

Eat with: Low-sodium, potassium-rich foods such as cantaloupe.

Strengths: Chicken noodle soup has a little iron, a bit of protein, and is low in fat and calories. Eating soup can help put the brakes on your appetite, so it's one way to avoid gaining weight.

A Better Idea: Make your own low-sodium soup by adding diced chicken and noodles to chicken stock and simmering until done.

CHICKEN NOODLE, *Ready-to-Serve*

Serving: 1 cup

Calories	175
Fat	6.0 g
Saturated	1.4 g
Monounsaturated	2.7 g
Polyunsaturated	1.5 g
Calories from fat	31%
Cholesterol	19 mg
Sodium	850 mg
Protein	12.7 g
Carbohydrate	17.0 g
Dietary fiber	3.8 g

CHIEF NUTRIENTS

NUTRIENT	AMOUNT	% RDA
Niacin	4.3 mg	23
Vitamin B$_{12}$	0.3 mcg	16
Iron	1.4 mg	14
Vitamin A	122.4 RE	12
Riboflavin	0.2 mg	10
Zinc	1.0 mg	6

Strengths: More nutritious than condensed or dehydrated soup prepared with water, ready-to-serve chicken noodle soup is a very respectable source of B vitamins for healthy skin and nerves. It's also a good source of iron to help ward off anemia and vitamin A to help protect against cancer. It has a small amount of zinc to help boost immunity and a good serving of protein. Eating soup before a meal can suppress your appetite somewhat, so dieters often go for it as an appetizer. Also, if you have a head cold, chicken soup can help relieve a stuffy nose.

Caution: Although this ready-to-serve soup is lower in sodium than condensed or dehydrated, it's not recommended if you're on a low-salt diet to control your blood pressure. *Allergy alert:* Read labels. May contain eggs, wheat, or other common food allergens.

A Better Idea: To reduce sodium intake, purchase low-salt chicken noodle soup. Or make your own by adding diced chicken and noodles to chicken stock and simmering until done.

CHICKEN RICE, Condensed

Serving: 1 cup, made w/water

Calories	60
Fat	1.9 g
Saturated	0.5 g
Monounsaturated	0.9 g
Polyunsaturated	0.4 g
Calories from fat	28%
Cholesterol	7 mg
Sodium	815 mg
Protein	3.5 g
Carbohydrate	7.2 g
Dietary fiber	N/A

CHIEF NUTRIENTS

NUTRIENT	AMOUNT	% RDA
Iron	0.8 mg	8
Vitamin A	65.1 RE	7
Vitamin B$_{12}$	0.1 mcg	7
Niacin	1.1 mg	6

Strengths: Includes small amounts of B vitamins, with a little iron and protein. Eating soup can help suppress your appetite, preventing weight gain. Also, there really is some home-remedy value in chicken soup: If you have a head cold, it can help relieve a stuffy nose.

Eat with: Low-sodium, high-fiber foods such as whole grain bread and a hearty salad. These additions round out your meal when soup is the main course.

Caution: Any condensed soup is fairly high in sodium—not recommended if you're on a low-salt diet to control blood pressure. *Allergy alert:* Read labels. May contain monosodium glutamate or other ingredients that bother some people.

A Better Idea: Look for low-sodium soup. You get the same nutrients with just a fraction of the sodium. Or make your own soup by simmering diced chicken and rice in chicken stock.

CHICKEN RICE, Dehydrated

Serving: 1 cup, made w/water

Calories	61
Fat	1.4 g
Saturated	0.3 g
Monounsaturated	0.6 g
Polyunsaturated	0.4 g
Calories from fat	21%
Cholesterol	3 mg
Sodium	981 mg
Protein	2.5 g
Carbohydrate	9.3 g
Dietary fiber	0.8 g

CHIEF NUTRIENTS

(None of the nutrients meet or exceed 5% of the RDA.)

Caution: With nearly 1,000 mg of sodium per serving, dehydrated soup isn't for those who are avoiding salty foods to control blood pressure. Avoid chicken soup if you have gout, since it contains purines. *Allergy alert:* Dehydrated soups have some monosodium glutamate (MSG), which is contained in hydrolyzed vegetable protein; MSG may cause headaches in sensitive individuals.

Eat with: Low-sodium, potassium-rich foods such as fruit.

Strengths: Low in calories, with barely any fat or cholesterol. Eating soup can help suppress your appetite, preventing weight gain. Also, if you have a head cold, chicken soup can help relieve a stuffy nose.

A Better Idea: Look for low-sodium canned soup. Or make your own low-sodium version by adding diced chicken and rice to chicken stock and simmering until done.

CHICKEN RICE, *Ready-to-Serve*

Serving: 1 cup

Calories	127
Fat	3.2 g
Saturated	1.0 g
Monounsaturated	1.4 g
Polyunsaturated	0.7 g
Calories from fat	23%
Cholesterol	12 mg
Sodium	888 mg
Protein	12.3 g
Carbohydrate	13.0 g
Dietary fiber	1.0 g

CHIEF NUTRIENTS

NUTRIENT	AMOUNT	% RDA
Vitamin A	585.6 RE	59
Niacin	4.1 mg	22
Iron	1.9 mg	19
Vitamin B$_{12}$	0.3 mcg	16
Riboflavin	0.1 mg	6
Vitamin C	3.8 mg	6
Zinc	1.0 mg	6

Strengths: The ready-to-serve soup is more nutritious than condensed or dehydrated versions prepared with water. It offers a fair amount of protein and is an excellent source of vitamin A, which helps protect against cancer. Chicken rice soup also offers healthy amounts of niacin and iron for rich, red blood. Some vitamin C and zinc help fend off illness. This soup has a moderate amount of calories, with very little fat and cholesterol. And, if you have a head cold, chicken soup can help relieve a stuffy nose.

Eat with: A yogurt fruit shake or other food rich in potassium and calcium.

Caution: Even the ready-to-serve soup is quite high in sodium, which is a potential problem if you're prone to high blood pressure. Chicken contains purines, so avoid chicken soup if you have gout. *Allergy alert:* Read labels. May contain ingredients that trigger troublesome reactions in certain individuals.

A Better Idea: Look for low-sodium soup: You get the same nutrients with a fraction of the salt. Or make your own by adding diced chicken and rice to chicken stock and simmering until done.

CHICKEN VEGETABLE, *Condensed*

Serving: 1 cup, made w/water

Calories	75
Fat	2.8 g
Saturated	0.8 g
Monounsaturated	1.3 g
Polyunsaturated	0.6 g
Calories from fat	34%
Cholesterol	10 mg
Sodium	945 mg
Protein	4.0 g
Carbohydrate	8.6 g
Dietary fiber	1.0 g

CHIEF NUTRIENTS

NUTRIENT	AMOUNT	% RDA
Vitamin A	265.1 RE	27
Iron	0.9 mg	9
Niacin	1.2 mg	6
Vitamin B$_{12}$	0.1 mcg	6

Strengths: Here's a convenient way to get a little extra iron and some B vitamins in your diet—along with a noticeable dollop of resistance-boosting vitamin A. If you're trying to avoid overeating at meals, a bowl of chicken vegetable soup before the entrée can help suppress your appetite, preventing weight gain. Also, if you have a head cold, chicken soup can help relieve a stuffy nose.

Eat with: Whole grain bread and a hearty salad to make a filling, fiber-rich meal.

Caution: With nearly 1,000 mg of sodium per serving, this isn't a wise choice if you've been advised to avoid salty foods to control blood pressure—or to guard against stroke and kidney disease. Since chicken contains purines, you'll want to avoid this soup if you have gout. *Allergy alert:* Read labels. May contain ingredients that trigger bothersome reactions in sensitive individuals.

CHICKEN VEGETABLE, *Dehydrated*

Serving: 1 cup, made w/water

Calories	50
Fat	0.8 g
Saturated	trace
Monounsaturated	trace
Polyunsaturated	trace
Calories from fat	14%
Cholesterol	3 mg
Sodium	807 mg
Protein	2.7 g
Carbohydrate	7.8 g
Dietary fiber	N/A

CHIEF NUTRIENTS

(None of the nutrients meet or exceed 5% of the RDA.)

Caution: Prohibitively high in sodium: Steer clear of all dehydrated soups if you've been advised to avoid salty food. *Allergy alert:* Read labels. May include ingredients, such as wheat, that trigger reactions in some individuals. If you're susceptible to gout, avoid chicken soup since it contains purines. Dehydrated soups may contain monosodium glutamate (MSG), which is contained in hydrolyzed vegetable protein; MSG is a common cause of headaches in sensitive individuals.

Eat with: Fruit salad, a baked potato, or other low-sodium, potassium-rich foods.

Strengths: Low in calories, chicken vegetable soup has barely any fat or cholesterol. Eating soup can help suppress your appetite, preventing weight gain. Also, if you have a head cold and your nose is stuffed up, eating chicken soup can help you breathe more freely.

A Better Idea: Look for low-sodium canned soup. Or make your own low-sodium version.

CHICKEN VEGETABLE, *Ready-to-Serve*

Serving: 1 cup

Calories	166
Fat	4.8 g
Saturated	1.4 g
Monounsaturated	2.2 g
Polyunsaturated	1.0 g
Calories from fat	26%
Cholesterol	17 mg
Sodium	1,068 mg
Protein	12.3 g
Carbohydrate	18.9 g
Dietary fiber	N/A

CHIEF NUTRIENTS

NUTRIENT	AMOUNT	% RDA
Vitamin A	600.0 RE	60
Niacin	3.3 mg	17
Iron	1.5 mg	15
Zinc	2.2 mg	14
Vitamin B_{12}	0.2 mcg	12
Potassium	367.2 mg	10
Riboflavin	0.2 mg	10
Vitamin C	5.5 mg	9
Folate	12.0 mcg	6

Strengths: One of the most nutrient-dense soups available, chicken vegetable soup is loaded with vitamin A to boost disease resistance and help protect against cancer. It also supplies appreciable amounts of B vitamins, with enough iron to help build healthier blood. Its vitamin C and zinc help fight infection and promote wound healing. The ready-to-serve soup offers a decent amount of protein, although the exact value varies from product to product.

Eat with: Whole wheat bread, to get some fiber along with soup.

Caution: With more than 1,000 mg of sodium per serving, this ready-to-serve soup should be avoided if you're on a low-salt diet to control blood pressure or avoid stroke and kidney disease. Chicken soup contains purines, so steer clear if you're prone to gout. *Allergy alert:* Read labels. May contain wheat or other common food allergens.

A Better Idea: Make your own low-sodium version with chunks of chicken, potatoes, carrots, onions, and tomatoes in a chicken stock.

CLAM CHOWDER, *Manhattan, Condensed*

Serving: 1 cup, made w/water

Calories	78
Fat	2.2 g
Saturated	0.4 g
Monounsaturated	0.4 g
Polyunsaturated	1.3 g
Calories from fat	26%
Cholesterol	2 mg
Sodium	578 mg
Protein	2.2 g
Carbohydrate	12.2 g
Dietary fiber	N/A

CHIEF NUTRIENTS

NUTRIENT	AMOUNT	% RDA
Vitamin B$_{12}$	4.1 mcg	203
Iron	1.6 mg	16
Vitamin A	97.6 RE	10
Vitamin C	3.9 mg	7
Zinc	1.0 mg	7
Potassium	187.9 mg	5

Strengths: An excellent low-fat, low-cholesterol source of blood-building vitamin B$_{12}$. A good source of iron, Manhattan clam chowder may also help you lose weight by dulling your appetite for a calorie-laden main course.

Caution: Commercially prepared soups are frequently loaded with sodium and monosodium glutamate (MSG). People who have high blood pressure or who are sensitive to MSG may want to avoid canned chowder. *Allergy alert:* May contain milk, wheat, and tomatoes—all of which can occasionally cause an allergic reaction.

Serve: With bread that is brushed very lightly with olive oil, rubbed with garlic, and topped with a grated low-salt, low-fat cheese. Broil until the cheese is melted.

A Better Idea: Make it from scratch and use low-sodium canned tomatoes.

CLAM CHOWDER, *New England, Condensed*

Serving: 1 cup, made w/milk & water

Calories	95
Fat	2.9 g
Saturated	0.4 g
Monounsaturated	1.2 g
Polyunsaturated	1.1 g
Calories from fat	27%
Cholesterol	5 mg
Sodium	915 mg
Protein	4.8 g
Carbohydrate	12.4 g
Dietary fiber	N/A

CHIEF NUTRIENTS

NUTRIENT	AMOUNT	% RDA
Vitamin B$_{12}$	8.0 mcg	400
Iron	1.5 mg	15
Calcium	43.9 mg	5
Niacin	1.0 mg	5

Strengths: With a rich, creamy taste, clam chowder is a fabulous source of blood-building vitamin B$_{12}$ and has plentiful iron. When you eat it as an appetizer, plan on a light entrée afterward: Chowder may aid in weight loss by reducing your appetite for the main (calorie-laden) course.

Caution: Commercially prepared soups are frequently loaded with sodium and monosodium glutamate (MSG). People who have high blood pressure or who are sensitive to MSG may want to avoid canned New England clam chowder. *Allergy alert:* May contain milk or wheat—both of which can occasionally cause an allergic reaction.

Description: The traditional New England clam chowder is made from clams, white potatoes, onions, black pepper, and hot milk, and it includes heavy cream and a cube of diced salt pork.

A Better Idea: Make your own chowder and use a low-fat milk. The only thing you'll lose is fat and calories.

CRAB, *Ready-to-Serve*

Serving: 1 cup

Calories	76
Fat	1.5 g
Saturated	0.4 g
Monounsaturated	0.7 g
Polyunsaturated	0.9 g
Calories from fat	18%
Cholesterol	10 mg
Sodium	1,235 mg
Protein	5.5 g
Carbohydrate	10.3 g
Dietary fiber	N/A

CHIEF NUTRIENTS

NUTRIENT	AMOUNT	% RDA
Thiamine	0.2 mg	13
Iron	1.2 mg	12
Vitamin B_{12}	0.2 mcg	10
Zinc	1.5 mg	10
Potassium	327.0 mg	9
Calcium	65.9 mg	8
Folate	14.6 mcg	7
Niacin	1.3 mg	7
Vitamin B_6	0.1 mg	6
Vitamin A	51.2 RE	5

Strengths: A taste of the Old South, crab soup has been elevated to an art form by southern cooks. It is a good source of blood-building vitamin B_{12} and iron. The canned variety is low in cholesterol and fat.

Caution: Commercially prepared soups are frequently loaded with sodium and monosodium glutamate (MSG). People who have high blood pressure or who are sensitive to MSG may want to avoid them. (MSG may also be listed on labels as "hydrolyzed plant protein," "hydrolyzed vegetable protein," or "natural flavor.") Since crabs are high in purines, people prone to gout may want to avoid this soup. *Allergy alert:* May contain milk and wheat—both of which can occasionally cause an allergic reaction.

A Better Idea: Make your own. You can significantly reduce the sodium and MSG without detracting from the flavor.

CREAM OF ASPARAGUS, *Condensed*

Serving: 1 cup, made w/water

Calories	85
Fat	4.1 g
Saturated	1.1 g
Monounsaturated	1.0 g
Polyunsaturated	1.8 g
Calories from fat	43%
Cholesterol	5 mg
Sodium	981 mg
Protein	2.3 g
Carbohydrate	10.7 g
Dietary fiber	0.7 g

CHIEF NUTRIENTS

NUTRIENT	AMOUNT	% RDA
Folate	22.0 mcg	11
Iron	0.8 mg	8
Zinc	0.9 mg	6

Strengths: This soup is a good source of folate, a B vitamin that's essential for women who are pregnant or taking birth control pills. It also offers a bit of iron and zinc to reinforce immunity. Supplies some fiber, thanks to the asparagus. Eating soup can help suppress your appetite, preventing weight gain.

Caution: A salt mine of sodium, this condensed soup is probably out of the question if you're on a low-salt diet to control your blood pressure. Asparagus is high in purines and should be eaten sparingly by anyone with gout. *Allergy alert:* Read labels. May contain milk, wheat, or other common food allergens.

A Better Idea: If you want a boost of bone-building calcium, prepare with skim milk instead of water. Or make homemade asparagus soup without salt and avoid unwanted sodium.

CREAM OF ASPARAGUS, *Dehydrated*

Serving: 1 cup, made w/water

Calories	58
Fat	1.7 g
Saturated	0.1 g
Monounsaturated	0.7 g
Polyunsaturated	0.3 g
Calories from fat	27%
Cholesterol	0
Sodium	800 mg
Protein	2.2 g
Carbohydrate	9.0 g
Dietary fiber	N/A

CHIEF NUTRIENTS

NUTRIENT	AMOUNT	% RDA
Iron	0.5 mg	5

Caution: Slightly lower in sodium than condensed asparagus soup, but still much too high if you've been advised to avoid excessive amounts of salt in your diet. Overall, the dehydrated cream of asparagus soup is less nutritious than the canned version. Asparagus is high in purines and should be eaten sparingly by anyone with gout. *Allergy alert*: Read labels. May contain milk, wheat, or other common food allergens. Dehydrated soups may contain monosodium glutamate (MSG), which is contained in hydrolyzed vegetable protein; MSG is a common cause of headaches in sensitive individuals.

Strengths: Low in fat and calories, this soup may offer some fiber, thanks to the asparagus. Soup seems to suppress appetite, so it may have some weight-loss benefits.

A Better Idea: Mixed with skim milk instead of water, dehydrated soup supplies calcium for stronger bones. Or you can reduce the sodium content by making your own asparagus soup without salt.

CREAM OF CELERY, *Condensed*

Serving: 1 cup, made w/water

Calories	90
Fat	5.6 g
Saturated	1.4 g
Monounsaturated	1.3 g
Polyunsaturated	2.5 g
Calories from fat	56%
Cholesterol	15 mg
Sodium	949 mg
Protein	1.7 g
Carbohydrate	8.8 g
Dietary fiber	N/A

CHIEF NUTRIENTS

NUTRIENT	AMOUNT	% RDA
Vitamin B$_{12}$	0.2 mcg	12
Iron	0.6 mg	6

Strengths: Cream of celery soup is a good source of vitamin B$_{12}$ for healthy nerves and rich, red blood. Its iron helps fight infections and contributes to healthier blood.

Eat with: Foods rich in vitamin C to boost absorption of the nonheme iron.

Caution: A bit higher in fat and calories than cream of asparagus or cream of potato soup, cream of celery has a comparably high amount of sodium. It's not a prudent choice if you're trying to control your blood pressure by adhering to a low-salt diet.

A Better Idea: If you're watching the sodium in your diet, make homemade cream of celery soup and omit the salt.

CREAM OF CELERY, *Dehydrated*

Serving: 1 cup, made w/water

Calories	64
Fat	1.6 g
Saturated	0.3 g
Monounsaturated	0.7 g
Polyunsaturated	0.6 g
Calories from fat	23%
Cholesterol	0
Sodium	838 mg
Protein	2.6 g
Carbohydrate	9.8 g
Dietary fiber	0.8 g

CHIEF NUTRIENTS

NUTRIENT	AMOUNT	% RDA
Iron	0.5 mg	5

Caution: Like other dehydrated and condensed soups, this is a high-sodium menu item. Not recommended if you're trying to control your blood pressure by staying on a low-salt diet. *Allergy alert*: Read labels. May contain milk, wheat, or other common food allergens. Dehydrated soups contain monosodium glutamate (MSG), contained in hydrolyzed vegetable protein; MSG is a common cause of headaches in sensitive individuals.

Eat with: A turkey sandwich or other high-protein, potassium-rich foods.

Strengths: With a bit of iron, cream of celery soup is low in fat if you make it with water instead of whole milk. Eating soup can help suppress your appetite, preventing weight gain.

A Better Idea: To avoid an excessive amount of sodium, make your own cream of celery soup and skip the salt.

CREAM OF CHICKEN, *Condensed*

Serving: 1 cup, made w/water

Calories	117
Fat	7.4 g
Saturated	2.1 g
Monounsaturated	3.3 g
Polyunsaturated	1.5 g
Calories from fat	57%
Cholesterol	10 mg
Sodium	986 mg
Protein	3.4 g
Carbohydrate	9.3 g
Dietary fiber	0.2 g

CHIEF NUTRIENTS

NUTRIENT	AMOUNT	% RDA
Iron	0.6 mg	6
Vitamin A	56.1 RE	6

Strengths: Contains a little protein, with some resistance-boosting iron and vitamin A.

Caution: Condensed cream of chicken soup is a bit high in fat compared to other soups. It's also high in sodium—so this is a poor choice for anyone who's been advised to avoid salty foods as a means of controlling elevated blood pressure. Chicken soups contain purines and should be eaten sparingly by anyone with gout. *Allergy alert:* Read labels: May contain milk, wheat, or other common food allergens.

A Better Idea: Make homemade cream of chicken soup without adding salt, and avoid unwanted sodium altogether.

CREAM OF CHICKEN, *Dehydrated*

Serving: 1 cup, made w/water

Calories	107
Fat	5.3 g
Saturated	3.4 g
Monounsaturated	1.2 g
Polyunsaturated	0.4 g
Calories from fat	45%
Cholesterol	3 mg
Sodium	1,185 mg
Protein	1.8 g
Carbohydrate	13.3 g
Dietary fiber	0.3 g

CHIEF NUTRIENTS

NUTRIENT	AMOUNT	% RDA
Niacin	2.6 mg	14
Vitamin B$_{12}$	0.3 mcg	13
Riboflavin	0.2 mg	12
Vitamin A	122.7 RE	12
Zinc	1.6 mg	10
Calcium	75.7 mg	9
Thiamine	0.1 mg	7
Potassium	214.1 mg	6

Caution: Cream of chicken soup is higher in saturated fat and calories than noncreamy dehydrated soups. And it's one of the saltiest soups on the market, with nearly 1,200 mg of sodium in just one serving. Obviously, this is not the best choice for anyone who needs to avoid excess salt. *Allergy alert:* Read labels. May contain milk, wheat, or other common food allergens. Like most dehydrated soups, this kind contains monosodium glutamate (MSG), which is contained in hydrolyzed vegetable protein; MSG is a common cause of headaches in sensitive individuals.

Eat with: Low-sodium, low-fat, potassium-rich foods such as cantaloupe and other fresh fruit.

Strengths: An array of B vitamins helps improve the health of skin and nerves. There's some vitamin A, which is essential for healthy skin, along with calcium for bone building. Cream of chicken soup also has immunity-boosting zinc.

A Better Idea: To avoid excess sodium, make homemade cream of chicken soup and skip the salt. Prepare with skim milk instead of water to increase calcium.

CREAM OF MUSHROOM, *Condensed*

Serving: 1 cup, made w/water

Calories	129
Fat	9.0 g
Saturated	2.4 g
Monounsaturated	1.7 g
Polyunsaturated	4.2 g
Calories from fat	62%
Cholesterol	2 mg
Sodium	1,032 mg
Protein	2.3 g
Carbohydrate	9.3 g
Dietary fiber	0.5 g

CHIEF NUTRIENTS

NUTRIENT	AMOUNT	% RDA
Calcium	46.4 mg	6
Iron	0.5 mg	5
Riboflavin	0.1 mg	5

Caution: High in fat and sodium, this soup is hard to justify if you're on a low-fat, low-sodium, heart-healthy diet. It's also higher in calories than noncreamy soups. Mushrooms have purines and should be eaten sparingly by anyone who has gout. *Allergy alert:* Read labels. May contain milk, wheat, or other common food allergens.

Eat with: A hearty salad of beans, greens, and other low-sodium, low-fat, potassium-rich foods.

Strengths: A little protein and calcium contribute to healthy nerves. Cream of mushroom soup also contains riboflavin, a B vitamin that helps produce healthy red blood cells, and iron, a mineral that helps convert beta-carotene to vitamin A.

A Better Idea: To avoid excess sodium and fat, make your own cream of mushroom soup with skim milk and no added salt. Skim milk has the added benefit of bone-building calcium.

CREAM OF POTATO, *Condensed*

Serving: 1 cup, made w/water

Calories	73
Fat	2.4 g
Saturated	1.2 g
Monounsaturated	0.6 g
Polyunsaturated	0.4 g
Calories from fat	29%
Cholesterol	5 mg
Sodium	1,000 mg
Protein	1.8 g
Carbohydrate	11.5 g
Dietary fiber	0.5 g

CHIEF NUTRIENTS

(None of the nutrients meet or exceed 5% of the RDA.)

Strengths: Cream of potato is not quite as fatty as cream of chicken, mushroom, or celery soup. Eating a starchy potato soup before a meal can reduce your appetite for fatty foods, which can be helpful if you're trying to lose weight.

Eat with: Salads chock-full of chopped greens and vegetables for vitamins, minerals, and fiber.

Caution: Condensed potato soup has a whopping amount of sodium, making this off-limits for many trying to cut down on salty foods for medical reasons. *Allergy alert:* Read labels. May contain milk, wheat, or other common food allergens.

Curiosity: A French version of this soup known as *potage Parmentier* is named after an 18th-century Frenchman, Antoine-Auguste Parmentier. Parmentier shunned potatoes, believing they were a food fit only for livestock, until he was taken prisoner by the Germans and had to live on potatoes for a year. Then he changed his tune—and created the *potage* named after him.

GAZPACHO, *Ready-to-Serve*

Serving: 1 cup

Calories	56
Fat	2.2 g
Saturated	0.3 g
Monounsaturated	0.5 g
Polyunsaturated	1.3 g
Calories from fat	36%
Cholesterol	0
Sodium	1,183 mg
Protein	8.7 g
Carbohydrate	0.8 g
Dietary fiber	3.7 g

CHIEF NUTRIENTS

NUTRIENT	AMOUNT	% RDA
Iron	1.0 mg	10
Vitamin B_6	0.2 mg	8
Potassium	224.5 mg	6
Vitamin C	3.2 mg	5

Caution: Gazpacho is one of the saltiest soups around. If you're on a low-sodium diet, you're probably better off having "just a taste" of this soup rather than a full serving.

Strengths: Moderate in fat and low in calories, gazpacho has a very good amount of fiber and a decent amount of protein. A good amount of nonheme iron contributes to healthier blood, and nice amounts of vitamins C and B_6 help fight infection. It has a little potassium, a mineral that's essential for carbohydrate and protein metabolism. Eating soup can help suppress your appetite, preventing weight gain.

Origin: This soup was first concocted in the Andalusia region of southern Spain.

Description: A refreshingly chilled, uncooked soup of pureed tomatoes, peppers, onions, celery, cucumbers, garlic, olive oil, vinegar, and sometimes lemon juice.

Serve: Garnished with scallions, and with slices of hot, crusty, whole wheat bread.

A Better Idea: To avoid exorbitantly high levels of sodium, make your own gazpacho.

GREEN PEA, *Condensed*

Serving: 1 cup, made w/water

Calories	165
Fat	2.9 g
Saturated	1.4 g
Monounsaturated	1.0 g
Polyunsaturated	0.4 g
Calories from fat	16%
Cholesterol	0
Sodium	988 mg
Protein	8.6 g
Carbohydrate	26.5 g
Dietary fiber	2.8 g

CHIEF NUTRIENTS

NUTRIENT	AMOUNT	% RDA
Iron	2.0 mg	20
Magnesium	40.0 mg	11
Zinc	1.7 mg	11
Niacin	1.2 mg	7
Thiamine	0.1 mg	7
Potassium	190.0 mg	5

Strengths: A hearty, low-fat soup, green pea offers a substantial amount of protein. Respectable amounts of nonheme iron and zinc help to boost immunity, and a significant amount of magnesium can promote resistance to tooth decay. Eating soup is a terrific way to put the brakes on your appetite at the beginning of a meal so you don't overdo it on the entrée.

Caution: Quite high in sodium. Not recommended if you're on a low-sodium diet to control your blood pressure. Soup containing peas and other legumes is high in purines and should be eaten sparingly by anyone with gout. *Allergy alert:* Read labels. May contain milk, wheat, or other common food allergens.

Serve: With salad and whole grain bread.

A Better Idea: If you're trying to avoid sodium, make your own pea soup with peas, onions, and chicken or vegetable stock.

LENTIL, *with Ham, Ready-to-Serve*

Serving: 1 cup

Calories	139
Fat	2.8 g
Saturated	1.1 g
Monounsaturated	1.3 g
Polyunsaturated	0.3 g
Calories from fat	18%
Cholesterol	7 mg
Sodium	1,319 mg
Protein	9.3 g
Carbohydrate	20.2 g
Dietary fiber	N/A

CHIEF NUTRIENTS

NUTRIENT	AMOUNT	% RDA
Iron	2.7 mg	27
Folate	49.6 mcg	25
Vitamin B$_{12}$	0.3 mcg	15
Thiamine	0.2 mg	11
Vitamin B$_6$	0.2 mg	11
Potassium	357.1 mg	10
Niacin	1.4 mg	7
Vitamin C	4.2 mg	7

Strengths: Low in fat, lentil soup with ham is a fairly decent source of protein. Worthwhile amounts of folate and nonheme iron provide plenty of assistance for healthy blood. Good amounts of vitamin B$_{12}$, thiamine, and vitamin B$_6$ provide a boost for healthy skin and nerves. This soup also has a little resistance-boosting vitamin C. Lentils in general are a decent source of soluble and insoluble fiber. Eating starchy soup containing legumes before a meal can reduce appetite for fatty foods, helping to cut overall fat intake.

Eat With: Fruit salad or other vitamin C–rich foods to enhance absorption of nonheme iron.

Caution: Like most canned soups, this one is brimming with sodium. If you're on a sodium-restricted diet, steer clear. Also, it contains legumes and pork, which are high in purines: Eat sparingly if you're susceptible to gout. *Allergy alert:* Read labels. May contain wheat or other common food allergens.

A Better Idea: Look for low-sodium versions of lentil soup.

MINESTRONE, *Condensed*

Serving: 1 cup, made w/water

Calories	82
Fat	2.5 g
Saturated	0.6 g
Monounsaturated	0.7 g
Polyunsaturated	1.1 g
Calories from fat	28%
Cholesterol	2 mg
Sodium	911 mg
Protein	4.3 g
Carbohydrate	11.2 g
Dietary fiber	1.0 g

CHIEF NUTRIENTS

NUTRIENT	AMOUNT	% RDA
Vitamin A	233.8 RE	23
Iron	0.9 mg	9
Folate	16.2 mcg	8
Potassium	313.3 mg	8

Strengths: Low in fat, a cup of minestrone is a good way to sneak an extra serving of vegetables into your diet. Also, eating soup that contains beans and pasta or beans and rice can help suppress your appetite, preventing weight gain.

Caution: This condensed soup is loaded with sodium, so it's not a wise choice for anyone trying to avoid salt. *Allergy alert:* Read labels. May contain wheat or other food allergens.

Description: Made with a variety of good-for-you vegetables—like celery, onions, beans, cabbage, tomatoes, turnips—plus pasta or rice. Minestrone, which means ''big soup'' in Italian, makes a meal in itself.

Serve: With slices of hot, crusty, whole wheat bread.

A Better Idea: Look for low-sodium versions of minestrone. Or make your own, adding any variety of vegetables you have on hand. (Minestrone is one of those soups that taste best a day after being made.)

MINESTRONE, *Dehydrated*

Serving: 1 cup, made w/water

Calories	79
Fat	1.7 g
Saturated	0.8 g
Monounsaturated	0.7 g
Polyunsaturated	0.1 g
Calories from fat	20%
Cholesterol	3 mg
Sodium	1,026 mg
Protein	4.4 g
Carbohydrate	11.9 g
Dietary fiber	N/A

CHIEF NUTRIENTS

NUTRIENT	AMOUNT	% RDA
Iron	1.0 mg	10
Folate	17.8 mcg	9
Potassium	340.2 mg	9
Niacin	1.0 mg	5
Thiamine	0.1 mg	5
Vitamin B_6	0.1 mg	5
Zinc	0.8 mg	5

Caution: Brimming with sodium, dehydrated minestrone lacks the vitamin A and the iron that are present in the ready-to-serve version. *Allergy alert:* Read labels. May contain wheat or other common food allergens. Dehydrated soups have some monosodium glutamate (MSG), which is contained in hydrolyzed vegetable protein; MSG is a common cause of headaches in sensitive individuals.

Strengths: Low in fat, minestrone has a bit of zinc and iron for better resistance, folate for healthy blood, and other useful B vitamins. Since a bowl of soup consumed before a meal can help suppress your appetite, it may be a way to reduce fat in your diet.

Eat with: Low-sodium, low-fat, potassium-rich foods.

A Better Idea: Look for low-sodium versions of minestrone. Or make your own with fresh and frozen vegetables. The soup tastes best when it's served the day after it's made.

MINESTRONE, *Ready-to-Serve*

Serving: 1 cup

Calories	127
Fat	2.8 g
Saturated	1.5 g
Monounsaturated	0.9 g
Polyunsaturated	0.3 g
Calories from fat	20%
Cholesterol	5 mg
Sodium	864 mg
Protein	5.1 g
Carbohydrate	20.7 g
Dietary fiber	N/A

CHIEF NUTRIENTS

NUTRIENT	AMOUNT	% RDA
Vitamin A	434.4 RE	43
Iron	1.8 mg	18
Folate	31.2 mcg	16
Potassium	612.0 mg	16
Vitamin B_6	0.2 mg	12
Zinc	1.4 mg	10
Calcium	60.0 mg	8
Vitamin C	4.8 mg	8
Riboflavin	0.1 mg	7
Niacin	1.2 mg	6

Strengths: Richer in nutrients than dehydrated or condensed minestrone, the ready-to-serve soup is a bountiful source of vitamin A, which boosts disease resistance and helps protect against cancer. Minestrone also has good amounts of other resistance-enhancing nutrients—notably iron, folate, zinc, and a bit of vitamin C—with useful amounts of B vitamins for healthy skin and nerves. It's low in fat—a good way to dampen an overzealous appetite before the entrée.

Eat with: High-protein, potassium-rich foods such as a turkey sandwich on whole wheat bread with sliced tomatoes.

Caution: Sodium is too high for many people who are on sodium-restricted diets. *Allergy alert:* Read labels. May contain wheat or other common food allergens.

Description: Filled with a variety of nutrient-rich vegetables, a good serving of minestrone makes a meal in itself.

A Better Idea: Look for low-sodium versions of minestrone, or make your own with a potpourri of vegetables.

MUSHROOM, *Dehydrated*

Serving: 1 cup, made w/water

Calories	96
Fat	4.9 g
Saturated	0.8 g
Monounsaturated	2.3 g
Polyunsaturated	1.5 g
Calories from fat	45%
Cholesterol	0
Sodium	1,020 mg
Protein	2.2 g
Carbohydrate	11.1 g
Dietary fiber	0.8 g

CHIEF NUTRIENTS

NUTRIENT	AMOUNT	% RDA
Thiamine	0.3 mg	19
Vitamin B_{12}	0.3 mcg	13
Calcium	65.8 mg	8
Riboflavin	0.1 mg	6
Iron	0.5 mg	5
Potassium	199.9 mg	5

Caution: High in sodium, dehydrated mushroom soup is not an advisable choice for those on salt-restricted diets. Mushrooms are high in purines and should be eaten sparingly by anyone with gout. *Allergy alert:* Dehydrated soups often contain monosodium glutamate (MSG), also known as hydrolyzed vegetable protein, a common cause of headaches in sensitive individuals.

Eat with: Foods high in potassium, such as carrot and raisin salad, to help offset some of the potentially harmful effects of sodium.

Strengths: This soup has respectable amounts of thiamine, a B vitamin that helps convert carbohydrates into energy, and vitamin B_{12}, which helps to produce mature red blood cells. Also contains some calcium and potassium. There's a little resistance-boosting iron, too.

A Better Idea: For lean, low-calorie mushroom soup without excess sodium, make your own with fresh mushrooms and homemade beef or chicken stock.

MUSHROOM BARLEY, Condensed

Serving: 1 cup, made w/water

Calories	73
Fat	2.3 g
Saturated	0.4 g
Monounsaturated	1.0 g
Polyunsaturated	0.7 g
Calories from fat	28%
Cholesterol	0
Sodium	891 mg
Protein	1.9 g
Carbohydrate	11.7 g
Dietary fiber	0.5 g

CHIEF NUTRIENTS

NUTRIENT	AMOUNT	% RDA
Vitamin B$_6$	0.2 mg	9
Iron	0.5 mg	5
Riboflavin	0.1 mg	5

Strengths: Low in fat and moderately low in calories, mushroom barley soup has a nice helping of resistance-boosting vitamin B$_6$. It also contains a smidgen of riboflavin and iron, two nutrients essential for healthy red blood cells. This is a tasty way to get a serving of barley, a high-fiber grain. Eating starchy soups like mushroom barley before a meal can curb your appetite, helping to cut overall fat intake.

Eat with: A salad of tomatoes and other potassium-rich vegetables. The potassium helps improve blood pressure.

Caution: Watch that sodium content! With nearly 900 mg per serving, this soup may be off-limits if you've been advised to limit your salt intake. Mushrooms are high in purines and should be eaten sparingly by anyone with gout.

A Better Idea: To avoid excess sodium, make your own mushroom barley soup with fresh mushrooms, carrots, celery, garlic, chicken stock, and medium-size pearl barley.

ONION, Condensed

Serving: 1 cup, made w/water

Calories	58
Fat	1.7 g
Saturated	0.3 g
Monounsaturated	0.8 g
Polyunsaturated	0.7 g
Calories from fat	27%
Cholesterol	0
Sodium	1,053 mg
Protein	3.8 g
Carbohydrate	8.2 g
Dietary fiber	1.0 g

CHIEF NUTRIENTS

NUTRIENT	AMOUNT	% RDA
Folate	15.2 mcg	8
Iron	0.7 mg	7

Strengths: Onions are allium vegetables, which have been found to offer some protection from cancer. Onion soup (also called French onion soup) is a low-calorie first course with barely any fat, as long as it's served without cheese. Often, people who have soup before the main course seem to eat less food—so eating soup can help fill you up without filling you out. The condensed version of onion soup offers a little resistance-boosting folate and iron.

Eat with: Baked potatoes or other potassium-rich foods to offset some of the potentially harmful effects of sodium on blood pressure.

Caution: Since condensed soup is excessively high in sodium, it's a poor choice for anyone who's been advised to cut out salty foods.

A Better Idea: If you would rather not have a lot of sodium in your diet, make your own onion soup with mild white onions and low-sodium chicken or beef stock.

ONION, *Dehydrated*

Serving: 1 cup, made w/water

Calories	27
Fat	0.6 g
Saturated	trace
Monounsaturated	trace
Polyunsaturated	trace
Calories from fat	19%
Cholesterol	0
Sodium	849 mg
Protein	1.1 g
Carbohydrate	5.1 g
Dietary fiber	1.0 g

CHIEF NUTRIENTS

(None of the nutrients meet or exceed 5% of the RDA.)

Caution: High in sodium, onion (also called French onion) soup is sparse in health-building nutrients when it comes in dehydrated form. *Allergy alert:* Dehydrated soups have some monosodium glutamate (MSG), which is contained in hydrolyzed vegetable protein; MSG can trigger migraines in sensitive individuals.

Eat with: A hearty salad of tomatoes, avocado, and other potassium-rich foods to help offset some of the negative effects of sodium.

Strengths: Low in fat and calories, a bowl of soup before a meal can dampen your appetite so you don't overeat the main course. That's a plus for dieters.

A Better Idea: To enjoy the soup with far less sodium, make your own with mild white onions and low-sodium chicken or beef stock.

SCOTCH BROTH, *Condensed*

Serving: 1 cup, made w/water

Calories	80
Fat	2.6 g
Saturated	1.1 g
Monounsaturated	0.8 g
Polyunsaturated	0.6 g
Calories from fat	30%
Cholesterol	5 mg
Sodium	1,012 mg
Protein	5.0 g
Carbohydrate	9.5 g
Dietary fiber	1.2 g

CHIEF NUTRIENTS

NUTRIENT	AMOUNT	% RDA
Vitamin A	216.9 RE	22
Vitamin B$_{12}$	0.3 mcg	14
Zinc	1.6 mg	11
Iron	0.8 mg	8
Niacin	1.2 mg	6

Strengths: A good source of cancer-protecting vitamin A, Scotch broth also provides vitamin B$_{12}$ to help maintain healthy blood and nerves. It has immunity-boosting zinc, with some blood-building iron and niacin. The barley in the soup supplies a bit of fiber to boost digestive health.

Eat with: A fresh salad of tomatoes, mushrooms, beans, and other potassium-rich foods.

Caution: Like most prepared soups, Scotch broth yields an overabundance of sodium—so it's not a good choice if you're on a sodium-restricted diet.

Curiosity: Barley—which is used in Scotch broth—is the main ingredient in traditional ''barley water,'' an old-fashioned restorative for invalids.

Description: The traditional broth is a Scottish soup made with lamb or mutton, barley, carrots, turnips, or other vegetables. Scotch broth is also known as barley broth.

A Better Idea: If you've been advised to avoid salty foods, make your own Scotch broth with lamb shoulder, carrots, celery, onions, pearl barley, parsley, and low-sodium stock.

SPLIT PEA with Ham, Condensed

Serving: 1 cup, made w/water

Calories	190
Fat	4.4 g
Saturated	1.8 g
Monounsaturated	1.8 g
Polyunsaturated	0.6 g
Calories from fat	21%
Cholesterol	8 mg
Sodium	1,007 mg
Protein	10.3 g
Carbohydrate	28.0 g
Dietary fiber	N/A

CHIEF NUTRIENTS

NUTRIENT	AMOUNT	% RDA
Iron	2.3 mg	23
Magnesium	48.1 mg	14
Vitamin B_{12}	0.3 mcg	13
Potassium	399.7 mg	11
Thiamine	0.2 mg	10
Zinc	1.3 mg	9
Niacin	1.5 mg	8

Strengths: Relatively low in fat, split pea soup is a good source of protein. Peas have some fiber, and they also contain useful amounts of health-building nutrients. This soup has resistance-boosting iron and zinc, along with healthy amounts of magnesium and potassium, minerals that help keep blood pressure under control. As a bonus, you get a little niacin, thiamine, and a kick of vitamin B_{12}—vitamins that help keep skin and nerves healthy.

Caution: This condensed soup is much too high in sodium for anyone who needs to avoid salty foods. *Allergy alert:* Read labels. May contain pork, wheat, or other common food allergens.

A Better Idea: Look for low-sodium split pea soup. Or make your own, but leave out the salt and salt pork that are in some recipes. To prepare homemade soup, soak the peas overnight, then drain and simmer with onions, celery, stock, and black pepper for 2 hours. Puree before serving.

SPLIT PEA, Dehydrated

Serving: 1 cup, made w/water

Calories	133
Fat	1.6 g
Saturated	0.4 g
Monounsaturated	0.7 g
Polyunsaturated	0.3 g
Calories from fat	11%
Cholesterol	3 mg
Sodium	1,220 mg
Protein	7.7 g
Carbohydrate	22.7 g
Dietary fiber	N/A

CHIEF NUTRIENTS

NUTRIENT	AMOUNT	% RDA
Folate	42.0 mcg	21
Thiamine	0.2 mg	15
Vitamin B_{12}	0.3 mcg	14
Magnesium	46.1 mg	13
Iron	1.0 mg	10
Riboflavin	0.2 mg	9
Niacin	1.3 mg	7
Potassium	238.5 mg	6

Strengths: Low in fat with a substantial amount of protein, this soup is a good source of iron and folate for healthy blood. It also supplies a decent amount of magnesium and a bit of potassium, both of which help keep nerves and muscles in tip-top condition. Eating soup can help suppress your appetite, preventing weight gain.

Caution: Way too high in sodium to be considered acceptable if you're on a salt-restricted diet. Soup containing peas or pork (or both) is high in purines and should be eaten sparingly by anyone with gout. *Allergy alert*: Read labels. May contain wheat or other common food allergens. Dehydrated soups may contain monosodium glutamate (MSG), also known as hydrolyzed vegetable protein, a common cause of headaches in sensitive individuals.

Serve: With salad and whole wheat bread.

A Better Idea: If you're trying to avoid sodium, make your own split pea soup with dried peas, lean low-sodium ham, carrots, and onions.

SPLIT PEA with Ham, Ready-to-Serve

Serving: 1 cup

Calories	185
Fat	4.0 g
Saturated	1.6 g
Monounsaturated	1.6 g
Polyunsaturated	0.6 g
Calories from fat	19%
Cholesterol	7 mg
Sodium	965 mg
Protein	11.1 g
Carbohydrate	26.8 g
Dietary fiber	4.1 g

CHIEF NUTRIENTS

NUTRIENT	AMOUNT	% RDA
Vitamin A	487.2 RE	49
Iron	2.1 mg	21
Zinc	3.1 mg	21
Niacin	2.5 mg	13
Vitamin B_{12}	0.2 mcg	12
Vitamin C	7.0 mg	12
Magnesium	38.4 mg	11
Vitamin B_6	0.2 mg	11

Strengths: Rich in vitamin A, this soup also has a good amount of vitamin C—so you'll be getting two nutrients that help protect against cancer. It offers an appreciable amount of anemia-fighting iron, and fair to good amounts of various B vitamins to help keep skin and nerves healthy. Split pea soup with ham is also a good source of immunity-boosting zinc. Low in fat, this soup has the benefit of being high in protein.

Caution: While this is one of the most nutritious soups you can choose, keep in mind that it's high in sodium; that's a definite drawback if you've been advised to reduce your sodium intake to avoid kidney disease or stroke. *Allergy alert*: If you're allergic to pork, wheat, or other common food allergens, read the label. You might have to avoid some brands of this soup.

A Better Idea: Buy or make your own low-sodium split pea soup, leaving out the salt or salt pork if they're in the recipe.

TOMATO, Condensed

Serving: 1 cup, made w/water

Calories	85
Fat	1.9 g
Saturated	0.4 g
Monounsaturated	0.4 g
Polyunsaturated	1.0 g
Calories from fat	20%
Cholesterol	0
Sodium	871 mg
Protein	2.1 g
Carbohydrate	16.6 g
Dietary fiber	N/A

CHIEF NUTRIENTS

NUTRIENT	AMOUNT	% RDA
Vitamin C	66.4 mg	111
Iron	1.8 mg	18
Folate	14.6 mcg	7
Niacin	1.4 mg	7
Potassium	263.5 mg	7
Vitamin A	68.3 RE	7
Thiamine	0.1 mg	6

Strengths: Tomato soup outstrips all others as a great source of vitamin C. In addition, it has a respectable amount of anemia-fighting iron, as well as useful amounts of folate and various other B vitamins for healthy skin and nerves. Its potassium is a mineral that is essential for normal functioning of nerves and muscles. Because tomato soup is low in fat yet filling, it's often recommended for weight watchers.

Eat with: A turkey sandwich or other low-sodium, high-protein, potassium-rich foods.

Caution: A tad less salty than some other soups, but still too high in sodium, condensed tomato soup is not advisable if you're on a salt-restricted diet. *Allergy alert:* Read labels. May contain wheat or other common food allergens.

A Better Idea: Look for low-sodium tomato soup. If you prepare it with skim milk instead of water, you'll get a boost of bone-building calcium, but the milk does contribute some sodium.

TOMATO, *Dehydrated*

Serving: 1 cup, made w/water

Calories	103
Fat	2.4 g
Saturated	1.1 g
Monounsaturated	0.9 g
Polyunsaturated	0.2 g
Calories from fat	21%
Cholesterol	0
Sodium	943 mg
Protein	2.5 g
Carbohydrate	19.4 g
Dietary fiber	0.5 g

CHIEF NUTRIENTS

NUTRIENT	AMOUNT	% RDA
Potassium	294.2 mg	8
Vitamin A	82.2 mg	8
Vitamin C	4.5 mg	8
Calcium	53.0 mg	7
Vitamin B_6	0.1 mg	5

Caution: Dehydrated tomato soup is brimming with sodium, and its vitamin C is a bare fraction of what you get from the condensed soup. *Allergy alert:* Read labels. May contain wheat or other common food allergens. Dehydrated soups have some monosodium glutamate (MSG), which is contained in hydrolyzed vegetable protein; MSG is a common migraine trigger in sensitive individuals.

Eat with: A chicken sandwich or other low-sodium, high-protein, potassium-rich foods.

Strengths: After the drying process, the powdered soup still contains small amounts of vitamins A and C, two resistance-building nutrients. You'll also get a smidgen of bone-building calcium and nerve-enhancing potassium. People who eat more soup seem to eat less food—so a bowl of soup is probably a good choice if you're on a weight-reduction diet.

A Better Idea: Look for low-sodium tomato soup.

TOMATO RICE, *Condensed*

Serving: 1 cup, made w/water

Calories	119
Fat	2.7 g
Saturated	0.5 g
Monounsaturated	0.6 g
Polyunsaturated	1.3 g
Calories from fat	21%
Cholesterol	2 mg
Sodium	815 mg
Protein	2.1 g
Carbohydrate	21.9 g
Dietary fiber	1.5 g

CHIEF NUTRIENTS

NUTRIENT	AMOUNT	% RDA
Vitamin C	14.8 mg	25
Potassium	331.0 mg	9
Iron	0.8 mg	8
Vitamin A	76.6 mg	8
Folate	13.6 mcg	7
Niacin	1.1 mg	6

Strengths: Contains a pretty decent amount of immunity-boosting vitamin C. From tomato rice soup you also get a fair amount of anemia-fighting iron and some potassium, a mineral that can improve nerve and muscle cells. Rice lends a little fiber, which is good for overall digestive health. Studies suggest that eating starchy soups before a meal can help reduce appetite, helping to cut overall fat intake.

Eat with: A tuna salad sandwich and a glass of skim milk: The potassium and calcium may help counteract some of the less desirable effects of high sodium.

Caution: Watch out for the sodium content if you've been warned to stay away from salty food. *Allergy alert:* Read labels. May contain wheat or other common food allergens.

A Better Idea: Look for low-sodium tomato rice soup—or make your own from fresh ingredients.

TURKEY NOODLE, *Condensed*   ⬇ 🏛

Serving: 1 cup, made w/water

Calories	68
Fat	2.0 g
Saturated	0.6 g
Monounsaturated	0.8 g
Polyunsaturated	0.5 g
Calories from fat	26%
Cholesterol	5 mg
Sodium	815 mg
Protein	3.9 g
Carbohydrate	8.6 g
Dietary fiber	1.5 g

CHIEF NUTRIENTS

NUTRIENT	AMOUNT	% RDA
Iron	1.0 mg	10
Vitamin B_{12}	0.2 mcg	8
Niacin	1.4 mg	7

Strengths: Often a favorite among children, turkey noodle soup is a fair source of iron, a mineral that helps prevent anemia. It also provides some niacin and vitamin B_{12}, nutrients that assist in keeping skin and nerves in tip-top condition. This soup is relatively low in calories, with very little fat and some protein. Eating a starchy noodle soup before a meal can curb your appetite for fatty foods—so there's a weight-loss benefit.
Eat with: An apple, banana, or orange. These high-potassium foods help keep blood pressure moderated.
Caution: Brimming with sodium, this condensed soup could pose problems if you're on a low-salt diet. Turkey and turkey soups contain purines and should be eaten sparingly by anyone with gout. *Allergy alert:* Read labels. May contain wheat or other common food allergens.
A Better Idea: Look for low-sodium turkey noodle soup. Or make your own with homemade stock, turkey breast, and noodles—with no added salt.

TURKEY VEGETABLE, *Condensed*   © ✳ ⬇ 🏛

Serving: 1 cup, made w/water

Calories	72
Fat	3.0 g
Saturated	0.9 g
Monounsaturated	1.3 g
Polyunsaturated	0.7 g
Calories from fat	38%
Cholesterol	2 mg
Sodium	906 mg
Protein	3.1 g
Carbohydrate	8.6 g
Dietary fiber	0.5 g

CHIEF NUTRIENTS

NUTRIENT	AMOUNT	% RDA
Vitamin	243.4 RE	24
Vitamin B_{12}	0.2 mcg	9
Iron	0.8 mg	8
Niacin	1.0 mg	5
Potassium	175.9 mg	5

Strengths: On a par with chicken vegetable soup, the turkey variety has a respectable quantity of vitamin A, nice amounts of B vitamins, and a little anemia-fighting iron. It also contributes a bit of potassium, a mineral that might help to hold your blood pressure in check—provided you aren't sensitive to all the sodium in this condensed product.
Eat with: Whole wheat bread and fruit or other potassium-rich foods.
Caution: Because of its high salt content, this soup should be avoided if you've been advised to eliminate high-sodium foods from your diet. Turkey and turkey soups contain purines and should be eaten sparingly by anyone with gout. *Allergy alert:* Read labels. May contain wheat or other common food allergens.
A Better Idea: Make your own turkey vegetable soup from turkey and fresh vegetables, with no added salt.

VEGETABLE BEEF, *Condensed*

Serving: 1 cup, made w/water

Calories	78
Fat	1.9 g
Saturated	0.9 g
Monounsaturated	0.8 g
Polyunsaturated	0.1 g
Calories from fat	22%
Cholesterol	5 mg
Sodium	956 mg
Protein	5.6 g
Carbohydrate	10.2 g
Dietary fiber	0.5 g

CHIEF NUTRIENTS

NUTRIENT	AMOUNT	% RDA
Vitamin A	190.3 RE	19
Vitamin B_{12}	0.3 mcg	16
Iron	1.1 mg	11
Zinc	1.5 mg	10
Folate	10.5 mcg	5
Niacin	1.0 mg	5

Strengths: Fair to good amounts of folate, vitamin A, iron, and zinc help to strengthen your body's resistance. Vegetable beef soup contributes a useful amount of protein, along with some niacin, a B vitamin that helps build healthy blood. Weight watchers note: A bowl of soup before a meal can suppress your appetite so you don't overeat.

Caution: High in sodium, this condensed soup is a poor choice if you need to restrict your sodium consumption. Beef broth contains purines and should be eaten sparingly by anyone with gout. *Allergy alert:* Soups may contain wheat and other common food allergens. Read labels.

Serve: With a sandwich on whole wheat bread, a glass of skim milk, and some fruit.

A Better Idea: Buy reduced-sodium or low-sodium versions: You get the same nutrients, with considerably less sodium. Or make your own low-sodium vegetable beef soup with homemade stock, lean beef, and chopped vegetables.

VEGETABLE BEEF, *Dehydrated*

Serving: 1 cup, made w/water

Calories	53
Fat	1.1 g
Saturated	0.6 g
Monounsaturated	0.5 g
Polyunsaturated	trace
Calories from fat	19%
Cholesterol	0
Sodium	1,002 mg
Protein	2.9 g
Carbohydrate	8.0 g
Dietary fiber	0.5 g

CHIEF NUTRIENTS

NUTRIENT	AMOUNT	% RDA
Vitamin B_{12}	0.3 mcg	13
Iron	0.9 mg	9
Magnesium	22.8 mg	7

Caution: Avoid this dehydrated soup if you've been advised to minimize sodium intake to control high blood pressure. Even the condensed version is higher in nutrients and lower in sodium. Beef-based soups have purines, and should be eaten sparingly by anyone with gout. *Allergy alert:* Dehydrated soups may contain monosodium glutamate (MSG), also known as hydrolyzed vegetable protein, a common migraine trigger in sensitive individuals. Read the label if you're sensitive to wheat or other common allergens.

Eat with: A tuna sandwich, green salad, or other foods that have potassium to help lower blood pressure.

Strengths: Low in fat and calories, even the dehydrated vegetable beef soup has an appreciable supply of vitamin B_{12}.

A Better Idea: To avoid an overdose of sodium, make your own vegetable beef soup.

VEGETARIAN VEGETABLE, *Condensed*

Serving: 1 cup, made w/water

Calories	72
Fat	1.9 g
Saturated	0.3 g
Monounsaturated	0.8 g
Polyunsaturated	0.7 g
Calories from fat	24%
Cholesterol	0
Sodium	822 mg
Protein	2.1 g
Carbohydrate	12.0 g
Dietary fiber	N/A

CHIEF NUTRIENTS

NUTRIENT	AMOUNT	% RDA
Vitamin A	301.3 RE	30
Iron	1.1 mg	11
Potassium	209.7 mg	6
Folate	10.6 mcg	5

Strengths: As many parents have discovered, kids will often eat vegetable soups even if they shun vegetables. The condensed form is rich in vitamin A to help strengthen immunity and protect against cancer, and it has a good amount of iron to help fight anemia. Weight watchers have found that sipping soup before a meal can take the edge off their appetite.

Eat with: Low-sodium foods that are rich in calcium and potassium, such as a glass of skim milk and a sandwich on whole wheat bread.

Caution: This high sodium soup is not recommended for those on low-salt diets. Vegetarians, take note: Some vegetable soups are made with beef broth. *Allergy alert:* Read labels. May contain milk, wheat, or other common food allergens.

A Better Idea: To avoid sodium overload, make your own vegetable soup with no added salt.

STEWS

BEEF, *Ready-to-Serve*

Serving: 1 cup

Calories	194
Fat	7.6 g
Saturated	2.5 g
Monounsaturated	N/A
Polyunsaturated	N/A
Calories from fat	35%
Cholesterol	34 mg
Sodium	1,007 mg
Protein	14.2 g
Carbohydrate	17.4 g
Dietary fiber	N/A

CHIEF NUTRIENTS

NUTRIENT	AMOUNT	% RDA
Iron	2.2 mg	22
Niacin	2.5 mg	13
Vitamin C	7.4 mg	12
Potassium	426.3 mg	11
Riboflavin	0.1 mg	7

Caution: This canned stew is high in sodium and best avoided if you're on a sodium-restricted diet. Also, like other meat soups, beef stew is high in purines and should be eaten sparingly by those with gout.

Strengths: Beef stew is a good source of iron and vitamin C, which are essential for collagen production for healthier skin and bones. Potassium and essential B vitamins enhance normal functioning of nerves and muscles.

A Better Idea: You can spare yourself a considerable amount of sodium if you make your own beef stew using a low-salt or no-salt recipe. And homemade stew is higher in nearly all nutrients.

BEEF, *Homemade*

Serving: 1 cup

Calories	218
Fat	10.5 g
Saturated	5.0 g
Monounsaturated	N/A
Polyunsaturated	N/A
Calories from fat	44%
Cholesterol	64 mg
Sodium	91 mg
Protein	15.7 g
Carbohydrate	15.2 g
Dietary fiber	N/A

CHIEF NUTRIENTS

NUTRIENT	AMOUNT	% RDA
Iron	2.9 mg	29
Vitamin C	17.2 mg	29
Niacin	4.7 mg	25
Potassium	612.5 mg	16
Riboflavin	0.2 mg	10
Thiamine	0.2 mg	10

Strengths: Homemade beef stew is an excellent way to incorporate leaner but less tender cuts of beef into your menu. It's a great source of protein and has far less sodium than canned stew, as long as you simmer the beef in water rather than preparing the stew with a high-sodium stock. (And, of course, don't add salt.) Plus, you get significant amounts of immunity-boosting iron and vitamin C. This tasty stew has plentiful niacin, along with good amounts of riboflavin and thiamine, B vitamins that help keep skin and nerves healthy.

Caution: Homemade beef stew often has more fat than canned beef stew. Like other meat soups, beef stew is high in purines and should be eaten sparingly by those with gout.

Description: Beef stew consists of cubed lean chuck roast and vegetables, which are simmered until the natural juices thicken into a souplike broth.

Preparation: Trim stew meat of all visible fat and brown it in a thin film of unsaturated oil (such as canola) instead of salt pork, rendered fat, or shortening.

OYSTER, *Condensed*

Serving: 1 cup, made w/water

Calories	58
Fat	3.8 g
Saturated	2.5 g
Monounsaturated	0.9 g
Polyunsaturated	0.1 g
Calories from fat	60%
Cholesterol	14 mg
Sodium	981 mg
Protein	2.1 g
Carbohydrate	4.1 g
Dietary fiber	N/A

CHIEF NUTRIENTS

NUTRIENT	AMOUNT	% RDA
Vitamin B_{12}	2.2 mcg	110
Zinc	10.3 mg	69
Iron	1.0 mg	10
Vitamin C	3.1 mg	5

Strengths: Studies indicate that an oyster-rich diet may lower your cholesterol and triglyceride levels. Canned oyster stew is a great low-cholesterol source of vitamin B_{12} and zinc to power your immune system.

Caution: Commercially prepared soups are frequently loaded with sodium and monosodium glutamate (MSG). People who have high blood pressure or who are sensitive to MSG may want to avoid them. (MSG is also listed on labels as "hydrolyzed plant protein," "hydrolyzed vegetable protein," or "natural flavor.") People prone to gout may wish to avoid oysters because of their high purine content. *Allergy alert:* May contain milk and wheat—both of which can occasionally cause an allergic reaction.

Preparation: Virtually none. Add water, heat, and serve.

OYSTER, *Homemade*

Serving: 1 cup

Calories	233
Fat	15.4 g
Saturated	7.2 g
Moncunsaturated	N/A
Polyunsaturated	N/A
Calories from fat	59%
Cholesterol	86 mg
Sodium	814 mg
Protein	12.5 g
Carbohydrate	10.8 g
Dietary fiber	N/A

CHIEF NUTRIENTS

NUTRIENT	AMOUNT	% RDA
Iron	4.6 mg	46
Calcium	273.6 mg	34
Riboflavin	0.4 mg	25
Niacin	2.2 mg	11
Potassium	319.2 mg	9
Thiamine	0.2 mg	9

Strengths: A single serving of homemade oyster stew has nearly half a day's supply of iron, plus a good dose of riboflavin and bone-building calcium. Studies indicate that an oyster-rich diet may lower your cholesterol and triglyceride levels.

Caution: Buy commercially harvested oysters for your stew. Since oysters are susceptible to "red tide" microorganisms, check with your local public health department before you harvest your own. Also, people with gout may want to avoid oysters, which are high in purines. *Allergy alert:* Fish occasionally causes allergic reactions.

Curiosity: Scientists have discovered that when oysters spawn during the summer months, they produce an excessive amount of glycogen, which gives them an undesirable taste and texture. Traditionally, people avoid eating oysters during May, June, July, and August.

Preparation: A classic oyster stew calls for a pint or so of undrained oysters, 1 cup of milk, and ½ cup of cream. Traditionally, makers of this heavy-style stew will add a few tablespoons of butter, a dash of paprika, and a stalk of chopped cooked celery.

A Better Idea: Replace the cream with milk and skip the butter.

Vegetables

ALFALFA SPROUTS

Serving: $\frac{1}{2}$ *cup, raw*

Calories	5
Fat	0.1 g
Saturated	trace
Monounsaturated	trace
Polyunsaturated	trace
Calories from fat	22%
Cholesterol	0
Sodium	1 mg
Protein	0.7 g
Carbohydrate	0.6 g
Dietary fiber	N/A

CHIEF NUTRIENTS

NUTRIENT	AMOUNT	% RDA
Folate	5.9 mcg	3

Strengths: Low in fat and calories, with a little folate.

Caution: Raw legumes like alfalfa contain a substance that inhibits trypsin, a digestive enzyme needed to digest protein. Also, uncooked alfalfa contains a factor that counteracts vitamin E and may damage the liver or muscles. Because of these potentially harmful factors, you should eat only alfalfa that has been sprouted or cooked. (If you're eating a large quantity of sprouts, simmer or steam them for about 3 minutes.)

Description: Crisp, tender, delicate sprouts of alfalfa seeds.

Serve: In soups, salads, and sandwiches. They may also be sautéed, steamed, or served in stir-fries.

AMARANTH LEAVES

Serving: $\frac{1}{2}$ *cup, boiled*

Calories	14
Fat	0.1 g
Saturated	trace
Monounsaturated	trace
Polyunsaturated	trace
Calories from fat	8%
Cholesterol	0
Sodium	14 mg
Protein	1.4 g
Carbohydrate	2.7 g
Dietary fiber	N/A

CHIEF NUTRIENTS

NUTRIENT	AMOUNT	% RDA
Vitamin C	27.1 mg	45
Folate	37.5 mcg	19
Vitamin A	182.8 RE	18
Calcium	137.9 mg	17
Iron	1.5 mg	15
Potassium	423.1 mg	11
Magnesium	36.2 mg	10
Vitamin B$_6$	0.1 mg	6
Riboflavin	0.1 mg	5

Strengths: Amaranth is a powerhouse of low-fat, low-calorie nutrition: $\frac{1}{2}$ cup of cooked greens provides almost one-half the RDA for infection-fighting vitamin C and almost one-fifth the RDA for calcium and wound-healing vitamin A. Contains fair to good amounts of folate, riboflavin, vitamin B$_6$, and iron, all blood builders; also magnesium and potassium, both helpful in lowering blood pressure.

Curiosity: Once considered a simple weed in the U.S.; grows along roadsides.

History: Cultivated in the Mediterranean as a vegetable, but after 1700 used only for pig food; hence its other name: pigweed.

Description: Delicious, slightly sweet flavor.

Storage: The leaves wilt quickly. Soak in ice water to keep crisp—and use as soon as possible.

Preparation: Stir-fry or steam. To preserve nutrients, cook quickly with as little water as possible.

Serve: With baked ham, lamb, pork chops, or fish, along with tiny spring potatoes, just as you would turnip or spinach greens.

ARTICHOKE HEARTS

Serving: ¹/₂ cup, boiled

Calories	42
Fat	0.1 g
Saturated	trace
Monounsaturated	trace
Polyunsaturated	trace
Calories from fat	3%
Cholesterol	0
Sodium	80 mg
Protein	2.9 g
Carbohydrate	9.4 g
Dietary fiber	4.4 g

CHIEF NUTRIENTS

NUTRIENT	AMOUNT	% RDA
Folate	42.8 mcg	21
Magnesium	50.4 mg	14
Vitamin C	8.4 mg	14
Iron	1.1 mg	11
Potassium	297.4 mg	8

Strengths: A good source of folate, vitamin C, and iron, all of which help boost immunity. Artichokes are an excellent choice for women who take birth control pills, which rob the body of folate and vitamin C. The vegetable is also a decent source of magnesium and has a small amount of potassium, two minerals that help control blood pressure. It has few calories and an infinitesimal amount of fat—as long as you don't drench the soft, fleshy leaves in oil, butter, or rich sauce.

Caution: Canned or bottled artichoke hearts may have added salt. If you're on a low-sodium diet, rinse before serving to flush away salt.

Curiosity: Brought to the Americas by French and Spanish explorers. Castroville, California, is known as the Artichoke Capital of the World.

Serve: As an appetizer or side dish with fish, meat, or poultry. Good hot or cold, with lemon juice or other seasoning.

ASPARAGUS

Serving: ¹/₂ cup, boiled

Calories	23
Fat	0.3 g
Saturated	trace
Monounsaturated	trace
Polyunsaturated	trace
Calories from fat	11%
Cholesterol	0
Sodium	4 mg
Protein	2.3 g
Carbohydrate	4.0 g
Dietary fiber	1.5 g

CHIEF NUTRIENTS

NUTRIENT	AMOUNT	% RDA
Folate	88.3 mcg	44
Vitamin C	24.4 mg	41
Potassium	279.0 mg	7
Vitamin A	74.7 RE	7
Vitamin B$_6$	0.1 mg	7
Iron	0.6 mg	6
Riboflavin	0.1 mg	6
Thiamine	0.1 mg	6

Strengths: An excellent source of blood-building folate and infection-fighting vitamin C. Also supplies small amounts of iron, potassium, and the B vitamin riboflavin, which helps the absorption of other B vitamins and is vital for mental balance. Naturally low in sodium. Also low in fat and calories. Great for weight watchers—if you skip the butter and hollandaise sauce.

Caution: Asparagus contains purines, substances that may contribute to gout. Anyone susceptible to gout attacks should eat asparagus sparingly. Some people notice a pungent odor when they urinate within a few hours of eating asparagus; this is due to the high sulfur content.

Curiosity: A member of the Lily family, asparagus was a favorite of the French monarch Louis XIV.

Selection: Most abundant in spring, from April through June. Choose stalks with fresh, firm, compact tips.

Serve: With steamed mushrooms and lemon juice. Blanched and chilled asparagus is delicious in salads. It's also good in omelets or quiche, or added to stir-fries. Makes a satisfying, velvety soup!

BAMBOO SHOOTS, Canned

Serving: 1 cup slices, drained

Calories	25
Fat	0.5 g
Saturated	trace
Monounsaturated	trace
Polyunsaturated	trace
Calories from fat	19%
Cholesterol	0
Sodium	9 mg
Protein	2.3 g
Carbohydrate	4.2 g
Dietary fiber	3.9 g

CHIEF NUTRIENTS

NUTRIENT	AMOUNT	% RDA
Vitamin B$_6$	0.2 mg	9
Zinc	0.9 mg	6

Strengths: Bamboo shoots are low in calories, fiber-filled, and have practically no fat. They supply some zinc, which has many benefits: It is necessary for speedy wound healing, improves proper blood clotting, and increases resistance to infection. The vitamin B$_6$ in bamboo shoots is one of the B vitamins needed to maintain health of the skin and nervous system.

Caution: Raw bamboo shoots contain poisonous amounts of cyanogens, which cause cyanide poisoning. Therefore, they must be thoroughly cooked before eating. Canned shoots should be safe, as they have already been cooked to dissipate the poison.

Description: These tender, crisp, sweet shoots of a number of different edible species of bamboo are usually canned. However, fresh shoots are sometimes found in Asian grocery stores.

Serve: In stir-fries, with chopped green peppers and other vegetables. (Note: Serving size is usually smaller than ½ cup.)

BEETS

Serving: ½ cup slices, boiled

Calories	26
Fat	trace
Saturated	trace
Monounsaturated	trace
Polyunsaturated	trace
Calories from fat	1%
Cholesterol	0
Sodium	42 mg
Protein	0.9 g
Carbohydrate	5.7 g
Dietary fiber	1.4 g

CHIEF NUTRIENTS

NUTRIENT	AMOUNT	% RDA
Folate	45.2 mcg	23
Magnesium	31.5 mg	9
Vitamin C	4.7 mg	8
Potassium	265.2 mg	7
Iron	0.5 mg	5

Strengths: Beet lovers get a healthy dollop of folate, which helps prevent fatigue and depression. Beets also supply fair amounts of magnesium and potassium for better blood pressure, along with some vitamin C and iron for healthy blood and stronger immunity. Because beets contain vitamin C, the iron in them is readily absorbed. And their fiber—1.4 g per serving—lowers cholesterol and prevents constipation. Despite their red pigment, beet roots aren't high in beta-carotene as the greens are.

Caution: People who eat a lot of beets may be in for a surprise—their urine may turn red. This condition, called beeturia, is harmless.

Selection: Whether you cook them fresh or get beets from the can, nutrient values are nearly the same.

Serve: As a side dish, pickled, or as baby food. Borscht, the classic Russian beet soup, may be served hot or cold—but serve it with plain nonfat yogurt instead of sour cream to avoid fat while adding a little bone-building calcium to the dish.

BROCCOLI

Serving: ½ cup, chopped, boiled

Calories	22
Fat	0.3 g
Saturated	trace
Monounsaturated	trace
Polyunsaturated	trace
Calories from fat	11%
Cholesterol	0
Sodium	20 mg
Protein	2.3 g
Carbohydrate	4.0 g
Dietary fiber	2.0 g

CHIEF NUTRIENTS

NUTRIENT	AMOUNT	% RDA
Vitamin C	58.2 mg	97
Folate	39.0 mcg	20
Vitamin A	108.4 RE	11
Iron	0.7 mg	7
Potassium	227.8 mg	6
Vitamin B$_6$	0.1 mg	6
Magnesium	18.8 mg	5
Riboflavin	0.1 mg	5

Strengths: Broccoli is a star member of the cruciferous family, which is believed to help protect against cancer. It's a super source of vitamin C that's needed for healthy blood and stronger immunity. Respectable amounts of other immune-boosting nutrients are also present. A good choice for women on oral contraceptives, broccoli helps replenish folate and vitamin B$_6$. It's also a great fiber source, for lower cholesterol, control of diabetes, and better digestion.

Curiosity: Broccoli's name comes from the Latin word *brachium,* for arm or branch.

Storage: Keep broccoli cool to prevent yellowing.

Serve: With browned garlic, sliced red peppers, or pimientos. Or serve raw, chilled, with yogurt dip. Goes well with ocean fish and seafood and is delicious in soups, salads, omelets, and side dishes.

A Better Idea: Microwave cooking saves nutrients.

BRUSSELS SPROUTS

Serving: ½ cup, boiled

Calories	30
Fat	0.4 g
Saturated	trace
Monounsaturated	trace
Polyunsaturated	trace
Calories from fat	12%
Cholesterol	0
Sodium	16 mg
Protein	2.0 g
Carbohydrate	6.8 g
Dietary fiber	3.4 g

CHIEF NUTRIENTS

NUTRIENT	AMOUNT	% RDA
Vitamin C	48.4 mg	81
Folate	46.8 mcg	23
Iron	0.9 mg	9
Potassium	247.3 mg	7
Vitamin B$_6$	0.1 mg	7
Vitamin A	56.2 RE	6
Thiamine	0.1 mg	5

Strengths: Brussels sprouts are a rich source of vitamin C, with a good amount of folate; both these ingredients are important for strengthening immunity. And sprouts belong to the cancer-fighting cruciferous family. With high fiber, they help control diabetes, lower cholesterol levels, and contribute to weight loss. A serving of brussels sprouts has as much or more fiber than two slices of whole grain bread!

Caution: Brussels sprouts give some people gas.

Origin: Developed near their namesake city of Brussels, Belgium.

Selection: Look for firm, compact, bright green buds.

Serve: With ocean fish and seafood. But skip the traditional, high-fat sauces like hollandaise, and season instead with basil, caraway seed, dill, mustard seed, sage, or thyme.

CABBAGE, *Celery*

Serving: ½ cup, shredded, boiled

Calories	8
Fat	0.1 g
Saturated	trace
Monounsaturated	trace
Polyunsaturated	trace
Calories from fat	11%
Cholesterol	0
Sodium	5 mg
Protein	0.9 g
Carbohydrate	1.4 g
Dietary fiber	1.0 g

CHIEF NUTRIENTS

NUTRIENT	AMOUNT	% RDA
Folate	31.8 mcg	16
Vitamin C	9.4 mg	16
Vitamin A	57.7 RE	6
Vitamin B$_6$	0.1 mg	5

Strengths: This close relative of Chinese cabbage is a good source of folate and vitamin C. Women who take birth control pills may need more folate, and pregnant women require folate for healthy babies. Vitamin C performs a dozen or more valuable functions, including bolstering immunity, fighting infections, and speeding wound healing. Celery cabbage, like other cruciferous vegetables, may protect against some types of cancer.

Curiosity: Also known as *pe-tsai*.

Description: Resembles Romaine lettuce.

Selection: Choose cabbage heads that look fresh and green.

Preparation: Cut the cabbage into wedges before cooking. Or shred raw cabbage to make into slaw.

CABBAGE, *Chinese*

Serving: ½ cup, shredded, boiled

Calories	10
Fat	0.1 g
Saturated	trace
Monounsaturated	trace
Polyunsaturated	trace
Calories from fat	12%
Cholesterol	0
Sodium	29 mg
Protein	1.3 g
Carbohydrate	1.5 g
Dietary fiber	1.4 g

CHIEF NUTRIENTS

NUTRIENT	AMOUNT	% RDA
Vitamin C	22.1 mg	37
Vitamin A	218.5 RE	22
Folate	34.5 mcg	17
Calcium	79.1 mg	10
Iron	0.9 mg	9
Potassium	315.4 mg	8
Vitamin B$_6$	0.1 mg	7

Strengths: With Chinese cabbage in a meal, you're getting a good supply of calcium for strong bones, some iron for healthy blood, and some potassium. A respectable source of dietary fiber, Chinese cabbage also has a good supply of vitamin A for better night vision and folate, which is essential to immune system function. A member of the cruciferous family, it may protect against some types of cancer. It has virtually no fat or calories. Dieters can eat as much as they want!

Eat with: Beans and grain, to enhance absorption of their plant iron.

Curiosity: Also known as *bok choy* or *pak-choi*.

Description: The leaf resembles spinach and the stalk looks like celery. Milder in flavor than other cabbages.

Storage: Wilts more rapidly than head cabbage; use quickly.

Preparation: Wash the leaves and stem and cut into pieces to add to soup, stew, or stir-fries.

Serve: Chopped and stir-fried with other vegetables, meats, poultry, or seafood.

CABBAGE, *Common*

Serving: ½ cup, shredded, boiled

Calories	16
Fat	0.2 g
Saturated	trace
Monounsaturated	trace
Polyunsaturated	trace
Calories from fat	11%
Cholesterol	0
Sodium	14 mg
Protein	0.7 g
Carbohydrate	3.6 g
Dietary fiber	1.8 g

CHIEF NUTRIENTS

NUTRIENT	AMOUNT	% RDA
Vitamin C	18.2 mg	30
Folate	15.2 mcg	8

Strengths: An honored member of the cruciferous (mustard) family, praised for its ability to protect against certain types of cancer. Cabbage is also a rich source of vitamin C, which plays a role against cancer, as well as boosts immunity in general. This vegetable boasts a fair amount of folate, a B vitamin needed during pregnancy. And cabbage contributes a respectable amount of fiber—more than the average store-bought bran muffin.

Eat with: Beans and grains: Vitamin C in cabbage enhances absorption of nonheme iron in legumes.

Caution: Cabbage may cause intestinal gas.

Curiosity: Grows well in cold climates. Popular in Austria, Germany, Poland, and Russia.

Selection: Look for hard, firm heads.

Storage: One great advantage of close-head–type cabbage is that it will keep for months if refrigerated.

Preparation: May be microwaved, steamed, or stir-fried.

Serve: In salads (like coleslaw).

CABBAGE, *Red*

Serving: ½ cup, shredded, boiled

Calories	16
Fat	0.2 g
Saturated	trace
Monounsaturated	trace
Polyunsaturated	trace
Calories from fat	9%
Cholesterol	0
Sodium	6 mg
Protein	0.8 g
Carbohydrate	3.5 g
Dietary fiber	1.8 g

CHIEF NUTRIENTS

NUTRIENT	AMOUNT	% RDA
Vitamin C	25.8 mg	43
Vitamin B$_6$	0.1 mg	6

Strengths: Like common green cabbage, its close relative, red cabbage seems to defend against some types of cancer. And it's even richer in vitamin C than its pale green cousin. Red cabbage also supplies some vitamin B$_6$.

Eat with: Beans and grains: The vitamin C in cabbage enhances absorption of plant iron in legumes.

Caution: *Allergy alert:* Reactions to cabbage are rare, but some evidence suggests cabbage allergy may be on the rise.

Curiosity: The odor of cooking cabbage intensifies if it's overcooked.

Storage: Keeps for months under refrigeration.

Serve: Baked, with sliced apples. When it's shredded or chopped, raw red cabbage makes an attractive addition to green salads.

CABBAGE, *Savoy*

Serving: ½ cup, shredded, boiled

Calories	18
Fat	0.1 g
Saturated	trace
Monounsaturated	trace
Polyunsaturated	trace
Calories from fat	4%
Cholesterol	0
Sodium	18 mg
Protein	1.3 g
Carbohydrate	4.0 g
Dietary fiber	2.3 g

CHIEF NUTRIENTS

NUTRIENT	AMOUNT	% RDA
Vitamin C	12.4 mg	21
Folate	33.8 mcg	17
Vitamin A	65.0 RE	7
Vitamin B_6	0.1 mg	6

Strengths: A fancy-looking member of the cruciferous family, savoy is among the vegetables that seem to protect against certain types of cancer. It's a great source of fiber, which relieves constipation, helps lower cholesterol levels and control blood sugar, and also protects against colon cancer. The folate and vitamins A and C team up for rich, red blood and stronger immunity. Savoy also lends some vitamin B_6 to the diet, which may help relieve some cases of carpal tunnel syndrome. It supplies complex carbohydrates, recommended for dieters and diabetics.

Eat with: Plant sources of iron, such as beans and grains, to enhance absorption of the mineral.

Caution: May cause gas in some people.

Preparation: Chop fine, then boil or steam.

CARROT

Serving: 1 (about 2½ oz.), raw

Calories	31
Fat	0.1 g
Saturated	trace
Monounsaturated	trace
Polyunsaturated	trace
Calories from fat	4%
Cholesterol	0
Sodium	25 mg
Protein	0.7 g
Carbohydrate	7.3 g
Dietary fiber	2.3 g

CHIEF NUTRIENTS

NUTRIENT	AMOUNT	% RDA
Vitamin A	2,025.4 RE	203
Vitamin C	6.7 mg	11
Potassium	232.6 mg	6
Vitamin B_6	0.1 mg	6
Folate	10.1 mcg	5

Strengths: Just about the best source of vitamin A you'll find, the carrot helps protect against cancer and boost immunity. Its vitamin A also plays a vital role in improving night vision. A carrot supplies enough fiber to help lower cholesterol, control diabetes, speed weight loss, and help fight digestive system cancers. The extra fiber can reduce the risk of hemorrhoids. The iron in a carrot is fairly well absorbed by the body—which is unusual for iron supplied by vegetables. Few nutrients are lost in cooking, and cooked carrot is easier to digest than raw.

Caution: *Allergy alert:* If you're allergic to birch pollen, carrots may make your mouth itch.

Curiosity: Carrots were originally purple. Europeans developed orange varieties in the early 17th century. During World War II, the British developed high beta-carotene carrots to help their aviators see better at night.

Serve: With chopped apples, raisins, and raw cabbage; in salads, stews, or carrot cake.

CASSAVA

Serving: 3½ oz., raw

Calories	120
Fat	0.4 g
Saturated	trace
Monounsaturated	trace
Polyunsaturated	trace
Calories from fat	3%
Cholesterol	0
Sodium	8 mg
Protein	3.1 g
Carbohydrate	26.9 g
Dietary fiber	N/A

CHIEF NUTRIENTS

NUTRIENT	AMOUNT	% RDA
Vitamin C	48.2 mg	80
Iron	3.6 mg	36
Potassium	764.0 mg	20
Magnesium	66.0 mg	19
Thiamine	0.2 mg	15
Vitamin B$_6$	0.3 mg	15
Calcium	91.0 mg	11
Folate	22.1 mcg	11
Niacin	1.4 mg	7
Riboflavin	0.1 mg	6

Strengths: Better known as tapioca, cassava is a rich source of iron, which converts beta-carotene into vitamin A. The super-high level of vitamin C in this root vegetable enhances iron absorption, boosts immunity, heals wounds, fights infections, and helps resist cancer. Since cassava has good supplies of calcium, magnesium, and potassium, it's also helpful in keeping blood pressure within healthy limits. In addition, cassava is a decent source of calcium for people who avoid dairy products.
Curiosity: Also known as manioc or yucca.
Origin: South America.
Description: The starchy tuber of a tropical shrub. It looks like a long, narrow sweet potato.
Serve: In tapioca pudding made with skim milk to contribute bone-building calcium. It's also good with picante sauce.

CAULIFLOWER

Serving: 3 florets (2 oz.), raw

Calories	13
Fat	0.1 g
Saturated	trace
Monounsaturated	trace
Polyunsaturated	trace
Calories from fat	4%
Cholesterol	0
Sodium	8 mg
Protein	1.1 g
Carbohydrate	2.8 g
Dietary fiber	1.3 g

CHIEF NUTRIENTS

NUTRIENT	AMOUNT	% RDA
Vitamin C	40.0 mg	67
Folate	37.0 mcg	19
Vitamin B$_6$	0.1 mg	7
Potassium	198.8 mg	5

Strengths: A pricey member of the cancer-fighting cabbage family, cauliflower is an excellent source of vitamin C, the nutrient that helps fight infection, improve iron absorption, and prevent anemia. It's strong in blood-building folate, too. The less common purple varieties contain some vitamin A, which helps maintain nerve cell sheaths. Because it's extremely low in calories and sodium and has decent amounts of fiber, cauliflower is a dieter's delight. Cooking hardly depletes the nutrients at all.
Caution: Cauliflower contains purines and should be eaten sparingly by those with gout.
Selection: Buy this delicacy on sale or at farmer's markets. Avoid spotted, speckled, or bruised heads.
Preparation: Steaming or boiling cauliflower and other cruciferous vegetables may destroy some of the cancer-fighting substances they contain, so eating cauliflower raw is a prudent idea.

CELERIAC ⬇ 🏛

Serving: 3¹/₂ oz., boiled

Calories	25
Fat	0.2 g
Saturated	N/A
Monounsaturated	N/A
Polyunsaturated	N/A
Calories from fat	7%
Cholesterol	0
Sodium	61 mg
Protein	1.0 g
Carbohydrate	5.9 g
Dietary fiber	1.8 g

CHIEF NUTRIENTS

NUTRIENT	AMOUNT	% RDA
Vitamin C	3.6 mg	6

Strengths: Celeriac's chief virtue is its fair vitamin C content, helpful for boosting resistance and building healthier blood. Vitamin C also acts as an antioxidant, blocking harmful oxidation in cellular substances. For dieters, celeriac offers a nice low-calorie, fat-free change of pace; in fact, anyone who wants to put a little pizzazz in their menu may want to try this vegetable. With some fiber, celeriac is filling but not fattening, and it's an excellent source of complex carbohydrates for blood sugar control.

Curiosity: Celeriac is sometimes called celery root, knob celery, or German celery.

Description: A parsley-scented type of celery characterized by a bulbous, gnarly, edible root.

Preparation: Peel the tuber, then cut into strips or slices and boil, bake, or stew. Blanching in saltwater and lemon juice may be necessary to counteract celeriac's slightly bitter taste.

Serve: Pureed. Adds body to mashed potatoes, pea soup, and stews. Or grate raw into salads. It's also excellent in poultry stuffing.

CELERY ⬇ 🏛

Serving: ¹/₂ cup, diced, raw

Calories	10
Fat	0.1 g
Saturated	trace
Monounsaturated	trace
Polyunsaturated	trace
Calories from fat	8%
Cholesterol	0
Sodium	52 mg
Protein	0.5 g
Carbohydrate	2.2 g
Dietary fiber	1.0 g

CHIEF NUTRIENTS

NUTRIENT	AMOUNT	% RDA
Folate	16.8 mcg	8
Vitamin C	4.2 mg	7

Strengths: Celery is a fair source of folate, which is important for the breakdown of amino acids, and it also contains some vitamin C. The vegetable is practically calorie-free, and has a fair amount of fiber to help prevent constipation, lower cholesterol, and control blood sugar.

Caution: *Allergy alert:* Celery can cause anaphylactic shock in some people. People who are allergic to mugwort, birch pollen, or fennel may also react to celery. Symptoms may range from itching in the mouth to hives, swelling of the larynx, runny nose, watery eyes, and asthma. Some people may develop a rash if they go out in the sun after eating celery, especially if the stalks are old or diseased.

Origin: Developed in the Mediterranean from wild celery.

Serve: With stewed tomatoes or stir-fried in vegetable-and-poultry combos. For a snack or side dish, try stuffing the uncooked stalk with peanut butter or pureed cooked beans. Use as a garnish for vegetable juice. Stalks, nutrient-rich leaves, and seeds also make tasty additions to the soup pot.

CHICORY

Serving: ½ cup, chopped, raw

Calories	21
Fat	0.3 g
Saturated	trace
Monounsaturated	trace
Polyunsaturated	trace
Calories from fat	12%
Cholesterol	0
Sodium	41 mg
Protein	1.5 g
Carbohydrate	4.2 g
Dietary fiber	3.6 g

CHIEF NUTRIENTS

NUTRIENT	AMOUNT	% RDA
Folate	98.6 mcg	49
Vitamin A	360.0 RE	36
Vitamin C	21.6 mg	36
Calcium	90.0 mg	11
Potassium	378.0 mg	10
Iron	0.8 mg	8
Magnesium	27.0 mg	8
Riboflavin	0.1 mg	5

Strengths: Chicory is a very good source of folate—which is a plus if you take aspirin, certain antacids, or lipid-lowering drugs which can deplete stores of that B vitamin. It's also rich in resistance-building vitamins A and C. Chicory is a good source of calcium for building strong bones and teeth, and it has some potassium to help lower blood pressure. Its magnesium works with potassium as a muscle relaxant. This is a low-calorie addition to weight-loss diets.

Curiosity: Related to the pricier leaf vegetables like radicchio and Belgian endive. Roasted and ground, chicory roots are sometimes used as a coffee substitute—a popular beverage in Creole kitchens.

Serve: Combined with other milder-tasting forms of lettuce; also with chopped apples or with diced beets, and in bean soups or minestrone. Can be used instead of collards or other greens.

CHIVES

Serving: 1 Tbsp., chopped, raw

Calories	1
Fat	trace
Saturated	trace
Monounsaturated	trace
Polyunsaturated	trace
Calories from fat	24%
Cholesterol	0
Sodium	0.2 mg
Protein	0.1 g
Carbohydrate	0.1 g
Dietary fiber	0.1 g

CHIEF NUTRIENTS

(None of the nutrients meet or exceed 5% of the RDA.)

Strengths: A member of the genus *Allium*, chives are among the vegetables that may guard against stomach cancer. A single tablespoon contains a very small amount of vitamin C, which helps improve the health of bones, teeth, skin, and blood vessels.

Curiosity: Chives can be found growing wild as far north as the Arctic Circle.

Description: Bright green chives resemble super-skinny scallions, to which they're related. They have a delicate onion flavor.

Storage: Fresh chives should be used soon after cutting, as they'll only keep for a few days in the refrigerator. If you want to store them longer, chop the leaves, place them in a plastic bag, and freeze immediately.

Serve: With low-fat cottage cheese or plain nonfat yogurt. Chives go well in potato salad and omelets; they're also good with low-fat cheese spread and salad dressing.

CORN, *Sweet Yellow*

Serving: kernels from 1 ear, boiled

Calories	83
Fat	1.0 g
Saturated	0.2 g
Monounsaturated	0.3 g
Polyunsaturated	0.5 g
Calories from fat	11%
Cholesterol	0
Sodium	13 mg
Protein	2.6 g
Carbohydrate	19.3 g
Dietary fiber	2.9 g

CHIEF NUTRIENTS

NUTRIENT	AMOUNT	% RDA
Folate	35.7 mcg	18
Thiamine	0.2 mg	11
Vitamin C	4.8 mg	8
Magnesium	24.6 mg	7
Niacin	1.2 mg	7
Potassium	191.7 mg	5

Strengths: Corn is a good source of folate, which you may need if you're on certain medications—including oral contraceptives, aspirin, colestipol, and some antacids. It also offers good thiamine, which might be deficient if you're drinking several cups of coffee a day. Corn has good fiber and contributes a little niacin, magnesium, and vitamin C as well. Nutrient values for white and yellow corn are equal; fiber values may differ.
Eat with: Beans and other legumes. Like wheat, rice, oats, and barley, corn is a grain and forms a complete protein when eaten with legumes.
Caution: *Allergy alert:* Corn is a common food allergen and shows up in many prepared foods. Read labels.
Curiosity: One ear of corn may contain up to 1,000 kernels, which always grow in even-numbered rows. (Count 'em.)

CUCUMBER

Serving: ½ (about 5 oz.), raw

Calories	20
Fat	0.2 g
Saturated	trace
Monounsaturated	trace
Polyunsaturated	trace
Calories from fat	9%
Cholesterol	0
Sodium	3 mg
Protein	0.8 g
Carbohydrate	4.4 g
Dietary fiber	1.5 g

CHIEF NUTRIENTS

NUTRIENT	AMOUNT	% RDA
Vitamin C	7.1 mg	12
Folate	20.9 mcg	10
Potassium	224.3 mg	6

Strengths: Low in calories, with practically no fat, cucumbers have a bit of fiber that can help reduce the risk of hemorrhoids and may assist in protecting against colon cancer.
Caution: Cucumbers that have been waxed for shipment or storage should be peeled before they're eaten. *Allergy alert:* Like other members of the melon family, raw cucumbers have ingredients that might make your mouth itch if you're allergic to pollen. And cukes may trigger a reaction in those who are allergic to aspirin.
Selection: Look for cucumbers that are firm, fresh, bright, and medium- to dark-green in color. Shriveled cukes are usually tough and bitter.
Serve: With the skin on. Top with acidic dressings like vinegar, or low-fat mayonnaise or yogurt.

EGGPLANT ⬇ 🗑

Serving: ½ cup, cubed, boiled

Calories	13
Fat	0.1 g
Saturated	trace
Monounsaturated	trace
Polyunsaturated	trace
Calories from fat	7%
Cholesterol	0
Sodium	1 mg
Protein	0.4 g
Carbohydrate	3.2 g
Dietary fiber	1.2 g

CHIEF NUTRIENTS

NUTRIENT	AMOUNT	% RDA
Folate	13.8 mcg	7
Potassium	238.1 mg	6

Strengths: Like celeriac, eggplant is filling but not fattening, with only a small percentage of calories from fat. It's a staple in countries where meat is used sparingly. Eggplant is a nice source of potassium, which tends to flush excess sodium from the body, thereby helping to lower blood pressure over the long haul. (Potassium is also essential for healthy nerves and muscles.) Eggplant contributes some folate to the diet, which is helpful if you're pregnant or taking birth control pills.

Origin: India. Later domesticated in China, then taken to Africa by Arabs and later introduced to Italy.

Selection: Flesh should be springy. Avoid soft, flabby, shriveled eggplant—it may taste bitter.

Preparation: Eggplant tends to soak up oil like a sponge, so avoid sautéing in oil. You're better off steaming or baking this vegetable to avoid adding fat and calories to your diet. Cook thoroughly.

Serve: With Mediterranean meats, vegetables, and seasonings. Goes well with lamb, zucchini, tomatoes, chick-peas, garlic, onions, and parsley. It is used in moussaka, curries, eggplant parmigiana, and ratatouille.

ENDIVE ⬇ 🗑

Serving: ½ cup, chopped, raw

Calories	4
Fat	0.1 g
Saturated	trace
Monounsaturated	N/A
Polyunsaturated	trace
Calories from fat	11%
Cholesterol	0
Sodium	6 mg
Protein	0.3 g
Carbohydrate	0.8 g
Dietary fiber	0.5 g

CHIEF NUTRIENTS

NUTRIENT	AMOUNT	% RDA
Folate	35.5 mcg	18
Vitamin A	51.3 RE	5

Strengths: This leafy green is a good source of folate (consuming too little of this B vitamin weakens immunity). Endive contains some beta-carotene, which the body converts into resistance-building vitamin A. Also low in calories—great for people watching their weight.

Caution: *Allergy alert*: For some people, handling endive may cause a skin rash, but this is rare.

Curiosity: Related to chicory and escarole. Old-time herbalists believed endive's slightly bitter taste stimulated the flow of digestive juices, but little research has been done on the subject.

Description: There are two varieties, broad-leafed and curly, with yellow or light green hearts. Stronger tasting than lettuce.

Storage: Wilted heads can be freshened in cold water.

Serve: Raw in sandwiches or in salads with other greens. Add to stir-fries, soups, or stews. For use in a side dish, it can be steamed and seasoned with lemon juice.

GARDEN CRESS

Serving: ½ cup, raw

Calories	8
Fat	0.2 g
Saturated	trace
Monounsaturated	trace
Polyunsaturated	trace
Calories from fat	20%
Cholesterol	0
Sodium	4 mg
Protein	0.7 g
Carbohydrate	1.4 g
Dietary fiber	0.3 g

CHIEF NUTRIENTS

NUTRIENT	AMOUNT	% RDA
Vitamin C	17.3 mg	29
Vitamin A	232.5 RE	23
Folate	20.1 mcg	10

Strengths: A respectable source of vitamins C and A. Vitamin C helps fight infections, heals wounds, and may lower the risk of stomach cancer. Vitamin A maintains healthy mucous membranes that line the lungs, stomach, and urinary tract. (People who take cholesterol-lowering drugs may need more vitamin A, and individuals who take aspirin may need to replenish vitamin C.) Garden cress also contains small amounts of some B vitamins and minerals. Like other leafy greens, it is low in calories. As a cabbage relative, cress may slow the spread of cancer.

Curiosity: Sometimes called field cress. Similar to watercress, a marsh plant.

Serve: Like alfalfa sprouts to add distinctive flavor to salads and sandwiches. May also be used to flavor soups, or added to sauces for fish dishes. Use on ocean fish and seafood as a vitamin C–rich garnish.

GARLIC

Serving: 1 clove, raw

Calories	4
Fat	trace
Saturated	trace
Monounsaturated	trace
Polyunsaturated	trace
Calories from fat	3%
Cholesterol	0
Sodium	0.5 mg
Protein	0.2 g
Carbohydrate	1.0 g
Dietary fiber	trace

CHIEF NUTRIENTS

(None of the nutrients meet or exceed 5% of the RDA.)

Strengths: Garlic supplies a trace of vitamin B_6, valuable for the manufacture of heme, the iron-containing part of blood hemoglobin. But garlic's chief virtue, according to clinical studies, is its ability to lower elevated cholesterol levels. A natural blood thinner, it may help protect against clogged arteries. Garlic also seems to act as a mild antibiotic and digestive aid. Some research indicates that people who relish garlic may get less stomach and colorectal cancer.

Caution: Because garlic thins the blood, people who take aspirin or other blood-thinning drugs daily, or who have a blood disorder, should not use garlic capsules. Also, some people experience gastric discomfort if they go too heavy on the garlic.

Curiosity: Related to onions, scallions, and leeks.

Serve: Minced into salads, in pasta sauce, or on garlic bread. Suggestion: Rinse your mouth with fresh water after eating garlic to wash out the pungent oils. Or chew peppermint or wintergreen gum to keep your breath socially acceptable.

GINGER ROOT

Serving: 1 Tbsp., raw

Calories	4
Fat	0.1 g
Saturated	trace
Monounsaturated	trace
Polyunsaturated	trace
Calories from fat	10%
Cholesterol	0
Sodium	1 mg
Protein	0.1 g
Carbohydrate	0.9 g
Dietary fiber	0.1 g

CHIEF NUTRIENTS

(None of the nutrients meet or exceed 5% of the RDA.)

Strengths: Although ginger contains negligible amounts of nutrients per se, it's valuable as a flavor enhancer. By grating fresh ginger on top, you'll add pizzazz to chicken, fish, and vegetables (mainstays of a healthful diet). Ginger seems to relieve motion sickness: As little as 1 gram can reduce motion-induced nausea and vomiting, according to one scientific study. Ginger may also help relieve heartburn and get rid of stomach gas.

Curiosity: Powdered ginger is spicy and intense, with a slightly musty flavor, while the fresh root is sweet and tangy.

Description: A tan, knobby root with yellowish flesh. Sold in the produce section.

Selection: Choose fresh ginger; avoid shriveled-looking roots.

Preparation: Peel and grate, mash or slice thinly.

Serve: In stir-fries. By adding ginger, you cut down on the need for salty condiments like soy sauce. (To maximize flavor, stir-fry ginger in well-heated oil for 1 minute before adding other ingredients.)

GREENS, *Beet*

Serving: 1/2 cup, boiled

Calories	19
Fat	0.1 g
Saturated	trace
Monounsaturated	trace
Polyunsaturated	trace
Calories from fat	6%
Cholesterol	0
Sodium	174 mg
Protein	1.9 g
Carbohydrate	3.9 g
Dietary fiber	2.1 g

CHIEF NUTRIENTS

NUTRIENT	AMOUNT	% RDA
Vitamin A	367.2 RE	37
Vitamin C	17.9 mg	30
Potassium	654.5 mg	17
Iron	1.4 mg	14
Magnesium	49.0 mg	14
Riboflavin	0.2 mg	12
Calcium	82.1 mg	10
Folate	10.3 mcg	5
Thiamine	0.1 mg	5

Strengths: A treasure trove of valuable nutrients, beet greens are an unusually rich source of vitamin A, which may reduce the risk of cancer of the breast, lung, and colon. Also a very good source of vitamin C, associated with enhanced immunity and reduced risk of stomach cancer. They have appreciable amounts of the blood pressure–lowering minerals magnesium and potassium, and their good dose of iron helps prevent anemia. As a bonus, you also get some folate, fiber, and plentiful riboflavin in a food that has very few calories.

Eat with: Foods high in vitamin B_6, like carrot juice. B_6 reduces the amount of oxalic acid excreted into the urine, helping to prevent new kidney stones from forming.

Caution: People who tend to develop oxalate-containing kidney stones should eat beet greens sparingly, as they contain higher-than-average amounts of oxalates, a nonnutrient substance.

Serve: In soups. Or add uncooked beet greens to salads.

A Better Idea: If you're looking for iron, eat beet roots instead of greens: The iron from the root is absorbed better.

GREENS, *Collard*

Serving: ½ cup, chopped, boiled

Calories	17
Fat	0.1 g
Saturated	N/A
Monounsaturated	N/A
Polyunsaturated	N/A
Calories from fat	6%
Cholesterol	0
Sodium	10 mg
Protein	0.9 g
Carbohydrate	3.9 g
Dietary fiber	1.3 g

CHIEF NUTRIENTS

NUTRIENT	AMOUNT	% RDA
Vitamin A	174.7 RE	17
Vitamin C	7.7 mg	13

Strengths: Collards are a healthy source of vitamin A, which reduces susceptibility to infections. They also have a good supply of vitamin C, which may be depleted in those who take aspirin, tetracycline, or oral contraceptives. Like their close relative, kale, collards qualify as a cancer-fighting vegetable. And they have practically no calories!

Curiosity: The ancient Romans erroneously believed that if they ate collards before a banquet, they wouldn't get befuddled by wine.

Serve: With ocean fish or seafood. Collard greens are also good with black-eyed peas, sweet potatoes, and cornbread. They can also be used as a soup green or a side dish.

GREENS, *Dandelion*

Serving: ½ cup, chopped, boiled

Calories	17
Fat	0.3 g
Saturated	N/A
Monounsaturated	N/A
Polyunsaturated	N/A
Calories from fat	16%
Cholesterol	0
Sodium	23 mg
Protein	1.0 g
Carbohydrate	3.3 g
Dietary fiber	1.5 g

CHIEF NUTRIENTS

NUTRIENT	AMOUNT	% RDA
Vitamin A	608.4 RE	61
Vitamin C	9.4 mg	16
Calcium	72.8 mg	9
Iron	0.9 mg	9
Riboflavin	0.1 mg	5

Strengths: Dandelion greens are a superb source of beta-carotene, the plant form of vitamin A that promotes good night vision, keeps cell membranes healthy, and assists in immune reactions and bone remodeling. In addition, beta-carotene may reduce the risk of breast cancer. Since they're a good source of vitamin C, dandelion greens work as an infection-fighter and wound-healer. They contribute iron, an essential component of hemoglobin in red blood cells, and calcium, which is associated with healthier blood pressure.

Caution: Don't be tempted to pluck dandelion greens growing along a roadway; they may contain dangerously high levels of lead from auto exhaust. Also, don't harvest dandelions from your lawn if it has been sprayed with herbicides.

Curiosity: Dandelion is derived from the French phrase *dent de lion,* which means ''lion's tooth''—an apt term considering the plant's saw-toothed leaves.

Selection: Look for fresh, young, tender, green leaves.

Serve: In soups. Or eat raw in salads, tartly flavored with acidic dressings such as lemon juice or vinaigrette.

GREENS, *Mustard*

Serving: ½ cup, chopped, boiled

Calories	11
Fat	0.2 g
Saturated	trace
Monounsaturated	trace
Polyunsaturated	trace
Calories from fat	15%
Cholesterol	0
Sodium	11 mg
Protein	1.6 g
Carbohydrate	1.5 g
Dietary fiber	1.4 g

CHIEF NUTRIENTS

NUTRIENT	AMOUNT	% RDA
Vitamin C	17.7 mg	30
Folate	51.4 mcg	26
Vitamin A	212.1 RE	21
Calcium	51.8 mg	6

Strengths: A very respectable source of vitamins C and A and folate, making this a good choice for those who want to build up their resistance. Mustard greens also supply a bit of calcium, which helps ward off osteoporosis. Another member of the family of cancer-fighting cruciferous vegetables.

Eat with: Beans and other vegetable sources of iron, to boost absorption. Mustard greens and black-eyed peas combine to make a high-fiber, high-nutrition combination.

Description: May be dark or light, smooth or curly (resembling collards). Most mustard greens sold in the U.S. have brilliant green oval leaves, with frilly edges and long stems. They taste sharp and pungent, much like prepared mustard.

Serve: In casseroles, soups, or stews; as a side dish with chopped onions; or raw in salads. Also good with ocean fish and seafood.

GREENS, *Turnip*

Serving: ½ cup, chopped, boiled

Calories	14
Fat	0.2 g
Saturated	trace
Monounsaturated	trace
Polyunsaturated	trace
Calories from fat	11%
Cholesterol	0
Sodium	21 mg
Protein	0.8 g
Carbohydrate	3.1 g
Dietary fiber	2.2 g

CHIEF NUTRIENTS

NUTRIENT	AMOUNT	% RDA
Folate	85.3 mcg	43
Vitamin A	396.0 RE	40
Vitamin C	19.7 mg	33
Calcium	98.6 mg	12
Vitamin B_6	0.1 mg	7
Iron	0.6 mg	6

Strengths: Like other leafy greens, turnip greens are chock-full of immunity-boosting vitamins (A, C, folate), with some iron and vitamin B_6. A superb source of dietary fiber, they're great for people who want to lose weight, prevent constipation, reduce risk of hemorrhoids, and lower cholesterol. In fact, cooked turnip greens contain more fiber per serving than some brands of breakfast cereal. And the extra helping of calcium helps to build strong bones. A member of the cruciferous family, turnip greens contain substances which seem to protect against some cancers.

Eat with: Beans or grains—their iron is more easily absorbed in the presence of vitamin C, which is readily supplied by turnip greens.

Selection: You can buy turnip greens with or without the roots.

Serve: In soups or as a side dish. Raw greens may be chopped finely and used in salads.

JERUSALEM ARTICHOKES

Serving: ½ cup slices, raw

Calories	57
Fat	trace
Saturated	trace
Monounsaturated	trace
Polyunsaturated	trace
Calories from fat	0.2%
Cholesterol	0
Sodium	3 mg
Protein	1.5 g
Carbohydrate	13.1 g
Dietary fiber	1.2 g

CHIEF NUTRIENTS

NUTRIENT	AMOUNT	% RDA
Iron	2.6 mg	26
Thiamine	0.2 mg	10
Potassium	321.8 mg	9
Folate	10.1 mcg	5
Niacin	1.0 mg	5

Strengths: One of the few vegetable sources of iron, a vital mineral that helps produce infection-fighting antibodies, synthesize collagen, and turn beta-carotene into vitamin A. (Symptoms of iron deficiency include fatigue, irritability, headaches, and tingling in the hands and feet.) Also a good source of thiamine, a B vitamin that helps convert carbohydrates into energy. Has fair amounts of folate and niacin. A lifelong intake of potassium-yielding foods like this can help control blood pressure. It has fewer calories than potatoes.

Eat with: Vitamin C–rich foods, like cauliflower, peppers, or citrus fruit, to enhance iron absorption.

Origin: Native to North America. "Jerusalem" is a corruption of *girasol,* Spanish for "turning toward the sun."

Serve: Grated or thinly sliced in salads. May be cooked, pureed, and served like mashed potatoes. Dried and ground, this tuber is used in breads and noodles; look for Jerusalem artichoke flour if you are allergic to wheat or other grains.

KALE

Serving: ½ cup, chopped, boiled

Calories	18
Fat	0.3 g
Saturated	trace
Monounsaturated	trace
Polyunsaturated	trace
Calories from fat	13%
Cholesterol	0
Sodium	29 mg
Protein	1.2 g
Carbohydrate	3.7 g
Dietary fiber	1.3 g

CHIEF NUTRIENTS

NUTRIENT	AMOUNT	% RDA
Vitamin C	34.1 mg	57
Iron	1.3 mg	13
Vitamin A	129.4 RE	13
Calcium	85.8 mg	11
Magnesium	37.1 mg	11
Potassium	178.1 mg	5

Strengths: May guard against cancer in several ways. Contains a bonus of vitamin C (over half the daily requirement) and a good serving of vitamin A. Like cabbage, it contains cancer-combating indoles. Kale is lower in oxalates than spinach, which is an advantage because the calcium it supplies is readily absorbed. Kale is also a good plant source of magnesium (which may be depleted in those taking thiazide diuretics) and iron.

Eat with: Meats, poultry, and fish to enhance iron absorption.

Curiosity: One of the first cabbages to be cultivated for food. Sometimes grown as an ornamental.

Serve: Steamed and seasoned with caraway or fennel seeds; stir-fried with pork, ginger, and garlic; or in soups with beans, barley, or potatoes. It can also be blanched and layered with pasta, as in lasagna.

KELP

Serving: 1 oz., raw

Calories	12
Fat	0.2 g
Saturated	trace
Monounsaturated	trace
Polyunsaturated	trace
Calories from fat	12%
Cholesterol	0
Sodium	65 mg
Protein	0.5 g
Carbohydrate	2.7 g
Dietary fiber	0.1 g

CHIEF NUTRIENTS

NUTRIENT	AMOUNT	% RDA
Folate	50.4 mcg	25
Magnesium	33.9 mg	10
Iron	0.8 mg	8
Calcium	47.0 mg	6

Strengths: A low-fat, no-cholesterol source of folate, kelp is a type of seaweed that assists in the breakdown of protein and the regeneration of red blood cells. It's also an excellent natural source of iodine, which is essential for the thyroid to produce hormones that regulate growth and development. Kelp contains small amounts of calcium, along with a good amount of magnesium, which holds calcium in the tooth enamel and aids in the prevention of tooth decay.

Curiosity: Almost a quarter of all food in Japan contains some variety of seaweed.

History: Through the ages, seaweed has been used to treat everything from constipation (and other gastric problems) to goiter.

Description: Kelp, the most common of all seaweeds, is brown in color and used to flavor Japanese soup stocks. Dried kelp, called kombu, is black, and can be wrapped around sushi or steeped to make a tea.

Preparation: All dried seaweeds should be soaked before use.

KOHLRABI

Serving: ½ cup slices, boiled

Calories	24
Fat	0.1 g
Saturated	trace
Monounsaturated	trace
Polyunsaturated	trace
Calories from fat	3%
Cholesterol	0
Sodium	17 mg
Protein	1.5 g
Carbohydrate	5.5 g
Dietary fiber	0.9 g

CHIEF NUTRIENTS

NUTRIENT	AMOUNT	% RDA
Vitamin C	44.3 mg	74
Potassium	278.8 mg	7
Vitamin B$_6$	0.1 mg	7

Strengths: Yet another superb source of vitamin C, which heals wounds, fights infections, and helps to form collagen, a major component of skin, teeth, and cartilage. Vitamin C also aids absorption of calcium. Kohlrabi supplies useful amounts of potassium, which is valuable for nerve and muscle function, and vitamin B$_6$. (Women who take birth control pills need to assure themselves of adequate supplies of B$_6$.) It's a member of the cancer-fighting cruciferous family.

Eat with: Black-eyed peas for a classic southern dish that's high in fiber and rich in vitamin C.

Description: The edible part of the kohlrabi is a fleshy stem that resembles a turnip but tastes sweeter and juicier.

Selection: Look for small or medium-sized kohlrabi; try to find bulbs that are smooth and free of cracks.

Serve: In stew or soup. May also be eaten raw in salad. The leaves are also edible.

LAMB'S-QUARTERS

Serving: ½ cup, chopped, boiled

Calories	29
Fat	0.6 g
Saturated	trace
Monounsaturated	trace
Polyunsaturated	trace
Calories from fat	20%
Cholesterol	0
Sodium	26 mg
Protein	2.9 g
Carbohydrate	4.5 g
Dietary fiber	1.9 g

CHIEF NUTRIENTS

NUTRIENT	AMOUNT	% RDA
Vitamin A	873.0 RE	87
Vitamin C	33.3 mg	56
Calcium	232.2 mg	29
Riboflavin	0.2 mg	14
Vitamin B$_6$	0.2 mg	8
Potassium	259.2 mg	7
Folate	12.2 mcg	6
Iron	0.6 mg	6
Magnesium	20.7 mg	6
Thiamine	0.1 mg	6

Strengths: A nutritious but little-used leafy green, lamb's-quarters are decidedly rich in vitamin A, which assists the immune system in various ways and may reduce the risk of breast cancer. They're also an excellent source of vitamin C (which enhances the immune system) and a fair to good source of hard-to-get B vitamins. In addition to riboflavin, lamb's-quarters provide a bit of iron, magnesium, and potassium for healthy blood pressure.

Eat with: Foods high in heme iron, such as lean beef, to maximize absorption of nonheme iron (the plant form of iron).

Caution: Like their cousin, spinach, lamb's-quarters boast a higher-than-average amount of calcium. But also like spinach, they're high in oxalates, substances that bind to calcium and prevent its absorption—so people who take medication for oxalate-containing kidney stones should eat the vegetable sparingly.

Curiosity: Also called goosefoot or pigweed.

LAVER

Serving: 1 oz., raw

Calories	10
Fat	0.1 g
Saturated	trace
Monounsaturated	trace
Polyunsaturated	trace
Calories from fat	7%
Cholesterol	0
Sodium	13 mg
Protein	1.6 g
Carbohydrate	1.4 g
Dietary fiber	N/A

CHIEF NUTRIENTS

NUTRIENT	AMOUNT	% RDA
Folate	41.0 mcg	20
Vitamin C	10.9 mg	18
Vitamin A	145.6 RE	15
Riboflavin	0.1 mg	7
Iron	0.5 mg	5

Strengths: This nutritious variety of seaweed is rich in folate. It also has good amounts of vitamins A and C, which are important to the body's immune system. Beta-carotene, the vitamin A precursor in laver, may reduce the risk of several kinds of cancer; vitamin C helps in the absorption of calcium and iron. In addition, laver is low in sodium, calories, fat, and cholesterol.

Caution: Laver is abundant in iodine, which could be harmful in high doses over an extended period.

Description: A red seaweed, closely related to nori, that comes in dried, pressed, paper-thin sheets.

Preparation: Soak in cold water before using in soups or wrapping around rice for sushi. Toasted laver or nori can be used as a garnish or eaten as a snack. Plain laver can be toasted by passing it over an open flame a few times.

LEEKS

Serving: ½ cup, chopped, boiled

Calories	16
Fat	0.1 g
Saturated	trace
Monounsaturated	trace
Polyunsaturated	trace
Calories from fat	6%
Cholesterol	0
Sodium	5 mg
Protein	0.4 g
Carbohydrate	4.0 g
Dietary fiber	1.3 g

CHIEF NUTRIENTS

NUTRIENT	AMOUNT	% RDA
Iron	0.6 mg	6
Folate	12.6 mcg	6

Strengths: Leeks are a member of the genus *Allium*. A report in the *Journal of the National Cancer Institute* suggests allium vegetables may help protect against stomach cancer. Leeks supply some fiber, which may help prevent colon cancer. And like onions, leeks may have a cholesterol-lowering effect.
Curiosity: This veggie is the national emblem of Wales.
Description: A leek roughly resembles an oversize scallion, but it tastes milder. Its stalk is about an inch in diameter, with flat rather than tubular leaves.
Preparation: Split lengthwise and wash well to rinse away grit that collects within the leaves and stalk.
Serve: In quiche or potato soup, or as a side dish. Raw leeks may be chopped and mixed with low-fat cottage cheese.

LETTUCE, *Butterhead*

Serving: 1 cup, shredded, raw

Calories	5
Fat	0.1 g
Saturated	trace
Monounsaturated	trace
Polyunsaturated	trace
Calories from fat	15%
Cholesterol	0
Sodium	2 mg
Protein	0.5 g
Carbohydrate	1.0 g
Dietary fiber	0.4 g

CHIEF NUTRIENTS

NUTRIENT	AMOUNT	% RDA
Folate	29.9 mcg	15
Vitamin C	3.3 mg	5
Vitamin A	39.5 RE	4

Strengths: Butterhead lettuce supplies a respectable amount of the B vitamin folate and some vitamin C for stronger immunity. It has a little fiber, but like most lettuce, butterhead's chief virtue is what it doesn't have—fat, sodium, or calories.
Eat with: Nutrient-rich vegetables like carrots, leafy greens, raw broccoli or cauliflower, and red cabbage for a salad bowl chock full of healing nutrients. Top with a yogurt dressing for some bone-building calcium.
Curiosity: A member of the sunflower family.
History: Etchings on tombs indicate that ancient Egyptians raised lettuce. But lettuce production didn't boom until the development of refrigerated railroad cars and storage facilities in the early 1900s.
Description: Less sturdy than iceberg lettuce, butterhead has loose, soft leaves.
Selection: Often sold as Bibb or Boston lettuce.

LETTUCE, *Iceberg*

Serving: 1 cup, shredded, raw

Calories	18
Fat	0.3 g
Saturated	trace
Monounsaturated	trace
Polyunsaturated	trace
Calories from fat	13%
Cholesterol	0
Sodium	12 mg
Protein	1.4 g
Carbohydrate	2.8 g
Dietary fiber	1.4 g

CHIEF NUTRIENTS

NUTRIENT	AMOUNT	% RDA
Folate	75.5 mcg	38
Vitamin C	5.3 mg	9
Iron	0.7 mg	7
Potassium	213.0 mg	6

Strengths: This sturdy salad green supplies a rich amount of folate, which can help bolster your immune system. It also has some vitamin C, iron, and potassium. But keep in mind that you have to eat the equivalent of a cupful (shredded) of this fairly dense vegetable to reap any nutritional benefits. One or two leaves don't have much to offer.

Caution: *Allergy alert:* Gardeners, beware! Handling lettuce sometimes triggers hives in sensitive people.

Selection: Look for clean, crisp, firm heads, with no brown spots or streaks (leaf rust). Iceberg is also known as crisphead lettuce.

Storage: Refrigerate in the crisper drawer or in a tightly closed plastic bag.

Preparation: Iceberg lettuce is a good choice for a Greek salad, which can be served as an entrée.

LETTUCE, *Looseleaf*

Serving: 1 cup, shredded, raw

Calories	10
Fat	0.2
Saturated	trace
Monounsaturated	trace
Polyunsaturated	trace
Calories from fat	14%
Cholesterol	0
Sodium	5 mg
Protein	0.7 g
Carbohydrate	2.0 g
Dietary fiber	1.0 g

CHIEF NUTRIENTS

NUTRIENT	AMOUNT	% RDA
Vitamin C	10.1 mg	17
Folate	27.9 mcg	14
Vitamin A	106.0 RE	11
Iron	0.8 mg	8

Strengths: Looseleaf boasts good amounts of resistance-building vitamins A and C and folate, along with a fair amount of iron. Considering its benefits, this attractive leafy green is a better choice for small salads than iceberg. Looseleaf lettuce is succulent and filling, but not fattening as long as you use a low- or reduced-fat dressing.

Selection: Often sold as oakleaf (red or green) or salad bowl lettuce. The darker the leaves, the higher the vitamin A content. Avoid bunches that are discolored or otherwise defective.

Preparation: For best results, rinse just before using. If leaves are dried in a salad spinner, dressing clings better, so a little bit goes further and helps to keep calories down.

Serve: Braised, or added to soups.

LETTUCE, *Romaine*

Serving: 1 cup, shredded, raw

Calories	9
Fat	0.1 g
Saturated	trace
Monounsaturated	trace
Polyunsaturated	trace
Calories from fat	12%
Cholesterol	0
Sodium	4 mg
Protein	0.9 g
Carbohydrate	1.3 g
Dietary fiber	1.0 g

CHIEF NUTRIENTS

NUTRIENT	AMOUNT	% RDA
Folate	76.0 mcg	38
Vitamin C	13.4 mg	22
Vitamin A	145.6 RE	15
Iron	0.6 mg	6

Strengths: Romaine is the reigning monarch of salad greens when it comes to nutrition. It's very high in folate (eating too little folate can cause fatigue, depression, confusion, anemia, and low resistance to infections). In addition, it has plentiful vitamin C. And romaine boasts ample amounts of beta-carotene, which may reduce the risk of lung and colon cancer. So next time you find yourself dining in a smoky restaurant, order a Caesar salad with its abundance of Romaine!

Curiosity: Also called cos lettuce. Probably the oldest form of cultivated lettuce. The Roman emperor Augustus Caesar, who believed that this vegetable rescued him from illness, erected a statue in its honor.

Preparation: Chop and cook like Swiss chard.

Serve: With low-fat or fat-free dressing.

MUSHROOMS

Serving: ½ cup pieces, boiled

Calories	21
Fat	0.4 g
Saturated	trace
Monounsaturated	trace
Polyunsaturated	trace
Calories from fat	16%
Cholesterol	0
Sodium	2 mg
Protein	1.7 g
Carbohydrate	4.0 g
Dietary fiber	1.7 g

CHIEF NUTRIENTS

NUTRIENT	AMOUNT	% RDA
Niacin	3.5 mg	18
Iron	1.0 mg	14
Riboflavin	0.2 mg	14
Folate	14.2 mcg	7
Potassium	277.7 mg	7
Vitamin C	3.1 mg	5

Strengths: Homely but nutritious, mushrooms supply respectable amounts of fiber and complex carbohydrates—so they're a super choice for dieters, diabetics, and anyone else who wants to add a healthy bonus to a meal. Order your pizza with mushrooms, for example, and you get a good dose of iron and niacin for healthy red blood cells, and riboflavin (associated with a sharp memory). Mushrooms also contain nice amounts of resistance-building folate and vitamin C, with some blood pressure–lowering potassium. If you eat the oriental favorite, shiitake mushrooms, you'll get some zinc and vitamin B_6. Mushrooms may also combat illness and heart disease: Like garlic, they may have some powerful antiviral, antitumor, and anticholesterol powers.

Eat with: Garlic, for increased disease resistance.

Caution: Mushrooms contain purines and should be eaten sparingly by people with gout. Also, don't eat wild mushrooms. Many innocent-looking fungi resemble cultivated mushrooms but are highly poisonous.

Curiosity: There are about 38,000 species of mushrooms.

OKRA

Serving: ½ cup slices, boiled

Calories	26
Fat	0.1 g
Saturated	trace
Monounsaturated	trace
Polyunsaturated	trace
Calories from fat	5%
Cholesterol	0
Sodium	4 mg
Protein	1.5 g
Carbohydrate	5.8 g
Dietary fiber	2.6 g

CHIEF NUTRIENTS

NUTRIENT	AMOUNT	% RDA
Vitamin C	13.0 mg	22
Folate	36.6 mcg	18
Magnesium	45.6 mg	13
Vitamin B$_6$	0.2 mg	8
Potassium	257.6 mg	7
Thiamine	0.1 mg	7
Calcium	50.4 mg	6

Strengths: A substantial source of dietary fiber. Pregnant women and women on birth control pills would benefit from the extra folate supplied by this vegetable. Also a decent source of vitamin B$_6$, which helps your body process cholesterol and may help relieve some cases of carpal tunnel syndrome. Provides a significant amount of magnesium, along with some calcium and potassium to help keep blood pressure under control.
Curiosity: Also called lady's finger. Thrives in the southern U.S. and in tropical areas.
Selection: Look for young, tender, fresh pods.
Preparation: Cook in stainless steel or enamel. Brass, copper, or iron utensils may discolor the vegetable (although this isn't harmful).
Serve: With stewed tomatoes. In vegetable soup or in stews with meat, fish, poultry, and/or crab (a combination called gumbo).

ONIONS

Serving: ½ cup, chopped, boiled

Calories	30
Fat	0.1 g
Saturated	trace
Monounsaturated	trace
Polyunsaturated	trace
Calories from fat	4%
Cholesterol	0
Sodium	2 mg
Protein	0.9 g
Carbohydrate	6.9 g
Dietary fiber	1.3 g

CHIEF NUTRIENTS

NUTRIENT	AMOUNT	% RDA
Vitamin C	5.1 mg	9
Folate	15.3 mcg	8

Strengths: Even a ½-cup serving supplies some fiber to help relieve constipation, prevent hemorrhoids, and stave off colon cancer. Onions, like other allium vegetables, may also help protect against stomach cancer. And they contribute useful amounts of folate, which is beneficial for pregnant women. Researchers have reason to think that eating fresh onions (and garlic) regularly may protect against heart attack.
Caution: In some people, the stomach lining may become sensitive to certain substances in onions, causing gastric discomfort.
Serve: Sautéed with other vegetables, or raw in salads.

PARSLEY

☉ ⚝ ⚚ ⬇ ▥

Serving: ½ cup, chopped, raw

Calories	10
Fat	0.1 g
Saturated	N/A
Monounsaturated	N/A
Polyunsaturated	N/A
Calories from fat	8%
Cholesterol	0
Sodium	12 mg
Protein	0.7 g
Carbohydrate	2.1 g
Dietary fiber	1.3 g

CHIEF NUTRIENTS

NUTRIENT	AMOUNT	% RDA
Vitamin C	27.0 mg	45
Folate	54.9 mcg	27
Iron	1.9 mg	19
Vitamin A	156.0 RE	16

Strengths: One of the most nutritious garnishes, parsley can help you meet the daily requirement for vitamin C. It's definitely recommended for anyone taking tetracycline, which can deplete vitamin C. Parsley supplies ample amounts of folate, iron, and vitamin A (for immunity), and it also acts as a breath freshener.

Caution: *Allergy alert:* May cause photodermatitis in sensitive individuals.

Curiosity: There are two varieties, curly and flat-leafed.

Origin: Related to carrots, celery, dill, fennel, parsnips, and caraway, parsley is native to the Mediterranean.

Selection: Look for fresh, bright-green sprigs, free of dirt or yellow leaves.

Storage: Slightly wilted parsley can be revived in cold water. Dried parsley is not as flavorful as fresh.

Serve: Finely chopped and added as a garnish to casseroles, cheese spreads, quiches, stir-fried dishes, salads, soups, stews, poultry stuffing, and fish or meat sauces. (Parsley is a common ingredient in *fines herbes,* a mix of herbs used in omelets and sauces.)

PARSNIPS

☉ ⚝ ⚅ ⚚ ⬇ ▥

Serving: ½ cup slices, boiled

Calories	63
Fat	0.2 g
Saturated	trace
Monounsaturated	trace
Polyunsaturated	trace
Calories from fat	3%
Cholesterol	0
Sodium	8 mg
Protein	1.0 g
Carbohydrate	15.2 g
Dietary fiber	3.8 g

CHIEF NUTRIENTS

NUTRIENT	AMOUNT	% RDA
Folate	45.4 mcg	23
Vitamin C	10.1 mg	17
Potassium	286.3 mg	8
Magnesium	22.6 mg	6

Strengths: A very good supply of complex carbohydrates and fiber makes this vegetable an excellent choice for diabetics. Eating parsnips can also help you lose weight, lower your cholesterol, and improve your digestive health. Parsnips are a good source of folate (a plus for pregnant women) and vitamin C (which acts as an antioxidant and cancer preventive). Smaller amounts of magnesium and potassium contribute to good blood pressure control.

Caution: *Allergy alert*: Like celery, parsnips may trigger a photosensitive reaction in some people due to photosensitizing substances (called psoralens) that are naturally present in the plant. Handling moist parsnips, followed by exposure to sunlight, can trigger a rash.

Curiosity: Related to celery and parsley.

Serve: In soups and stews. May also be pureed and whipped with skim milk (for a bonus serving of bone-building calcium).

PEPPERS, *Chili*

Serving: 1 Tbsp., raw

Calories	4
Fat	trace
Saturated	trace
Monounsaturated	trace
Polyunsaturated	trace
Calories from fat	3%
Cholesterol	0
Sodium	0.7 mg
Protein	0.2 g
Carbohydrate	0.9 g
Dietary fiber	0.1 g

CHIEF NUTRIENTS

NUTRIENT	AMOUNT	% RDA
Vitamin C	22.7 mg	38
Vitamin A	100.8 RE	10

Strengths: Chili peppers help increase metabolism, burn calories, and speed weight loss. They also block the formation of cancer-causing compounds in cured meats. And if you want to ease pain caused by shingles, just munch on a few chili peppers. The super healing powers of chili peppers are attributable to capsaicin, the substance that makes hot peppers hot. (They contain a good amount of vitamin C, too.) Of course, they're too hot to eat in significant quantities, but why not make them a regular addition to spice up entrées?

Eat with: Milk, yogurt, rice, bread, or hard candy to cool down inflamed taste buds. Oil-based foods and starches also have a soothing after-effect. (Water doesn't work!)

Caution: Wash your hands with soap and water after handling hot chili peppers to avoid getting capsaicin in your eyes. Or wear rubber gloves.

Origin: Mexico.

Serve: With Mexican food. Chili is the key constituent of Tabasco sauce, cayenne pepper, red pepper flakes, chili powder, and chili sauce.

PEPPERS, *Green Bell*

Serving: ½ cup, chopped, raw

Calories	14
Fat	0.1 g
Saturated	trace
Monounsaturated	trace
Polyunsaturated	trace
Calories from fat	7%
Cholesterol	0
Sodium	1 mg
Protein	0.5 g
Carbohydrate	3.2 g
Dietary fiber	0.8 g

CHIEF NUTRIENTS

NUTRIENT	AMOUNT	% RDA
Vitamin C	44.7 mg	74
Folate	11.0 mcg	6
Vitamin B$_6$	0.1 mg	6

Strengths: A fantastic source of vitamin C, which enhances iron absorption when you eat other plant foods like beans and grains that have nonheme iron. Vitamin C also boosts calcium absorption, a big plus for men and women prone to weak bones. Along with vitamin C, green bell peppers supply some folate and vitamin B$_6$. For women taking birth control pills, this vegetable resupplies some of the B vitamins that may be depleted by oral contraceptive use. And peppers have some fiber to keep bowel movements regular, reduce the risk of hemorrhoids, and stave off digestive woes like diverticulosis.

Eat with: Iron-rich vegetables and legumes.

Curiosity: May also be red, yellow, orange, purple, or gold.

Selection: Avoid soft, shriveled, or limp peppers.

Preparation: For an entrée, stuff peppers with rice or bread crumbs and meat, poultry, or mashed lentils, then bake.

PEPPERS, *Red Bell*

Serving: ½ cup, chopped, raw

Calories	14
Fat	0.1 g
Saturated	trace
Moncunsaturated	trace
Polyunsaturated	trace
Calories from fat	7%
Cholesterol	0
Sodium	1 mg
Protein	0.5 g
Carbohydrate	3.2 g
Dietary fiber	0.8 g

CHIEF NUTRIENTS

NUTRIENT	AMOUNT	% RDA
Vitamin C	95.0 mg	158
Vitamin A	285.0 RE	29
Folate	11.0 mcg	6
Vitamin B$_6$	0.1 mg	6

Strengths: The caviar of vegetables, sweet red peppers are even higher in vitamin C than the green variety. And they're a good source of vitamin A, which speeds wound healing and helps boost resistance in various ways. Like the green kind, red peppers contribute a respectable amount of folate and vitamin B$_6$. Easy on the waistline, too!

Eat with: Iron-rich vegetables and legumes.

Curiosity: In 1494, the ship's doctor sailing with Columbus alerted Spanish authorities to the potential of this New World vegetable. The Spanish called it ''pimienta.''

Selection: Look for firm, shiny, bright-colored peppers.

Serve: In salads and on crudité platters with yogurt dip for a bit of calcium. Cooking doesn't seem to deplete peppers' nutrients, so you can use sweet peppers generously in stir-fries and on pizza. Or dice them and cook with corn or rice.

POTATO

Serving: 1 (about 7 oz.), baked

Calories	220
Fat	0.2 g
Saturated	trace
Monounsaturated	trace
Polyunsaturated	trace
Calories from fat	1%
Cholesterol	0
Sodium	16 mg
Protein	4.7 g
Carbohydrate	51.0 g
Dietary fiber	2.2 g

CHIEF NUTRIENTS

NUTRIENT	AMOUNT	% RDA
Vitamin C	26.1 mg	43
Vitamin B$_6$	0.7 mg	35
Iron	2.8 mg	28
Potassium	844.4 mg	23
Niacin	3.3 mg	18
Magnesium	54.5 mg	16
Thiamine	0.2 mg	15
Folate	22.2 mcg	11

Strengths: The high iron in potatoes is fairly well absorbed when this nutrient-rich vegetable is part of your meal. A potato has almost twice as much potassium as a banana, good news for anyone on thiazide diuretics. And they're a terrific source of complex carbohydrates and fiber, which help control blood sugar in diabetics. (A baked potato with skin has as much fiber as a ⅓-cup serving of oat bran.) Potatoes make a satisfying side dish for dieters—as long as you don't top them with gobs of butter, sour cream, or cheese sauce. And of course, try to resist french fries or scalloped potatoes, which tend to be higher in fat, sodium, and calories.

Caution: Pare away green areas, and discard heavily sprouted spuds—they can make you sick. *Allergy alert:* Some people with asthma are severely allergic to dried potatoes, hash browns, or french fries treated with sulfites.

Preparation: Microwaved potatoes retain nutrients well. And contrary to popular belief, peeling potatoes does not strip away their vitamin C content.

Serve: Topped with salsa, plain nonfat yogurt, low-fat cottage cheese, or herbs and butter-flavor granules.

PUMPKIN, *Canned*

Serving: ½ cup

Calories	41
Fat	0.3 g
Saturated	trace
Monounsaturated	trace
Polyunsaturated	trace
Calories from fat	7%
Cholesterol	0
Sodium	6 mg
Protein	1.3 g
Carbohydrate	9.9 g
Dietary fiber	3.4 g

CHIEF NUTRIENTS

NUTRIENT	AMOUNT	% RDA
Vitamin A	2,691.3 RE	269
Iron	1.7 mg	17
Vitamin C	5.1 mg	9
Folate	15.0 mcg	8
Magnesium	28.1 mg	8
Potassium	251.3 mg	7

Strengths: Pumpkin is one of the most abundant sources of beta-carotene, the vegetable pigment that the body turns into resistance-building vitamin A. This member of the gourd family also offers a little bit of protein, a hefty portion of fiber, and a healthy amount of iron. There's enough vitamin C to help boost iron absorption.

Eat with: Foods like turkey that contain heme iron to boost absorption of pumpkin's nonheme iron (the form supplied by plant foods).

Caution: Pumpkin is generally used in pie, which will inflate the calorie and fat content of each serving. When making pumpkin pie, use egg substitute and evaporated skim milk instead of whole eggs and cream to minimize fat and calorie count.

Description: Stronger flavored than squashes, which are its close relatives in the gourd family.

Preparation: In pumpkin bread, use whole grain flour and raisins for added fiber. Or make pumpkin soup, topped with yogurt for a little bone-building calcium.

PURSLANE

Serving: ½ cup, boiled

Calories	10
Fat	0.1 g
Saturated	N/A
Monounsaturated	N/A
Polyunsaturated	N/A
Calories from fat	10%
Cholesterol	0
Sodium	26 mg
Protein	0.9 g
Carbohydrate	2.0 g
Dietary fiber	N/A

CHIEF NUTRIENTS

NUTRIENT	AMOUNT	% RDA
Magnesium	38.9 mg	11
Vitamin A	107.3 RE	11
Vitamin C	6.1 mg	10
Potassium	283.1 mg	8
Calcium	45.2 mg	6

Strengths: Anyone who's looking for a new source of resistance-building, infection-fighting, wound-healing vitamins A and C should try purslane. This sturdy, cloverlike green is also a good source of magnesium, with a fair amount of calcium and potassium, which are valuable blood pressure–lowering minerals.

Caution: People who tend to develop oxalate-containing kidney stones should eat purslane sparingly, as it contains higher-than-average amounts of oxalates, a non-nutrient substance.

Origin: Greece and India.

Description: A fleshy plant with paddle-shaped leaves.

Preparation: Cook lightly like spinach.

Serve: With mashed potatoes. Like okra, purslane adds body to soups and stews. May also be eaten raw, in salads.

RADISHES

Serving: ½ cup slices, raw

Calories	10
Fat	0.3 g
Saturated	trace
Moncunsaturated	trace
Polyunsaturated	trace
Calories from fat	28%
Cholesterol	0
Sodium	14 mg
Protein	0.4 g
Carbohydrate	2.0 g
Dietary fiber	1.3 g

CHIEF NUTRIENTS

NUTRIENT	AMOUNT	% RDA
Vitamin C	13.2 mg	22
Folate	15.7 mcg	8

Strengths: More than just a pretty face on the relish tray, radishes are a member of the cancer-fighting cruciferous family. Above all, they are a good source of vitamin C to help fight colds and protect against environmental stress, along with some blood-nurturing folate. Since radishes are a fairly decent source of complex carbohydrates (with a barely noticeable number of calories and a harmless trace of fat), they are popular with dieters who get the munchies before dinner.

Curiosity: The largest radishes may grow up to 50 pounds in size.

Selection: If you want the crisp kind with a mild-flavored taste, look for radishes that are smooth, well-formed, and firm.

Serve: In stir-fries with oriental vegetables. Radish slices are a great stand-in for water chestnuts, which are usually canned and therefore high in sodium.

RHUBARB

Serving: 1 cup, diced, raw

Calories	26
Fat	0.2 g
Saturated	N/A
Monounsaturated	N/A
Polyunsaturated	N/A
Calories from fat	8%
Cholesterol	0
Sodium	5 mg
Protein	1.1 g
Carbohydrate	5.5 g
Dietary fiber	2.2 g

CHIEF NUTRIENTS

NUTRIENT	AMOUNT	% RDA
Vitamin C	10.0 mg	16
Calcium	105.0 mg	13
Potassium	351.4 mg	9

Strengths: The ruddy, fibrous stalks of this leafy plant contain a respectable amount of vitamin C—enough to help boost resistance—and a good amount of fiber for digestive health. Rhubarb also supplies a bit of potassium, useful for keeping blood pressure at a healthy level.

Caution: Despite its attributes, rhubarb has two strikes against it. Rhubarb is so tart that most people eat it with some kind of sweetener (usually sugar), which inflates the calorie count considerably. Also, the calcium it contains is in the form of calcium oxalate, which blocks absorption of calcium not only from rhubarb itself but also from any other food you eat at the same time. Rhubarb should also be avoided by people who tend to develop oxalate-containing kidney stones: The stalks contain extremely high amounts of oxalates. Rhubarb leaves contain even higher amounts—enough to be fatal, sometimes—and should *never* be consumed in any amount.

Serve: In pie, jam, or compote. The redder the stalk, the less sweetener you'll need in preparation.

RUTABAGAS

Serving: ¹/₂ cup cubes, boiled

Calories	29
Fat	0.2 g
Saturated	trace
Monounsaturated	trace
Polyunsaturated	trace
Calories from fat	5%
Cholesterol	0
Sodium	15 mg
Protein	0.9 g
Carbohydrate	6.6 g
Dietary fiber	2.0 g

CHIEF NUTRIENTS

NUTRIENT	AMOUNT	% RDA
Vitamin C	18.6 mg	31
Folate	13.2 mcg	7
Potassium	244.0 mg	7
Magnesium	17.9 mg	5

Strengths: A member of the cancer-fighting cruciferous family, rutabagas are rich in vitamin C, which aids calcium and iron absorption, boosts immunity, fights infections, and heals wounds. If you are taking aspirin for a chronic condition, or you have been prescribed a course of tetracycline, you may benefit from the vegetable, since both drugs may deplete the body of vitamin C. Rutabagas also provide a little folate (valuable to pregnant women). They have potassium and magnesium, both essential for healthy muscle and nerve tissue.

Caution: Be sure to remove any paraffin or wax that appears on the root.

Origin: A hybrid of turnips and kale.

Description: Large, yellow-fleshed tubers with a slightly sweet taste.

Serve: Mashed, like potatoes. (A mixture of mashed rutabaga and potato has fewer calories than mashed potatoes alone.) Good with fish and seafood.

SALSIFY

Serving: ¹/₂ cup slices, boiled

Calories	46
Fat	0.1 g
Saturated	N/A
Monounsaturated	N/A
Polyunsaturated	N/A
Calories from fat	2%
Cholesterol	0
Sodium	11 mg
Protein	1.9 g
Carbohydrate	10.5 g
Dietary fiber	2.1 g

CHIEF NUTRIENTS

NUTRIENT	AMOUNT	% RDA
Vitamin B$_6$	0.2 mg	8
Riboflavin	0.1 mg	7
Folate	10.3 mcg	5
Potassium	192.4 mg	5
Vitamin C	3.1 mg	5

Strengths: A nice way to add a little extra fiber, potassium, vitamin C, or B vitamins to your diet. A good source of complex carbohydrates for diabetics. Low in sodium, with a little protein.

Caution: May cause gas in some people. Start with a small amount.

Curiosity: Also known as oysterplant, salsify is related to dandelions, endive, and chicory. According to an old wives' tale, the juice of this plant is an antidote for snakebite.

Description: A taproot which may reach a foot or more in length. Tastes a little like oysters.

Preparation: Best if scrubbed whole before cooking, then peeled. Boil and eat like potatoes.

Serve: Steamed and prepared like carrots or parsnips. Salsify is a good addition to soups or stews. May also be sliced and eaten raw in salads, or with yogurt dip (for a bit of calcium.) The young, tender leaves are also edible.

SCALLIONS

Serving: ¹/₂ cup, chopped, raw

Calories	16
Fat	0.1 g
Saturated	trace
Monounsaturated	trace
Polyunsaturated	trace
Calories from fat	6%
Cholesterol	0
Sodium	8 mg
Protein	0.9 g
Carbohydrate	3.7 g
Dietary fiber	1.2 g

CHIEF NUTRIENTS

NUTRIENT	AMOUNT	% RDA
Folate	32.0 mcg	16
Vitamin C	9.4 mg	16
Iron	0.7 mg	7

Strengths: Higher in nutrients than leeks and onions, scallions are a good source of folate, which is essential during pregnancy, and resistance-building vitamin C. Research suggests that scallions—like onions, garlic, and other allium vegetables—may help protect against stomach cancer. And scallions contain some fiber to aid in digestive health.

Eat with: Dried beans, whole grains, and other vegetables that are sources of nonheme iron; the C in scallions should boost absorption of the plant iron.

Origin: The name comes from the Latin *Ascalonia,* for Ascalon, in ancient Judea, where onions were first grown.

Description: Scallions are actually young, undeveloped onions. They're sometimes called spring onions or green onions.

Serve: Chopped, in salads, soups, and stir-fries. Or add to non-fat yogurt to make a tasty dip for raw veggies like broccoli, cauliflower, cherry tomatoes, and mushrooms.

SHALLOTS

Serving: 1 Tbsp., chopped, raw

Calories	7
Fat	trace
Saturated	trace
Monounsaturated	trace
Polyunsaturated	trace
Calories from fat	1%
Cholesterol	0
Sodium	1 mg
Protein	0.3 g
Carbohydrate	1.7 g
Dietary fiber	0.2 g

CHIEF NUTRIENTS

NUTRIENT	AMOUNT	% RDA
Vitamin A	124.8 RE	12

Strengths: Even though shallots are usually eaten in small quantities, they contain a respectable amount of resistance-building vitamin A. And research suggests that allium vegetables like shallots may help protect against stomach cancer.

Description: A cluster of bulbs covered in brown, yellowish-brown, or reddish-brown parchmentlike skin, shallots are similar to garlic cloves. Their distinctive flavor is a cross between garlic and onions.

Selection: Look for firm, dry, unsprouted bulbs. Avoid soft, spongy specimens.

Preparation: Shallots should never be browned, as they turn bitter. Instead, simmer or sauté them gently.

Serve: Minced, as a superb low-sodium flavor enhancer for sauces, soups, and salad dressings.

SPINACH

Serving: ½ cup, boiled

Calories	21
Fat	0.2 g
Saturated	trace
Monounsaturated	trace
Polyunsaturated	trace
Calories from fat	10%
Cholesterol	0
Sodium	63 mg
Protein	2.7 g
Carbohydrate	3.4 g
Dietary fiber	2.0 g

CHIEF NUTRIENTS

NUTRIENT	AMOUNT	% RDA
Vitamin A	737.1 RE	74
Folate	131.2 mcg	66
Iron	3.2 mg	32
Magnesium	78.3 mg	22
Calcium	122.4 mg	15
Vitamin C	8.8 mg	15
Riboflavin	0.2 mg	12
Potassium	419.0 mg	11
Vitamin B_6	0.2 mg	11

Strengths: Spinach is considered a good source of iron, but less than 2% of it is bioavailable. That means that up to 98% of the iron in a serving can't be readily absorbed by the body. Spinach does contain useful amounts of magnesium, though, which may help prevent migraine headaches. And of course, spinach is a superb source of vitamin A and folate, with a decent amount of vitamin C and other B vitamins—all of which help maximize immunity. This leafy green also contains a good amount of blood pressure–lowering calcium and potassium, and ample fiber.

Caution: People with gout should eat spinach sparingly as it contains purines. And people who tend to develop oxalate-containing kidney stones should avoid spinach as it contains higher-than-average amounts of oxalates.

Preparation: Uncooked spinach needs to be washed thoroughly to remove grit. Delicious with grapefruit sections and sliced mushrooms in a tossed salad.

SQUASH, *Acorn*

Serving: ½ cup cubes, baked

Calories	57
Fat	0.1 g
Saturated	trace
Monounsaturated	trace
Polyunsaturated	trace
Calories from fat	2%
Cholesterol	0
Sodium	4 mg
Protein	1.1 g
Carbohydrate	14.9 g
Dietary fiber	2.9 g

CHIEF NUTRIENTS

NUTRIENT	AMOUNT	% RDA
Vitamin C	11.0 mg	18
Magnesium	43.9 mg	13
Potassium	445.7 mg	12
Thiamine	0.2 mg	11
Folate	19.1 mcg	10
Iron	1.0 mg	10
Vitamin B_6	0.2 mg	10
Calcium	44.9 mg	6

Strengths: Like other dark yellow vegetables, acorn squash is a good source of cancer-fighters like vitamin C and beta-carotene, the plant form of vitamin A. And it has a useful amount of iron for healthy blood. In addition, this squash has plenty of fiber and ample complex carbohydrates, the preferred source of calories for diabetics and weight watchers. Virtually free of fat and sodium, acorn squash supplies decent amounts of potassium and magnesium, which help control blood pressure and prevent strokes. Each serving has appreciable amounts of folate, thiamine, and vitamin B_6. Considering all these benefits, it's the perfect food for peak mental and physical health.

Curiosity: Related to pumpkins (but sweeter tasting).

Preparation: Can be cooked speedily in a microwave oven.

SQUASH, *Butternut*

Serving: ¹/₂ cup cubes, baked

Calories	41
Fat	0.1 g
Saturated	trace
Monounsaturated	trace
Polyunsaturated	trace
Calories from fat	2%
Cholesterol	0
Sodium	4 mg
Protein	0.9 g
Carbohydrate	10.7 g
Dietary fiber	2.9 g

CHIEF NUTRIENTS

NUTRIENT	AMOUNT	% RDA
Vitamin A	714.0 RE	71
Vitamin C	15.0 mg	26
Folate	20.0 mcg	10
Magnesium	30.0 mg	9
Potassium	290.0 mg	8
Vitamin B$_6$	0.1 mg	7
Iron	0.6 mg	6
Calcium	42.0 mg	5
Niacin	1.0 mg	5

Strengths: Roughly equal to acorn squash in nutritional content, the fiber-rich butternut variety has somewhat fewer calories. Merits a place on the menu for anyone who wants to boost intake of resistance-building vitamins A and C. Also contains a bit of iron, which teams with vitamin C to build healthy blood. Butternut squash supplies a little niacin, a B vitamin that plays a vital role in nutrient metabolism.
Origin: Mexico and Central America.
Description: Club-shaped, with buff-colored skin.
Preparation: Peel and cube. May also be cooked and eaten unpeeled. Microwaving can cut cooking time by two-thirds.
Serve: Topped with butter-flavor granules or maple syrup instead of butter or margarine. (This will keep fat intake under control.)

SQUASH, *Crookneck or Straightneck*

Serving: ¹/₂ cup slices, boiled

Calories	18
Fat	0.3 g
Saturated	trace
Monounsaturated	trace
Polyunsaturated	trace
Calories from fat	14%
Cholesterol	0
Sodium	0.9 mg
Protein	0.8 g
Carbohydrate	3.9 g
Dietary fiber	1.3 g

CHIEF NUTRIENTS

NUTRIENT	AMOUNT	% RDA
Folate	18.0 mcg	9
Vitamin C	5.0 mg	8
Magnesium	22.0 mg	6

Strengths: These summer squash are even lower in calories than winter squash (like acorn or butternut), with insignificant amounts of fat and sodium. In general, however, summer squash isn't nearly as rich in nutrients as winter squash. It has nice levels of folate: That's good news for women who are pregnant or taking oral contraceptives, because they need a good supply of this B vitamin. Summer squash contributes some vitamin C (which can help enhance absorption of iron from other foods) and magnesium (needed to convert vitamin D into its active form). And it also contains fiber, which plays a role in lowering cholesterol.
Description: Summer squash has soft rinds, and it's more watery than winter squash. May be yellow or green.
Serve: In soups and stir-fries. To boost the nutrient content, stuff with bread crumbs and lean meat and serve as an entrée.

SQUASH, Hubbard

Serving: ½ cup cubes, baked

Calories	51
Fat	0.6 g
Saturated	trace
Monounsaturated	trace
Polyunsaturated	trace
Calories from fat	11%
Cholesterol	0
Sodium	8 mg
Protein	3.0 g
Carbohydrate	11.0 g
Dietary fiber	2.9 g

CHIEF NUTRIENTS

NUTRIENT	AMOUNT	% RDA
Vitamin A	616.0 RE	62
Vitamin C	10.0 mg	16
Potassium	365.0 mg	10
Vitamin B$_6$	0.2 mg	9
Folate	17.0 mcg	8
Magnesium	22.0 mg	6
Thiamine	0.1 mg	5

Strengths: If you're dieting to lose weight and control your cholesterol—or if you're watching your blood sugar—consider hubbard squash for your menu. It's a useful source of fiber and complex carbohydrates. Like other orange-hued winter squash, hubbard is a super source of vitamin A, one of the premier cancer-fighting nutrients, and it boasts a healthy supply of vitamin C to help fight infections and heal wounds. This squash has nice amounts of folate, thiamine, and vitamin B$_6$. (B vitamins are essential to healthy skin and high energy levels.) Hubbard also supplies some magnesium and potassium, mighty minerals that help keep your blood flowing freely.

Curiosity: Related to cucumber and watermelon.

Description: Has a tough, warty, gourdlike rind.

Storage: Fresh hubbard squash keeps for several weeks. Once cooked, it should be refrigerated and used within several days.

Preparation: You may need a heavy hammer and meat cleaver to cut open this monster. Do not peel. Cut in wedges and steam, bake, or microwave. Microwaving saves preparation time.

SQUASH, Spaghetti

Serving: 1 cup, boiled/baked

Calories	45
Fat	0.4 g
Saturated	trace
Monounsaturated	trace
Polyunsaturated	trace
Calories from fat	8%
Cholesterol	0
Sodium	28 mg
Protein	1.0 g
Carbohydrate	10.0 g
Dietary fiber	2.2 g

CHIEF NUTRIENTS

NUTRIENT	AMOUNT	% RDA
Vitamin C	5.4 mg	9
Vitamin B$_6$	0.2 mg	8
Niacin	1.3 mg	7
Folate	12.4 mcg	6
Iron	0.5 mg	5
Potassium	181.4 mg	5

Strengths: Spaghetti squash is praiseworthy for what it doesn't have—fat, sodium, or calories. That makes it a filling but not fattening change of pace for dieters and diabetics. It's also a convenient pasta alternative for people who are allergic to wheat products.

Curiosity: Sometimes called vegetable spaghetti.

Description: Pale yellow and football-shaped. Bland in flavor compared to other winter squash. The golden strands of pulp look very much like spaghetti or angel hair pasta.

Storage: Uncut squash can be stored for weeks at room temperature. Once it has been cooked, use within 2 or 3 days.

Preparation: Cut in half lengthwise. Steam, bake, or microwave, then pick out the strands with a fork.

Serve: With pesto sauce or with garlic and herbs. Or top with tomato sauce to boost the levels of resistance-building vitamin C.

SQUASH, *Zucchini*

Serving: 1/2 cup slices, boiled

Calories	14
Fat	0.1 g
Saturated	trace
Monounsaturated	trace
Polyunsaturated	trace
Calories from fat	3%
Cholesterol	0
Sodium	3 mg
Protein	0.6 g
Carbohydrate	3.5 g
Dietary fiber	1.3 g

CHIEF NUTRIENTS

NUTRIENT	AMOUNT	% RDA
Folate	15.1 mcg	8
Vitamin C	4.1 mg	7
Magnesium	19.8 mg	6
Potassium	227.7 mg	6

Strengths: Zucchini is a summer squash with only a fraction of the nutritional value of the dark orange and yellow winter varieties such as butternut squash. It holds the honor of being one of the lowest-calorie foods in existence. Its respectable fiber content makes zucchini a great staple on weight watchers' menus. Also contributes some potassium and magnesium. And it has a little vitamin C and folate to help boost resistance.

Eat with: A stuffing of meat, bread crumbs, and seasonings, like chopped onions, garlic, and parsley. All these ingredients boost the nutritional value.

Selection: Look for small zucchini. Anything larger than a foot long will be watery and full of huge seeds.

Preparation: Do not peel (the skin is edible). And don't overcook zucchini; you'll find that it turns to mush.

Serve: Sliced into rounds or matchstick-shaped strips and added to stir-fries.

SWEET POTATO

Serving: 1 (about 4 oz.), baked

Calories	117
Fat	0.1 g
Saturated	trace
Monounsaturated	trace
Polyunsaturated	trace
Calories from fat	1%
Cholesterol	0
Sodium	11 mg
Protein	2.0 g
Carbohydrate	27.7 g
Dietary fiber	3.4 g

CHIEF NUTRIENTS

NUTRIENT	AMOUNT	% RDA
Vitamin A	2,488.0 RE	249
Vitamin C	28.0 mg	47
Vitamin B_6	0.3 mg	14
Folate	25.8 mcg	13
Potassium	396.7 mg	11
Riboflavin	0.1 mg	8
Magnesium	22.9 mg	7
Iron	0.5 mg	5
Thiamine	0.1 mg	5

Strengths: When it comes to vitamin A content, the sweet potato certainly beats most other vegetables. And it's also very high in vitamin C, supplying nearly half the daily requirement in just one serving. Canned sweet potatoes are nearly as high in nutrition as those cooked fresh. Both versions contain fairly respectable amounts of B vitamins, although fresh-cooked have more riboflavin, which is important for healthy blood cells. Both contain a respectable amount of protein. Best of all, a sweet potato is rich in fiber and complex carbohydrates, a combination that will help regulate blood sugar levels and sate the appetite of weight watchers. (If eating less than a whole sweet potato, count on proportionately fewer nutrients.)

Caution: *Drug Interaction:* People taking levodopa, an anti-parkinsonism drug, should not eat excess amounts of foods rich in pyridoxine (vitamin B_6). That includes sweet potatoes.

Curiosity: Not related to white potatoes.

Serve: Baked with apple slices or pineapple rings. Or mash like white potatoes.

SWISS CHARD

Serving: ½ cup, chopped, boiled

Calories	18
Fat	0.1 g
Saturated	N/A
Monounsaturated	N/A
Polyunsaturated	N/A
Calories from fat	4%
Cholesterol	0
Sodium	158 mg
Protein	1.7 g
Carbohydrate	3.6 g
Dietary fiber	1.8 g

CHIEF NUTRIENTS

NUTRIENT	AMOUNT	% RDA
Vitamin A	276.3 RE	28
Vitamin C	15.8 mg	26
Magnesium	75.7 mg	22
Iron	2.0 mg	20
Potassium	483.1 mg	13
Calcium	51.0 mg	6

Strengths: Quite a good source of vitamins A and C, star defenders against cancer. Also a good source of minerals like iron (to boost immunity and prevent fatigue) and calcium, magnesium, and potassium (valuable for good blood pressure control). As with spinach, the minerals in chard may not be readily absorbed unless you eat it with vitamin C–rich foods.

Eat with: Peppers or cauliflower, which have vitamin C to enhance iron absorption.

Caution: People who tend to develop oxalate-containing kidney stones should eat Swiss chard greens sparingly, as they contain higher-than-average amounts of oxalates, a nonnutrient substance.

Curiosity: A beet relative. No one is sure why this green is called *Swiss* chard.

Description: Looks like spinach, tastes like beet greens.

Selection: Look for fresh, turgid stalks with tender leaves. Avoid discolored stalks or yellowed leaves. (Chard is often available frozen like spinach.)

Serve: Combined with garlic and onions, or added to soups.

TOMATO

Serving: 1 (about 4 oz.), raw

Calories	26
Fat	0.4 g
Saturated	trace
Monounsaturated	trace
Polyunsaturated	trace
Calories from fat	14%
Cholesterol	0
Sodium	11 mg
Protein	1.1 g
Carbohydrate	5.7 g
Dietary fiber	1.6 g

CHIEF NUTRIENTS

NUTRIENT	AMOUNT	% RDA
Vitamin C	23.5 mg	39
Folate	18.5 mcg	9
Vitamin A	76.3 RE	8
Potassium	273.1 mg	7
Iron	0.6 mg	6

Strengths: Although the iron content of tomatoes does not seem impressive, absorption of the mineral is enhanced because of the generous amounts of vitamin C also supplied. Tomatoes also provide some folate, which women need during pregnancy, and the fiber in tomatoes helps relieve constipation and prevent hemorrhoids.

Eat with: Dried beans, whole grains, and other vegetable sources of nonheme iron, as the C in tomatoes will boost absorption of the mineral.

Caution: *Allergy alert:* Tomatoes are a common food allergen and may trigger hives, headaches, mouth itching, or other reactions in sensitive people. People who are allergic to aspirin might want to avoid tomatoes, which contain salicylate.

Origin: Tomatoes appear to have evolved from the cherry tomato in Peru and Ecuador. Their name is derived from the Aztec word *tomatl*.

Selection: Look for firm, plump, red tomatoes, free of soft spots, bruises, or signs of overripeness.

TURNIPS

Serving: ½ cup cubes, boiled

Calories	14
Fat	0.1 g
Saturated	trace
Monounsaturated	N/A
Polyunsaturated	trace
Calories from fat	4%
Cholesterol	0
Sodium	39 mg
Protein	0.6 g
Carbohydrate	3.8 g
Dietary fiber	1.6 g

CHIEF NUTRIENTS

NUTRIENT	AMOUNT	% RDA
Vitamin C	9.1 mg	15

Strengths: Although less nutritious than its cousin the rutabaga, the turnip is a good source of vitamin C, with a fair amount of fiber as a bonus. Virtually free of fat and calories, this vegetable is a dieter's dream. And it supplies the kind of complex carbohydrates that can keep blood sugar levels on an even keel in diabetics. A member of the cruciferous family, turnips contain substances that seem to protect against some cancers.

Eat with: Plant sources of nonheme iron such as beans and grains: The vitamin C in turnips will enhance the absorption of the mineral.

Selection: Select turnip roots that are smooth, firm, and heavy for their size.

Serve: Mashed and seasoned, like potatoes, or in soups and stews.

WAKAME

Serving: 1 oz., raw

Calories	13
Fat	0.2 g
Saturated	trace
Monounsaturated	trace
Polyunsaturated	trace
Calories from fat	12%
Cholesterol	0
Sodium	244 mg
Protein	0.8 g
Carbohydrate	2.5 g
Dietary fiber	0.1 g

CHIEF NUTRIENTS

NUTRIENT	AMOUNT	% RDA
Magnesium	30.0 mg	9
Iron	0.6 mg	6
Calcium	42.0 mg	5

Strengths: Low in fat and calories, with no cholesterol, this form of seaweed contains small amounts of calcium to help guard against osteoporosis; iron for healthier blood; and magnesium, which can protect against high blood pressure. Wakame, like other seaweeds, provides an ample amount of iodine, which the thyroid needs to produce hormones that regulate growth and development.

Caution: High in sodium.

Description: Dried wakame comes in long, curly strands and normally is used in soups.

Preparation: Soak the seaweed in warm water for 20 minutes or so, then remove and discard the tough center vein. To preserve its vitamins and minerals, simmer for no longer than a minute.

WATER CHESTNUTS, *Canned*

Serving: ½ cup slices

Calories	35
Fat	trace
Saturated	N/A
Monounsaturated	N/A
Polyunsaturated	N/A
Calories from fat	1%
Cholesterol	0
Sodium	6 mg
Protein	0.6 g
Carbohydrate	8.7 g
Dietary fiber	1.5 g

CHIEF NUTRIENTS

NUTRIENT	AMOUNT	% RDA
Iron	0.6 mg	6
Vitamin B$_6$	0.1 mg	6

Strengths: With negligible amounts of fat, sodium, and calories, Chinese water chestnuts are a great "extender" for stir-fries. They get a passing grade for iron, which helps prevent fatigue, headaches, and irritability. Water chestnuts also contribute some vitamin B$_6$. (Too little B$_6$ in the diet can trigger dermatitis and other skin problems.)

Eat with: Vitamin C–rich foods like peppers and broccoli to boost iron absorption.

Curiosity: Not related to chestnuts from the chestnut tree. Water chestnut is the edible tuber of a sedge that grows in mud on the banks of ponds, lakes, and marshes.

Description: Crisp and white-fleshed, like tulip bulbs.

Storage: After opening canned chestnuts, you may keep them in a glass jar of water, refrigerated. If you change the water daily, they'll keep for a month.

WATERCRESS

Serving: ½ cup, chopped, raw

Calories	2
Fat	trace
Saturated	trace
Monounsaturated	trace
Polyunsaturated	trace
Calories from fat	10%
Cholesterol	0
Sodium	7 mg
Protein	0.4 g
Carbohydrate	0.2 g
Dietary fiber	0.4 g

CHIEF NUTRIENTS

NUTRIENT	AMOUNT	% RDA
Vitamin C	7.3 mg	12
Vitamin A	79.9 RE	8

Strengths: Diets rich in cruciferous vegetables such as watercress have been linked to a lower incidence of various digestive system cancers. The protective effect may be due to chemical compounds called indole glucosinolates, present in significant quantities in these vegetables. But cooking may break down glucosinolates, so it's a good idea to eat watercress raw. Watercress supplies a respectable amount of vitamin C and some vitamin A—nutrients that also help protect against cancer. The fiber in this delicate, dark-green, leafy vegetable helps prevent constipation and regulate blood sugar.

Serve: In salads and sandwiches; fresh watercress is a crunchy, tasty addition. Use instead of lettuce.

YAMS

Serving: ½ cup cubes, boiled/baked

Calories	79
Fat	0.1 g
Saturated	trace
Monounsaturated	trace
Polyunsaturated	trace
Calories from fat	1%
Cholesterol	0
Sodium	5 mg
Protein	1.0 g
Carbohydrate	18.7 g
Dietary fiber	2.7 g

CHIEF NUTRIENTS

NUTRIENT	AMOUNT	% RDA
Vitamin C	8.2 mg	14
Potassium	455.6 mg	12
Vitamin B_6	0.2 mg	8
Folate	10.9 mcg	5

Strengths: This is just the sort of low-fat, high-fiber, starchy carbohydrate that's the best source of calories for people watching their weight or blood sugar levels. Since yams contribute a good amount of potassium, they are a good choice if you're trying to keep your blood pressure under control. They supply a respectable amount of resistance-building vitamin C, along with some folate. Another valuable nutrient is vitamin B_6, which can help some cases of carpal tunnel syndrome. The fiber might help prevent constipation.

Curiosity: "Yam" comes from the Senegalese word *nyami,* which means "to eat."

Description: Yams are not related to sweet potatoes, which they closely resemble.

Preparation: As with other tubers, cooking yams in the microwave saves time. They can also be baked, boiled, or roasted like Irish potatoes.

Nutritional Reference Tables

Top-Ranked Food Sources of Vitamins and Minerals

Every time we shop at the supermarket, sit down at a restaurant, munch cereal at the kitchen counter, or pull up to the drive-in window of a fast food chow house, we're making choices our bodies have to live with.

Of course, it would be wonderful if we could always grab bites that were great for our health, or exquisitely balance our grocery carts to serve our dietary needs.

But more often than not, we have a somewhat foggy notion of the best and the brightest in the nutritional galaxy. ("What *is* the tuna that's sky-high in Omega-3's?" "What did I hear about a new cereal that's low in sodium *and* super in fiber?" "Did someone mention a nutrition-packed *pizza*?") Well, to jog your memory with some hard data, here's an entire section of nutritional reference tables that will help you weigh the heard-abouts against the hardly-knowns, and compare them with your personal favorites.

First, you'll find tables listing the *top* sources of vitamins and minerals drawn from the 1,000 food entries in this book. (Before you make a selection from among these champs, however, be sure to read the **Caution** in the individual entries. Foods that are loaded with one kind of nutrient may *not* be the best for your all-around health.) You'll also find tables with nutritional information about hundreds of other items, from frozen entrées and cheeses to fast food specials, alcoholic beverages, and frozen yogurts.

So, whether you're browsing, grazing, or shopping, don't miss a tour of these tables to compare your choices before you go on.

Vitamin A: Grade A for All-Around Health

As nutrients go, vitamin A is without a doubt one of nature's star performers. Among other basic functions, it plays a vital role in good vision, keeps skin healthy, assists the body's immune system, and helps manufacture red blood cells. As if that weren't enough, foods high in Vitamin A—like dark-green and deep-orange fruits and vegetables—seem to protect against certain kinds of cancer. One caveat: Stay away from cholesterol-laden foods such as liver and giblets. They may be high in vitamin A, but they come at a high price nutritionally and may contain other substances that you don't want in your diet. Be sure to check the individual entries in the book.

Food	Portion	% RDA
Beef liver, braised	3 oz.	901
Veal liver, braised	3 oz.	684
Lamb liver, braised	3 oz.	637
Carrot juice, canned	6 oz.	474
Liverwurst, fresh	3 slices (app. 2 oz.)	471
Pork liver, braised	3 oz.	459
Chicken liver, simmered	3 oz.	418
Turkey liver, simmered	3 oz.	318
Pumpkin, canned	½ cup	269
Sweet potato, baked	1 (app. 4 oz.)	249
Braunschweiger	2 oz.	239
Carrot, raw	1	203
Chicken giblets, simmered	3 oz.	190
Eel, cooked, dry heat	3 oz.	97
Lamb's-quarters, chopped, boiled	½ cup	87
Spinach, boiled	½ cup	74
Butternut squash, baked cubes	½ cup	71
Tuna, fresh, cooked, dry heat	3 oz.	64
Hubbard squash, baked cubes	½ cup	62
Dandelion greens, chopped, boiled	½ cup	61
Cantaloupe, cubed	1 cup	52
Mango	½ (app. 3½ oz.)	40
Turnip greens, chopped, boiled	½ cup	40
Beet greens, boiled	½ cup	37
Persimmon	1 (6 oz.)	36

Thiamine: Converting Carbohydrates to Power

Strictly speaking, almost any cut of pork is high in thiamine, a B vitamin that converts carbohydrates into energy. But some cuts of pork are too fatty to be consumed regularly. (Any diet high in saturated fat increases the likelihood of heart disease.) So your best bet is to skip fatty pork products, such as sausage and salami, in favor of other, leaner sources of thiamine like pork tenderloin, wheat germ, and pasta. Detailed nutritional profiles of these and other thiamine-rich foods appear throughout the book.

Food	Portion	% RDA
Pork tenderloin, roasted	3 oz.	53
Pork center loin, roasted	3 oz.	51
Pork sirloin, roasted	3 oz.	45
Sunflower seeds, dried	1 oz.	43
Ham, cured, roasted, boneless	3 oz.	42
Florida pompano, cooked, dry heat	3 oz.	39
Ham, fresh, roasted	3 oz.	39
Pork center rib, roasted	3 oz.	36
Pork salami	3 slices (app. 2 oz.)	35
Pork blade loin roast, roasted	3 oz.	33
Pork picnic shoulder arm, roasted	3 oz.	33
Pork shoulder blade, roasted	3 oz.	33
Pork sausage, smoked	1 link (app. 2½ oz.)	32
Squab (pigeon), uncooked	1	32
Wheat germ, toasted	¼ cup (1 oz.)	31
Bratwurst, fresh	1 link (app. 3 oz.)	29
Italian sausage, fresh, cooked	1 link (app. 2½ oz.)	28
Spinach noodles, enriched, cooked	1 cup	26
Canadian bacon, grilled	2 slices (app. 1½ oz.)	25
Chorizo, dried	1 link (app. 2 oz.)	25
Piñon pine nuts, dried	1 oz.	23
Egg noodles, enriched, cooked	1 cup	20
Fresh pasta, cooked	1 cup	20
Rice bran, raw	2 Tbsp.	19

Riboflavin: Going for Energy

Like its sister nutrients niacin and thiamine, the B vitamin riboflavin plays a critical role in converting the food we eat into energy. So you want to incorporate high-riboflavin foods into your daily diet. The problem is, the same foods that are extraordinarily high in riboflavin—beef liver and lamb liver, for instance—are also extraordinarily high in cholesterol, a substance that can sabotage the health of your heart and arteries. The prudent strategy is to stick with lean, readily available sources of riboflavin like low-fat or nonfat yogurt, milk, mackerel, and, occasionally, modest portions of Swiss cheese. For more specific nutrient data, see individual entries for foods listed here.

Food	Portion	% RDA
Beef liver, braised	3 oz.	205
Lamb liver, braised	3 oz.	202
Pork liver, braised	3 oz.	110
Veal liver, braised	3 oz.	97
Chicken liver, simmered	3 oz.	88
Turkey liver, simmered	3 oz.	71
Braunschweiger	2 oz.	51
Chicken giblets, simmered	3 oz.	48
Caribou, roasted	3 oz.	45
Bear, simmered	3 oz.	41
Roe, uncooked	3 oz.	37
Liverwurst, fresh	3 slices (app. 2 oz.)	34
Nonfat yogurt	1 cup	31
Tilsit cheese	1 oz.	30
Low-fat yogurt	1 cup	29
Skim milk	8 fl. oz.	28
Cocoa	8 fl. oz.	26
Duck, roasted	3 oz.	24
King mackerel, uncooked	3 oz.	24
Low-fat (1%) milk	8 fl. oz.	24
Low-fat (2%) milk	8 fl. oz.	24
Swiss cheese	1 oz.	24
Gjetost cheese	1 oz.	23
Squid, fried	3 oz.	23
Whole milk	8 fl. oz.	23

Niacin: A Supercharge for Cells

Niacin works like millions of little hydroelectric power plants in your body—it helps living cells generate energy from food. You can be sure you're getting an adequate amount of this important B vitamin by regularly including niacin-rich foods, such as chicken and turkey breast, tuna, veal, rice bran, and certain fish, in your diet. (As with folate, organ meats like liver are also rich in niacin, but they're too high in saturated fat and cholesterol to recommend.) You may wish to consult individual entries to determine which of the foods listed in the table are best for you.

Food	Portion	% RDA
Chicken breast, roasted	½ (app. 3 oz.)	62
Squab (pigeon), uncooked	1	61
Light meat tuna, canned in water	3 oz.	55
Lamb liver, braised	3 oz.	54
Swordfish, cooked, dry heat	3 oz.	53
Sturgeon, smoked	3 oz.	50
Beef liver, braised	3 oz.	48
Fresh tuna, cooked, dry heat	3 oz.	47
Veal leg, roasted	3 oz.	45
Veal loin, roasted	3 oz.	42
Veal sirloin, roasted	3 oz.	42
Capon, roasted	3 oz.	40
Quail, uncooked	1	40
American shad, uncooked	3 oz.	38
King mackerel, uncooked	3 oz.	38
Pork liver, braised	3 oz.	38
Veal liver, braised	3 oz.	38
Veal arm roast, roasted	3 oz.	37
Ground veal, broiled	3 oz.	36
Veal rib roast, roasted	3 oz.	34
Halibut, cooked, dry heat	3 oz.	32
Atlantic mackerel, cooked, dry heat	3 oz.	31
Rainbow trout, cooked, dry heat	3 oz.	31
Turkey white meat, roasted	3 oz.	31

Folate: The Chockablock Blood Booster

Animal organs—liver and giblets—tend to accumulate awesome amounts of this important B vitamin, which the body uses to manufacture new blood cells. The problem is, organ meats are also sky-high in cholesterol. So your best bet is to bypass those delicacies in favor of plant sources of folate, which are still highly respectable treasure troves of this immunity-boosting, blood-building nutrient.

Food	Portion	% RDA
Chicken liver, simmered	3 oz.	327
Veal liver, braised	3 oz.	323
Turkey liver, simmered	3 oz.	283
Chicken giblets, simmered	3 oz.	160
Beef liver, braised	3 oz.	92
Cowpeas, boiled	½ cup	89
Lentils, boiled	½ cup	89
Mung beans, boiled	½ cup	80
Pinto beans, boiled	½ cup	73
Pink beans, boiled	½ cup	71
Adzuki beans, boiled	½ cup	70
Pork liver, braised	3 oz.	69
Baby lima beans, boiled	½ cup	68
Spinach, boiled	½ cup	66
Black beans, boiled	½ cup	64
Navy beans, boiled	½ cup	64
Yard-long beans, boiled	½ cup	63
Kidney beans, boiled	½ cup	57
Wheat germ, toasted	¼ cup (1 oz.)	50
Chicory, chopped, raw	½ cup	49
Great Northern beans, boiled	½ cup	45
Asparagus, boiled	½ cup	44
Broadbeans, boiled	½ cup	44
Turnip greens, chopped, boiled	½ cup	43
Chickpeas, canned	½ cup	40

Vitamin B$_6$: To Fight Off the Blahs

Dietary sources of this B vitamin are diverse, ranging from rice bran and squab to carrot juice and trout. (And that's good, because people who have a serious deficiency of vitamin B$_6$ run the risk of skin rashes, depression, anemia, and other symptoms.) The trick is to concentrate on the low-fat foods in which B$_6$ is readily available—such as potatoes, bananas, chicken breast, and prune juice—while forgoing high-cholesterol foods like liver or fatty foods like goose. For more specific data on the fat content of these foods, please refer to individual entries.

Food	Portion	% RDA
Squab (pigeon), uncooked	1	45
Beef liver, braised	3 oz.	39
Potato, baked	1 (app. 7 oz.)	35
Banana	1 (app. 4 oz.)	33
Pheasant, uncooked	3 oz.	31
Chickpeas, canned	½ cup	29
Prune juice, canned	8 fl. oz.	28
Whelks, cooked, moist heat	3 oz.	28
Pork sausage, fresh	4 links (app. 2 oz.)	27
Chicken breast, roasted	½ (app. 3 oz.)	26
Chicken liver, simmered	3 oz.	25
Quail, uncooked	1	25
Beef top round steak, broiled	3 oz.	24
Pork liver, braised	3 oz.	24
Fresh tuna, cooked, dry heat	3 oz.	23
Turkey white meat, roasted	3 oz.	23
Turkey liver, simmered	3 oz.	22
Lamb liver, braised	3 oz.	21
Rice bran, raw	2 Tbsp.	21
Striped mullet, cooked, dry heat	3 oz.	21
Veal liver, braised	3 oz.	21
Carrot juice, canned	6 fl. oz.	20
Goose, roasted	3 oz.	20
Rainbow trout, cooked, dry heat	3 oz.	20

Vitamin B$_{12}$: Sheathe Your Nerve Cells

This vitamin is unique in that it helps produce myelin, the protective coating that surrounds delicate nerve tissue (much like the insulation around electrical wiring). If your diet provides too little vitamin B$_{12}$, or if you absorb it poorly, you may experience memory loss, poor appetite, and low energy levels. Luckily, you don't have to rely on organ meats like liver to supply B$_{12}$. Dietary deficiencies of B$_{12}$ are rare, and this vitamin shows up in significant amounts in many kinds of meat, seafood, and dairy products. (For specific amounts, see individual entries throughout this book.)

Food	Portion	% RDA
Clams, steamed	20 small (app. 3 oz.)	4,450
Lamb liver, braised	3 oz.	3,252
Beef liver, braised	3 oz.	3,018
Turkey liver, simmered	3 oz.	2,019
Veal liver, braised	3 oz.	1,552
Blue mussels, cooked, moist heat	3 oz.	1,020
Octopus, uncooked	3 oz.	850
Chicken liver, simmered	3 oz.	824
Atlantic mackerel, cooked, dry heat	3 oz.	808
Eastern oysters, steamed	6 medium (1½ oz.)	804
Pork liver, braised	3 oz.	794
Whelks, cooked, moist heat	3 oz.	771
King mackerel, uncooked	3 oz.	663
Braunschweiger	2 oz.	570
Atlantic herring, cooked, dry heat	3 oz.	559
Alaskan king crab, steamed	3 oz.	489
Fresh tuna, cooked, dry heat	3 oz.	463
Chicken giblets, simmered	3 oz.	431
Roe, uncooked	3 oz.	425
Liverwurst, fresh	3 slices (app. 2 oz.)	382
Blue crab, cooked, moist heat	3 oz.	311
Caribou, roasted	3 oz.	282
Rabbit, roasted	3 oz.	277
Sockeye salmon, canned, drained	3 oz.	247
Bluefish, uncooked	3 oz.	229

Vitamin C: The Star Infection Fighter

Most people think of orange juice and other citrus drinks when they think of vitamin C. And citrus fruit certainly is a rich and reliable source of this vitamin, which is critical in fighting off infections. But did you know that when it comes to vitamin C, tropical fruits like guava and kiwi give oranges a run for their money? Crunchy, flavorful vegetables like broccoli and red peppers are also super sources. The following table is a guide to the winners of the highest vitamin C ratings. As it shows, your best bet is to include vitamin C–rich fruits, vegetables, and juices in your diet.

Food	Portion	% RDA
Acerola cherries	3 (app. ½ oz.)	403
Black currants	1 cup	338
Guava	1 (app. 3 oz.)	275
Orange juice	8 fl. oz.	207
Pummelos, sectioned	1 cup	193
Red bell peppers, raw, chopped	½ cup	158
Papaya	½ (app. 5½ oz.)	157
Cranberry juice cocktail	8 fl. oz.	149
Strawberries	1 cup	141
Kiwifruit	1 (app. 2½ oz.)	124
Grapefruit juice	8 fl. oz.	120
Orange	1 (app. 4½ oz.)	116
Cantaloupe, cubed	1 cup	113
Citrus fruit juice drink	8 fl. oz.	112
Broccoli, chopped, boiled	½ cup	97
Longans	20 (app. 2 oz.)	90
Dried litchis	1 oz.	85
Vegetable juice cocktail	6 fl. oz.	84
Brussels sprouts, boiled	½ cup	81
Cassava, raw	3½ oz.	80
Pink or red grapefruit	½ (app. 4 oz.)	78
Red or white currants	1 cup	77
Green bell peppers, raw, chopped	½ cup	74
Kohlrabi, sliced, boiled	½ cup	74
Honeydew, cubed	1 cup	70

Vitamin E: Bracing Your Body to Battle Disease

It's nice to know oils have some bona fide nutritional value, since at more than 100 calories per tablespoon, they're calorically dense. Like wheat germ, vegetable oils are among the richest food sources of vitamin E, a natural immune-system booster. Vitamin E also shows up in dressings and spreads made with vegetable oil, such as mayonnaise, margarine, and blue cheese dressing. The problem is that those items also contribute a fair amount of saturated fat, which most people nowadays need to avoid. So stick with the unsaturated sources of this nutrient. In addition to the ''top 25'' in the following table, other good sources of E include nuts, almonds, whole grains, and sunflower seeds.

Food	Portion	% RDA
Hazelnut oil	1 Tbsp.	64
Sunflower oil	1 Tbsp.	61
Almond oil	1 Tbsp.	53
Cottonseed oil	1 Tbsp.	48
Safflower oil	1 Tbsp.	46
Grapeseed oil	1 Tbsp.	39
Wheat germ, toasted	¼ cup (1 oz.)	39
Mayonnaise	1 Tbsp.	29
French vinaigrette dressing	2 Tbsp.	22
Corn oil	1 Tbsp.	19
Blue cheese or Roquefort dressing, regular	2 Tbsp.	18
Russian dressing, regular	2 Tbsp.	18
French dressing, regular	2 Tbsp.	16
Olive oil	1 Tbsp.	16
Peanut oil	1 Tbsp.	16
Italian dressing, regular	2 Tbsp.	15
Soybean oil	1 Tbsp.	15
Mango	½ (app. 3½ oz.)	12
Oil and vinegar dressing	2 Tbsp.	12
Thousand Island dressing, regular	2 Tbsp.	12
Black currants	1 cup	11
Blackberries	1 cup	9
Stick margarine	2 tsp.	9
Apple	1 (app. 5 oz.)	8
Soft margarine	2 tsp.	7

Calcium: The Essence of Better Bones

One of the nicest things about the new math of reduced-fat dairy products is that when you subtract the fat, you come up with more calcium per serving. As a result, nonfat and low-fat yogurt, skim milk, and part-skim ricotta cheese are among the richest sources of dietary calcium, needed for strong, fracture-resistant bones and comfortable blood pressure levels. One important note: Both lamb's-quarters and spinach, which are listed in this book, appear to contribute calcium to the diet, but neither has it in an easy-to-absorb form. So dairy products are, by far, your best bet.

Food	Portion	% RDA
Nonfat yogurt	1 cup	56
Low-fat yogurt	1 cup	52
Skim milk	8 fl. oz.	44
Part-skim ricotta cheese	½ cup	42
Low-fat fruit-flavored yogurt	1 cup	39
Low-fat (1%) milk	8 fl. oz.	38
Malted milk beverage from powder	8 fl. oz.	38
Cocoa	8 fl. oz.	37
Low-fat (2%) milk	8 fl. oz.	37
Buttermilk	8 fl. oz.	36
Carob beverage from powder	8 fl. oz.	36
Whole milk	8 fl. oz.	36
Chocolate milk	8 fl. oz.	35
Gruyère cheese	1 oz.	35
Swiss cheese	1 oz.	34
Whole-milk yogurt	1 cup	34
Whole-milk ricotta cheese	½ cup	32
Lamb's-quarters, chopped, boiled	½ cup	29
Edam cheese	1 oz.	26
Monterey Jack cheese	1 oz.	26
Provolone cheese	1 oz.	26
Cheddar cheese	1 oz.	25
Muenster cheese	1 oz.	25
Tilsit cheese	1 oz.	25

Iron: The Power to Build Red Blood

Without sufficient iron, the body can't manufacture enough new blood cells packed with hemoglobin, the red-cell protein that transports oxygen in the blood. Meat is the best source of iron because it contains heme iron, which is readily absorbed by the body. Other food sources of iron—like liver—are especially rich in heme iron, but they're too high in saturated fat and cholesterol to be included on a heart-smart diet. Be sure to check other entries to find lean meats that are good sources of iron.

Plant foods, such as carrots, potatoes, beets, pumpkin, broccoli, tomatoes, cauliflower, cabbage, turnips, and sauerkraut, are also good sources of iron, although it's in a less readily absorbed form called nonheme iron. Absorption of the nonheme iron in legumes (like soybeans) and iron-fortified cereals can be enhanced by combining them with either meat or vitamin C.

Food	Portion	% RDA
Clams, steamed	20 small (app. 3 oz.)	252
Pork liver, braised	3 oz.	152
Cream of Wheat cereal, cooked	¾ cup	103
Bear, simmered	3 oz.	91
Whelks, cooked, moist heat	3 oz.	86
Natto	½ cup	76
Squab (pigeon), uncooked	1	76
Chicken liver, simmered	3 oz.	72
Lamb liver, braised	3 oz.	70
Turkey liver, simmered	3 oz.	66
Tofu, regular	¼ block (app. 4 oz.)	62
Eastern oysters, steamed	6 medium (1½ oz.)	61
Beef liver, braised	3 oz.	58
Blue mussels, cooked, moist heat	3 oz.	57
Chicken giblets, simmered	3 oz.	55
Braunschweiger	2 oz.	53
Caribou, roasted	3 oz.	52
Turkey gizzard, simmered	3 oz.	48
Octopus, uncooked	3 oz.	45
Soybeans, boiled	½ cup	44
Pumpkin seeds, hulled, dried	1 oz.	43
Quail, uncooked	1	42
Quinoa	¼ cup	39

Magnesium: General Guardian of Good Health

Even mild deficiencies of magnesium may provoke major health problems, and adequate amounts of this mineral may help protect your bones, heart, arteries, and blood pressure. So be sure to incorporate at least one good food source of this "forgotten mineral" into your diet every day. In general, whole grains, whole grain cereal, nuts, seeds, and beans tend to be the most reliable sources, although a few fish and fruits also contribute significant amounts.

Food	Portion	% RDA
Rice bran, raw	⅓ cup	63
Pumpkin seeds, hulled, dried	1 oz.	43
Whelks, cooked, moist heat	3 oz.	42
Amaranth seeds, raw	¼ cup	37
Tofu, regular	¼ block (app. 4 oz.)	34
Natto	½ cup	29
Sunflower seeds, dried	1 oz.	29
Halibut, cooked, dry heat	3 oz.	26
Quinoa	¼ cup	26
Wheat germ, toasted	¼ cup (1 oz.)	26
Prickly pear	1 (app. 3½ oz.)	25
Spinach spaghetti, cooked	1 cup	25
Almonds, dried, unblanched	1 oz.	24
Atlantic mackerel, cooked, dry heat	3 oz.	24
Yard-long beans, boiled	½ cup	24
Filberts, dried, unblanched	1 oz.	23
Rice bran, raw	2 Tbsp.	23
Spinach, boiled	½ cup	22
Swiss chard, chopped, boiled	½ cup	22
Cashew nuts, dry-roasted, unsalted	1 oz.	21
Soybeans, boiled	½ cup	21
Cassava, raw	3½ oz.	19
Pignolia pine nuts, dried	1 oz.	19
Piñon pine nuts, dried	1 oz.	19
Brazil nuts, dried, unblanched	1 oz.	18

Potassium: One Way to Say No to High Blood Pressure

Yams, bananas, raisins, and avocados—among other kinds of fruits and vegetables—are fairly good sources of potassium. But certain fish, beans, and dairy products also contain plentiful amounts of this mineral, which can often help control blood pressure without drugs. There is no official RDA for potassium, but if you eat several servings of a variety of these foods per day, you'll get about 3,500 milligrams, which is considered an adequate amount.

Food	Portion	% RDA*
Potato, baked	1 (app. 7 oz.)	24
Cassava, raw	3½ oz.	22
Prune juice	8 fl. oz.	20
Peaches, dried	5 halves (app. 2 oz.)	18
Zante currants, dried	½ cup	18
Adzuki beans, boiled	½ cup	17
Avocado	½ (app. 3 oz.)	17
Nonfat yogurt	1 cup	17
Whelks, cooked, moist heat	3 oz.	17
Clams, steamed	20 small (app. 3 oz.)	16
Seedless raisins	½ cup	16
Carrot juice	6 fl. oz.	15
Florida pompano, cooked, dry heat	3 oz.	15
Golden seedless raisins	½ cup	15
Low-fat yogurt	1 cup	15
Rainbow trout, cooked, dry heat	3 oz.	15
Apricots, dried	10 halves	14
Cantaloupe, cubed	1 cup	14
Lima beans, boiled	½ cup	14
Orange juice	8 fl. oz.	14
Banana	1 (app. 4 oz.)	13
Coho salmon, cooked, moist heat	3 oz.	13
Honeydew melon, cubed	1 cup	13
Pears, dried	5 halves	13
Yams, cubed, boiled/baked	½ cup	12

*Using a recommendation of 3,500 milligrams as an unofficial RDA.

Zinc: To Help You Mend

Zinc is essential for proper wound healing and healthy skin. And research has shown that *low* blood levels of zinc are associated with loss of taste and smell. Women will want to note that the need for zinc increases during pregnancy. With a few exceptions, meat tends to be the richest source of zinc. So you have to pick and choose your zinc sources carefully to be sure you're not inadvertently loading up on saturated fat in the course of ensuring adequate zinc intake. Stick with lean cuts, such as beef top round, and consume liver and fatty steaks or roasts in moderation. (Consult individual entries for the specific fat content of the foods that are listed below.)

Food	Portion	% RDA
Eastern oysters, steamed	6 medium (1½ oz.)	509
Beef shank cross cuts, simmered	3 oz.	60
Beef blade roast, braised	3 oz.	58
Veal liver, braised	3 oz.	54
Beef arm pot roast, braised	3 oz.	49
Lamb foreshank, braised	3 oz.	49
Lamb liver, braised	3 oz.	45
Beef short ribs, braised	3 oz.	44
Alaskan king crab, steamed	3 oz.	43
Beef tip round steak, roasted	3 oz.	40
Rib eye steak, broiled	3 oz.	40
Beef brisket, braised	3 oz.	39
Pork liver, braised	3 oz.	38
Beef wedge bone sirloin, broiled	3 oz.	37
Lamb blade roast, roasted	3 oz.	37
Beef liver, braised	3 oz.	34
Lamb shoulder, roasted	3 oz.	34
Lamb stew/kabob meat, broiled	3 oz.	33
Beef top round steak, broiled	3 oz.	32
Filet mignon, broiled	3 oz.	32
Veal blade roast, roasted	3 oz.	32
Wheat germ, toasted	¼ cup (1 oz.)	32
Beef bottom round roast, braised	3 oz.	31
Miso	½ cup	31
Porterhouse steak, broiled	3 oz.	31

Comparison Tables

Choices, choices.

Do you ever wonder *which* brands of cereal are really the highest in fiber . . . which cuts of meat are the leanest . . . or which cheeses are the lowest in sodium?

The comparison tables in this section are designed to give you quick answers to those questions and many more. With these tables, you'll be able to keep an eye on fiber, fat, calories, cholesterol, sodium, sugar, and omega-3's in brand name and fast food items as well as ordinary foods.

For instance, if you're considering a glass of wine and a few slices of cheese before dinner, first have a look at "Calorie Content of Alcoholic Beverages" and "Fat, Calorie, Cholesterol, and Sodium Content of Cheeses". You'll discover that a glass of white table wine (3½ fluid ounces) has 70 calories, and 1 ounce of regular cheddar cheese has 9 grams of fat. A quick glance at the tables will tell you how white wine stacks up, calorie-wise, against other alcoholic beverages, and how regular cheddar compares to some of the reduced-fat versions.

Going out for pizza? You can look up your favorites, with different toppings, listed by brand name in "Fat, Calorie, Cholesterol, and Sodium Content of Pizzas". Compare frozen yogurts by brand-name in "Fat, Calorie, and Cholesterol Content of Frozen Yogurt". Or look up your favorite breakfast cereal under "Nutrient Content of Cereals".

So use these tables to help you make comparisons when shopping at the restaurant, supermarket, health food store, or even the liquor store. And when you're cooking up a storm, use the "Healthy Substitutes" table to help create leaner, healthier dishes.

Healthy Substitutes

Small changes in your daily eating habits can add up to big health rewards in the long run. When you have a recipe that calls for some ingredient listed on the left side of this table, look for a substitute ingredient listed on the right. You can slash fat, salt, and calories from your diet with the heart-smart, diet-wise substitutes listed in this table.

Replace . . .	With . . .
Butter (as a spread)	• Fruit preserves, jam, or jelly • Margarine made from liquid vegetable oil • Reduced-calorie margarine
Butter, lard, or shortening (for cooking)	• Nonstick vegetable spray • Polyunsaturated oil or margarine made with corn, safflower, sesame seed, soybean, or sunflower oil • Monounsaturated oil such as olive, peanut, or canola
Cream	• Evaporated skim milk • Skim milk • Low-fat (1%) milk
Cream cheese	• Pureed low-fat (1%) cottage cheese • Yogurt cheese
Creamed cottage cheese (4%)	• Nonfat cottage cheese • Low-fat (1%) cottage cheese
Cheese, regular	• Low-fat cheese • Part-skim mozzarella or ricotta
Mayonnaise	• Nonfat mayonnaise • Reduced-calorie mayonnaise • Plain nonfat or low-fat yogurt • Pureed low-fat (1%) cottage cheese
Oil or margarine	• Applesauce (for muffins or quick breads) • Half the amount
Sour cream	• Plain nonfat or low-fat yogurt • Pureed low-fat (1%) cottage cheese with a little lemon juice added
Whipped cream	• Whipped evaporated skim milk
Whole eggs	• Egg substitute • Egg whites
Whole milk	• Evaporated skim milk • Skim milk • Low-fat (1%) milk • Reconstituted nonfat dry milk

Calorie Content of Alcoholic Beverages

You might call alcohol the ultimate junk food—but let's start with the pluses. It's true that some alcoholic beverages, like Bloody Marys and screwdrivers, contain a smidgen of vitamins. However, if you drink a lot of alcohol, there's the potential risk of alcohol-induced heart or liver damage. So, if it's nutrition you're after, the better sources are plain tomato juice and orange juice rather than Bloody Marys and screwdrivers.

Alcohol consumption, even in moderation, is linked with high blood pressure, an enlarged heart, lower immunity, memory loss, and cancer—and it frequently leads to weight gain. What's more, men who drink beer are particularly susceptible to colon cancer. Also, overconsumption of alcohol causes cirrhosis of the liver and can lead to deficiencies of vitamins A, B$_6$, folate, thiamine, and zinc.

Combining alcohol with prescription or over-the-counter medicine increases the risk of adverse effects. Some alcoholic beverages may trigger reactions if you're sensitive to grain, yeast, or tyramine (a substance that causes migraine in some people). Others can cause cause headaches, palpitations, nausea, vomiting, and dangerous increases in blood pressure if combined with monoamine oxidase inhibitors (MAOs). And drinking alcohol during pregnancy can lead to birth defects.

The bottom line? if you like to drink, know the risks. For some people, nonalcoholic beverages may be a smarter choice.

The following chart shows the percentage of alcohol by volume and calorie content of some drinks. Note that serving sizes vary.

Beverage	Alcohol (% by volume)	Calories
Beer (12 fl. oz.)		
Ale	5.0–5.7	148
Light	4	99
Porter	5–7.5	N/A
Regular	5	146
Stout	5.8	192
Liquor (1.5 fl. oz.)		
Bourbon	40–50	107
Gin	45	110
Rum	40	97
Scotch	40–43	107
Vodka	40	97
Whiskey	43	105

Beverage	Alcohol (% by volume)	Calories
Mixed Drinks		
Bloody Mary (5 fl. oz.)	11.7	115
Bourbon and soda (4 fl. oz.)	16.1	104
Daquiri (2 fl. oz.)	28.3	112
Gin and tonic (7.5 fl. oz.)	8.8	171
Manhattan (2 fl. oz.)	36.9	128
Martini (2.5 fl. oz.)	38.4	156
Piña colada (4.5 fl. oz.)	12.3	262
Rum cooler (12 fl. oz.)	4.0	205
Screwdriver (7 fl. oz.)	8.2	175
Tequila sunrise (5.5 fl. oz.)	13.5	189
Tom Collins (7.5 fl. oz.)	9.0	122
Whiskey sour (3 fl. oz.)	20.6	122
Wine		
Champagne (3 fl. oz.)	12	75
Cooler (12 fl. oz.)	5.1	180
Port (2 fl. oz.)	18–19	82–96
Red table wine (3.5 fl. oz.)	11.5	74
Rosé (3.5 fl. oz.)	11.5	73
Sherry (2 fl. oz.)	18–19	79
Vermouth, dry (2 fl. oz.)	17	66
Vermouth, sweet (2 fl. oz.)	17	88
White table wine (3.5 fl. oz.)	11.5	70

Fat, Calorie, Cholesterol, and Sodium Content of Cheeses

People have been eating cheese for over 4,000 years, but for most of those millennia they didn't fret about fat. In recent years, however, food scientists have come up with a number of reduced-fat cheeses to help consumers who are watching their calories, fat, or cholesterol enjoy cheese without sabotaging their low-fat diet. Keep in mind, however, that "low-fat" doesn't mean "no fat." Most reduced-fat brands still get half their calories from fat, and most contribute a fair amount of sodium to the diet. So your best bet is to look for cheese with 5 grams of fat or less per ounce—and limit yourself to two slices or 1-inch cubes.

The following table shows calories, fat, percent of calories from fat, cholesterol, and sodium for seven different types of regular cheese. Various brands of low-calorie and low-fat cheese are given for comparison.

Cheese	Portion	Calories
American		
Regular American*	1 oz.	105
Alpine Lace American Flavor	1 oz.	60
Borden Lite Line American Flavor	⅔ oz. or 1 slice	35
Kraft Light n' Lively American Flavor	1 oz.	70
Kraft Light Singles American Flavor	1 oz.	70
Weight Watchers Low-Sodium American Flavor	1 oz.	50
Cheddar		
Regular Cheddar*	1 oz.	113
Borden Lite-Line Sharp Cheddar Flavor	⅔ oz. or 1 slice	35
Cracker Barrel Light Sharp White Cheddar	1 oz.	90
Dorman's Light Reduced-Fat Cheddar	1 oz.	80
Kraft Light Naturals Mild Reduced-Fat Cheddar	1 oz.	80
Kraft Light Naturals Sharp Reduced-Fat Cheddar	1 oz.	90
Weight Watchers Natural Sharp Cheddar	1 oz.	80
Colby		
Regular Colby*	1 oz.	110
Alpine Lace Colbi-Lo	1 oz.	80
Kraft Light Naturals Reduced-Fat Colby	1 oz.	80

*Based on rounded values

Fat (g)	Calories from fat (%)	Cholesterol (mg)	Sodium (mg)
9	77	26	401
4	60	20	200
2	51	10	280
4	51	15	420
4	51	15	410
2	36	5	120
9	72	29	174
2	51	5	300
5	50	20	200
5	56	20	140
5	56	20	210
5	50	20	200
5	56	15	150
9	73	27	169
5	56	20	85
5	56	20	150

(continued)

Fat, Calorie, Cholesterol, and Sodium Content of Cheeses—Continued

Cheese	Portion	Calories
Monterey Jack		
Regular Monterey Jack*	1 oz.	105
Alpine Lace Monti-Jack Lo	1 oz.	80
Kraft Light Naturals Reduced-Fat Monterey Jack	1 oz.	80
Mozzarella		
Regular Mozzarella*	1 oz.	79
Borden Lite-Line Mozzarella Flavor	⅔ oz. or 1 slice	35
Frigo Truly Lite Mozzarella	1 oz.	60
Muenster		
Regular Muenster*	1 oz.	103
Alpine Lace Muenster Low Sodium	1 oz.	100
Swiss		
Regular Swiss*	1 oz.	105
Borden Lite-Line Swiss Flavor	⅔ oz. or 1 slice	35
Dorman's Reduced-Fat Swiss	1 oz.	90
Kraft Light Naturals Reduced-Fat Swiss	1 oz.	90
Kraft Light Singles Swiss Flavor	1 oz.	70
Main Streeet Swiss Lite	1 oz.	100
Weight Watchers Natural Swiss Flavor	1 oz.	90

Fat (g)	Calories from fat (%)	Cholesterol (mg)	Sodium (mg)
8	69	25	150
5	56	15	75
6	68	20	160
6	68	22	104
2	51	10	230
2	30	8	140
8	70	27	176
9	81	25	85
8	69	26	73
2	51	10	260
5	50	15	60
5	50	20	55
3	39	15	350
8	72	25	70
5	50	15	40

Fat, Calorie, Cholesterol, and Sodium Content of Frozen Entrées

Frozen meals have come a long way since the TV dinners and potpies of a generation ago. Today you can buy freezer-to-oven versions of many tasty entrées that break out of the meat-and-mashed-potatoes mold, including Italian, Chinese, and Mexican alternatives to the traditional TV-table standards.

To trim excess fat from your diet, look for entrées with less than 30 percent of calories from fat—the uppermost level recommended by the American Heart Association. And if you're watching your salt consumption, stick to products with less than 850 milligrams of sodium.

Even the best frozen entrées don't make a complete meal. So you should supplement frozen dinners with fresh vegetables, fruit, and whole grain bread to help ensure that you get plenty of vitamins, minerals, and fiber.

Entrée	Weight	Calories
Budget Gourmet Light		
Lasagna with Meat Sauce	9⅖ oz.	290
Orange Glazed Chicken	9 oz.	280
Sirloin of Beef in Herb Sauce	9½ oz.	270
Sirloin Salisbury Steak	8½ oz.	260
Healthy Choice		
Beef Pepper Steak	9½ oz.	250
Cheese Manicotti	9¼ oz.	230
Cheese Ravioli	9 oz.	240
Chicken Chow Mein	8½ oz.	220
Fettucini Alfredo	8 oz.	240
Glazed Chicken	8½ oz.	220
Lasagna with Meat Sauce	9 oz.	250
Roasted Turkey and Mushrooms in Gravy	8½ oz.	200
Seafood Newburg	8 oz.	200
Spaghetti with Meat Sauce	10 oz.	280
Kraft Eating Right		
Chicken Breast Parmesan	9 oz.	300
Glazed Chicken Breast	10 oz.	240
Lasagna with Meat Sauce	10 oz.	270
Macaroni and Cheese	9 oz.	270
Sliced Turkey Breast	9 oz.	250

Fat (g)	Calories from fat (%)	Cholesterol (mg)	Sodium (mg)
11	34	40	760
3	10	35	1,050
10	33	60	720
13	45	65	700
4	14	40	340
4	16	20	450
2	8	30	460
3	12	45	440
7	26	45	370
3	12	45	390
4	14	20	420
3	14	25	310
3	14	55	440
6	19	20	480
10	30	50	540
4	15	35	560
7	23	30	440
8	27	15	590
7	25	50	560

(continued)

Fat, Calorie, Cholesterol, and Sodium Content of Frozen Entrées—Continued

Entrée	Weight	Calories
Lean Cuisine		
Breast of Chicken Parmesan	10 oz.	240
Cheese Ravioli	8½ oz.	240
Chicken Chow Mein	11¼ oz.	250
Glazed Chicken	8½ oz.	270
Macaroni and Beef	10 oz.	240
Oriental Beef	8⅝ oz.	250
Salisbury Steak	9½ oz.	260
Sliced Turkey Breast	7⅞ oz.	240
Spaghetti with Beef and Mushroom Sauce	11½ oz.	270
Stuffed Cabbage	10¾ oz.	220
Zucchini Lasagna	11 oz.	260
Weight Watchers		
Beef Sirloin Tips and Mushrooms in Wine Sauce	7½ oz.	220
Cheese Manicotti	9¼ oz.	280
Chicken Fajitas	6¾ oz.	230
Fettucini Alfredo	9 oz.	210
Garden Lasagna	11 oz.	290
Italian Cheese Lasagna	11 oz.	350
Lasagna with Meat Sauce	11 oz.	320
London Broil in Mushroom Sauce	7⅖ oz.	140
Spaghetti with Meat Sauce	10½ oz.	280
Stuffed Turkey Breast	8½ oz.	260

Fat (g)	Calories from fat (%)	Cholesterol (mg)	Sodium (mg)
7	26	70	870
8	30	55	590
5	18	35	980
8	27	60	810
6	23	40	590
7	25	45	900
9	31	85	800
6	23	45	720
6	20	30	940
10	41	55	930
7	24	25	950
7	29	50	220
8	26	75	490
5	20	30	590
8	34	35	600
7	22	20	670
12	31	30	690
10	28	45	630
3	19	40	510
7	23	25	610
10	35	80	910

Fat, Calorie, Cholesterol, and Sodium Content of Pizzas

Zesty, cheesy pizza ranks right up there with hot fudge sundaes when it comes to foods people crave most.

When they think of pizza, most health-conscious people assume, reasonably enough, that pizza is loaded with fat, calories, and sodium. And generally, it contributes a fair amount of all three. But as you can see from the following table, pizza's nutritional profile varies somewhat from brand to brand (and topping to topping).

Obviously, it's nearly impossible to find a low-sodium pizza. But you can keep fat levels from getting stratospheric by choosing certain brands of pizza over others. (While you're making that selection, keep in mind that authorities urge people to limit their total fat intake to somewhere between 37 grams (if you're consuming 1,300 calories) and 78 grams (if you're consuming 2,800 calories) per day. A ferociously fatty pizza can deliver up to 32 grams of fat, nearly half an entire day's allotment for some people. Toppings make a difference, too. Extra cheese, sausage, and pepperoni are the fattiest. Plain or veggie-topped pizzas tend to be much lower in fat, but don't add extra cheese!

Type	Portion	Calories
Fast Food Pizza		
Domino's		
Cheese	2 slices	376
Deluxe (sausage, onions, pepperoni, green peppers, mushrooms)	2 slices	498
Pepperoni	2 slices	460
Sausage and mushroom	2 slices	430
Veggie (mushrooms, onions, green peppers, double cheese, olives)	2 slices	498
Little Caesar's		
Cheese	2 slices	340
Pepperoni, green peppers, onions, mushrooms	2 slices	380
Pizza Hut		
Hand-tossed cheese	2 slices	518
Hand-tossed pepperoni	2 slices	500
Hand-tossed supreme	2 slices	540
Pan cheese	2 slices	492
Pan pepperoni	2 slices	540
Pan supreme	2 slices	589
Personal Pan pepperoni	1 pizza	675
Personal Pan supreme	1 pizza	647
Thin'N Crispy cheese	2 slices	398
Thin'N Crispy pepperoni	2 slices	413
Thin'N Crispy supreme	2 slices	459

Fat (g)	Calories from fat (%)	Cholesterol (mg)	Sodium (mg)
10	24	19	483
20	36	39	954
18	35	28	825
16	33	26	552
19	34	36	1,035
12	32	20	570
14	33	30	680
20	35	55	1,276
23	41	50	1,267
26	43	55	1,470
18	33	34	940
22	37	42	1,127
30	46	48	1,363
29	39	53	1,335
28	39	49	1,313
17	38	33	867
20	44	46	986
22	43	42	1,328

(continued)

Fat, Calorie, Cholesterol, and Sodium Content of Pizzas—Continued

Type	Portion	Calories
Frozen Pizza		
Celeste Pizza for One		
Cheese	1 pizza	500
Deluxe	1 pizza	580
Pepperoni	1 pizza	545
Sausage	1 pizza	570
Vegetable	1 pizza	490
Lean Cuisine		
French Bread cheese	1 piece	320
French Bread pepperoni	1 piece	340
French Bread sausage	1 piece	350
Pillsbury Microwave		
Cheese	½ pizza	240
Pepperoni	½ pizza	300
Sausage	½ pizza	280
French Bread cheese	1 piece	370
French Bread pepperoni	1 piece	430
French Bread sausage	1 piece	410
Stouffer's		
French Bread Canadian-style bacon	1 piece	360
French Bread cheese	1 piece	340
French Bread hamburger	1 piece	410
French Bread pepperoni	1 piece	410
French Bread sausage	1 piece	420
French Bread vegetable deluxe	1 piece	420
Weight Watchers		
Deluxe (sausage, mushrooms, onions, peppers)	1 pizza	330
Pepperoni and cheese	1 pizza	320

Fat (g)	Calories from fat (%)	Cholesterol (mg)	Sodium (mg)
25	45	40	1,070
32	50	20	1,290
30	50	20	1,290
32	51	20	1,300
26	48	N/A	1,200
10	28	15	700
12	32	30	970
11	28	40	960
10	38	N/A	540
15	45	N/A	790
13	42	N/A	680
15	36	N/A	680
19	40	N/A	940
16	35	N/A	860
14	35	N/A	960
13	34	N/A	840
19	42	N/A	1,010
20	44	N/A	1,120
20	43	N/A	1,110
20	43	N/A	830
10	27	25	650
10	28	35	710

Fat Content of Cuts of Meat

All meat is not created equal. Fat and calorie values for beef, lamb, and poultry, for instance, cover a wide range—and the type of meat you choose can make a big difference in the amount of these substances you consume. What's more, fat and calories of each type of meat vary considerably from cut to cut. Three ounces of ground beef, for example, have three times as much fat as the same size portion of beef top round, and over 50 additional calories. Pork spareribs are even worse, with more than *five* times as much fat as an equal portion of pork tenderloin—and nearly an additional 200 calories per serving.

The following table ranks various cuts of meat in order of increasing fattiness. Using this table, you can plan a diet that derives less than 30 percent of calories from fat, as recommended by the American Heart Association. Also, choosing leaner cuts of meat automatically cuts calories, helping to keep off unwanted pounds.

Meat	Calories from fat (%)*	Fat (g)	Calories
Beef			
Top round steak	25	4.2	153
Eye round roast	26	4.2	143
Shank cross cuts	28	5.4	171
Tip round steak	34	5.9	157
Bottom round roast	35	7.0	178
Arm pot roast	35	7.1	184
Top loin steak	41	8.0	176
Filet mignon	43	8.5	179
Flank steak	44	8.6	176
T-bone steak	44	8.8	182
Porterhouse steak	45	9.2	185
Rib eye steak	47	10.0	191
Brisket	47	10.9	206
Blade roast	47	11.1	213
Ground beef	57	13.9	218
Lamb			
Foreshank	29	5.1	159
Shank	33	5.7	153
Leg	36	6.6	162
Sirloin	40	7.8	173
Arm roast	43	7.9	163
Loin roast	44	8.3	172

*Rounded

Meat	Calories from fat (%)*	Fat (g)	Calories
Lamb—Continued			
Shoulder	47	9.2	173
Blade roast	50	9.8	178
Rib roast	52	11.3	197
Ground lamb	62	16.7	241
Pork			
Tenderloin	26	4.1	141
Ham, cured	42	6.5	140
Ham, fresh	45	9.4	187
Picnic shoulder arm	50	10.7	194
Sirloin	50	11.2	201
Center rib	51	11.7	208
Top loin	51	11.7	208
Shoulder blade	59	14.3	218
Blade loin roast	62	16.4	237
Spareribs	69	25.8	337
Poultry			
Turkey, white meat	18	2.7	133
Chicken breast	19	3.1	142
Turkey, dark meat	35	6.1	159
Chicken leg	40	8.0	181
Capon	46	9.9	195
Goose	48	10.8	202
Veal			
Leg	20	2.9	128
Arm roast	32	4.9	139
Sirloin	33	5.3	143
Blade roast	36	5.9	145
Loin	36	5.9	149
Rib roast	38	6.3	150
Ground veal	40	6.4	146

Fat, Calorie, and Cholesterol Content of Frozen Yogurt

Most frozen yogurt is lower in fat than ice cream. But you have to read labels on each brand of frozen yogurt to find out exactly how much fat you're getting.

As you can see from the following table, fat, percentage of calories from fat, cholesterol, and calories vary from product to product. In general, when you're reading labels, look for frozen yogurt that has less than 1 gram of fat per ounce. Since most labels indicate a 4-ounce serving, that means less than 4 grams of fat *per serving*. In comparison, most premium ice cream contains between 8 and 17 grams of fat per 4-ounce serving. In the tables all values are per 4-ounce serving.

Yogurt	Calories	Fat (g)*	Calories from fat (%)†	Cholesterol (mg)
Breyer's All-Natural Low-Fat Vanilla	120	1	8	15
Colombo Nonfat Flavors	100	0	0	0
Crowley Silver Premium Raspberry	107	1	11	7
Häagen-Dazs Vanilla Almond Crunch	200	7	30	48
I Can't Believe It's Yogurt Nonfat Flavors	80	0	0	0
Kemp's Low-Fat Vanilla	107	3	22	7
Kemp's Nonfat Vanilla	93	0	0	0
Sealtest Fat-Free Nonfat Vanilla	100	0	0	0
Simple Pleasures Chocolate	140	<1	<6	10
T & W Temptations Free Nonfat Raspberry	93	0	0	0
TCBY Nonfat Flavors	110	<1	<8	<5
Yoplait Soft Vanilla	120	3	20	7

*Rounded
†Based on unrounded values

Food Sources of Fiber

When we say a food is ''high in fiber,'' we may be talking about two different kinds of fiber—but both are good for us. Insoluble fiber is often called roughage or bulk. This is the fiber that helps relieve constipation and may also help prevent hemorrhoids. And research has shown that insoluble fiber may also help prevent colorectal cancer and diverticular disease. Insoluble fiber is found in most foods that contain fiber, but it's particularly concentrated in the peels, skins, stalks, and husks of fruits, vegetables, and whole grains.

Soluble fiber is the more digestible kind that can help control weight by reducing appetite. This kind of fiber forms bulk in the stomach, creating a sense of fullness. Naturally, when we feel full, we tend to eat less—with the net result that we put on less weight. Soluble fiber is found in fruits, vegetables, legumes, and whole grains.

Experts recommend that we consume approximately 20 to 35 grams of soluble and insoluble fiber per day—which is just about twice as much as most of us get. The foods on this list are the top sources of fiber included in this book.

Food	Portion	Dietary Fiber (g)
Pears, dried	5 halves	11.5
Cherimoya	½ (app. 10 oz.)	8.7
Triticale, raw	½ cup	8.7
Cowpeas, boiled	½ cup	8.3
Corn bran, raw	2 Tbsp.	7.9
Amaranth seeds, raw	¼ cup	7.5
Miso	½ cup	7.5
Blackberries	1 cup	7.2
Chickpeas, canned	½ cup	7.0
Kidney beans, boiled	½ cup	6.9
Lima beans, boiled	½ cup	6.8
Refried beans, canned	½ cup	6.7
Baby lima beans, boiled	½ cup	6.6
Mamey	¼ (app. 7½ oz.)	6.3
Black beans, boiled	½ cup	6.1
Ralston cereal, cooked	¾ cup	6.0
Raspberries	1 cup	6.0
Apples, dried	10 rings (app. 2 oz.)	5.6
Whole wheat spaghetti, cooked	1 cup	5.4
Peaches, dried	5 halves (app. 2 oz.)	5.3
Figs, dried	3 (app. 2 oz.)	5.2
Lentils, boiled	½ cup	5.2
Succotash	½ cup	5.2
Figs	3 (app. 5 oz.)	5.0
Guava	1 (app. 3 oz.)	4.9
Navy beans, boiled	½ cup	4.9
Zante currants, dried	½ cup	4.9

Nutrient Content of Cereals

Cereal is an excellent source of complex carbohydrates, iron, and fiber. But with over 100 cereals to choose from, it's easy to get confused. The challenge is to find a cereal you like that also has plenty of fiber and nutrients, but not too much fat, calories, sugar, or sodium. Your best bet is to look for adult cereals on the top shelf in most supermarket aisles. Some of the better buys, nutritionally speaking, are shown below.

Note that serving sizes vary somewhat, depending on the brand and the kind of cereal. While most products quote a serving size of 1 ounce, you might be in the habit of eating much more. On the one hand, larger portions bring you closer to the daily total fiber intake of 20 to 35 grams per day recommended by the National Cancer Institute. On the other, if you're on a low-fat, heart-healthy diet, remember that larger portions will mean an increase in your consumption of calories, fat, sugar, and sodium.

Research suggests that starting the day with a bowl of high-fiber cereal may help curb calorie intake for the rest of the day. With the following chart, you can tell at a glance how some selected brands and types of cereal measure up to each other in fiber, calories, fat, sugar, and sodium.

Cereal	Portion	Fiber (g)
General Mills		
Cheerios	1 oz. (1¼ cups)	2
Fiber One	1 oz. (½ cup)	13
Raisin Nut Bran	1 oz. (½ cup)	3
Total	1 oz. (1 cup)	3
Wheaties	1 oz. (1 cup)	3
Health Valley		
Amaranth Flakes	1 oz. (app. ½ cup)	3
Fruit & Fitness	2 oz. (¾ cup)	11
Oat Bran O's	1 oz. (app. ½ cup)	4
Kellogg's		
All-Bran	1 oz. (app. ⅓ cup)	10
Bran Flakes	1 oz. (app. ⅔ cup)	5
Common Sense Oat Bran	1 oz. (app. ½ cup)	3
Corn Flakes	1 oz. (app. 1 cup)	1
Cracklin' Oat Bran	1 oz. (app. ½ cup)	4
Fiberwise Crunchy Flakes	1 oz. (app. 1 cup)	6
Frosted Mini-Wheats Biscuits	1 oz. (app. 4)	3
Fruitful Bran	1.3 oz. (app. ⅔ cup)	5
Kenmei Rice Bran	1 oz. (app. ¾ cup)	1
Mueslix Five Grain Muesli	1.5 oz. (app. ½ cup)	4
Nut & Honey Crunch	1 oz. (app. ⅔ cup)	trace
Nutri-Grain Raisin Bran	1.4 oz. (app. 1 cup)	5

Calories	Fat (g)	Sugar (g)	Sodium (mg)
110	2	1	290
60	1	0	140
110	3	8	140
100	1	3	200
100	1	3	200
90	<1	N/A	5
220	4	N/A	5
90	<1	N/A	0
70	1	5	260
90	0	5	220
100	1	5	270
100	0	2	290
110	4	7	150
100	0	6	0
110	0	11	230
90	1	5	125
110	1	4	230
140	1	11	55
110	1	9	200
130	1	9	200

(continued)

Nutrient Content of Cereals—Continued

Cereal	Portion	Fiber (g)
Kellogg's—Continued		
Oatbake Raisin Nut	1 oz. (app. ⅓ cup)	2
Product 19	1 oz. (app. 1 cup)	1
Raisin Bran	1.4 oz. (app. ¾ cup)	5
Rice Krispies	1 oz. (app. 1 cup)	trace
Special K	1 oz. (app. 1 cup)	trace
Nabisco		
Fruit Wheats Oat Bran Raisin	1 oz. (app. ½ cup)	3
100% Bran	1 oz. (½ cup)	10
Shredded Wheat	⅚ oz. (1 biscuit)	3
Post		
Fruit & Fiber with Dates, Raisins, and Walnuts	1.3 oz. (app. ⅔ cup)	5
Grape Nuts	1 oz. (app. ¼ cup)	3
Raisin Bran	1.4 oz. (app. ⅔ cup)	6
Quaker		
Crunchy Rice Bran	1 oz. (⅔ cup)	2
Puffed Rice	½ oz. (1 heaping cup)	0
Puffed Wheat	½ oz. (1 heaping cup)	1
Ralston		
Corn Chex	1 oz. (1 cup)	N/A
Fruit Muesli	1.5 oz. (app. ½ cup)	3
Multi Bran Chex	1 oz. (⅔ cup)	4
Rice Bran Options	1.1 oz. (app. 1 cup)	3
Rice Chex	1 oz. (1⅛ cups)	N/A
Wheat Chex	1 oz. (⅔ cup)	2

Calories	Fat (g)	Sugar (g)	Sodium (mg)
110	3	8	190
100	0	3	320
120	1	13	230
110	0	3	290
110	0	3	230
90	0	5	0
70	1	6	190
80	<1	0	0
120	2	6	170
110	0	3	170
120	1	7	200
100	1	6	250
50	0	0	0
50	0	0	0
110	0	3	310
140	2	4	95
90	0	6	200
120	2	7	140
110	0	2	280
100	0	3	230

Fast Food Comparisons

With a little comparison shopping, you can enjoy the convenience of roadside dining while keeping a lid on fat and cholesterol. A grilled chicken sandwich at Hardee's, for example, has just 9 grams of fat, compared with 40 grams (an entire day's allotment for some people) for a chicken sandwich at Burger King. Similarly, a Junior Hamburger at Wendy's has 9 grams of fat and 34 milligrams of cholesterol (both considered moderate levels) compared with 33 grams of fat and 90 milligrams of cholesterol for a Big Classic at the same chain.

One important caveat: Menu items that are lower in fat also tend to be lower in sodium and cholesterol. But, as a whole, fast food tends to be high in sodium, so if you're trying to follow a reduced-sodium diet, your best bet is probably to order a salad without the dressing.

Using the following table, you'll be able to compare some of the lowest-fat, lowest-calorie items with some of the highest-fat and -calorie offerings. For comparison, we've listed a variety of burgers, chicken, fish, roast beef, and salads offered by a number of fast food chains.

Food	Calories	Fat (g)	Calories from fat (%)	Sodium (mg)	Cholesterol (mg)
Chicken					
Arby's					
Grilled Chicken Barbeque	378	14	33	1,059	44
Chicken Cordon Bleu	658	37	51	1,824	65
Burger King					
Chicken Tenders (6 pcs.)	236	13	50	541	46
Chicken Sandwich	685	40	53	1,417	82
Carl's Jr.					
Charbroiler BBQ Chicken Sandwich	310	6	17	680	30
Charbroiler Chicken Club Sandwich	570	29	46	1,160	60
Chick-fil-A					
Chargrilled Chicken (no bun)	128	2	14	698	32
Chicken Salad Sandwich	449	26	52	888	50
Hardee's					
Grilled Chicken Sandwich	310	9	26	890	60
Chicken Stix (9 pcs.)	310	14	41	1,020	55
Kentucky Fried Chicken					
Lite N' Crispy Drumstick	121	7	52	196	61
Extra Tasty Crispy Thigh	406	30	67	688	129
Long John Silver's					
Baked Chicken Sandwich (without sandwich sauce)	320	8	23	960	70
Batter-Dipped Chicken Sandwich (2 pcs., without sandwich sauce)	440	17	35	1,280	95

Food	Calories	Fat (g)	Calories from fat (%)	Sodium (mg)	Cholesterol (mg)
Chicken—Continued					
McDonald's					
Chicken McNuggets (6 pcs.)	270	15	50	580	56
McChicken	490	29	53	780	43
Roy Rogers					
Chicken Leg	117	7	54	162	64
Chicken Thigh	282	20	64	505	89
Wendy's					
Chicken Club Sandwich	506	25	44	930	70
Grilled Chicken Fillet	100	3	27	330	55
Fish					
Arby's					
Fish Fillet	537	29	49	994	79
Burger King					
Ocean Catch Fish Fillet	495	25	45	879	57
Carl's Jr.					
Catch Fish Sandwich	560	30	48	1,220	5
Hardee's					
Fisherman's Fillet	500	24	43	1,030	70
Long John Silver's					
Homestyle Fish (1 pc.)	125	7	50	200	20
Batter-Dipped Fish (1 pc.)	210	12	51	570	30
McDonald's					
Filet-O-Fish	440	26	53	1,030	50
Wendy's					
Fish Fillet Sandwich	460	25	49	780	55
Hamburgers					
Burger King					
Hamburger	272	11	36	505	37
Double Whopper with Cheese	935	61	59	1,245	194

(continued)

Fast Food Comparisons—Continued

Food	Calories	Fat (g)	Calories from fat (%)	Sodium (mg)	Cholesterol (mg)
Hamburgers—Continued					
Carl's Jr.					
Hamburger	320	14	39	590	35
Double Western Bacon Cheeseburger	1,030	63	55	1,810	145
Hardee's					
Hamburger	270	10	33	490	20
Bacon Cheeseburger	610	39	58	1,030	80
McDonald's					
Hamburger	260	10	35	500	37
McD.L.T.	580	37	57	990	109
Roy Rogers					
Hamburger	456	28	55	495	73
RR Bar Burger	611	39	57	1,826	115
Wendy's					
Jr. Hamburger	260	9	31	570	34
Big Classic	570	33	52	1,085	90
Roast Beef					
Arby's					
Junior Roast Beef	218	11	45	345	23
Bac N'Cheddar Deluxe	532	33	56	1,672	83
Carl's Jr.					
Roast Beef Deluxe Sandwich	540	26	43	1,340	40
Roast Beef Club Sandwich	620	34	49	1,950	45
Hardee's					
Regular Roast Beef	260	9	31	730	35
Big Roast Beef	300	11	33	880	45
Roy Rogers					
Roast Beef Sandwich	317	10	28	785	55
Large Roast Beef Sandwich with Cheese	467	21	40	1,953	95

Food	Calories	Fat (g)	Calories from fat (%)	Sodium (mg)	Cholesterol (mg)
Salads					
Arby's					
Side Salad	25	0	0	30	0
Cashew Chicken Salad	590	37	56	1,140	65
Burger King					
Side Salad	25	0	0	27	0
Chef's Salad	178	9	46	568	103
Carl's Jr.					
Garden Salad-to-Go	50	3	54	75	5
Chicken Salad-to-Go	200	8	36	300	70
Chick-fil-A					
Tossed Salad (without dressing)	21	0.3	13	19	0
Chicken Salad Plate	579	45	70	980	387
Hardee's					
Side Salad	20	<1	<45	15	0
Chef's Salad	240	15	56	930	115
Long John Silver's					
Small Salad (without dressing or crackers)	11	0	0	5	0
Seafood Salad	230	5	20	580	90
McDonald's					
Side Salad	60	3	45	85	41
Chef's Salad	230	13	51	490	128
Subway					
Roast Beef Salad (small)	222	10	41	775	38
Tuna Salad (small)	430	38	80	380	43
Wendy's					
Garden Salad	102	5	44	110	0
Taco Salad	660	37	50	1,110	35

Omega-3 Fatty Acids in Fish

While consuming too much fat is blamed for the development of heart disease, fish contains a type of fat that's actually good for your heart: omega-3 fatty acids. Researchers have shown that, when eaten in sufficient quantity, omega-3 fatty acids can slash high blood levels of triglycerides and possibly reduce cholesterol as well. And omega-3's seem to reduce the tendency of blood to form clots, the major cause of heart attacks. The following table shows what quantity (in grams) of omega-3's you get by eating 3 ounces of certain finfish and shellfish. It includes a description of how the fish is cooked (in some cases the figures are for raw fish and shellfish). Note that deep-frying seafood destroys these beneficial fatty acids.

Fish or Shellfish	Omega-3's (g)
Atlantic herring, cooked, dry heat	1.82
Anchovies, canned in olive oil	1.76
Pink salmon, canned	1.45
Bluefin tuna, cooked, dry heat	1.28
Atlantic sardines, canned in oil	1.26
Atlantic mackerel, cooked, dry heat	1.12
Sockeye salmon, cooked, dry heat	1.10
Swordfish, cooked, dry heat	0.90
Eastern oysters, steamed	0.81
Rainbow smelts, cooked, dry heat	0.81
Tilefish, cooked, dry heat	0.77
Rainbow trout, cooked, dry heat	0.75
Whiting, cooked, dry heat	0.75
Shark, uncooked	0.74
Blue mussels, cooked, moist heat	0.70
Carp, cooked, dry heat	0.68
Albacore tuna, canned in water	0.65
Sea bass, cooked, dry heat	0.65
Striped bass, uncooked	0.65
Eel, cooked, dry heat	0.63
Squid, fried	0.55
Halibut, cooked, dry heat	0.47
Blue crab, cooked, moist heat	0.42
Pollack, cooked, dry heat	0.40
Atlantic ocean perch, cooked, dry heat	0.38
Chinook salmon (lox), smoked	0.38
Sturgeon, smoked	0.25
Haddock, cooked, dry heat	0.20

Index of Foods and Dishes

Note: Page references in **boldface** indicate tables.